British Plant Communities I

British Plant Communities

VOLUME I

WOODLANDS AND SCRUB

J. S. Rodwell (editor)
C. D. Pigott, D. A. Ratcliffe
A. J. C. Malloch, H. J. B. Birks
M. C. F. Proctor, D. W. Shimwell
J. P. Huntley, E. Radford
M. J. Wigginton, P. Wilkins

for the
Nature Conservancy Council

CAMBRIDGE
UNIVERSITY PRESS

CAMBRIDGE UNIVERSITY PRESS
Cambridge, New York, Melbourne, Madrid, Cape Town, Singapore, São Paulo

Cambridge University Press
The Edinburgh Building, Cambridge CB2 2RU, UK

Published in the United States of America by Cambridge University Press, New York

www.cambridge.org
Information on this title: www.cambridge.org/9780521235587

First published 1991
First paperback edition 1998
Reprinted 2003

A catalogue record for this publication is available from the British Library

Library of Congress Cataloguing in Publication data
British plant communities / J. S. Rodwell, editor: C. D. Pigott . . . [et al.] for the Nature Conservancy Council.
    p. cm.
Includes bibliographical references.
Contents. v. 1. Woodlands and scrub
ISBN 0-521-23558-8 (v. 1)
1. Plant communities–Great Britain.   2. Vegetation
classification–Great Britain.   I. Rodwell, J. S.   II. Nature Conservancy Council (Great Britain)
QK306.B857   1991
581.5′247′0941–dc20   90-1300   CIP

ISBN-13 978-0-521-23558-7 hardback
ISBN-10 0-521-23558-8 hardback

ISBN-13 978-0-521-62721-4 paperback
ISBN-10 0-521-62721-4 paperback

Transferred to digital printing 2005

CONTENTS

FIGURES

FOREWORD

This work is a landmark in the historical journey of British botany. It will provide a source of reliable information about vegetation for research workers, conservation managers and dedicated amateurs for many years to come.

It is the culmination of fifteen years' detailed survey and analysis of British vegetation. It began in 1975 with a contract from the Nature Conservancy Council to Lancaster University and involved leading ecologists in this and other universities contributing their particular skills in a coordinated team. During the course of the project, many other botanists and researchers have willingly supplied their data and ideas, which have enhanced the final work greatly.

The published work will be in five volumes – woodlands and scrub; mires and heaths; grasslands and montane vegetation; aquatic communities, swamps and tall-herb fens; maritime and weed communities.

The aim of the project is to describe British vegetation as a series of plant communities, to understand how these relate to one another and to set them in a wider European context. The scheme is systematic and analytical and it takes into account the influence of environmental factors and management practices, an understanding of which is vital for the conservation of vegetation.

We who are the inheritors of the good work of the Nature Conservancy Council will use the work also in measuring our own progress in conserving and studying the rich variety of British plant communities.

Professor Sir Frederick Holliday CBE DL FRSE
Chairman Designate
Joint Nature Conservation Committee

PREFACE AND ACKNOWLEDGEMENTS

The publication of this the first of five volumes of *British Plant Communities* sees the work of the National Vegetation Classification into its final stage. When the research team first came together in 1975, we none of us thought that the task ahead would be so laborious as to be fifteen years in the completion. It has been a long haul for all of us and, particularly in these closing years, when the responsibility for bringing the work to a good end has fallen mainly on me, the continuing interest of all the participants, and the encouragement and expectation of many others, have been of enormous importance.

As coordinator of the project and editor of the volumes, I know the extent of the debt which we all owe to the originators of the proposal that here comes to fruition. Among our team, Donald Pigott provided a firm conviction that an understanding of plant communities is of inestimable value in ecology and conservation, and his own perception has helped set the style for the kind of vegetation descriptions we have aimed to produce. Andrew Malloch's concern to see the vision realised has been just as constant: his own studies were a model for our work and, from the start, he has given that firm support with the everyday working of the project that was essential for its success.

The Nature Conservancy had first accepted the need for such a project in 1971, in research proposals by Derek Ratcliffe, who, as Chief Scientist of its successor, the Nature Conservancy Council, welcomed the proposal by Donald Pigott and was later instrumental in placing the contract for the work and helping to launch it. His own achievements provided a vital inspiration for our work, and his continuing faith in its value helped sustain us in the long task of bringing its results to light. The Nature Conservancy Council has maintained its funding for the research throughout and, without this commitment, the project would have foundered. Tim Bines, who became NCC nominated officer to the project part way through, brought an understanding which was greatly valued in the difficult middle years, and Lynne Farrell, who succeeded him towards the

close, has helped see the enterprise through with vigour.

The description of woodland and scrub communities, which are the subject of this volume, was just one part of the enterprise, but it was an especially difficult one. These kinds of vegetation are among the most complex of which we had to give an account and there were few existing data on which we could draw. I am more than ever grateful to my fellow members of the research team, then, for the industry with which they searched out and sampled woodlands and scrub along with all the other vegetation types they had to survey in their regions. The bulk of the sampling here was carried out by the four research assistants, Jacqueline Huntley (née Paice), Elaine Radford (née Grindey), Martin Wigginton, Paul Wilkins, and myself. Donald Pigott and the research supervisors, John Birks, Andrew Malloch, Michael Proctor and David Shimwell, also provided some data and, while maintaining all their responsibilities in their university posts, controlled the direction of the whole project and monitored its progress. Cooperation was the watchword here, sampling and data review being very much team efforts, though our common purpose did not preclude a diversity of individual contributions nor some entertaining exchanges. Indeed, when we were first beginning to work together, one of our meetings to discuss the scope of the task before us took us to a Cambridgeshire wood, where we had a lively discussion about the vegetation patterns and their meaning. It is one of my chief hopes that, among the sheer grind of our labour, we have managed to preserve some of that early freshness of observation and enquiry.

A number of other workers and organisations helped us at this stage of our research into woodlands and scrub. Katherine Hearn, seconded from the NCC for a single year, was to be a very welcome addition to our team, providing much needed samples from southern Scotland, while Messrs Eric Birse and James Robertson of the Macaulay Institute in Aberdeen kindly allowed us access to all their data from Scotland. Drs Bryan Wheeler and Mary Edwards and the Rev. Gordon

Graham generously let us incorporate samples which they had collected and many NCC regional staff, members of conservation and natural history societies, and flora recorders helped us spread the overall geographical and florisitic coverage by directing us to representative sites for sampling.

Among the NVC team, John Birks and Jacqueline Huntley had particular responsibility for the analysis of what became the largest data set ever assembled from British woodlands and scrub, and they in turn were kindly assisted by Dr Brian Huntley in the development of computerised classification techniques at Cambridge. Jacqueline generously continued working with John Birks beyond the end of her contract with the project and I am greatly indebted to them both for the firm foundation which they laid for our woodland scheme, and for their continuing concern for the welfare of the whole project.

During my own further processing of the material which they provided and in the writing of the community accounts, the major debt I owe is to Donald Pigott whose encouragement and criticism, reflecting a deep appreciation of woodland ecology, were a continual challenge to me. Professor Roy Clapham also provided helpful comments on early drafts and I have benefited at various stages in the writing from discussions with Drs Margaret Atherden, Susan Barker, Peter Grubb, Martin Hermy, Daniel Kelly, Oliver Rackham, Ulriche Sachse, Bryan Wheeler and Bogdan Zamanek, Messrs Jack Lavin and Geoffrey Wilmore of the West Yorkshire Data Bank and various staff of the Forestry Commission. Undergraduate students at Lancaster University and course members attending Juniper Hall Field Study Centre have also provided valuable field tests of the classification.

As our scheme began to take shape, it was very heartening to us that the NCC was able to look forward to integrating our proposals with the invaluable historical perspective on woodlands developed by Dr George Peterken. I have been especially encouraged by Dr Keith Kirby who has been providing a searching commentary on our classification for some time, and building on our early experience of training NCC staff in using the NVC approach to the description of woodland vegetation. Among these staff, Ms Jane MacKintosh and Mr Richard Tidswell pioneered the use of the classification in Scotland with exemplary industry and initiative,

while Dr Tony Whitbread and Mr Gavin Saunders have helped to produce some highly informative guides to the scheme.

Other people have made an important contribution to the progress of the whole project at various stages. In the early years, Mr Philip Harper provided technical help and Miss Frances Rake assistance with data handling while, at various times, Mrs Beryl Fletcher, Mrs Sylvia Peglar, Mrs Mary Pettit, Mrs Margaret Pigott and Mrs Mary West coped cheerfully with the tedium of data coding. Then, secretarial help was provided by Mrs Jennie Ford, Mrs Claire Ashworth and, over the final seven years, by Mrs Carol Barlow, to whom I am especially grateful for her outstanding efficiency in typing virtually the entire text, tables and indices for all five volumes. In each of the universities involved in the work, Cambridge, Exeter, Manchester and particularly Lancaster, which was the major contractor, many other staff have helped provide a productive home for the project, servicing its numerous calls on administration, computing facilities and libraries.

In our discussions about publication, I have been much helped by Mr Philip Oswald of the NCC and by various staff at the Cambridge University Press who have been enthusiastic from the beginning at the prospect of handling this task. Mr Martin Walters dealt with the early negotiations for the Press and continued to provide encouragement as the manuscript was being prepared. More recently, Dr Alan Crowden has faced with great confidence the formidable task of transforming the material into books. At this stage, too, as we look forward to the increasing application of the work in a challenging time for conservation, I have been enormously cheered by the enthusiasm of Dr Peter Bridgewater, who succeeded Dr Ratcliffe as Chief Scientist of the NCC.

Finally, this kind of enterprise takes a toll which cannot be counted in time and money, and the sharing of which it is more difficult to bear or to acknowledge. A few have taken this strain with us, bound to the task by their loyalty to us. For my part, I want to thank Rosemary, my wife, who, over fifteen years of living with this work, has known something of its real cost.

John Rodwell

Lancaster

PREAMBLE

GENERAL INTRODUCTION

The background to the work

It is a tribute to the insight of our early ecologists that we can still return with profit to *Types of British Vegetation* which Tansley (1911) edited for the British Vegetation Committee as the first coordinated attempt to recognise and describe different kinds of plant community in this country. The contributors there wrote practically all they knew and a good deal that they guessed, as Tansley himself put it, but they were, on their own admission, far from comprehensive in their coverage. It was to provide this greater breadth, and much more detailed description of the structure and development of plant communities, that Tansley (1939) drew together the wealth of subsequent work in *The British Islands and their Vegetation*, and there must be few ecologists of the generations following who have not been inspired and challenged by the vision of this magisterial book.

Yet, partly because of its greater scope and the uneven understanding of different kinds of vegetation at the time, this is a less systematic work than *Types* in some respects: its narrative thread of explication is authoritative and engaging, but it lacks the light-handed framework of classification which made the earlier volume so very attractive, and within which the plant communities might be related one to another, and to the environmental variables which influence their composition and distribution. Indeed, for the most part, there is a rather self-conscious avoidance of the kind of rigorous taxonomy of vegetation types that had been developing for some time elsewhere in Europe, particularly under the leadership of Braun-Blanquet (1928) and Tüxen (1937). The difference in the scientific temperament of British ecologists that this reflected, their interest in how vegetation works, rather than in exactly what distinguishes plant communities from one another, though refreshing in itself, has been a lasting hindrance to the emergence in this country of any consensus as to how vegetation ought to be described, and whether it ought to be classified at all.

In fact, an impressive demonstration of the value of the traditional phytosociological approach to the description of plant communities in the British Isles was published in German after an international excursion to Ireland in 1949 (Braun-Blanquet & Tüxen 1952), but more immediately productive was a critical test of the techniques among a range of Scottish mountain vegetation by Poore (1955a, b, c). From this, it seemed that the really valuable element in the phytosociological method might be not so much the hierarchical definition of plant associations, as the meticulous sampling of homogeneous stands of vegetation on which this was based, and the possibility of using this to provide a multidimensional framework for the presentation and study of ecological problems. Poore & McVean's (1957) subsequent exercise in the description and mapping of communities defined using this more flexible approach then proved just a prelude to the survey of huge tracts of mountain vegetation by McVean & Ratcliffe (1962), work sponsored and published by the Nature Conservancy (as it then was) as *Plant Communities of the Scottish Highlands*. Here, for the first time, was the application of a systematised sampling technique across the vegetation cover of an extensive and varied landscape in mainland Britain, with assemblages defined in a standard fashion from full floristic data, and interpreted in relation to a complex of climatic, edaphic and biotic factors. The opportunity was taken, too, to relate the classification to other European traditions of vegetation description, particularly that developed in Scandinavia (Nordhagen 1943, Dahl 1956).

McVean & Ratcliffe's study was to prove a continual stimulus to the academic investigation of our mountain vegetation and of abiding value to the development of conservation policy, but their methods were not extended to other parts of the country in any ambitious sponsored surveys in the years immediately following. Despite renewed attempts to commend traditional phytosociology, too (Moore 1962), the attraction of this whole approach was overwhelmed for many by the heated debates that preoccupied British plant ecologists in the 1960s, on the issues of objectivity in the sampling and sorting of data, and the respective values of classification or ordination as analytical techniques. Others, though, found it perfectly possible to integrate multivariate analysis into phytosociological survey, and demonstrated the advantage of computers for the display and interpretation of ecological data, rather than the simple testing of methodologies (Ivimey-Cook & Proctor 1966). New generations of research students also began to draw inspiration from the Scottish and Irish initiatives by applying phytosociology to the solving of particular descriptive and interpretive problems, such as variation among British calcicolous grasslands (Shimwell 1968*a*), heaths (Bridgewater 1970), rich fens (Wheeler 1975) and salt-marshes (Adam 1976), the vegetation of Skye (Birks 1969), and Cornish cliffs (Malloch 1970) and Upper Teesdale (Bradshaw & Jones 1976). Meanwhile, too, workers at the Macaulay Institute in Aberdeen had been extending the survey of Scottish vegetation to the lowlands and the Southern Uplands (Birse & Robertson 1976, Birse 1980, 1984).

With an accumulating volume of such data and the appearance of uncoordinated phytosociological perspectives on different kinds of British vegetation, the need for an overall framework of classification became ever more pressing. For some, it was also an increasingly urgent concern that it still proved impossible to integrate a wide variety of ecological research on plants within a generally accepted understanding of their vegetational context in this country. Dr Derek Ratcliffe, as Scientific Assessor of the Nature Conservancy's Reserves Review from the end of 1966, had encountered the problem of the lack of any comprehensive classification of British vegetation types on which to base a systematic selection of habitats for conservation. This same limitation was recognised by Professor Sir Harry Godwin, Professor Donald Pigott and Dr John Phillipson who, as members of the Nature Conservancy, had been asked to read and comment on the Reserves Review. The published version, *A Nature Conservation Review* (Ratcliffe 1977), was able to base the description of only the lowland and upland grasslands and heaths on a phytosociological treatment. In 1971, Dr Ratcliffe, then Deputy Director (Scientific) of the Nature Conservancy, in proposals for development of its research programme, drew attention to 'the need for a national and systematic phytosociological treatment of British vegetation, using standard methods in the field and in analysis/classification of the data'. The intention of setting up a group to examine the issue lapsed through the splitting of the Conservancy which was announced by the Government in 1972. Meanwhile after discussions with Dr Ratcliffe, Professor Donald Pigott of the University of Lancaster proposed to the Nature Conservancy a programme of research to provide a systematic and comprehensive classification of British plant communities. The new Nature Conservancy Council included it as a priority

item within its proposed commissioned research programme. At its meeting on 24 March 1974, the Council of the British Ecological Society welcomed the proposal. Professor Pigott and Dr Andrew Malloch submitted specific plans for the project and a contract was awarded to Lancaster University, with sub-contractual arrangements with the Universities of Cambridge, Exeter and Manchester, with whom it was intended to share the early stages of the work. A coordinating panel was set up, jointly chaired by Professor Pigott and Dr Ratcliffe, and with research supervisors from the academic staff of the four universities, Drs John Birks, Michael Proctor and David Shimwell joining Dr Malloch. At a later stage, Dr Tim Bines replaced Dr Ratcliffe as nominated officer for the NCC, and Miss Lynne Farrell succeeded him in 1985.

With the appointment of Dr John Rodwell as full-time coordinator of the project, based at Lancaster, the National Vegetation Classification began its work officially in August 1975. Shortly afterwards, four full-time research assistants took up their posts, one based at each of the universities: Mr Martin Wigginton, Miss Jacqueline Paice (later Huntley), Mr Paul Wilkins and Dr Elaine Grindey (later Radford). These remained with the project until the close of the first stage of the work in 1980, sharing with the coordinator the tasks of data collection and analysis in different regions of the country, and beginning to prepare preliminary accounts of the major vegetation types. Drs Michael Lock and Hilary Birks and Miss Katherine Hearn were also able to join the research team for short periods of time. After the departure of the research assistants, the supervisors supplied Dr Rodwell with material for writing the final accounts of the plant communities and their integration within an overall framework. With the completion of this charge in 1989, the handover of the manuscript for publication by the Cambridge University Press began.

The scope and methods of data collection

The contract brief required the production of a classification with standardised descriptions of named and systematically arranged vegetation types and, from the beginning, this was conceived as something much more than an annotated list of interesting and unusual plant communities. It was to be comprehensive in its coverage, taking in the whole of Great Britain but not Northern Ireland, and including vegetation from all natural, semi-natural and major artificial habitats. Around the maritime fringe, interest was to extend to the start of the truly marine zone, and from there to the tops of our remotest mountains, covering virtually all terrestrial plant communities and those of brackish and fresh waters, except where non-vascular plants were the dominants. Only short-term leys were specifically excluded, and, though care was to be taken to sample more pristine and long-established kinds of vegetation, no undue attention was to be given to assemblages of rare plants or to especially rich and varied sites. Thus widespread and dull communities from improved pastures, plantations, run-down mires and neglected heaths were to be extensively sampled, together with the vegetation of paths, verges and recreational swards, walls, man-made waterways and industrial and urban wasteland.

For some vegetation types, we hoped that we might be able to make use, from early on, of existing studies, where these had produced data compatible in style and quality with the requirements of the project. The contract envisaged the abstraction and collation of such material from both published and unpublished sources, and discussions with other workers involved in vegetation survey, so that we could ascertain the precise extent and character of existing coverage and plan our own sampling accordingly. Systematic searches of the literature and research reports revealed many data that we could use in some way and, with scarcely a single exception, the originators of such material allowed us unhindered access to it. Apart from the very few classic phytosociological accounts, the most important sources proved to be postgraduate theses, some of which had already amassed very comprehensive sets of samples of certain kinds of vegetation or from particular areas, and these we were generously permitted to incorporate directly.

Figure 1. Standard NVC sample card.

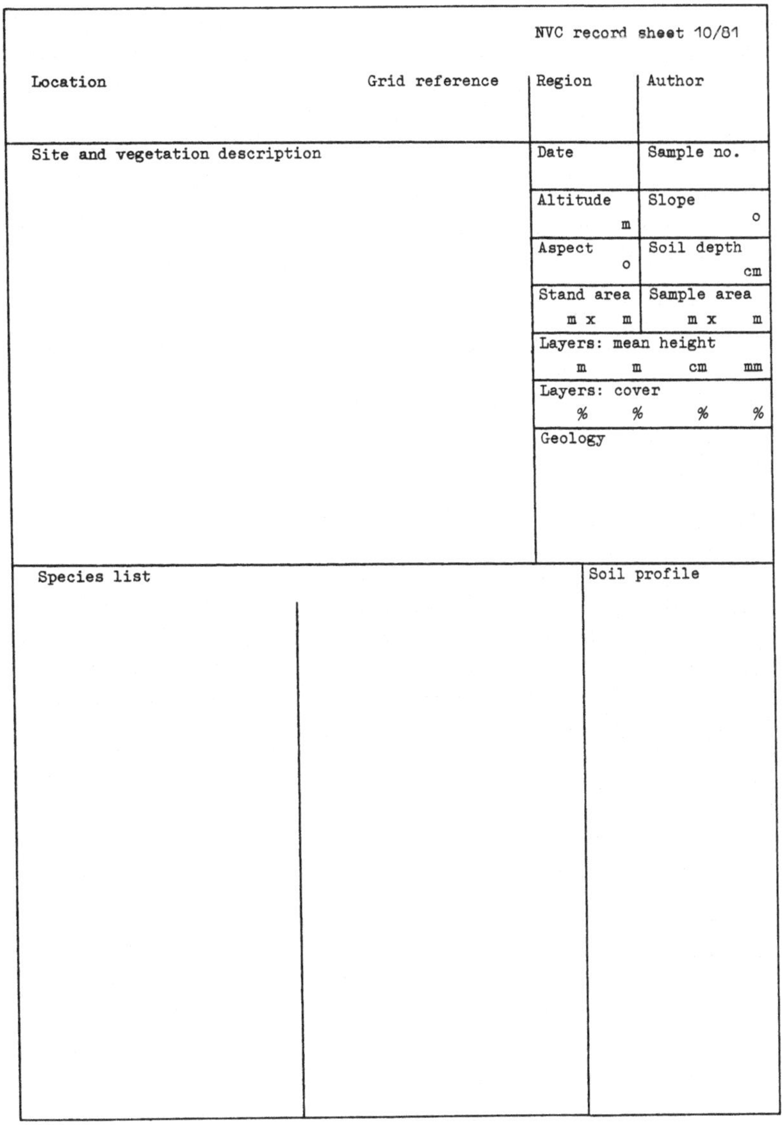
NVC record sheet 10/81
Location
Grid reference
Region
Author
Site and vegetation description
Date
Sample no.
Altitude m
Slope o
Aspect o
Soil depth cm
Stand area m x m
Sample area m x m
Layers: mean height m m cm mm
Layers: cover % % % %
Geology
Species list
Soil profile

Then, from the NCC and some other government agencies, or from individuals who had been engaged in earlier contracts for them, there were some generally smaller bodies of data, occasionally from reports of extensive surveys, more usually from investigations of localised areas. Published papers on particular localities, vegetation types or individual species also provided small numbers of samples. In addition to these sources, the project was able to benefit from and influence ongoing studies by institutions and individuals, and itself to stimulate new work with a similar kind of approach among university researchers, NCC surveyors, local flora recorders and a few suitably qualified amateurs. An initial assessment and annual monitoring of floristic and geographical coverage were designed to ensure that the accumulating data were fairly evenly spread, fully representative of - the range of British vegetation, and of a consistently high quality. Full details of the sources of the material, and our acknowledgements of help, are given in the preface and introduction to each volume.

Our own approach to data collection was simple and pragmatic, and a brief period of training at the outset ensured standardisation among the team of five staff who were to carry out the bulk of the sampling for the project in the field seasons of the first four years, 1976–9. The thrust of the approach was phytosociological in its emphasis on the systematic recording of floristic information from stands of vegetation, though these were chosen solely on the basis of their relative homogeneity in composition and structure. Such selection took a little practice, but it was not nearly so difficult as some critics of this approach imply, even in complex vegetation, and not at all mysterious. Thus, crucial guidelines were to avoid obvious vegetation boundaries or unrepresentative floristic or physiognomic features. No prior judgements were necessary about the identity of the vegetation type, nor were stands ever selected because of the presence of species thought characteristic for one reason or another, nor by virtue of any observed uniformity of the environmental context.

From within such homogeneous stands of vegetation, the data were recorded in quadrats, generally square unless the peculiar shape of stands dictated otherwise. A relatively small number of possible sample sizes was used, determined not by any calculation of minimal areas, but by the experienced assessment of their appropriateness to the range of structural scale found among our plant communities. Thus plots of 2×2 m were used for most short, herbaceous vegetation and dwarf-shrub heaths, 4×4 m for taller or more open herb communities, sub-shrub heaths and low woodland field layers, 10×10 m for species-poor or very tall herbaceous vegetation or woodland field layers and dense scrub, and 50×50 m for sparse scrub, and woodland canopy and understorey. Linear vegetation, like that in streams and ditches, on walls or from hedgerow field layers, was sampled in 10 m strips, with 30 m strips for hedgerow shrubs and trees. Quadrats of 1×1 m were rejected as being generally inadequate for representative sampling, although some bodies of existing data were used where this, or other sizes different from our own, had been employed. Stands smaller than the relevant sample size were recorded in their entirety, and mosaics were treated as a single vegetation type where they were repeatedly encountered in the same form, or where their scale made it quite impossible to sample their elements separately.

Samples from all different kinds of vegetation were recorded on identical sheets (Figure 1). Priority was always given to the accurate scoring of all vascular plants, bryophytes and macrolichens (*sensu* Dahl 1968), a task which often required assiduous searching in dense and complex vegetation, and the determination of difficult plants in the laboratory or with the help of referees. Critical taxa were treated in as much detail as possible though, with the urgency of sampling, certain groups, like the brambles, hawkweeds, eyebrights and dandelions, often defeated us, and some awkward bryophytes and crusts of lichen squamules had to be referred to just a genus. It is more than likely, too, that some very diminutive mosses and especially hepatics escaped notice in the field and, with much sampling taking place in summer, winter annuals and vernal perennials might have been missed on occasion. In general, nomenclature for vascular plants follows *Flora Europaea* (Tutin *et al.* 1964 *et seq.*) with Corley & Hill (1981) providing the authority for bryophytes and Dahl (1968) for lichens. Any exceptions to this, and details of any difficulties with sampling or identifying particular plants, are given in the introductions to each of the major vegetation types.

A quantitative measure of the abundance of every taxon was recorded using the Domin scale (*sensu* Dahl & Hadač 1941), cover being assessed by eye as a vertical projection on to the ground of all the live, above-ground parts of the plants in the quadrat. On this scale:

Cover of 91–100% is recorded as Domin	10
76–90%	9
51–75%	8
34–50%	7
26–33%	6
11–25%	5
4–10%	4
<4% { with many individuals	3
with several individuals	2
with few individuals	1

In heaths, and more especially in woodlands, where the vegetation was obviously layered, the species in the different elements were listed separately as part of the same sample, and any different generations of seedling or saplings distinguished. A record was made of the total cover and height of the layers, together with the cover of any bare soil, litter, bare rock or open water. Where existing data had been collected using percentage cover or the Braun-Blanquet scale (Braun-Blanquet 1928), it was possible to convert the abundance values to the Domin scale, but we had to reject all samples where DAFOR scoring had been used, because of the inherent confusion within this scale of abundance and frequency.

Each sample was numbered and its location noted using a site name and full grid reference. Altitude was estimated in metres from the Ordnance Survey 1:50 000 series maps, slope estimated by eye or measured using a hand level to the nearest degree, and aspect measured to the nearest degree using a compass. For terrestrial samples, soil depth was measured in centimetres using a probe, and in many cases a soil pit was dug sufficient to allocate the profile to a major soil group (*sensu* Avery 1980). From such profiles, a superficial soil sample was removed for pH determination as soon as possible thereafter using an electric meter on a 1:5 soil:water paste. With aquatic vegetation, water depth was measured in centimetres wherever possible, and some indication of the character of the bottom noted. Details of bedrock and superficial geology were obtained from Geological Survey maps and by field observation.

This basic information was supplemented by notes, with sketches and diagrams where appropriate, on any aspects of the vegetation and the habitat thought likely to help with interpretation of the data. In many cases, for example, the quantitative records for the species were filled out by details of the growth form and patterns of dominance among the plants, and an indication of how they related structurally one to another in finely organised layers, mosaics or phenological sequences within the vegetation. Then, there was often valuable information about the environment to be gained by simple observation of the gross landscape or microrelief, the drainage pattern, signs of erosion or deposition and patterning among rock outcrops, talus slopes or stony soils. Often, too, there were indications of biotic effects including treatments of the vegetation by man, with evidence of grazing or browsing, trampling, dunging, mowing, timber extraction or amenity use. Sometimes, it was possible to detect obvious signs of ongoing change in the vegetation, natural cycles of senescence and regeneration among the plants, or successional shifts consequent upon invasion or particular environmental impacts. In many cases, also, the spatial relationships between the stand and neighbouring vegetation types were highly informative and, where a number of samples were taken from an especially varied or complex site, it often proved useful to draw a map indicating how the various elements in the pattern were interrelated.

The approach to data analysis

At the close of the programme of data collection, we had assembled, through the efforts of the survey team and by the generosity of others, a total of about 35 000 samples of the same basic type, originating from more than 80% of the 10 × 10 km grid squares of the British mainland and many islands (Figure 2). Thereafter began a coordinated phase of data processing, with each of the four universities taking responsibility for producing preliminary analyses from data sets crudely separated into major vegetation types – mires, calcicolous grasslands, sand-dunes and so on – and liaising with the others where there was a shared interest. We were briefed in the contract to produce accounts of discrete plant communities which could be named and mapped, so our attention was naturally concentrated on techniques of multivariate classification, with the help of computers to sort the very numerous and often complex samples on the basis of their similarity. We were concerned to employ reputable methods of analysis, but the considerable experience of the team in this kind of work led us to resolve at the outset to concentrate on the ecological integrity of the results, rather than on the minutiae of mathematical technique. In fact, each centre was free to

Figure 2. Distribution of samples available for analysis.

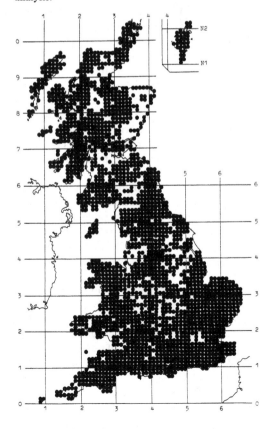

some extent to make its own contribution to the development of computer programs for the task, Exeter concentrating on Association and Information Analysis (Ivimey-Cook *et al.* 1975), Cambridge and Manchester on cluster analysis (Huntley *et al.* 1981), Lancaster on Indicator Species Analysis, later Twinspan (Hill *et al.* 1975, Hill 1979), a technique which came to form the core of the VESPAN package, designed, using the experience of the project, to be particularly appropriate for this kind of vegetation survey (Malloch 1988).

Throughout this phase of the work, however, we had some important guiding principles. First, this was to be a new classification, and not an attempt to employ computational analysis to fit groups of samples to some existing scheme, whether phytosociological or otherwise. Second, we were to produce a classification of vegetation types, not of habitats, so only the quantitative floristic records were used to test for similarity between the samples, and not any of the environmental information: this would be reserved, rather, to provide one valuable correlative check on the ecological meaning of the sample groups. Third, no samples were to be rejected at the outset because they appeared nondescript or troublesome, nor removed during the course of analysis or data presentation where they seemed to confuse an otherwise crisply-defined result. Fourth, though, there was to be no slavish adherence to the products of single analyses using arbitrary cut-off points when convenient numbers of end-groups had been produced. In fact, the whole scheme was to be the outcome of many rounds of sorting, with data being pooled and reanalysed repeatedly until optimum stability and sense were achieved within each of the major vegetation types. An important part of the coordination at this stage was to ensure roughly comparable scales of definition among the emerging classifications and to mesh together the work of the separate centres so as to avoid any omissions in the processing or wasteful overlaps.

With the departure from the team of the four research assistants in 1980, the academic supervisors were left to continue the preparation of the preliminary accounts of the vegetation types for the coordinator to bring to completion and integrate into a coherent whole. Throughout the periods of field work and data analysis, we had all been conscious of the charge in the contract that the whole project must gain wide support among ecologists with different attitudes to the descriptive analysis of vegetation. Great efforts were therefore made to establish a regular exchange of information and ideas through the production of progress reports, which gained a wide circulation in Britain and overseas, via contacts with NCC staff and those of other research agencies, and the giving of papers at scientific meetings. This meant that, as we approached the presentation of

the results of the project, we were well informed about the needs of prospective users, and in a good position to offer that balance of concise terminology and broadly-based description that the NCC considered would commend the work, not only to their own personnel, but to others engaged in the assessment and management of vegetation, to plant and animal ecologists in universities and colleges, and to those concerned with land use and planning.

The style of presentation

The presentation of our results gives priority to the definition of the vegetation types, rather than to the construction of a hierarchical classification. We have striven to characterise the basic units of the scheme on roughly the same scale as a Braun-Blanquet association, but these have been ordered finally not by any rigid adherence to the higher phytosociological categories of alliance, order and class, but in sections akin to the formations long familiar to British ecologists. In some respects, this is a more untidy arrangement, and even those who find the general approach congenial may be surprised to discover what they have always considered to be, say, a heath, grouped here among the mires, or to search in vain for what they are used to calling 'marsh'. The five volumes of the work gather the major vegetation types into what seem like sensible combinations and provide introductions to the range of communities included: aquatic vegetation, swamps and tall-herb fens; grasslands and montane vegetation; heaths and mires; woodlands and scrub; salt-marsh, sand-dune and sea-cliff communities and weed vegetation. The order of appearance of the volumes, however, reflects more the exigencies of publishing than any ecological viewpoint.

The bulk of the material in the volumes comprises the descriptions of the vegetation types. After much consideration, we decided to call the basic units of the scheme by the rather non-committal term 'community', using 'sub-community' for the first-order sub-groups which could often be distinguished within these, and 'variant' in those very exceptional cases where we have defined a further tier of variation below this. We have also refrained from erecting any novel scheme of complicated nomenclature for the vegetation types, invoking existing names where there is an undisputed phytosociological synonym already in widespread use, but generally using the latin names of one, two or occasionally three of the most frequent species. Among the mesotrophic swards, for example, we have distinguished a *Centaurea nigra-Cynosurus cristatus* grassland, which is fairly obviously identical to what Braun-Blanquet & Tüxen (1952) called *Centaureo-Cynosuretum cristati*, and within which, from our data, we have characterised three sub-communities. For the convenience of shorthand description and mapping, every vegetation type has been given a code letter and number, so that *Centaurea-Cynosurus* grassland for example is MG5, MG referring to its place among the mesotrophic grasslands. The *Galium verum* sub-community of this vegetation type, the second to be distinguished within the description, is thus MG5b.

Vegetation being as variable as it is, it is sometimes expedient to allocate a sample to a community even though the name species are themselves absent. What defines a community as unique are rarely just the plants used to name it, but the particular combination of frequency and abundance values for all the species found in the samples. It is this information which is presented in summary form in the floristic tables for each of the communities in the scheme. Figure 3, for example, shows such a table for MG5 *Centaurea-Cynosurus* grassland. Like all the tables in the volumes, it includes such vascular plants, bryophytes and lichens as occur with a frequency of 5% or more in any one of the sub-communities (or, for vegetation types with no sub-communities, in the community as a whole). Early tests showed that records of species below this level of frequency could be largely considered as noise, but cutting off at any higher level meant that valuable floristic information was lost. The vascular species are not separated from the cryptogams on the table though, for woodlands and scrub, the vegetation is sufficiently complex for it to be sensible to tabulate the species in a way which reflects the layered structure.

Every table has the frequency and abundance values arranged in columns for the species. Here, 'frequency' refers to how often a plant is found on moving from one sample of the vegetation to the next, irrespective of how much of that species is present in each sample. This is summarised in the tables as classes denoted by the Roman numerals I to V: 1–20% frequency (that is, up to one sample in five) = I, 21–40% = II, 41–60% = III, 61–80% = IV and 81–100% = V. We have followed the usual phytosociological convention of referring to species of frequency classes IV and V in a particular community as its constants, and in the text usually refer to those of class III as common or frequent species, of class II as occasional and of class I as scarce. The term 'abundance' on the other hand, is used to describe how much of a plant is present in a sample, irrespective of how frequent or rare it is among the samples, and it is summarised on the tables as bracketed numbers for the Domin ranges, and denoted in the text using terms such as dominant, abundant, plentiful and sparse. Where there are sub-communities, as in this case, the data for these are listed first, with a final column summarising the records for the community as a whole.

The species are arranged in blocks according to their pattern of occurrence among the different sub-communities and within these blocks are generally ordered by decreasing frequency. The first group, *Festuca rubra* to *Trifolium pratense* in this case, is made up of the community constants, that is those species which have an overall frequency IV or V. Generally speaking, such plants tend to maintain their high frequency in each of the sub-communities, though there may be some measure of variation in their representation from one to the next: here, for example, *Plantago lanceolata* is somewhat less common in the last sub-community than the first two, with *Holcus lanatus* and a number of others showing the reverse pattern. More often, there are considerable differences in the abundance of these most frequent species: many of the constants can have very high covers, while others are more consistently sparse, and plants which are not constant can sometimes be numbered among the dominants.

The last group of species on a table, *Ranunculus acris* to *Festuca arundinacea* here, lists the general associates of the community, sometimes referred to as companions. These are plants which occur in the community as a whole with frequencies of III or less, though sometimes they rise to constancy in one or other of the sub-communities, as with *R. acris* in this vegetation. Certain of the companions are consistently common overall like *Rumex acetosa*, some are more occasional throughout as with *Rhinanthus minor*, some are always scarce, for example *Calliergon cuspidatum*. Others, though, are more unevenly represented, like *R. acris*, *Heracleum sphondylium* or *Poa trivialis*, though they do not show any marked affiliation to any particular sub-community. Again, there can be marked variation in the abundance of these associates: *Rumex acetosa*, for example, though quite frequent, is usually of low cover, while *Arrhenatherum elatius* and some of the bryophytes, though more occasional, can be patchily abundant; *Alchemilla xanthochlora* is both uncommon among the samples and sparse within them.

The intervening blocks comprise those species which are distinctly more frequent within one or more of the sub-communities than the others, plants which are referred to as preferential, or differential where their affiliation is more exclusive. For example, the group *Lolium perenne* to *Juncus inflexus* is particularly characteristic of the first sub-community of *Centaurea-Cynosurus* grassland, although some species, like *Leucanthemum vulgare* and, even more so, *Lathyrus pratensis*, are more strongly preferential than others, such as *Lolium*, which continues to be frequent in the second sub-community. Even uncommon plants can be good preferentials, as with *Festuca pratensis* here: it is not often found in *Centaurea-Cynosurus* grassland but, when it does occur, it is generally in this first sub-type.

The species group *Galium verum* to *Festuca ovina* helps to distinguish the second sub-community from the first, though again there is some variation in the strength

Floristic table MG5

	a	b	c	MG5
Festuca rubra	V (1–8)	V (2–8)	V (2–7)	V (1–8)
Cynosurus cristatus	V (1–8)	V (1–7)	V (1–7)	V (1–8)
Lotus corniculatus	V (1–7)	V (1–5)	V (2–4)	V (1–7)
Plantago lanceolata	V (1–5)	V (1–5)	IV (1–4)	V (1–7)
Holcus lanatus	IV (1–6)	IV (1–6)	V (1–5)	IV (1–6)
Dactylis glomerata	IV (1–7)	IV (1–6)	V (1–6)	IV (1–7)
Trifolium repens	IV (1–9)	IV (1–6)	V (1–4)	IV (1–9)
Centaurea nigra	IV (1–5)	IV (1–4)	V (2–4)	IV (1–5)
Agrostis capillaris	IV (1–7)	IV (1–7)	V (3–8)	IV (1–8)
Anthoxanthum odoratum	IV (1–7)	IV (1–8)	IV (1–8)	IV (1–8)
Trifolium pratense	IV (1–5)	IV (1–4)	IV (1–3)	IV (1–5)
Lolium perenne	IV (1–8)	III (1–7)	I (2–3)	III (1–8)
Bellis perennis	III (1–7)	II (1–7)	I (4)	II (1–7)
Lathyrus pratensis	III (1–5)	I (1–3)	I (1)	II (1–5)
Leucanthemum vulgare	III (1–3)	I (1–3)	II (1–3)	II (1–3)
Festuca pratensis	II (1–5)	I (2–5)	I (1)	I (1–5)
Knautia arvensis	I (4)			I (4)
Juncus inflexus	I (3–5)			I (3–5)
Galium verum	I (1–6)	V (1–6)		II (1–6)
Trisetum flavescens	II (1–6)	IV (1–6)	II (1–3)	III (1–6)
Achillea millefolium	III (1–6)	V (1–4)	III (1–4)	III (1–6)
Carex flacca	I (1–4)	II (1–4)	I (1)	I (1–4)
Sanguisorba minor	I (4)	II (3–5)		I (3–5)
Koeleria macrantha	I (1)	II (1–6)		I (1–6)
Agrostis stolonifera	I (1–7)	II (1–6)	I (6)	I (1–7)
Festuca ovina		II (1–6)		I (1–6)
Prunella vulgaris	III (1–4)	III (1–4)	IV (1–3)	III (1–4)
Leontodon autumnalis	II (1–5)	II (1–3)	IV (1–4)	III (1–5)
Luzula campestris	II (1–4)	II (1–6)	IV (1–4)	III (1–6)
Danthonia decumbens	I (2–5)	I (1–3)	V (2–5)	I (1–5)
Potentilla erecta	I (1–4)	I (3)	V (1–4)	I (1–4)
Succisa pratensis	I (1–4)	I (1–5)	V (1–4)	I (1–5)
Pimpinella saxifraga	I (1–4)	I (1–4)	III (1–4)	I (1–4)
Stachys betonica	I (1–5)	I (1–4)	III (1–4)	I (1–5)
Carex caryophyllea	I (1–4)	I (1–3)	II (1–2)	I (1–4)
Conopodium majus	I (1–4)	I (1–5)	II (2–3)	I (1–5)
Ranunculus acris	IV (1–4)	II (1–4)	IV (2–4)	III (1–4)
Rumex acetosa	III (1–4)	III (1–4)	III (1–3)	III (1–4)
Hypochoeris radicata	III (1–5)	II (2–4)	III (1–4)	III (1–5)
Ranunculus bulbosus	III (1–7)	II (1–5)	III (1–2)	III (1–7)
Taraxacum officinale agg.	III (1–4)	III (1–4)	III (1–3)	III (1–4)
Brachythecium rutabulum	II (1–6)	III (1–4)	II (2)	III (1–6)
Cerastium fontanum	III (1–3)	II (1–3)	II (1–3)	II (1–3)
Leontodon hispidus	II (1–6)	III (2–4)	III (1–5)	II (1–6)
Rhinanthus minor	II (1–5)	II (1–4)	II (1–3)	II (1–5)
Briza media	II (1–6)	III (1–4)	III (2–3)	II (1–6)
Heracleum sphondylium	II (1–5)	II (1–3)	III (1–3)	II (1–5)
Trifolium dubium	II (1–8)	II (1–5)	I (2)	II (1–8)
Primula veris	II (1–4)	II (2–4)	I (2)	II (1–4)
Arrhenatherum elatius	II (1–6)	II (1–7)	I (3–4)	II (1–7)
Cirsium arvense	II (1–3)	II (1–4)	I (1)	II (1–4)
Eurhynchium praelongum	II (1–5)	II (1–4)	I (1–2)	II (1–5)
Rhytidiadelphus squarrosus	II (1–7)	II (1–5)	III (1–4)	II (1–7)
Poa pratensis	II (1–6)	II (2–5)		II (1–6)
Poa trivialis	II (1–8)	I (1–3)	I (1–2)	II (1–8)
Veronica chamaedrys	II (1–4)	I (1–4)	I (1)	II (1–4)
Alopecurus pratensis	I (1–6)	I (1–4)	I (1)	I (1–6)
Cardamine pratensis	I (1–3)	I (1)	I (3)	I (1–3)
Vicia cracca	I (1–4)	I (1–3)	I (1–2)	I (1–4)
Bromus hordeaceus hordeaceus	I (1–6)	I (2–3)	I (3)	I (1–6)
Phleum pratense pratense	I (1–6)	I (1–5)	I (1)	I (1–6)
Juncus effusus	I (2–3)	I (3)	I (1–2)	I (1–3)
Phleum pratense bertolonii	I (1–3)	I (1–3)	I (1)	I (1–3)
Calliergon cuspidatum	I (1–5)	I (2–4)	I (3)	I (1–5)
Ranunculus repens	II (1–7)	I (2)	II (1–4)	I (1–7)
Pseudoscleropodium purum	I (1–5)	I (3–4)	II (2)	I (1–5)
Ophioglossum vulgatum	I (1–5)	I (1)		I (1–5)
Silaum silaus	I (1–5)	I (1–3)		I (1–5)
Agrimonia eupatoria	I (1–5)	I (1–3)		I (1–5)
Avenula pubescens	I (1–3)	I (2–5)		I (1–5)
Plantago media	I (1–4)	I (1–4)		I (1–4)
Alchemilla glabra	I (2)	I (3)		I (2–3)
Alchemilla filicaulis vestita	I (1–3)	I (3)		I (1–3)
Alchemilla xanthochlora	I (1–3)	I (2)		I (1–3)
Carex panicea	I (1–4)	I (2–4)		I (1–4)
Colchicum autumnale	I (3–4)	I (1–3)		I (1–4)
Crepis capillaris	I (1–5)	I (3)		I (1–5)
Festuca arundinacea	I (1–5)	I (3–5)		I (1–5)

Figure 3. Floristic table for NVC community MG5 *Centaurea nigra-Cynosurus cristatus* grassland.

of association between these preferentials and the vegetation type, with *Achillea millefolium* being less markedly diagnostic than *Trisetum flavescens* and, particularly, *G. verum*. There are also important negative
features, too, because, although some plants typical of
the first and third sub-communities, such as *Lolium* and
Prunella vulgaris, remain quite common here, the disappearance of others, like *Lathyrus pratensis*, *Danthonia
decumbens*, *Potentilla erecta* and *Succisa pratensis* is
strongly diagnostic. Similarly, with the third sub-
community, there is that same mixture of positive and
negative characteristics, and there is, among all the
groups of preferentials, that same variation in abundance as is found among the constants and companions.
Thus, some plants which can be very marked preferentials are always of rather low cover, as with *Prunella*,
whereas others, like *Agrostis stolonifera*, though diagnostic at low frequency, can be locally plentiful.

For the naming of the sub-communities, we have
generally used the most strongly preferential species, not
necessarily those most frequent in the vegetation type.
Sometimes, sub-communities are characterised by no
floristic features over and above those of the community
as a whole, in which case there will be no block of
preferentials on the table. Usually, such vegetation types
have been called Typical, although we have tried to
avoid this epithet where the sub-community has a very
restricted or eccentric distribution.

The tables organise and summarise the floristic variation which we encountered in the vegetation sampled:
the text of the community accounts attempts to expound
and interpret it in a standardised descriptive format. For
each community, there is first a synonymy section which
lists those names applied to that particular kind of
vegetation where it has figured in some form or another
in previous surveys, together with the name of the
author and the date of ascription. The list is arranged
chronologically, and it includes references to important
unpublished studies and to accounts of Irish and Continental associations where these are obviously very similar. It is important to realise that very many synonyms
are inexact, our communities corresponding to just part
of a previously described vegetation type, in which case
the initials *p.p.* (for *pro parte*) follow the name, or being
subsumed within an older, more broadly-defined unit.
Despite this complexity, however, we hope that this
section, together with that on the affinities of the vegetation (see below), will help readers translate our scheme
into terms with which they may have been long familiar.
A special attempt has been made to indicate correspondence with popular existing schemes and to make sense
of venerable but ill-defined terms like 'herb-rich meadow', 'oakwood' or 'general salt-marsh'.

There then follow a list of the constant species of the
community, and a list of the rare vascular plants,

bryophytes and lichens which have been encountered in
the particular vegetation type, or which are reliably
known to occur in it. In this context, 'rare' means, for
vascular plants, an A rating in the *Atlas of the British
Flora* (Perring & Walters 1962), where scarcity is measured by occurrence in vice-counties, or inclusion on
lists compiled by the NCC of plants found in less than
100 10 × 10 km squares. For bryophytes, recorded presence in under 20 vice-counties has been used as a
criterion (Corley & Hill 1981), with a necessarily more
subjective estimate for lichens.

The first substantial section of text in each community
description is an account of the physiognomy, which
attempts to communicate the feel of the vegetation in a
way which a tabulation of data can never do. Thus, the
patterns of frequency and abundance of the different
species which characterise the community are here filled
out by details of the appearance and structure, variation
in dominance and the growth form of the prominent
elements of the vegetation, the physiognomic contribution of subordinate plants, and how all these components relate to one another. There is information, too,
on important phenological changes that can affect the
vegetation through the seasons and an indication of the
structural and floristic implications of the progress of
the life cycle of the dominants, any patterns of regeneration within the community or obvious signs of competitive interaction between plants. Much of this material
is based on observations made during sampling, but it
has often been possible to incorporate insights from
previous studies, sometimes as brief interpretive notes,
in other cases as extended treatments of, say, the biology
of particular species such as *Phragmites australis* or
Ammophila arenaria, the phenology of winter annuals or
the demography of turf perennials. We trust that this
will help demonstrate the value of this kind of descriptive classification as a framework for integrating all
manner of autecological studies (Pigott 1984).

Some indication of the range of floristic and structural
variation within each community is given in the discussion of general physiognomy, but where distinct sub-
communities have been recognised these are each given a
descriptive section of their own. The sub-community
name is followed by any synonyms from previous studies, and by a text which concentrates on pointing up the
particular features of composition and organisation
which distinguish it from the other sub-communities.

Passing reference is often made in these portions of
the community accounts to the ways in which the nature
of the vegetation reflects the influence of environmental
factors upon it, but extended treatment of this is
reserved for a section devoted to the habitat. An opening
paragraph here attempts to summarise the typical conditions which favour the development and maintenance
of the vegetation types, and the major factors which

control floristic and structural variation within it. This is followed by as much detail as we have at the present time about the impact of particular climatic, edaphic and biotic variables on the community, or as we suppose to be important to its essential character and distribution. With climate, for example, reference is very frequently made to the influence on the vegetation of the amount and disposition of rainfall through the year, the variation in temperature season by season, differences in cloud cover and sunshine, and how these factors interact in the maintenance of regimes of humidity, drought or frosts. Then, there can be notes of effects attributable to the extent and duration of snow-lie or to the direction and strength of winds, especially where these are icy or salt-laden. In each of these cases, we have tried to draw upon reputable sources of data for interpretation, and to be fully sensitive to the complex operation of topographic climates, where features like aspect and altitude can be of great importance, and of regional patterns, where concepts like continental, oceanic, montane and maritime climates can be of enormous help in understanding vegetation patterns.

Commonly, too, there are interactions between climate and geology that are best perceived in terms of variations in soils. Here again, we have tried to give full weight to the impact of the character of the landscape and its rocks and superficials, their lithology and the ways in which they weather and erode in the processes of pedogenesis. As far as possible, we have employed standardised terminology in the description of soils, trying at least to distinguish the major profile types with which each community is associated, and to draw attention to the influence on its floristics and structure of processes like leaching and podzolisation, gleying and waterlogging, parching, freeze-thaw and solifluction, and inundation by fresh- or salt-waters.

With very many of the communities we have distinguished, it is combinations of climatic and edaphic factors that determine the general character and possible range of the vegetation, but we have often also been able to discern biotic influences, such as the effects of wild herbivores or agents of dispersal, and there are very few instances where the impact of man cannot be seen in the present composition and distribution of the plant communities. Thus, there is frequent reference to the role which treatments such as grazing, mowing and burning have on the floristics and physiognomy of the vegetation, to the influence of manuring and other kinds of eutrophication, of draining and re-seeding for agriculture, of the cropping and planting of trees, of trampling or other disturbance, and of various kinds of recreation.

The amount and quality of the environmental information on which we have been able to draw for interpreting such effects has been very variable. Our own sampling provided just a spare outline of the physical and edaphic conditions at each location, data which we have summarised where appropriate at the foot of the floristic tables; existing sources of samples sometimes offered next to nothing, in other cases very full soil analyses or precise specifications of treatments. In general, we have used what we had, at the risk of great unevenness of understanding, but have tried to bring some shape to the accounts by dealing with the environmental variables in what seems to be their order of importance, irrespective of the amount of detail available, and by pointing up what can already be identified as environmental threats. We have also benefited by being able to draw on the substantial literature on the physiology and reproductive biology of individual species, on the taxonomy and demography of plants, on vegetation history and on farming and forestry techniques. Sometimes, this information provides little more than a provisional substantiation of what must remain for the moment an interpretive hunch. In other cases, it has enabled us to incorporate what amount to small essays on, for example, the past and present role of *Tilia cordata* in our woodlands with variation in climate, the diverse effects of dunging by rabbit, sheep and cattle on calcicolous swards, or the impact of burning on *Calluna-Arctostaphylos* heath on different soils in a boreal climate. Debts of this kind are always acknowledged in the text and, for our part, we hope that the accounts indicate the benefits of being able to locate experimental and historical studies on vegetation within the context of an understanding of plant communities (Pigott 1982).

Mention is often made in the discussion of the habitat of the ways in which stands of communities can show signs of variation in relation to spatial environmental differences, or the beginnings of a response to temporal changes in conditions. Fuller discussion of zonations to other vegetation types follows, with a detailed indication of how shifts in soil, microclimate or treatment affect the composition and structure of each community, and descriptions of the commonest patterns and particularly distinctive ecotones, mosaics and site types in which it and any sub-communities are found. It has also often been possible to give some fuller and more ordered account of the ways in which vegetation types can change through time, with invasion of newly available ground, the progression of communities to maturity, and their regeneration and replacement. Some attempt has been made to identify climax vegetation types and major lines of succession, but we have always been wary of the temptation to extrapolate from spatial patterns to temporal sequences. Once more, we have tried to incorporate the results of existing observational and experimental studies, including some of the classic accounts of patterns and processes among British vegetation, and to point up the great advantages of a reliable scheme of classification as a basis for the monitoring and management of plant communities (Pigott 1977).

Throughout the accounts, we have referred to particular sites and regions wherever we could, many of these visited and sampled by the team, some the location of previous surveys, the results of which we have now been able to redescribe in the terms of the classification we have erected. In this way, we hope that we have begun to make real a scheme which might otherwise remain abstract. We have also tried in the habitat section to provide some indications of how the overall ranges of the vegetation types are determined by environmental conditions. A separate paragraph on distribution summarises what we know of the ranges of the communities and sub-communities, then maps show the location, on the 10×10 km national grid, of the samples that are available to us for each. Much ground, of course, has been thinly covered, and sometimes a dense clustering of samples can reflect intensive sampling rather than locally high frequency of a vegetation type. However, we believe that all the maps we have included are accurate in their general indication of distributions, and we hope that this exercise might encourage the production of a comprehensive atlas of British plant communities.

The last section of each community description considers the floristic affinities of the vegetation types in the scheme, and expands on any particular problems of synonymy with previously described assemblages. Here, too, reference is often given to the equivalent or most closely-related association in continental phytosociological classifications and an attempt made to locate each community in an existing alliance. Where the fuller account of British vegetation that we have been able to provide necessitates a revision of the perspective on European plant communities as a whole, some suggestions are made as to how this might be achieved.

Meanwhile, each reader will bring his or her own needs and commitment to this scheme and perhaps be dismayed by its sheer size and apparent complexity. For those requiring some guidance as to the scope of each volume and the shape of that part of the classification with which it deals, the introductions to the major vegetation types will provide an outline of the variation and how it has been treated. The contents page will then give directions to the particular communities of interest. For readers less sure of the identity of the vegetation types with which they are dealing, a key is provided to each major group of communities which should enable a set of similar samples organised into a constancy table to be taken through a series of questions to a reasonably secure diagnosis. The keys, though, are not infallible short cuts to identification and must be used in conjunction with the floristic tables and community descriptions. An alternative entry to the scheme is provided by the species index which lists the occurrences of all taxa in the communities in which we have recorded them. There is also an index of synonyms which should help readers find the equivalents in our classification of vegetation types already familiar to them.

Finally, we hope that whatever the needs, commitments or even prejudices of those who open these volumes, there will be something here to inform and challenge everyone with an interest in vegetation. We never thought of this work as providing the last word on the classification of British plant communities: indeed, with the limited resources at our disposal, we knew it could offer little more than a first approximation. However, we do feel able to commend the scheme as essentially reliable. We hope that the broad outlines will find wide acceptance and stand the test of time, and that our approach will contribute to setting new standards of vegetation description. At the same time, we have tried to be honest about admitting deficiencies of coverage and recognising much unexplained floristic variation, attempting to make the accounts sufficiently open-textured that new data might be readily incorporated and ecological puzzles clearly seen and pursued. For the classification is meant to be not a static edifice, but a working tool for the description, assessment and study of vegetation. We hope that we have acquitted ourselves of the responsibilities of the contract brief and the expectations of all those who have encouraged us in the task, such that the work might be thought worthy of standing in the tradition of British ecology. Most of all, we trust that our efforts do justice to the vegetation which, for its own sake, deserves understanding and care.

WOODLANDS AND SCRUB

INTRODUCTION TO WOODLANDS AND SCRUB

The sampling of woodland vegetation

Sampling woodland vegetation poses some particular problems. In the first place, woods are often a prominent feature of the landscape and tend to impress themselves upon us in their entirety. Some have a long-established integrity, enshrined in a name and witnessing to a complex history of economic and social recognition and use, but even those woods which are the product of more recent afforestation or neglect can demand attention as a whole. In fact, of course, individual stretches of wooded vegetation are often very varied internally. They can consist of a number of widely-differing woodland communities, and have scrub around their margins, on the edges of clearings and along rides, quite apart from including stands of bracken, heath and grassland, mires and swamps, and aquatic vegetation in pools and flooded ruts. It is very important, in the description and assessment of sites, to be sensitive to the frequently intimate ecological and historical relationships between the vegetation types that make up particular woods, but our priority here is to characterise woodland communities in the strict sense. This volume therefore includes just vegetation in which trees and/or shrubs are dominant, together with closely related underscrub, and readers will have to consult other parts of the work for full accounts of grasslands and heaths and so on which they might encounter within the boundaries of woods. At the same time, frequent reference is made here to such zonations and successions between the various vegetation types that are found in and around woods, and to the different kinds of wooded landscapes characteristic of Britain.

Within woodland communities, a second difficulty is that the range of size among the plants represented is very great, from hepatics, mosses and lichens that can be tiny, through herbs and ferns of various proportions, to shrubs and sometimes enormous trees. On entering a stretch of woodland, it may be clear enough that the plants are organised in a stratified arrangement of one sort or another, approximating to ground and field layers, understorey and canopy, or some variation on this theme, but it is often difficult to take in the extremes of this structural complexity at a single glance, or do ready justice to all its subtleties. Indeed, it is something of a commonplace that ecologists tend to concentrate on looking either upwards or down in woodlands, and to be inclined to give precedence in their understanding of this kind of vegetation to either the field layer or the woody plants: as often, smaller cryptogams get neglected.

More seriously, reasons are sometimes advanced for at least wishing to treat the various structural elements of woodland vegetation separately in exercises of description or classification, because they may show different responses to the environmental conditions at work among them. In particular, ground and field layers are generally supposed to reflect the continuing influence of site conditions like climate and soil, whereas the shrubs and trees have often been more directly affected by sylvicultural treatments. In our view, the relationships between the patterns of floristic and physiognomic variation that we see in the different layers of woodland vegetation, even in more natural situations, are complex, and we know very little, in fact, about the ways in which treatments modify or disrupt any shared responses among the woodland flora to soils and climate. Indeed, we have not so far had any comprehensive indication of the actual range of variation in the different structural elements of woodlands throughout Britain, so we are hardly in a position to assess with authority the meaning of any coincidence of patterning, or lack of it, between the layers. In attempting to provide such information and to begin to make sense of it, we therefore resolved to sample all the layers together.

The problem then is that the eye needs to accommodate to different scales of floristic and physiognomic variation within the layers. Detecting homogeneity among trees and shrubs is not actually all that difficult: it can be much easier than in some grasslands or mires, for example, where the patterning is very fine. However, a lot of ground might have to be covered to assess the extent of diversity within a single tract of woodland canopy and understorey, and samples need to be large enough to represent adequately the scale of uniformity in the selected stands. Beneath such a homogeneous stretch of shrubs and trees, though, the field and ground layers may be much less coarsely structured. In contrast to many continental phytosociologists, therefore, we adopted not a single moderately-sized quadrat for sampling all the layers, but a large plot of 50 × 50 m for the canopy and understorey, and within this either a 4 × 4 m or 10 × 10 m plot for the field and ground layers, depending on their own scale of organisation, the records then being combined to constitute a single sample. Where the field and ground layers were quite diverse beneath a more or less uniform tract of shrubs and trees, the smaller quadrats might be taken from two or more different homogeneous patches, each being combined with the same set of records for the canopy and understorey in the large plot (Figure 4). As usual, if stands were a peculiar shape, as where woodland vegetation occurred in narrow strips in valley bogs or on small alluvial terraces, or in irregular shelterbelts, differently-shaped quadrats of the same area were employed. And, where uniform tracts of trees and shrubs extended over areas of less than 50 × 50 m, as was often the case in wooded flushes, the stand was treated as the sample.

Clearly, in any one sample, tree and shrub species might figure several times in the list, being recorded as constituents of a canopy and in an understorey, as saplings and as seedlings. Climbers and lianes might also appear twice, or even three times where they formed a tangle among the herbs or a ground carpet. To interpret

Figure 4. Sampling from a woodland.
Four homogeneous stands of canopy and understorey have been distinguished and a sample plot laid out in each, 50 × 50 m where possible (A and B) or an identical area of different shape (C) or, with a small patch, the entire stand (D). In each plot, homogeneous areas of field and ground vegetation are delineated, as in B enlarged below, with samples of either 4 × 4 m (a) or 10 × 10 m (b and c).

data such as these, and to convey the full complexity of the woodland organisation, it was often highly informative, too, to supplement the numerical records with observations on the detailed physiognomy of the woody plants and their contribution to the layers, on the occurrence of subsidiary tiers to the canopy or understorey, and on the apparent age-structure and patterns of regeneration, if any. Notes could also be made on the precise relationships of variation in the field and ground layers to that among the trees and shrubs, on such phenological changes as might be evident from single visits, and on zonations to neighbouring vegetation types or apparent successions (Figure 5).

As throughout the survey, therefore, relative homogeneity was the only criterion used to locate samples. More particularly, we did not bias our sampling in favour of ancient woodlands and never used floristic or structural indicators of antiquity to select stands. We were briefed in the contract to look at all British vegetation as it exists at the present time, and younger secondary woods and plantations, including many with non-native broadleaves and conifers, therefore figured prominently. At the same time, every effort was made to be sensitive to the impact of historical processes upon the vegetation being sampled. Thus, the examination of the woodland structure involved noting signs of treatments like coppicing, pollarding and different regimes of timber extraction and planting, or of the neglect of such management; of the impact of grazing and browsing, mowing or burning; of the effects of draining or peat extraction, and so on. Also, though there was not time to consult documentary evidence, the presence of obvious field features of historical or archaeological interest, such as boundary banks and old settlements, was recorded and an attempt made to locate the samples within their landscape context, whether this was semi-natural high forest or old coppice, wood-pasture or heathy common land, small remnants or plantations in an intensive agricultural context or huge conifer forests. In this way, we hoped to be able to assess the effects of treatments against the influence of such natural vari-

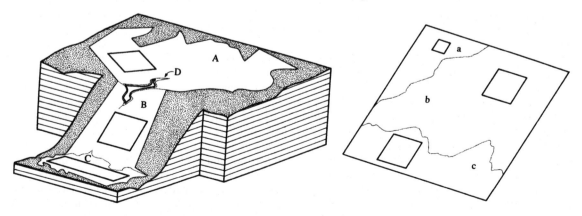

ables, like climate and soils, about which we had some information.

By including scrub and transitions to herbaceous vegetation in the sampling programme, we hoped too that we might bring some order to the understanding of natural successional processes and regeneration, as a background to the discussion of the effects of treatments of woodlands and their seral precursors. Stands of scrub and underscrub were selected as usual by homogeneity, and sampled using 10 × 10 m quadrats for all layers, except where the cover of bushes and saplings was more open, when the usual 50 × 50 m plot was used for the woody plants, and a smaller quadrat for the field and ground layers. We also sampled woodland and scrub where they occurred in the rather particular situation of hedges. Here, to give some uniformity of approach with previous surveys, we used a strip 30 m long for recording the trees and shrubs, and within this a strip 10 m long for the field and ground layer. Special attention was then given in the vegetation description to the way in which the hedge related to adjacent underscrub, grasslands and other communities disposed over the bank, ditch and verge, which might be sampled separately (Figure 6).

Figure 5. Three completed sample cards from woodlands.

NVC record sheet 10/81

Location	Grid reference	Region	Author
Johnny's Wood, Cumbria	NY(35) 250145	NW	MJW

Site and vegetation description: Uneven boulder-strewn slope with rather open oak-birch canopy and scattered rowan, uneven-topped. Understorey thin and patchy with occasional hazel bushes and a few second generation oak and birch, but rowan is regenerating well and sometimes plentiful. Much of ground covered by block scree, some fragments up to 3m across, with extensive carpet of bryophytes on tops & sides. Species marked # make up bulk of cover on bigger stable rocks on block tops, with occasional Deschampsia tufts & Oxalis. Those marked @ can form patches on stone sides & hang down in mats, with others sometimes over steeper and barer surface & in sometimes very humid crevices between boulders. Ferns, sometimes luxuriant, occur on deeper soil between & provide floristic continuity with grassy field layer downslope (See sample 169)

Date	3.9.1977
Sample no.	168
Altitude	155 m
Slope	15°
Aspect	40°
Soil depth	2-10 cm
Stand area	60 m × 100 m
Sample area	50/10 m × 70/10 m
Layers: mean height	14 m, 4 m, 40 cm, 50 mm
Layers: cover	60%, 20%, 60%, 60%
Geology	Borrowdale Volcanics

Species list

Canopy in 50×50m
Betula pubescens 6
Quercus petraea 3
Sorbus aucuparia 4
Understorey in 50×50m
Sorbus aucuparia 1
Betula pubescens 1
Quercus petraea 1
Corylus avellana 3
Field layer in 10×10m
Deschampsia flexuosa 7
Agrostis canina montana 2
Dryopteris filix-mas 5
Dryopteris dilatata 4
Dryopteris borreri 2
Oxalis acetosella 2
Hymenophyllum wilsonii 2
Ground layer in 10×10m
Rhytidiadelphus loreus # 5
Dicranum majus # 5
Hypnum cupressiforme # 4
Hylocomium splendens # 4
Mnium hornum

Isothecium myosuroides #@+ 4
Polytrichum formosum # 2
Pleurozium schreberi # 2
Plagiothecium undulatum # 2
Thuidium tamariscinum # 3
Thuidium delicatulum #@ 2
Hylocomium umbratum @ 2
Diplophyllum albicans @ 3
Lepidozia reptans @ 3
Plagiochila spinulosa @ 2
Scapania gracilis @ 2
Dicranodontium denudatum 3
Plagiochila punctata # 1
Scapania umbrosa 1
Jamesoniella autumnalis @ 2
Bazzania trilobata @ 3
Scapanylum micans # 1
Sphagnum quinquefarium 4
Ptilium crista-castrensis 3
Hypnum callichroum 2
Heterocladium heteropterum 2
Isopterygium elegans 1
Rhizomnium punctatum 1
Bazzania tricrenata 1

Soil profile: Shallow humic ranker over boulders with deeper pockets between. pH 0-2cm = 3.4

Calypogeia fissa 1
Calypogeia neelelorum 1
Lepidozia pearsonii 1
Lophozia ventricosa 1
Mylia taylori 2
Cephalozia media 2
Cephaloziella sp. 1
Marsupella emarginata 1
Cladonia squamosa 1
Cladonia polydactyla 1
Cladonia uncialis 1
bare soil 2
bare litter 2
bare rock 4

NVC record sheet 10/81

Location	Grid reference	Region	Author
Sixpenny Handley, Cranborne Chase, Dorset	ST(31) 976173	SW	MCFP

Site and vegetation description: Neglected stretch of hazel coppice with oak standards & a few other trees and shrubs, including some big ash emergent from the underwood, also scattered maple & willow. Hazel bushes fairly regular in distribution & size but underwood is scrubby in places with dense bramble and there is a scatter of associates including young maple, spindle and, nearby, dogwood. Field layer quite rich and somewhat varied from place to place, but differences are mainly shifts in proportions of Mercurialis, Hyacinthoides & Anemone over the somewhat uneven ground, last tending to thicken up in gentle hollows. Fairly rich epiphytic bryophyte flora on ash including Neckera complanata, N. pumila, Orthotrichum lyellii, O. striatum & O. affine with profuse growth of lichens on upper branches. Area of similar vegetation across road recently coppiced : no standards there but big birch left standing.

Date	2.5.1976
Sample no.	260
Altitude	97 m
Slope	0°
Aspect	-°
Soil depth	>30 cm
Stand area	50 m × 50 m
Sample area	50/50 m × 10/10 m
Layers: mean height	20 m, 4 m, 20 cm, 30 mm
Layers: cover	30%, 20%, 90%, 10%
Geology	Cretaceous + drift. Chalk with drift of Clay-with-Flints

Species list

Canopy in 50×50m
Quercus robur 6
Fraxinus excelsior 2
Acer campestre 1
Salix caprea 1
Understorey in 50×50m
Corylus avellana 8
Crataegus monogyna 4
Rosa arvensis 2
Rosa canina agg. 4
Acer campestre sapling 2
Euonymus europaeus 2
Sorbus aria 1
Field layer in 10×10m
Hyacinthoides non-scripta 4
Mercurialis perennis 7
Deschampsia cespitosa 7
Anemone nemorosa 4
Viola reichenbachiana 3
Viola riviniana 3
Potentilla sterilis 3
Fragaria vesca 2
Conopodium majus 5
Sanicula europaea 3
Lamiastrum galeobdolon 2

Bromus ramosus agg. 5
Melica uniflora 5
Ajuga reptans 3
Brachypodium sylvaticum 2
Poa trivialis 2
Veronica chamaedrys 2
Glechoma hederacea 2
Carex sylvatica 3
Arum maculatum 2
Hedera helix 3
Geum urbanum 2
Luzula pilosa 2
Festuca gigantea 1
Acer campestre seedling 1
Rumex sanguineus 1
Primula vulgaris 2
Ranunculus ficaria 2
Melampyrum pratense 1
Arctium minus agg. 1
Fagus sylvatica seedling 1
Euphorbia amygdaloides 2
Vicia sepium 1
Ground layer in 10×10m
Eurhynchium striatum 3
Brachythecium rutabulum 3
Fissidens taxifolius 2
Eurhynchium praelongum 2

Thuidium tamariscinum 2
Plagiomnium undulatum 1
Atrichum undulatum 1
Cirriphyllum piliferum 1
Thamnium alopecurum 1

Soil profile: Brown Earth, somewhat clayey above but becoming very stony below 20cm and with obvious Chalk fragments below 25cm. pH 0-5cm = 6.7

NVC record sheet 10/81

Location	Grid reference	Region	Author
Pont Burn Woods, Durham	NZ(45) 155653	NE	JR

Site and vegetation description: Irregular-shaped stand of wet woodland below a flush-line with uneven-topped & somewhat open cover of Alnus, many suckering, with occasional Fraxinus & Quercus and, a little lower, scattered sickly Betula. Understorey quite dense with Corylus predominating as large uncoppiced bushes and quite numerous saplings up to 3-4m, then below a scattered cover of Rosa, Rose, Salix and Viburnum. Field layer as here in central wettest area with patchy dominance of Juncus, Carices & Equisetum hyemale. Gives ground to drier Deschampsia cespitosa field layer under some canopy (sample 544), then to oak-birch woodland with Holcus mollis field layer (546). Patches of drier ground around planted with Fagus & occasional conifers. Coal Measure valleys about grits-shale junctions. No obvious signs of grazing or other recent treatments.

Date	3.8.1977
Sample no.	543
Altitude	60 m
Slope	10
Aspect	270°
Soil depth	70 cm
Stand area	30 m × 70 m
Sample area	30/30 m × 4/4 m
Layers: mean height	15 m, 3 m, 70 cm, 25 mm
Layers: cover	85%, 25%, 95%, 20%
Geology	Carboniferous Coal Measure sandstones & shales

Species list

Canopy in 30×70m
Alnus glutinosa 7
Betula pubescens 3
Fraxinus excelsior 6
Betula pendula 2
Quercus petraea 2
Quercus hybrid 4
Understorey in 30×70m
Corylus avellana 5
Alnus glutinosa sapling 2
Crataegus monogyna 2
Fraxinus excelsior sapling 2
Acer pseudoplatanus sapling 2
Prunus padus 2
Rosa canina agg. 2
Rubus idaeus 2
Quercus robur sapling 1
Ribes rubrum 1
Salix cinerea 1
Viburnum opulus 1
Lonicera periclymenum 1
Field layer in 4×4m
Equisetum sylvaticum 4
Athyrium filix-femina 5
Cirsium palustre 3

Equisetum hyemale 1
Filipendula ulmaria 2
Juncus effusus 6
Allium ursinum 4
Viola riviniana 1
Lysimachia nemorum 1
Galium aparine 1
Angelica sylvestris 1
Ribes rubrum seedling 1
Ranunculus repens 5
Deschampsia cespitosa 5
Bromus ramosus 2
Carex remota 4
Carex laevigata 3
Festuca gigantea 1
Ground layer in 4×4m
Mnium hornum 1
Plagiomnium rostratum 1
Plagiomnium undulatum 5
Eurhynchium praelongum 5
Lophocolea bidentata 2
Cirriphyllum piliferum 1
Brachythecium rutabulum 2
Eurhynchium striatum 1
Thuidium tamariscinum 1
bare soil & litter 3

Soil profile: Stagnogley, with deep upper horizon of wet rather structureless silt in centre of flush, becoming clayey below with shale fragments, whole profile tending to slump downslope. pH 0-5cm = 5.7

In addition to the samples provided by our own extensive field work, we were able to make use of some existing data, though less so than for most other vegetation types because little previous work had produced compatible material. However, along with the samples included from the Scottish Highlands by McVean & Ratcliffe (1962), we were especially grateful to have access to data from many parts of the Southern Uplands and Scottish lowlands collected at the Macaulay Institute (Birse & Robertson 1976, Birse 1980, 1982, 1984). Then, studies like those of Birks (1973) on Skye, Graham (1971, 1988) in Durham and Edwards in North Wales (Edwards & Birks 1986) yielded valuable data from more restricted localities, with Wheeler (1975, 1978, 1980c) providing data from lowland mire woodlands. Small sets of samples were sometimes also found in NCC reports (e.g. Meres Report 1980). Whenever incorporating such material, we were careful to check that, as with our own sampling, vernal herbs had not been missed, and that attention had been given to

Figure 6. Sampling from hedges.
Samples are taken from homogeneous stretches of the hedgerow core, with trees and shrubs recorded in a 30 m strip (1a in each example), and of the associated field and ground vegetation, recording in a 10 m strip (1b). Vegetation on adjacent banks, ditches and verges is separately recorded in samples of relevant size (2,3,4).

taxonomic difficulties like distinguishing the two species and their hybrids in oak and birch, something which really is important and informative, and identifying bryophytes and macrolichens, the former in particular being diagnostic of certain woodland types. Even with our own samples, however, we always recorded brambles as *Rubus fruticosus* agg. and very rarely distinguished microspecies of roses. Indeterminate vegetative material of *Viola riviniana* and *V. reichenbachiana* was recorded as *Viola* sp., *Salix cinerea* included ssp. *oleifolia* (formerly *S. atrocinerea*) and *Betula pubescens* included ssp. *carpatica* (formerly ssp. *odorata*), though the presence of this taxon is usually indicated in the text. External data generally recorded Hieracia simply as a group and *Sphagnum capillifolium* sometimes included what is now separated as *S. quinquefarium*. *Lophocolea bidentata sensu lato* may include some records for infertile *L. cuspidata*. In total, about 2800 samples were available for analysis and floristic and geographical coverage across much of the country was good (Figure 7), although scrub vegetation and upland conifer forests are certainly under-represented in relation to their extent and variety.

Data analysis and the description of woodland communities

As usual in our data analysis, only the floristic records for the samples were used to characterise the vegetation

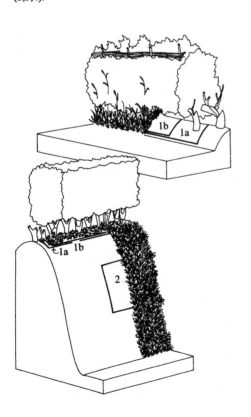

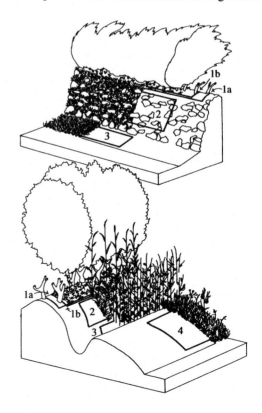

types. The quantitative scores for all vascular plants, bryophytes and lichens were employed and no special weighting was given in the analysis to either trees and shrubs on the one hand, or field- and ground-layer species on the other, or to reputed ancient woodland indicators and other kinds of character species. In general, in the classification that emerged, the communities are defined by assemblages of plants from all the structural elements and the most obvious distinctions that occur among them are not concerned with antiquity. Almost all the communities are sufficiently varied internally to warrant sub-division, but in some cases the sub-communities are distinguished more by differences in the trees and shrubs, while in others this element remains more constant and there is greater variation among the herbs and cryptogams. This can seem rather confusing at first, but we consider it to be an accurate reflection of the complexity of our woodland vegetation.

The floristic tables are presented in the usual style except that, for ease of reading, canopy trees and shrubs are listed first, then those of the understorey, then the field- and ground-layer plants together, different generations of woody plants being distinguished by the suffixes 'sapling' and 'seedling'. For each structural element, the constants, companions and preferentials are then displayed on the tables, and the complexities of composition and physiognomy explained in the text.

In all, 25 communities of woodland and scrub have been characterised. There are six types of mixed deciduous and oak-birch woodland, three kinds of *Fagus sylvatica* woodland, communities dominated by *Taxus baccata*, *Pinus sylvestris* and *Juniperus communis* ssp. *communis* or Arctic-Alpine willows, and seven types of wetter woodland with various mixtures of *Alnus glutinosa*, *Betula pubescens* and Salices. Many communities survive as both ancient and younger stands, and plantations of introduced broadleaf trees and conifers have in all cases been considered as modified forms of these woodland types.

Many of the communities also contain stands which, though young, are still dominated by much the same mixtures of trees and shrubs as are characteristic of mature woodlands, and some, of course, include coppiced versions where shrubs figure prominently in an underwood. However, three separate vegetation types have been characterised from scrubs in which the dominants are species like *Crataegus monogyna*, *Prunus spinosa*, *Sambucus nigra*, Rubi and roses, small calcicolous trees and shrubs, *Ulex europaeus* or *Cytisus scoparius*. Many stands are obviously seral, others occurring in hedges and along woodland margins have become stabilised. Finally, two underscrubs with abundant *Rubus fruticosus* agg. and/or *Pteridium aquilinum* are best included here, though the bulk of our less mesophytic bracken vegetation is described in the volume on grass-

lands and montane communities. One other distinctive kind of scrub with *Hippophae rhamnoides* is included among the sand-dune vegetation.

Mixed deciduous and oak-birch woodlands

Major floristic trends in relation to soils

The most important lines of floristic variation which we detected among the six communities of mixed deciduous and oak-birch woodland are best understood initially in relation to differences in soils. Two kinds of woodland are associated with rendzinas and brown calcareous earths, not always of high surface pH, but certainly base-rich and calcareous below, and with a tendency to develop mull humus except where the profiles are maintained in a permanently immature state. To the opposite extreme is a pair of communities characteristic of acid rankers, brown podzolic soils and podzols with mor humus. Between, there are two types of woodland typically found on brown earths of low base-status, frequently with moder. A significant complicating factor is that many of the intermediate and base-rich brown soils are subject to gleying, sometimes because of inherently poor drainage where the profiles are heavy-

Figure 7. Distribution of samples available from woodlands.

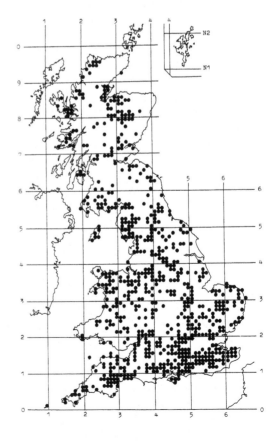

textured, in other cases by virtue of flushing from springs or streams (Figure 8).

Floristic responses to this pattern of edaphic variation are complex: they differ somewhat in the lowland south and east and the sub-montane north and west and, particularly among the woody species, are often masked by the effects of sylvicultural treatments. However, the fundamental outline is clear enough. To the more base-rich extreme, the *Fraxinus-Acer-Mercurialis* (W8) and *Fraxinus-Sorbus-Mercurialis* (W9) woodlands have *Mercurialis perennis* as a strongly preferential herb with other calcicoles like *Geum urbanum*, *Circaea lutetiana*, *Sanicula europaea*, *Viola reichenbachiana*, *Arum maculatum* and *Brachypodium sylvaticum* among the frequent associates of the field layer, and *Fissidens taxifolius*, *Ctenidium molluscum*, *Eurhynchium striatum* and *Thamnium alopecurum* as common bryophytes. Quite often, though, the soils here are moist, being derived in many cases from calcareous argillaceous rocks or clayey superficials, or being flushed with base-rich waters. Then, a more strictly calcicolous element in the field layer can be masked by an abundance of such species as *Hyacinthoides non-scripta*, *Anemone nemorosa*, *Primula vulgaris* (or the rare *P. elatior* in parts of East Anglia), *Glechoma hederacea*, *Ajuga reptans* and *Lamiastrum galeobdolon*. The woody component in these communities is among the richest and most diverse of any kind of British woodland. The oaks are very frequent, *Tilia cordata* and *Carpinus betulus* can figure prominently and very many stands have been treated as hazel coppice, but

the best preferential species are *Fraxinus excelsior*, elms (both *Ulmus glabra* and suckering taxa of the *procera* and *carpinifolia* sections) and, especially in southern Britain, *Acer campestre*, *Cornus sanguinea*, *Rhamnus catharticus* and *Viburnum lantana*.

With a shift on to more base-poor brown earths, most of the more calcicolous plants characteristic of the above two communities fade in importance. In the *Quercus-Pteridium-Rubus* (W10) and *Quercus-Betula-Oxalis* (W11) woodlands, *T. cordata*, *Carpinus* and also *Castanea sativa* can be important local dominants in southern Britain, but overall it is the oaks and birches which provide the most consistent components of the canopy. *Corylus avellana* and the hawthorns remain prominent in the understorey, together with *Ilex aquifolium* and *Sorbus aucuparia*. The typical vernal dominant in the field layer is *Hyacinthoides* (or *Anemone* on spring-waterlogged soils and in regions with a more boreal climate) but, by mid-summer, these woodlands often have a thick cover of *Pteridium aquilinum*, dense tangles of *Rubus fruticosus* agg. and *Lonicera periclymenum* or, where grazing is important, a grassy carpet. Indeed, *Holcus mollis* is distinctive throughout, and other characteristic associates include *Silene dioica*, *Stellaria holostea*, *Teucrium scorodonia*, *Luzula pilosa* and *Digitalis purpurea* with, among the bryophytes, *Rhytidiadelphus squarrosus*, *Dicranum scoparium*, *Hylocomium splendens* and *Hypnum cupressiforme s.l.*

With a continuing reduction in the contribution from virtually all other canopy trees, the oaks and birches further increase their prominence in the *Quercus-Betula-Deschampsia* (W16) and *Quercus-Betula-Dicranum* (W17) woodlands, the two communities of the base-

Figure 8. The six mixed deciduous and oak-birch woodlands in relation to soils and climate.

W9 *Fraxinus-Sorbus-Mercurialis* woodland	W11 *Quercus-Betula-Oxalis* woodland	W17 *Quercus-Betula-Dicranum* woodland	COOL & WET NORTH-WESTERN SUB-MONTANE ZONE
W8 *Fraxinus-Acer-Mercurialis* woodland	W10 *Quercus-Pteridium-Rubus* woodland	W16 *Quercus-Betula-Deschampsia* woodland	WARM & DRY SOUTH-EASTERN LOWLAND ZONE
RENDZINAS & BROWN CALCAREOUS EARTHS	BROWN EARTHS OF LOW BASE STATUS	RANKERS, BROWN PODZOLIC SOILS & PODZOLS	

poor edaphic extreme, *Ilex* and *S. aucuparia* become the commonest understorey species, and though *Pteridium* can still be a prominent plant by summer, *Rubus* and, especially, *Hyacinthoides* are much reduced. Typical herbs here are *Deschampsia flexuosa*, *Potentilla erecta*, *Galium saxatile* and *Melampyrum pratense*, but often more prominent are the ericoid sub-shrubs *Vaccinium myrtillus* and, in less shady places, *Calluna vulgaris* which give a distinctly heathy stamp to the vegetation. The bryophyte flora is often very rich and plentiful, especially in the humid north-west and, throughout, more calcifuge species like *Polytrichum formosum*, *Dicranum majus*, *Pleurozium schreberi* and *Rhytidiadelphus loreus* are good preferentials.

Climatic contrasts between south-east and north-west Britain

With each of the three pairs of mixed deciduous and oak-birch woodlands, there is a striking pattern of regional replacement of one community by the other in moving from the lowland south and east of the country to the sub-montane north-west (Figure 9). Although the regional boundaries are roughly coincident, the floristic differences between the suites of south-eastern and

Figure 9. Regional patterns among the mixed-deciduous and oak-birch woodlands.

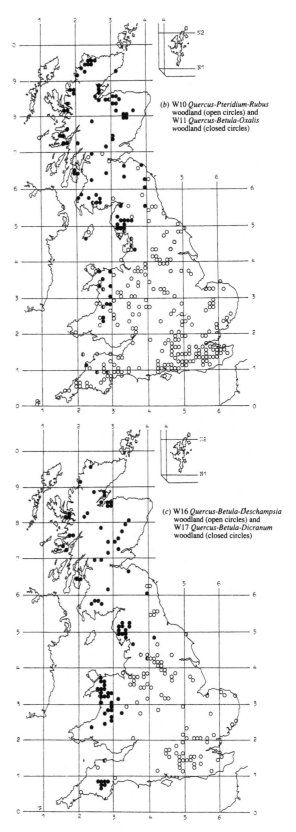

(b) W10 *Quercus-Pteridium-Rubus* woodland (open circles) and W11 *Quercus-Betula-Oxalis* woodland (closed circles)

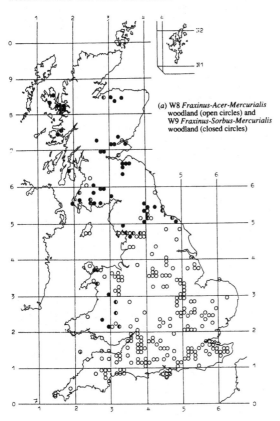

(a) W8 *Fraxinus-Acer-Mercurialis* woodland (open circles) and W9 *Fraxinus-Sorbus-Mercurialis* woodland (closed circles)

(c) W16 *Quercus-Betula-Deschampsia* woodland (open circles) and W17 *Quercus-Betula-Dicranum* woodland (closed circles)

north-western woodland types are in fact quite varied
and of diverse cause, but by and large it is direct and
indirect effects of climate which produce this contrast.

First, there is the influence of temperature. The floristic divide between the regions corresponds crudely with
the 26 °C mean annual maximum isotherm (Conolly &
Dahl 1970), and the cooler, cloudier and shorter summers of the north and west adversely affect the sexual
reproduction of a number of important species, such
that Continental and Continental Southern elements in
the flora of these woodlands (Matthews 1955) are largely
confined to the communities of the south and east. This
is especially important towards the calcicolous end of
the spectrum, where such plants as *A. campestre*, *Cornus*, *V. lantana*, *Rhamnus*, *Euonymus europaeus*, *Arum*
and *Viola reichenbachiana* all help to distinguish the
Fraxinus-Acer-Mercurialis woodland from its north-
western analogue, the *Fraxinus-Sorbus-Mercurialis*
woodland. Other Continental species are a little more
catholic in their soil preferences, being confined largely
to the south and east, but occurring there in both the
Fraxinus-Acer-Mercurialis woodland and its counterpart on less base-rich brown earths, the *Quercus-Pteridium-Rubus* woodland: *T. cordata* and *Carpinus* fall into
this category and, among the herbs, *Lamiastrum* and
Euphorbia amygdaloides. The occurrence of these plants,
and the abundance in many stands of *Hyacinthoides*,
emphasise the affinity of both these communities with
the Carpinion woodlands of mainland Europe (Figure
16). To the north and west, however, most of the thermophilous species decline in importance, although some
Continental Northern and Northern Montane plants
help give a positive definition to the woodlands there.
Again, these are more numerous towards the calcicolous
extreme, where *Prunus padus*, *Rubus saxatilis*, *Actaea
spicata*, *Trollius europaeus*, *Crepis paludosa* and *Cirsium
helenioides* can characterise the *Fraxinus-Sorbus-Mercurialis* woodland, but *Trientalis europaeus* is also very
characteristic of some more calcifuge woodlands in
northern Britain.

Second among the climatic factors, there is rainfall. In
broad terms, the boundaries between the suites of
communities coincide with the 1000 mm mean annual
isohyet (*Climatological Atlas* 1952), the higher rainfall
to the north and west making for an increased tendency
to leaching, the impact of which is all the greater because
of the fortuitous abundance there of pervious, arenaceous bedrocks. This contrast plays a major part in the
switch from *Quercus robur* and *Betula pendula* as the
predominant oak and birch among the south-eastern
woodlands, to *Q. petraea* and *B. pubescens* in the north-
west, though this distinction becomes less obvious the
more acidic the soils and, particularly with the oaks, has
been much affected by treatments.

One other effect of the prevalence of leaching to the

north and west is the marked transgression of more
calcifuge herbs and bryophytes on to the middle ground
of woodland variation. Plants like *P. erecta*, *G. saxatile*,
D. flexuosa, *H. splendens*, *Pleurozium schreberi* and
Dicranum majus, for example, are generally rare in the
south-eastern *Quercus-Pteridium-Rubus* woodland, but
their extension into much *Quercus-Betula-Oxalis* woodland in the north-west emphasises its floristic continuity
with the communities of the more acidic edaphic
extreme. In phytosociological terms, then, this woodland could be placed with the *Quercus-Betula-Deschampsia* and *Quercus-Betula-Dicranum* types in the
Quercion alliance. The most calcicolous of the north-
western woodlands, the *Fraxinus-Sorbus-Mercurialis*
community, thus tends to be maintained somewhat
against the climatic odds, and to occur very locally,
often marking out places where flushing with base-rich
waters offsets the effects of leaching. The resulting
prominence of moisture-tolerant plants brings it close to
more calcicolous damp woodlands of the Alno-Ulmion
alliance and sharpens up the contrasts with much south-
eastern *Fraxinus-Acer-Mercurialis* woodland.

Even where the soils under north-western woodlands
are more free-draining and leached, the higher rainfall
helps keep the profiles generally moist throughout the
year: in particular, they are free from marked droughting in summer, something which can affect even the
more ill-draining of our lowland soils in the very dry
parts of the south-east. One good indication of this is the
occurrence through much of the range of north-western
woodlands of plants like *Corylus*, *Viola riviniana* and
Oxalis acetosella which, in the south-east, are more
confined to heavier and more base-rich soils. Another is
the increasing prominence, in moving to the north and
west, of *Acer pseudoplatanus*: although a very widely
planted introduction in Britain, this species becomes
much more frequent in our semi-natural woodlands in
exactly the same climatic conditions as it favours in its
natural European range, where it is essentially a tree of
cool, humid ravines.

The consistency of surface and atmospheric humidity
in the north-west, where there are rarely fewer than 160
wet days yr^{-1} (Ratcliffe 1968), also encourages a greater
profusion of ferns and bryophytes in all the woodland
types, an abundance often accentuated by the rockier
terrain of north-western landscapes with a profusion of
sheltered niches. Different groups of species can characterise each of the woodlands in this part of the country.
Among the ferns for example, *Dryopteris dilatata* tends
to increase its contribution throughout but, towards the
calcicolous extreme, *Phyllitis scolopendrium* and *Polystichum* spp. are very distinctive, with *Blechnum spicant*
becoming preferential at the opposite pole. With the
bryophytes, the regional difference is best seen in the
more calcifuge communities, where an abundance of

mosses and liverworts provide what is the major distinction between the *Quercus-Betula-Dicranum* woodland and its analogue in the less humid south-east, the *Quercus-Betula-Deschampsia* woodland.

Other floristic patterns related to climate

Major contrasts in these climatic factors, then, are responsible for some of the most obvious floristic variations between the woodlands of different parts of the country. In fact, though, the shifts in the patterns of temperature, rainfall and humidity are more or less continuous in moving from the warmer and drier south-east to the cooler and wetter north-west. This means that, within the woodlands of both regions, it is often possible to characterise sub-communities which represent transitional types. In both the *Fraxinus-Acer-Mercurialis* and *Quercus-Pteridium-Rubus* woodlands, for example, there are sub-communities that, towards the Welsh borders and along the Pennine fringes, begin to show floristic features that become more fully expressed further west and north. Similarly, in Wales, the Lake District and southern Scotland, the *Quercus-Betula-Oxalis* and *Quercus-Betula-Dicranum* woodlands have sub-communities which still show remnants of essentially south-eastern elements, though transitions among our more acidophilous woodlands in northern England are confused by the demise of diagnostic bryophytes as a result of long-maintained atmospheric pollution.

Other sub-communities among the mixed deciduous and oak-birch woodlands are related to further climatic effects. First, on soils which are not too base-poor or droughty, there is a marked increase in the frequency and abundance of ground-growing *Hedera helix* in moving from East Anglia towards the south-west (Figure 10). This probably represents a response to increasing mildness of climate as measured, say, by minimum winter temperatures or the incidence of frost. It helps characterise sub-communities in the *Fraxinus-Acer-Mercurialis* and *Quercus-Pteridium-Rubus* woodlands and it complicates the value of an ivy carpet as an indicator of undisturbed growth of canopy or underwood in these vegetation types, such as occurs in young plantations and neglected coppice. Ivy can also be grazed out by deer, although grazing and browsing are of much more widespread significance in reducing the abundance of another useful indicator of oceanic climate, *Ilex aquifolium*. Variation in the occurrence of holly is thus generally not consistent enough to help define sub-communities, but it can show marked local abundance in woodlands towards the south-west of England, as in the *Quercus-Pteridium-Rubus* and *Quercus-Betula-Deschampsia* woodlands of the New Forest (and, incidentally, their beech analogues: see below).

The impact of oceanicity of climate is also felt elsewhere. Along the far north-western seaboard of Scotland, for example, the cool but equable conditions, combined with the very heavy and consistent rainfall and humidity, encourages the development of striking richness and variety among Atlantic bryophytes. This is most obvious in the more calcifuge *Quercus-Betula-Dicranum* woodland, already often well endowed with mosses and hepatics, but it can be seen too in both the *Quercus-Betula-Oxalis* and *Fraxinus-Sorbus-Mercurialis* woodlands, and therefore results in something of a floristic convergence among these communities, well seen on Skye. The similarity between the woodlands of this region is further enhanced by the prevalence throughout of *B. pubescens* ssp. *carpatica* and *S. aucuparia* (and, more locally, of *Corylus*) in low, scrubby covers. The absence of big canopy trees here is partly due to widespread timber extraction in the past, but the very windy climate also contributes to the development of a distinctive physiognomy (Figure 11).

A third effect is visible in the contrast between the kinds of *Quercus-Betula-Dicranum* and *Quercus-Betula-Oxalis* woodlands in these more oceanic western parts of Scotland and in the distinctly continental east, as in Aberdeenshire. There, lower rainfall and humidity,

Figure 10. Distribution of samples of mixed-deciduous and oak-birch woodlands with a ground carpet of *Hedera helix*.

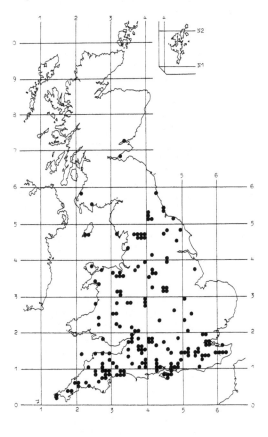

combined with a greater temperature range through the year, are reflected in a switch from *Betula pubescens* back to *B. pendula*, in the replacement of *Hyacinthoides* by *Anemone* on less base-poor soils and in the appearance of *Trientalis* as a frequent associate; there is also a striking rise in the occurrence of *Rhytidiadelphus trique-trus* in the moss carpet. Such features help define eastern Scottish sub-communities in these woodlands, and are also prominent in associated pine and juniper woods (see below) in what is the nearest approach to boreal forest in Britain (Figure 12).

The effects of sylvicultural treatments

Differences in soil and climate provide the best basis for understanding the major floristic variations among our mixed deciduous and oak-birch woodlands but, very often, this diversity is overlain by the effects of sylvicultural treatments. Treatments operate within the general constraints that edaphic and climatic conditions impose, though they do not always work in the same direction. Sometimes, treatments sharpen up patterns of floristic variation related to natural environmental factors;

often, they work against them, blurring the variation. Most obviously, treatments such as the removal or planting of timber trees, or the selective coppicing of underwood crops, can result in a great diversity of tree and shrub covers in what is essentially the same kind of woodland. Conversely, the same sort of treatment can produce identical stands of timber or underwood in what are actually different woodland communities. Figure 13 shows the great variety of treatment-derived dominance in the six communities of mixed deciduous and oak-birch woodland and indicates how our classification of these vegetation types (the columns) cuts across schemes based primarily on the contribution of trees and shrubs (the rows). The sections of the community accounts on synonymy and affinities give particular attention to this question, especially as it touches upon the proposals of Rackham (1980) and Peterken (1981), and every effort has been made throughout the text to point up areas of integration of our own approach with the highly informative historical perspectives on woodland that these authors have developed.

Figure 11. Distribution of samples of mixed-deciduous and oak-birch woodlands with rich Atlantic bryophyte flora under a scrubby canopy.

Figure 12. Distribution of samples of mixed-deciduous and oak-birch woodlands with a Boreal element among their flora.

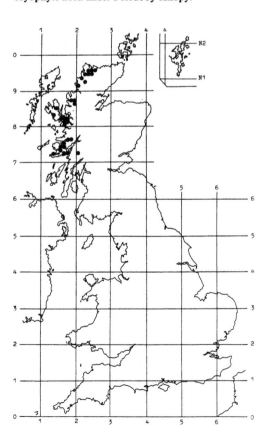

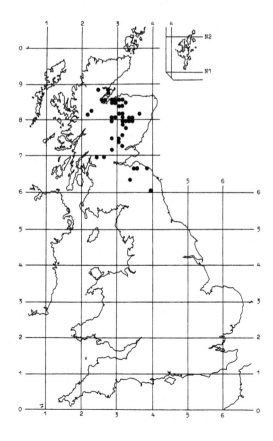

Although the effects of treatments on the floristics of the woodlands are rarely sufficiently obvious to characterise sub-communities, their impact is dealt with in detail in the sections on habitat. Special attention has been given to features like the artificially low cover of oak in many kinds of coppiced woodland (particularly in the *Fraxinus-Acer-Mercurialis* and *Quercus-Pteridium-Rubus* woodlands) and to the part that planting has played in local balances between *Quercus robur* and *Q. petraea* (especially noticeable in the *Quercus-Pteridium-Rubus* woodland and in eastern Scottish stands of the *Quercus-Betula-Oxalis* and *Quercus-Betula-Dicranum* woodlands); the impact of coppicing for hazel, ash, lime, hornbeam and chestnut (very obvious in the *Fraxinus-Acer-Mercurialis* and *Quercus-Pteridium-Rubus* woodlands) and for oak (important in the *Quercus-Betula-Oxalis* and *Quercus-Betula-Dicranum* woodlands); the distinctive post-coppice floras which develop in the different communities; the results of coppice neglect; and the effects of replanting or the establishment of new plantations with either the natural woody dominants or other broadleaf trees or conifers.

One other important sort of treatment in woodlands (or incidental biotic effect where herbivores get into

Figure 13. Dominance and its relation to treatment among the major trees and shrubs of the mixed-deciduous and oak-birch woodlands.
The size of the hollow circle gives a crude indication of the relative contribution of each species to the communities. The solid circle shows roughly how often the species dominates in coppiced stands.

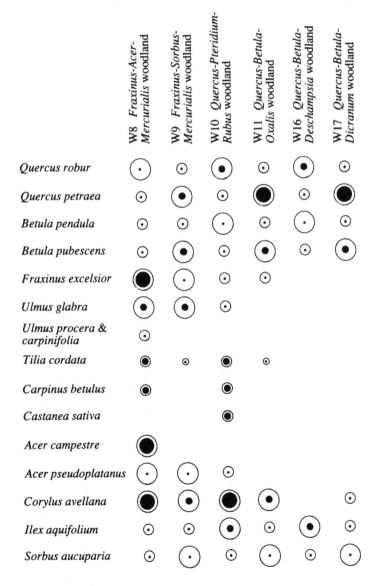

unenclosed sites) is grazing and browsing by stock and wild animals. The effects of such predation on the floristics, structure and regeneration of the vegetation are widespread and various, but of particular importance in the lowland wood-pasture landscapes of south-eastern commons, parks and ancient hunting forests (where much *Quercus-Pteridium-Rubus* and *Quercus-Betula-Deschampsia* woodland is found) and more consistently obvious throughout the north-west of Britain where many woods persist as declining remnants within the grazings of the upland fringes. There, for example, grazing can mediate a floristic convergence among *Quercus-Betula-Oxalis* and *Quercus-Betula-Dicranum* woodlands on transitional soils, by eliminating sensitive preferentials and favouring a spread throughout of tolerant grasses and dicotyledons. Such responses help reinforce the contrast between these two communities and their south-eastern equivalents.

Beech and yew woodlands

The classification recognises a distinct suite of three woodland communities in which *Fagus sylvatica* is the overwhelming canopy dominant. In general terms, these can be considered as beech analogues of the three south-eastern mixed deciduous and oak-birch woodlands, one being strongly calcicolous in character, one markedly calcifugous and one in between, so the scheme preserves, by and large, the traditional English view of these vegetation types, as well as the subdivisions of the Fagion alliance (Figure 14).

However, the communities have a more restricted distribution than their mixed deciduous and oak-birch counterparts, being concentrated in what appears to be the native British range of *Fagus* and represented only locally elsewhere by beech plantations (which are, incidentally, often floristically indistinguishable). Moreover, the formidable shading power of the tree and the intense root competition that it exerts often produce an impoverished version of the field and ground layers than might be expected given the particular soil conditions. Indeed, in extreme situations, there can be considerable difficulty in telling dense stands of the different kinds of beech woodland apart until gaps or clearings appear. This convergence is also accentuated by the increase throughout the communities of more shade-tolerant woody associates like *Ilex* and, to a lesser extent, *Taxus baccata*.

The proportion of *Fagus* in our more southerly woodlands is very variable, so the dividing line between the beech woodlands and their analogues can sometimes look very ill-defined. Towards the calcicolous end of the spectrum, separation is a little easier to comprehend because, though the woodlands are floristically quite similar, they often show clear environmental preferences. Essentially, the *Fagus-Mercurialis* woodland

(W12) is an edaphic replacement for the *Fraxinus-Acer-Mercurialis* woodland on more free-draining calcareous soils within the natural range of beech, most notably on the scarps of the southern Chalk. On moving on to more base-poor brown earths, the distinction between the communities is harder to understand in simple edaphic terms, although the *Fagus-Rubus* woodland (W14) still tends to prefer more permeable profiles, the *Quercus-Pteridium-Rubus* woodland heavier and less freely-draining soils. On the most acidic profiles, however, separation of the communities is much more difficult and almost infinitely variable mixtures of beech, oak and birch can be found over very similar field and ground layers in the *Fagus-Deschampsia* (W15) and *Quercus-Betula-Deschampsia* woodlands.

Successional interpretations of these kinds of contrasts and gradations should be undertaken with caution. There is no doubt that each of the mixed deciduous and oak-birch woodlands can function as seral precursors to their beech analogues; also that *Fagus*, once it becomes well established is an especially uncompromising canopy dominant. But whether the different kinds of beech woodland are inevitable climax forests is not so clear: beech fruits erratically and spreads with some difficulty when unaided. So, in circumstances where edaphic conditions are equally favourable to woodlands with and without *Fagus*, there may be some quite natural cyclical replacements through time, one type repeatedly giving way to the other.

Even within the native British range of beech, the communities that it dominates are often of planted origin or severely affected by treatments. Coppice is rarely found (though it was probably more common in the past), but timber extraction is still widespread. In the Chilterns, many stands still bear signs of their management under a selection system, though cropping is now usually on a clear-fall or shelterwood regime, and replanting often involves the use of beech-conifer mixtures. Planted stands beyond the natural range of *Fagus* can still generally be accommodated within the relevant beech community, though a few far-flung (but long-established) Scottish plantations are best placed among the *Quercus-Betula-Oxalis* woodlands.

Taxus is a characteristic associate of *Fagus* in all three kinds of beech woodland, often forming part of a second tier of shade-tolerant trees and being especially prominent in areas like the New Forest and on the Chalk of south-east England. In certain circumstances, however, especially on warm and sunny south-facing slopes over shallow limestone soils, it can pre-empt beech and become dominant in a distinctive *Taxus* woodland (W13). Floristically, such vegetation can be seen as a yew analogue of the calcicolous woodlands of south-eastern Britain (Figure 14) and part of the Fagion alliance, although extremely dense shade and the very inhospitable edaphic environment make this community one of the most species-poor of all our woodlands. Nonetheless, these yew woods are among the finest in the whole of Europe, and they, with transitions to related beechwoods, provide the usual context in this country for what are probably native populations of *Buxus sempervirens*.

Taxus grows very well far to the north of the main area of distribution of our yew woods, and scattered trees are a perfectly natural feature of many stands of

Fraxinus-Acer-Mercurialis woodland on Carboniferous Limestone in Wales, Derbyshire, the Pennines and around Morecambe Bay. Indeed, in some places, yew attains local abundance in such vegetation and, though stands have generally been retained as part of the mixed deciduous woodland, they could probably equally well be seen as northern outliers of the *Taxus* community.

Pine and juniper woodlands and montane willow scrub

Planted *Pinus sylvestris* figures as a replacement canopy tree in a variety of mixed deciduous and oak-birch woodlands, in both the south-east and the north-west,

Figure 14. The beech and yew woodlands in relation to soils and climate.

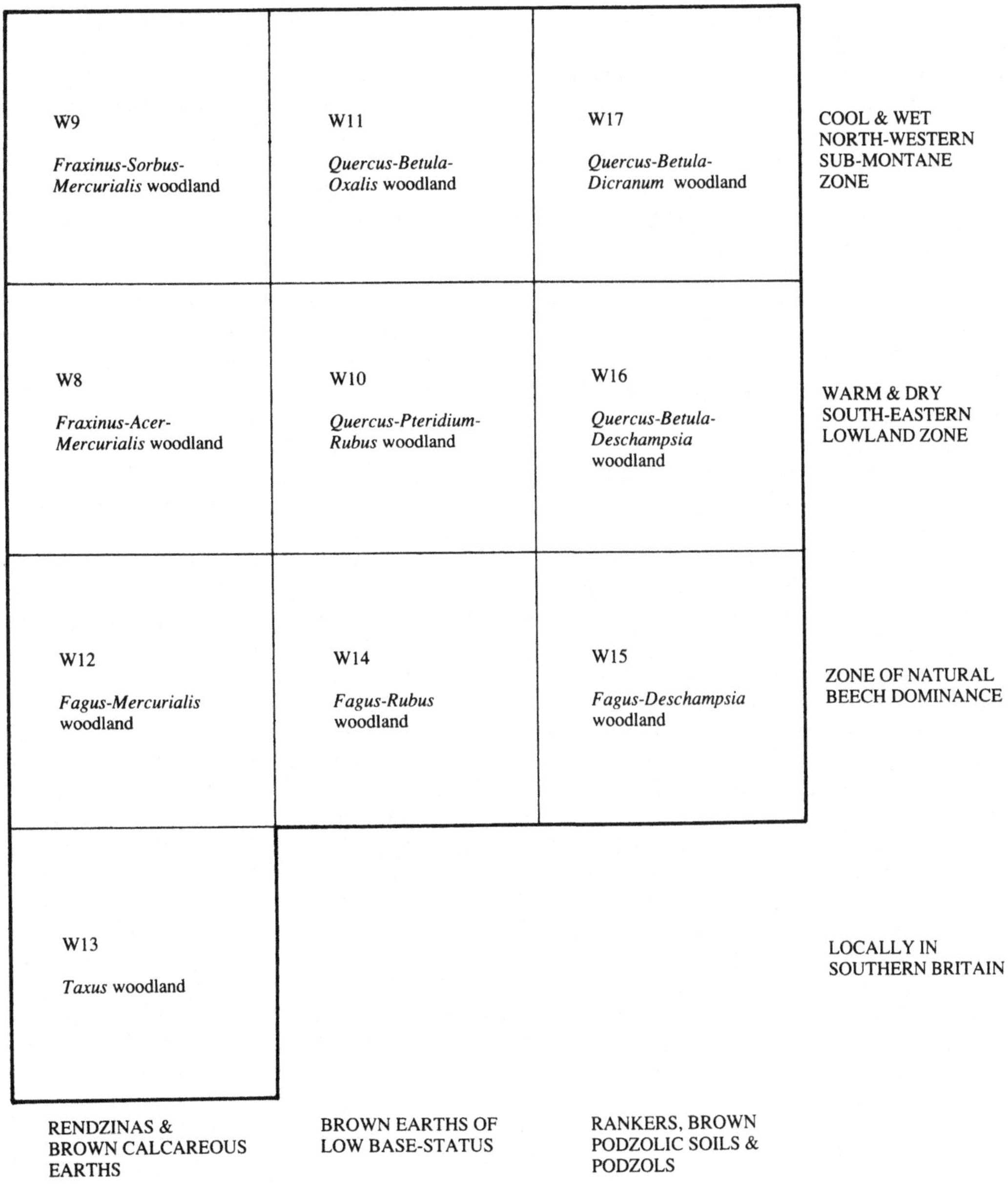

W9 *Fraxinus-Sorbus-Mercurialis* woodland	W11 *Quercus-Betula-Oxalis* woodland	W17 *Quercus-Betula-Dicranum* woodland	COOL & WET NORTH-WESTERN SUB-MONTANE ZONE
W8 *Fraxinus-Acer-Mercurialis* woodland	W10 *Quercus-Pteridium-Rubus* woodland	W16 *Quercus-Betula-Deschampsia* woodland	WARM & DRY SOUTH-EASTERN LOWLAND ZONE
W12 *Fagus-Mercurialis* woodland	W14 *Fagus-Rubus* woodland	W15 *Fagus-Deschampsia* woodland	ZONE OF NATURAL BEECH DOMINANCE
W13 *Taxus* woodland			LOCALLY IN SOUTHERN BRITAIN
RENDZINAS & BROWN CALCAREOUS EARTHS	BROWN EARTHS OF LOW BASE-STATUS	RANKERS, BROWN PODZOLIC SOILS & PODZOLS	

and it has also been widely planted within the range of our beech woodlands. In Scotland, however, within what is presumed to be the natural distribution of pine, the tree dominates in a distinctive *Pinus-Hylocomium* woodland (W18). In many ways, this community can be seen as an analogue, on similarly base-poor soils, of the *Quercus-Betula-Dicranum* woodland. The same ericoid sub-shrubs and calcifuge bryophytes are prominent, though here they are often accompanied by *Vaccinium vitis-idaea* and *Ptilium crista-castrensis* and a striking contingent of rare Northern Montane and Continental Northern plants such as *Goodyera repens*, *Listera cordata*, *Linnaea borealis* and the various wintergreens. The community extends a little way on to more fertile soils, too, now generally occupied in this part of Britain by the *Quercus-Betula-Oxalis* woodland and, in wetter areas, it has a prominent *Sphagnum* component in the ground layer, which brings it close floristically to the *Betula-Molinia* mire woodland (see below). Although pine plantations in some regions of Scotland can acquire many elements of the typical *Pinus-Hylocomium* flora, there seems little doubt that the remaining distribution of more natural stands represents a fragmentary inheritance of the prevalence of pine in the cooler parts of the country through the post-Glacial period (Figure 15).

Evidence suggests that, in the past, canopies of our native pine forests were more mixed, with birch and *Juniperus communis* ssp. *communis* occurring much more often as integral components of the woody cover. These species can still figure in close association with the *Pinus-Hylocomium* woodland, and juniper is also found in the *Quercus-Betula-Oxalis* and *Quercus-Betula-Dicranum* woodlands of eastern Scotland. Very often now, though, a distinct *Juniperus-Oxalis* scrubby woodland (W19) persists in this region as a seral community which, for one reason or another, does not progress to mature forest. Overall, this community is analogous to the *Quercus-Betula-Oxalis* woodland, although one sub-community extends its occurrence on to markedly acidic soils, another on to flushed profiles that can be quite base-rich. Also, centred on the more continental parts of eastern Scotland, *Trientalis*, *Anemone* and *Rhytidiadelphus triquetrus* are common and there can be records for some of the Continental Northern and Northern Montane rarities. These features help to locate the vegetation phytosociologically with the *Pinus-Hylocomium* woodland in the Dicrano-Pinion alliance, and emphasise the very different character of the community from the seral scrub in which juniper can occur in the south-eastern lowlands of Britain (Figure 16).

The *Juniperus-Oxalis* scrubby woodland extends southwards into northern England but it is typically a high-altitude community and, in some places, may represent a sub-alpine climax vegetation. However, at scattered localities on higher ground through the southern and central Highlands of Scotland, it is replaced on moister, mesotrophic soils by a *Salix-Luzula* scrub (W20), provided there is sufficient freedom from grazing to allow the development of a low canopy of Arctic-Alpine willows among tall herbs. These mixtures of *Salix lapponum*, *S. myrsinites*, *S. lanata*, *S. arbuscula* and *S. reticulata*, with *Luzula sylvatica*, lush dicotyledons, ferns and ericoid sub-shrubs, are thus now restricted to isolated and inaccessible banks and ledges and they represent our nearest approach to the kind of Salicion arbusculae vegetation seen in Scandinavia (Figure 16).

Wet woodlands with alder, birch and willows

Seven types of woodland are characterised by the dominance in the canopy of various mixtures of *Alnus glutinosa*, *Betula pubescens* (only very rarely *B. pendula*) and *Salix* spp. in situations where wetness of the ground is (or until recently has been) the overriding element in the environment. For the most part, floristic variation among these communities can be understood in terms of interactions between the amount of soil moisture, the degree of base-richness of the soils and waters and the trophic state of the system. Often, the combination of particular conditions favoured by the different woodland types is associated with a distinctive kind of mire or flush, and the distribution of the communities is often a reflection of the occurrence of these characteristic site types within regions of broadly suitable climate. In the early stages of colonisation around water margins and on strongly waterlogged ground, there is often a chance element in the assortment of woody plants that invade. Then, as the woodland matures, the composition of the canopy tends to equilibrate and its increasing closure, together with the advancing terrestrialisation of the ground, exert effects on the herb and cryptogam flora inherited from the preceding swamp or mire.

We have tried wherever possible to draw attention to the continuing influence of these dynamic processes among the communities and to sketch possible lines of succession. Again, too, we have attempted to assess the impact of sylvicultural treatments, though these are usually not so important here. Osier-dominated vegetation is still cropped commercially in some areas, and drier stands of a number of the communities have sometimes been coppiced, notably for alder, but generally human influence has been felt more through activities such as draining, reclamation and eutrophication which have altered and fragmented many stretches of wetter woodland. More locally, neglect of mowing for marsh crops like reed, sedge and litter has allowed some secondary spread of certain communities.

Two of the woodlands are especially associated with the primary colonisation of swampy vegetation in open-water transitions. Where the waters are fairly base-rich

Figure 15. The pine and juniper woodlands and
montane willow scrub in relation to soils and climate.

W20 *Salix-Luzula* scrub	W19 *Juniperus-Oxalis* woodland	W18 *Pinus-Hylocomium* woodland	COLD NORTHERN UPLANDS & SUB-ALPINE ZONE
W9 *Fraxinus-Sorbus-Mercurialis* woodland	W11 *Quercus-Betula-Oxalis* woodland	W17 *Quercus-Betula-Dicranum* woodland	COOL & WET NORTH-WESTERN SUB-MONTANE ZONE
W8 *Fraxinus-Acer-Mercurialis* woodland	W10 *Quercus-Pteridium-Rubus* woodland	W16 *Quercus-Betula-Deschampsia* woodland	WARM & DRY SOUTH-EASTERN LOWLAND ZONE
W12 *Fagus-Mercurialis* woodland	W14 *Fagus-Rubus* woodland	W15 *Fagus-Deschampsia* woodland	ZONE OF NATURAL BEECH DOMINANCE
W13 *Taxus* woodland			LOCALLY IN SOUTHERN BRITAIN
RENDZINAS & BROWN CALCAREOUS EARTHS	BROWN EARTHS OF LOW BASE-STATUS	RANKERS, BROWN PODZOLIC SOILS & PODZOLS	

Figure 16. Possible phytosociological affinities of drier
British woodlands.

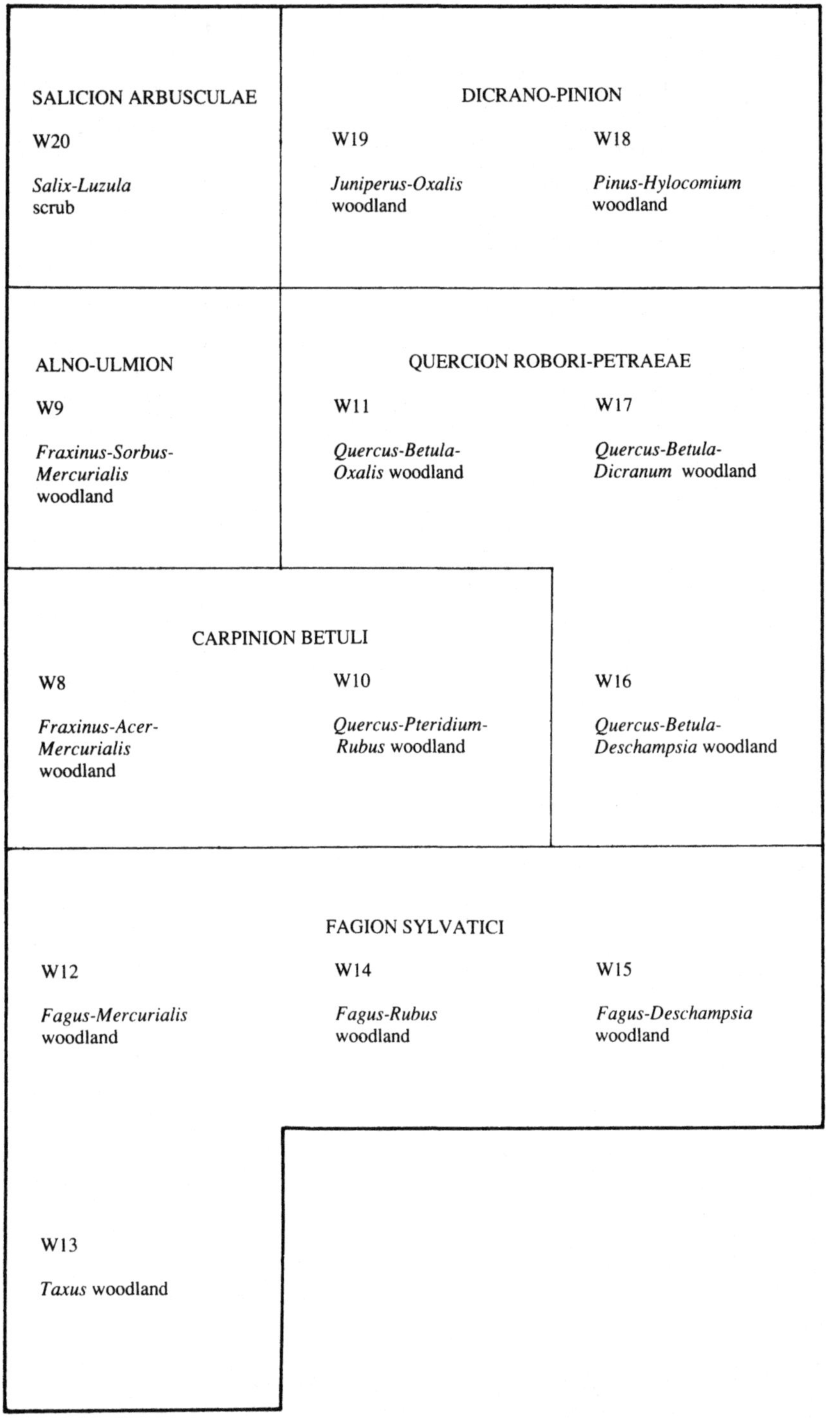

SALICION ARBUSCULAE

W20

Salix-Luzula
scrub

DICRANO-PINION

W19

Juniperus-Oxalis
woodland

W18

Pinus-Hylocomium
woodland

ALNO-ULMION

W9

Fraxinus-Sorbus-
Mercurialis
woodland

QUERCION ROBORI-PETRAEAE

W11

Quercus-Betula-
Oxalis woodland

W17

Quercus-Betula-
Dicranum woodland

CARPINION BETULI

W8

Fraxinus-Acer-
Mercurialis
woodland

W10

Quercus-Pteridium-
Rubus woodland

W16

Quercus-Betula-
Deschampsia woodland

FAGION SYLVATICI

W12

Fagus-Mercurialis
woodland

W14

Fagus-Rubus
woodland

W15

Fagus-Deschampsia
woodland

W13

Taxus woodland

and eutrophic, favouring the accumulation of fen peat in topogenous mires and valley fens, the *Alnus-Carex paniculata* woodland (W5) is very characteristic. This is our major Alnion glutinosae community in Britain, representing a more oceanic replacement of Continental swamp forests and having a rich inherited flora with bulky sedges like *C. paniculata*, *C. acutiformis* and *C. elata*, herbs such as *Filipendula ulmaria*, *Eupatorium cannabinum*, *Iris pseudacorus*, *Valeriana officinalis*, *Angelica sylvestris*, *Lycopus europaeus*, *Lythrum salicaria*, *Lysimachia vulgaris* and, mostly now in Broadland, *Peucedanum palustre*, together with ferns including *Athyrium filix-femina*, *Thelypteris palustris* and *Osmunda regalis*. The *Alnus-Carex* woodland has a fairly widespread distribution on mire remnants throughout the British lowlands, forming classic swamp carr, though there is some doubt as to whether it always progresses to drier forest types, perhaps being prone to a cyclical process of decay back to swamp.

The counterpart of the *Alnus-Carex* woodland in very wet poor-fen systems, where there is a somewhat different balance of base-status and trophic level, is the *Salix-Carex rostrata* woodland (W3). This is largely a community of northern Britain, its distribution reflecting the local occurrence of basin mires where the waters are not too acidic or oligotrophic. Typically, there, it colonises floating rafts of such plants as *C. rostrata*, *Menyanthes trifoliata*, *Potentilla palustris*, *Filipendula*, *Angelica*, *Valeriana* spp., *Geum rivale*, *Crepis paludosa*, *Cardamine pratensis*, *Caltha palustris* and larger *Calliergon* spp. and Mniaceae. *Alnus* has often not penetrated to these isolated sites, and the stands are usually dominated by promiscuous mixtures of willows, with the Continental Northern *S. pentandra* especially distinctive in the cool climate.

Base-rich and moderately eutrophic systems where the influence of the ground water is still strong, but not sufficient to maintain swamp vegetation, typically have the *Salix-Betula-Phragmites* woodland (W2), the classic fen carr of terrestrialising mires with fen peat, though often now found as a secondarily developed cover on neglected mowing-marsh. It shares many species with the *Alnus-Carex* woodland, though here the commonest monocotyledon is *Phragmites australis* rather than one of the bulky sedges and, in secondary stands, the composition of the field layer often reflects the lasting influence of different cutting regimes in the preceding fen. Again, the isolation of stands can have a marked effect on the pattern of colonisation by the woody plants, such that species which are usually no more than occasional can attain marked local prominence (as with *Frangula alnus* and *Rhamnus* at Wicken Fen, for example).

In phytosociological terms, both the *Salix-Carex* and *Salix-Betula-Phragmites* woodlands would be considered as part of the Salicion cinereae alliance. In more markedly eutrophic systems, both they and the *Alnus-Carex* woodland are replaced by what Continental workers would call Salicion albae forest, all our examples of which have been grouped in a single compendious *Alnus-Urtica* woodland (W6). This can develop in fen peat systems where there has been some degree of enrichment through drainage and disturbance or by eutrophication of the waters, but it also includes woodlands which appear naturally on rich alluvium accumulating in more mature river valleys and around silting water-bodies. Although *Alnus* and *S. cinerea* can be prominent invaders in such situations, a variety of other Salices can come to dominate including, in the early stages and in osier beds, *S. viminalis*, *S. triandra* and *S. purpurea* and, on wet river terraces, *S. fragilis*. The field layer is much less species-rich than in typical swamp and fen woodland, being dominated by patchy mixtures of more nutrient-demanding herbs, like *Urtica dioica*, *Galium aparine* and *Epilobium hirsutum*.

There are some clear floristic transitions between these communities and drier mixed deciduous woodlands but, in the early stages of development of the *Salix-Betula-Phragmites* woodland (and, to a lesser extent, the *Salix-Carex* woodland), there is evidence of a successional divergence to more calcifuge vegetation. This is seen in a shift away from *Alnus* and *Salix* dominance towards a prevalence of *Betula pubescens* and an appearance of less demanding Sphagna like *S. recurvum*, *S. palustre*, *S. squarrosum* and *S. fimbriatum*, in what seem to be ombrogenous nuclei isolated from the influence of the ground water table. Such a floristic trend continues into the *Betula-Molinia* woodland (W4), where these Sphagna, along with *Molinia caerulea*, become a more or less consistent feature beneath rather open and often moribund canopies of birch on moist and moderately acid peaty soils. As well as occurring on raised areas within flood-plain and basin mires, this community also extends on to the drier fringes of blanket bogs and valley mires. In all these situations, the same general balance of soil conditions is marked by a poverty of rich-fen herbs and the presence of associates more typical of wet heath or mires with small sedges and rushes. In phytosociological schemes, the heathy affinities of this kind of vegetation have been recognised in its location alongside pine woodlands in a Betulion pubescentis alliance.

Certain types of *Betula-Molinia* woodland extend into markedly soligenous situations with nutrient-poor peaty gleys but, in more base-rich flushes with strongly-gleyed and often mesotrophic mineral soils, it is replaced by the *Alnus-Fraxinus-Lysimachia* woodland (W7). In this community, *Alnus*, *S. cinerea* and *B. pubescens* all retain some prominence but *Fraxinus* appears as a distinctive constant and the field and ground layers both show the influence of flushing, with such plants as

Lysimachia nemorum, Ranunculus repens, Chrysosplenium oppositifolium, Brachythecium rivulare and *Pellia epiphylla*, and bulky dominants like *Juncus effusus, Deschampsia cespitosa, Carex pendula, C. remota* and *C. laevigata*. This community has a strongly western and northern distribution, reflecting the occurrence of suitable flushes on slopes around the upland fringes, and there it can come close in its floristics to the north-western *Fraxinus-Sorbus-Mercurialis* woodland, being best placed with it in the Alno-Ulmion alliance.

Exposed situations along the northern and western coasts of Britain also provide some of our most striking examples of *Salix-Galium palustre* woodland (W1), where *S. cinerea* is the sole woody dominant over mixtures of plants like *J. effusus, G. palustre* and *Mentha aquatica*, generally without any big swamp or fen mono-cotyledons. This vegetation develops there on cliff-top flushes, but it can also be seen on waterlogged mineral soils of only moderate base-status in other parts of the country, around dune-slacks and in damp field hollows where, for one reason or another, alder invasion has not occurred. In exposed and isolated situations in more equable climates, it may attain some measure of stability as a climax Salicion scrub.

Scrub and underscrub communities

The bulk of our sub-climax woody vegetation on circumneutral and base-rich soils in the lowlands has been classified in the *Crataegus-Hedera* scrub (W21). Many stabilised wood margins and hedges (the latter often planted) can be included here but more extensive stands usually develop by the invasion of neglected bare ground, pasture or meadow or where woodland has been degraded, and the floristics and physiognomy of the community often reflect its transitional and unstable character. Although variation in climate and soils has obvious influences on the vegetation, the previous treatment of the land and the availability of seed-parents and agents of dispersal play important roles in the early stages of development. Spinose plants with fruits attractive to birds and other animals are very prominent here, with *Crataegus monogyna, Prunus spinosa*, Rubi and roses very common, but tree saplings can be frequent from the outset, *Fraxinus, Acer pseudoplatanus, Quercus robur* (on heavier soils) and, more locally, *Fagus* all being characteristic. *Sambucus nigra* is an abundant shrub on more eutrophic soils and on shallow, base-rich profiles in the warmer south of Britain, there is a very striking enrichment with a suite of thermophilous calcicoles like *Cornus sanguinea, Viburnum lantana, Ligustrum vulgare* and *Clematis vitalba*. The southern stations of *Juniperus communis* ssp. *communis* are also concentrated in this kind of scrub and *Taxus* is a preferential invader. The field layer in denser stands of *Crataegus-*

Hedera scrub is typically species-poor, a ground carpet of ivy bearing testimony to the deepening shade, *Urtica* and *Galium aparine* indicating richer soils and *Brachypodium sylvaticum* becoming frequent on disturbed but more impoverished ground. However, younger open mosaics of this sort of scrub among weedy vegetation, pasture or meadow can retain very rich and diverse assemblages of herbs and develop a complex fringe of underscrub and rank grassland which some rarities seem to favour. More long-established stands slowly acquire a woodland field layer of the Carpinion type as herbs spread in and an opening canopy admits more light..

In certain situations, notably on deep, moist and fairly nutrient-rich soils and where exposure prevents the establishment of many other trees and shrubs, *Prunus spinosa* comes to dominate in a dense suckering canopy to *Prunus-Rubus* scrub (W22). To some extent, floristic variation in the field layer parallels that found in the *Crataegus-Hedera* scrub, though with a shift towards more mesophytic herbs. The community also includes some sea-cliff scrub where maritime plants can figure. *Pteridium aquilinum* and grasses, prominent here in more open stands, continue to be strongly represented in the third community in this group, the *Ulex europaeus-Rubus* scrub (W23) which takes in most of our vegetation with the common gorse and brambles, or *Cytisus scoparius*. Generally, however, this is a community of more acidic soils and the high frequency of species such as *Agrostis capillaris, Galium saxatile* and *Rhytidiadelphus squarrosus* provides a floristic link with heathland precursors of our lowland Quercion oak-birch woodlands.

Finally, there are two types of what we have called underscrub in which brambles are prominent, the *Rubus-Holcus lanatus* (W24) and *Pteridium-Rubus* (W25) communities. These are very widespread and common vegetation types of woodland and scrub margins, of clearings and rides, hedgebanks and new colonised open ground or grassland where trees and shrubs have not yet established dominance or from which they have been cleared. The *Rubus-Holcus* underscrub can have a strong weedy or mesotrophic grassland element inherited from the invaded swards or waste ground, while the *Pteridium-Rubus* community frequently has a relic woodland field layer in sites where bracken has expanded its cover to replace a cleared canopy.

In phytosociological terms, all these scrubs and underscrubs could be accommodated in various subdivisions of the order Prunetalia, but this is one area where further sampling and interpretation might help to clarify the range of variation represented in Britain and better understand its significance in successions.

KEY TO WOODLANDS AND SCRUB

With something as complex and variable as vegetation, no key can pretend to offer an infallible short cut to diagnosis. The following should therefore be seen as simply a crude guide to identifying the types of woodland and scrub in the scheme and must always be used in conjunction with the data tables and community descriptions. It relies on floristic (and, to a lesser extent, physiognomic) features of the vegetation and demands a knowledge of the British vascular flora and, at certain points, of some bryophytes. It does not make primary use of any habitat features, though these can provide a valuable confirmation of a diagnosis.

Because the major distinctions between the vegetation types in the classification are based on inter-stand frequency, the key works best when sufficient samples of similar composition are available to construct a constancy table. It is the frequency values in this (and, to a lesser extent, the ranges of abundance) which are then subject to interrogation with the key. Most of the questions are dichotomous and notes are provided at particularly difficult points and where confusing zonations are likely to be found.

As in the construction of the scheme, samples should be of 50 × 50 m for woodland canopy and shrub layer or sparse scrub, 10 × 10 m for dense scrub canopy, with either 4 × 4 m or 10 × 10 m for the field and ground layer.

---

1   Low scrub dominated by *Salix lapponum*, sometimes with *S. lanata*, *S. myrsinites* or *S. reticulata*, with luxuriant mixtures of *Vaccinium myrtillus*, *V. vitis-idaea* and *Empetrum nigrum* ssp. *hermaphroditum*, *Luzula sylvatica*, *Deschampsia cespitosa*, tall dicotyledons and bryophytes; a rare community of montane crags and ledges

      W20 *Salix lapponum-Luzula sylvatica* scrub

Scrub or underscrub with one or more of *Crataegus monogyna*, *Prunus spinosa*, *Sambucus nigra*, *Ulex europaeus*, *Cytisus scoparius*, *Rosa canina* agg. and *Rubus fruticosus* agg. frequent and often abundant; saplings of taller trees sometimes numerous but never forming an overtopping canopy   57

High forest or coppice in which *Crataegus monogyna*, *Prunus spinosa*, *Sambucus nigra*, *Rosa canina* agg. and *Rubus fruticosus* agg. can all occur frequently but not as dominants and very often as a minor component with taller trees overtopping or in underwood with other species abundant   2

In floristics and structure, woodland and scrub grade continuously one into the other, so it is hard to devise a foolproof separation between the two: if in doubt, work right through the key starting at 2

---

2   Canopy dominated by *Salix cinerea* or *Betula pubescens* or mixtures of these two species, sometimes with *Salix pentandra* but all other woody species usually infrequent   3

Not as above   8

3   Single tree/shrub layer dominated by mixtures of *Salix cinerea* and *S. pentandra* with occasional *Betula pubescens*; field layer swampy and dominated by mixtures of *Carex rostrata*, *Menyanthes trifoliata* and *Equisetum fluviatile* with five or more of *Filipendula ulmaria*, *Cardamine pratensis*, *Galium palustre*, *Caltha palustris*, *Angelica sylvestris*, *Valeriana dioica*, *Geum rivale* and *Crepis paludosa*; bryophyte mat often extensive with *Calliergon cuspidatum*, *Rhizomnium punctatum*, *Mnium hornum*, *Eurhynchium praelongum*, *Climacium dendroides* and occasional patches of *Sphagnum squarrosum*, *S. recurvum* and/or *S. palustre*.

W3 *Salix pentandra-Carex rostrata* woodland

Salix pentandra and combinations of the above field-
and ground-layer species absent   4

4   *Salix cinerea* and *Betula pubescens* both frequent
and often co-dominant in single tree/shrub layer; field
layer with *Phragmites australis* constant and often abundant with *Carex paniculata*, *Lythrum salicaria* and *Lysimachia vulgaris* infrequent

W2 *Salix cinerea-Betula pubescens-Phragmites
australis* woodland   5

Either *Salix cinerea* or *Betula pubescens* more frequent
and abundant; *Phragmites australis* absent or rare in
field layer and never dominant   6

5   Field layer dominated by mixtures of *Phragmites
australis*, *Filipendula ulmaria*, *Urtica dioica* and *Eupatorium cannabinum*; Sphagna rare in the ground layer

W2 *Salix cinerea-Betula pubescens-Phragmites
australis* woodland
Alnus glutinosa-Filipendula ulmaria sub-
community

Alnus glutinosa is a ready invader of this woodland type and there is a complete gradation
between it and the Typical sub-community of the
Alnus-Urtica woodland; the higher frequency of
A. glutinosa there and its tendency to be overwhelmingly dominant in a distinct tree canopy
can usually be used as a separation

Field layer not dominated by mixtures of the above
species and a prominent carpet of mixtures of *Sphagnum
squarrosum*, *S. fimbriatum*, *S. recurvum* and *S. palustre*
always present

W2 *Salix cinerea-Betula pubescens-Phragmites
australis* woodland
Sphagnum sub-community

A carpet of these Sphagna is also characteristic of
the *Sphagnum* sub-community of the *Betula-
Molinia* woodland but, although *M. caerulea*
occurs occasionally here, *Phragmites australis* is
rare in that woodland type

---

6   *Salix cinerea* constant and generally dominant in
single tree/shrub canopy with occasional *Betula pubescens*; field layer somewhat varied but *Mentha aquatica*
and *Galium palustre* frequent and *Molinia caerulea* rare

W1 *Salix cinerea-Galium palustre* woodland

Betula pubescens constant and generally dominant with
occasional *Salix cinerea* in a usually well-defined though
quite often open and somewhat moribund tree canopy,
occasionally with saplings of both species beneath;
Molinia caerulea constant and often abundant in the
field layer

W4 *Betula pubescens-Molinia caerulea* woodland
                                                          7

7   Prominent carpet of mixtures of *Sphagnum fimbriatum*, *S. recurvum*, *S. palustre*, *S. squarrosum* and *S.
papillosum* present with *Molinia caerulea* generally the
only prominent grass

W4 *Betula pubescens-Molinia caerulea* woodland
Sphagnum sub-community

Sphagnum recurvum and, less frequently, *S. palustre* may
be prominent but field layer has, in addition to *Molinia
caerulea*, frequent and often abundant *Holcus mollis*, *H.
lanatus* and *Deschampsia cespitosa*

W4 *Betula pubescens-Molinia caerulea* woodland
Juncus effusus sub-community

Sphagna rare beneath a field layer with frequent and
often abundant *Rubus fruticosus* agg., *Dryopteris dilatata* and *Lonicera periclymenum*

W4 *Betula pubescens-Molinia caerulea* woodland
Dryopteris dilatata-Rubus fruticosus sub-
community

These field-layer species can also be abundant
under canopies rich in *Betula pubescens* in the
Betula sub-community of the *Alnus-Urtica* woodland but the frequency of *U. dioica* and absence of
Molinia caerulea there should effect a separation

---

8   *Alnus glutinosa* and/or *Salix fragilis* constant and
often prominent in a tree canopy or one or more of *Salix
triandra*, *S. viminalis*, *S. purpurea* or hybrids in scrubby
vegetation or osier beds   9

All these species absent or, if present, then only as rare
scattered trees or shrubs within woodland dominated by
other species   19

9   *Carex paniculata* (or locally *Scirpus sylvaticus*)
constant and almost always prominent in the field layer
with some of *Rubus fruticosus* agg., *Cirsium palustre*,
Galium palustre, *Filipendula ulmaria*, *Carex acutiformis*,
Eupatorium cannabinum, *Urtica dioica*, *Dryopteris dilatata*, *Mentha aquatica*, *Valeriana officinalis*

W5 *Alnus glutinosa-Carex paniculata* woodland
10

Carex paniculata and combinations of the above species absent from the field layer
12

10   Canopy often purely of *Alnus glutinosa* with *Chrysosplenium oppositifolium* and *Pellia epiphylla* constant and often prominent as a carpet around the other species

W5 *Alnus glutinosa-Carex paniculata* woodland
Chrysosplenium oppositifolium sub-community

Fraxinus excelsior, Betula pubescens and *Salix cinerea* frequent canopy associates and sometimes locally dominant; *Chrysosplenium oppositifolium* and *Pellia epiphylla* rare and never abundant
11

11   *Frangula alnus* a frequent associate; field layer a rich mixture of sedges (*C. paniculata, C. acutiformis, C. remota*), rich-fen dicotyledons (*Lysimachia vulgaris, Lythrum salicaria, Lycopus europaeus, Peucedanum palustre*) and ferns (*Thelypteris palustris, Osmunda regalis*)

W5 *Alnus glutinosa-Carex paniculata* woodland
Lysimachia vulgaris sub-community

Frangula alnus rare in canopy; rich-fen dicotyledons and ferns rare in field layer which is generally dominated by *Carex paniculata* and/or *C. acutiformis* or locally *Scirpus sylvaticus* with *Filipendula ulmaria* and *Phragmites australis* frequent

W5 *Alnus glutinosa-Carex paniculata* woodland
Phragmites australis sub-community

12   Field layer with *Urtica dioica* constant and often abundant with two of *Galium aparine, Poa trivialis, Rubus fruticosus* agg. and *Dryopteris dilatata* but *Chrysosplenium oppositifolium, Lysimachia nemorum* and *Athyrium filix-femina* rare

W6 *Alnus glutinosa-Urtica dioica* woodland   13

Urtica dioica absent or, if present, then also with *Chrysosplenium oppositifolium* and *Lysimachia nemorum* or *Athyrium filix-femina*
16

13   *Rubus fruticosus* agg. constant and often abundant, frequently with *Dryopteris dilatata* and *Lonicera periclymenum*
14

Above combination of species rare
15

14   *Sambucus nigra* and *Salix cinerea* frequent in the shrub layer; *Dryopteris filix-mas* and *Hedera helix* fre-

quent in the field layer

W6 *Alnus glutinosa-Urtica dioica* woodland
Sambucus nigra sub-community

Canopy often with *Betula pubescens* more frequent than *Alnus glutinosa* or sometimes with planted or seeding-in *Pinus sylvestris*; *Sambucus nigra* and *Salix cinerea* infrequent in shrub layer; *Dryopteris filix-mas* and *Hedera helix* rare in field layer

W6 *Alnus glutinosa-Urtica dioica* woodland
Betula pubescens sub-community

15   *Salix fragilis* dominant or co-dominant with *Alnus glutinosa* in the canopy; *Sambucus nigra* frequent in the shrub layer; *Solanum dulcamara* constant as a sprawler

W6 *Alnus glutinosa-Urtica dioica* woodland
Salix fragilis sub-community

Scrubby woodland or managed or derelict osier beds dominated by *Salix triandra, S. viminalis, S. purpurea* or hybrids

W6 *Alnus glutinosa-Urtica dioica* woodland
Salix viminalis/triandra sub-community

S. fragilis and osier-bed Salices rare in woodland dominated by *Alnus glutinosa* with *Salix cinerea* frequent in the shrub layer

W6 *Alnus glutinosa-Urtica dioica* woodland
Typical sub-community

16   *Fraxinus excelsior* and/or *Betula pubescens* frequent with *Alnus glutinosa* in an often open tree canopy; shrub layer comprising mixtures of *Salix cinerea, Corylus avellana* and *Crataegus monogyna*; field layer with four or more of *Filipendula ulmaria, Lysimachia nemorum, Athyrium filix-femina, Poa trivialis, Holcus mollis, Ranunculus repens* and *Chrysosplenium oppositifolium*

W7 *Alnus glutinosa-Fraxinus excelsior-Lysimachia nemorum* woodland
17

Stands of this woodland are often found in small flushes on slopes and the proportions of the canopy and shrub-layer species are somewhat variable

Not as above

W2 *Salix cinerea-Betula pubescens-Phragmites australis* woodland
Alnus glutinosa-Filipendula ulmaria sub-community

17   *Ranunculus repens* and *Chrysosplenium oppositi-folium* constant as a ground carpet; field layer often dominated by *Urtica dioica* and *Galium aparine* or, less frequently, *Phalaris arundinacea*

> **W7** *Alnus glutinosa-Fraxinus excelsior-Lysimachia nemorum* woodland
> *Urtica dioica* sub-community

Chrysosplenium oppositifolium absent or, if present, not with *Urtica dioica, Galium aparine* or *Phalaris arundinacea*   18

18   *Corylus avellana* and *Crataegus monogyna* the most frequent and abundant shrubs; field layer with constant and often abundant *Deschampsia cespitosa, Dryopteris dilatata* and *Oxalis acetosella*; *Mnium hornum* and *Atrichum undulatum* frequent in ground layer

> **W7** *Alnus glutinosa-Fraxinus excelsior-Lysimachia nemorum* woodland
> *Deschampsia cespitosa* sub-community

This woodland type grades floristically to mixed deciduous *Fraxinus-Sorbus-Mercurialis* and *Quercus-Betula-Oxalis* woodlands around flushes on the upland margins of the north and west and it is sometimes difficult to partition samples between the communities: in general, the abundance of *Alnus* in the canopy here is a good criterion but many stands are small and the cover of even this tree may be low on occasion

---

Cirsium palustre and *Valeriana officinalis* frequent in the field layer with above herbaceous species and bryophytes infrequent and rarely abundant; field layer with a variety of possible dominants: most frequently *Carex remota* but sometimes *C. pendula* (often with *Equisetum telmateia*), *C. laevigata* or *Juncus effusus*; bryophytes often abundant with *Brachythecium rivulare, Calliergon cuspidatum, Chiloscyphus polyanthos* and *Cratoneuron commutatum* preferential

> **W7** *Alnus glutinosa-Fraxinus excelsior-Lysimachia nemorum* woodland
> *Carex remota-Cirsium palustre* sub-community

19   *Fagus sylvatica* constant and dominant in the tree canopy   20

Fagus sylvatica absent or, if present, then only as rare scattered trees within a canopy dominated by other species   26

Many woods where *Fagus* is represented show a continuous gradation in its prominence, so this distinction can be difficult to make: in particular, regeneration gaps in beech-dominated woodland types are often temporarily occupied by other communities, thus producing complex mosaics

---

20   *Quercus robur* an infrequent associate in the tree canopy; shrub layer with *Fagus sylvatica* saplings and *Ilex aquifolium* constant, the latter often abundant but all other shrubs and tree saplings rare; field layer often sparse but with *Rubus fruticosus* agg., *Pteridium aquilinum* and *F. sylvatica* seedlings constant and with various species showing characteristically patchy local dominance: *Milium effusum, Melica uniflora, Deschampsia cespitosa, Hedera helix, Ruscus aculeatus*

> **W14** *Fagus sylvatica-Rubus fruticosus* woodland

It may be difficult to separate samples of the above woodland type with a very sparse field layer from the *Fagus* sub-community of the *Fagus-Deschampsia* woodland but *Ilex aquifolium* is rare in the shrub layer there; and regeneration cores often show a transition to the *Quercus-Pteridium-Rubus* woodland with an increase in *Quercus robur* and/or *Betula pendula*

Ilex aquifolium rare in the shrub layer or, if present, then either *Mercurialis perennis* and *Hedera helix* or *Deschampsia flexuosa, Vaccinium myrtillus* or *Calluna vulgaris* present in the field layer   21

21   Two or more of the following trees or shrubs present: *Aesculus hippocastanum, Fraxinus excelsior, Acer pseudoplatanus, Taxus baccata, Sorbus aria, Corylus avellana, Crataegus monogyna*; field layer with *Mercurialis perennis* and/or *Hedera helix* (or rather sparse if *T. baccata* is prominent in the canopy)

> **W12** *Fagus sylvatica-Mercurialis perennis* woodland   22

Two or more of the following trees or shrubs present: *Quercus robur, Q. petraea, Betula pendula, Ilex aquifolium, Fagus sylvatica* saplings and *B. pendula* saplings; field layer with *Deschampsia flexuosa, Vaccinium myrtillus* or *Calluna vulgaris*

> **W15** *Fagus sylvatica-Deschampsia flexuosa* woodland   24

22   *Taxus baccata* constant as a canopy associate with occasional *Sorbus aria* and *Buxus sempervirens* but *Corylus avellana* and *Crataegus monogyna* rare; field layer often a sparse mixture of *Mercurialis perennis* and *Rubus fruticosus* agg.

W14 *Fagus sylvatica-Mercurialis perennis* woodland
Taxus baccata sub-community

Samples of this woodland type may be difficult to separate from the *Sorbus* sub-community of the *Taxus* woodland but *Fagus* is rare there and never a canopy dominant

Taxus baccata rare and never co-dominant in the tree canopy; *Corylus avellana* and *Crataegus monogyna* frequent and sometimes abundant in the shrub layer; field layer with *Hedera helix* constant   23

23   *Fraxinus excelsior* a constant associate in the tree canopy and *Acer pseudoplatanus* frequent; field layer generally dominated by mixtures of *Mercurialis perennis*, *Hedera helix*, *Brachypodium sylvaticum* and *Rubus fruticosus* agg.

W14 *Fagus sylvatica-Mercurialis perennis* woodland
Mercurialis perennis sub-community

Fraxinus excelsior and *Acer pseudoplatanus* rare in the tree canopy; field layer with two or more of *Sanicula europaea*, *Mycelis muralis*, *Melica uniflora*, *Poa nemoralis*, *Carex flacca*, *Ligustrum vulgare*; *Rubus fruticosus* agg. infrequent

W14 *Fagus sylvatica-Mercurialis perennis* woodland
Sanicula europaea sub-community

Regeneration cores within these kinds of *Fagus* woodland, particularly the *Mercurialis* subcommunity, often show a local dominance of *Fraxinus* and/or *A. pseudoplatanus* and are best considered as *Fraxinus-Acer-Mercurialis* woodland

24   *Fagus sylvatica* often the sole canopy tree with no shrubs and very sparse field layer with just *F. sylvatica* seedlings and scattered patches of *Eurhynchium praelongum*, *Mnium hornum* and *Dicranella heteromalla*

W15 *Fagus sylvatica-Deschampsia flexuosa* woodland
Fagus sylvatica sub-community

Quercus robur, *Q. petraea* or *Betula pendula* often present as canopy associates; *Deschampsia flexuosa* or mixtures of *Vaccinium myrtillus*, *Calluna vulgaris* and *Pteridium aquilinum* prominent in sometimes sparse field layer   25

Regeneration cores within this kind of *Fagus sylvatica* woodland often show a local dominance of either *Quercus* spp. or *Betula pendula*; such stands are best considered as within the *Quercus-Betula-Deschampsia* woodland

25   Field layer dominated by open or closed carpet of *Deschampsia flexuosa* without *Vaccinium myrtillus* or *Calluna vulgaris*

W15 *Fagus sylvatica-Deschampsia flexuosa* woodland
Deschampsia flexuosa sub-community

Field layer with *Vaccinium myrtillus* constant in the absence of *Calluna vulgaris*

W15 *Fagus sylvatica-Deschampsia flexuosa* woodland
Vaccinium myrtillus sub-community

Field layer with *Calluna vulgaris* constant with or without *Vaccinium myrtillus*

W15 *Fagus sylvatica-Deschampsia flexuosa* woodland
Calluna vulgaris sub-community

26   *Taxus baccata* constant and dominant in the tree canopy

W13 *Taxus baccata* woodland   27

Taxus baccata absent or, if present, then only as rare scattered trees within a canopy dominated by other species   28

27.   *Sorbus aria* a frequent canopy associate; shrub layer absent or with occasional *Sambucus nigra*; field and ground layers often totally absent

W13 *Taxus baccata* woodland
Sorbus aria sub-community

Sorbus aria rare; *Sambucus nigra* frequent in shrub layer; field layer often sparse but with constant *Mercurialis perennis*

W13 *Taxus baccata* woodland
Mercurialis perennis sub-community

28   *Pinus sylvestris* constant and dominant in the tree canopy; field layer with two or more of *Deschampsia flexuosa*, *Calluna vulgaris*, *Vaccinium myrtillus*, *V. vitisidaea*; ground layer always well developed with *Hylocomium splendens*, *Pleurozium schreberi*, *Dicranum scopar-*

ium and two or more of *Rhytidiadelphus triquetrus,*
Pseudoscleropodium purum, Plagiothecium undulatum,
Rhytidiadelphus loreus, and *Ptilium crista-castrensis,*
Hypnum jutlandicum and *Lophocolea bidentata s. l.*

> **W18** *Pinus   sylvestris-Hylocomium   splendens*
> woodland   29

Pinus sylvestris absent or rare or, if constant and dominant, then never with combinations of the above bryo-
phytes   33

Strictly speaking, the *Pinus-Hylocomium* woodland comprises native pinewoods or pine plantations within the range of native *P. sylvestris,* but
long-established plantations elsewhere may
approximate to its floristic composition

---

29   *Rhytidiadelphus triquetrus* constant and *Pseudos-*
cleropodium purum frequent in the ground layer   30

Above bryophytes absent or rare in ground layer but
Sphagnum capillifolium and *Dicranum majus* frequent
                                                                                                        32

30   *Goodyera repens* and *Erica cinerea* constant in the
field layer in the absence of *Vaccinium myrtillus* and *V.*
vitis-idaea; field layer with *Plagiothecium undulatum,*
Rhytidiadelphus loreus and *Ptilium crista-castrensis* rare
or absent

> **W18** *Pinus   sylvestris-Hylocomium   splendens*
> woodland
> *Erica cinerea-Goodyera repens* sub-community

Goodyera repens and *Erica cinerea* absent or, if present,
then also with the listed ericoids and bryophytes   31

31   *Luzula pilosa* constant and *Galium saxatile* and
Oxalis acetosella frequent in field layer

> **W18** *Pinus   sylvestris-Hylocomium   splendens*
> woodland
> *Luzula pilosa* sub-community

Above species absent from field layer

> **W18** *Pinus   sylvestris-Hylocomium   splendens*
> woodland
> *Vaccinium   myrtillus-V.   vitis-idaea*   sub-
> community

32   Ground layer with *Scapania gracilis* constant and
Thuidium tamariscinum and *Diplophyllum albicans*
frequent

> **W18** *Pinus   sylvestris-Hylocomium   splendens*
> woodland
> *Scapania gracilis* sub-community

Erica tetralix frequent in field layer; above bryophytes
absent or rare but *Sphagnum capillifolium* constant in
ground layer

> **W18** *Pinus   sylvestris-Hylocomium   splendens*
> woodland
> *Sphagnum   capillifolium-Erica   tetralix*   sub-
> community

33   *Juniperus communis* ssp. *communis* constant and
dominant in an often open canopy, occasionally with
overtopping *Betula pubescens;* field layer with *Vacci-*
nium myrtillus, Oxalis acetosella, Galium saxatile, Luz-
ula pilosa, Agrostis capillaris, A. canina montana constant; ground layer always well developed with
Hylocomium splendens and *Thuidium tamariscinum*
constant

> **W19** *Juniperus   communis-Oxalis   acetosella*
> woodland   34

Juniperus communis ssp. *communis* absent or, if present,
then as scattered shrubs within a woodland dominated
by other canopy species   35

Where *Juniperus* attains some local dominance in
the understoreys of the *Quercus-Betula-Oxalis* or
Quercus-Betula-Dicranum woodlands (a quite
widespread phenomenon in east-central Scotland, for example), this separation can be difficult
to make

---

34   Field layer with *Vaccinium vitis-idaea, Calluna*
vulgaris and *Deschampsia flexuosa* constant; ground
layer with four or more of *Dicranum scoparium, Pla-*
giothecium undulatum, Hypnum cupressiforme, Rhytidia-
delphus loreus, Pleurozium schreberi and *Dicranum*
majus

> **W19** *Juniperus   communis-Oxalis   acetosella*
> woodland
> *Vaccinium vitis-idaea-Deschampsia flexuosa* sub-
> community

Field layer somewhat variable but with combinations of
the above species lacking, with *Viola riviniana* and
Anemone nemorosa constant and with *Festuca rubra, F.*
ovina, Campanula rotundifolia and *Cardamine flexuosa*
frequent; *Rhytidiadelphus triquetrus* constant in ground
layer

> **W19** *Juniperus   communis-Oxalis   acetosella*
> woodland
> *Viola   riviniana-Anemone   nemorosa*   sub-
> community

Juniperus can invade a variety of heaths and grasslands and fragments of these often form a complex mosaic among denser stretches of Juniperus-Oxalis woodland

35   Well-defined tree canopy or thicket-like tree/shrub layer or coppice regrowth with Quercus petraea (rarely Q. robur) and/or Betula pubescens constant and dominant and Sorbus aucuparia, Corylus avellana and Ilex aquifolium the most frequent associates; field layer sometimes sparse but with Deschampsia flexuosa constant and Pteridium aquilinum, Vaccinium myrtillus and Oxalis acetosella frequent; ground layer always well-developed with six or more of the following present: Rhytidiadelphus loreus, Polytrichum formosum, Dicranum majus, Hylocomium splendens, Pleurozium schreberi, Plagiothecium undulatum, Dicranum scoparium, Thuidium tamariscinum

   W17 Quercus petraea-Betula pubescens-Dicranum majus woodland   36

Above combination of canopy, field-layer and ground-layer species absent   39

In areas of past or present atmospheric pollution, for example in the southern Pennines, it may be difficult to partition samples on the basis of bryophyte richness

36   Quercus petraea and Corylus avellana infrequent and canopy often a single tree/shrub layer with mixtures of Betula pubescens and Sorbus aucuparia dominant; Rhytidiadelphus triquetrus and Pseudoscleropodium purum frequent in ground layer; Trientalis europaea an occasional preferential

   W17 Quercus petraea-Betula pubescens-Dicranum majus woodland
   Rhytidiadelphus triquetrus sub-community

Quercus petraea (rarely Q. robur) constant in the canopy; Corylus avellana occasional to frequent; above bryophytes absent or rare in the ground layer   37

37   Isothecium myosuroides and Diplophyllum albicans constant in ground layer with four or more of Hypnum cupressiforme, Lepidozia reptans, Thuidium delicatulum, Leucobryum glaucum, Campylopus paradoxus, Plagiochila spinulosa, Scapania gracilis and Bazzania trilobata

   W17 Quercus petraea-Betula pubescens-Dicranum majus woodland

Isothecium myosuroides-Diplophyllum albicans sub-community

Combinations of the above bryophytes absent   38

38   Galium saxatile, Anthoxanthum odoratum, Agrostis capillaris and Holcus mollis constant in a grassy field layer with Vaccinium myrtillus very much reduced in frequency and abundance

   W17 Quercus petraea-Betula pubescens-Dicranum majus woodland
   Anthoxanthum odoratum-Agrostis capillaris sub-community

In grazed woodlands to the west and north, it may sometimes be difficult to partition samples between this woodland type and the Quercus-Betula-Oxalis woodland but the bryophytes typical of the former are less frequent and abundant there, Hyacinthoides non-scripta is more frequent and there is more often an underscrub of Rubus fruticosus agg. and dryopteroid ferns

Above grasses and Galium saxatile infrequent or absent

   W17 Quercus petraea-Betula pubescens-Dicranum majus woodland
   Typical sub-community

39   Well-defined tree canopy or thicket-like tree/shrub canopy or coppice regrowth with one or more of Quercus petraea, Q. robur, Betula pubescens and B. pendula constant and dominant or plantations of Pinus sylvestris, Pinus nigra var. maritima, Pseudotsuga menziesii or Larix spp.; Sorbus aucuparia occasional to frequent but Corylus avellana rare; field layer with Deschampsia flexuosa and Pteridium aquilinum constant with Oxalis acetosella, Anthoxanthum odoratum, Agrostis capillaris and Holcus mollis all rare

   W16 Quercus-Betula-Deschampsia flexuosa woodland   40

Above combination of canopy and field-layer species absent   41

40   Canopy generally dominated by Betula pendula and/or Quercus robur or plantation conifers; field layer often very species-poor with only Deschampsia flexuosa and Pteridium aquilinum at all frequent

   W16 Quercus-Betula-Deschampsia flexuosa woodland
   Quercus robur sub-community

Canopy generally dominated by Quercus petraea, Betula spp. and Sorbus aucuparia; Vaccinium myrtillus and

Dryopteris dilatata frequent in field layer; ground layer often sparse but occasionally with *Dicranella heteromalla*, *Hypnum cupressiforme*, *Isopterygium elegans*, *Mnium hornum* and *Lepidozia reptans* as scattered patches

> **W16** *Quercus-Betula-Deschampsia*   *flexuosa*
> woodland
> *Vaccinium*   *myrtillus-Dryopteris*   *dilatata*
> sub-community

41   Well-defined tree canopy or thicket-like tree/ shrub layer or coppice regrowth with one or more of *Betula pubescens*, *B. pendula*, *Quercus petraea* and *Quercus* hybrids (only occasionally *Q. robur*) constant and dominant with *Sorbus aucuparia* and *Corylus avellana* the most frequent associates; field layer with six or more of the following constant: *Oxalis acetosella*, *Anthoxanthum odoratum*, *Agrostis capillaris*, *Deschampsia flexuosa*, *Holcus mollis*, *Galium saxatile*, *Viola riviniana*, *Potentilla erecta*, *Pteridium aquilinum*; *Hyacinthoides non-scripta* or *Anemone nemorosa* often conspicuous in spring; ground layer sometimes extensive with two or more of *Rhytidiadelphus squarrosus*, *Thuidium tamariscinum*, *Pseudoscleropodium purum* and *Hylocomium splendens*

> **W11** *Quercus petraea-Betula pubescens-Oxalis*
> *acetosella* woodland   42

Above combinations of canopy, field-layer and ground-layer species absent   45

42   *Quercus petraea* constant in the canopy and often co-dominant with *Betula pubescens*; *Corylus avellana* frequent in a shrub layer; field layer with frequent vernal *Hyacinthoides non-scripta* and often with an underscrub of *Rubus fruticosus* agg., *Lonicera periclymenum*, *Dryopteris dilatata* and *D. borreri*; *Potentilla erecta* and *Hylocomium splendens* rare

> **W11** *Quercus petraea-Betula pubescens-Oxalis*
> *acetosella* woodland
> *Dryopteris dilatata* sub-community

On more calcareous and base-rich substrates to the west and north, this woodland type often grades to the Typical sub-community of the *Fraxinus-Sorbus-Mercurialis* woodland but *Fraxinus excelsior*, *Ulmus glabra* and *Acer pseudoplatanus* almost always exceed *Quercus petraea* in frequency and abundance there, the grasses characteristic here are all rare there and *Mercurialis perennis* becomes increasingly prominent

---

Quercus petraea more occasional in the canopy which is sometimes dominated by *Betula pubescens* and/or *B. pendula* in a single thicket-like tree/shrub layer; *Corylus avellana* only rare or occasional; *Hyacinthoides non-scripta* may be frequent and abundant in the vernal field layer but underscrub of *Rubus fruticosus* agg. and dryopteroids rare; *Potentilla erecta* and *Hylocomium splendens* constant   43

43   Canopy sometimes a thicket-like mixture of *Betula pubescens* and *Sorbus aucuparia*; *Blechnum spicant*, *Thelypteris limbosperma*, *Primula vulgaris* frequent in the field layer with *Hyacinthoides non-scripta* often prominent in spring; ground layer usually extensive with mixtures of *Pleurozium schreberi*, *Dicranum majus*, *Polytrichum formosum*, *Hypnum cupressiforme* and *Rhytidiadelphus loreus* present in addition to bryophytes of the community

> **W11** *Quercus petraea-Betula pubescens-Oxalis*
> *acetosella* woodland
> *Blechnum spicant* sub-community

On neutral to acidic soils in north-western Britain, it is often difficult to distinguish this woodland type from the *Anthoxanthum-Agrostis* sub-community of the *Quercus-Betula-Dicranum* woodland, particularly if grazing is heavy (as it often is) and sampling is too late in the year to catch any *Hyacinthoides*: under such circumstances, the abundance and diversity of calcifuge bryophytes can serve to partition samples but problems with diagnosis are a real reflection of the convergence of these two woodland types with particular kinds of treatment

---

Canopy usually a thicket-like mixture of *Betula pubescens* and *B. pendula* sometimes with *Quercus robur*, *Q. petraea* or *Quercus* hybrids; above field-layer and ground-layer species infrequent but *Rhytidiadelphus triquetrus* constant   44

44   *Luzula pilosa* and *Trientalis europaea* constant with *Anemone nemorosa* a common vernal plant

> **W11** *Quercus petraea-Betula pubescens-Oxalis*
> *acetosella* woodland
> *Anemone nemorosa* sub-community

Stellaria holostea, *Hypericum pulchrum*, *Luzula multiflora*, *Ajuga reptans*, *Festuca rubra*, *Veronica officinalis* and *Cerastium fontanum* frequent in field layer; *Plagiomnium undulatum* and *Lophocolea bidentata s.l.* frequent in the ground layer

W11 *Quercus petraea-Betula pubescens-Oxalis acetosella* woodland
Stellaria holostea-Hypericum pulchrum sub-community

45 High forest or coppice with constant *Quercus robur* (locally *Q. petraea*) and frequent *Betula pendula*, and occasional (sometimes locally dominant) *Carpinus betulus, Tilia cordata* or *Castanea sativa; Fraxinus excelsior, Acer pseudoplatanus* and *Ulmus glabra* generally infrequent, though more common to the north-west; *Corylus avellana* and *Crataegus monogyna* the most frequent shrubs, the former very commonly a coppice dominant; field layer generally dominated by diverse mixtures of *Rubus fruticosus* agg., *Pteridium aquilinum* and *Lonicera periclymenum* with *Hyacinthoides non-scripta* a common vernal dominant; *Mercurialis perennis* and other calcicolous herbs rare

OR Plantations of *Pinus sylvestris, Pinus nigra* var. *maritima, Pseudotsuga menziesii* or *Larix* spp. with the same kind of field layer

    W10 *Quercus robur-Pteridium aquilinum-Rubus fruticosus* woodland   46

Above combinations of canopy, shrub, field- and ground-layer species absent   50

46 *Anemone nemorosa* largely replacing *Hyacinthoides non-scripta* as the vernal dominant and *Castanea sativa* often abundant in the canopy

    W10 *Quercus robur-Pteridium aquilinum-Rubus fruticosus* woodland
    Anemone nemorosa sub-community

Anemone nemorosa infrequent, though *Castanea sativa* sometimes a canopy dominant   47

47 *Quercus robur* and/or *Betula pendula* frequent and dominant in a well-defined canopy or as thicket tree/shrub layer or conifer plantations of *Pinus sylvestris, Pinus nigra* var. *maritima, Pseudotsuga menziesii* or *Larix* spp.; *Corylus avellana* rare; field layer with *Holcus lanatus* constant and often abundant but *H. mollis* and *Hyacinthoides non-scripta* rare

    W10 *Quercus robur-Pteridium aquilinum-Rubus fruticosus* woodland
    Holcus lanatus sub-community

Corylus avellana frequent and often abundant with *Holcus lanatus* rare in the field layer   48

48 *Quercus robur* usually the dominant in a well-defined tree canopy; *Hedera helix* constant in the field layer and often forming a prominent carpet

    W10 *Quercus robur-Pteridium aquilinum-Rubus fruticosus* woodland
    Hedera helix sub-community

Towards the south-west of England, this woodland type grades to the *Hedera helix* sub-community of *Fraxinus-Acer-Mercurialis* woodland; although a *Hedera* carpet and an under-scrub of *Rubus, Pteridium* and *Lonicera* can occur there, *Mercurialis perennis* is almost always present and often abundant in the field layer, *Fraxinus excelsior* is an important component of the canopy and *Acer campestre* is frequent in either the canopy or shrub layer

Quercus robur may be dominant but *Hedera helix* rare in the field layer   49

49 *Fraxinus excelsior* and *Acer pseudoplatanus* frequent and sometimes locally prominent in the canopy and as saplings in the shrub layer; *Oxalis acetosella* and *Dryopteris dilatata* frequent in the field layer

    W10 *Quercus robur-Pteridium aquilinum-Rubus fruticosus* woodland
    Acer pseudoplatanus-Oxalis acetosella sub-community

To the north-east of England, it may be difficult to partition samples between this woodland type and the *Dryopteris* sub-community of its northern counterpart, the *Quercus-Betula-Oxalis* woodland; in that woodland, there is a general switch from *Quercus robur* and *Betula pendula* to *Q. petraea* and *B. pubescens* but the best separator is probably the constancy there of *Anthoxanthum odoratum, Agrostis capillaris* and *Deschampsia flexuosa* in the field layer

Fraxinus excelsior and *Acer pseudoplatanus* never more than occasional in the canopy or shrub layer; *Oxalis acetosella* rare in the field layer

    W10 *Quercus robur-Pteridium aquilinum-Rubus fruticosus* woodland
    Typical sub-community

50 High forest or coppice with constant *Fraxinus excelsior* and *Quercus robur* or *Q. petraea* and occasional (sometimes locally dominant) *Carpinus betulus, Tilia cordata* or suckering elms; *Corylus avellana* and *Crataegus monogyna* frequent in the shrub layer, the former

often a coppice dominant; *Acer campestre* frequent in canopy or understorey with occasional *Cornus sanguinea* and *Euonymus europaeus* but *Sorbus aucuparia* scarce; field layer often showing strong vernal dominance of *Mercurialis perennis, Hyacinthoides nonscripta, Anemone nemorosa* or *Allium ursinum* (or in parts of East Anglia, *Primula elatior*); ferns and bryophytes often sparse

> **W8** *Fraxinus excelsior-Acer campestre-Mercurialis perennis* woodland   52

High forest (sometimes scrubby) or coppice with frequent *Fraxinus excelsior, Sorbus aucuparia, Betula pubescens* and occasional *Quercus petraea*; *Tilia cordata, Carpinus betulus*, suckering elms, *Acer campestre, Cornus sanguinea* and *Euonymus europaeus* all very scarce; field layer often without a uniform dominant but with *Oxalis acetosella* and *Viola riviniana* frequent, together with *Athyrium filix-femina* but only rare *Arum maculatum* and *Viola reichenbachiana*; bryophytes often abundant with *Thuidium tamariscinum, Eurhynchium striatum* and *Plagiomnium undulatum* common

> **W9** *Fraxinus excelsior-Sorbus aucuparia-Mercurialis perennis* woodland   51

51   Canopy often a scrubby and sometimes open mixture of *Betula pubescens, Sorbus aucuparia* and *Corylus avellana* with frequent *Fraxinus excelsior* and locally abundant *Prunus padus*; field layer often luxuriant with five or more of *Filipendula ulmaria, Conopodium majus, Crepis paludosa, Deschampsia cespitosa, Arrhenatherum elatius, Geum rivale, Cirsium helenioides, Geranium sylvaticum, Rumex acetosa, Vicia sepium*

> **W9** *Fraxinus excelsior-Sorbus aucuparia-Mercurialis perennis* woodland
> *Crepis paludosa* sub-community

Usually tall high forest or coppice dominated by mixtures of *Fraxinus excelsior, Ulmus glabra* and *Acer pseudoplatanus* with occasional *Quercus petraea; Crataegus monogyna* frequent in shrub layer; above herbs scarce and field layer usually a shorter, more open cover with frequent *Mercurialis perennis, Geum urbanum, Circaea lutetiana* and *Potentilla sterilis*

> **W9** *Fraxinus excelsior-Sorbus aucuparia-Mercurialis perennis* woodland
> Typical sub-community

52   Well-defined tree canopy or coppice regrowth with *Fraxinus excelsior, Ulmus glabra, Acer pseudoplatanus* or mixtures of these species constant and often dominant and *Quercus robur* rare   53

Ulmus glabra and *Acer pseudoplatanus* rare but *Quercus robur* frequent and sometimes dominant   55

53   Prominent and varied shrub layer with *Cornus sanguinea, Viburnum opulus, Sorbus aucuparia* and *Rhamnus catharticus* all frequent with *Corylus avellana, Crataegus monogyna* and *Acer campestre*; field layer often a complex mosaic but with five or more of the following: *Deschampsia cespitosa, Teucrium scorodonia, Campanula latifolia, C. trachelium, Angelica sylvestris, Melica uniflora, Arrhenatherum elatius, Sanicula europaea, Viola riviniana, Polystichum aculeatum; Melica nutans* and *Rubus caesius* differential at low frequency

> **W8** *Fraxinus excelsior-Acer campestre-Mercurialis perennis* woodland
> *Teucrium scorodonia* sub-community

Above combinations of shrub and field-layer species absent   54

54   *Allium ursinum* a constant and often dominant vernal species in the field layer, frequently with some *Urtica dioica* and *Galium aparine*

> **W8** *Fraxinus excelsior-Acer campestre-Mercurialis perennis* woodland
> *Allium ursinum* sub-community

Allium ursinum infrequent and never a vernal field-layer dominant; field layer usually dominated by mixtures of *Mercurialis perennis, Brachypodium sylvaticum, Hedera helix, Geranium robertianum, Urtica dioica, Galium aparine*, sometimes with an underscrub of *Rubus fruticosus* agg.

> **W8** *Fraxinus excelsior-Acer campestre-Mercurialis perennis* woodland
> *Geranium robertianum* sub-community

55   *Deschampsia cespitosa* constant and often dominant in the field layer with occasional *Mercurialis perennis, Potentilla sterilis* and *Filipendula ulmaria* but *Primula vulgaris, P. elatior* and their hybrids and *Glechoma hederacea* rare

> **W8** *Fraxinus excelsior-Acer campestre-Mercurialis perennis* woodland
> *Deschampsia cespitosa* sub-community

Deschampsia cespitosa rare and never dominant in the field layer, but *Hedera helix* constant and often abundant as a ground carpet with frequent *Brachypodium sylvaticum; Mercurialis perennis* constant but often sparse and other listed associates scarce

W8 *Fraxinus excelsior-Acer campestre-Mercurialis perennis* woodland
Hedera helix sub-community

Deschampsia cespitosa rare and *Hedera helix* only occasional and neither abundant in the field layer, but *Primula vulgaris*, *Glechoma hederacea* and *Poa trivialis* frequent 56

56 *Anemone nemorosa* and *Ranunculus ficaria* constant and often dominant vernal species

W8 *Fraxinus excelsior-Acer campestre-Mercurialis perennis* woodland
Anemone nemorosa sub-community

Anemone nemorosa and *Ranunculus ficaria* rare

W8 *Fraxinus excelsior-Acer campestre-Mercurialis perennis* woodland
Primula vulgaris-Glechoma hederacea sub-community

57 Underscrub dominated by *Rubus fruticosus* agg. with other woody species infrequent and not abundant 58

Rubus fruticosus agg. often a frequent and prominent component of the vegetation but one or more of *Crataegus monogyna*, *Prunus spinosa*, *Ulex europaeus* and *Cytisus scoparius* also present and commonly dominant, often in taller vegetation 61

58 *Pteridium aquilinum* a constant and often abundant companion to *Rubus fruticosus* agg.

W25 *Rubus fruticosus* agg.-*Pteridium aquilinum* underscrub 59

Pteridium aquilinum absent

W24 *Rubus fruticosus* agg.-*Holcus lanatus* underscrub 60

59 *Hyacinthoides non-scripta* frequent and often abundant in spring with *Urtica dioica*, *Galium aparine* and *Holcus mollis* common

W25 *Rubus fruticosus* agg.-*Pteridium aquilinum* underscrub
Hyacinthoides non-scripta sub-community

Above species infrequent but *Teucrium scorodonia* and *Holcus lanatus* common

W25 *Rubus fruticosus* agg.-*Pteridium aquilinum* underscrub
Teucrium scorodonia sub-community

60 *Cirsium arvense*, *C. vulgare* and *Agrostis stolonifera* frequent in rank and weedy vegetation with patchy local abundance of a wide variety of species such as *Epilobium angustifolium*, *E. hirsutum* and *Urtica dioica*

W24 *Rubus fruticosus* agg.-*Holcus lanatus* underscrub
Cirsium arvense-Cirsium vulgare sub-community

Cirsium spp. and *Agrostis stolonifera* scarce but *Arrhenatherum elatius*, *Festuca rubra*, *Dactylis glomerata*, *Heracleum sphondylium*, *Anthriscus sylvestris*, *Chaerophyllum temulentum* all common in rank grassy vegetation

W24 *Rubus fruticosus* agg.-*Holcus lanatus* underscrub
Arrhenatherum elatius-Heracleum sphondylium sub-community

61 *Ulex europaeus* and/or *Cytisus scoparius* frequent and often co-dominant with *Rubus fruticosus* agg. but other woody species scarce

W23 *Ulex europaeus-Rubus fruticosus* agg. scrub 62

Ulex europaeus and *Cytisus scoparius* occasionally present but as a minor component in scrub dominated by other woody species 63

62 *Teucrium scorodonia* constant; *Agrostis capillaris*, *Holcus lanatus*, *Galium saxatile* infrequent

W23 *Ulex europaeus-Rubus fruticosus* agg. scrub
Teucrium scorodonia sub-community

Teucrium scorodonia infrequent but *Rumex acetosella* and *Hypochoeris radicata* constant, *Senecio jacobaea* and *Plantago lanceolata* common, *Crepis capillaris* and *Jasione montana* distinctive occasionals

W23 *Ulex europaeus-Rubus fruticosus* agg. scrub
Rumex acetosella sub-community

All the above species infrequent but *Anthoxanthum odoratum* and *Potentilla erecta* frequent

W23 *Ulex europaeus-Rubus fruticosus* agg. scrub
Anthoxanthum odoratum sub-community

63 *Prunus spinosa* constant and typically the sole dominant in often very dense scrub with other woody plants scarce

W22 *Prunus spinosa-Rubus fruticosus* agg. scrub 64

Prunus spinosa a frequent and sometimes locally prominent species but always accompanied by and usually exceeded in cover by a variety of other woody species

> **W21** *Crataegus monogyna-Hedera helix* scrub 65

64   *Hedera helix* and *Silene dioica* frequent

> **W22** *Prunus spinosa-Rubus fruticosus* agg. scrub
> *Hedera helix-Silene dioica* sub-community

Viola riviniana and *Veronica chamaedrys* frequent

> **W22** *Prunus spinosa-Rubus fruticosus* agg. scrub
> *Viola riviniana-Veronica chamaedrys* sub-community

Dactylis glomerata constant with two or more of *Brachypodium sylvaticum*, *Festuca rubra*, *Holcus lanatus*, *Agrostis capillaris*, *Rumex acetosa*, sometimes with maritime herbs

> **W22** *Prunus spinosa-Rubus fruticosus* agg. scrub
> *Dactylis glomerata* sub-community

65   *Ligustrum vulgare*, *Viburnum lantana* and *Cornus sanguinea* constant with frequent scrambling *Clematis vitalba* and *Tamus communis*

> **W21** *Crataegus monogyna-Hedera helix* scrub
> *Viburnum lantana* sub-community

Above species no more than very occasional   66

66   *Sambucus nigra*, *Urtica dioica* and *Galium aparine* very common with only scarce *Brachypodium sylvaticum*
67

Brachypodium sylvaticum frequent, sometimes with *Sambucus nigra*, *Urtica dioica* and *Galium aparine*

> **W21** *Crataegus monogyna-Hedera helix* scrub
> *Brachypodium sylvaticum* sub-community

67   *Mercurialis perennis* and *Arum maculatum* very common, with occasional *Glechoma hederacea*, *Hyacinthoides non-scripta*, *Poa trivialis*

> **W21** *Crataegus monogyna-Hedera helix* scrub
> *Mercurialis perennis* sub-community

Above species rare but field layer often with an untidy look with occasional *Arrhenatherum elatius*, *Holcus lanatus*, *Heracleum sphondylium*, *Silene dioica*

> **W21** *Crataegus monogyna-Hedera helix* scrub
> *Hedera helix-Urtica dioica* sub-community

COMMUNITY DESCRIPTIONS

W1
Salix cinerea-Galium palustre woodland

Synonymy

Salix carr   Willis & Jefferies 1959 *p.p.; Salix cinerea*
carr Wheeler 1980*c p.p.;* Woodland plot type
32 Bunce 1982; *Scutellaria galericulata-Alnus glutinosa* Association   Birse 1982 *p.p.*

Constant species

Salix cinerea, Galium palustre.

Rare species

Lysimachia thyrsiflora.

Physiognomy

The *Salix cinerea-Galium palustre* woodland has a
canopy dominated by *S. cinerea* but stands vary considerably in their overall appearance. Where invasion is
more recent, there can be a confused mass of bushes of
varying height and density but older stands have a more
even look with usually a single tier of sallows forming a
canopy 4–8 m high. Here, there can be an abundance of
standing dead wood beneath, where thickly-set colonising bushes have been shaded out by the developing
survivors, but long-established sallows, especially multi-
stemmed individuals which form broadly-spreading
crowns, usually cast a light shade.

The commonest woody associate, though it is still no
more than occasional, is *Betula pubescens* and scattered
trees of this species, together with scarcer *Alnus glutinosa, Quercus robur* and *Betula pendula,* sometimes
break the sallow canopy, reaching 10–15 m. Other
Salices are uncommon (there are sometimes records for
the osiers, *S. viminalis* and *S. purpurea*) but there can be
scattered bushes of *Crataegus monogyna, Corylus avellana* and *Frangula alnus.* Saplings of *Betula pubescens,
Quercus robur* and *Alnus* occur very occasionally.

The field layer varies in its cover and composition but
the general appearance is of an open scatter of herbs
with different species attaining occasional local prominence. Quite commonly there are mosaics developed in
relation to canopy gaps and over undulations of wetter
and drier ground. The commonest species throughout is
Galium palustre but *Mentha aquatica* and *Juncus effusus*
are also frequent, the latter sometimes forming large
tussocks in lighter shade. Then, there are scattered
plants of *Angelica sylvestris, Lycopus europaeus, Ranunculus flammula, R. repens, Epilobium palustre, Equisetum
fluviatile, Filipendula ulmaria, Cirsium palustre* and
Rumex sanguineus. Less common, though sometimes
abundant, are *Caltha palustris, Hydrocotyle vulgaris,
Potentilla palustris* and *Iris pseudacorus* and, in North
Yorkshire and central Scotland, the national rarity
Lysimachia thyrsiflora occurs (e.g. Birse 1982). In some
stands, there may be much *Rubus fruticosus* agg. and
Solanum dulcamara and, where these sprawl over the
living and dead sallows, the vegetation can be virtually
impenetrable. *Hedera helix,* too, can be abundant over
the ground and among the bushes. In other cases, the
field layer may have a more grassy look with *Holcus
lanatus* and, especially where stock graze, sheets of
Agrostis canina ssp. *canina* or *A. stolonifera* (e.g. Willis &
Jefferies 1959). In general, swamp and fen dominants are
not a consistent feature of this community but occasional stands can be found with an abundance of *Carex
paniculata, C. riparia, C. vesicaria* (e.g. Bunce 1982) or
sparse *Phragmites australis.*

There is quite often a considerable amount of bare
ground among the herbs, particularly where stock have
trampled, and over this, and on fallen twigs and over the
bush bases, there is frequently a patchy cover of bryophytes. *Eurhynchium praelongum* is the most frequent
species but *Chiloscyphus polyanthos, Calliergon cuspidatum, C. cordifolium, Brachythecium rutabulum* and *Rhytidiadelphus squarrosus* can also be found. In more
sheltered situations, and especially towards the south-
west, there may be a striking profusion of epiphytic
lichens over the lightly-shaded sallow branches.
Usneion barbatae communities seem especially characteristic and, at Slapton Ley in Devon, Hawksworth
(1972) recorded *Ramalinetum fastigiatae, Usneetum sub-
floridanae* and *Usneetum articulato-floridae* var. *cerati-
nae* assemblages (see also James *et al.* 1977).

Habitat

The *Salix-Galium* woodland is essentially a community of wet mineral soils on the margins of standing or slow-moving open waters and in moist hollows, mainly in the lowlands. It occurs, often as a narrow fringe or scattered fragments, around ponds, lakes and dune slacks and along ditches, canals and sluggish streams and rivers. Less often, it can be found on silty peats in the lagg of basin mires and on flood-plain mires.

Salix cinerea can invade moist ground in a wide variety of situations, provided there is some period of freedom from surface standing water in early summer when the seeds germinate simultaneously (Ellis 1965). Here, such conditions are provided by the seasonal fall of the water-table after the winter inundation of lake and stream margins or by the gradual terrestrialisation of open waters by silting. Once established in congenial surroundings, it can grow very rapidly and reproduce not only by prolific fruiting but also from detached shoots. The soils are often strongly gleyed and *S. cinerea* seems to tolerate a fair degree of winter waterlogging. Stands may have residual pools of water between the bushes and, in exceptional circumstances, whole stands may be flooded but prolonged raising of the water-table probably kills the species (e.g. Hawksworth 1972).

Where suitable open waters or wet ground occur within agricultural land, freedom from grazing is another essential for the development of the community: in the Gordano valley in Somerset, for example, Willis & Jefferies (1959) considered that occasional grazing might be sufficient to set back the spread of the scattered *S. cinerea* bushes. In wet meadows and road-side ditches, regular cutting can also restrict the spread of sallows.

The field layer of the community is, generally speaking, a shaded version of the kind of poorly-developed swamp understorey characteristic of the limits of winter-flooding of mineral soils where *Galium palustre*, *Mentha aquatica* and scattered *Juncus effusus* are typically prominent. Much of the variation within and between stands reflects the pattern of light penetration beneath the developing canopy, differences in the amount of soil moisture over the often uneven ground and, where stock have access to established stands, the extent of grazing beneath. Some, however, may represent inherited variation from the diverse kinds of vegetation which *S. cinerea* can invade.

Zonation and succession

On the margins of open waters, the *Salix-Galium* woodland often forms a patchy fringe behind some kind of swamp vegetation growing on the wetter mineral substrates, e.g. the *Phragmitetum australis*, *Typhetum latifoliae*, *Sparganietum erecti*, *Caricetum ripariae* or *Caricetum vesicariae*. In more extensive transitions, there can

be an intervening zone of *Phragmites-Eupatorium* or *Phragmites-Urtica* fen or complex mosaics of woodland, fen and swamp which reflect the extent and pattern of colonisation by woody vegetation: such patchworks are a common feature of silted pools, disused ponds and canals and larger loops of sluggish streams. On occasion, the community can pass, towards drier ground, to some type of *Alnus-Urtica* woodland but, very frequently, agricultural improvement almost to the margins of open waters means that stands of the *Salix-Galium* woodland give way to modified herbaceous vegetation. Such transitions may be gradual, as in ill-drained farmland around field ponds, where the community may pass to grazed *Holco-Juncetum* or *Lolio-Cynosuretum*. In other cases, sharp treatment-related boundaries terminate the sequence more abruptly and the woodland remains isolated within an intensive agricultural landscape.

Some other particular contexts of the *Salix-Galium* woodland deserve special mention. In roadside ditches, the community can occur in linear stands which on occasion function as hedges and here it is usually fronted by some kind of mown *Arrhenatheretum*. In dune slacks, it can form a zone between the *Salix repens-Holcus lanatus* community on wetter ground and some kind of grassland or heath on the drier fixed sand around. Where the community occurs in the laggs of raised mires, it gives way towards the mire centre to herbaceous bog vegetation or, where trees have colonised the drying peat surface, to the *Betula-Molinia* woodland.

Seral developments of the *Salix-Galium* woodland have not been followed but, in its usual position in topogenous wetlands with mineral soils, it would seem most likely that the community would be replaced by *Alnus-Urtica* woodland with increased terrestrialisation. Extension of an *Alnus glutinosa* or *Betula pubescens* canopy above the sallows and an increase in *Urtica dioica*, *Rubus fruticosus* agg. and dryopteroid ferns in the field layer would result in a composition akin to that found in the *Sambucus* or *Betula* sub-communities of that woodland type. *Salix-Galium* woodland as defined here is not a precursor to the richer kinds of carr on fen peats. In more exposed localities, as around the coasts of Cornwall, west Wales and western Scotland, the community may represent the end-point of colonisation by woody species on wetter mineral soils.

Distribution

The community occurs in scattered localities throughout the lowlands.

Affinities

Most descriptions of scrub or woodland dominated by *S. cinerea* (e.g. Pearsall 1918, Tansley 1939, Willis & Jefferies 1959, Wheeler 1978, Birse 1982, 1984) involve vegetation which, at least in part, is best considered as immature forms of carr on fen peat. This is described elsewhere in this scheme and the present unit is retained for those stands dominated by *S. cinerea* which lack field layers with such swamp and fen dominants as *Phragmites australis, Carex paniculata, C. acutiformis* or *C. rostrata*. Although somewhat poorly defined in floristic terms and often present in fragmentary form, the community seems to have an ecological integrity as a pioneer woody vegetation of wet mineral soils. It cannot easily be incorporated into the *Alnus-Urtica* woodland (unlike scrubs dominated by the osiers *S. viminalis* and *S. purpurea*) but it is probably best considered as part of the Alnion glutinosae rather than the Salicion cinereae (Frangulo-Salicion auritae of Ellenberg 1978).

Floristic table W1

Salix cinerea	V (4–10)	Dryopteris carthusiana	I (4–6)
Betula pubescens	II (3–7)	Rumex acetosa	I (1–4)
Quercus robur	I (4–5)	Cardamine flexuosa	I (1–3)
Betula pendula	I (4–7)	Lotus uliginosus	I (3–4)
Crataegus monogyna	I (1–5)	Carex remota	I (1–5)
Alnus glutinosa	I (4)	Rhytidiadelphus squarrosus	I (1–3)
Frangula alnus	I (4–8)	Calliergon cuspidatum	I (1–4)
Corylus avellana	I (3–5)	Carex paniculata	I (1–8)
Salix viminalis	I (6)	Brachythecium rutabulum	I (3–5)
Salix purpurea	I (5)	Lychnis flos-cuculi	I (4–5)
		Lonicera periclymenum	I (2–4)
Betula pubescens sapling	I (2–4)	Juncus acutiflorus	I (3–8)
Quercus robur sapling	I (1–3)	Plagiomnium undulatum	I (1–4)
Alnus glutinosa sapling	I (3–9)	Thuidium tamariscinum	I (1–5)
		Lophocolea bidentata s.l.	I (2–4)
Galium palustre	IV (1–4)	Callitriche stagnalis	I (2–6)
		Equisetum palustre	I (2–5)
Juncus effusus	III (1–6)	Calliergon cordifolium	I (3–4)
Mentha aquatica	III (1–6)	Galium uliginosum	I (3–5)
Holcus lanatus	II (3–6)	Valeriana officinalis	I (3)
Eurhynchium praelongum	II (1–6)	Agrostis capillaris	I (2–4)
Angelica sylvestris	II (2–5)	Dactylis glomerata	I (3–4)
Rubus fruticosus agg.	II (1–6)	Myosotis scorpioides	I (2–3)
Ranunculus flammula	II (1–6)	Carex riparia	I (1–8)
Solanum dulcamara	II (1–7)	Phalaris arundinacea	I (5–6)
Lycopus europaeus	II (1–6)	Lemna minor	I (3–4)
Ranunculus repens	II (1–6)	Apium nodiflorum	I (4–5)
Equisetum fluviatile	II (1–4)	Glyceria fluitans	I (1–4)
Hedera helix	II (1–6)	Galium aparine	I (2–3)
Epilobium palustre	II (2–4)	Carex nigra	I (2–4)
Agrostis canina canina	I (3–7)		
Filipendula ulmaria	I (1–8)	Number of samples	38
Cirsium palustre	I (3–4)	Number of species/sample	17 (3–32)
Agrostis stolonifera	I (1–7)		
Rumex sanguineus	I (1–4)	Tree/shrub height (m)	6 (1–15)
Hydrocotyle vulgaris	I (3–4)	Tree/shrub cover (%)	82 (40–100)
Potentilla palustris	I (1–5)	Herb height (cm)	56 (5–200)
Chiloscyphus polyanthos	I (1–4)	Herb cover (%)	82 (10–100)
Caltha palustris	I (4–8)	Ground height (mm)	18 (10–30)
Phragmites australis	I (2–4)	Ground cover (%)	24 (0–100)
Molinia caerulea	I (4–8)		
Iris pseudacorus	I (1–4)	Altitude (m)	102 (1–390)

W1 *Salix cinerea-Galium palustre* woodland

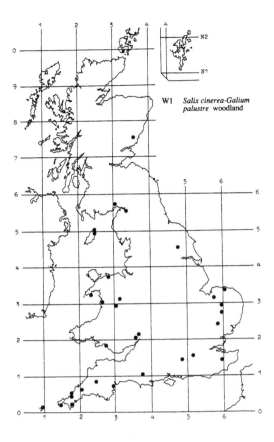

W1   *Salix cinerea-Galium palustre* woodland

W2
Salix cinerea-Betula pubescens-Phragmites australis woodland

Synonymy

Fen carr Pallis 1911, Godwin & Turner 1933, Godwin 1936, Tansley 1939, Lambert 1951; Closed carr Pearsall 1918 *p.p.; Rhamnetum, Franguletum, Rhamno-franguletum* Godwin & Tansley 1929, Godwin 1936, 1943*a, b; Betuletum* Godwin & Turner 1933; Semiswamp and pseudo-swamp carrs Lambert 1951 *p.p.; Calamagrostis* sere Poore 1956*b p.p.;* Alder woodland type 2c McVean 1956*b p.p.; Betulo-Dryopteridetum cristatae* Wheeler 1978; *Osmundo-Alnetum* (Klötzli 1970) Wheeler 1980*c p.p.; Frangula alnus* and *Rhamnus catharticus* sociations Wheeler 1980*c p.p.; Betulo-Myricetum gale* Wheeler 1980*c p.p.; Alnus-Salix-Betula* woodland Meres Report 1980 *p.p.;* Fen Woodland Fitter *et al.* 1980 *p.p.*

Constant species

Betula pubescens, Salix cinerea, Phragmites australis.

Rare species

Carex elongata, Dryopteris cristata, D. × uliginosa, Peucedanum palustre, Pyrola rotundifolia, Thelypteris palustris.

Physiognomy

The *Salix cinerea-Betula pubescens-Phragmites australis* woodland is rather variable in the floristics and physiognomy of its woody component. *Salix cinerea* and *Betula pubescens* are the most frequent species throughout and, together with some *Alnus glutinosa*, they usually form the bulk of the canopy. These major colonists show some differences in their preference for the various habitat conditions characteristic of the community (see below) but the chance availability of seed-parents and vagaries of propagule dispersal clearly play a large part in determining the composition and structure of the canopy, especially in the early stage of development (e.g. Godwin 1936). Moreover, these species show some ability to 'leapfrog' over one another in the colonisation process, so that, though *S. cinerea* is a common early invader (e.g. Pearsall 1918, Lambert 1946), it may be overtaken by waves of *Alnus* (e.g. Pigott & Wilson 1978) or *B. pubescens*. Without extensive detailed studies on a large number of sites, it is difficult to say whether there are consistent sequences of invasion which are reflected in the subsequent woodland structure. Young canopies of this community are certainly very heterogeneous internally and very different, one from another, comprising mixtures of these three species in various proportions, with individuals of different ages, forming a cover that varies from low, open scrub or very dense thickets to taller woodland with some stratification (e.g. Pallis 1911, Godwin & Turner 1933, Lambert 1951). Furthermore, in some stands, other woody species, which are generally no more than occasional through the community as a whole, can attain a striking prominence in the early stages of colonisation and exert a long-lasting, though by no means simple and unchanging, influence on the vegetation. The most renowned example of this is at Wicken Fen in Cambridgeshire where *Frangula alnus* and *Rhamnus catharticus* were, and over much of the fen remain, major components of the canopy (Godwin 1936, Godwin *et al.* 1974) but such local dominance, especially of *Frangula*, can be encountered elsewhere.

The composition and structure of the field layer is strongly influenced by the character of the preceding fen vegetation and the extent to which its various elements can survive in the changing environment. Since both the floristics and physiognomy of the invaded fen and the cover and density of the colonising canopy which increasingly affects it are very variable, the field layer of the community as a whole has few species that provide a strong, consistent core throughout. Of those bulky monocotyledons which commonly dominate the kinds of fen over which this woodland usually develops, only *Phragmites australis* is sufficiently persistent to remain a frequent species overall and, even then, its cover is very variable. In some stands it is very abundant, forming dense patches, a metre or more tall, which dominate the field layer; elsewhere it is much more sparse and, rarely, it is absent altogether. Other fen dominants, like *Carex acutiformis*, are more sporadic or are generally confined to younger and more open canopies, e.g. *Cladium mariscus* (Godwin 1936, Godwin *et al.* 1974), *Calamagrostis canescens* (Lambert 1951) or, at Woodwalton Fen in Cambridgeshire, *C. epigejos* (Poore 1956b). *Carex paniculata* is another species which can be very persistent but it is scarce in the community and much more typical of the *Alnus-Carex* woodland. However, despite their patchy representation, the gross physiognomy of these species means that, when they do occur, they can give stands a striking individuality.

No member of the associated flora of tall-herb fens remains frequent throughout the community but *Thelypteris palustris*, particularly towards the heart of its range in East Anglia, is a distinctive occasional. Indeed, its occurrence in the fen systems there has been associated (Wheeler 1978) with the development of young carr but it remains decidedly local in both the sub-communities, being very abundant in some stands and quite absent from others. The tall dicotyledons characteristic of fens are likewise somewhat patchily distributed. The commonest species throughout are *Filipendula ulmaria* and *Eupatorium cannabinum* with, less frequently, *Lysimachia vulgaris*, *Lythrum salicaria* and the rare *Peucedanum palustre*. Again, these seem to be more frequently found where stands of this woodland have developed among richer fens, such as those of Broadland, but certain of them are also more consistently associated with the *Alnus-Filipendula* sub-community. It is possible that variation in this component, like that in the amount of *Myrica gale* and *Molinia caerulea*, which are preferential to the *Sphagnum* sub-community, represents in part differences inherited from the preceding fen vegetation.

Other general floristic features of the field layer are weakly developed but three elements deserve comment, especially in the light of the possible seral developments of the community (see below). First, there is quite commonly an undershrub layer. *Rubus fruticosus* agg. and *Rosa canina* agg. are its most frequent components but, less often, there can be some *Rubus idaeus* (or *R. caesius* at Wicken: Godwin 1936, Godwin *et al.* 1974), *Ribes rubrum* or *R. nigrum*. Second, *Dryopteris dilatata*, generally speaking an uncommon fern of fens, except where it gains a hold on drier sedge tussocks, makes an appearance. And, third, over the sometimes very extensive areas of bare ground, there can be loose mats of *Poa trivialis* and wefts of *Eurhynchium praelongum*.

Sub-communities

***Alnus glutinosa-Filipendula ulmaria* sub-community:** Fen carr Pallis 1911, Godwin 1936, Lambert 1951; *Rhamnetum, Franguletum, Rhamno-franguletum* Godwin & Tansley 1929, Godwin 1936, 1943 *a, b*; Semi-swamp and pseudo-swamp carrs Lambert 1951 *p.p.*; *Calamagrostis* sere Poore 1956b; Alder woodland type 2c McVean 1956b *p.p.*; *Osmundo-Alnetum* (Klötzli 1970) Wheeler 1980c *p.p.*; *Frangula alnus* and *Rhamnus catharticus* sociations Wheeler 1980c *p.p.*; Fen Woodland Fitter *et al.* 1980 *p.p.* In its typical and more mature forms, the canopy in this sub-community is more species-rich, varied and structurally complex than that of the *Sphagnum* sub-community. *Alnus glutinosa* is preferentially frequent here and it can exceed *S. cinerea* and *B. pubescens* in abundance. Well-grown trees, together with occasional *Fraxinus excelsior* and scarcer *Quercus robur*, sometimes form an upper tier to the canopy, reaching 15 m, with somewhat shorter *S. cinerea* and *B. pubescens* beneath. Commonly, however, such stratification is indistinct, even in longer-established stands, and there is often a complete gradation from taller trees and well-grown shrubs to shorter individuals. Among the smaller woody associates, the most frequent are *Crataegus monogyna*, *Frangula alnus* and, especially distinctive of this sub-community, *Viburnum opulus* and *Rhamnus catharticus*. Under more mature canopies, the commonest saplings are of *Alnus* and *Fraxinus*.

Two features can increase the physiognomic complexity of this woody component. First, competition between the developing canopy elements can leave many overtopped shrubs (notably smaller individuals of *S. cinerea* and *Frangula*) dying or dead and their remains can choke the understorey in places and leave an abundance of brushwood litter over the ground. Second, on substrates which are less firm, as for example where there is unconsolidated silt below peat, the weight of the developing canopy may cause some subsidence or surface breakdown of the substrate so that the trees and shrubs sink somewhat or lean over. As a rule, such leaning is not so pronounced as in the *Alnus-Carex* woodland where trees are often rooted in *Carex paniculata* tussocks which can eventually roll under the strain, but something approaching swampy conditions can develop locally (as in the situations described as semi-swamp and pseudo-swamp carr by Lambert 1951). Where *S. cinerea* and *V. opulus* topple over, their branches may take root and show a renewed flush of growth in their displaced positions, further increasing the tangle of woody material. Sometimes, this process is repeated producing a 'banyan' physiognomy (e.g. Godwin 1936, Godwin *et al.* 1974). As well as complicating the appearance of the canopy, such developments may have direct consequences for the patterning in the field layer by creating new patchworks of wetter or less-shaded ground.

In general, the field layer under more mature canopies is dominated by mixtures of *Phragmites australis*, occasionally accompanied or sometimes replaced by *Carex acutiformis*, and a variety of tall herbs. Among these latter, the most frequent species are *Filipendula ulmaria*, *Eupatorium cannabinum* and *Urtica dioica* which can each occur as scattered individuals or in locally prominent patches. More occasional and usually less abundant are *Cirsium palustre*, *Angelica sylvestris*, *Epilobium hirsutum*, *Phalaris arundinacea*, *Berula erecta*, *Lycopus europaeus*, *Epilobium palustre*, *Iris pseudacorus*, *Lythrum salicaria*, *Lysimachia vulgaris* and *Peucedanum palustre*. The vegetative vigour and flowering of these species, and of the persistent fen dominants, often show some depression in areas of deeper shade and a resurgence where the canopy opens up with the ageing or death of trees and shrubs (e.g. Godwin 1936, Haslam 1971*b*, 1972).

Although ferns are not so numerous or varied as in the *Sphagnum* sub-community, the two community species, *Thelypteris palustris* and, on drier knolls, *Dryopteris dilatata*, occur occasionally. Quite often, too, the vegetation is tangled with underscrub, especially of *Rubus fruticosus* agg., and with sprawlers and climbers: *Galium palustre*, *G. aparine* and *Solanum dulcamara* can all be prominent and, among the woody branches, there may be some *Calystegia sepium*, *Tamus communis*, *Humulus lupulus* or *Lonicera periclymenum*. *Carex elongata* survives in vegetation of this sub-community at Askham Bog in North Yorkshire (Fitter *et al.* 1980).

Smaller herbaceous species are, generally speaking, not numerous and, apart from *Poa trivialis* which can grow in large patches, their cover is low, with just scattered plants of, for example, *Mentha aquatica* or *Caltha palustris*. Indeed, in areas of denser shade, there can be large stretches of peat and litter which are totally bare of vegetation apart from bryophytes. These can form an extensive cover, though the species involved are generally few. *Eurhynchium praelongum* and, preferential here, *Brachythecium rutabulum* and *Plagiothecium undulatum* are by far the commonest species, though *Plagiothecium sylvaticum*, *Amblystegium riparium*, *A. serpens*, *Calliergon cuspidatum* and *Lophocolea heterophylla* can also be found. Sphagna are characteristically very scarce in this sub-community.

The woodlands described from Wicken Fen (Godwin & Tansley 1929, Godwin 1936, Godwin *et al.* 1974) are best considered, for the most part, as a distinctive variant of this sub-community. At this site, secondary fen was quickly colonised, on the cessation of mowing, by *Frangula alnus* and smaller amounts of *Rhamnus catharticus*, *Salix cinerea*, *Crataegus monogyna*, *Viburnum opulus* and *Prunus spinosa* to produce a mosaic of young canopies of varying composition and density. Over the first 20 years or so of the invasion, there was a gradual demise, under the thicker scrub, of the once-dominant *Cladium mariscus* (and, to a lesser extent, of

Phragmites australis) and its mown-fen associates *Salix repens* and *Molinia caerulea*. These were replaced by an open patchwork of the tall herbs and underscrub characteristic of this sub-community, some species, such as *Eupatorium cannabinum*, *Filipendula ulmaria* and *Angelica sylvestris*, surviving from the preceding fen vegetation and showing an expansion, others, like *Thelypteris palustris*, *Urtica dioica* and *Rubus caesius* invading afresh. Where the scrub remained more open, the fen dominants persisted longer to produce a more mixed field layer. Changes in the developing canopy, due to internal competition between its expanding components and markedly affected by early die-back of *Frangula* after infection by a *Fusarium* sp. and *Nectria cinnabarina*, superimposed on this general trend a fine scale of variation in canopy shade. This was reflected in the field layer by waves of waxing and waning of the tall herbs or, under the initially thinner canopies, of the surviving fen dominants. After some 50 years (Godwin *et al.* 1974), there were signs at Wicken of a reversion to the kind of canopy more typical of this sub-community with the appearance in various parts of the site of *Alnus*, *Betula* spp., *Fraxinus* and *Quercus robur*. However, the field layer of the most closely monitored of the original plots (Reserve A), though it retained a cover of *Rhamnus* and *Frangula*, seemed to have progressed somewhat beyond the limits of this vegetation type. With a reduction in *Phragmites* and tall fen herbs, an increase in *Urtica dioica* and *Poa trivialis* and the appearance of *Glechoma hederacea*, it resembled more the field layer of the *Alnus-Urtica* woodland (see below).

The young scrub described by Poore (1956*b*) from Woodwalton Fen in Cambridgeshire illustrates a different variation on the same theme. Here the major coloniser was *Salix cinerea* with smaller amounts of *Betula* spp. (unusually, mostly *B. pendula*), *Crataegus monogyna*, *Quercus robur*, *Viburnum opulus* and local *Alnus*. *Rhamnus* was rare, *Frangula* limited to a single (perhaps planted) individual and *Fraxinus* absent. The field layer had many of the typical features of this sub-community with frequent, though sparse, *Phragmites*, a strong contingent of tall herbs and well-developed underscrub but intermixed with these were prominent remnants of the rather peculiar secondary fen vegetation typical of this much-modified site: *Calamagrostis canescens*, *C. epigejos*, *Symphytum officinale*, *Lotus uliginosus*, *Thalictrum flavum* and *Vicia cracca*. As at Wicken, these various components came and went somewhat with the closure and thinning of the developing canopy.

***Sphagnum* ssp. sub-community:** *Betuletum* Godwin & Turner 1933; Fen carr Godwin *et al.* 1974 *p.p.*; *Betulo-Dryopteridetum cristatae* Wheeler 1978; *Betulo-Myricetum peucedanetosum* Wheeler 1980*c p.p.*; *Alnus-Salix-Betula* woodland XXIii Meres Report 1980.

Betula pubescens is almost always the most abundant woody species here, though its stature is very variable. Sometimes it forms the basis of small islands of open birch scrub with saplings only a metre or so high; elsewhere, the cover is more extensive, denser and taller. There is commonly a little *Salix cinerea* but both *Alnus* and *Fraxinus* are markedly less prominent than in the previous sub-community, often occurring just as occasional saplings. *Frangula* can be locally abundant and there is sometimes a little *Salix aurita* and some young *Quercus robur* but *Rhamnus* and *Viburnum opulus* are characteristically absent. Occasionally, bushes of *Myrica gale* and *Salix repens* form a patchy lower tier to the canopy and there can be tangles of *Rubus fruticosus* agg., *Rosa canina* agg. and *Lonicera periclymenum*.

Phragmites remains frequent in the field layer and it can be abundant but other fen dominants are generally sparse, even in younger stands: there is occasionally some *Calamagrostis canescens* and, less commonly, *Cladium mariscus, Carex elata* or *C. appropinquata* can persist. *C. paniculata*, however, is rare and *C. acutiformis* absent. Tall fen herbs are also rather patchy here compared with the previous sub-community. *Filipendula ulmaria* remains occasional and it can be accompanied, especially among richer fen systems, by *Eupatorium cannabinum, Lysimachia vulgaris, Lythrum salicaria* and *Peucedanum palustre* but *Urtica dioica* is noticeably very uncommon. Quite often, the field layer is marked by the patchy abundance of various grasses, including *Holcus lanatus, Molinia caerulea, Agrostis canina* spp. *canina, A. stolonifera* and *Poa trivialis*, and there can be scattered tussocks of *Juncus effusus*. *Potentilla erecta* and *Hydrocotyle vulgaris* are occasional associates and, in Broadland, there are rare records for *Pyrola rotundifolia* and *Drosera rotundifolia*.

Two further features are especially characteristic of this vegetation. The first is the abundance and variety of ferns. As well as *Thelypteris palustris, Dryopteris dilatata* and *Athyrium filix-femina, Dryopteris carthusiana* is more frequent here than in the previous sub-community and of particular note in Broadland is *D. cristata*. This last species seems to thrive best in a more open shrubby cover and its distinctive clusters of bright green, upright fronds often form a striking fringe around the margins of stands or in glades (Wheeler 1978, Page 1982). In denser shade, sparse sterile fronds can sometimes be found. It hybridises with *D. carthusiana* and the large and vigorous *D.* × *uliginosa* occurs in some stands. Other ferns encountered rarely are *Thelypteris phegopteris, T. limbosperma, Osmunda regalis* and *Ophioglossum vulgatum*.

The second distinctive characteristic is the abundance of Sphagna which can occur as scattered patches around the bases of the *Phragmites* or bush stools or sometimes as a virtually continuous cover. *Sphagnum fimbriatum*,

S. squarrosum, S. palustre and *S. recurvum* are all very frequent, *S. subnitens* somewhat less so. Other preferential bryophytes here are *Mnium hornum, Plagiothecium denticulatum, Aulacomnium palustre, Rhizomnium pseudopunctatum* and *Calypogeia fissa*.

Habitat

The *Salix-Betula-Phragmites* woodland is typically a community of topogenous fen peats. It is especially characteristic of flood-plain mires but can also be found on the terraces of some valley mires and, rarely, in basin mires. It can develop as a primary woodland cover by the direct invasion of herbaceous fen but very many stands represent a secondary succession on abandoned mowing-marsh and these often show floristic and structural peculiarities related to the complex treatment histories of individual sites. Drier stands can show signs of past coppicing.

In primary hydrarch successions, this community is able to develop when the accumulation of consolidated litter eventually raises the surface of the peat mat above the limit of the winter flood (Godwin & Bharucha 1932, Lambert 1951, Poore 1956*b*). However, quite delicate balances between the timing of the spring fall in the water-table, the amount of residual surface moisture and the germination requirements of the different pioneer species may exert some control over which of these shrubs and trees are able to establish themselves initially. A late fall in the water-table may, for example, inhibit some invaders and not others and it has been suggested (White 1932) that very marked differences between winter and summer water-levels might hinder any kind of colonisation. We know that certain of the important early invaders, such as *Frangula* (Godwin 1943*b*) and *Alnus* (McVean 1953) have generally high moisture requirements during and after germination but there are no systematic data available on the comparative ability of the shrubs and trees of this community to establish themselves under different conditions.

Other physical features may be important for establishment too. Some species, *Frangula* for example (Kinzel 1926), will not germinate without light. *Rhamnus*, on the other hand, can germinate in darkness (Kinzel 1926) and so can *Alnus*, though it is unable to put up its cotyledons through even a thin covering of soil or litter and needs high light intensities during early seedling growth (McVean 1953), as does *Salix cinerea*. The physiognomy of the existing herbaceous fen vegetation is therefore likely to exert a strong influence on colonisation because, even where abundant seeds drop on to suitably moist surfaces, thick standing dead material (as in *Phragmites*: Haslam 1971*a, b*), dense winter-green foliage (as in *Cladium*: Lambert 1951), lush early spring growth (as in *Glyceria maxima*: Lambert 1946) or matted litter (Godwin & Tansley 1929) may hinder or prevent establishment. Certainly, in general, bush and tree growth tends to be more marked where structural variation in the fen vegetation has provided a patchwork of more open areas where seeds could lodge and germinate. The tops of *Carex paniculata* tussocks fulfil such requirements very well but fens in which this species is prominent are invaded early and the derived woodlands preserve a richer and rather different balance of herbaceous species than is usual here. Although *C. paniculata* is an occasional in this community, woodlands in which it remains an important component are included in this scheme in the *Alnus-Carex* woodland.

Within these limitations imposed by the habitat, chance plays a considerable part in determining which woody species invade and in what order. Clearly, seed-parents must be available. With species such as *Salix cinerea* and *Betula pubescens*, which produce vast quantities of very light fruits that can be carried far by wind, the immediate proximity of existing individuals is less important. With *Alnus* the situation is not so simple because, though its fruits can be wind-dispersed, they seldom seem to be borne far by this means (McVean 1953). Water-dispersal, on the other hand, may be very effective here, as with other species whose fruits float readily, and the winter flood can leave a tide-mark of such disseminules towards its upper limit. For *Frangula* and *Rhamnus*, bird-dispersal is important, though the birds may not roam far and the distribution of excreted seeds may be closely linked to the occurrence of perching and roosting sites (Godwin 1936, 1943a, b). The fruits of *Frangula* are eaten also by fieldmice (*Apodemus* spp.) and seed may germinate from the fruit stores to produce dense tufts of seedlings (Godwin 1936, 1943b). For both these shrubs, simple falling to the ground of the heavy fruits may be a major means of dispersal (Godwin 1943a, b). However, no matter how effective these various mechanisms are, initial discrepancies in the numbers of the different woody species can have a long-lasting influence on the composition of the new canopies produced by colonisation. At Wicken, for example, the striking disparity between the amounts of *Frangula* and *Rhamnus* in the young scrub and the absence of *Alnus* in the early years seemed to be attributable mainly to differences in the numbers of seed-parents when invasion began (Godwin 1936). Even after 50 years, the influence of such differences was still very apparent (Godwin *et al.* 1974).

Human activities can have a marked effect on each of these variables and so influence the rate of colonisation and the composition and structure of the developing canopies. This is true even in primary successions in natural herbaceous fen where interference with surrounding land can affect the progress of invasion but it is seen especially clearly in those mires which themselves have a long history of human interference. Here, the community has developed over surfaces and from vegetation which have been disturbed or modified to varying degrees and preserves within its canopies the effects of such treatments.

In the first place, the very widespread draining of flood-plain mires has resulted in an artificial lowering of the water-table over many of those remnants of fen that have escaped conversion to intensive agricultural use. Often, such a fall has been so marked that development of this community has been rapidly overtaken by seral progressions to other woodland types, where, that is, woodland has been allowed to grow up at all. In other cases, the substrate has remained sufficiently moist for this community to appear, though the balance of the invading shrubs and trees has been affected. This seems to have happened at Woodwalton (Poore 1956b) where *Crataegus monogyna* and *Quercus robur* were unusually frequent and *Betula pubescens* partly replaced by *B. pendula*. Such anomalous canopies may take many years to equilibrate: indeed, they may never do so before further successional changes ensue.

Sometimes, peat extraction has complicated the influence of the water-table and modified the pattern of invasion by woody species. On a grand scale, of course, digging has created new areas, like the Norfolk Broads, in which the whole process of primary succession has proceeded afresh to leave us with some of our most extensive and natural stands of this community (Lambert 1951, 1965). Elsewhere, it has exposed previously hidden deposits of fen peat on which the later stages of invasion have taken place, as seems to have happened at Woodwalton (Poore 1956b), or left complex surface mosaics of wetter and drier areas over which colonisation has been uneven. At Wicken, for example, scrub development over peat-cut surfaces was confined in the early years to remnant baulks of peat between the excavated furrows which remained too wet for invasion to occur (Godwin 1936). There is evidence, too, that the rather special water regimes within shallow peat-diggings may play some part in determining the floristic differences between the two sub-communities.

The disposition of seed-parents from which invasion can begin has also been affected by human activity. Systematic drainage helped destroy and fragment any existing woodland on some flood-plain mires, such that remnant fens could be left isolated in largely treeless pastoral and arable landscapes. In such cases, any subsequent colonisation has depended on the survival of occasional shrubs and trees in hedges, along ditches and streams and, where the fens were mown, in neglected compartments. Which species remained in surviving mires such as Woodwalton and Wicken may have been entirely a matter of chance. In Broadland, things may have been rather different. Here, the upper reaches of the flood-plain mires remained more closely under the influence of the natural river drainage and human activity actually increased the movement of water between and over the fens. It is possible, too, that a mixed marsh economy with exploitation of both herbaceous and woodland vegetation persisted longer than elsewhere so that more balanced mixtures of potential invaders remained closer at hand. Certainly, stands of the community in this area have a more uniform canopy composition than those in more isolated stretches of fen. Finally, on the positive side here, it is possible that shrubs and trees were deliberately brought within colonising distance of herbaceous vegetation. Rods of *S. cinerea*, for example, were often used to mark out the limits of peat-holdings and could quickly take root. Woody species were also sometimes planted to stabilize the banks of dikes. Once established, such individuals could provide a local source of seed or a centre of vegetative expansion.

Where suitable soils lie within the range of colonising shrubs and trees, the most important variable influencing the establishment of the community is the harvesting of herbaceous marsh crops which has deflected the natural succession to this woodland type, maintaining herbaceous fen in a wide variety of secondary forms. With the almost total demise of such treatments, extensive stands of the community have developed by secondary succession. As already indicated, the physiognomy of the preceding fen vegetation can exert an influence on the actual process of colonisation by shrubs and trees. Its floristics also make an important contribution to the field layer of the developing woodland, especially in the early stages. Since both the structure and the composition of the herbaceous fen are strongly affected by the mowing regime, the kind of treatment it received can have a long-lasting influence on the appearance of the subsequent woodland vegetation.

In essence, the general characteristics of the field layer of younger stands of the community, whether these arise by primary or secondary succession, are a product of the effects of decreasing light on the invaded fen flora. The rate at which light declines under the woody cover will depend largely on the speed with which the canopy closes, though this is itself much influenced by the initial density of the invading shrubs and trees (e.g. Godwin 1936) and subject to local and temporary reversal in the general trend as canopy components die or open up with ageing. The effects of increasing shade are usually first seen among the one-time fen dominants which, almost without exception, decline. The most persistent of these is *Phragmites* but its continuing high frequency throughout the community is, in part, a reflection of its almost universal prominence in the kinds of fen invaded. It is certainly sensitive to shading, showing a reduction in shoot density, shoot height and flowering (Haslam 1971*b*, 1972), though its rhizome displays some adap-

tation to lower light levels and the continued growth of its terminal branches may make it very persistent at low abundance (Haslam 1965). Under lighter canopies, it often remains proportionately important and it may even extend its cover to fill gaps left by less tolerant species. Where high light levels are restored beneath canopy gaps, it may show some recovery of a more vigorous growth form (Haslam 1971*b*). *Calamagrostis canescens* can also be quite persistent, though it is much less uniformly distributed in the invaded fens, so its presence in this community is patchy. When mowing ceases (and this species is very much a plant of drier secondary fens), it often adopts a markedly tussocky habit and, though these tussocks become less vigorous in shade, they can remain prominent for some time (Lambert 1951). At Wicken, *C. canescens* actually increased in some older *Frangula* woodland, perhaps spreading in some areas in response to burning (Godwin *et al.* 1974; see also Luck 1964). *Cladium mariscus* is somewhat different again. It is more shade-sensitive than either *Phragmites* or *C. canescens* (Conway 1942), but, like the former, it can be initially very abundant and last for many years as drawn-up individuals before finally disappearing (Godwin 1936). Finally, among this group, there are *Carex paniculata* and *C. acutiformis*. In contrast to the rest of the dominants, both these are shade-tolerant: indeed, the latter may show a marked spread under developing woodland. They are, however, not widely distributed among the fens from which this woodland develops and remain, at most, occasional in the field layer.

Among the associates of the invaded fen vegetation, very few retain a high frequency with substantial increase in shade. The most persistent element comprises certain of the taller dicotyledons, notably *Filipendula ulmaria* and *Eupatorium cannabinum*, and these may increase their cover at first as the dominants fade or even reappear where particular mowing regimes had eradicated them. Again, however, their vigour and flowering show a decline as the canopy thickens up and very commonly the most prominent feature of really dense stands of this community is the extent of bare ground. Then, only very shade-tolerant species such as *Poa trivialis* and bryophytes with scattered *Iris pseudacorus* may be able to survive.

The floristic differences between the two sub-communities appear to reflect variation in the base-richness and calcium levels in the peat which, in the topogenous mires of which this woodland is characteristic, are largely dependent on the height and movement of the ground water. The *Alnus-Filipendula* sub-community is, in general, typical of fen peats which remain under the comparatively close influence of the fluctuating water-table. In Broadland, such waters have dissolved calcium levels of 60–120 mg l^{-1} (Wheeler 1983) and the surface pH under this vegetation remains generally high, between 6.5 and 7.5, very much like that beneath the preceding fen vegetation. The *Sphagnum* sub-community, by contrast, is characteristically developed where, for some reason or other, the influence of the base-rich and calcareous ground water is not so great. Sometimes, it is found on areas which are at a slightly higher level than the surrounding peats and/or somewhat further removed from freely-circulating waters in open lakes or dike systems. This was the case at Calthorpe Broad in Norfolk where Godwin & Turner (1933) noted that the *Alnus-Filipendula* sub-community gave way to the *Sphagnum* sub-community with increasing height above and distance from the broad. In other places, the *Sphagnum* sub-community occurs on floating rafts of peat, as in the Ant valley in Broadland (Wheeler 1978, 1980c) and here the rising of the substrate with the waters prevents frequent inundation of the surface layers of the peat. Some of the finest stands in this area occur within fens in shallow nineteenth-century peat-cuttings, like those on the Catfield Fens (Lambert 1965, Wheeler 1978), and here conditions may be virtually stagnant. Whatever its particular situation, the pH beneath the *Sphagnum* sub-community is typically lower than that beneath the *Alnus-Filipendula* sub-community, though not always markedly so, being generally between 5.5 and 6.5.

Differences in trophic levels may be involved here too. When the surface of the more base-rich and calcareous peat under the *Alnus-Filipendula* sub-community dries somewhat in summer, there may be some oxidation and the release of a flush of nutrients (e.g. Haslam 1965). Then, litter turnover may be enhanced and the gradual development of peaty mull soil initiated (e.g. Poore 1956b). In this respect, the markedly preferential abundance of the nutrient-demanding *Urtica dioica* in this sub-community is especially interesting. Under the *Sphagnum* sub-community, by contrast, the greater acidity of the surface layers of the peat may prevent such developments, even when they dry out. The greater isolation from any nutrients in the circulating waters further hinders any tendency towards eutrophication.

Zonation and succession

In the more extensive flood-plain mires of the Broadland river valleys, it is still possible to see complete and fairly straightforward zonations from open-water vegetation, through swamp and primary fen, to the *Salix-Betula-Phragmites* woodland, over fen peats which are increasingly free from inundation by the seasonally-fluctuating waters (e.g. Pallis 1911, Lambert & Jennings 1951, Wheeler 1980c, 1983). Where the base-status and calcareous nature of the peat are maintained, such sequences involve the *Alnus-Filipendula* sub-community which replaces, beyond the upper limit of the winter flood, the *Peucedano-Phragmitetum*. Local variations in the pat-

tern of water fluctuation and uneven invasion of shrubs and trees frequently blur this boundary so that a hazy zone of scrubby vegetation lies between the fen and the woodland proper. There is variation, too, in the particular kind of *Peucedano-Phragmitetum* to be found in these zonations: along the Bure valley, for example, it is generally the Typical sub-community, dominated by *Phragmites* or less often and perhaps where conditions are more oligotrophic, *Cladium*; in the more eutrophic Yare fens, drier forms of the *Glyceria* sub-community, usually dominated by *Phragmites*, are often involved.

Outside Broadland, where flood-plain mires have been much fragmented and extensively drained, intact zonations of this kind are rare and such sequences as do survive generally have the less species-rich vegetation of the *Phragmites-Eupatorium* fen (most often its *Phragmites* or *Cladium* sub-communities) replacing the *Peucedano-Phragmitetum*. Less commonly, the *Alnus-Filipendula* sub-community may pass directly to the rather dry swamp vegetation of the *Galium* sub-community of the *Phragmitetum*. Transitions of this kind can be seen around some open waters in the Shropshire and Cheshire meres (Sinker 1962, Meres Report 1980) and in more base-rich valley mires, such as those in Breckland (e.g. Haslam 1965).

Stratigraphical studies around the Norfolk Broads (Lambert & Jennings 1951, 1965, Jennings & Lambert 1951, Lambert 1951, Lambert *et al.* 1960, 1965) have shown that zonations of this kind can represent the progress of primary hydrarch successions in which the *Alnus-Filipendula* sub-community is the natural product of invasion of a variety of herbaceous fen types. This sub-community corresponds most closely with Lambert's 'fen carr' but it also includes some of what she would have called 'semi-swamp carr' and 'pseudo-swamp carr', woodlands which differ largely in their physiognomy and whose floristics can be comfortably subsumed within this single vegetation type. Although it has not been confirmed by stratigraphical analysis or long observation elsewhere, it seems likely that this line of development represents the major succession on our more base-rich topogenous mires, except where *Carex paniculata* figures prominently in the primary fens. Its rate of progress is difficult to assess but comparisons of maps led Lambert & Jennings (1951) to suggest that some of the swamp around the Norfolk Broads could have progressed to woodland over 150 years and Haslam (1965) adduced a similar process to have taken some 50 years in a Breckland valley fen.

In many areas, including Broadland where exploitation of the fens was formerly very intensive, such primary succession was deflected by regular mowing for herbaceous crops and many of the present stands of the *Alnus-Filipendula* sub-community are found as the product of secondary invasion among complex patchworks of vegetation types, variation among which is more closely related to the previous mowing regime and the date of abandonment than to natural differences in the water-level. Often, such patterns have been further complicated by other activities. Where there has been peat-digging, for example, secondary fen and woodland on the drier baulks may be intermingled with primary herbaceous vegetation and woodland developing afresh in flooded workings (e.g. Lambert & Jennings 1951, Wheeler 1978). On more isolated mire fragments, where drainage and even cultivation have occurred over the peat surface, stands of the *Alnus-Filipendula* sub-community can be found developing from fen remnants alongside suites of other rather different vegetation types on even drier and more disturbed ground or isolated within intensive agricultural landscapes.

The *Sphagnum* sub-community is more local than the *Alnus-Filipendula* sub-community and its development has not been studied but it seems to be the product of a divergent succession from the kind described above. Quite small differences in the water regime, in surface pH and nutrient status may be responsible for determining which of the two sub-communities ultimately develops and such variation may be initiated early in the succession, even before the appearance of any woodland cover. Indeed, the *Sphagnum* sub-community is occasionally to be found in some basin and valley mires where the whole character of the habitat is shifted somewhat towards being less base-rich and calcareous. Usually, however, it is the occurrence of such conditions on a more local scale that seems to be the prelude to its appearance. In some cases, it seems possible that the remnants of a former, more extensive cover of acid peat may provide a congenial surface (Wheeler 1978). Ombrogenous peats are known or strongly presumed to have formed within some of the East Anglian floodplain mires in the past (Godwin & Clifford 1938, Poore 1956*b*, Walker 1970) and, though these have been largely stripped away, fragments may remain. Often, however, the typically small stands of the *Sphagnum* sub-community give every indication of representing new ombrogenous nuclei in which surface layers of more acid peat are accumulating afresh beneath a *Sphagnum* cover (Wheeler 1978, 1980*c*, 1983). Small domes of Sphagna can sometimes be found within herbaceous fens (e.g. Pallis 1911, Tansley 1939, Wheeler 1983) and these may form the basis of subsequent stands of the *Sphagnum* sub-community, being preferentially invaded from the start by *Betula pubescens*. Such a development seems to be particularly associated with floating mats of fen vegetation like those found in various parts of the Ant valley, most distinctively in shallow peat-cuttings as on the Catfield fens (Wheeler 1978). Here, rich *Phragmitetum* appears to progress to a striking kind of *Peucedano-Phragmitetum* dominated by *Carex lasiocarpa*. Within this, small islands of Sphagna form and these then act as centres for *B. pubescens* invasion (Giller

1982, Wheeler 1983). In these fens, and among other kinds developed on solid peat, certain kinds of mowing regime may help accentuate surface impoverishment and acidification so that, once treatment ceases, conditions are inimical to the development of the *Alnus-Filipendula* sub-community.

In other cases, it is possible that the *Sphagnum* sub-community makes a late appearance in the succession, developing from the *Alnus-Filipendula* sub-community as conditions change. Such a process may be natural, as where the accumulation of litter raises the peat surface above the level of close influence by the ground water (e.g. Godwin & Turner 1933) or, as seems to have been the case at Wicken in recent years, it may be precipitated by the artificial prevention of flooding (Godwin *et al.* 1974). Only close monitoring of a variety of particular stretches of fen over long periods could provide some indication of the importance of these various possibilities.

In neither of the two divergent lines of succession is it clear what the final vegetation types might be. It has been suggested on a number of occasions (e.g. Godwin & Turner 1933, Tansley 1939, Lambert 1951) that some kind of oakwood is the natural development from the *Alnus-Filipendula* sub-community. In some sites, transitions between these two woodland types can be found but extrapolating from such sequences to seral developments is dubious because the oak woodland frequently lies on the drier land surrounding the mire where the influence of its ground water had probably never been great. A more likely immediate successor to the *Alnus-Filipendula* sub-community is some kind of drier *Alnus-Urtica* woodland. The virtually complete eclipse of fen herbs in this latter community, the prominence of *Urtica dioica* and Rubi, the appearance of *Sambucus nigra* and the occurrence of a canopy in which *Alnus* generally predominates over *Salix cinerea* and *Betula pubescens* could be seen as natural developments over a peat surface which was being slowly converted to a humose mull with a high nutrient turnover. This kind of process seems to be in train, at least within the field layer, in Reserve A at Wicken (Godwin *et al.* 1974) and within the Askham woods (Fitter *et al.* 1980) and perhaps accounts for some of the later stages of the succession in the Breckland valley mires (Haslam 1965).

The *Sphagnum* sub-community, on the other hand, seems more likely to progress to the *Betula-Molinia* woodland, in certain kinds of which Sphagna remain very prominent, along with an increased cover of *Molinia caerulea*, under an often moribund canopy of *B. pubescens*. Some of the islands of the *Sphagnum* sub-community described from the Ant valley have a core of this kind of woodland in their drier centres (Wheeler 1978, 1980*c*) and such a development may also have occurred towards the back of Esthwaite North Fen in Cumbria (Pearsall 1918, Tansley 1939, Pigott & Wilson

1978). It is possible that such vegetation might eventually give way to ombrogenous mire with the disappearance of all tree cover, recapitulating the process that seems to have occurred in the past, even on those of our lowland mires which experience a drier climate (Pearsall 1918, Godwin & Turner 1933, Walker 1970, Wheeler 1978, 1983).

The various vegetation patterns which result from the operation of these successions can be further complicated by the close proximity of the products of other seral developments within mires. In some cases, it is variation within the herbaceous vegetation that is being invaded that deflects the succession away from the formation of the *Salix-Betula-Phragmites* woodland. Where *Carex paniculata* is a prominent feature of swamp or primary fen, for example, the invasion of shrubs and trees is often early and results instead in the development of swampy *Alnus-Carex* woodland. Stands of this woodland can be found developing in both flood-plain and valley mires, sometimes between the *Salix-Betula-Phragmites* woodland and open water, sometimes, marking the position of old channels and pools, behind it (e.g. Pallis 1911, Lambert & Jennings 1951). Usually, it is the *Alnus-Filipendula* sub-community that is found in this kind of zonation but, in the Ant valley, some of the larger stands of the *Sphagnum* sub-community have a core of *Alnus-Carex* woodland (Wheeler 1978).

In other cases, variation in the pattern of sediment accumulation may lead to the development of other woodland types alongside the *Salix-Betula-Phragmites* woodland. Where there is deposition of alluvium, for example, more eutrophic fens and woody vegetation can occur among the sequences of communities on the peat. Such patterns are especially characteristic of some valley mires, where levees build up along the water's edge as suspended sediments are dropped in the flood. Then the *Salix-Betula-Phragmites* woodland behind gives way to a fringe of such herbaceous vegetation as the *Phragmites-Urtica* fen or the *Phalaridetum* progressing to wetter kinds of *Alnus-Urtica* woodland, dominated by *Salix purpurea*, *S. triandra*, *S. viminalis* or *S. fragilis*. This kind of transition is a marked feature of the enriched zone alongside the Black Beck in Esthwaite North Fen (Pearsall 1918, Tansley 1939, Pigott & Wilson 1978).

Finally, significant differences in the water regime and in the pH, calcium concentration and nutrient status may create conditions which locally prevent the development of tall-herb fen and subsequent *Salix-Betula-Phragmites* woodland. This seems to happen along the margins of certain of the Broadland flood-plains, where the fens are isolated from the moving waters and in some places influenced by soligenous seepage, and also in some shallow, stagnant peat-cuttings (Wheeler 1978, 1980a, 1983). Here, stretches of the *Salix-Betula-Phrag-mites* woodland give way to small-sedge mires. Similar habitat variation may also play some part in the complex pattern of vegetation types seen in Esthwaite North Fen (Pearsall 1918, Tansley 1939, Pigott & Wilson 1978) where the *Salix-Betula-Phragmites* woodland lies behind a zone of the *Potentillo-Caricetum* fen which is developing into the *Salix-Carex* woodland.

Interesting as these seral developments are, the formation of the *Salix-Betula-Phragmites* woodland from a range of primary and secondary fens inevitably results in a loss of diversity among the vegetation. With the demise of mowing, extensive stretches of often species-rich herbaceous vegetation, together with its particular associations of other biota, notably invertebrates (e.g. Ellis 1965), have been lost and the agricultural record, which the pattern of fen compartments, dikes and droves preserved, has been obscured. Only in a very few sites does mowing of fens continue (e.g. Godwin 1978) and attempts to preserve heterogeneity by woodland clearance are laborious and costly. However, where such clearance has been undertaken, at Woodwalton for example (Duffey 1971), the effects have been judged worthwhile. At this particular site, there has been the additional interest of the reappearance, after many years, of *Viola persicifolia* and the spread of *Luzula pallescens* in the open vegetation.

Distribution

More extensive and undisturbed tracts of the *Salix-Betula-Phragmites* woodland are now largely confined to East Anglia (in Broadland and the valley mires of the Chalk) and some of the Cheshire and Shropshire meres with fragmentary stands scattered elsewhere on remnant fens throughout the lowlands. The *Sphagnum* sub-community is the more local of the two but particularly good stands occur in the Ant valley in Broadland (Wheeler 1978: see Figure 2) and here the spread of this vegetation is of especial significance for the distribution of *Dryopteris cristata* which seems to have increased in recent years (c.f. Petch & Swann 1968, Jermy *et al.* 1978 and Page 1982).

Affinities

This community takes in most of the woodland recognised as 'fen carr' in British descriptive accounts including more unusual canopies dominated by species such as *Frangula alnus* and *Rhamnus catharticus*. It also includes those *Sphagnum*-rich types which were noted early but not fully described until Wheeler (1978) characterised his *Betulo-Dryopteridetum cristatae*. Although this kind of woodland is very distinctive, especially in its Broadland form on which Wheeler concentrated, it seems best to unite it with the *Alnus-Filipendula* type to form a single woodland community with a more or less common origin. Both sub-communities have counterparts in Continental schemes and fall fairly clearly into the Salicion cinereae. This alliance has been placed by some authors (e.g. Westhoff & den Held 1969, Wheeler 1980*a*) in the Franguletea, the class of successional woodlands of minerotrophic fens, by others (e.g. Ellenberg 1978) among the possibly climax communities of the Alnetea glutinosae.

Two problems attend the definition of the community. The first is to separate it from the fen vegetation from which it is derived and to which it frequently grades through a scrubby fringe. Sometimes, vegetation equivalent to that included here has been grouped within primarily herbaceous fen, as in the *Pallavicinio-Sphagnetum* Meltzer 1945, in which Westhoff & den Held (1969) placed woodlands like those of the *Sphagnum* sub-community. Clearly, when invasion of woody species is gradual and sparse, it will be difficult to separate stands of the *Salix-Betula-Phragmites* woodland from fens like the *Peucedano-Phragmitetum*. Here, the rule of thumb is that, where the developing canopy begins to

modify the invaded fen, the vegetation is best considered as young woodland.

The second difficulty concerns the distinction between this community and other woodlands developing on mires. The *Alnus-Carex* woodland is generally considerably more species-rich than the *Salix-Betula-Phragmites* woodland and usually of different physiognomy with much of its canopy rooted in ultimately unstable *Carex paniculata* tussocks. Although stands with composition and structure intermediate between the two communities can sometimes be found, the traditional separation between 'swamp carr' and 'fen carr' made by Pallis (1911) and confirmed by Lambert (1951) seems basically a sound one.

In their later stages of development, both sub-communities show floristic transitions to other woodland types, the *Alnus-Filipendula* type to the drier woodlands of the Alno-Ulmion alliance, the *Sphagnum* type to the mire forests of the Betulion pubescentis.

Floristic table W2

	a	b	2
Betula pubescens	III (3–8)	V (6–9)	IV (3–9)
Salix cinerea	III (3–9)	IV (2–6)	IV (2–9)
Frangula alnus	I (1–9)	II (4–6)	I (1–9)
Quercus robur	I (6)	I (1)	I (1–6)
Salix aurita	I (4)	I (3)	I (3–4)
Alnus glutinosa	III (4–10)	I (1–4)	II (1–10)
Fraxinus excelsior	II (3–6)	I (1)	I (1–6)
Crataegus monogyna	II (2–5)	I (1–2)	I (1–5)
Viburnum opulus	II (1–6)		I (1–6)
Salix fragilis	I (5–8)		I (5–8)
Rhamnus catharticus	I (3–5)		I (3–5)
Alnus glutinosa sapling	I (3–4)	I (5)	I (3–5)
Fraxinus excelsior sapling	I (1–5)	I (1–7)	I (1–7)
Betula pendula sapling		II (4–6)	I (4–6)
Phragmites australis	V (2–9)	IV (2–8)	IV (2–9)
Filipendula ulmaria	IV (1–7)	II (3–4)	III (1–7)
Brachythecium rutabulum	IV (2–7)	II (2)	III (2–7)
Urtica dioica	III (2–7)	I (1)	II (1–7)
Eupatorium cannabinum	III (3–6)	I (1–3)	II (1–6)
Plagiomnium undulatum	II (2–5)	I (2)	I (2–5)
Galium palustre	II (1–4)	I (1–3)	I (1–4)
Cirsium palustre	II (1–4)	I (2)	I (1–4)
Carex acutiformis	II (2–9)		I (2–9)
Epilobium hirsutum	II (1–4)		I (1–4)
Galium aparine	II (2–5)		I (2–5)
Angelica sylvestris	II (1–4)		I (1–4)

Floristic table W2 *(cont.)*

	a	b	2
Mentha aquatica	II (3–5)		I (3–5)
Solanum dulcamara	II (1–5)		I (1–5)
Caltha palustris	I (2–4)		I (2–4)
Phalaris arundinacea	I (3–7)		I (3–7)
Stellaria media	I (3)		I (3)
Lycopus europaeus	I (2–4)		I (2–4)
Calystegia sepium	I (4–5)		I (4–5)
Carex acuta	I (4–8)		I (4–8)
Epilobium palustre	I (2–3)		I (2–3)
Hedera helix	I (1–5)		I (1–5)
Plagiothecium sylvaticum	I (2–3)		I (2–3)
Amblystegium riparium	I (2–3)		I (2–3)
Symphytum officinale	I (2)		I (2)
Fraxinus excelsior seedling	I (2–3)		I (2–3)
Tamus communis	I (3–4)		I (3–4)
Amblystegium serpens	I (2–3)		I (2–3)
Geranium robertianum	I (1–4)		I (1–4)
Deschampsia cespitosa	I (3–4)		I (3–4)
Glechoma hederacea	I (2–4)		I (2–4)
Humulus lupulus	I (3–5)		I (3–5)
Sphagnum squarrosum	I (4)	V (2–7)	III (2–7)
Sphagnum fimbriatum	I (5)	V (4–7)	III (4–7)
Sphagnum recurvum	I (3)	IV (2–6)	III (2–6)
Sphagnum palustre		IV (3–8)	III (3–8)
Lonicera periclymenum	I (3–5)	III (1–6)	II (1–6)
Mnium hornum	I (1–3)	III (1–3)	II (1–3)
Plagiothecium denticulatum	I (1)	III (1–3)	II (1–3)
Holcus lanatus	I (1–5)	III (2–5)	II (1–5)
Juncus effusus	I (2–3)	III (2–5)	II (2–5)
Dryopteris carthusiana	I (3–4)	II (1–3)	I (1–4)
Hydrocotyle vulgaris	I (4)	II (3–4)	I (3–4)
Molinia caerulea	I (5)	II (2–6)	I (2–6)
Potentilla erecta	I (3)	II (2–4)	I (2–4)
Calypogeia fissa	I (2–3)	II (1–2)	I (1–3)
Myrica gale		II (5)	I (5)
Dryopteris cristata		II (1–3)	I (1–3)
Aulacomnium palustre		II (2)	I (2)
Rhizomnium pseudopunctatum		II (2–3)	I (2–3)
Agrostis canina canina		II (3–7)	I (3–7)
Agrostis stolonifera		II (1–4)	I (1–4)
Carex vesicaria		I (4–6)	I (4–6)
Calliergon giganteum		I (3–7)	I (3–7)
Deschampsia flexuosa		I (4)	I (4)
Sphagnum subnitens		I (1–4)	I (1–4)
Thelypteris phegopteris		I (3–8)	I (3–8)
Carex nigra		I (2)	I (2)
Thelypteris limbosperma		I (3–4)	I (3–4)

Orthodontium lineare		I (2–3)	I (2–3)
Quercus petraea seedling		I (1)	I (1)
Menyanthes trifoliata		I (2–3)	I (2–3)
Eurhynchium praelongum	III (2–6)	III (2–5)	III (2–6)
Dryopteris dilatata	II (1–4)	II (2)	II (1–4)
Poa trivialis	II (2–7)	II (3–4)	II (2–7)
Rubus fruticosus agg.	II (2–8)	II (2–4)	II (2–8)
Thelypteris palustris	II (4–5)	II (1–5)	II (1–5)
Ajuga reptans	I (3–4)	II (2–4)	I (2–4)
Lotus uliginosus	I (3–4)	II (3)	I (3–4)
Rosa canina agg.	I (2–4)	II (2–4)	I (2–4)
Athyrium filix-femina	I (1–4)	I (2–4)	I (1–4)
Berula erecta	I (4)	I (3)	I (3–4)
Carex paniculata	I (3–7)	I (1)	I (1–7)
Carex remota	I (1–3)	I (1–3)	I (1–3)
Cladium mariscus	I (2–4)	I (1–3)	I (1–4)
Equisetum palustre	I (1–4)	I (2–3)	I (1–4)
Peucedanum palustre	I (3)	I (1–3)	I (1–3)
Lythrum salicaria	I (3–4)	I (1–3)	I (1–4)
Lysimachia vulgaris	I (2)	I (1–3)	I (1–3)
Juncus subnodulosus	I (4)	I (1–3)	I (1–4)
Glyceria maxima	I (1–2)	I (3)	I (1–3)
Calamagrostis canescens	I (2–3)	I (1–4)	I (1–4)
Rubus idaeus	I (4)	I (2)	I (2–4)
Pellia epiphylla	I (2)	I (2)	I (2)
Lophocolea heterophylla	I (1–3)	I (2)	I (1–3)
Scutellaria galericulata	I (3)	I (2)	I (2–3)
Lophocolea bidentata s.l.	I (4)	I (2–3)	I (2–4)
Valeriana officinalis	I (2–4)	I (1)	I (1–4)
Calliergon cuspidatum	I (4–5)	I (1–3)	I (1–5)
Atrichum undulatum	I (3)	I (1)	I (1–3)
Campylopus paradoxus	I (3)	I (2)	I (2–3)
Hypnum cupressiforme	I (3)	I (2)	I (2–3)
Pohlia nutans	I.(4)	I (3)	I (3–4)
Number of samples	33	11	44
Number of species/sample	18 (7–27)	23 (15–30)	19 (7–30)
Tree height (m)	9 (5–11)	8 (6–10)	9 (5–11)
Tree cover (%)	74 (5–100)	67 (50–85)	72 (5–100)
Shrub height (m)	4 (3–5)	5 (3–6)	4 (3–6)
Shrub cover (%)	8 (0–100)	15 (0–60)	10 (0–100)
Herb height (cm)	86 (10–200)	80 (10–200)	85 (10–200)
Herb cover (%)	86 (25–100)	86 (50–100)	86 (25–100)
Ground height (mm)	16 (10–30)	8 (5–10)	14 (5–30)
Ground cover (%)	8 (0–85)	63 (50–80)	42 (0–80)
Altitude (m)	23 (1–45)	59 (30–76)	36 (1–76)

a *Alnus glutinosa-Filipendula ulmaria* sub-community

b *Sphagnum* sub-community

2 *Salix cinerea-Betula pubescens-Phragmites australis* woodland (total)

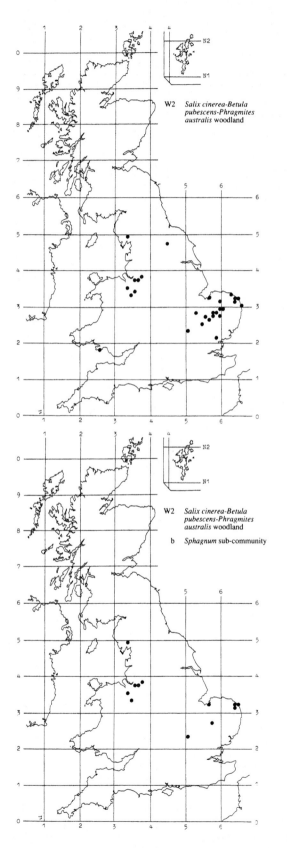

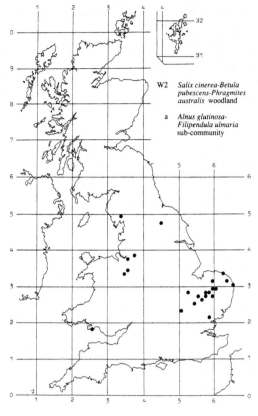

W3
Salix pentandra-Carex rostrata woodland

Synonymy
Alder wood Rankin 1911*b*; Open carr Pearsall 1918; Closed carr Pearsall 1918 *p.p.*; *Salix* carr Ingram *et al.* 1959; Fen carr Proctor 1974; Willow carr Adam *et al.* 1975 *p.p.*; *Crepido-Salicetum pentandrae* Wheeler 1980*c*; *Sphagno-Salicetum atrocinereae* Birse 1984 *p.p.*

Constant species
Salix cinerea, S. pentandra, Angelica sylvestris, Cardamine pratensis, Carex rostrata, Caltha palustris, Filipendula ulmaria, Galium palustre, Geum rivale, Valeriana dioica, Calliergon cuspidatum, Mnium hornum, Rhizomnium punctatum.

Rare species
Salix nigricans, Carex appropinquata, Carex diandra, Corallorhiza trifida, Lysimachia thyrsiflora, Pyrola rotundifolia.

Physiognomy
The *Salix pentandra-Carex rostrata* woodland is a very distinctive woodland type, fairly constant in its overall composition and structure and with clear northern European affinities in its flora. The canopy is always low, though often uneven-topped and invariably dominated by bushy Salices, most commonly *Salix pentandra* and/or *S. cinerea*. Although the former Northern Montane willow can be found occasionally in other kinds of wet woodland in northern Britain, only in this community does it make more than a local contribution to the canopy. When it is abundant here, its typical spreading habit, with branches reaching down to the ground and often re-rooting, gives stands a very characteristic appearance. Mature individuals of *S. cinerea* can open up with age, too, so that, even where the cover of bushes of these species is dense, the canopy can be somewhat open. Other Salices are rare, though they can be locally abundant. Most distinctive among these are two other Northern Montane species, *S. nigricans* and *S. phyli-*

cifolia which, when they occur together, frequently hybridise to produce a perplexing range of intermediates (*S.* × *tetrapla*: Meikle 1975) among which the identity of the parents can be lost (e.g. Proctor 1974, Adam *et al.* 1975, Lock & Rodwell 1981). *S. aurita* also occurs in some stands and, more rarely, the osiers *S. viminalis* and *S. purpurea* and these, too, may hybridise among one another and with the other Salices.

The only other woody species to occur with any frequency in the community is *Betula pubescens* and even this is only occasional. *Alnus glutinosa* is rare and the associates of the canopy of southern fen woods like *Frangula alnus, Rhamnus catharticus* and *Viburnum opulus* are typically absent. Neither does an underscrub of Rubi or *Ribes* spp. ever play a prominent part here.

In the field layer, there is no consistent pattern of dominance. Indeed, many stands have no single herbaceous dominant at all but the assemblage of species typical of the community is nonetheless very distinctive. The most prominent component generally comprises taller dicotyledons, though the mixture of species here is rather different from that characteristic of southern fen woods. The commonest species are *Filipendula ulmaria, Angelica sylvestris, Valeriana dioica, V. officinalis, Geum rivale* and *Cirsium palustre*. Most of these can be locally abundant and, under the frequently light shade of the canopy, they often flower. *Equisetum fluviatile* is common too, its stems growing tall and often branching. *Urtica dioica* is notably infrequent and species such as *Eupatorium cannabinum, Lysimachia vulgaris, Lythrum salicaria* and *Iris pseudacorus* are usually absent.

Herbs of shorter stature form a patchy lower tier to the vegetation. The most notable among these are *Cardamine pratensis* and the Northern Montane *Crepis paludosa* but also frequent are *Caltha palustris, Mentha aquatica, Lychnis flos-cuculi, Ranunculus repens, Poa trivialis, Dactylorhiza fuchsii* and *Equisetum palustre. Menyanthes trifoliata* and *Potentilla palustris* also occur commonly and are especially conspicuous in wetter areas where they can form large patches. Then there is almost always some *Galium palustre* trailing over the ground and other plants. Ferns are not a prominent feature but *Dryopteris dilatata* is occasionally found on drier areas. Three rare Northern Montane species which have been recorded in this woodland type are *Lysimachia thyrsiflora* (in Scotland: Birse 1984), the saprophytic orchid *Corallorhiza trifida* (Wheeler 1980c, Birse 1984) and *Pyrola rotundifolia*, which seems to favour the more open margins of the stands where it occurs (Wheeler 1975).

Bulkier monocotyledons are very variable in their abundance, though a number can assume local dominance. The commonest of these is *Carex rostrata*, which is constant and which can exceptionally form an extensive cover in the field layer; usually, though, it occurs as rather sparse scattered shoots. Other Carices recorded less frequently are *C. diandra*, *C. lasiocarpa*, *C. appropinquata*, *C. paniculata*, *C. laevigata*, *C. vesicaria*, *C. nigra* and *C. curta*. *C. acutiformis* is not found and *Phragmites australis* occurs only very occasionally and rarely with any abundance. There are sometimes prominent tussocks of *Juncus acutiflorus* or *J. effusus*.

The remaining striking feature of the community is the abundance of bryophytes which, in some stands, form a virtually complete carpet over the ground. Most conspicuous and distinctive of the commoner species are *Calliergon cuspidatum*, *Climacium dendroides* and various of the larger Mniaceae, usually *Rhizomnium punctatum* but, in some stands, *Plagiomnium affine*, *P. ellipticum*, *P. rostratum* or *P. elatum*. There is frequently a little *Mnium hornum* too, especially over the drier tree bases, and also some *Eurhynchium praelongum*. More occasionally, *Brachythecium rutabulum* and *Hypnum cupressiforme* can be found and some stands have locally prominent patches of Sphagna, usually *S. recurvum*, *S. squarrosum* or *S. palustre*. *Drepanocladus uncinatus*, *Ptilidium pulcherrimum*, *Douinia ovata* and *Nowellia curvifolia* have been recorded in this woodland on shrub boles, branches and decaying wood and there can be an abundance of Usneion barbatae lichens festooning the shrub bark.

Habitat

The *Salix-Carex* woodland is most typical of peat soils kept moist by moderately base-rich and calcareous ground water in open-water transitions and basin mires in northern Britain. It can develop as a primary woodland cover with the increased terrestrialisation of such sites or by secondary colonisation of peat-cut surfaces. In drier situations, grazing can be an important factor inhibiting its development.

The general geographical limits of the community are undoubtedly influenced by climate. Shrubs such as *Salix pentandra*, *S. nigricans* and *S. phylicifolia* and herbs like *Crepis paludosa*, *Menyanthes trifoliata*, *Potentilla palus-*

tris and the rarer *Carex diandra*, *C. lasiocarpa*, *C. appropinquata*, *Lysimachia thyrsiflora*, *Corallorhiza trifida* and *Pyrola rotundifolia* are all species which, to varying degrees, are confined to the cooler, more northerly parts of Britain or are commoner there (Matthews 1955). Conversely, a number of woody and herbaceous species prominent in communities like the *Salix-Betula-Phragmites* woodland and the *Alnus-Carex* woodland become very much sparser towards the north or increasingly restricted to the coast.

Within this general range, the *Salix-Carex* woodland probably has a potentially widespread distribution within the sub-montane zone: the altitudinal range of the available samples is from 45–370 m (Ingram *et al.* 1959, Proctor 1974, Adam *et al.* 1975, Wheeler 1975, Birse 1984). It is, however, a distinctly local vegetation type because of the scarcity of suitably base-rich and calcareous situations within the northern uplands. Such conditions are most extensively developed in the region within often fairly small and isolated basins set in calcareous bedrocks or drift, as at Semerwater (Ingram *et al.* 1959) and Malham Tarn (Proctor 1974, Adam *et al.* 1975) in North Yorkshire, in the Eden valley in Cumbria, in southern Scotland (Wheeler 1975, 1980c, 1983) and north-east Scotland (Birse 1980, 1984). Other basins, such as Esthwaite in Cumbria (Pearsall 1918, Tansley 1939, Pigott & Wilson 1978) and Crag Lough in Northumberland (Lock & Rodwell 1981), show a more localised development of similar conditions. It is in such sites that the most well-developed stands of the *Salix-Carex* woodland are to be found, growing on fen peats or peaty gleys, irrigated with waters with a pH of 5–7 and dissolved calcium levels in the order of 10–90 mg l^{-1} (Proctor 1974, Wheeler 1983).

In general, the community occupies tracts of such substrates which are sufficiently free from surface waterlogging to allow shrub colonisation to take place. On solid peats, such a condition will arise where autochthonous accumulation of litter raises the surface above the usual upper limit of the flood. In the rather cool climate of the region and with the often low level of nutrients in the waters, such accumulation may be slow (e.g. Spence 1964). Furthermore, some of the basins in which the community is typically found are prone to sudden and dramatic unseasonal fluctuations in water-level after heavy rain (e.g. Ingram *et al.* 1959, Lock & Rodwell 1981) and this may be inimical to the permanent establishment of a shrub cover on the marginal fens. Where expanses of such vegetation develop as a floating mat, a not uncommon feature of these basins, the effect of this may be obviated: then the raft of vegetation, bound together below by the stout, interweaving rhizomes of species such as *Carex rostrata*, *Menyanthes trifoliata* and *Potentilla palustris*, will rise and fall with the waters, keeping the surface free from inundation (e.g. Lock & Rodwell 1981). Where drier baulks of peat remain after cutting of the surface, these may provide centres for initial invasion (Proctor 1974, Wheeler 1983).

Occasionally, stands of the *Salix-Carex* woodland can be found in other situations where this balance of base-richness and moisture level is maintained. It may occur, for example, in the lagg of raised mires, where marginal seepage ameliorates the ombrotrophic conditions developed on the higher peats (e.g. Rankin 1911*b*). It is also probably a potential woodland cover on soligenous mires and small flushes where seepage of base-rich and calcareous waters results in the development of humose gleys. Almost invariably, however, such areas are set within tracts of upland pasture and then grazing repeatedly sets back any shrub colonisation. Even on the drier parts of basin mires, such grazing (usually by sheep or deer) may be an important factor in delaying invasion: at Crag Lough, for example, a stand of the community is very sharply delimited from fen vegetation by a fence which has prevented access by stock (Lock & Rodwell 1981).

The isolation of many of the sites on which this woodland can develop may have some influence on the composition of the canopy. Although all the common Salices of the community produce large quantities of light, readily wind-dispersed fruits, the relative proximity of seed-parents of the different species may play some part in the rather striking site-specific variation sometimes found in the canopy. The isolation of the established stands will also influence the complex and varied patterns of hybridisation that occur among the Salices at different sites. It may play a part, too, in the scarcity of *Alnus glutinosa* in the community: although the climatic and edaphic conditions characteristic of the *Salix-Carex* woodland are generally within the range favourable to this species, it may not be able to migrate readily into sites which are distant from seed-parents (McVean 1953, 1956*b*). Interestingly, it does play some part in colonisation at Esthwaite North Fen where it is common in adjacent woodlands (Pigott & Wilson 1978).

The composition of the field layer of this woodland is closely related to that of the preceding herbaceous vegetation. Quite a wide range of rich and poor fens seem to be potential precursors to the *Salix-Carex* woodland and some of these are marked by considerable physiognomic variety. Thus, even young stands of the community often lack the fairly standard patterns of dominance found in the field layers to communities like the *Salix-Betula-Phragmites* woodland and the *Alnus-Carex* woodland. Many of the potential dominants of the fens are also somewhat shade-tolerant, so, under the relatively open canopy, they do not show such a striking sequential demise as is typical of the *Salix-Betula-Phragmites* woodland. This is true also of the dicotyle-

donous component of the field layer, most of the species of which persist from the invaded fens. Indeed, there is very little about the herbaceous element of the *Salix-Carex* woodland that is peculiar to it alone though, with increasing age, many species become more confined to the more open and wetter parts of the woodland, the drier areas around the tree bases having but a sparse cover of *Dryopteris dilatata* and patches of *Mnium hornum*.

Zonation and succession

At some sites, the *Salix-Carex* woodland occupies a clear position in open-water transitions running from the woodland on the driest peats through fen and swamp to aquatic vegetation. At Crag Lough, for example, the community has a fringing front of very wet *Potentillo-Caricetum* fen which gives way, in turn, to the *Caricetum rostratae*, the *Equisetetum fluviatile* or related societies of *Potentilla palustris* and *Menyanthes trifoliata* (Lock & Rodwell 1981). A similar pattern is found in Esthwaite North Fen, though here, the *Phragmitetum* and *Scirpetum lacustris* also play a part in the swamp sequence (Pearsall 1918, Tansley 1939, Pigott & Wilson 1978).

In more fully terrestrialised basin mires, the community characteristically occurs in complex patchworks of vegetation that involve a variety of rich and poor fens related to local variations in seepage and through-put of waters of different pH and calcium content. Again, the *Potentillo-Caricetum* can play a part in such mosaics but, more often, it is some kind of *Carex rostrata-Calliergon* mire that provides the bulk of the herbaceous vegetation around the *Salix-Carex* woodland with a more local development of poor fens like the *Carex rostrata-Sphagnum squarrosum* community. If there is some deposition of alluvium alongside inflow streams, *Filipendulion* communities may complicate the pattern. This kind of mosaic is seen very clearly at Malham Tarn (Proctor 1974, Adam *et al.* 1975; Figure 17).

At this site and in some other basin mires, the *Salix-Carex* woodland can also be found forming a marginal belt around areas of *Sphagnum*-dominated mires with patchy *Betula-Molinia* woodland. Such situations approach those found in raised mires where, likewise, the community can occur in the lagg giving way sharply towards the mire centre to acidophilous bog and woodland (e.g. Rankin 1911*b*).

The small amount of stratigraphical data available from mires such as these (Lock & Rodwell 1981, Webb & Moore 1982) suggests that the *Salix-Carex* woodland has developed from fens such as the *Potentillo-Caricetum* and the *Carex-Calliergon* mire though the exact relationship of such successions to natural changes in the habitat has not been monitored. Nor do we know how rapidly such developments proceed under normal circumstances, though photographic evidence from

Figure 17. Simple and complex mosaics in mires with
Salix-Carex woodland.

(*a*) Crag Lough, Northumberland

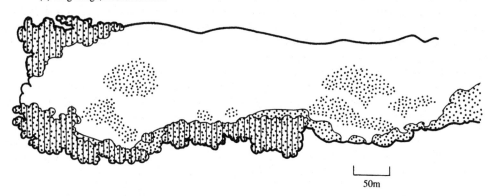

50m

(*b*) Malham Tarn Fen, North Yorkshire (after Proctor 1974)

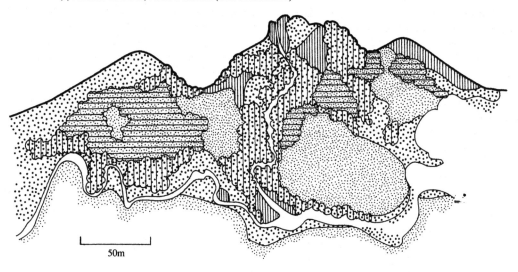

50m

 W3 *Salix-Carex* woodland

 S27 *Carex-Potentilla*
           & related fens

 S9 *Carex rostrata* &
         S10 *Equisetum fluviatile*
          swamps

 W4 *Betula-Molinia* woodland

 M19 *Calluna-Eriophorum* &
        M25 *Molinia-Potentilla* mires

 M27 *Filipendula-Angelica* mire

Crag Lough (Lock & Rodwell 1981) suggests that a good cover of this woodland had formed there within a century or so. At Esthwaite, too, colonisation has been quite rapid since the early surveys (Tansley 1939, Pigott & Wilson 1978). But both these sites have moderately eutrophic waters and, in more impoverished basins, change may be much slower. The eventual fate of such successions is unknown but it seems unlikely that the *Salix-Carex* woodland progresses to, say, *Alnus-Urtica* woodland except where stands are drained and disturbed. It is possible, too, that sinking of the fen mat under the weight of the developing canopy sets back further succession in some cases.

One other possibility is that the community is a prelude to the development of more ombrotrophic vegetation as terrestrialisation isolates the mire surface from the ground-water. In this respect, the occasional occurrence of patches of Sphagna within the field layer of the *Salix-Carex* woodland is of interest. Further sampling may permit the characterisation of a distinct sub-community of this woodland in which Sphagna are more prominent, with a decreased representation of some of the dicotyledons and certain Carices, as in the southern *Salix-Betula-Phragmites* woodland. The progression of such vegetation to herbaceous bog (perhaps with an intervening stage with the *Betula-Molinia* woodland) would then leave the *Salix-Carex* woodland persisting only on the marginal fen peat (Walker 1970, Proctor 1974, Wheeler 1983). As on southern flood-plain mires, the conditions for such a development might well be initiated within the herbaceous fen before any shrub invasion had taken place.

Distribution
The *Salix-Carex* woodland occurs locally throughout the sub-montane zone of northern Britain. Although no stands have been encountered in Wales, its natural range probably extends there and further sampling may reveal its presence.

Affinities
The distinctive floristic features of northern British fen woods were noted by Proctor (1974) and this vegetation type given association status by Wheeler (1975, 1980c) as the *Crepido-Salicetum pentandrae*. Although woodlands with *Salix pentandra* have been described from parts of Europe (e.g. Passarge 1961, Westhoff & den Held 1969), their overall composition is very different to that of the *Salix-Carex* woodland. Its nearest relative among Continental communities seems to be the montane scrub described by Dahl (1956) as the *Mnieto-Salicetum phylicifoliae*.

The community can be considered as the northern counterpart of the *Salix-Betula-Phragmites* woodland within the Salicion cinereae (Wheeler 1975, 1980c) and its closest affinities outside woodlands are with the rich fens of the Caricion davallianae.

Floristic table W3

Salix pentandra	IV (4–10)		*Menyanthes trifoliata*	III (1–5)
Salix cinerea	IV (3–8)		*Equisetum fluviatile*	III (1–3)
Betula pubescens	II (2–6)		*Eurhynchium praelongum*	III (1–4)
Alnus glutinosa	I (3–6)		*Poa trivialis*	III (1–4)
Salix nigricans	I (4–9)		*Crepis paludosa*	III (1–3)
Salix phylicifolia	I (4–9)		*Equisetum palustre*	III (1–3)
Salix aurita	I (9)		*Climacium dendroides*	III (1–5)
Salix viminalis	I (4)		*Valeriana officinalis*	III (1–5)
			Mentha aquatica	III (1–5)
Cardamine pratensis	V (1–4)		*Potentilla palustris*	III (1–5)
Galium palustre	V (1–6)		*Ranunculus repens*	II (1–7)
Caltha palustris	V (1–8)		*Lychnis flos-cuculi*	II (1)
Filipendula ulmaria	V (1–6)		*Ranunculus acris*	II (1–3)
Calliergon cuspidatum	V (1–5)		*Cirsium palustre*	II (1–3)
Angelica sylvestris	IV (1–2)		*Brachythecium rutabulum*	II (1–4)
Valeriana dioica	IV (1–5)		*Ajuga reptans*	II (1–4)
Rhizomnium punctatum	IV (1–7)		*Holcus lanatus*	II (1–3)
Mnium hornum	IV (1–4)		*Phragmites australis*	II (1–4)
Carex rostrata	IV (1–6)		*Molinia caerulea*	II (1–3)
Geum rivale	IV (1–6)		*Dactylorhiza fuchsii*	II (1–3)

Floristic table W3 *(cont.)*

Plagiomnium affine	II (1–3)	*Scutellaria galericulata*	I (2)
Dryopteris dilatata	II (1–3)	*Chiloscyphus polyanthos*	I (1)
Juncus acutiflorus	II (1–6)	*Deschampsia cespitosa*	I (1)
Urtica dioica	II (1–5)	*Eupatorium cannabinum*	I (4)
Juncus effusus	I (1–5)	*Veronica scutellata*	I (1)
Carex paniculata	I (3–4)	*Crataegus monogyna* seedling	I (3)
Sphagnum recurvum	I (1–3)	*Pellia endiviifolia*	I (1)
Hypnum cupressiforme	I (1–4)	*Plagiochila asplenoides*	I (1)
Sphagnum squarrosum	I (4–9)	*Chiloscyphus pallescens*	I (1)
Viola palustris	I (1–2)	*Fissidens adianthoides*	I (1)
Plagiomnium rostratum	I (1–6)	*Aulacomnium androgynum*	I (2)
Phalaris arundinacea	I (1–4)	*Plagiomnium elatum*	I (8)
Sphagnum palustre	I (1–4)	*Carex appropinquata*	I (4)
Chrysosplenium oppositifolium	I (1–8)	*Parnassia palustris*	I (1)
Lophocolea bidentata s.l.	I (1–3)	*Trollius europaeus*	I (1)
Hedera helix	I (2–3)	*Conocephalum conicum*	I (2)
Galium uliginosum	I (1–3)	*Dicranoweissia cirrata*	I (1)
Galium aparine	I (1–3)	*Carex echinata*	I (1)
Carex nigra	I (2–7)	*Potamogeton polygonifolius*	I (2)
Ranunculus flammula	I (1–2)	*Alnus glutinosa* seedling	I (1)
Epilobium palustre	I (1–2)	*Salix cinerea* seedling	I (1)
Agrostis stolonifera	I (2–6)	*Carex laevigata*	I (4)
Succisa pratensis	I (1–3)	*Saccogyna viticulosa*	I (3)
Lycopus europaeus	I (1–3)	*Marchantia polymorpha*	I (1)
Myosotis scorpioides	I (1–2)	*Cephalozia bicuspidata*	I (1)
Lophocolea cuspidata	I (1–3)	*Jungermannia obovata*	I (2)
Drepanocladus uncinatus	I (2–3)	*Calliergon giganteum*	I (1)
Lophocolea heterophylla	I (1–3)	*Douinia ovata*	I (2)
Plagiomnium ellipticum	I (4–8)	*Carex vesicaria*	I (4)
Solanum dulcamara	I (1)	*Eurhynchium speciosum*	I (1)
Cardamine flexuosa	I (1–3)	*Pyrola rotundifolia*	I (1)
Hydrocotyle vulgaris	I (1)		
Galeopsis tetrahit	I (1)	Number of samples	18
Lysimachia vulgaris	I (3)	Number of species/sample	28 (20–38)
Callitriche stagnalis	I (4)		
Lemna minor	I (1)	Shrub height (m)	5 (3.5–7)
Rubus fruticosus agg.	I (1)	Shrub cover (%)	93 (80–100)
Veronica beccabunga	I (3)	Herb height (cm)	52 (20–100)
Calliergon cordifolium	I (4)	Herb cover (%)	60 (5–90)
Athyrium filix-femina	I (1)	Ground height (mm)	15 (10–35)
Myosotis laxa caespitosa	I (4)	Ground cover (%)	42 (10–90)
Oenanthe crocata	I (2)		
		Altitude (m)	267 (45–380)

W3 *Salix pentandra-Carex rostrata* woodland

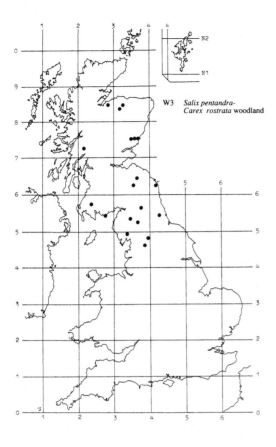

W3   *Salix pentandra-*
Carex rostrata woodland

W4
Betula pubescens-Molina caerulea woodland

Synonymy

Betuletum pubescentis (Hueck 1929) R.Tx. 1955 *p.p.*; Alder woodland types 2a & 3a   McVean 1956*b*; Woodwalton *Molinia* sere   Poore 1956*b p.p.*; Woodwalton Birch wood   Poore 1956*b*; Waveney/Ouse fens Community R   Bellamy & Rose 1961; Clarepool Moss woodland   Sinker 1962; Malham Tarn Birchwoods Proctor 1974, Adam *et al.* 1975; Woodland *Molinia* nodum   Daniels 1978; *Betula-Sphagnum flexuosum* nodum   Daniels 1978; *Betulo-Myricetum molinietosum*   Wheeler 1980*c*; *Osmundo-Alnetum sphagnetosum*   Wheeler 1980*c*; *Betula pubescens* woodland Meres Report 1980; *Sphagnum palustre-Betula pubescens* Community   Birse 1982; *Sphagno-Salicetum atrocinereae*   Birse 1984 *p.p.*

Constant species

Betula pubescens, Molinia caerulea, Sphagnum recurvum/palustre.

Rare species

Dryopteris cristata.

Physiognomy

The general floristic and physiognomic features of the *Betula pubescens-Molinia caerulea* woodland are very simple. *Betula pubescens* is the only constant woody species and it is almost invariably the dominant, though the canopy it forms is often rather open, with well-spaced individuals. Quite commonly, the trees have a moribund look: infestation with *Piptoporus betulinus* is frequent and its large fruiting bodies can often be seen on birches that have fallen over or are obviously dying upright. *B. pendula* is typically very scarce though it is sometimes to be found in local abundance invading drier stands (as at Woodwalton: Poore 1956*b*). Indeed, no other tree is even occasional throughout: *Alnus glutinosa* comes a poor second to *B. pubescens* (though it is a little more common in the *Juncus* sub-community and locally dominant there) and oaks are typically infrequent. In marked contrast to woodlands on more base-rich soils,

Fraxinus excelsior is extremely uncommon here.

Smaller woody species are also few in number and the understorey is characteristically sparse. *Salix cinerea* is the most frequent shrub and sometimes it grows sufficiently tall to break the birch canopy making stratification indistinct. *S. caprea, S. pentandra* and *S. aurita* are also sometimes found and very occasionally *Corylus avellana, Crataegus monogyna* and *Ilex aquifolium* occur. *B. pubescens* saplings are quite common though they rarely form the dense thickets so typical of much birch invasion. Young *Alnus* and oaks occur infrequently.

The most distinctive feature of the field layer here is the consistent presence and often great abundance of *Molinia caerulea*, which often seems to form an even grassy sward but which, on close inspection, is found in its characteristic tussocky form with systems of litter-lined runnels in which the vascular associates and bryophytes are disposed in mosaics. The herbaceous element in this kind of woodland is, in fact, rather variable and no species apart from *Molinia* attains any great frequency overall, but, except in the driest stands, there are usually some Sphagna, most typically throughout *S. palustre* and/or *S. recurvum* with, much less commonly *S. subnitens*, forming a patchy cover, sometimes a continuous carpet, between the tussocks with occasional *Aulacomnium palustre*. *Eurhynchium praelongum* and *Pseudoscleropodium purum* can also sometimes be found among the *Molinia* shoots and eroding tussock tops occasionally have a prominent cover of *Polytrichum commune*.

Sub-communities

***Dryopteris dilatata-Rubus fruticosus* sub-community:** Woodwalton Birch wood   Poore 1956*b*; Malham Tarn Birchwood   Proctor 1974 *p.p.*; *Sphagnum palustre-Betula pubescens* Community   Birse 1982 *p.p. B. pubescens* is almost always the dominant here in a canopy that is usually taller and denser than in the other sub-communities. Also, there is a little more variety in the woody cover than is typically the case in this community: *B. pendula* occasionally augments (and rarely replaces) *B. pubescens* but, much more obviously, there is frequently some *Sorbus aucuparia*; oaks are somewhat more common here, too, and *Pinus sylvestris* is sometimes found. Among the smaller species, *Salix cinerea* is very frequent and, with young birch and more occasional oaks, it forms an understorey that can be quite dense in parts. Then, there are sparse records for *Corylus avellana*, *Crataegus monogyna* and *Ilex aquifolium*.

In the field layer, *Molinia* remains very frequent and is often abundant, though here its cover is often masked by an underscrub of *Rubus fruticosus* agg. and trailing *Lonicera periclymenum*. *Dryopteris dilatata* is a very good preferential, too, its large crowns emerging through the bramble cover. Occasionally, there can be some patchy invasion of *Pteridium aquilinum* or, especially where there has been disturbance of the substrate, clumps of *Epilobium angustifolium*. Another feature of note in some stands is the occurrence of patches of *Deschampsia flexuosa* with scattered *Galium saxatile*: these are often centred around the tree bases, forming a mosaic with the intervening *Molinia*-dominated areas, a pattern which is particularly prominent where the peat between the trees is shrinking downwards.

On the usually drier soils here, Sphagna are sparser than in the other sub-communities, being generally found as small tufts (mostly of *S. palustre*) among the *Molinia* litter. By contrast, there is an increase in species such as *Mnium hornum*, *Hypnum cupressiforme*, *Polytrichum commune*, *P. formosum*, *Plagiothecium denticulatum* and *Isopterygium elegans*.

Juncus effusus sub-community: Alder woodland type 2a McVean 1956*b; Sphagnum palustre-Betula pubescens* Community Birse 1982 *p.p.; Sphagno-Salicetum atrocinereae* Birse 1984 *p.p.* Again, *B. pubescens* is the usual dominant here but it typically forms an open and rather low canopy. *Alnus glutinosa* is a little more frequent than elsewhere in the community and occasionally it exceeds *B. pubescens* in its abundance but the trees are usually poorly grown and often have irregular spreading crowns. Bushes of *Salix cinerea* are frequent and *B. pubescens* saplings occasional but they do not usually compose a distinct understorey, occurring rather in scattered patches among the more open areas of trees. Where this kind of woodland occurs in valleyside flushes, substrate instability can result in slumping and produce further irregularity in the woody cover.

The general appearance of the field layer here is of a grassy ground with scattered tussocks of rushes and sedges. *Molinia* remains very common and abundant but it is frequently accompanied by *Holcus mollis, H.*

lanatus, Deschampsia cespitosa and, less commonly, by *Anthoxanthum odoratum* and *Agrostis stolonifera*. *Juncus effusus* is constant and often very prominent; *J. acutiflorus* is more occasional, but it can replace *J. effusus*, perhaps where there is more obvious water seepage. In other stands, *Carex laevigata* is abundant and sometimes *C. nigra* though, in contrast to similar field layers in the *Alnus-Fraxinus-Lysimachia* woodland, *C. remota* is typically absent here. *Myrica gale* occurs occasionally, in some cases forming a quite thick cover of leggy bushes. Where stands are grazed (a fairly common occurrence), this structural heterogeneity is accentuated by patchy break-up of the vegetation cover with trampling.

Scattered through the field layer is a variety of distinctive herbs. *Potentilla erecta, Hydrocotyle vulgaris, Viola palustris, Cirsium palustre* and *Lotus uliginosus* are the most common but also preferential at lower frequencies are *Galium uliginosum, Angelica sylvestris* and *Valeriana officinalis*.

Sphagnum recurvum and, less frequently, *S. palustre* form a sometimes extensive carpet between the tussocks of grasses, rushes and sedges, and *Calliergon cuspidatum* appears as a good preferential bryophyte.

Sphagnum spp. sub-community: Alder woodland type 3a McVean 1956*b*; Waveney/Ouse fens Community R Bellamy & Rose 1961; Clarepool Moss woodland Sinker 1962; Malham Tarn Birchwoods Proctor 1974 *p.p.*, Adam *et al.* 1975; *Betula pubescens* woodland Meres Report 1980; *Betulo-Myricetum molinietosum* Wheeler 1980*c; Osmundo-Alnetum sphagnetosum* Wheeler 1980*c. B. pubescens* is always dominant here though, as in the *Juncus* sub-community, its cover is usually open and low. Other woody species are poorly represented: there is very occasionally some *Alnus glutinosa* and even *Salix cinerea* is rather sparse.

The field layer, too, is characteristically species-poor, though it can show considerable variety from stand to stand. The constant feature is again *Molinia* and this is usually dominant. *Juncus effusus* occurs occasionally, but the diversity and abundance of grasses typical of the *Juncus* sub-community is absent. More often, here, the vegetation preserves elements of the wet heath or mire from which the woodland has developed: in some cases, *Calluna vulgaris* and *Erica tetralix* occur in a patchy cover; in others, there are tussocks of *Eriophorum vaginatum* and lawns or pools with *E. angustifolium, Carex nigra* and *Vaccinium oxycoccus*; or there can be very wet areas with *Carex rostrata, Menyanthes trifoliata* and *Potentilla palustris*.

The other generally distinctive feature of the vegetation is the prominence of Sphagna which are more varied and abundant than in other kinds of *Betula-Molinia* woodland. The community species *S. recurvum* and *S. palustre* are both very frequent but preferential here are *S. fimbriatum*, *S. squarrosum* and *S. papillosum*; *S. subnitens* and *S. cuspidatum* can also be found less commonly. Typically, these form extensive patches in the wet runnels between the *Molinia* tussocks with mosses like *Pohlia nutans* and *Polytrichum commune* picking out drier areas of exposed peat.

Habitat

The *Betula-Molinia* woodland is typically a community of moist, moderately acid, though not necessarily highly oligotrophic, peaty soils in a variety of mire types. It is especially characteristic of thin or drying ombrogenous peats around the margins of blanket mires and topogenous deposits that have become elevated above or isolated from the influence of more base-rich and eutrophic ground waters; but it also occurs on soligenous peats in valley mires and on flushed peaty gleys where there is irrigation by rather base- and nutrient-poor waters. Various human activities may play an important part in the development of the community, both in the spread of *Molinia* prior to woodland development and in subsequent birch invasion, and grazing by stock and deer is probably important in maintaining the distinctive character of some stands.

The different sub-communities show fairly clear relationships to particular conditions within this broad environmental definition. The *Sphagnum* sub-community is most typical of wetter and deeper peats where, as on degraded blanket bogs or in the centre of basin mires, the water-table is sufficiently low for birch to become well established but where there is enough surface moisture to sustain a fairly extensive *Sphagnum* cover, though with a clear shift away from the suites of hummock and pool species typical of the intact and active bog surface (e.g. Sinker 1962, Meres Report 1980). In such situations, the mire waters are generally base- and nutrient-poor but the *Sphagnum* sub-community is also occasionally found in more eutrophic and base-rich topogenous fen systems, in areas where the peat deposits have become isolated from the movement of the ground waters, as, for example, at the landward edge of some Broadland fens (e.g. Wheeler 1975, 1978, 1980c) and towards the back of Esthwaite North Fen (e.g. Pearsall 1918, Tansley 1939, Pigott & Wilson 1978). In many of these cases, human interference, such as the grazing, draining and burning of the mires, or peat-cutting and the abandonment of exposed and drying peat surface, may act as an encouragement for the spread of *Molinia* and *B. pubescens*.

The *Dryopteris-Rubus* sub-community often appears to be a longer-established and drier form of *Betula-Molinia* woodland than the above, occurring in similar situations but always where the peat is naturally thinner (as on the eroding fringes of blanket mires) or has become much better drained, either through natural

growth above a ground water table or as a result of human activity. Here the prominence of the Sphagna is much less obvious and many stands show a clear similarity to oak-birch woodland of dry, acid soils. Disturbance and surface enrichment of the peat is quite common here, too, and sometimes the vegetation approaches that of the *Betula* sub-community of the *Alnus-Urtica* woodland. *Urtica dioica* itself, however, is rare here, *Epilobium angustifolium* being the more characteristic marker of eutrophication in this community.

The *Juncus* sub-community is rather more specialised in its environmental associations, being typically associated with areas of soligenous influence of fairly acid, though not always cation-poor, waters. It can occur over peats, as for example in the lagg of basin mires and down the central zone of some more base-poor valley mires but it is also found on peaty gleys, as where flushes emerge over slopes of impermeable argillaceous rocks or non-calcareous superficials. In such situations, it can be seen as the more base-poor equivalent of the *Alnus-Fraxinus-Lysimachia* woodland: in both these communities, *Carex laevigata* is a very characteristic indicator of such edaphic conditions.

Zonation and succession

Very often, the *Betula-Molinia* woodland occurs in zonations and mosaics with other vegetation types which reflect the progress of birch invasion on mire surfaces. On ombrogenous and topogenous mires, it is typically found in association with Erico-Sphagnion and Ericion tetralicis bogs and wet heaths and fragments of such vegetation types persisting within the field layer of the *Sphagnum* sub-community can give some clue as to its origin. On raised mires and in basin mires where the waters are not base-rich and calcareous, this kind of *Betula-Molinia* woodland may represent a fairly natural first development in the progression to woodland with increased elevation of the peats above the ground water level and, in some places, crudely concentric zonations can be seen which approximate to the classic hydroseral pattern as in some Shropshire meres (Sinker 1962) and at Malham Tarn (e.g. Proctor 1974). At certain sites, as at Malham again, the zonation continues to the drier *Dryopteris-Rubus* sub-community which may represent the continuation of a succession towards acid oak-birch woodlands. Very often, however, patterns are not so clear as this and fragments of these two sub-communities frequently occur intermixed with variously modified remnants of mire vegetation on peat surfaces that have been subject to complex histories of draining, burning and cutting (as at Woodwalton: Poore 1956b). Similar treatments may also have encouraged the development of the *Betula-Molinia* woodland where it occurs on the margins of run-down blanket mires.

Other kinds of complexities arise where there has been woodland development over mire surfaces with varying water and nutrient regimes. In flood-plain mires, for example, the *Betula-Molinia* woodland can mark out ombrogenous nuclei which have formed within tall-herb or *Carex*-dominated fens. In the Ant valley, the *Sphagnum* sub-community forms a core to some of the striking islands of *Betula-Salix-Phragmites* woodland that have grown up among the *Peucedano-Phragmitetum* fen (Wheeler 1978, 1980*c*) and, at Esthwaite, it has developed behind a front of *Potentillo-Caricetum* fen which is progressing to the *Salix-Carex* woodland (Pearsall 1918, Tansley 1939, Pigott & Wilson 1978). Similar, though more complex, patterns occur around peat islands in the Malham fens (Proctor 1974, Adam *et al.* 1975).

The *Juncus* sub-community can also be found in association with mire communities. In the rands of basin mires, for example, it often occurs amongst rush- or sedge-dominated vegetation of the Caricion curto-nigrae or Junco-Molinion and these would seem to be the normal seral precursors to this kind of woodland in such situations. In valley mires, and in soligenous soak-ways on some ombrogenous mires, it can be found in mosaics with Ericion tetralicis vegetation and may develop from this by birch invasion. Where this kind of *Betula-Molinia* woodland marks out slope flushes with peaty mineral soils, rather different patterns are encountered. Usually here, the *Juncus* sub-community occurs as small stands within more extensive tracts of drier woodland, very often at junctions between acid oak-birch woodlands on permeable sandstones above the flush line, and less acidophilous Carpinion woodland on the shales or drift below.

Distribution

The community is widespread but local throughout the lowlands and on the upland fringes of Britain. The *Sphagnum* sub-community and, especially now with increased mire drainage, the *Dryopteris-Rubus* sub-community are the commoner types; the *Juncus* sub-community has been much less frequently encountered.

Affinities

This is a quite distinct woodland type though it has usually been recognised only on a local basis as part of particular complexes of mire vegetation rather than as a nationally-distributed component of the series of wetter woodlands in Britain. It figured in McVean's (1956*b*) scheme for British alderwoods though, as defined here, *Alnus* is a rare component; and it is also represented in Birse's accounts of Scottish woodlands (Birse 1982, 1984), though his samples are split into two separate communities. In phytosociological terms, the *Betula-Molinia* woodland is clearly very similar to part of the *Betuletum pubescentis* described from various parts of the Continent (e.g. Schwickerath 1944, LeBrun *et al.* 1949, Tüxen 1955, Matuszkiewicz 1963, Westhoff & den Held 1969), though that association has a good representation of heath dwarf-shrubs which are, at most, occasional here. The *Betuletum pubescentis* is located within the Vaccinio-Picetea: such vegetation is seen in Britain in the *Pinus-Hylocomium* and *Juniperus-Oxalis* woodlands but the most similar birch-dominated community is the *Quercus-Betula-Oxalis* woodland. The affinities of the *Betula-Molinia* woodland are perhaps more obviously with the Alnion glutinosae.

Floristic table W4

	a	b	c	4
Betula pubescens	V (6–10)	V (5–9)	V (5–10)	V (5–10)
Alnus glutinosa	I (1–8)	II (5–8)	I (6–8)	I (1–8)
Quercus robur	I (1–5)	I (2–5)	I (5)	I (1–5)
Betula pendula	I (1)	I (4)	I (1–4)	I (1–4)
Salix caprea	I (4)	I (3)	I (2)	I (2–4)
Pinus sylvestris	I (1–6)		I (1–5)	I (1–6)
Quercus petraea	I (2–4)	I (3–4)		I (2–4)
Salix pentandra	I (1–2)	I (1)		I (1–2)
Betula hybrids	I (3)	I (3)		I (3)
Fraxinus excelsior	I (7)	I (1)		I (1–7)
Salix cinerea	III (1–7)	III (1–7)	II (1–9)	III (1–9)
Betula pubescens sapling	III (2–7)	II (2–3)	I (4–5)	II (2–7)
Sorbus aucuparia	III (1–6)	I (1)	I (1)	II (1–6)
Quercus robur sapling	II (1–4)			I (1–4)
Alnus glutinosa sapling	I (1)	II (1–2)	I (4)	I (1–4)
Quercus petraea sapling	I (2–4)	I (1)	I (1)	I (1–4)
Quercus hybrids sapling	I (2)	I (4)	I (4)	I (2–4)

Floristic table W4 *(cont.)*

	a	b	c	4
Crataegus monogyna	I (1–3)	I (1–2)		I (1–3)
Corylus avellana	I (3–5)	I (1)		I (1–5)
Ilex aquifolium	I (2–4)	I (1)		I (1–4)
Salix aurita		I (7)	I (3)	I (3–7)
Molinia caerulea	IV (1–10)	IV (3–8)	IV (1–9)	IV (1–10)
Dryopteris dilatata	IV (4–9)	I (1)	I (1–3)	II (1–9)
Rubus fruticosus agg.	IV (1–10)			II (1–10)
Lonicera periclymenum	III (1–4)			I (1–4)
Mnium hornum	III (1–4)		I (1–2)	I (1–4)
Deschampsia flexuosa	II (2–6)	I (2–4)	I (3–6)	I (2–6)
Epilobium angustifolium	II (2–5)	I (1)	I (3)	I (1–5)
Galium saxatile	II (1–2)	I (1–2)	I (1)	I (1–2)
Lophocolea bidentata s.l.	II (1–3)	I (2)	I (1–3)	I (1–3)
Pteridium aquilinum	II (1–5)	I (2–5)		I (1–5)
Hypnum cupressiforme	II (2–6)		I (1–4)	I (1–6)
Hedera helix	I (2–7)			I (2–7)
Plagiothecium denticulatum	I (1–2)			I (1–2)
Isopterygium elegans	I (1–2)			I (1–2)
Polytrichum formosum	I (2–3)			I (2–3)
Juncus effusus	I (1–4)	V (1–8)	II (1–5)	II (1–8)
Potentilla erecta	I (1)	IV (3)	I (1–3)	II (1–3)
Holcus mollis	I (1–8)	IV (1–5)		II (1–8)
Deschampsia cespitosa	I (1–8)	IV (1–5)		I (1–8)
Holcus lanatus	I (3–7)	III (1–4)		I (1–7)
Hydrocotyle vulgaris		III (2–4)	I (3–4)	I (2–4)
Viola palustris		III (1–3)	I (2–4)	I (1–4)
Cirsium palustre		III (1–4)		I (1–4)
Lotus uliginosus		III (1–3)		I (1–3)
Calliergon cuspidatum		III (1–4)		I (1–4)
Myrica gale	I (3–4)	II (4–6)	I (3–7)	I (3–7)
Juncus acutiflorus		II (2–6)	I (3)	I (2–6)
Succisa pratensis		II (2)	I (2)	I (2)
Galium uliginosum		II (3–4)		I (3–4)
Carex laevigata		II (4–6)		I (4–6)
Angelica sylvestris		II (2)		I (2)
Valeriana officinalis		II (1)		I (1)
Sphagnum fimbriatum	I (2–7)		III (1–8)	II (1–8)
Sphagnum squarrosum	I (3–5)	I (5)	II (1–8)	I (1–8)
Sphagnum papillosum	I (1–5)		II (2–4)	I (1–5)
Pohlia nutans		I (3)	II (1–3)	I (1–3)
Eriophorum angustifolium			II (1–6)	I (1–6)
Calluna vulgaris			II (1–7)	I (1–7)
Eriophorum vaginatum			II (1–5)	I (1–5)
Erica tetralix			II (1–5)	I (1–5)
Menyanthes trifoliata			I (5–6)	I (5–6)

Potentilla palustris			I (5–7)	I (5–7)
Carex rostrata			I (1–4)	I (1–4)
Vaccinium oxycoccus			I (1–4)	I (1–4)
Drepanocladus fluitans			I (2–5)	I (2–5)
Equisetum fluviatile			I (4)	I (4)
Calliergon stramineum			I (3)	I (3)
Sphagnum cuspidatum			I (4)	I (4)
Sphagnum recurvum	I (5–8)	IV (1–7)	IV (1–8)	III (1–8)
Sphagnum palustre	II (1–9)	II (2–5)	III (2–9)	II (1–9)
Eurhynchium praelongum	II (2–6)	II (1–3)	I (2)	I (1–6)
Pseudoscleropodium purum	II (2–4)	II (1–2)	I (3–8)	I (1–8)
Aulacomnium palustre	I (1–3)	II (1–4)	II (1–4)	I (1–4)
Carex nigra	I (1)	II (1–5)	II (1–3)	I (1–5)
Polytrichum commune	II (1–7)	I (1)	II (1–7)	I (1–7)
Rhytidiadelphus squarrosus		II (2–3)	I (1)	I (1–3)
Agrostis canina canina	I (1–5)	I (2)	I (1–7)	I (1–7)
Dryopteris carthusiana	I (3–6)	I (2)	I (2)	I (2–6)
Rumex acetosa	I (1)	I (2–3)	I (1)	I (1–3)
Sphagnum subnitens	I (1–6)	I (2)	I (3–5)	I (1–6)
Pleurozium schreberi	I (2)	I (1)	I (1–4)	I (1–4)
Lophocolea cuspidata	I (1–2)	I (1)	I (3)	I (1–3)
Betula pubescens seedling	I (2)	I (2)	I (2–3)	I (2–3)
Anthoxanthum odoratum	I (1–2)	I (3–5)		I (1–5)
Lophocolea heterophylla	I (1–2)	I (1)		I (1–2)
Agrostis capillaris	I (1–7)	I (4)		I (1–7)
Agrostis stolonifera	I (2–3)	I (3–5)		I (2–5)
Oxalis acetosella	I (2–4)	I (4)		I (2–4)
Athyrium filix-femina	I (1–3)	I (1)		I (1–3)
Blechnum spicant	I (1)	I (1)		I (1)
Chiloscyphus polyanthos	I (2)	I (1)		I (1–2)
Filipendula ulmaria	I (1)	I (2)		I (1–2)
Brachythecium rutabulum	I (1)	I (3)		I (1–3)
Vaccinium myrtillus	I (1–5)		I (2–6)	I (1–6)
Quercus robur seedling	I (1–3)		I (2–3)	I (1–3)
Dicranum scoparium	I (1–4)		I (2–3)	I (1–4)
Hypnum jutlandicum	I (2–3)		I (2–4)	I (2–4)
Poa trivialis	I (1–2)		I (2)	I (1–2)
Calypogeia fissa	I (2)		I (1–2)	I (1–2)
Dryopteris cristata	I (4)		I (1–5)	I (1–5)
Equisetum palustre	I (2)		I (2)	I (2)
Campylopus paradoxus	I (1–2)		I (2)	I (1–2)
Dicranella heteromalla	I (1–3)		I (2)	I (1–3)
Leucobryum glaucum	I (2)		I (2)	I (2)
Orthodontium lineare	I (1)		I (1)	I (1)
Galium palustre		I (4)	I (2)	I (2–4)
Sphagnum capillifolium		I (5)	I (6)	I (5–6)
Epilobium palustre		I (3)	I (2)	I (2–3)
Number of samples	35	9	28	72
Number of species/sample	17 (7–30)	28 (15–52)	13 (2–23)	17 (2–52)

Floristic table W4 *(cont.)*

	a	b	c	4
Tree height (m)	13 (8–25)	8 (5–14)	7 (5–15)	10 (5–25)
Tree cover (%)	90 (20–100)	59 (10–85)	70 (20–100)	81 (10–100)
Shrub height (m)	3 (1–6)	5 (3–8)	3 (2–5)	3 (1–8)
Shrub cover (%)	17 (0–75)	10 (0–55)	5 (0–15)	10 (0–75)
Herb height (cm)	47 (15–120)	66 (30–140)	58 (30–150)	54 (15–150)
Herb cover (%)	81 (25–100)	86 (70–100)	69 (15–100)	77 (15–100)
Ground height (mm)	31 (10–110)	57 (10–100)	42 (10–110)	44 (10–110)
Ground cover (%)	28 (0–90)	29 (1–90)	66 (7–100)	42 (0–100)
Altitude (m)	78 (5–270)	100 (20–260)	100 (1–210)	90 (1–270)
Slope (°)	1 (0–15)	2 (0–5)	1 (0–2)	1 (0–15)

a *Dryopteris dilatata-Rubus fruticosus* sub-community
b *Juncus effusus* sub-community
c *Sphagnum* sub-community
4 *Betula pubescens-Molinia caerulea* woodland (total)

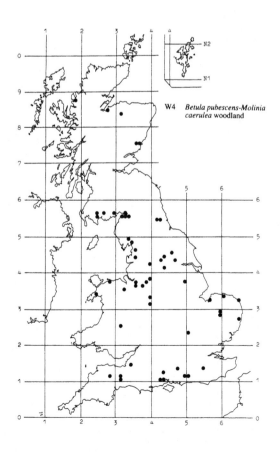

W4   *Betula pubescens-Molinia caerulea* woodland

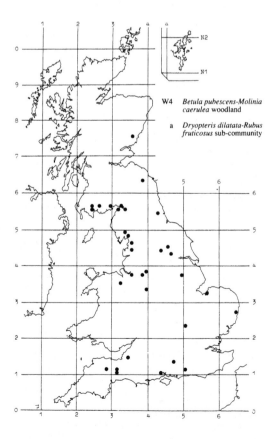

W4   *Betula pubescens-Molinia caerulea* woodland

a   *Dryopteris dilatata-Rubus fruticosus* sub-community

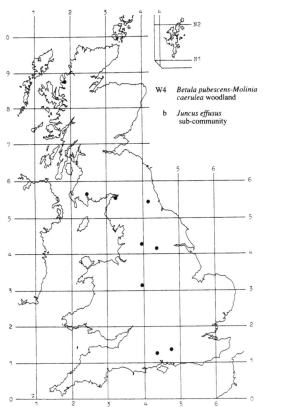

W4   *Betula pubescens-Molinia
         caerulea* woodland

b   *Juncus effusus*
      sub-community

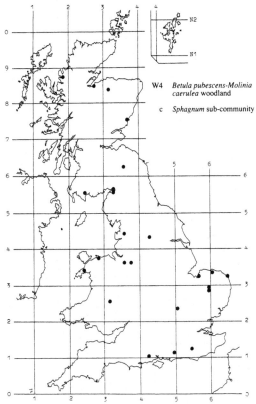

W4   *Betula pubescens-Molinia
         caerulea* woodland

c   *Sphagnum* sub-community

W5
Alnus glutinosa-Carex paniculata woodland

Synonymy

Swamp carr Pallis 1911; Alder thicket Rankin 1911*b* *p.p.*; Valley fen woods Farrow 1915; Alder wood Clapham 1940; Swamp carr and semi-swamp carr Lambert 1951; Alder woodland types 2c *p.p.* & 3c McVean 1956*b*; Alder carr Sinker 1962; Valley fen alderwoods Haslam 1965; *Osmundo-Alnetum glutinosae* (Klötzli 1970) Wheeler 1980*c p.p.*; *Alnus-Salix* woodland XXiii Meres Report 1980; Alder stand types 7Ba and 7Bb Peterken 1981; *Scutellaria galericulata-Alnus glutinosa* Association Birse 1982 *p.p.*; Woodland plot type 14 Bunce 1982 *p.p.*

Constant species

Alnus glutinosa, Carex paniculata, Galium palustre, Rubus fruticosus agg., *Brachythecium rutabulum, Eurhynchium praelongum.*

Rare species

Carex appropinquata, C. elongata, Cicuta virosa, Dryopteris cristata, Peucedanum palustre, Thelypteris palustris.

Physiognomy

The canopy of the *Alnus glutinosa-Carex paniculata* woodland is characterised by the high frequency and often the great abundance of *Alnus* but its detailed floristics and physiognomy vary considerably according to the age of the stand and the nature of the substrate. *Alnus* and *Salix cinerea* are the most frequent invaders of the kinds of swamp and fen from which this woodland is derived and, in the early stages of colonisation, their proportions are very much a reflection of the chance availability of propagules and the frequency and disposition of sites where seedlings can gain a hold, very often here the tops and sides of *Carex paniculata* tussocks. In general, though, young stands are characterised by the co-dominance of these two species in low, uneven and open canopies and the general trend with ageing is for *S. cinerea* to be relegated to an understorey or, where the *Alnus* grows up to cast a deep shade, to be completely extinguished. *Betula pubescens* can also appear early in invasion but it is decidedly patchy and its overall frequency is no more than occasional. Once the canopy has thickened up, light quickly becomes insufficient for its establishment. In drier situations, on initially firmer substrates or in sites where the fen mat consolidates with the passage of time, *Fraxinus excelsior* can become frequent but it generally makes no more than a local contribution to the canopy. *Quercus robur* can also occur in such situations though it remains rare through the community as a whole.

In young stands, *Salix cinerea* is frequently the only shrub present and, in general, it remains the commonest species in whatever understorey remains as the trees grow up to form a distinct, tall canopy. On drier substrates, *Crataegus monogyna* occurs occasionally (rarely *C. laevigata*) and there are sparse records, too, for *Ilex aquifolium, Sorbus aucuparia, Rhamnus catharticus* and *Salix aurita. Viburnum opulus* and *Frangula alnus* also occur throughout, though both these species, and especially the latter, are preferential for one particular sub-community. *Frangula* particularly can invade quite early and dense patches of saplings may be gradually grown over by the expanding woodland cover. In general, tree saplings are markedly rare in older stands, with only *Alnus* regenerating with any frequency under intact canopies.

In richer, well-established stands, the usual appearance of this woodland is of a clearly-defined canopy of well-grown *Alnus*, often multi-stemmed and up to 15 m tall, with scattered associated trees, over a distinct understorey of varying density, though sometimes very tangled and thick. Very often, however, the general pattern of development and structure is complicated by the local effects of substrate instability, exacerbated by the substantial weight of the trees which slowly depress the frequently thin fen mat into the amorphous sediments beneath. As the water-level rises above the tree bases, the *Alnus* can become moribund and die and the emergent tops of dead trees are a common feature of stands in such situations (Lambert 1951, 1965, McVean 1956*b*). Where, as is often the case, the alders are rooted in *Carex paniculata* tussocks, these may eventually roll over so that the trees, though they may survive, tilt at crazy angles (Lambert 1951, 1965). In both these cases, gaps may be opened up in the tree canopy, allowing other woody species, notably the quick-growing *Salix cinerea*, *Betula pubescens* and *Frangula*, to attain a local prominence, provided, of course, that the substrate beneath is not too wet. Where there is a more extensive breakdown of the fen mat with substantial surface waterlogging, a seral regression may occur (see below).

Although the constants of the field layer are few, the general character of the herbaceous vegetation in this woodland is very distinctive. In essence, it represents a modified flora of the preceding fen with but a small, strictly woodland component and structurally the plants are disposed according to the pattern of light and shade and the amount of surface moisture. Of the fen element, the most conspicuous species are usually large sedges, particularly the tussock-forming *Carex paniculata*, whose individuals can go on growing beneath the canopy to attain enormous size, and which often form a structural framework to the field layer. *C. acutiformis* is somewhat less frequent but it, too, can be very abundant and, being shade-tolerant, can extend its cover in the developing woodland by the spread of its far-creeping rhizomes. Either of these species may dominate and often both occur together. Other bulky fen sedges recorded less frequently, and then most often in one of the sub-communities, are *C. elata*, *C. appropinquata*, *C. riparia* and *C. pseudocyperus*. *Phragmites australis*, which is a prominent feature in many of the invaded fens, persists very patchily: it is quite common in one sub-community though, even there, it is only really abundant under more open areas of the canopy.

The other fen element which remains generally prominent comprises tall herbs. Of these, the most frequent throughout are *Urtica dioica*, *Filipendula ulmaria*, *Eupatorium cannabinum*, *Cirsium palustre*, *Valeriana officinalis* and *Iris pseudacorus* with, less commonly, *Angelica sylvestris* and *Valeriana dioica*. Where the community has developed from richer fens, *Lysimachia vulgaris* and *Lythrum salicaria* often occur and, in Broadland, *Peucedanum palustre*. Typically, as in the preceding fen, many of these species are rooted on the sides of or atop the *Carex paniculata* tussocks though *Iris* is habitually found between them and the others often spread on to the surface of the peat mat as the cover of prominent fen species such as *Phragmites* and *Calamagrostis canescens* declines. Here, they may attain patchy local prominence, giving a very variegated effect to the field layer, though where light is further reduced they, too, may show a decline or, as in *Iris*, stop flowering. Among these

taller species, woodland plants like *Geranium robertianum* and *Circaea lutetiana* make a very occasional appearance.

Generally speaking, smaller herbaceous species are rather few in number and rarely prominent. *Mentha aquatica* and *Poa trivialis* are by far the most frequent throughout but others recorded (sometimes with preferential frequency in the different sub-communities) include *Ranunculus repens*, *R. flammula*, *Viola palustris*, *Hydrocotyle vulgaris*, *Caltha palustris* and *Myosotis laxa* ssp. *cespitosa*. In contrast to the tall herbs, these plants are often more common between the sedge tussocks where, even on quite wet substrates, they can form a patchy ground cover. Where the substrate sinks to any marked degree, however, these plants, too, may disappear, leaving a network of sloppy peat surfaces between the tussocks, virtually bare apart from a little *Lemna minor* and treacherous to the unwary.

Three other elements of the field layer are distinctive and each can be conspicuous. The first comprises ferns, among which *Dryopteris dilatata* is the commonest species throughout, often growing epiphytically on the sedge tussocks or around the tree bases. *Athyrium filix-femina* occurs occasionally, tolerating somewhat wetter conditions and both *Dryopteris carthusiana* and the rare *D. cristata* have also been recorded. Much more strictly confined to the richer woodlands of one of the sub-communities are *Thelypteris palustris* and *Osmunda regalis*.

The second element consists of sprawlers. *Galium palustre* is a constant of the community and *Solanum dulcamara* is frequent and the latter, especially in one of the sub-communities, can be extremely abundant. The relative scarcity of *Galium aparine* and *Calystegia sepium* here is in contrast to the more eutrophic *Alnus-Urtica* woodland. Third, adding to the tangled field layer, there are usually some undershrubs. *Rubus fruticosus* agg. is very frequent, especially on drier ground where, with the more occasional *Lonicera periclymenum*, it can cover large areas. *Ribes nigrum*, *R. rubrum* and *Rosa canina* agg. can also occur.

Bryophytes are rather variable in their total cover but a number of species occur frequently and these can show a distinct patterning over the various available surfaces (e.g. Clapham 1940). The commonest mosses are *Eurhynchium praelongum*, *Brachythecium rutabulum*, *Plagiomnium undulatum* and *Rhizomnium punctatum* and each of these, together with the occasional *Lophocolea bidentata s.l.*, can be found on damp ground around the *Carex paniculata* tussocks or, where the sedge is growing healthily, on the sides or tops of the tussocks. *Calliergon cuspidatum* is rather more patchily distributed and, usually, it is more common over the bare, wet peat or on the stumpy wet remains of dead sedge. *Mnium hornum* is common here, too, but it is very much a plant of drier and somewhat more acid situations, among the decaying leaves of ageing tussocks and on tree bark, around bole bases or on fallen trunks and branches. In general, Sphagna are rare in this community, but small tufts of *S. palustre* or *S. squarrosum* may be found where there is seepage of more base-poor waters. A more frequent marker of such conditions here, however, and characteristic of one particular sub-community, is *Pellia epiphylla*.

Sub-communities

***Phragmites australis* sub-community:** *Osmundo-Alnetum glutinosae hydrocotyletosum* Klötzli 1970 *p.p.*; *Osmundo-Alnetum glutinosae typicum* (Klötzli 1970) Wheeler 1980c *p.p.*; Alder stand type 7Ba Peterken 1981 *p.p.*; *Scutellaria galericulata-Alnus glutinosa* Association Birse 1982 *p.p.* *Alnus* is always the dominant in mature tracts of this sub-community though young stands may have much *Salix cinerea* in the canopy and, in older woodlands, there is frequently some *Fraxinus* and *Betula pubescens*. *Sorbus torminalis* and *Populus tremula* occur rarely in less swampy situations.

The shrub layer is variable in total cover but there are usually some bushes of *S. cinerea* and sometimes these are sufficiently tall as to make stratification indistinct. *Crataegus monogyna* occurs occasionally but *Viburnum opulus* and *Frangula alnus* are rare.

The field layer is more species-poor here than in the two other sub-communities but, among the dominants of the preceding fen vegetation, *Phragmites* remains somewhat more frequent than elsewhere, though it is only abundant where the canopy has not yet closed or where developing gaps allow it to expand its cover (Haslam 1971b, 1972). The more usual dominants are *Carex paniculata* or *C. acutiformis* (rarely *C. elata*). Locally, *Scirpus sylvaticus* may totally replace these and its unusual abundance can give the field layer a very distinctive look.

Tall herbs are often prominent but there are no preferentially frequent species among this component, the commonest plants being those characteristic of the community as a whole: *Urtica dioica*, *Filipendula ulmaria*, *Eupatorium cannabinum*, *Iris pseudacorus*, *Cirsium palustre* and, less frequently, *Lycopus europaeus*, *Valeriana officinalis*, *V. dioica* and *Angelica sylvestris*. *Lysimachia vulgaris* and *Lythrum salicaria* are both scarce. Ferns, too, though they may be abundant, are not distinctive with only *Dryopteris dilatata* and *Athyrium filix-femina* occurring with any frequency. Smaller dicotyledons are also few in number, although *Mentha aquatica* occurs frequently. *Poa trivialis* is occasional and *Equisetum palustre* is weakly preferential here, showing a slight tendency to be more consistently associated with sites where *Scirpus sylvaticus* is dominant.

Scramblers and underscrub species are often very prominent. *Solanum dulcamara* is more frequent here than in the other sub-communities and it sometimes forms a virtually impenetrable mass of thin woody shoots. *Galium aparine* is a little commoner, too, than elsewhere in the community. In drier stands, *Rubus fruticosus* agg. can be very abundant.

The bryophytes are those characteristic of the community as a whole.

***Lysimachia vulgaris* sub-community:** *Osmundo-Alnetum glutinosae lycopetosum* Klötzli 1970 *p.p.*; Alder stand type 7Ba Peterken 1981 *p.p.* The tree canopy here retains the general floristic features of the previous sub-community and shows the same physiognomic variation with age as is characteristic of the whole community. The shrub layer, however, is somewhat more varied than usual: in addition to frequent bushes of *Salix cinerea* and the occasional *Crataegus monogyna*, *Frangula alnus* and *Viburnum opulus* occur with preferentially high frequency here and the former can be locally abundant.

However, the most distinctive feature of this vegetation is the field layer, where virtually all the components show greater richness and variety than elsewhere in the community. Among the sedges, both *C. paniculata* and *C. acutiformis* remain frequent here and either or both may dominate. Occasionally, they are supplemented or replaced by *C. elata* (as at Esthwaite North Fen and in ditches at Askham Bog). Other swamp and fen sedges persisting less commonly, and generally in smaller amounts, are *C. appropinquata*, *C. pseudocyperus* and *C. riparia*. Then, *C. remota* is frequent here, though it is rarely so abundant as to be worthy of calling a dominant. Some stands outside Broadland also provide the open and wet, but not continuously waterlogged, conditions under which *C. elongata* survives in Britain (David 1978, Wheeler 1980c, 1983). This is *the* characteristic sedge of this kind of woodland on the Continental mainland, though in this country it also occurs in the *Salix-Betula-Phragmites* woodland (as at Askham which has probably the largest British colony: David 1978, Fitter *et al.* 1980) and on open fen.

The tall herbs, too, are striking in their number with the community species all being common and frequently joined by *Lysimachia vulgaris*, *Lythrum salicaria* and *Lycopus europaeus* and, in Broadland, *Peucedanum palustre* and, less commonly, *Impatiens capensis* and *Thalictrum flavum*. The fern component is frequently enriched by *Thelypteris palustris*, the fronds of which can be very abundant, and, more occasionally, by *Osmunda regalis*, with stools of sometimes massive size. This latter species was used by Klötzli (1970) to name this whole community but it is rather strictly confined to this particular type of *Alnus-Carex* woodland and it also occurs in Britain in other kinds of wet woodlands, though more rarely. Among the smaller dicotyledons, *Myosotis laxa* ssp. *cespitosa* is strongly preferential and *Cardamine pratensis* becomes frequent. *Scutellaria galericulata*, *Viola palustris*, *Ranunculus repens* and *R. flammula* are slightly more frequent here than elsewhere.

Solanum dulcamara remains common in this sub-community and there is occasionally some climbing *Humulus lupulus*. Underscrub can be dense with much *Rubus fruticosus* agg. especially in drier stands, and occasionally prominent *Ribes nigrum* and *R. rubrum*.

Again, the bryophytes are not distinctive.

Chrysosplenium oppositifolium sub-community: *Osmundo-Alnetum glutinosae chrysosplenietosum* (Klötzli 1970) Wheeler 1980c; Alder stand type 7Bb Peterken 1981. The tree canopy here is simpler and less rich than in the other sub-communities and *Alnus* is often the only species: large *Salix cinerea* bushes can occasionally break the canopy but both *Fraxinus* and *Betula pubescens* are rare. As this vegetation is not usually developed on mats of fen peat in topogenous mires, there is not the typical pattern of flooding and subsidence of the substrate here, though older *Alnus* may still tumble over from the tops of sedge tussocks. The shrub layer is poorer in species and often less extensive than in the other sub-communities: *Salix cinerea* is occasional but *Crataegus monogyna* is absent and other species are rare.

In the field layer, *C. paniculata* is almost always the dominant with *C. acutiformis* reduced in frequency and rarely abundant. Other sedges are rare or absent and *Phragmites* is very uncommon. As in the first sub-community, the tall herbs here are usually just those of the community and even some of these (notably *Iris pseudacorus* and *Eupatorium cannabinum*) are reduced in frequency. The one distinctive species in this element, however, is *Oenanthe crocata* which is strongly preferential here and a good marker of these more oceanic alderwoods from their European mainland counterparts. *Epilobium hirsutum* and *Rumex sanguineus* occur occasionally. The ferns are not distinctive.

However, very striking here are the small-herb and bryophyte components which form a patchy carpet between the sedge tussocks. *Chrysosplenium oppositifolium* is strongly preferential and it can be very abundant with frequent *Cardamine pratensis*, *Ajuga reptans* and *Caltha palustris*. Among the bryophytes, *Pellia epiphylla* becomes very common and *Rhizomnium punctatum* is somewhat more frequent than in the other sub-communities.

Habitat

The *Alnus-Carex* woodland is most characteristic of wet to waterlogged organic soils, base-rich and moderately eutrophic, in topogenous and soligenous mires. It is especially associated with fen peats in open water transi-

tions and flood-plain mires where there is a strong influence of calcareous ground water and periodic deposition of allochthonous mineral material in winter-flooding but it can occur, too, in basin mires where there is a local influence of more base-rich water from marginal springs and in soligenous mires below seepage lines where very wet mineral soils develop a humose topsoil or thin surface peat. It is essentially a primary woodland, developing naturally from certain kinds of fen in hydrarch successions but drier, older stands have been subject to treatments such as coppicing and grazing.

As with the other main British carr community, the *Salix-Betula-Phragmites* woodland, the floristics and structure of the developing canopy and understorey are very much dependent on the availability of seed-parents and on propagule dispersal and the composition of the field layer is strongly influenced by the nature of the invaded fen vegetation. Here, though, the frequent presence of upstanding tussocks of *Carex paniculata* provide sites for early colonisation well before the accumulation of peat has raised the general fen surface above the limit of the winter flood. Very commonly, therefore, this community has the character of a swamp woodland and, in its early stages at least, is free from the complicating effects of human activity that are such a marked feature of many stands of the *Salix-Betula-Phragmites* woodland. Where *C. acutiformis* has been the more prominent sedge in the preceding vegetation, invasion of woody species is delayed until the substrate surface begins to rise above the limit of the inundating waters so that the woodland is initially less swampy. Though the fen mat may be depressed somewhat as the woodland ages, most of the trees and shrubs are rooted in the substrate itself and they do not show the same tendency to topple over. Such differences formed the basis of Lambert's separation between 'swamp' and 'semi-swamp' carrs (Lambert 1951, 1965) but it is important to realise that this is an essentially physiognomic distinction that may be of value in understanding the seral development of the woodland and its ultimate structure but which is not reflected in general floristic differences, as indeed Lambert herself acknowledged.

The floristic richness of this kind of woodland is partly related to the richness of the preceding swamp and fen. This is usually primary vegetation without the overwhelming dominance of densely-packed monocotyledonous herbage favoured by mowing. Even where bulky sedges are abundant, the cover is somewhat more open and, especially where *C. paniculata* tussocks predominate, amply provided with a variety of niches where the different associates can gain a hold. The high shade-tolerance of both *C. paniculata* and *C. acutiformis* also means that they persist beneath the developing canopy so there is often a strong floristic and structural continuity between the woodland and the herbaceous vegetation that is being invaded.

The *Lysimachia* sub-community includes the most species-rich *Alnus-Carex* woodlands that were the subject of the early classic accounts from Broadland (e.g. Pallis 1911, Lambert 1951). It is still a prominent feature there in those topogenous mires which remain under the close influence of the calcareous and eutrophic river waters, forming sometimes large stands on waterlogged fen peats around the Broad margins and over larger turf ponds and occurring as narrower strips alongside dykes and sluggish stretches of the rivers. Many stands remain swampy throughout the year, especially where colonisation is recent or where the substrate has begun to sink under the weight of larger trees. In those river systems which experience tidally-related water fluctuations, there is a more frequent pattern of inundation on the ground between the trees. This sub-community is also found occasionally outside Broadland where there is a similar combination of base-richness and moderate eutrophy in the waters and soils, on terraces in some river valleys (as in the Breck: Haslam 1965) and in a few of the Shropshire and Cheshire basin mires (Clapham in Tansley 1939, Sinker 1962). It is also found as a central strip of less acidophilous vegetation in a few soligenous mires, as in certain New Forest valleys (e.g. Rankin 1911b, Rose 1950, McVean 1956b).

The habitat of the more common and more widespread *Phragmites* sub-community is very similar and part of the floristic difference between this vegetation and that of the *Lysimachia* sub-community is phytogeographical in that more Continental and Continental Northern wetland species, such as *Thelypteris palustris*, *Peucedanum palustre* and *Carex appropinquata*, become rarer outside the eastern lowlands. But there may also have been some impoverishment due to human disturbance because many stands of this sub-community survive on fragments of mire which have been much interfered with and which now remain isolated within agriculturally-improved landscapes. There is also the possibility that the *Phragmites* sub-community is associated with less peaty soils or, at least, ones which have a more pronounced influence of minerotrophic waters. This is certainly the case with the distinctive *Scirpus sylvaticus*-dominated vegetation included here: it is typically found on substrates which below have the consistency and colour of thick tomato soup, indicating iron enrichment.

The habitat of the *Chrysosplenium* sub-community is much more obviously distinctive. This is typically a woodland of springs and seepage lines where the emergence of somewhat less base-rich waters than is usual for the community keeps the soils very wet but not surface-flooded. The soils often have a mineral base but, over the impervious drift or bedrock substrate, peat can accumu-

late in the waterlogged conditions. Calcicolous species are not so prominent here and, in the carpet of ground vegetation that typically forms on the wet soil between the sedge tussocks, plants like *Chrysosplenium oppositifolium* and *Pellia epiphylla* (regarded by Klötzli (1970) as more typical of his *Pellio-Alnetum*) become very frequent. This is a local woodland type found in often small incised valleys cut into less calcareous bedrocks, as in the Weald.

Zonation and succession

Around standing and slow-moving open waters, the *Lysimachia* and *Phragmites* sub-communities are often found in zonations running from woodland, through swamp, to aquatic vegetation, as around some of the Norfolk Broads (e.g. Pallis 1911, Lambert & Jennings 1951, Lambert 1951) and at Sweat Mere (Clapham in Tansley 1939, Sinker 1962). Most frequently, it is *Carex paniculata* swamp that forms an immediate front to the *Alnus-Carex* woodland but sometimes invasion of the developing sedge tussocks is so rapid that there is no distinct sedge swamp but simply a belt of helophyte swamp dominated by *Phragmites australis* or *Typha angustifolia* between the young woodland fringe and the open water. Where invasion has been substantially delayed, a belt of sedge-dominated fen (the *Peucedano-Phragmitetum* or *Phragmites-Eupatorium* fen) may supervene between the woodland and the swamp. On drier ground, the *Lysimachia* and *Phragmites* sub-communities can pass to drier carr of the *Salix-Betula-Phragmites* woodland, still over peat, or directly to some kind of mixed deciduous woodland, usually in these southern lowlands, the *Quercus-Pteridium-Rubus* woodland. However, in the latter case, quite gradual transitions in both woody species and field layer may, in fact, mask discontinuities in substrate and soils, as where mire deposits give way to solid valley sides around.

On more extensive flood-plain mires with a long history of exploitation, various kinds of human disturbance can complicate this basic pattern. Peat-digging, for example, can create new hollows which, when abandoned and flooded and colonised with swamp and swamp-woodland, produce local reversals in the general trend (well seen in some of the profiles of Lambert & Jennings 1951). Then, mowing for fen-crops can create extensive stretches of secondary herbaceous vegetation behind the *Alnus-Carex* woodland, though in most places such mowing-marshes have been abandoned now and colonised by the *Salix-Betula-Phragmites* woodland or, where there has been much disturbance and enrichment of the dry peat surface, the *Alnus-Urtica* woodland. Along the water's edge, too, zonations have been modified, as where boat-traffic and the depredations of coypu have fretted away the fringe of swamp and left rather moribund stands of *Alnus-Carex* woodland abutting directly on to open water. Eutrophication of open waters has also encouraged the development of *Phragmites-Urtica* fen along the front of some tracts.

There is a continuous series of zonations between the more extensive sequences of vegetation types described above and the much narrower belts of communities found on the peaty terraces of small slow-moving rivers. Here, the *Alnus-Carex* woodland (usually, in these sites, the *Phragmites* sub-community) can be found occupying the flat and winter-flooded ground of the terrace with fragmentary swamp along the water's edge and a very sharp transition at the valley side to mixed deciduous woodland. Quite commonly, the swamp-woodland is separated from the moving river waters by a narrow belt of the *Alnus-Urtica* woodland on levees of rich alluvium, behind which conditions may be virtually stagnant. These kinds of patterns now often survive isolated in intensive agricultural or afforested landscapes where the valley sides have been cleared for pasture or a tree crop.

In valley mires where there is some persisting influence of more base-rich and eutrophic waters along the central axis, the *Lysimachia* or *Phragmites* subcommunities can occur as a central strip of woodland sandwiched between parallel belts of very wet poor-fen with Sphagna (especially *S. recurvum*) and dominants such as *Carex rostrata*, *Equisetum fluviatile* and *Juncus acutiflorus* or Junco-Molinion vegetation, which in turn passes to *Sphagnum* pools and then to wet and dry heath. This is the classic kind of zonation described from some valley mires in the New Forest (Rankin 1911*b*, Rose 1950, McVean 1956*b*, Wheeler 1983). Similar, though often compressed and fragmented, patterns can be seen around the margins of some basin mires where there is a local influence of more base-rich soligenous waters in the lagg (as at Rhos-goch Common: Bartley 1960).

The characteristic association of the *Chrysosplenium* sub-community with seepage zones means that it is often found as small stands isolated within stretches of drier, usually mixed deciduous woodland. In such situations, it can be seen as the less base-rich analogue of the *Alnus-Fraxinus-Lysimachia* woodland: like that vegetation it usually interrupts stretches of bluebell or mercury woodland or straddles transitions between the two over junctions of less and more base-rich soils. Where seepage is sufficient to create the semblance of a small valley mire, it may form more extensive sinuous stands following the line of water movement.

Stratigraphical studies in Broadland (Jennings & Lambert 1951, Lambert & Jennings 1951, Lambert 1951, 1965) have shown that the more intact zonations from open water to *Alnus-Carex* woodland represent a primary hydrosere in which aquatic vegetation, then *Typha*, *Phragmites* and *Carex* swamps are naturally succeeded by invasion of woody species. More fragmen-

tary studies at Sweat Mere in Shropshire (e.g. Sinker 1962) suggest that a similar process has occurred there. Although both sedge-swamp and swamp-carr were poorly represented in the numerous profiles examined in Walker's (1970) survey, it seems likely that this succession is the normal one around more base-rich and eutrophic open waters in the British lowlands (Wheeler 1978, 1980*c*, 1983). Despite the primeval appearance of many stands of this woodland, it is clear that it can develop from open water within little more than a century (Lambert 1951, 1965, Sinker 1962). Some stands, though, may be considerably older because this appears to be a fairly stable kind of vegetation and there is no firm evidence that it habitually progresses to drier carr, say of the *Salix-Betula-Phragmites* type. This may happen in some more isolated parts of flood-plain mires or in those valley mires where there is substantial deposition of alluvium but often the woodland becomes moribund before the general peat surface has become markedly dry, trees collapsing and sedge tussocks decaying to initiate the cycle of development over again (e.g. McVean 1956*b*). In some situations, marked enrichment of the waters and substrates (say, through sewage or fertiliser contamination or massive deposition of mineral material), may precipitate a succession to wetter kinds of the *Alnus-Urtica* woodland.

Although some drier stands of the community may have been cleared in the past for the extension of mowing-marshes (e.g. Lambert 1965), the frequently intractable topography of the *Alnus-Carex* woodland has given it some protection against conversion to secondary fen. However, where clearance has occurred, the community is not able to re-develop on the abandonment of mowing unless there is substantial inundation by base-rich and moderately eutrophic waters, in marked contrast to the *Salix-Betula-Phragmites* woodland which is very often found as a secondary woodland over old mowing-marsh.

Where cattle have access to drier stands, they can graze down the tall herbs and physically damage the sedge tussocks which can lead to a gradual run-down of this woodland type. Light burning of stands among heathland in the New Forest does not totally destroy *Alnus* which can produce abundant basal sprouts (McVean 1956*b*).

Where the local influence of base-rich waters is ultimately insufficient to offset the development of more acid peat, as in some valley mires and basin mires, the *Alnus-Carex* woodland may give way eventually to *Sphagnum*-dominated mire. Such a development was suggested by Rose (1950) for some New Forest valley mires and it may be happening, too, around the ombrogenous nuclei found in some Broadland fens (as in the Ant valley: Wheeler 1978, 1983). Walker (1970) considered that *Sphagnum*-mire might be a natural climax of swamp-carr in small inland basins in the British lowlands.

Distribution

The *Alnus-Carex* woodland is now a fairly local, though quite widespread, community throughout the English lowlands, with a very few localities in Wales and Scotland. The *Phragmites* sub-community is the commonest and most widely distributed type, being especially well represented in the Shropshire and Cheshire meres. The *Lysimachia* sub-community is mostly confined to Broadland. The *Chrysosplenium* sub-community is much rarer but very good stands survive in the Wealden woods with outliers further north.

Affinities

This community has long been recognised, especially in its richer forms with striking swamp physiognomy, as a distinct kind of wet woodland in Britain, comprising, with the *Salix-Betula-Phragmites* fen carr, most of our remaining woodland cover over lowland fen peats and more base-rich peaty alluvial soils. The two communities are very close floristically and, in some stands, it is only the relative prominence of either bulky sedges or *Phragmites* that provides a distinction. Klötzli (1970), in fact, does not have an equivalent to the *Salix-Betula-Phragmites* woodland and neither does Wheeler (1978, 1980c, 1983) who expanded Klötzli's geographical coverage and defined a larger number of sub-communi-

ties. In so far as they have sampled *Salix-Betula-Phragmites* woodland, their stands have been placed within a single association, the *Osmundo-Alnetum*. This, however, produced a very variable and bulky vegetation type and it seems better to maintain two distinct communities, especially as such a division is supported by seral and environmental differences. Apart from the separation of samples lacking larger sedges, the treatment of *Alnus-Carex* woodland here largely follows the expanded account of the *Osmundo-Alnetum* given by Wheeler (1980c), though his *sphagnetosum* is much better considered as part of the *Betula-Molinia* woodland.

Klötzli (1970) conceived the *Osmundo-Alnetum* as an oceanic replacement along the the western European seaboard of the *Carici elongatae-Alnetum* and the two communities are generally very similar. *Osmunda regalis* was considered a good character species but it has not been used for naming here because of its quite severe restriction to the richer stands of the *Lysimachia* sub-community. *Oenanthe crocata* is another good distinct species of the British examples of this kind of woodland but it, too, is largely confined to a single sub-community. Both these species (and indeed the much rarer *C. elongata*) also occur in other wet woodlands in Britain.

In the more oceanic parts of Britain, the *Alnus-Carex* woodland is replaced by the *Alnus-Fraxinus-Lysimachia* woodland and the *Chrysosplenium* sub-communities of each grade floristically one into the other.

Floristic table W5

	a	b	c	5
Alnus glutinosa	V (1–10)	V (1–8)	V (4–6)	V (1–10)
Salix cinerea	III (3–9)	III (1–6)	II (1–8)	III (1–9)
Fraxinus excelsior	III (1–4)	III (1–4)	I (1)	III (1–4)
Betula pubescens	II (1–7)	III (1–6)	I (1)	II (1–7)
Quercus robur	I (4)	I (1–4)	I (1)	I (1–4)
Salix cinerea	III (2–10)	III (1–6)	II (1–5)	III (1–10)
Crataegus monogyna	II (1–4)	II (1–3)		II (1–4)
Alnus glutinosa sapling	I (1–3)	I (1–3)	I (1–3)	I (1–3)
Ilex aquifolium	I (3)	I (4)	I (1–3)	I (1–4)
Salix aurita	I (4)		I (4)	I (4)
Rhamnus catharticus	I (1–3)	I (1–3)		I (1–3)
Sorbus aucuparia		I (1–3)	I (1–3)	I (1–3)
Frangula alnus	I (1–3)	III (1–4)	I (1)	II (1–4)
Viburnum opulus	I (1–3)	II (1–3)	I (1–3)	I (1–3)
Eurhynchium praelongum	IV (1–3)	V (1–4)	V (1–4)	IV (1–4)
Carex paniculata	IV (1–8)	IV (1–5)	V (1–5)	IV (1–8)
Brachythecium rutabulum	IV (1–4)	III (1–4)	V (1–4)	IV (1–4)

Rubus fruticosus agg.	III (1–6)	V (1–4)	V (1–3)	IV (1–6)
Galium palustre	III (1–3)	V (1–4)	IV (1–3)	IV (1–4)
Solanum dulcamara	IV (1–4)	III (1–4)	II (1–3)	III (1–4)
Phragmites australis	III (1–5)	II (1–4)	I (1–5)	III (1–5)
Galium aparine	II (1–3)	I (1–3)	I (1–3)	I (1–3)
Equisetum palustre	II (1–3)	I (1–3)		I (1–3)
Scirpus sylvaticus	II (1–6)			I (1–6)
Lophocolea heterophylla	II (1–3)			I (1–3)
Plagiothecium denticulatum	II (1–2)			I (1–2)
Callitriche stagnalis	I (2–4)			I (2–4)
Symphytum officinale	I (1–3)			I (1–3)
Dryopteris cristata	I (3–4)			I (3–4)
Holcus lanatus	I (3–4)			I (3–4)
Hedera helix	I (2–4)			I (2–4)
Mnium hornum	III (1–3)	IV (1–3)	II (1–3)	III (1–3)
Lysimachia vulgaris	I (1–3)	IV (1–3)	I (1–3)	II (1–3)
Poa trivialis	II (1–4)	III (1–4)	II (1–3)	II (1–4)
Lycopus europaeus	II (1–3)	III (1–3)	I (1–3)	II (1–3)
Thelypteris palustris	I (1–4)	III (1–6)	I (1)	II (1–6)
Lythrum salicaria	I (1–3)	III (1–3)	I (1–3)	II (1–3)
Myosotis laxa caespitosa	I (1–6)	III (1–4)	I (1–3)	II (1–6)
Carex remota	I (2–7)	III (1–4)	I (1–4)	II (1–7)
Ribes nigrum	I (1–2)	II (1–4)	I (1)	I (1–4)
Carex elata	I (6–7)	II (1–8)		I (1–8)
Humulus lupulus	I (1)	II (1–4)		I (1–4)
Osmunda regalis		II (1–8)	I (1)	I (1–8)
Peucedanum palustre		II (1–3)		I (1–3)
Carex pseudocyperus		II (1–4)		I (1–4)
Carex appropinquata		I (1–3)		I (1–3)
Campylium stellatum		I (1–3)		I (1–3)
Impatiens capensis		I (1–3)		I (1–3)
Carex riparia		I (1–4)		I (1–4)
Thalictrum flavum		I (1–3)		I (1–3)
Pellia endiviifolia		I (1–3)		I (1–3)
Rumex hydrolapathum		I (1–3)		I (1–3)
Carex elongata		I (1–2)		I (1–2)
Bryum pseudotriquetrum		I (1)		I (1)
Cardamine pratensis	I (1–3)	III (1–3)	IV (1–3)	III (1–3)
Chrysosplenium oppositifolium	I (1)		V (1–6)	II (1–6)
Pellia epiphylla		II (1)	V (1–3)	II (1–3)
Ajuga reptans	II (1–3)		IV (1–4)	II (1–4)
Caltha palustris	I (1–4)	II (1–3)	III (1–3)	II (1–4)
Rhizomnium punctatum	I (2–3)	II (1–3)	III (1–4)	II (1–4)
Lonicera periclymenum	I (1–3)	II (1–3)	III (1–3)	II (1–3)
Oenanthe crocata	I (2–7)	I (1)	III (1–6)	II (1–7)
Ranunculus acris	I (1)	I (1–3)	II (1–3)	I (1–3)
Epilobium hirsutum	I (1–3)		II (1–3)	I (1–3)
Rumex sanguineus	I (2–3)		II (1–3)	I (1–3)
Atrichum undulatum			II (1–3)	I (1–3)

Floristic table W5 *(cont.)*

	a	b	c	5
Dryopteris dilatata	III (1–4)	IV (1–4)	III (1–4)	III (1–4)
Cirsium palustre	III (1–3)	III (1–3)	IV (1–3)	III (1–3)
Mentha aquatica	III (1–4)	III (1–3)	III (1–3)	III (1–4)
Urtica dioica	III (1–3)	III (1–4)	III (1–3)	III (1–4)
Filipendula ulmaria	III (1–4)	III (1–5)	III (1–4)	III (1–5)
Carex acutiformis	III (1–8)	III (1–5)	II (1–6)	III (1–8)
Eupatorium cannabinum	III (1–4)	III (1–4)	II (1–4)	III (1–4)
Iris pseudacorus	III (1–4)	III (1–4)	I (1–3)	III (1–4)
Valeriana officinalis	II (1–3)	III (1–3)	III (1–3)	III (1–3)
Calliergon cuspidatum	I (1–3)	III (1–4)	III (1–4)	II (1–4)
Plagiomnium undulatum	II (1–4)	II (1–3)	III (1–4)	II (1–4)
Angelica sylvestris	II (1–3)	II (1–3)	III (1–4)	II (1–4)
Athyrium filix-femina	II (1–3)	II (1–4)	II (1–3)	II (1–4)
Valeriana dioica	II (1–4)	II (1–3)	I (1–4)	II (1–4)
Geranium robertianum	I (1–3)	II (1–3)	II (1–3)	I (1–3)
Phalaris arundinacea	I (1–6)	II (1–4)	I (1)	I (1–6)
Scutellaria galericulata	I (1)	II (1–3)	I (1)	I (1–3)
Viola palustris	I (3)	II (1–3)	I (1)	I (1–3)
Ranunculus repens	I (2–6)	II (1–3)	I (1)	I (1–3)
Lophocolea bidentata s.l.	I (3)	II (1–3)	I (1)	I (1–3)
Ranunculus flammula	I (1)	II (1–3)	I (1)	I (1–3)
Juncus effusus	I (1–7)	II (1–3)	I (1)	I (1–3)
Sphagnum palustre	I (1–6)	II (1–3)	I (1–3)	I (1–6)
Dryopteris filix-mas	I (1–3)	I (1–3)	I (1–3)	I (1–3)
Circaea lutetiana	I (1–5)	I (1–3)	I (1–3)	I (1–5)
Rosa canina agg.	I (1–3)	I (1–3)	I (1)	I (1–3)
Berula erecta	I (2–4)	I (1–3)	I (1)	I (1–4)
Glyceria maxima	I (1–3)	I (1–3)	I (1–3)	I (1–3)
Scrophularia aquatica	I (1–3)	I (1)	I (1)	I (1–3)
Equisetum fluviatile	I (1–3)	I (1–3)	I (1–3)	I (1–3)
Calliergon cordifolium	I (1)	I (1–3)	I (1)	I (1–3)
Sphagnum squarrosum	I (1)	I (1–3)	I (1)	I (1–3)
Hydrocotyle vulgaris	I (1)	I (1)	I (1)	I (1)
Dryopteris carthusiana	I (1)	I (1)	I (1)	I (1)
Deschampsia cespitosa	I (1–3)		I (1–3)	I (1–3)
Ligustrum vulgare	I (1–3)		I (1)	I (1–3)
Hypericum tetrapterum	I (1–3)		I (1)	I (1–3)
Equisetum telmateia	I (1)		I (1)	I (1)
Amblystegium serpens	I (1–3)	I (1–3)		I (1–3)
Myosotis scorpioides	I (1–3)	I (1–3)		I (1–3)
Calamagrostis canescens	I (7)	I (1–6)		I (1–7)
Calystegia sepium	I (1–3)	I (1–3)		I (1–3)
Hypnum cupressiforme	I (1–3)	I (1)		I (1–3)
Lychnis flos-cuculi	I (1)	I (1–3)		I (1–3)
Cicuta virosa	I (1–7)	I (1–3)		I (1–7)
Agrostis stolonifera	I (1–3)	I (1)		I (1–3)
Ranunculus lingua	I (1)	I (1)		I (1)

Ribes rubrum	I (1)	I (1–3)		I (1–3)
Sparganium erectum	I (1–3)	I (1–3)		I (1–3)
Plagiomnium rostratum	I (1)	I (1)		I (1)
Cardamine amara		I (1–4)	I (1–3)	I (1–4)
Sphagnum fimbriatum		I (1–3)	I (1)	I (1–3)
Epilobium parviflorum		I (1)	I (1)	I (1)
Number of samples	46	46	15	107
Number of species/sample	20 (4–44)	29 (21–45)	26 (16–42)	24 (4–45)

a *Phragmites australis* sub-community

b *Lysimachia vulgaris* sub-community

c *Chrysosplenium oppositifolium* sub-community

5 *Alnus glutinosa-Carex paniculata* woodland (total)

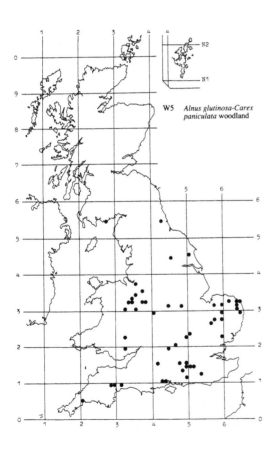

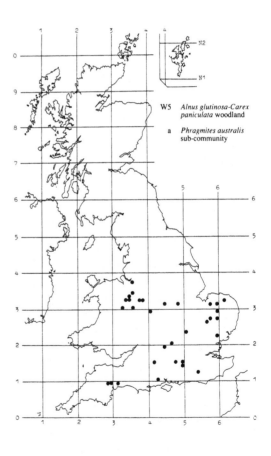

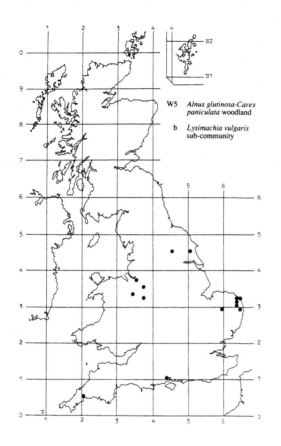

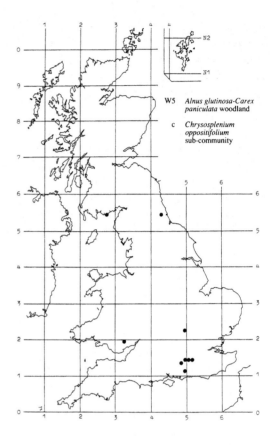

W6
Alnus glutinosa-Urtica dioica woodland

Synonymy

Valley fen woods   Farrow 1915 *p.p.*; *Betulo-Alnetum* Clapham in Tansley 1939; Woodwalton Birch wood Poore 1956*b p.p.*; Valley fen alderwoods   Haslam 1965 *p.p.*; *Alnus-Salix* woodland XXi & XXii   Meres Report 1980;   *Alnus-Salix-Betula*   woodland   XXi Meres Report 1980; Fen Woodlands B, C & D   Fitter *et al.* 1980; Alder stand types 7Aa & 7Ab   Peterken 1981.

Constant species

Alnus glutinosa, Urtica dioica.

Physiognomy

The *Alnus glutinosa-Urtica dioica* woodland is a rather ill-defined community which brings together a variety of canopies dominated by *Alnus glutinosa, Salix* spp. and *Betula pubescens* beneath which the rich assemblages of swamp and fen herbs characteristic of many of our wetter woods are replaced by a species-poor, though quite distinctive, field layer. There is considerable floristic and physiognomic diversity among the woodland types included here and, at first sight, it is often the peculiarities of stands which impress the visitor more than their underlying similarities. Nonetheless, there are sound ecological reasons for both the general species-poverty of these woodlands and for what little they have in common and it seems best to treat them within a single, rather disparate, group.

Alnus glutinosa is by far the commonest tree throughout and it remains frequent in all but the driest stands. In the wetter woodlands included here, it is often an overwhelming dominant, forming an even-topped and usually closed canopy of well-grown, usually multi-stemmed trees. In one sub-community, it is replaced as the most abundant tree by *Salix fragilis*. In the drier types of woodland within the community, *Betula pubescens* becomes increasingly frequent and locally dominant and *Pinus sylvestris* is an important invader or planted canopy replacement. Other tree species are generally uncommon but *Populus nigra* var. *betulifolia* is a very distinctive associate in some stands and it can attain a grand stature here with its black bossed trunk and irregular branches arching downwards. This tree is probably native in southern England and, as on the Continent, this community perhaps provides its natural woodland locus. There is occasionally some *Acer pseudoplatanus* or *Fraxinus excelsior* and sometimes a little *Quercus robur*.

Mature woodlands of these types usually have a distinct, though generally open and often rather patchy, understorey. Except where the substrate is dry, *Salix cinerea* is the leading shrub with, on drier ground, *Sambucus nigra*. *Crataegus monogyna* is occasional throughout and there are sparse records for *Salix caprea, Ilex aquifolium, Corylus avellana, Viburnum opulus* and *Prunus spinosa*. Exceptionally, as at Wicken Fen (Godwin *et al.* 1974), this kind of woodland has developed beneath a canopy dominated by *Frangula alnus* and *Rhamnus catharticus*. Tree saplings are quite common with occasional young *A. pseudoplatanus, Alnus, Fraxinus* and *S. fragilis*.

The other group of woody species which can attain prominence in these woodlands are the osiers. *Salix viminalis* and, less commonly, *S. triandra* and *S. purpurea*, are occasionally found as shrubs or small trees in the understorey but it is sensible to include in this community scrubby vegetation in which these species dominate, usually with only scattered *Alnus*, over the kind of field layer typical here. Planted osier beds can be seen as but a more ordered and managed version of such vegetation and their major weeds are the characteristic herbs of the community.

What distinguishes the field layer of this community from the herbaceous component of its closest relatives, the *Salix-Betula-Phragmites* and *Alnus-Carex* woodlands, is the very poor representation here of bulkier swamp dominants and tall rich-fen dicotyledons. Species such as *Phragmites australis, Carex paniculata* and *C. acutiformis* are, at most, occasional in the *Alnus-*

Urtica woodland and they do not usually form extensive and vigorous patches. Dicotyledons such as *Lysimachia vulgaris*, *Lythrum salicaria*, *Valeriana officinalis*, *V. dioica* and even *Eupatorium cannabinum*, *Filipendula ulmaria* and *Angelica sylvestris*, are likewise rather uncommon, occurring usually as only sparse scattered individuals. And, where this community comes close in its floristic composition to the *Alnus-Fraxinus-Lysimachia* woodland, there is only very rarely the sort of ground cover of *Chrysosplenium oppositifolium* and *Caltha palustris* so characteristic there of trickling surface water.

On the positive side, the really typical herb here is *Urtica dioica*, the sole constant of the field layer throughout and often very prominent, sometimes in a virtually continuous cover, in other cases as conspicuous patches. One good measure of the general species-poverty here is that, when the luxuriant *Urtica* litter dies down quickly at the end of the growing season, there are very few remains of other perennial herbs to be seen. The vernal aspect of these woodlands, which generally lack such plants as *Mercurialis perennis*, *Hyacinthoides non-scripta*, *Ranunculus ficaria* or *Anemone nemorosa*, is thus often decidedly bare.

The commoner field-layer associates of the community as a whole are few but they form an ill-defined series running from the wetter to the drier habitats. Where the soils remain moist towards the surface (as in the Typical, *Salix fragilis* and *Salix viminalis/triandra* sub-communities), *Poa trivialis* and *Galium aparine* are frequent and sometimes abundant and there is often some *Solanum dulcamara* scrambling through the shrubs. It is in these kinds of *Alnus-Urtica* woodlands that such swamp and fen species as occur in the community are best represented with small clumps of *Phragmites*, *Carex acutiformis*, *Phalaris arundinacea* and *Epilobium hirsutum* scattered among usually dense *Urtica* with occasional *Filipendula ulmaria* and *Iris pseudacorus*. On drier substrates, by contrast (as in the *Sambucus* and *Betula* sub-communities), the importance of all these species fades and there is an increasing prominence of *Lonicera periclymenum*, *Dryopteris dilatata* and *Rubus fruticosus* agg. among an *Urtica* cover that can be much thinner and patchier.

Other herbs present throughout at low frequencies include *Arrhenatherum elatius*, *Heracleum sphondylium*, *Ranunculus repens*, *Cardamine flexuosa*, *Glechoma hederacea*, *Angelica sylvestris* and *Cirsium palustre*. Towards the south and west, *Oenanthe crocata* is a scarce but locally abundant associate and, especially in the industrial north, *Impatiens glandulifera* has become prominent in some stands. This considerable variety among the field layer is also often accompanied by a generally untidy and run-down appearance: wetter stands are often choked with brushwood and litter after the winter flood and can have interesting artefacts like fertiliser bags and dead fish. Drier woodlands frequently show signs of disturbance.

Bryophytes are very variable in their abundance but, because of the often low cover of smaller vascular species, they frequently appear conspicuous over the soil surface and herb stools, especially in winter and spring when they can provide the only splashes of green. The species are few: *Eurhynchium praelongum* is by far the commonest but *Brachythecium rutabulum* occurs occasionally throughout and there are also sparse records for *Plagiothecium denticulatum*, *Plagiomnium undulatum*, *Rhizomnium punctatum* and, over bark and on twiggy litter, *Mnium hornum* and *Lophocolea heterophylla*. Species such as *Calliergon cuspidatum*, *Pellia epiphylla* and *Sphagna* are very rare, in contrast to other kinds of wet woodland.

Sub-communities

Typical sub-community: Fen Woodlands B, C & D Fitter *et al.* 1980. *Alnus* is almost always the woody dominant here, often growing as tall, multi-stemmed trees and forming an even-topped and virtually closed canopy. *Fraxinus* is occasional and there are sparse records also for *Acer pseudoplatanus* and *Quercus robur*. Shrub cover is generally thin with scattered bushes of *Salix cinerea* being the only frequent feature. *Crataegus monogyna* occurs occasionally and there is sometimes a little *Sambucus*, *Corylus*, *Ilex*, *Salix viminalis* and *Viburnum opulus*. *Frangula alnus* and *Rhamnus catharticus* are also sometimes encountered and they dominate in woodland of this kind in Reserve A at Wicken (Godwin *et al.* 1974). Saplings are usually rather sparse with scattered young *Fraxinus* and *Alnus*.

In the field layer, *Urtica* is usually very abundant and there is often much *Galium aparine*. Where the cover of these plants is somewhat patchier, there can be extensive mats of *Poa trivialis* over the ground surface and, less frequently, creeping *Ranunculus repens* and *Glechoma hederacea*. *Arrhenatherum elatius* and *Heracleum sphondylium* are found occasionally on drier areas and there can be sparse, but locally prominent, clumps of *Epilobium hirsutum* and *Phalaris arundinacea*. *Solanum dulcamara* and, less commonly, *Humulus lupulus* are sometimes found sprawling and climbing over the shrubs.

Against this general background, the field layer here can show floristic peculiarities which reflect its development from the herbaceous component of different kinds of swamp and fen. Tall dicotyledons such as *Filipendula ulmaria*, *Angelica sylvestris*, *Eupatorium cannabinum* and *Lysimachia vulgaris* are generally scarce in this vegetation but they can show local prominence. Bulkier monocotyledons too, can persist patchily, along dykes or streams for example or in wetter pools. *Phragmites* is the most frequent of these, though it is no more than occasional throughout, never attains more than moderate abundance and, even then, is generally found beneath gaps or along the margins of stands. *Carex acutiformis* and *C. paniculata* are less common though, being more shade-tolerant, tussocks of these sedges can remain prominent locally beneath intact canopies here. Where these species occur together in the same stand, it may be quite difficult to partition samples between the *Alnus-Urtica* woodland and either the *Salix-Betula-Phragmites* or *Alnus-Carex* woodlands, but, since both of these communities seem to be able to develop into *Alnus-Urtica* woodland, transitional stands should be expected (e.g. Godwin *et al.* 1974, Fitter *et al.* 1980).

Salix fragilis sub-community: Valley fen woods Farrow 1915 *p.p.*; *Alnus-Salix* woodland XXi & XXii Meres Report 1980. *Alnus* remains frequent here but it generally occurs as scattered trees in a canopy dominated by *Salix fragilis*. This willow can grow up to form tall individuals but its widely-spreading branches make for broad, irregular crowns so the canopy is often rather uneven-topped and somewhat open. Other trees are rare but, beneath gaps and in younger stands, shrubs can be large enough to make stratification indistinct. Mature woodlands of this kind usually have a low understorey with a patchy distribution of shrubs and saplings over mosaics of drier and wetter ground, a common feature of the habitat here. *Salix cinerea* and *Sambucus nigra* are the commonest species, the former thickening up in moister places, the latter more prominent, sometimes with a little *Crataegus monogyna*, in drier parts. Saplings can be numerous with young *S. fragilis*, *Alnus*, *Fraxinus* and *Acer pseudoplatanus*. As the branches of the canopy *S. fragilis* grow heavy with age, they readily crack off at their junctions and large limbs may crash down in high winds or with snow. Sometimes, these take root and sprout afresh but often they die, leaving the understorey choked with decaying wood. Winter-flooding also frequently washes in river drift and *Solanum* and *Humulus* can add to the tangle, making the vegetation almost impenetrable.

Urtica is again very frequent here and often very luxuriant, especially over marginal levees or over patches of alluvium deposited among the heaps of brushwood litter and around the tree bases. Here, too, there can be dense sprawls of *Galium aparine* and patches (or, on levees, strips) of *Phalaris arundinacea* or *Epilobium hirsutum* with occasional tussocks of *Arrhenatherum elatius* and, in the driest areas, *Dryopteris dilatata*. Where the cover is not so thick, *Poa trivialis* and *Ranunculus repens* can spread over the soil surface, together with mats of *Eurhynchium praelongum* and *Brachythecium rutabulum*. *Mnium hornum* and *Lophocolea heterophylla* are sometimes conspicuous on the

abundant decaying wood.

In wetter areas, these herbs and bryophytes become more sparse: there is occasionally some *Iris pseudacorus* and *Galium palustre* here but very often there are extensive bare stretches of sloppy mud.

Salix viminalis/triandra sub-community. This kind of vegetation has not been systematically sampled but included here are stands in which osiers dominate over the kind of field layer typical of the *Alnus-Urtica* woodland. In semi-natural situations, *Salix viminalis* and *S. triandra* seem to be the commonest species with *S. purpurea* somewhat less frequent and diverse mixtures of these species and their hybrids, together with occasional *S. cinerea* and crosses with this sallow, typically form a thicket-like cover, low and often very dense. There is sometimes a little *Sambucus nigra* and *Crataegus mongyna* and emergent *Alnus* and *Fraxinus* can also be found. The major osiers here have long been a source of rods for basket weaving and plantings, frequently using selected varieties with picturesque names like 'Black Maul', 'Glibskins', 'Champion Rod' and 'Mottled Spaniards', are probably the original source of much of the taxonomic diversity seen in the wild and also the direct forbears of many, now neglected, stands.

The field layer in such situations is usually similar to that of the Typical and *Salix fragilis* sub-communities with an abundance of *Urtica* and *Galium aparine* and a ground carpet of *Poa trivialis*. *Solanum dulcamara* can be very prominent and, more occasionally, *Humulus lupulus* and *Calystegia sepium*. Bulky herbs like *Phragmites*, *Carex riparia*, *Epilobium hirsutum* and *Phalaris arundinacea* can also be patchily abundant and there may be scattered plants of *Filipendula ulmaria*, *Angelica sylvestris* and *Rumex* spp. Often, too, there is that characteristic untidiness produced by deposition of flood detritus and rubbish and by the dropping of alluvial material among the osier stools. Where these Salices are colonising sand and shingle islands, there can be much greater floristic heterogeneity too, with stretches of woody vegetation intermixed with varied inundation communities on still-shifting areas of substrate.

In planted, coppiced stands, control of herbaceous vegetation among the osiers is of paramount importance and the most troublesome weeds of the crop are *Urtica*, *Galium aparine*, *Calystegia sepium* and Rumices.

Sambucus nigra sub-community: *Betulo-Alnetum* Clapham in Tansley 1939 *p.p.*; Woodwalton Birch wood Poore 1956*b p.p.*; *Alnus-Salix-Betula* woodland XXIi Meres Report 1980 *p.p.*; Alder stand type 7Ab Peterken 1981 *p.p.* *Alnus* is constant here and it is usually the dominant in a tall, more or less closed canopy, but *Betula pubescens* now becomes occasional and it can be locally abundant. *Fraxinus, Acer pseudoplatanus, Salix fragilis* and *Quercus robur* occur more sparsely. There is often a distinct understorey, though the cover is variable. *Salix cinerea* is still frequent but it is not usually abundant and a much more obvious feature here is the common presence of scattered bushes of *Sambucus nigra. Crataegus monogyna* occurs occasionally and *Prunus spinosa* makes an infrequent appearance. *Corylus* and *Ilex* are rare and saplings, too, are rather uncommon with sparse records for young *Fraxinus, A. pseudoplatanus* and *B. pubescens*.

In the field layer, both *Urtica* and *Galium aparine* remain frequent but they are not so consistently prominent here as in the Typical and *Salix fragilis* subcommunities and the first impression given by the vegetation is generally the abundance of *Rubus fruticosus* agg. which can form a thick underscrub with some trailing *Lonicera periclymenum* and occasional crowns of *Dryopteris dilatata. D. filix-mas*, generally speaking a rare species in the *Alnus-Urtica* woodland, also becomes frequent here and there is sometimes a little *D. borreri*. Another quite frequent preferential is *Hedera helix* which can form an extensive ground carpet beneath the herbs. Then, there are sparse records for *Rumex obtusifolius, Silene dioica, Cardamine flexuosa, Heracleum sphondylium, Glechoma hederacea, Holcus mollis, Ranunculus ficaria* and *Hyacinthoides non-scripta*. Where there is a little base-enrichment on the margins of small streams winding through this kind of woodland, such plants as *Geum urbanum, Circaea lutetiana* and *Mercurialis perennis* can be found but extensive vernal dominance by *Mercurialis, Hyacinthoides* or *Ranunculus ficaria* does not occur here, in contrast to similar kinds of field layers in the *Quercus-Pteridium-Rubus* and *Fraxinus-Acer-Mercurialis* woodlands. Some stands, though, are characterised by a local abundance of *Allium ursinum* in spring and early summer, which is then replaced as the field-layer dominant in mid- and late summer by *Petasites hybridus*, the umbrella-like leaves of which can reach an enormous size here.

***Betula pubescens* sub-community:** *Betulo-Alnetum* Clapham in Tansley 1939 *p.p.*; Alder stand type 7Aa Peterken 1981 *p.p. Alnus* is here reduced to an occasional canopy component and, even when it does occur, it is often subordinate in cover to *Betula pubescens* which rises to constancy. *Pinus sylvestris* is a frequent invader of these drier woodlands and pine plantations which have replaced cleared *Alnus* woodland on alluvial flats are best included here on the basis of their field-layer characteristics. Clearings are also occasionally colonised by *Acer pseudoplatanus* but only rarely is there any *Fraxinus* or *Quercus robur*. Shrubs, too, are rather few in number and *Salix cinerea* is conspicuously absent here: usually, there are just a few scattered bushes of *Sambucus* and *Crataegus monogyna* and scarce *Salix caprea*. Saplings are few and mostly of *A. pseudoplatanus*.

In the field layer, the tendency towards a reduction in the prominence of such species as *Galium aparine* and *Poa trivialis* continues here. Even *Urtica* is rather less common and, though it can still be patchily abundant, its cover is usually less extensive and luxuriant than in the other sub-communities. There are also no very frequent preferential herbs here, though *Epilobium angustifolium* and *Holcus lanatus* are good occasional markers of the disturbance that woodlands of this kind often suffer. The most obvious feature of the field layer is thus the underscrub of *Rubus* and *Lonicera* with scattered *Dryopteris dilatata* that is typical of drier *Alnus-Urtica* woodlands in general.

Habitat

This community is first and foremost a woodland of eutrophic moist soils. It is especially characteristic of sites where there is (or has been) substantial deposition of allochthonous mineral matter, as on alluvial terraces in more mature stretches of river valleys but it can occur, too, in open-water transitions and on flood-plain mires where strongly-enriched waters flood fen peats. In such situations, the *Alnus-Urtica* woodland can develop as a primary forest cover in natural hydrarch successions and persist for some time as the substrates become elevated and dry out somewhat. But it can also develop secondarily where there is eutrophication of substrates under other kinds of wet woodland and then it can even be found on disturbed and enriched acid peats in some basin mires.

It is this general tendency towards enrichment of soils that are becoming, at least patchily, dry towards the surface in summer that is marked here by the prominence of such species as *Sambucus nigra, Urtica dioica* and *Galium aparine*. Generally speaking, the substrates remain moist enough for *Alnus* and various Salices to maintain their prominence in the canopy of most of these woodlands but the trend towards terrestrialisation is marked in the community, among both the woody species and the herbs, by the beginnings of a move towards the flora of mixed deciduous woodland.

Where naturally eutrophic mineral soils are developing by the deposition of rich particulate matter in the slacker reaches of rivers and on flood plains, the Typical and *Salix fragilis* sub-communities are characteristic. Both these kinds of *Alnus-Urtica* woodland (the latter exclusively, though more locally) occur on raw aluvium on levees, small terraces on river bends and uncultivated flood plains. They can also be found around abandoned meanders and silting lakes and as a fringe to artificial water-bodies like ornamental pools and old mill-ponds. The native status of *S. fragilis* is, in fact, disputed (e.g. Meikle 1984) and it has certainly been widely planted in or close to such habitats, but there is no doubt that, like

Alnus, it is very much at home in these situations and, once established, can quickly come to dominate.

In riverside woodlands of this kind, the substrate can be repeatedly enriched with fresh alluvium for many years and, nowadays, fertiliser run-off and the discharge of sewage effluent rich in nitrates and orthophosphates adds further plentiful supplies of major nutrients. In this kind of habitat, the ground may be submerged for weeks on end in the winter floods and hollows can remain very wet throughout the summer, so that a swampy structure develops. But, with the increasing deposition of silt, the ground surface becomes dry enough for *Urtica* to play its prominent role in the field layer.

The *Alnus* sub-community can also occur over fen peats which are inundated by nutrient-rich waters, either in naturally eutrophic river systems like some of the Broadland valleys or, again, where there has been artificial enrichment. But it can be found, too, more deeply within flood-plain mires where the peats have begun to dry out and become surface-oxidised with the release of a flush of nutrients. This is a natural process attendant upon the gradual elevation of the fen surface but there is no doubt that it can be accentuated by physical disturbance. However such eutrophication occurs, gradual drying and enrichment are marked here by the characteristic waning of the fen dominants and the prominence of *Urtica* and other more eutrophic herbs.

The *Sambucus* and *Betula* sub-communities are usually found in drier situations than the Typical and *Salix fragilis* sub-communities and are commonest on brown alluvial soils or alluvial gleys on old river terraces, infilled pools and over peats in flood-plain and basin mires well removed from the influence of flooding waters. The *Sambucus* sub-community is perhaps characteristic of more eutrophic and slightly more base-rich situations than is the *Betula* sub-community but, in both cases, the increasing surface dryness is marked by the almost total absence of fen plants, the waning of the dominance of *Urtica* in the field layer and the development of a herbaceous element that is characteristic of more species-poor mixed deciduous woodlands. Accessible stands of these woodlands are often disturbed by various kinds of human activity (including the dumping of rubbish and use as shooting coverts) and, where they occur within sites that are being cleared for afforestation, the soils are dry and rich enough to support a good growth of conifers like *Pinus sylvestris*. In other cases, vegetation of the *Betula* sub-community has developed in semi-ornamental plantings on heavier gleyed soils or by the natural invasion of land disturbed by major construction work and opencast restoration.

In this range of habitats characteristic of the *Alnus-Urtica* woodland, the *Salix viminalis/triandra* sub-community is typically found in wetter situations like those preferred by the Typical and *Salix fragilis* sub-

communities. Sometimes, these are obviously natural, as where osiers have colonised river islands or fresh alluvium deposited along the slacker margins of moving waters, where repeated flooding maintains eutrophic conditions encouraging prolific canopy growth and, in areas of less dense shade, a luxuriant field layer. Even here, however, the osiers may ultimately originate from planted stock upstream: the native status of both *S. viminalis* and *S. triandra* is regarded as questionable and their widespread distributions are thought to have been much influenced by man (e.g. Meikle 1984). Other stands occur in situations which suggest a more obviously artificial provenance, around streams and in wetter fields near farms and settlements, where osiers were probably widely planted to supply local need.

Extensive commercial osier beds are now few in number and very local, though, in some areas, as around West Sedgemoor in Somerset, they still make a distinctive contribution to the landscape. Here, osiers grow extremely well on moist alluvial clays and silts over peat in a mild climate with fairly frost-free winters and warm summers. Traditionally, osier beds or 'holts' are spring-planted with close rows of 'sets', 30 cm lengths of first or second year shoots left with the top 10 cm protruding. The first crop of 'rods' from the one or two buds left exposed on the sets are usually of poor quality, crooked and often branched but, with repeated annual cutting, the crop builds in quality and quantity so that up to 20 rods can be obtained from each stool, giving yields of 800 000 or 15 tonnes ha^{-1}. Coppicing for 'buffs' (rods peeled after boiling) begins as the leaves start to fall in October to November with 'browns' (rods used with bark left on) being cut somewhat later; 'whites' (peeled, unboiled rods) are harvested either in March when renewed growth allows easy stripping of the bark or are produced from winter-cut rods which have been allowed to stand for some months in water (Troup 1966, Coate & Son undated).

Even with extensive preparation involving ploughing and harrowing and the application of residual herbicides, weed growth in the humid conditions of osier beds is often prolific and the close spacing of the rods usually necessitates repeated hand-hoeing. Sheep are sometimes turned into the crop in September to eat off any remaining herbs and there can be a further period of cattle-grazing in March and April to remove early shoots that may be damaged by frosts. With good tending, filling of blanks by new sets and fertilising where there is no winter-flooding with silt-laden waters, an osier bed can crop well for 25 years or more.

Zonation and succession

The *Alnus-Urtica* woodland is now rarely found as part of extensive zonations on alluvial soils because so many flood-plains have been extensively reclaimed for agriculture. Most often, the *Salix fragilis* and Typical sub-communities survive as small, isolated stands within active loops of slack lowland rivers or around abandoned but still wet pools. They sometimes pass directly to open water or are fringed by a belt of eutrophic herbaceous vegetation, such as the *Phragmites-Urtica* fen, or *Phalaris* fen over silt, or by a zone of inundation vegetation over less stable alluvium or river shingle. Landward boundaries are often abrupt, with a sharp transition to improved pasture or arable but more neglected fringes may have a belt of *Urtica* or *Epilobium hirsutum* or rank *Arrhenatheretum*; in other cases, a boundary ditch occurs with swamp vegetation. In narrower valleys, where small alluvial terraces abut directly on to surrounding slopes, stands of the *Salix fragilis* or Typical sub-communities can pass sharply to some kind of mixed deciduous woodland. Where slope-flushes run down on to the flats, the *Alnus-Fraxinus-Lysimachia* woodland may form a transitional zone between. On older drier terraces in this kind of situation, the *Sambucus* or *Betula* sub-communities occur on the stabilised alluvium, and there seems little doubt that these kinds of *Alnus-Urtica* woodland are a natural seral development from the wetter sub-communities on gradually accumulating mineral material. The succession might be expected to progress to *Quercus-Pteridium-Rubus* woodland with the gradual invasion of *Quercus* spp., *Pteridium aquilinum* and herbs like *Anemone nemorosa* and *Ranunculus ficaria*.

In valley mires and flood-plain mires, the wetter kinds of *Alnus-Urtica* woodland are sometimes found in zonations close to open water with stands of *Salix-Betula-Phragmites* or *Alnus-Carex* woodlands or the *Salix-Carex* woodland, but typically they mark areas of alluvial deposition behind which peat is accumulating, rather than forming an integral part of a single hydrarch succession. This is well seen alongside the Black Beck in Esthwaite North Fen (Pearsall 1918, Tansley 1939, Pigott & Wilson 1978). More usually on extensive peats, the *Alnus-Urtica* woodland occurs more deeply within the fen system forming complexes with other woodlands and herbaceous communities and here it seems to represent a secondary development attendant upon late eutrophication of the habitat. This may happen naturally where deep fen peats dry out superficially and become oxidised with a release of nutrients, but in many cases it has probably been assisted by draining, the surface disturbance of peat-digging and inwash of fertiliser run-off: complex histories of this kind seem to lie behind the development of the community in sites like Woodwalton (Poore 1956*b*), Reserve A at Wicken (Godwin *et al.* 1974) and Askham Bog (Fitter *et al.* 1980). In such situations, the usual precursor of the *Alnus-Urtica* woodland seems to be the *Salix-Betula-Phragmites* woodland which may progress fairly rapidly to the Typical or *Sambucus* sub-communities and then perhaps the *Betula* sub-community. In less base-rich situations, as in some basin mires, the *Betula* sub-community may develop more directly from the *Betula-Molinia* woodland on grossly-disturbed acid peats.

Distribution

The *Alnus-Urtica* woodland is a widespread but local community throughout the lowlands, its occurrence reflecting the distribution of active alluvial deposition on more mature rivers and the remnants of undrained flood-plains and eutrophicated mires.

Affinities

Although this is rather a cumbersome community, it provides a convenient location for a variety of woodland types which show general similarities in floristics and environmental relationships. Previously, these have been recognised in British descriptions only as locally-developed, enriched fragments of other carr communities but, if stands were not so isolated, we would probably acknowledge these woodlands as part of an important seral sequence on our more nutrient-rich flood plains. Phytosociologically, they clearly belong to the Salicion albae alliance in the Salicetea purpureae, colonising scrubs and woodlands in which a variety of willows play a prominent part. Associations like the *Saliceto-Populetum* (R.Tx. 1931) Meijer-Drees 1936 (Oberdorfer 1953, 1957) have a similar suite of herbs to the *Alnus-Urtica* woodland, show a corresponding variety of woody dominants and incorporate the same trend to drier mixed deciduous woodlands as seen here. Some authorities (e.g. Westhoff & den Held 1969) have divided this compendious community into smaller units, separating off the osier scrubs (e.g. *Salicetum triandrae* Malcuit 1929, *Salicetum triandro-viminalis* (Libbert 1931) R.Tx. 1951) from the woodlands dominated by larger willows (e.g. *Salicetum albo-fragilis* (Soo 1934) R.Tx. (1948) 1955) and further sampling might justify such a demarcation in Britain.

Floristic table W6

	a	b	d	e	6
Alnus glutinosa	V (7–10)	IV (4–8)	IV (4–10)	II (4–7)	IV (4–10)
Acer pseudoplatanus	I (3–4)		I (2–3)	II (4–5)	I (2–5)
Quercus robur	I (4)		I (3)	I (1)	I (1–4)
Fraxinus excelsior	II (3–5)		I (3–5)	I (4)	I (3–5)
Salix fragilis		V (6–10)	I (4–8)		II (4–10)
Betula pubescens			II (4–7)	V (5–9)	II (4–9)
Pinus sylvestris				III (3–9)	I (3–9)
Salix cinerea	III (4–8)	III (2–3)	III (1–9)		III (1–9)
Sambucus nigra	I (3–5)	III (1–4)	IV (2–5)	II (3–5)	III (1–5)
Crataegus monogyna	II (1–5)	II (3–4)	II (2–6)	II (3)	II (1–6)
Acer pseudoplatanus sapling		II (1)	I (1–4)	II (3–4)	I (1–4)
Alnus glutinosa sapling	I (1–3)	II (2)	I (1–4)	I (3)	I (1–4)
Fraxinus excelsior sapling	I (3–5)	I (3)	I (1–3)	I (3)	I (1–5)
Salix caprea		I (2)	I (5)	I (2–6)	I (2–6)
Salix viminalis	I (4–7)		I (5)	I (7)	I (4–7)
Ilex aquifolium	I (2)		I (2–3)	I (1–3)	I (1–3)
Corylus avellana	I (3)		I (2–5)		I (2–5)
Viburnum opulus	I (3)			I (4)	I (3–4)
Betula pubescens sapling		I (1)		I (3)	I (1–3)
Salix fragilis sapling		II (6)			I (6)
Prunus spinosa			II (3–9)		I (3–9)
Urtica dioica	V (2–9)	V (2–5)	IV (1–9)	III (3–9)	IV (1–9)
Poa trivialis	III (3–7)	III (1–4)	II (3–9)	I (2–4)	II (1–9)
Galium aparine	IV (3–7)	III (2–5)	III (2–6)		III (2–7)
Solanum dulcamara	II (1–4)	III (1–8)	I (1)	I (3)	II (1–8)
Lonicera periclymenum			II (2–5)	III (1–8)	II (1–8)
Dryopteris dilatata	I (1–4)	II (1)	II (1–7)	III (2–6)	II (1–7)
Rubus fruticosus agg.	I (8)	I (1)	V (1–9)	IV (1–8)	III (1–9)

Floristic table W6 *(cont.)*

	a	b	d	e	6
Filipendula ulmaria	II (3–6)	I (4)	I (6)	I (3)	I (3–6)
Phragmites australis	II (3–5)				I (3–5)
Carex acutiformis	II (3–6)				I (3–6)
Equisetum palustre	II (3–5)				I (3–5)
Phalaris arundinacea	I (3–4)	III (4–7)	I (3–4)	I (5–6)	I (3–7)
Galium palustre	I (3)	III (1–3)		I (3)	I (1–3)
Lophocolea heterophylla	I (2)	II (1–3)	I (3)		I (1–3)
Iris pseudacorus	I (3–4)	II (4–5)	I (4)		I (3–5)
Epilobium hirsutum	I (4–6)	II (1–6)			I (1–6)
Dryopteris filix-mas	I (4)		III (1–6)	I (2–5)	II (1–6)
Hedera helix			III (3–7)	I (4–8)	I (3–8)
Rumex obtusifolius	I (4)	I (1)	II (1–3)	I (3)	I (1–4)
Silene dioica	I (2–4)	I (4)	II (2–5)	I (3)	I (2–5)
Circaea lutetiana	I (1)		II (2–6)	I (1–3)	I (1–6)
Geum urbanum	I (2)		II (3–4)		I (2–4)
Allium ursinum			II (2–6)		I (2–6)
Petasites hybridus			II (5–8)		I (5–8)
Ranunculus ficaria			I (1–5)		I (1–5)
Dryopteris borreri			I (1–7)		I (1–7)
Epilobium angustifolium	III (2–5)	I (1)	I (3–5)	II (1–4)	I (1–5)
Holcus lanatus	II (1–4)	I (2)	I (1–8)	II (3–6)	I (1–8)
Eurhynchium praelongum	III (2–5)	III (1–3)	IV (2–9)	III (2–5)	III (1–9)
Brachythecium rutabulum	II (1–4)	I (2)	II (2–6)	II (3–6)	II (1–6)
Cardamine flexuosa	I (1)	II (2–3)	II (1–4)	I (3)	I (1–4)
Ranunculus repens	II (3–7)	II (2–3)	I (3–4)	I (4–10)	I (2–10)
Arrhenatherum elatius	II (4–5)	II (2–3)	I (4)	I (2–5)	I (2–5)
Heracleum sphondylium	II (2–5)	I (2)	II (1–4)		I (1–5)
Glechoma hederacea	II (2–4)		II (2–8)	I (4–6)	I (2–8)
Angelica sylvestris	I (2)	I (2)	I (2–6)	I (2–4)	I (2–6)
Plagiothecium denticulatum	I (1–3)	I (1)	I (2–4)	I (3)	I (1–4)

Species					
Cirsium palustre	I (3)		I (2)	I (2-3)	I (1-3)
Oenanthe crocata	I (6)	I (1)	I (4-5)		I (4-6)
Taraxacum officinale agg.	I (2)	I (4)	I (2)		I (2-4)
Lysimachia vulgaris	I (1-3)	I (4)	I (3)		I (1-3)
Mercurialis perennis	I (4-5)	I (2)	I (4-5)	I (3)	I (3-5)
Plagiomnium undulatum	I (1)		I (2-6)	I (1-3)	I (1-6)
Geranium robertianum	I (3)		I (2-4)	I (2-3)	I (2-4)
Stellaria media	I (3)		I (2-4)	I (3-5)	I (2-5)
Galium uliginosum	I (4)		I (3)	I (4)	I (3-4)
Dactylis glomerata	I (3)		I (3)	I (3)	I (3)
Digitalis purpurea		I (1)	I (2-3)	I (2)	I (1-3)
Holcus mollis		I (3)	I (4-5)	I (3)	I (3-5)
Rumex sanguineus	I (1-2)		I (1-4)		I (1-4)
Eupatorium cannabinum	I (3-5)		I (3)		I (3-5)
Impatiens glandulifera	I (2-6)		I (3-8)		I (3-8)
Humulus lupulus	I (4-5)		I (3)		I (3-5)
Chrysosplenium oppositifolium	I (4-7)		I (4)		I (4-7)
Mentha aquatica	I (4-5)		I (3)		I (3-5)
Stachys sylvatica	I (3)		I (4)		I (3-4)
Equisetum arvense	I (4)		I (3)		I (3-4)
Caltha palustris	I (2-4)	I (4)			I (2-4)
Carex paniculata	I (3-6)			I (4)	I (3-6)
Deschampsia cespitosa	I (1)			I (5-7)	I (1-7)
Epilobium montanum		I (1)	I (1-3)		I (1-3)
Epilobium palustre		I (1)	I (4)		I (1-4)
Veronica montana		I (1)	I (3)		I (1-3)
Rumex crispus		I (2)	I (3)		I (2-3)
Carex riparia		I (7)	I (4)		I (4-7)
Athyrium filix-femina			I (1-5)	I (1-3)	I (1-5)
Agrostis stolonifera			I (4)	I (2-4)	I (2-4)
Ranunculus acris			I (3)	I (2)	I (2-3)
Rubus idaeus			I (2-3)	I (3)	I (2-3)
Hyacinthoides non-scripta			I (4-9)	I (4)	I (4-9)
Plagiomnium affine			I (2-3)	I (3)	I (2-3)

Floristic table W6 (*cont.*)

	a	b	d	e	6
Number of samples	17	6	20	15	58
Number of species/sample	12 (4–23)	21 (15–23)	21 (10–40)	16 (10–25)	17 (4–40)
Tree height (m)	11 (7–15)	12 (8–18)	14 (6–22)	13 (8–20)	13 (6–22)
Tree cover (%)	92 (70–100)	76 (50–100)	92 (80–100)	84 (25–100)	88 (25–100)
Shrub height (m)	4 (2–5)	4 (2–5)	3 (1–4)	3 (2–6)	3 (1–6)
Shrub cover (%)	5 (0–30)	20 (0–35)	19 (0–100)	24 (10–70)	11 (0–100)
Herb height (cm)	75 (40–125)	97 (35–150)	66 (20–120)	77 (50–150)	75 (20–150)
Herb cover (%)	95 (60–100)	91 (70–100)	92 (60–100)	92 (60–100)	93 (60–100)
Ground height (mm)	3 (1–20)	5	12 (10–20)	25 (20–30)	12 (1–30)
Ground cover (%)	8 (0–75)	4 (0–20)	39 (1–100)	24 (5–40)	22 (0–100)
Altitude (m)	45 (4–140)	63 (30–115)	56 (8–160)	89 (25–121)	62 (4–160)

a Typical sub-community
b *Salix fragilis* sub-community
c *Salix viminalis/triandra* sub-community (not tabled)
d *Sambucus nigra* sub-community
e *Betula pubescens* sub-community
6 *Alnus glutinosa-Urtica dioica* woodland (total)

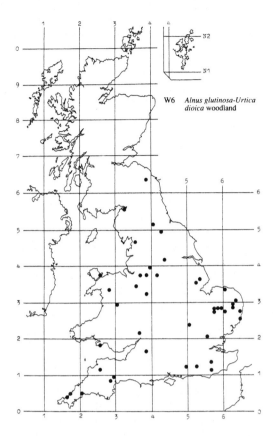

W6 *Alnus glutinosa-Urtica dioica* woodland

W7

Alnus glutinosa-Fraxinus excelsior-Lysimachia nemorum woodland

Synonymy

Alder woodland types 1b, 1c *p.p.* & 2b *p.p.* McVean 1956b; *Pellio-Alnetum* Klötzli 1970 *p.p.*; *Crepis paludosa-Alnus glutinosa* Association Birse 1980; Alder stand types 7Aa, 7Ab, 7Bc, 7D & 7Eb Peterken 1981; Woodland plot types 12, 13, 14 & 16 Bunce 1982.

Constant species

Alnus glutinosa, Filipendula ulmaria, Lysimachia nemorum, Eurhynchium praelongum.

Physiognomy

The *Alnus glutinosa-Fraxinus excelsior-Lysimachia nemorum* woodland typically has a somewhat open and, in slope-flush stands, rather irregular canopy of trees. *Alnus glutinosa* is the only woody constant throughout and, on more secure substrates, it can reach a grand stature with multiple stems and numerous small suckers and be an overwhelming dominant. Very frequently, however, there is some *Fraxinus excelsior* in the canopy and, rather less commonly, some *Betula pubescens* (only rarely *B. pendula*) and both of these can attain prominence in a more mixed cover of trees. *Acer pseudoplatanus* occurs occasionally where the ground is not permanently moist. Sometimes, larger specimens of *Salix cinerea* and *S. caprea* break the main canopy tier. Although edaphic conditions here are often suitable for *Q. robur*, this is predominantly a north-western woodland type and the species is rather scarce. *Q. petraea* can occur occasionally, though it is very much limited to the less strongly gleyed soils of the *Deschampsia* subcommunity. Slumping flushes carrying this woodland are sometimes overhung by trees rooted in the drier ground of the surrounding woodland.

There is usually an understorey in this community but its cover, height and composition are rather variable. *Corylus avellana, Crataegus monogyna* and *Salix cinerea* are the most frequent species overall and each can be abundant but the first two tend to be very poorly represented on periodically-flooded alluvial flats and confined to more stable and drier areas in slope flushes;

S. cinerea, by contrast, is commoner in wetter situations. *Sorbus aucuparia* and *Sambucus nigra* occur occasionally, the former tending to follow *Corylus* and *Crataegus*, the latter more often associated with *S. cinerea*. Other species that may be encountered are *Ilex aquifolium, Viburnum opulus, Prunus spinosa* and *P. padus*, the last sometimes with striking local abundance. Saplings of the main canopy trees are common with young *Fraxinus* and *A. pseudoplatanus* being especially frequent and abundant, *Alnus* and *Betula* rather less so. Where there is a tall tier of trees, these smaller woody elements may form a quite well defined understorey; in other stands, the layers tend to merge indistinctly.

The field layer here is somewhat variable with only two constant species, but the community has a number of associates which are frequent throughout and its general features are quite distinctive. The most consistent component is a low-growing cover of herbaceous dicotyledons and grasses, among which *Lysimachia nemorum, Ranunculus repens, Poa trivialis* and *Holcus mollis* are the commonest species. Scattered through this ground are frequent plants of *Filipendula ulmaria* and *Athyrium filix-femina* and, in the different sub-communities, this tall-herb and fern component is considerably enriched, so that a layered structure develops in the herbaceous vegetation. Often, though, this physiognomy is masked by the prominence of bulkier monocotyledons: *Juncus effusus* is very frequent in some sub-communities (essentially where there is strong gleying, rather than flooding) and it can be quite abundant; in other stands, *Carex remota* or, less commonly, *C. pendula* or *C. laevigata* occur in quantity. *C. paniculata* and *C. acutiformis*, however, are rare here, a good separation between this community and some similar kinds of *Alnus-Carex* woodland. And *Molinia caerulea* is likewise very uncommon which helps distinguish this vegetation from soligenous *Betula-Molinia* woodland. Finally, except in the wettest stands, there is frequently some *Rubus fruticosus* agg., though only on flush surrounds does it thicken up to form an underscrub.

Bryophyte cover in the community is somewhat

patchy but *Eurhynchium praelongum* and *Plagiomnium undulatum* are both very frequent and, in various of the sub-communities, *Lophocolea bidentata s.l.*, *Thuidium tamariscinum*, *Rhizomnium punctatum* and *Brachythecium rutabulum* occur occasionally with, in wetter stands, *B. rivulare* and *Chiloscyphus polyanthos* and, in drier situations, *Mnium hornum* and *Atrichum undulatum*. *Pellia epiphylla*, which Klötzli (1970) used to name this kind of woodland in Britain, is no more than occasional. Sphagna are characteristically rare.

Sub-communities

Urtica dioica **sub-community:** Alder woodland type 1c McVean 1956*b p.p.*; *Pellio-Alnetum* Klötzli 1970 *p.p.*; Alder stand types 7Ab & 7Bc Peterken 1981 *p.p.*; Woodland plot types 13 & 14 Bunce 1982 *p.p.* The canopy here is almost invariably dominated by *Alnus* with frequent and sometimes abundant *Fraxinus* but rather less *Betula pubescens* than is usual in the community. *Acer pseudoplatanus* is slightly preferential and can make very good growth in situations that are not too wet. Very occasionally, there is some *Salix fragilis* or *S. caprea*. The understorey is rather patchy but generally sparse and both *Corylus* and *Crataegus* are markedly uncommon, the usual cover being provided by scattered bushes of *Salix cinerea* and, preferential here, *Sambucus nigra*. Saplings are frequent, with *A. pseudoplatanus* being especially well represented, *Alnus* and *Fraxinus* more occasionally.

Among the smaller herbs of the field layer, *Chrysosplenium oppositifolium* is especially distinctive here and, where there is much water trickling over the surface, it can form extensive carpets or continuous strips in the narrow winding channels that carry flush waters down from nearby slopes. *C. alternifolium* can occur, too, though much less frequently and typically on the more stagnant areas between the channels. Occasionally, there may be some *Caltha palustris* in the moving waters and *Cardamine amara* along the channel banks and when all these are flowering in early spring before the tall-herb cover has fully grown up, they can give stands a very striking patterned appearance. In other cases, as on moist alluvium or colluvium, *Allium ursinum* may provide a distinctive vernal cover.

Later in the season, the ground is typically covered with a carpet of *C. oppositifolium* with *Ranunculus repens*, *Poa trivialis*, *Lysimachia nemorum* and, in drier places, *Holcus mollis*. Above this, among the scattered crowns of *Athyrium* and clumps of *Filipendula*, *Angelica sylvestris* is preferential and, even more distinctive here, *Urtica dioica*, which is sometimes present as scattered shoots but often as a patchily dense cover, among which there are thick sprawls of *Galium aparine*. *Rubus fruticosus* agg. is quite frequent but usually not abundant, especially where the ground is very wet. In some stands,

the tall-herb cover is enriched by clumps of *Phalaris arundinacea* or, less commonly but sometimes with local abundance and giving a distinctly Atlantic feel, *Oenanthe crocata*. In other places, *Impatiens glandulifera* has become prominent in this kind of woodland. In somewhat drier situations but with some base-enrichment, herbs such as *Mercurialis perennis*, *Geum urbanum*, *Geranium robertianum* and *Circaea lutetiana* can occur. *C. × intermedia* can also occur sparsely where flushing produces bare patches of sloppy ground.

Bryophyte cover is patchy in this sub-community, though sometimes quite extensive, with *Brachythecium rutabulum* (and, in wetter places, *B. rivulare*) frequently joining *Eurhynchium praelongum* and *Plagiomnium undulatum* to form lush wefts over bare soil and the bases of the herb stools. Quite commonly, after winter-flooding, there are patches of newly-deposited silt.

Carex remota-Cirsium palustre **sub-community:** Alder woodland type 2b McVean 1956*b p.p.*; *Pellio-Alnetum* Klötzli 1970 *p.p.*; *Crepis paludosa-Alnus glutinosa* Association Birse 1980 *p.p*; Alder stand types 7Bc, 7D & 7Eb Peterken 1981 *p.p.* *Alnus* is again the usual dominant in this sub-community but *Fraxinus* and *B. pubescens* are both quite frequent and, in the typically small and irregular stands developed over slumping slope flushes, the canopy can be very uneven-topped and rather open. In the gaps, *Salix cinerea* can attain a considerable size so that stratification becomes indistinct. Overall, this is the commonest of the smaller woody species but both *Corylus* and *Crataegus* become quite frequent here, rooted in patches of drier and more stable ground. *Ilex aquifolium* can also sometimes be found and *Prunus padus*, though not very common, can be locally abundant. Then there are scattered saplings of *Alnus*, *Fraxinus* and *B. pubescens*, but rarely of *A. pseudoplatanus* which is typically absent from the canopy here. Where the substrate is especially unstable, displaced or leaning trees and shrubs are frequent and much of the woody cover may then be provided by plants rooted on the drier surrounds to the flush.

Among the smaller herbs, *Ranunculus repens* remains very frequent here and with the community species, *Lysimachia nemorum*, it can form extensive ground carpets. *Chrysosplenium oppositifolium* occurs occasionally, too, though not with the consistently high frequency or abundance characteristic of the last sub-community. *Mentha aquatica* and *Ajuga reptans* are preferential among this element of the vegetation, though neither is very common. The tall-herb element is more distinctive: *Urtica* occurs only rarely but *Filipendula* is frequently joined here by *Cirsium palustre*, *Valeriana officinalis*, *Crepis paludosa* and, more occasionally, *Lychnis flos-cuculi*, *Eupatorium cannabinum* and *Succisa pratensis*.

Diverse and luxuriant mixtures of these species form a

background to the patchy dominance of a variety of bulky monocotyledons. *Juncus effusus* (rarely *J. acutiflorus*) is frequent and it can be abundant, but more peculiar to this sub-community is the common occurrence of large amounts of *Carex remota*. Sometimes this is accompanied or replaced by *C. laevigata* or *C. pendula*, the last quite often with *Equisetum telmateia* and some small patches of *Cratoneuron commutatum* on weak tufa, producing a very striking kind of vegetation that, with further sampling, could warrant recognition as a separate type of *Fraxinus-Alnus-Lysimachia* woodland. *Phalaris* and *Oenanthe crocata* occur rarely but sometimes with an abundance that lends a distinctive physiognomy. *Rubus* is very infrequent and typically has very sparse cover.

Bryophytes can form an important element of this vegetation though their cover is very variable and much dependent on the exposure of more open patches of slipping moist soil. In such situations, *Brachythecium rivulare* frequently joins *Eurhynchium praelongum* and *Plagiomnium undulatum* and each of these can be abundant; less commonly, there is some *Calliergon cuspidatum*, *Chiloscyphus polyanthos*, *Pellia epiphylla* and *Cirriphyllum piliferum* in addition to the community occasionals *Thuidium tamariscinum* and *Rhizomnium punctatum*. Rarely, there may be small tufts of less exacting Sphagna such as *S. palustre*, *S. recurvum* or *S. squarrosum* but these do not attain the prominence typical of certain similar kinds of *Betula-Molinia* woodland.

***Deschampsia cespitosa* sub-community:** Alder woodland type 1b McVean 1956b *p.p.*; *Pellio-Alnetum* Klötzli 1970 *p.p.*; *Crepis paludosa-Alnus glutinosa* Association Birse 1980 *p.p.*; Alder stand types 7Aa, 7Ab, 7D & 7E Peterken 1981 *p.p.*; Woodland plot type 16 Bunce 1982. The canopy in this vegetation, which often occurs as a transition from flushes to the surrounding woodland, is a little more varied than elsewhere in the community. *Alnus* has reduced frequency and quite commonly shares dominance with, not only *Fraxinus* and *B. pubescens*, but sometimes oak (generally *Q. petraea*) and occasionally *A. pseudoplatanus* and *Ulmus glabra* or large specimens of *Sorbus aucuparia*. The understorey is frequently well defined and quite extensive, with *Corylus* and *Crataegus* being often joined by *S. aucuparia* and, more rarely, by *Ilex*, *Prunus spinosa*, *P. padus* and *Viburnum opulus*. *Salix cinerea*, though, is much scarcer than in the other two sub-communities. Saplings are common with frequent young *Fraxinus, B. pubescens* and *A. pseudoplatanus* and, more occasionally, oaks and *Ulmus*; *Alnus* saplings, however, are rare.

Smaller dicotyledons are still prominent in the field layer but the species involved are somewhat different

from those in the other kinds of *Alnus-Fraxinus-Lysimachia* woodland. *Lysimachia* remains frequent but both *Ranunculus repens* and *Chrysosplenium oppositifolium* are very uncommon here and their place is taken by *Oxalis acetosella* (which can form extensive carpets), *Viola riviniana*, *Veronica montana*, *Potentilla sterilis*, *Fragaria vesca* and, occasionally in more southerly stands, *Lamiastrum galeobdolon*. Sometimes, too, there is a patchy cover of *Hedera helix*. Scattered through this ground are some tall herbs, though this component is generally not prominent here: only *Filipendula* is common and plants such as *Stachys sylvatica*, *Digitalis purpurea*, *Stellaria holostea* and *Teucrium scorodonia*, though preferential, are at most occasional.

More obvious and distinctive elements of the vegetation are provided by grasses and ferns. *Deschampsia cespitosa* is strongly preferential here and it can be very abundant, together with *Holcus mollis* and (somewhat reduced in frequency in this sub-community) *Poa trivialis*. Drier areas, especially where there is grazing, a common feature in the woodlands in which these flushes occur, may have some *Anthoxanthum odoratum* or *Agrostis capillaris* and, in more base-rich areas, there can be *Brachypodium sylvaticum*. Among the ferns, *Athyrium* is commonly accompanied by *Dryopteris dilatata* and, less frequently, by *D. filix-mas* or *D. borreri*. Rarely, a few sparse fronds of *Pteridium aquilinum* can be seen.

As in the *Carex-Cirsium* sub-community, *Juncus effusus* occurs frequently and sometimes in abundance but, in contrast to that vegetation type, there is often also a dense tangle of *Rubus*.

Bryophyte cover is again rather variable but it can be high and some distinctive species are involved. Along with *Eurhynchium praelongum* and *Plagiomnium undulatum*, *Mnium hornum* becomes a constant feature on twiggy litter and around tree bases and fern stools and there is frequently some *Atrichum undulatum* on exposed soil. *Plagiothecium denticulatum* and *Isopterygium elegans* occur more rarely.

Habitat

The *Alnus-Fraxinus-Lysimachia* woodland is typical of moist to very wet mineral soils, only moderately base-rich and usually only mesotrophic, in the wetter parts of Britain. Some stands are grazed and some have undoubtedly been a source of coppice-wood, mostly of alder, in the past.

The community takes much of its general floristic character from the fact that it occupies wetter mineral soils in which there is no great tendency for the accumulation of either fen peat, over which it is usually replaced by the *Alnus-Carex* woodland, or more acidic organic matter, where the *Betula-Molinia* woodland is typical, a feature recognised in the intermediate position accorded to this kind of vegetation in the scheme of alder woodlands proposed by McVean (1956*b*). In fact, the *Alnus-Fraxinus-Lysimachia* woodland can closely approach both of these other communities in its floristics, resembling the *Alnus-Carex* woodland in the *Urtica* and *Carex-Cirsium* sub-communities, and the *Betula-Molinia* woodland in the *Carex-Cirsium* and *Deschampsia* sub-communities; and the relationships of these various woodland types to the base-status and calcium content of the soils is probably quite delicate. In general, though, the *Alnus-Fraxinus-Lysimachia* woodland is marked off from the alder woodlands of fen peat by the scarcity here of *Carex paniculata* and *C. acutiformis* and the relative poverty of the tall-herb element; and it differs from the *Betula-Molinia* woodland in the rarity of *Molinia caerulea* and even the less exacting Sphagna. But the presence of some of the larger field-layer dominants, such as *Urtica dioica*, *Carex remota* and *Juncus effusus*, beneath canopies with much *Alnus* and *B. pubescens* can easily give a misleading first impression.

Typically, then, the *Alnus-Fraxinus-Lysimachia* woodland is found on various kinds of moister brown soils, such as brown alluvial soils, stagnogleyic brown earths or stagnogleys proper. Usually the superficial pH lies between 5 and 6, the combination of only moderate base-richness and high soil moisture being well marked by such herbs as *Athyrium filix-femina*, *Ranunculus repens*, *Juncus effusus* and *Chrysosplenium oppositifolium* and the liverwort *Pellia epiphylla*. Although the local influence of somewhat more base-rich waters seeping from limy partings or calcareous drift can allow the sporadic appearance of such plants as *Mercurialis*, *Geum urbanum* and *Circaea lutetiana*, drier patches of soil can show quite marked surface leaching, an indication of the generally high rainfall experienced by most sites: by and large, this is a community of those parts of Britain which have more than 800 mm annual rainfall (*Climatological Atlas* 1952). The more equable character of the climate in such regions is well indicated by the occasional presence of plants such as *Carex laevigata* and *Oenanthe crocata*.

Within these areas, the community is often associated with situations where there is some topogenous or soligenous movement of minerotrophic waters, though it can also occur where impervious substrates maintain a perched water-table on gentle slopes or plateaus. And the different sub-communities show fairly clear relationships with the extent of the waterlogging and the nature of the water supply and its movement. The *Urtica* sub-community is most typical of often light-textured brown alluvial soils over the flat or gently-sloping terraces of young river systems cut into arenaceous or argillaceous rocks, sometimes running up on to colluvium that has washed down from neighbouring flushed slopes. Such soils are generally free-draining but they are kept moist

throughout the year by the high water-table of the streams alongside which they occur or by constant flushing from above; and, on lower terraces, there may be some surface flooding in winter. Small depressions or areas receiving flush waters may remain permanently very wet; other raised patches can dry out somewhat in summer, when the soil surface shows a good, crumbly mull structure: such small-scale edaphic patterning is often reflected in the mosaic-like character of the field layer.

Also of importance here is the fact that the soils are quite eutrophic, and maintained in a moderately enriched state by the repeated deposition of allochthonous mineral material in flooding or flushing. *Urtica dioica* and *Galium aparine* show luxuriant growth and a rapid turnover of their remains: very little standing dead material persists in this vegetation until the start of the next growing season and, in early spring, the ground is often decidedly bare. The leaves of *Alnus* itself, which can attain grand stature here, are also rich in minerals and decay rapidly. Large woody nodules with nitrogen-fixing bacteria can also often be seen on *Alnus* roots exposed along the sides of small channels. Generally, however, conditions are not so enriched as to favour the development of the *Alnus-Urtica* woodland, though this community can often be found on more eutrophic terraces further downstream in the same river systems.

The *Carex-Cirsium* sub-community, in contrast to the *Urtica* sub-community, is very much a woodland of soligenous, minerotrophic flushes, being especially associated with springs or seepage lines where ground water emerges at impervious strata on shedding slopes, as where shaley partings occur within grits or sandstones or where such pervious deposits give way below to clays. The soils in such situations are kept permanently wet and, though they can be quite well structured below, on steeper slopes they approach undifferentiated sloppy masses of silt, clay and small rock fragments washed down by the trickling waters. The size and shape of the stand and the sharpness of its boundary are very much a reflection of the physical character of the flush. Where there is strong flushing over incompetent substrates, there can be massive slumping under this kind of woodland which produces a more open and unstable vegetation cover, often sharply marked off from the surrounding vegetation on firmer ground. Smaller and less vigorous flushes, whose waters percolate slowly over gentle slopes of stronger rocks, can have less well defined stands, grading with the neighbouring vegetation.

It is in such transitional situations that the *Deschampsia* sub-community is often found, over brown earths that show sometimes strong signs of gleying. But this kind of *Alnus-Fraxinus-Lysimachia* woodland is also characteristic of other situations where moderately base-rich soils are kept moist by drainage impedence, as, for example, over level-bedded argillaceous bedrocks or where there are smears of heavy-textured superficials over gentle slopes or flat ground. In both of these situations, the vegetation approaches the less calcifugous oak-birch woodlands, the *Quercus-Pteridium-Rubus* woodland towards the south-east, the *Quercus-Betula-Oxalis* woodland to the north-west, more commonly the latter since this kind of alder woodland is largely a community of the upland fringes. However, the reasonably high frequency of *Alnus* here and the abundance of such species as *Juncus effusus* and *Deschampsia cespitosa* provide a good separation. Like many of the north-western oak-birch woodlands, the *Deschampsia* sub-community is often grazed, usually by sheep and/or deer which can readily reach the firmer ground around flushes, and this helps favour the prominence of grasses.

Zonation and succession

Typically, this community occurs as fairly small stands within more extensive stretches of other kinds of woodland, marking out sites where a strong influence of ground water interrupts vegetation patterns determined in large measure by variations in the base-richness of the soils.

The most frequent context for the *Carex-Cirsium* sub-community is some kind of less calcifugous oak-birch woodland clothing the surrounding slopes of arenaceous or argillaceous bedrocks or fairly light-textured drift with more freely draining brown earths. Where the interruption of drainage of the ground water is very slight or localised, as where thin shaley partings occur within sandstone sequences or where superficials become more clayey, the flush woodland may form no more than a small (though often repeated) variation in a more or less uniform woodland cover. In other cases, the patterning is more pronounced, as where there are more vigorous flushes or where very young river valleys have cut through shales, where the concave slope above the outer swing of every meander can have an undercut, slumping flush occupied by the *Carex-Cirsium* sub-community, alternating with drier oak-birch woodland on the intervening bluffs. Some very good examples of this last kind of pattern can be seen in the Carboniferous sediments of the Pennines and in Northumberland. The particular kind of oak-birch woodland involved in such zonations depends on the regional locality of the flush. Over much of the range of the *Alnus-Fraxinus-Lysimachia* woodland, it is the north-western *Quercus-Betula-Oxalis* woodland, as in the Lake District, the northern Pennines and Scotland. Further south, it is the *Quercus-Pteridium-Rubus* woodland, often in the southern Pennines and Wales the *Acer-Oxalis* sub-community, in the occasional Wealden localities the *Anemone* or Typical sub-communities (Figure 18).

A quite frequent complication of this general pattern is produced where the impervious substrate giving rise to the flushing is more calcareous than the pervious deposit above. Then, the *Carex-Cirsium* sub-community can straddle the junction between two kinds of woodland, the flush biting back into the more calcifugous vegetation above and spilling down on to the more calcicolous cover below. This is very well seen in those Wealden woods where flushing occurs at the junction of the Hythe Beds above, carrying *Quercus-Pteridium-Rubus* woodland and the Atherfield Clay and Weald Clay below, with *Fraxinus-Acer-Mercurialis* woodland. A similar zonation occurs further north where more calcareous shales interrupt sequences of less calcareous grits, though where there is flushing with markedly base-rich waters in the north-west, the *Alnus-Fraxinus-Lysimachia* woodland is generally replaced by wetter kinds of *Fraxinus-Sorbus-Mercurialis* woodland.

In all these situations, the *Deschampsia* sub-community can be found in association with the *Carex-Cirsium* sub-community as a transition to the woodland on drier soils around. However, it frequently occurs without the *Carex-Cirsium* sub-community, marking out strongly-gleyed patches with the *Quercus-Pteridium-Rubus* and *Quercus-Betula-Oxalis* woodlands on less well drained gentle slopes and plateaus or where there is a switch to more heavy-textured drift. It can also spread along rides and paths within these woodlands where trampling consolidates heavy-textured soils. To the north-west, transitions between the *Deschampsia* sub-community and the surrounding woodland, which are already often very gentle, can be further blurred by grazing throughout.

The *Urtica* sub-community is often found in the same kinds of woodlands as the two other sub-communities and, where flush waters drain down through colluvium and on to stream-side flats, it can occur in close association within them, all three forming a complex suite. In other cases, it occurs alone on alluvium, passing more sharply to the woodland on the drier valley sides.

Where there is some variation in canopy cover, a frequent natural feature in less stable stands of the *Carex-Cirsium* sub-community, but also found where timber has been cleared, the *Alnus-Fraxinus-Lysimachia* woodland can form mosaics with Filipendulion tall-herb vegetation. Where the *Urtica* sub-community is opened up with some disturbance, patches of nitrophilous tall-herb vegetation are often found and clearance of the *Deschampsia* sub-community can result in a spread of *Rubus-Holcus* underscrub and the *Deschampsia-Holcus* grassland. Where clearance has been more extensive, the site of the woodland may be marked only by a small patch of flushed herbaceous vegetation in a stretch of improved grassland.

In general terms, it is probably some kind of Filipendulion vegetation which forms the natural precursor to

Figure 18. Different topographies and vegetation patterns with *Alnus-Fraxinus-Lysimachia* woodland.

W4b *Betula-Molinia* woodland, *Juncus* sub-community

W4c *Betula-Molinia* woodland, *Sphagnum* sub-community

W7a *Alnus-Fraxinus-Lysimachia* woodland, *Urtica* sub-community

W7b *Alnus-Fraxinus-Lysimachia* woodland, *Carex-Cirsium* sub-community

W7c *Alnus-Fraxinus-Lysimachia* woodland, *Deschampsia* sub-community

W10 *Quercus-Pteridium-Rubus* woodland

W11 *Quercus-Betula-Oxalis* woodland

W16 *Quercus-Betula-Deschampsia* woodland

W17 *Quercus-Betula-Dicranum* woodland

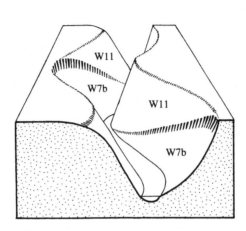

(*a*) Valley cut through shales in the upland fringes of north-west Britain with well-drained brows and slumping undercut hollows

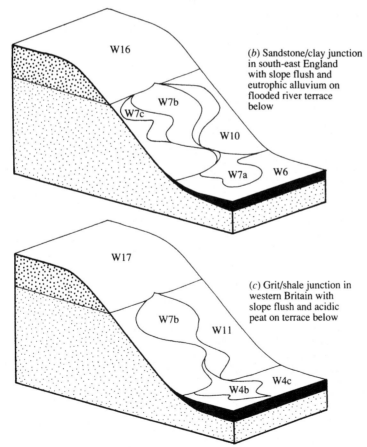

(*b*) Sandstone/clay junction in south-east England with slope flush and eutrophic alluvium on flooded river terrace below

(*c*) Grit/shale junction in western Britain with slope flush and acidic peat on terrace below the *Alnus-Fraxinus-Lysimachia* woodland, though succession has never been followed in detail. With continued flushing and gleying, the community can probably persist as a more or less permanent feature but, with increased drying of the soil, as, for example, on up-building alluvium beneath the *Urtica* sub-community, progression to the *Quercus-Pteridium-Rubus* or *Quercus-Betula-Oxalis* woodland might be expected, with a decline in nitrophilous herbs and *Alnus*, a spread of *Rubus* and an invasion of oak, *Hyacinthoides* and *Pteridium*.

Distribution
The community is widely, though locally, distributed throughout the upland fringes of the north and west, with outlying occurrences in the wetter parts of southern England, notably the Weald.

Affinities
Apart from the fragmentary accounts provided by McVean (1956*b*) and Klötzli (1970), the detailed but partial definition of Birse (1980) and the appearance of this community within a number of alder woodland stand types in Peterken (1981), the *Alnus-Fraxinus-Lysimachia* woodland has not figured prominently in descriptions of British woodlands, alder-rich stands of this kind often being subsumed under general headings such as 'damp oakwood' (e.g. Tansley 1939). However, despite its floristically intermediate character, in relation to such factors as base-richness, soil moisture and trophic state (e.g. McVean 1956*b*), it is a well-defined woodland type occupying a very distinctive suite of habitats.

Moreover, it has a very clear counterpart indeed, among the woodlands of mainland Europe, in the *Carici remotae-Fraxinetum* Koch 1926 and its various emendations. This community has been described from exactly comparable situations, sometimes showing very similar suites of sub-communities to those recognised here: from Germany and Austria (Oberdorfer 1953, 1957), France (Noirfalise 1952), Belgium (Dethioux 1955), The Netherlands (Westhoff & den Held 1969), Czechoslovakia (Neuhäuslova-Novotna 1977) and Poland (Matuszkiewicz 1981). Stands towards the north-west of Europe are distinctive in the presence of species such as *Carex laevigata* and *Oenanthe crocata* and, in northern Britain, *Crepis paludosa* adds a Continental Northern feel. With the *Fraxinus-Sorbus-Mercurialis* woodland of more base-rich wet mineral soils, the *Alnus-Fraxinus-Lysimachia* woodland represents the Alno-Ulmion in Britain.

Floristic table W7

	a	b	c	7
Alnus glutinosa	V (4–10)	V (1–10)	III (1–10)	IV (1–10)
Fraxinus excelsior	III (1–7)	II (1–6)	III (1–7)	III (1–7)
Betula pubescens	I (2–4)	II (2–10)	II (1–6)	II (1–10)
Salix caprea	I (1–3)	I (1)	I (1–4)	I (1–4)
Quercus robur	I (2)	I (2–6)	I (1–4)	I (1–6)
Acer pseudoplatanus	II (1–5)		II (1–5)	II (1–5)
Betula pendula	I (1–7)		I (1–8)	I (1–8)
Salix cinerea		II (1–9)	I (2–5)	I (1–9)
Quercus hybrids		I (1–4)	I (2–5)	I (1–5)
Betula hybrids		I (2–6)	I (5)	I (2–6)
Salix pentandra		I (1–6)		I (1–6)
Quercus petraea		I (1–4)	III (1–9)	II (1–9)
Ulmus glabra			I (1–4)	I (1–4)
Crataegus monogyna	I (1–4)	III (1–6)	III (1–5)	II (1–6)
Corylus avellana	I (1–7)	II (1–7)	III (3–8)	II (1–8)
Fraxinus excelsior sapling	II (2–8)	II (1–4)	II (2–5)	II (1–8)
Salix cinerea	II (1–6)	II (1–8)	I (1–4)	II (1–8)
Alnus glutinosa sapling	II (3–6)	II (1–4)	I (2–6)	II (1–6)
Betula pubescens sapling	I (2)	II (1–7)	II (1–7)	II (1–7)
Ilex aquifolium	I (1)	II (1–4)	I (1–4)	I (1–4)
Viburnum opulus	I (1)	I (1–3)	I (1–2)	I (1–3)

Prunus spinosa	I (1)	I (1–4)	I (1–3)	I (1–4)
Prunus padus		I (1–4)	I (2–5)	I (1–5)
Quercus robur sapling		I (1–3)	I (1–4)	I (1–4)
Betula hybrids sapling		I (1–2)		I (1–2)
Acer pseudoplatanus sapling	III (1–4)	I (1–3)	III (1–5)	II (1–5)
Sambucus nigra	II (1–4)	I (1–2)	I (1–2)	I (1–4)
Sorbus aucuparia	I (2–3)	I (1–4)	III (1–4)	II (1–4)
Ulmus glabra sapling			I (1–4)	I (1–4)
Quercus petraea sapling			I (1–2)	I (1–2)
Quercus hybrids sapling			I (2–5)	I (2–5)
Filipendula ulmaria	III (1–7)	IV (1–8)	III (1–7)	IV (1–8)
Lysimachia nemorum	II (1–3)	IV (1–5)	III (2–6)	IV (1–6)
Eurhynchium praelongum	III (2–7)	III (1–5)	IV (1–5)	IV (1–7)
Ranunculus repens	IV (1–8)	IV (1–5)	I (1–7)	III (1–8)
Chrysosplenium oppositifolium	IV (2–8)	II (1–6)	I (1–8)	II (1–8)
Urtica dioica	IV (1–8)	I (2–4)	I (1–2)	II (1–8)
Galium aparine	III (1–7)	I (1–3)	I (1–2)	II (1–7)
Angelica sylvestris	III (1–3)	II (1–6)	I (3)	II (1–6)
Brachythecium rutabulum	III (1–4)		I (1–5)	I (1–5)
Phalaris arundinacea	II (1–7)	I (2–6)		I (1–7)
Cirsium palustre	I (4)	III (1–4)	I (1–3)	II (1–4)
Brachythecium rivulare	II (1–5)	III (1–6)	I (1–4)	II (1–6)
Valeriana officinalis		III (1–4)	I (1–3)	II (1–4)
Carex remota		III (1–8)	I (1–6)	II (1–8)
Chiloscyphus polyanthos	I (2–4)	II (1–3)	I (1–4)	I (1–4)
Mentha aquatica	I (1)	II (1–4)		I (1–4)
Crepis paludosa		II (2–6)	I (1–4)	I (1–6)
Ajuga reptans		II (1–4)	I (1–3)	I (1–4)
Carex laevigata		II (1–7)	I (3)	I (1–7)
Calliergon cuspidatum		II (1–5)	I (1–3)	I (1–5)
Carex pendula		II (1–7)		I (1–7)
Eupatorium cannabinum		I (2–5)		I (2–5)
Juncus acutiflorus		I (1–6)		I (1–6)
Lychnis flos-cuculi		I (1–4)		I (1–4)
Succisa pratensis		I (1–2)		I (1–2)
Tussilago farfara		I (2–3)		I (2–3)
Cratoneuron commutatum		I (2–4)		I (2–4)
Achillea ptarmica		I (4–6)		I (4–6)
Cardamine pratensis		I (1–3)		I (1–3)
Carex nigra		I (1–2)		I (1–2)
Deschampsia cespitosa	I (1–5)	II (1–5)	IV (1–9)	III (1–9)
Mnium hornum	I (2–5)	II (1–4)	IV (1–7)	III (1–7)
Dryopteris dilatata	I (1–4)	II (1–3)	IV (1–6)	III (1–6)
Oxalis acetosella		II (1–7)	III (1–7)	II (1–7)
Atrichum undulatum		I (1–5)	III (1–4)	II (1–5)
Agrostis capillaris		I (3)	II (1–6)	I (1–6)
Viola riviniana		I (1–3)	II (1–4)	I (1–4)
Anthoxanthum odoratum		I (1–2)	II (1–5)	I (1–5)

Floristic table W7 *(cont.)*

	a	b	c	7
Hedera helix	I (1–4)		II (1–6)	I (1–6)
Dryopteris borreri	I (1)		II (1–6)	I (1–6)
Veronica montana	I (2)		II (1–3)	I (1–3)
Plagiothecium denticulatum	I (1–3)		II (1–3)	I (1–3)
Stachys sylvatica	I (1–6)		II (1–3)	I (1–6)
Potentilla sterilis			II (1–5)	I (1–5)
Brachypodium sylvaticum			II (1–9)	I (1–9)
Fragaria vesca			II (1–3)	I (1–3)
Dryopteris filix-mas			II (1–6)	I (1–6)
Digitalis purpurea			I (1–3)	I (1–3)
Galium saxatile			I (1–4)	I (1–4)
Isopterygium elegans			I (1–3)	I (1–3)
Pteridium aquilinum			I (1–4)	I (1–4)
Teucrium scorodonia			I (1–3)	I (1–3)
Stellaria holostea			I (2–6)	I (2–6)
Conopodium majus			I (1–5)	I (1–5)
Lamiastrum galeobdolon			I (3–4)	I (3–4)
Primula vulgaris			I (1–2)	I (1–2)
Athyrium filix-femina	III (1–4)	III (1–4)	III (1–6)	III (1–6)
Holcus mollis	III (1–7)	II (2–5)	III (2–8)	III (1–8)
Poa trivialis	III (1–7)	III (1–6)	II (1–3)	III (1–7)
Plagiomnium undulatum	II (1–4)	III (1–6)	III (1–4)	III (1–6)
Juncus effusus	I (1–4)	III (1–6)	III (1–9)	III (1–9)
Rubus fruticosus agg.	III (1–4)	I (1–4)	III (1–8)	II (1–8)
Cardamine flexuosa	II (1–4)	II (1–4)	I (1–3)	II (1–4)
Caltha palustris	II (3–7)	II (1–6)	I (6)	II (1–7)
Lophocolea bidentata s.l.	II (2–3)	I (1–2)	II (1–3)	II (1–3)
Galium palustre	I (3)	II (1–5)	I (1–3)	I (1–5)
Holcus lanatus	I (1–5)	II (1–7)	I (1–4)	I (1–7)
Lonicera periclymenum	I (1–4)	II (1–3)	II (1–6)	II (1–6)
Circaea lutetiana	II (1–3)		II (1–4)	II (1–4)
Pellia epiphylla		II (2–5)	II (1–5)	II (1–5)
Thuidium tamariscinum		II (1–4)	II (1–5)	II (1–5)
Rumex acetosa		II (1–4)	II (2–4)	I (1–4)
Rhizomnium punctatum		II (1–4)	II (1–4)	I (1–4)
Stellaria alsine	II (1–3)	I (2–4)	I (3)	I (1–4)
Geranium robertianum	II (1–4)	I (1–2)	I (1–5)	I (1–5)
Mercurialis perennis	II (1–5)	I (1–2)	I (1–6)	I (1–6)
Ranunculus ficaria	II (3–5)	I (1–4)	I (3–4)	I (1–5)
Hyacinthoides non-scripta	I (1–6)	I (1–2)	I (1–4)	I (1–6)
Equisetum sylvaticum	I (3–6)	I (1–5)	I (2)	I (1–6)
Ranunculus acris	I (3)	I (1–4)	I (1–4)	I (1–4)
Dactylis glomerata	I (1–6)	I (2–3)	I (1–3)	I (1–6)
Rosa canina agg.	I (1)	I (2–3)	I (1–3)	I (1–3)
Cirriphyllum piliferum	I (1–5)	I (1–6)	I (1–3)	I (1–6)
Glechoma hederacea	I (1–3)		I (1–2)	I (1–3)
Geum urbanum	I (1–2)		I (1–2)	I (1–2)

Silene dioica	I (1–4)		I (1–2)	I (1–4)
Iris pseudacorus	I (2–5)	I (3–7)		I (2–7)
Oenanthe crocata	I (1–6)	I (1–6)		I (1–6)
Cardamine amara	I (2–7)	I (5)		I (2–7)
Equisetum telmateia	I (5)	I (7–10)		I (5–10)
Equisetum arvense		I (1–3)	I (1)	I (1–3)
Geum rivale		I (2–4)	I (1–3)	I (1–4)
Viola palustris		I (1–5)	I (4)	I (1–5)
Eurhynchium striatum		I (1–2)	I (1–5)	I (1–5)
Eurhynchium swartzii		I (1–2)	I (1–2)	I (1–2)
Agrostis canina canina		I (1–8)	I (1–2)	I (1–8)
Plagiothecium succulentum		I (1–2)	I (1–2)	I (1–2)
Plagiochila asplenoides		I (1–3)	I (1)	I (1–3)
Fraxinus excelsior seedling		I (1–2)	I (1–3)	I (1–3)
Polytrichum commune		I (1–5)	I (1–6)	I (1–6)
Sphagnum palustre		I (1–4)	I (2–7)	I (1–7)
Anemone nemorosa		I (2)	I (3–7)	I (2–7)
Epilobium obscurum		I (1)	I (3)	I (1–3)
Equisetum fluviatile		I (1–2)	I (2)	I (1–2)
Glyceria fluitans		I (2–4)	I (3)	I (2–4)
Ranunculus flammula		I (1–4)	I (3)	I (1–4)
Conocephalum conicum		I (1–2)	I (3)	I (1–3)
Sphagnum recurvum		I (1–3)	I (3–4)	I (1–4)
Myosotis secunda		I (1–2)	I (3)	I (1–3)
Pellia endiviifolia		I (1–4)	I (1)	I (1–4)
Sphagnum squarrosum		I (2–5)	I (3)	I (2–5)
Number of samples	23	39	40	102
Number of species/sample	26 (7–53)	32 (19–46)	31 (9–53)	30 (7–53)
Tree height (m)	13 (10–20)	12 (7–27)	14 (6–30)	13 (6–30)
Tree cover (%)	77 (60–100)	70 (10–100)	71 (20–100)	72 (10–100)
Shrub height (m)	4 (2–5)	4 (2–7)	4 (1–7)	4 (1–7)
Shrub cover (%)	13 (0–55)	32 (1–90)	42 (5–95)	32 (0–95)
Herb height (cm)	57 (30–120)	48 (15–120)	42 (10–150)	48 (10–150)
Herb cover (%)	95 (80–100)	86 (40–100)	85 (15–100)	88 (15–100)
Ground height (mm)	7 (1–25)	26 (1–10)	27 (10–40)	22 (1–40)
Ground cover (%)	11 (0–20)	22 (3–100)	22 (2–85)	21 (0–100)
Altitude (m)	117 (60–300)	141 (2–340)	185 (8–366)	152 (2–366)
Slope (°)	5 (0–25)	10 (2–40)	16 (0–45)	11 (0–45)

a *Urtica dioica* sub-community

b *Carex remota-Cirsium palustre* sub-community

c *Deschampsia cespitosa* sub-community

7 *Alnus glutinosa-Fraxinus excelsior-Lysimachia nemorum* woodland (total)

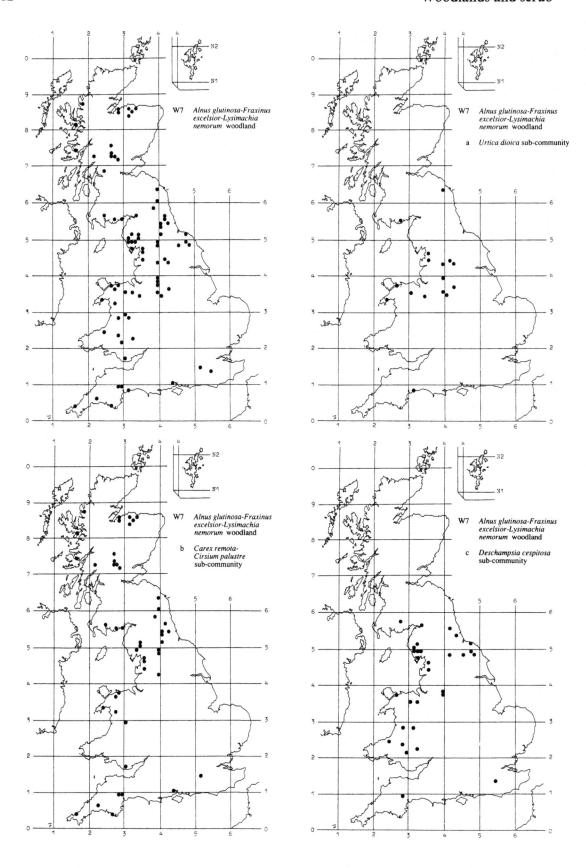

W7 *Alnus glutinosa-Fraxinus excelsior-Lysimachia nemorum* woodland

W7 *Alnus glutinosa-Fraxinus excelsior-Lysimachia nemorum* woodland

a *Urtica dioica* sub-community

W7 *Alnus glutinosa-Fraxinus excelsior-Lysimachia nemorum* woodland

b *Carex remota-Cirsium palustre* sub-community

W7 *Alnus glutinosa-Fraxinus excelsior-Lysimachia nemorum* woodland

c *Deschampsia cespitosa* sub-community

W8

Fraxinus excelsior-Acer campestre-Mercurialis perennis woodland

Synonymy

Oak-hazel woods Moss 1907; Damp oakwood association Moss *et al.* 1910; Ash-oakwood association Moss *et al.* 1910; Ashwood association Moss *et al.* 1910; *Quercetum roburis* Tansley 1911; *Fraxinetum excelsioris* Moss 1911; *Fraxinetum excelsioris calcareum* Tansley & Rankin 1911; *Quercetum roboris* Tansley 1939 *p.p.*; *Fraxinetum calcicolum* Tansley 1939; *Dryopterido dilatatae-Fraxinetum* Klötzli 1970 *p.p.*; *Querco-Fraxinetum* Klötzli 1970 *p.p.*; *Hyperico-Fraxinetum* Klötzli 1970 *p.p.*; Ash-maple-hazel woodlands Rackham 1980 *p.p.*; Hornbeam-woods Rackham 1980 *p.p.*; Limewoods Rackham 1980 *p.p.*; Elmwoods Rackham 1980 *p.p.*; Ash-wych elm woodland Peterken 1981 *p.p.*; Ash-maple woodland Peterken 1981; Hazel-ash woodland Peterken 1981 *p.p.*; Ash-lime woodland Peterken 1981 *p.p.*; Hornbeam woodland Peterken 1981 *p.p.*; Suckering elm woodland Peterken 1981; Woodland plot types 1, 2, 3, 4, 5, 6, 7, 8 & 12 Bunce 1982; *Querco-Ulmetum glabrae* Birse & Robertson 1976 *emend.* Birse 1984 *p.p.*

Constant species

Acer campestre, Corylus avellana, Fraxinus excelsior, Mercurialis perennis, Rubus fruticosus agg., *Eurhynchium praelongum.*

Rare species

Cardamine impatiens, Daphne mezereum, Festuca altissima, Ribes alpinum, Primula elatior, P. vulgaris × elatior, Tilia platyphyllos.

Physiognomy

Although individual kinds of *Fraxinus excelsior-Acer campestre-Mercurialis perennis* woodland are amongst our most distinctive and readily-recognisable woodland types, this community as a whole is very diverse, in both the floristics of the field layer and the composition and structure of the woody component. The patterns of variation in these separate elements are complex enough, but they can be systematised fairly readily into sets of either different field layers or different canopy types. The real problem here is in relating the two because, though field-layer and canopy differences are by no means independent, variation in the one is not fully consonant with variation in the other, largely because of the very great influence which different sylvicultural treatments have had on this kind of woodland, sometimes working with diversity related to natural factors like climate and soil, sometimes not. This makes it very difficult to devise a single satisfactory scheme for comprehending the variation included here and, with this community, more than ever, it should be remembered that the sub-communities represent a necessarily approximate summary of diversity which is complex and multi-directional. Exceptionally here, as an additional guide to variation within the community, the following general account has been sub-headed.

Major variations within the canopy

In essence, these are woodlands in which *Fraxinus excelsior, Acer campestre* and *Corylus avellana* provide the general diagnostic character of the woody component, though they are not always ash-maple-hazel woods in the sense of Rackham (1980) and Peterken (1981). Often, these three species do dominate in various proportions and then they frequently form the basis of some kind of coppiced underwood (though often now abandoned) but they are sometimes relegated to a minor role within coppice characterised by a local abundance of trees which are no more than occasional throughout the community as a whole, and they also occur widely here as elements of more diverse and structurally complex high forest. Some *Fraxinus* plantations, often with very little *Corylus* and *A. campestre*, or none, are also best included here on the basis of the general similarity of the vegetation.

The next most common tree throughout is oak, though there is a fairly sharp distinction among *Fraxinus-Acer-Mercurialis* woodlands as to which of the two species predominates. This is a reflection of edaphic variation and is expressed in geographical terms because the heavier kinds of base-rich soil favoured by *Quercus robur* are predominantly to the south-east of a line from the Severn to the Humber. In this region, *Q. robur* is a very frequent constituent of these woodlands, so much so that, in early descriptive accounts, they were usually described as some kind of 'oak woodland' (e.g. Moss 1907, Moss *et al.* 1910, Tansley 1911, 1939). In fact, the abundance of oak in individual stands is very variable and often low: frequently, *Q. robur* is present only as the occasional timber trees in coppice-with-standards and, even there, it is now often overshadowed by emergent maidens of *Fraxinus* or the uncut poles of large-coppice underwood. Only rarely is it truly a dominant: in some places (as in parts of Hatfield Forest in Essex), it persists in abundance in abandoned underwood and occasionally, as with *Fraxinus*, it predominates in plantation replacements of this community. Nonetheless, though it is rather unhelpful to refer to these *Fraxinus-Acer-Mercurialis* woodlands as 'oak woods', the high frequency of *Q. robur* is the best preferential feature of the canopy in one suite of sub-communities (a–d on the table) which show a predominantly south-eastern distribution, extending to the north-west only where the soils are locally heavy and moist.

In these kinds of *Fraxinus-Acer-Mercurialis* woodland, *Q. petraea* and hybrid oaks are, generally speaking, rare. Locally in the south-east, they can make an appearance where small stands of *Fraxinus-Acer-Mercurialis* woodland mark out somewhat more base-rich conditions, as on stream-side flats, within *Quercus-Pteridium-Rubus* woodland (where *Q. petraea* is frequent) but such situations are but a minor exception. To the north and west, by contrast, *Q. petraea* and sometimes the hybrids tend to predominate, even partially replacing *Q. robur* on moister soils and becoming the usual kind of oak in the remaining sub-communities (e–g on the table) which are characteristic of more free-draining, base-rich soils over permeable limestones, very much the habitat of the *Fraxinus-Acer-Mercurialis* woodlands in this region.

But it is a very striking feature of these woodlands to the north-west that the occurrence of oak is very patchy. In some places, as along the Welsh borders, around Morecambe Bay and in the North York Moors, *Q. petraea* shows high frequency: elsewhere, as in woodlands in Derbyshire and many Yorkshire Dales, oak is relegated almost to the status of a rarity. Perhaps, here, superimposed on to the basic pattern of oak distribution in relation to soil and climate, we can see the effects of localised clearance and regeneration of these woodlands in the north-west in which oak, once removed, has failed to re-establish itself as a canopy component at the high levels of frequency which it still shows in the south-east

(Pigott 1969, Merton 1970). Not that, even there, oak regeneration is a simple matter because saplings of *Q. robur* are now generally scarce and local in coppiced stands of the community (e.g. Watt 1919, Streeter 1974, Rackham 1980). In early accounts, the great emphasis on the diagnostic value of oak often led to the north-western woodlands in which its occurrence is sporadic being classified quite separately as some kind of '*Fraxinetum*' with various types of 'oak-ash woodland' as intermediates between this and 'oak woodland' proper (e.g. Moss 1911, Tansley & Rankin 1911, Tansley 1939).

This difference in the representation of different species of oak in these woodlands is overlain by a further marked variation among the trees of the community which gives the north-western sub-communities a much clearer positive characterisation than that provided by *Q. petraea*. This involves *Ulmus glabra* and *Acer pseudoplatanus* which, generally speaking, are rare in south-eastern *Fraxinus-Acer-Mercurialis* woodlands but, towards the north-west, become very frequent and often abundant as components of high-forest or large-coppice underwood and as regenerating saplings. However, this distinction, which can convincingly be related to the wetness of the climate, does not follow quite the same geographical line as that shown by the oaks. Across the Midlands, it makes roughly the same separation along a Humber–Severn line but, further south, *Fraxinus-Acer-Mercurialis* woodlands with much *U. glabra* and *A. pseudoplatanus* occur commonly in the wet south-west peninsula and, further east, pick out locally wet sites like the north-facing coombes cut into the downland of the Home Counties. Here, then, the geographical separation between the two suites of sub-communities becomes rather blurred, though the floristic distinction on *U. glabra* and *A. pseudoplatanus* remains a good one and southern stands can usually be allocated to one or the other group without difficulty. The value of these two trees in making a major separation among these woodlands was recognised by Klötzli (1970), though his *Dryopterido-Fraxinetum* (with *U. glabra* and *A. pseudoplatanus*) and *Querco-Fraxinetum* (generally without) do not exactly correspond with the two groups characterised here. *U. glabra* has also figured prominently in the diagnosis of distinctive western base-rich woods in Peterken's scheme (1981). In fact, of the two species, *A. pseudoplatanus* is the sharper indicator of the wetter climatic conditions, though it has often been neglected in classifications of British woodlands, perhaps because it is an introduction: it probably arrived here in late medieval times (Jones 1945). It is, however, very much at home now in the north-western *Fraxinus-Acer-Mercurialis* woodlands and shows exactly the same response to wetness of climate in Britain as it does in mainland Europe, where it is strikingly sparse in drier areas like the Paris Basin (Pigott 1984). With the demise of *U. glabra* as Dutch Elm Disease spreads through our northern woods, it might be expected to increase its representation even further. Indeed, there already are some areas, like certain of the Durham coastal denes, where, for some reason, *Fraxinus* and *A. pseudoplatanus* dominate to the virtual exclusion of *U. glabra*.

Other trees in the community

Of the other tree species which occur in this community, none is as generally common as those dealt with above and the patterns of occurrence, in terms of both frequency and abundance, are very variable. Some species are found occasionally throughout; others, though infrequent overall, show varying degrees of preference for different sub-communities or either of the two main suites of sub-communities. Some are characteristically solitary: others usually occur as scattered individuals but can, on occasion, be locally prominent; yet others typically occur with striking local abundance. Such variations are partly a reflection of responses to climatic and edaphic differences and partly related to the growth form of the particular trees; very often, though, they are also strongly influenced by different sylvicultural treatments. Hardly any of these trees show an exclusive association with the *Fraxinus-Acer-Mercurialis* woodland as a whole; it is, rather, their occurrence here with *Fraxinus*, *Corylus* and *A. campestre* (and with the characteristic field layer of the community) that is distinctive. The community is also defined, in a negative fashion, by the absence or great infrequency of certain tree species, though some of its floristic boundaries with other kinds of woodland are rather hazy.

Among the infrequent and generally sparse associates here, the most widespread is birch. Both *Betula pendula* and *B. pubescens* occur, though the former is a little commoner overall and not so much confined to the heavier soils typical of the south-eastern sub-communities. Although both species can be coppiced, they are typically found as maiden poles which have originated from wind-blown seed that has germinated after coppicing or in natural gaps. Quite commonly, now, birch trees of about 35–40 years old can be found, perhaps related to the widespread, though sporadic, coppicing that occurred in the cold winter of 1946/7, usually without any weeding afterwards. Given sufficient light, birch can attain a grand stature here but, where woods have been well managed, the trees characteristically remain as scattered individuals, in contrast to most kinds of *Quercus-Pteridium-Rubus* woodland where birch, especially *B. pendula*, is a very frequent canopy tree and the commonest coloniser of gaps. Even here, however, drastic clearance with subsequent neglect can give birch an opportunity to attain a disproportionate temporary prominence. It has also been noted as an important component of early seral stages in the development of *Fraxinus-Acer-Mercurialis* woodland in some places over the limestones of the north-west, as in Derbyshire, where its increase was especially noticeable after myxomatosis (Merton 1970).

Two other trees which frequently accompany birch in woodlands on less base-rich soils throughout Britain, *Ilex aquifolium* and *Sorbus aucuparia*, are, like birch, rather uncommon here, though both are a little more frequent in north-western *Fraxinus-Acer-Mercurialis* woodlands than to the south-east. *Ilex* is the commoner, occurring sometimes in the tree canopy, though more usually as scattered individuals in the understorey and its distribution in this community reflects its general avoidance of areas with a more continental climate. *Sorbus aucuparia* is generally less frequent and, among the north-western *Fraxinus-Acer-Mercurialis* woodlands, is strongly preferential to the distinctive *Teucrium* sub-community.

Much more strictly non-gregarious than these species are *Sorbus torminalis* and *Malus sylvestris*: indeed, these often occur so sparsely as to be missed if only single canopy samples, even very large ones, are taken from woods, though they often turn up if whole sites are searched. *S. torminalis* is a suckering tree which can grow to a considerable size; *M. sylvestris* is smaller, though it was rarely coppiced and can attain a venerable appearance with its sturdily-branched round crown. Feral *M. communis* can also seed into woodland of this kind. A third non-gregarious tree, the wild (though possibly introduced) pear, *Pyrus communis*, occurs much more rarely (e.g. Rackham 1975, 1980).

Similarly infrequent, though often showing a local abundance when they do occur, are *Populus tremula* and *Prunus avium*, both of which sucker to produce usually small clones. The former shows a distinct preference for the generally moister soils of the south-eastern *Fraxinus-Acer-Mercurialis* woodlands, particularly for those of the *Anemone* and *Deschampsia* sub-communities where waterlogging can continue well into late spring. *Prunus avium*, by contrast, seems to favour better-drained ground. It is not associated with any particular sub-community, but it does tend to be more obvious over wood-banks and perhaps also in more recent stands of this kind of woodland (Rackham 1980).

The remaining group of characteristic trees is rather different. These are species which, though infrequent in the community as a whole, show a distinctly preferential occurrence among the south-eastern sub-communities and which often occur there with a very marked local dominance, so much so that, in some schemes (e.g. Rackham 1980, Peterken 1981), they have been used as the diagnostic markers of quite separate kinds of woodland. This solution is not adopted here though, because of the importance accorded to these trees elsewhere and because of their particularly complex (and different) patterns of occurrence, some special notes on each have been appended at the end of this general account. Two of these species, *Tilia cordata* and *Carpinus betulus*, are not confined to *Fraxinus-Acer-Mercurialis* woodland. They occur in a number of other communities, most notably the *Quercus-Pteridium-Rubus* woodland, the counterpart of this community on somewhat less base-rich soils and, if anything, the stands they dominate there tend to be more extensive. Both species can be found as scattered individuals, when they have sometimes formed part of a mixed-coppice crop; isolated trees of *Tilia*, in particular, also often occur as relics of a previous woodland cover. But both these trees are gregarious and both cast dense shade and they are often found here as dominants (sometimes alone, sometimes mixed) and dominants of a very uncompromising kind. Very often, the structural uniformity of this kind of *Fraxinus-Acer-Mercurialis* woodland has been further accentuated by generations of cropping for large-coppice underwood. It should be noted that *Castanea sativa*, which is often found in association with *Tilia* and *Carpinus* in *Quercus-Pteridium-Rubus* woodland is typically very rare here.

The other trees which are important as local dominants in south-eastern *Fraxinus-Acer-Mercurialis* woodlands are the invasive elms referred to the *Ulmus carpinifolia* and *procera* sections. These are very aggressive trees which are able to expand into existing tracts of woodland and into previously unwooded ground by the suckering of clonal patches. They are not confined to this community, though their expansion seems very much to be centred on it and their invasion results in the development of a rather uniform and simple physiognomy, often in association with a distinctly eutrophic kind of field layer. Much more locally, this community can be dominated by a range of other elms which seem mostly to be hybrids between *U. carpinifolia* and *U. glabra* (Rackham 1980).

Finally, among the tree associates, there are those species whose scarcity helps to define the floristic limits of the *Fraxinus-Acer-Mercurialis* woodland and these are worth mentioning because some of these boundaries can be rather imprecise, depending on the particular area and situation in which this community is encountered. The first case concerns *Fagus sylvatica* and, to a lesser extent, *Taxus baccata*, for on lighter, base-rich soils in southern Britain, these species, and especially *Fagus*, become important competitors of the major trees of this community and often relegate it to the status of a seral stage in the development of high forest dominated by beech or, more locally, yew. As defined here, the *Fraxinus-Acer-Mercurialis* woodland takes in vegetation which some early authors classed as successional facies of these other woodland types or as treatment-derived woodland from which *Fagus* had been systematically removed (e.g. Adamson 1921, Watt 1923, 1924, 1925, 1926, 1934*a, b,* Tansley 1939). Both *Fagus* and *Taxus* occur here only very infrequently, as either sparse members of the tree canopy or, especially with *Fagus*, as scattered saplings, but it is important to realise that intermediate stands can occur in which either or both of these trees increase somewhat in abundance over a field layer which can remain virtually unchanged. Although *Fagus* has been widely planted beyond its supposed natural limit in Britain and can flourish much further north on soils favoured by *Fraxinus-Acer-Mercurialis* woodland, it generally dominates there in replacements for woodlands on more acid soils and so does not often pose such sharp problems of definition as in the south. *Taxus*, likewise, is only locally dominant in the north though, in a few places, as around Morecambe Bay and in Castle Eden Dene in Durham, it may be difficult to separate stands in which it has become abundant from the surrounding vegetation of this community.

Generally much less difficult but locally a problem around flushes or on very wet plateaus with perched gleys is the interface with various kinds of *Alnus glutinosa* woodland. *Alnus* is a rare tree in the *Fraxinus-Acer-Mercurialis* woodland, occurring only as scattered individuals on wetter soils of the *Anemone* and *Deschampsia* sub-communities, where it has often been included in a coppice crop, but it can become more abundant where stands of this community include areas of permanently waterlogged ground and then it may be difficult to discern a boundary. Also, when *Fraxinus-Acer-Mercurialis* woodland on wetter ground is coppiced, the more extreme kinds of *Deschampsia* field layer which develop then may become temporarily identical to the herbaceous element of the *Alnus-Fraxinus-Lysimachia* woodland.

Shrubs of the understorey or small-coppice underwood

Because of the frequent treatment of this kind of woodland as coppice, the typical canopy/understorey structure of high forest is often absent here, especially among the south-eastern sub-communities where coppicing has been more universal. There, the typical shrubs of the community have often provided the basis of small-coppice underwood cut on a variety of rotations, though, with the widespread abandonment of such treatment, very many stands have lost a well-defined coppice physiognomy. In the north-western sub-communities, where a high-forest structure is more common, the shrubs often occur as an understorey in the strict sense, forming a frequently rather open lower tier to the woodland. In fact, in both regions now, a whole range of structures between simple coppice or coppice-with-standards and high forest can be seen. And, superimposed on to this structural variation, there are, as with the trees, some distributional patterns among the shrubs which reflect edaphic and climatic differences between the regions.

As noted above, *Corylus avellana* is one of the most frequent woody species throughout this community and, in coppice, except where the crop consists of trees like *Tilia cordata* and *Carpinus betulus*, it is usually the most abundant component, and almost always the most prominent shrub, of the underwood. In such situations, its high frequency is natural but its great abundance is not: it reflects the ancient and widespread recognition of the general facility with which it responds to coppicing and an appreciation of the particular qualities of its young rods. The favouring of *Corylus* may also have extended to planting it in some woods to further increase its contribution to the underwood. Such a predominance is not limited to *Fraxinus-Acer-Mercurialis* woodland. It is widespread, too, in coppiced *Quercus-Pteridium-Rubus* woodland and, where uniform sylvicultural treatment has been imposed over boundaries between these two communities, an identical coppice composition and physiognomy may extend over the different field layers. Where the *Fraxinus-Acer-Mercurialis* woodland occurs as high forest, as it does more commonly to the north-west, *Corylus* remains frequent, though it is rarely so disproportionately abundant.

Apart from *Fraxinus* and *Acer campestre*, the most frequent woody companions of *Corylus*, though rarely rivalling it in cover in coppices, are the hawthorns. *Crataegus monogyna* is the more common species overall and, among the north-western sub-communities, it is the usual woodland hawthorn. *C. laevigata*, by contrast, is very much a preferential for the south-eastern sub-communities and, within this region, is much more obviously restricted to long-established stands of this community than is *C. monogyna*, which also occurs widely in younger tracts of *Fraxinus-Acer-Mercurialis* woodland, in plantation replacements of it and around its scrubby margins. When the two occur in close proximity, as they often do, their different habits and phenology can be well seen. *C. monogyna* is, strictly speaking, a small tree with a single bole, though it can be coppiced; *C. laevigata* is an untidy, multi-stemmed bush and lends itself better to cutting, under which treatment it can form stools of very considerable size (e.g. Rackham 1975). It also comes into leaf and flowers a little earlier than does *C. monogyna*. However, the two species hybridise very freely – hybrids are as common in the data as pure *laevigata* – and individual populations can show a complete range of variation between the parents (Bradshaw 1953). Both species, especially *C. laevigata*, are shade-tolerant and will persist in overgrown cop-

pice. Very big trees of *C. monogyna* can also occasionally be found growing up into the tree canopy of high forest *Fraxinus-Acer-Mercurialis* woodland.

Considerably less common than the hawthorns but occurring throughout and with patchy abundance are *Sambucus nigra* and *Prunus spinosa*. The former is very much a scrub of more eutrophic situations: it is more common and sometimes prominent where older stands of the community have been disturbed and enriched in some way and in younger woods where it is often associated with species-poor canopies of *Fraxinus* and *Crataegus monogyna* as in the *Hedera* sub-community. It is also somewhat more common in the north-western sub-communities where the free-draining soils may be naturally more eutrophic (see below). *Prunus spinosa* is less frequent than *Sambucus* but likewise rather sporadic in its occurrence: it is especially characteristic here of the more open post-coppice and ride vegetation of the *Deschampsia* sub-community where it can form dense suckering patches.

Cornus sanguinea is another associate which can attain local prominence in more open stands but it is also very much an indicator of more base-rich soil conditions, becoming more frequent where this community extends on to more calcareous superficials or on to rocky sites over the north-western limestones. *Euonymus europaeus* and *Ligustrum vulgare*, frequent companions of *Cornus* in scrub over our lowland limestones, also occur here, though less commonly than *Cornus*.

Finally here, though *Alnus glutinosa* is very uncommon, low frequencies of *Salix caprea* and, to a lesser extent, *S. cinerea*, are quite characteristic and their scattered bushes or trees are especially distinctive in early spring when the catkins appear before the canopy has come into leaf. Both species coppice very well, *S. cinerea* producing the most rapidly-growing spring of all the underwood species; when they are left, both can attain a substantial size. *Viburnum opulus*, a common associate of *S. cinerea* in fen carr, also occurs here at low frequencies and on moister soils.

Major variations in the field and ground layers

Like the woody component of this woodland, the field layer here is very variable but a good qualitative definition for the community as a whole is provided by the occurrence of *Mercurialis perennis* with mixtures of *Hyacinthoides non-scripta*, *Circaea lutetiana*, *Geum urbanum*, *Arum maculatum* and *Viola riviniana/reichenbachiana* (incompletely distinguished in the available data and difficult to separate when vegetative: Wigginton & Graham 1981, Phillips 1982). Less frequent overall, but also quite characteristic, are *Lamiastrum galeobdolon*, *Carex sylvatica*, *Sanicula europaea*, *Adoxa moschatellina* and *Conopodium majus*. *Hedera helix* and *Brachypodium sylvaticum* occur fairly commonly, too, though with rather more uneven frequency in the different sub-communities. Combinations of these species were recognised early as a distinctive kind of woodland field layer on fairly moist, base-rich soils in Britain (e.g. Tansley & Rankin 1911, Moss 1911, Tansley 1911, 1939) and they comprise the major group of herbs which unite the more calcicolous woodlands in the classifications of Klötzli (1970), Rackham (1980), Peterken (1981) and Bunce (1982). In this scheme, they can be found in other communities, but they always occur there either with a different canopy (as under *Fagus sylvatica* or *Taxus baccata*) or with frequent records for additional herbs which are typically scarce here (as in the upland *Fraxinus-Sorbus-Mercurialis* woodland).

Among these latter species, whose infrequency here helps distinguish the *Fraxinus-Acer-Mercurialis* woodland from its north-western upland counterpart, are ferns. No species occurs more than occasionally through the community as a whole here and most records are for *Dryopteris filix-mas* with, much less commonly, *D. borreri*, *D. dilatata*, *Polystichum setiferum* and *Athyrium filix-femina*. Towards the north and west, there is some increase in the variety and prominence of the fern flora but there are never here the consistently high frequencies for dryopteroid ferns and *A. filix-femina* characteristic of the much wetter climate and somewhat surface-leached soils of the *Fraxinus-Sorbus-Mercurialis* woodland. This difference was noted by Klötzli (1970) in his characterisation of a north-western *Dryopterido-Fraxinetum*.

Though ferns do not provide a consistent element in the structure of the field layer here, an underscrub does, though its prominence is very variable. *Rubus fruticosus* agg. is the most frequent component, occasionally with some *Rosa canina* agg., and much more sporadically, with *Rubus idaeus*, *R. caesius*, *Ribes rubrum* and *R. uva-crispa*. Ground-growing *Lonicera periclymenum* is quite common, too, adding to the tangle. Where such an underscrub becomes prominent here with lower covers of *Mercurialis* (quite a common situation), it may be difficult to distinguish this community from the *Quercus-Pteridium-Rubus* woodland, where *Rubus fruticosus* agg., *Lonicera* and *Hyacinthoides* are all common and where sylvicultural treatments can result in canopies very similar to some of those found here (as in oak-hazel, lime or hornbeam coppice). Usually, however, except in the densest shade, the presence of scattered plants of *Circaea lutetiana*, *Geum urbanum*, *Arum maculatum* and *Viola* spp., will help effect a separation. Additionally, *Pteridium aquilinum*, often very prominent in *Quercus-Pteridium-Rubus* woodland, is typically very rare here. This distinction, between what are essentially, on the one hand, dog's mercury woodlands and, on the other, bluebell woodlands has not always been maintained in

British schemes. Tansley (1939) united the two types in his broadly-defined *Quercetum roboris* and Klötzli's (1970) two major mixed deciduous woodlands also each include both types. More recent classifications which have used the dominance of such trees as *Tilia cordata* and *Carpinus betulus* (e.g. Rackham 1980, Peterken 1981) also relegate this distinction to sub-community level.

The bryophytes of the community as a whole are not especially distinctive, though they are quite commonly a prominent feature of the vegetation, especially in the winter and early spring. The most frequent species overall are *Eurhynchium praelongum*, *Brachythecium rutabulum* and *Plagiomnium undulatum* with, more occasionally throughout, *Atrichum undulatum*, *Mnium hornum*, *Eurhynchium swartzii*, *E. confertum*, *Thuidium tamariscinum*, *Hypnum cupressiforme* and *Lophocolea bidentata s.l.* Fallen and rotting timber can provide a considerably enriched bryophyte flora and, where stands occur in the more equable oceanic climate of the south-west, epiphytic species can become important.

These floristic features, then, provide a good general definition for the herbaceous element of the community as a whole, but there is a fairly sharp distinction among other field-layer species which corresponds to the major division noted among the shrubs and trees and which helps characterise more precisely the two suites of sub-communities. One group of these additional herbs follows *Quercus robur*, *Carpinus betulus*, *Tilia cordata*, the invasive elms and *Crataegus laevigata* in being more consistently associated with heavier, base-rich soils in the warmer and drier south-east. These are *Poa trivialis*, *Glechoma hederacea*, *Ajuga reptans* and *Primula vulgaris* (partly replaced in two areas in East Anglia by *P. elatior*). Among the more widespread community species, *Hyacinthoides non-scripta* and *Rosa canina* agg. also show a somewhat higher frequency here.

Except on moister soils along stream-sides or over pockets of heavy drift, all these species are distinctly less common in *Fraxinus-Acer-Mercurialis* woodlands towards the north and west. There, the increase in *Acer pseudoplatanus*, *Ulmus glabra* and, to a lesser extent, *Quercus petraea* and *Ilex aquifolium*, is matched in the field layer by consistently higher frequencies for *Urtica dioica*, *Galium aparine*, *Geranium robertianum* and *Phyllitis scolopendrium* and, among the bryophytes, *Eurhynchium striatum*, *Thamnobryum alopecurum* and, less commonly, *Ctenidium molluscum*. Among these preferentials, the increase in *U. dioica* and *G. aparine* (and perhaps also in *G. robertianum*) is perhaps a reflection of the generally more rapid nutrient turnover in the usually lighter base-rich soils characteristic of the north-west. These species do occur occasionally through the south-eastern sub-communities, but they are very closely related there to local enrichment of the soils, often resulting from human disturbance. The prominence of *Phyllitis scolopendrium*, by contrast, is a response to the milder and wetter climatic conditions towards the west of Britain. *Hedera helix*, another oceanic species, follows it somewhat in being more frequent in these sub-communities than it generally is to the south-east, but its distribution as an important field-layer species in this community is rather broader than that of *P. scolopendrium* and it also occurs in abundance across southern England under canopies which are of the south-eastern type. Its occurrence is also complicated by the fact that, with *Brachypodium sylvaticum*, it tends to become prominent in young stands of this community and in plantation replacements of it.

This major floristic distinction among the field layers is matched by a physiographic difference between the habitats characteristic of each of the suites of sub-communities. As described below, this has important implications for the kinds of vegetation that are found within each of the suites, but it also has a much more general effect on the appearance of the field layer. To the south-east, *Fraxinus-Acer-Mercurialis* woodlands are typically found over the gentle slopes and undulating plateaus developed over superficials and easily-weathered argillaceous rocks. Stands are often quite extensive (though much reduced now by clearance) and the field layer is frequently characterised by a physiognomic uniformity and gentle shifts in dominance, except where recent coppicing has introduced variation. The north-western sub-communities, on the other hand, are much more typical of the sharply-defined scenery of limestone bedrocks, with steep slopes and sometimes very intricate patterns of weathering of the exposures. Even where the field layer is floristically uniform, it can show a very complex structure, being disposed in a mosaic over the substrate surface. Such patterning can be seen in its most extreme form where *Fraxinus-Acer-Mercurialis* woodland occurs over Carboniferous Limestone pavement.

Other variations in the field and ground layers

The qualitative features described above serve to define the major floristic trends in the field layer of this community. Often, though, it is the very considerable quantitative variation among these widespread species that is the more striking here and the way in which species of fairly low overall frequency can come to dominate in field layers that seem to have little in common, one with another.

Much of this variation can be seen, in spring and summer, in terms of the differing prominence of *Mercurialis perennis* because, though this is the most characteristic herb here, its frequency and abundance are very variable. Overall, it is the commonest plant of the field layer in this community and the most frequent herba-

ceous dominant, often forming an extensive network of discrete patches or a virtually continuous carpet of coalesced clones, with sometimes upwards of 400 shoots m^{-2} (Wilson 1968, Hutchings & Barkham 1976). It is a long-lived perennial, propagating mainly by slow vegetative expansion and, though it spreads very gradually into young stands, once established it is very well adapted, in its patterns of rhizome and shoot growth, to maintaining itself as a dominant (Mukerji 1936, Wilson 1968, Hutchings & Barkham 1976). And, when it does grow well, there is every indication that it exerts a very strong influence on the composition and patterning of the field layer by virtue of the dense shade that it casts (Abeywickrama 1949, Wilson 1968, Martin 1968, Pigott 1977); it can even effect canopy regeneration by hindering the growth of tree seedlings (Wardle 1959). Its shoots emerge very early, usually from February onwards, and well in advance of most of the other possible vernal dominants here (Salisbury 1916, Wilson 1968). They can grow tall, up to almost half a metre, and quickly put out their decussate, horizontally-held leaves to form the dull green carpet that characterises these woodlands until late in the summer after which the senescing shoots finally succumb to frost and fungal infection in September and October (Wilson 1968).

In certain situations, however (and these are quite well defined ecologically), the cover of *Mercurialis* is reduced and it becomes increasingly scarce; in some kinds of woodland included in this community, it is totally absent. Minor reductions in *Mercurialis* are often marked simply by an increase in the abundance of species already present. *Hyacinthoides non-scripta* is often the most obvious of these and it commonly assumes dominance here on somewhat moister soils, where *Mercurialis* suffers from the reduced aeration (Abeywickrama 1949, Martin 1968). This can be seen within both the south-eastern and north-western kinds of *Fraxinus-Acer-Mercurialis* woodland, in the former on gentler slopes where surface water cannot easily percolate away, in the latter where colluvial soils are receiving water shed from the slopes above. Then, too, provided *Hyacinthoides* itself is not too overwhelmingly dominant (it can be), there is an increase in the cover of the other herbs and bryophytes characteristic of the community as a whole and of each of the regions. Much of this kind of quantitative variation can be contained within particular sub-communities: it is a common feature in the two mainstream kinds of *Fraxinus-Acer-Mercurialis* woodland, the *Primula-Glechoma* sub-community to the south-east and the *Geranium* sub-community to the north-west.

Sometimes, however, environmental conditions are such as to so reduce the cover of *Mercurialis*, that species which are generally scarce throughout the community as a whole are able to assume a marked local dominance because of the lack of competition. This more marked qualitative change in the floristics of the field layer helps define some other sub-communities in each of the regions. Very often, such differences are due to very pronounced waterlogging of the soils. Among the south-eastern sub-communities, the prominence of *Anemone nemorosa* and *Ranunculus ficaria* in this kind of situation on very wet plateau soils, defines the *Anemone* sub-community. To the north-west, moist conditions on free-draining colluvial soils are marked by vegetation placed here in an *Allium ursinum* sub-community.

A reduction in *Mercurialis* cover is also a feature of soils which become ill-aerated through trampling, often associated here with coppicing, which usually takes place in winter when the soils are more susceptible to poaching. This kind of variation, often accompanied by vegetative expansion of some existing shade-tolerant species and a flush of flowering, can be seen in various kinds of *Fraxinus-Acer-Mercurialis* woodland and is often best understood as a temporary perturbation of the more stable floristic patterns of the particular sub-community that has been coppiced. Sometimes, however, coppicing (or clearance) produces a more dramatic and qualitative shift in the composition and structure of the field layer. This kind of vegetation is included here in a *Deschampsia cespitosa* sub-community among the suite of south-eastern *Fraxinus-Acer-Mercurialis* woodlands.

The remaining variation within the field layers can be accommodated in two further sub-communities. The association of frequent and often abundant *Hedera helix* with the south-eastern type of canopy noted earlier produces a very distinctive kind of woodland common in the more oceanic parts of southern Britain. The herbaceous element in this *Hedera* sub-community is often rather impoverished and, though *Mercurialis* is itself quite frequent, this sub-community also includes younger woodlands on base-rich soils from throughout the south-east.

Finally, among the *Fraxinus-Acer-Mercurialis* woodlands of the north-west, there are some very striking stands in which unusually rich and uneven-aged mixtures of trees and shrubs occur over a field layer in which the typical herbs of the region are accompanied by a variety of species reflecting an often patchy woody cover, a complex rocky topography and climatic conditions approaching those characteristic of montane parts of Britain. In general in this community, sylvicultural treatments have been so extensively applied that more natural assemblages of woodland plants survive only under much modified canopies but, in this *Teucrium scorodonia* sub-community, we perhaps have a glimpse of what certain north-western *Fraxinus-Acer-*

Mercurialis woodlands were like before widespread alteration and clearance.

Tilia cordata in *Fraxinus-Acer-Mercurialis* woodland

In this community, as in the other major kind of woodland in which it is well represented, the *Quercus-Pteridium-Rubus* woodland, *Tilia cordata* has a low overall frequency but is preferentially common and locally abundant in the south-eastern suite of sub-communities, though, even here, it shows a markedly patchy geographical distribution, being more frequent in Essex, Suffolk, Northamptonshire, Lincolnshire, the West Midlands and Somerset, and decidedly scarce south of the Thames. In general, though, it is characteristic of that part of the country where the fall in summer temperatures since the period of its great prominence at the Forest Maximum has not been so critical as to reduce it to being a relic tree, able only very sporadically to set viable seed (Pigott & Huntley 1978, 1980, 1981). This is the status which it seems to have in the north-western sub-communities of this woodland and of the *Quercus-Pteridium-Rubus* woodland, and in the most southerly stands of their upland analogues which just reach down to what seems to have been the past geographical limit of *Tilia* in northern England (Pigott & Huntley 1980). Here, *Tilia* usually persists as what are obviously very ancient stools, often isolated on more intractable topographies, around which their original woodland context has come and gone. Locating such individuals phytosociologically is difficult now: quite apart from the fact that they have outlasted clearance and regeneration, many trees occupy cliff edges with their horizontal roots beneath one kind of woodland, their vertical roots in another below, or occur in ravines with complex mosaics of habitats and field layers. Then, their association is more strictly with a site type than with a particular vegetation type, though this does not make their conservation any the less pressing: indeed, as what are probably among some of the most venerable trees in the British countryside, they are especially deserving of care.

Isolated limes can often be found further south, too, well within the range of the tree's present ability to set viable seed (Pigott & Huntley 1981). Single poles, sometimes pollarded, or stools or their clonal remnants, again often of immense size, occur commonly in stands of the south-eastern sub-communities, often on wood-banks or territorial boundaries (see maps in Rackham 1980). *Tilia* can persist in this way in those stands of *Fraxinus-Acer-Mercurialis* woodland which have a known (or strongly-suspected) history of fairly recent clearance and regeneration or which are obviously quite young plantations, as well as in those less-disturbed woodlands where it has never been common for reasons quite unrelated to treatment. However, documentary evidence shows that where, among these south-eastern sub-communities, *Tilia* attains its highest frequencies and abundance, there has usually been a continuity of wood-land cover and composition for a very considerable period of time. There are, though, some documented exceptions to this general rule and it does not follow, of course, that an absence of *Tilia* indicates recent origin: many *Fraxinus-Acer-Mercurialis* woodlands with a proven ancient history have no *Tilia*.

Where *Tilia* is abundant in these woodlands, it has often clearly been a coppice crop, cut for timber, wood and bast, though it seems to have been unpopular as a charcoal source because of a tendency to reignite when the stacks are opened (Pigott & Huntley 1978). Usually, now, it dominates as large coppice poles, sprung from stools which are often of great size and which sometimes have decayed centres so that rings of apparently discrete trees remain. The poles grow vigorously and straight, this last feature being especially noticeable where the tree is intermixed with the wavy-boled *Carpinus betulus*, as it quite often is, both here and in the *Quercus-Pteridium-Rubus* woodland. In abandoned coppice, *Tilia* regrowth can overtop smaller standard oaks, though in fact most stands do have some scattered *Quercus robur* and often a little *Fraxinus*. There are also occasional trees of *Acer campestre* and *Betula pendula* but underwood associates are very sparse with only rare *Corylus* or *Crataegus*.

Among the south-eastern sub-communities, *Tilia* is most often encountered and generally shows the greatest local abundance in vegetation which resembles the *Primula-Glechoma* and *Anemone* sub-communities. Woodlands of this kind have been classified by Rackham (1980) as lime-ash woods and were included by Peterken (1981) within his Stand type 4Ba of lowland maple-ash-lime woods. However, there is a problem in deciding whether *Tilia* is more associated with one or the other of these field layers because, wherever *Tilia* is dominant, its very dense canopy eliminates all but the most shade-tolerant of the herbaceous associates and even these often occur so sparsely that it is difficult to discern what the floristic affinities of the vegetation are. Indeed, it is sometimes hard to be sure that the vegetation does not belong to some other sub-community apart from these two, or even whether it is a *Fraxinus-Acer-Mercurialis* woodland at all. *Tilia* characteristically produces a convergent impoverishment of the field layer of whatever kind of woodland it has become prominent in and where uniform sylvicultural treatment for producing crops of lime has been applied across boundaries between, say, this community and the *Quercus-Pteridium-Rubus* woodland, the poverty of the herbaceous associates means that there is often next to no sign that different vegetation types have been transformed.

In comparison with typical kinds of *Primula-Glechoma* and *Anemone* vegetation, only *Rubus fruticosus* agg., among the common associates, retains anything like its usual frequency here and it is often reduced to sparse trailing branches rather than forming a dense underscrub. Of the other community species, *Mercurialis perennis*, *Hyacinthoides non-scripta*, *Circaea lutetiana*, *Geum urbanum*, *Viola* spp. and *Arum maculatum*, all show depressed frequencies under dense *Tilia* and, among the sub-community preferentials, only *Ajuga reptans* occurs more than occasionally and even this is usually found as scattered individuals. The only really positive feature of the herbaceous vegetation in these woodlands is the high frequency of certain bryophytes: although more delicate species such as *Eurhynchium praelongum*, *Brachythecium rutabulum* and *Plagiomnium undulatum* decline, there is a marked increase in the occurrence of the more robust *Mnium hornum* and, especially, *Atrichum undulatum*.

Recently-coppiced *Tilia* has not been sampled but there seems little doubt that where stands on wetter soils are cut, the characteristic plants of the *Deschampsia* sub-community could attain the temporary prominence that they show beneath other kinds of young underwood. By contrast, the low frequency of *Tilia* in the *Hedera* sub-community seems to be a real reflection of the scarcity of this tree south of a Thames–Severn line, where this kind of *Fraxinus-Acer-Mercurialis* woodland is most common. This sub-community also includes some plantations in which *Tilia* is but an infrequent survivor of an earlier woodland cover.

In those much rarer situations where *Tilia* occurs as a coppice-underwood dominant in stands of north-western *Fraxinus-Acer-Mercurialis* woodlands, a similar physiognomy to that described above can be found, though here *Quercus petraea* partially replaces *Q. robur* and there can be occasional scattered trees of *Ulmus glabra* and *Acer pseudoplatanus*. Because of the dense shade, however, the distinctive features of the north-western kind of field layer are often obscured. Where *Tilia cordata* occurs as isolated relic trees in the north-west, it is occasionally found, sometimes in association with our other native lime, *T. platyphyllos*, as in the Wye valley and in Derbyshire, in the very distinctive *Teucrium* sub-community (Pigott 1969). North-western *Fraxinus-Acer-Mercurialis* woodlands with *Tilia cordata* were included in Peterken's (1981) Stand types 4Bb western maple-ash-lime woodlands and 4C sessile oak-ash-lime woods.

***Carpinus betulus* in *Fraxinus-Acer-Mercurialis* woodland**
Like *Tilia cordata*, *Carpinus betulus* occurs commonly as a woody dominant in both this community and the *Quercus-Pteridium-Rubus* woodland and it has sometimes been used to characterise various types of 'horn-

beam woodland' which subsume parts of both these communities, as by Salisbury (1916, 1918a), in Rackham's (1980) hornbeam-ash woods and in Peterken's (1981) Stand types 9A and 9B. It has been widely planted to the north of Britain (and, in a few sites, is established in some abundance) but, in general, it is much more strictly confined to the south-east than is *Tilia*, having a clear central European distribution (Matthews 1955). Furthermore, its floristic affinities are a little clearer because, though it is, like *Tilia*, a markedly gregarious tree, it does not seem to cast such a heavy shade, so the field layer usually survives in sufficient abundance as to indicate the general nature of the vegetation. And there is no doubt that, within both the woodland communities in which it is common, *Carpinus* shows a very frequent (though not exclusive) association with vegetation in which *Anemone nemorosa* and *Ranunculus ficaria* become abundant on spring-wet soils.

As with *Tilia*, the usual woody associates where *Carpinus* is abundant in *Fraxinus-Acer-Mercurialis* woodland are *Quercus robur* and *Fraxinus*, both of which, though usually the former, can be found as standards. *Acer campestre* is generally infrequent, except where smaller *Carpinus*-dominated stands occur within more orthodox stretches of mixed canopies (as in Rackham's (1980) maple-hornbeam woodlands) and *Corylus* and *Crataegus* spp. are scarce. Structural uniformity has often been enhanced by many generations of coppicing (hornbeam has long been valued as an excellent fuel) but it has been suggested (Rackham 1980) that more extreme dominance by *Carpinus* is maintained only by a long-rotation coppice cycle. If cropping is at less than, say, fifteen years and there are no hornbeam standards to provide seed, any *Corylus*, which can fruit on a shorter cycle, may gain an advantage. Also some stools of *Carpinus* usually die after coppicing so, where hornbean and lime grow together (quite a common occurrence), *Tilia* may increase its share of the canopy. Nowadays, with the abandonment of coppicing, *Carpinus* often grows to an impressive height, overtopping oak standards, though its stools are seldom as enormous as those of *Tilia*.

Except in the densest shade, most of the herbs and bryophytes of the south-eastern *Fraxinus-Acer-Mercurialis* woodlands retain high frequency. On somewhat drier soils *Mercurialis* or *Hyacinthoides* can be dominant in a field layer of the *Primula-Glechoma* type (equivalent to Salisbury's (1916, 1918a) *Mercurialis*, *Scilla* and *Nepeta* societies) but more often the herbaceous vegetation is that of the *Anemone* sub-community (as in Salisbury's (1916, 1918a) *Anemone* and *Ficaria* societies). Along paths and especially in more open areas, as in recently-cut panels, the *Deschampsia* sub-community can attain local prominence and post-coppice floras of

this kind can be more persistent here than beneath other kinds of underwood because *Carpinus* shows only slow early regrowth. Field layers of all these types were included in Rackham's (1980) maple-hornbeam and hornbeam-ash woods and in Peterken's (1981) Stand types 9Ab and 9Bb.

Unlike *Tilia*, which is generally a very rare coloniser in more recent spontaneous woodland, *Carpinus* is quite often found as the dominant in young extensions to older woods in which it is abundant. However, it does not seem to be an early invader but rather takes advantage of any clearance of woody pioneers and, even then, becomes prominent in the secondary canopy only slowly (Salisbury 1918b, Adamson 1932, Rackham 1980).

Invasive elms in *Fraxinus-Acer-Mercurialis* woodland
Invasion of existing woodland by suckering elms of the *Ulmus* sections *carpinifolia* and *procera* is widespread throughout lowland Britain, especially towards the south and east and most markedly of all on the heavy claylands of the Midlands and East Anglia (Rackham 1980, Peterken 1981). All available evidence suggests that the woodland types most frequently invaded are the south-eastern *Fraxinus-Acer-Mercurialis* woodlands, particularly those of the *Primula-Glechoma* sub-community, but also the wetter *Anemone* and *Deschampsia* sub-communities. These suckering elms also readily extend into neglected farmland and over abandoned settlements and such more recent woodland can have a field layer of the *Hedera* type. In certain parts of East Anglia, some other clonal, but apparently non-suckering, elms can also be encountered as dominants over these kinds of field layers (Rackham 1980). These seem to be hybrids between *U. carpinifolia* (rarely *U. procera*) and *U. glabra* but, unlike the suckering elms, these trees can also be found in association with *Tilia*, *Carpinus* and *Castanea sativa* in *Quercus-Pteridium-Rubus* woodland.

The suckering elms are aggressive trees but their invasion is characteristically local and it can be slow. Usually invasions start from established trees on wood margins, in hedges, on internal boundaries or around settlements and the expanding clones of suckers have a typically rounded outline, part-circular where the trees are invading from a wood edge, roughly circular within woods (see maps in Rackham 1975, 1980). Origin from seedlings may be a rare event but it is important in extending colonisation to new sites and in further increasing the already dismaying morphological variety that can be found between clones (e.g. Melville 1975, 1978, Rackham 1980, Richens 1983).

Although these invasive elms can fairly quickly become the tallest trees in stands of *Fraxinus-Acer-Mercurialis* woodland, where invasion is recent many of the typical floristic features persist. Mixtures of *Fraxi-*

nus, Corylus and *Acer campestre* with oak standards often form the context for elm-expansion and, beneath, all the characteristic south-eastern herbs can be frequent. Commonly, though, there is more *Urtica dioica* than usual and *Sambucus nigra* can be prominent in the underwood. This may be because these invasive elms spread more readily on soils which are already eutrophic (they are often very conspicuous over infilled moats and around old settlements and their gardens) or because the elm litter, with its high calcium levels (e.g. Martin & Pigott 1975), assists rapid nutrient turnover; perhaps both. However, it is not true to say (Rackham 1975, 1980) that either *Glechoma hederacea* or *Poa trivialis* are especially 'elm plants', except in so far as they are common in the kinds of wood invaded and persist under the increasing elm canopy as other herbs fade.

More advanced elm invasion of this kind also affects the existing trees and shrubs, as the oak standards and even any emergent *Fraxinus* are overtopped and shaded out as well as the underwood. Older elm clones thus often have a lot of dead wood and timber, with oak especially prominent among the latter, its killed standards being slow to fall and rot (Rackham 1975, 1980). With time, such elm woodland acquires a floristic and structural uniformity which makes it very different from the vegetation it has replaced, a point of some importance as elm invasion is one of the few major changes occurring in British woodlands at this time. Interestingly, its progress has not been markedly affected by the current epidemics of Dutch Elm Disease (*Ceratocystis ulmi*). This seems at present to attack mostly *procera* elms, the less common of the sections in woodlands, and *U. glabra* which is generally scarce in the south-east (Rackham 1980). Even where larger poles have succumbed, swarms of young suckers may persist.

Sub-communities
The seven sub-communities are described in the usual style, the four south-eastern types first, then the three north-western. In each suite, the account of the mainstream sub-community precedes the others.

***Primula vulgaris-Glechoma hederacea* sub-community:** Oak-hazel woods Moss 1907 *p.p.*; Damp oakwood association Moss *et al.* 1910; *Quercetum roburis* Tansley 1911 *p.p.*; Ash-oakwood Association Tansley 1911 *p.p.*; *Mercurialis* and *Fragaria* societies Adamson 1912; Oak-hazel woods Salisbury 1916 *p.p.*; *Quercus robur-Carpinus* woods, *Mercurialis* society Salisbury 1916; *Quercus sessiliflora-Carpinus* woods, *Mercurialis* and *Nepeta* societies Salisbury 1918*a*; *Quercetum roboris, Scilla, Mercurialis* and *Primula* societies Tansley 1939; Hayley Wood zones 5 & 6 Abeywickrama 1949; *Querco-Fraxinetum typicum* Klötzli 1970;

Ash-maple-hazel woodlands Rackham 1980 *p.p.*; Lime-ash woods Rackham 1980 *p.p.*; Maple-hornbeam woods and Hornbeam-ash woods *p.p.* Rackham 1980; Elmwoods Rackham 1980 *p.p.*; Ash-Wych elm stand type 1Bb Peterken 1981 *p.p.*; Ash-maple stand types 2A & 2B Peterken 1981 *p.p.*; Ash-lime stand type 4Ba Peterken 1981 *p.p.*; Hornbeam stand types 9Ab & 9Bb Peterken 1981 *p.p.*; Suckering elm woodland Peterken 1981; Woodland plot types 5 & 7 Bunce 1982. The three woody constants of the community, *Fraxinus, Corylus* and *Acer campestre*, are all frequent here, occurring in all possible combinations and very variable proportions. Very commonly, they form the basis of what has obviously been simple coppice or coppice-with-standards with variable numbers of standard timber trees, usually oak, and almost always here *Quercus robur*. Occasionally, *Fraxinus* can be found as standards or, more unusually, *A. campestre* and, rarely (as at Hatfield Forest in Essex), *Q. robur* occurs prominently in the underwood. In coppiced woodland of this kind, the composition and structure of the vegetation can still give strong clues as to the pattern of cropping for small, large or mixed coppice on various rotations in different panels. However, with the demise of coppicing, the present appearance of this kind of woodland is very varied, being strongly influenced not only by the original coppice physiognomy but also by the local environmental conditions that have affected regrowth and regeneration since the last cut.

Of the four common woody species, both *Fraxinus* and *A. campestre* are readily able to promote themselves to a tree canopy from the unchecked growth of their stools. *Fraxinus* is also the best regenerator from seed in the community as a whole: seedlings and saplings are very frequent and, provided the shade from trees and shrubs, or from *Mercurialis*, is not too intense, they can get away rapidly and form vigorous maiden poles. Emergent *Fraxinus* are thus very common here and they can shade out small standard oaks. *A. campestre* is, generally speaking, initially less abundant, springs up not quite so far and regenerates erratically from seed, so it is not so common an emergent, but it can tolerate *Fraxinus* shade and itself casts dense shade. Existing individuals can thus hold their own for a considerable time even though their proportional contribution to the canopy falls. A common pattern resulting from such interactions is for there now to be a quite well defined but uneven-topped canopy of trees, frequently extensive though usually not completely closed, made up of mixtures of any persisting oak standards, much coppice regrowth and maiden *Fraxinus* and usually smaller amounts of coppice regrowth of *A. campestre*. Less commonly, stands can be found in which *Fraxinus*, or very occasionally *A. campestre*, or rarely *Q. robur*, have

come to be sole dominants and it is possible that some of
the stands with much ash and oak represent long-
established plantations.

Older woodlands of this kind, whether treated as
coppice or approximating more to high forest, often
have occasional records for other trees: *Betula* spp., *Tilia
cordata*, *Carpinus betulus*, *Sorbus torminalis* and *Malus
sylvestris* occurring as scattered individuals, *Populus
tremula* and *Prunus avium* in small suckering patches.
Those species which can be coppiced have often been
included in a mixed-coppice crop but stands in which
Tilia or *Carpinus* predominate (forming 'lime-' or 'horn-
beam-woodland') also occur here with, when the shade
is not too dense, a typical *Primula-Glechoma* field layer.
Elm invasion is also quite common and tracts of this
sub-community can be found under *U. carpinifolia* or *U.
procera*.

Except where some large-coppice tree or older elm
suckers dominate or in plantation replacements for the
canopy, the most obvious feature of the underwood in
these woodlands is generally the persistent and often
very great abundance of *Corylus*. Where the shade cast
by the trees is not too dense, very large hazel bushes can
themselves be the dominant in old coppice and, even
where the tree canopy is closing, the bushes often grow
so tall as to make stratification indistinct. But, with
increasing shade, the cover of *Corylus* declines from
what are really artificially high levels in coppiced
versions of this woodland. The shoots are replaced more
slowly and fruiting becomes less prolific; when photo-
synthesising at a lower rate, *Corylus* may also be less able
to withstand the depredations of the *Armillaria mellea*
which very commonly lives on its roots (Rackham
1980). Individual bushes begin to die and the species
takes its place as one member, though usually still the
most abundant, of a more open and balanced shrub
cover.

Typical underwood companions of *Corylus* here are
the hawthorn species and their hybrids (*Crataegus
monogyna* often extending out as a prominent marginal
shrub, *C. laevigata* scarcer and more obviously associated with the older more intact parts of stands), with,
less commonly, *Sambucus nigra*, *Prunus spinosa*, *Salix
caprea*, *S. cinerea*, *Cornus sanguinea*, *Euonymus europaeus*, *Viburnum opulus*, *V. lantana*, *Ilex aquifolium* and
Sorbus aucuparia. Many of these would have been
included with hazel in a small-coppice crop; now, with
neglect, they often show a local abundance in relation to
particular environmental conditions.

The particular diagnostic feature of the field layer of
this sub-community as against other south-eastern types
of *Fraxinus-Acer-Mercurialis* woodland is the frequent
presence here of *Poa trivialis*, *Glechoma hederacea*,
Ajuga reptans and *Primula vulgaris* (or, in wetter stands

in certain parts of East Anglia, *P. elatior*) in the general
absence of *Anemone nemorosa* and *Ranunculus ficaria* or
Deschampsia cespitosa or *Hedera helix*. All of these
latter species are scarce here and never abundant within
uniform stands of the *Primula-Glechoma* sub-community: small patches of *Anemone* and *Ranunculus* can mark
out slight surface depressions where the ground is
undulant and *Deschampsia* often spreads along the
edges of paths but such areas are best regarded as
fragments of other sub-communities associated with
habitat heterogeneity.

Both *Poa trivialis* and, more especially under certain
circumstances (see below), *Glechoma hederacea* can be
quite abundant here with large patches of creeping leafy
shoots. They are also both evergreen and, with the
common bryophytes of the community, can give a
distinctive stamp to this vegetation in late winter and
early spring before either *Mercurialis* or *Hyacinthoides*,
the usual vernal dominants, have begun to put up their
shoots. Often, though, they occur in small amounts,
together with scattered plants of *Primula vulgaris* and
other species of wide occurrence throughout the
community: *Circaea lutetiana*, *Geum urbanum*, *Viola*
spp. and, more occasionally, *Arum maculatum*, *Carex
sylvatica*, *Lamiastrum galeobdolon* and *Sanicula europaea*. Each of these can also attain some measure of local
prominence (as in the 'societies' of early descriptive
accounts) but frequently they form a diverse mosaic
overlain by varying amounts of *Rubus fruticosus* agg.

In spring and summer, however, the most prominent
feature of the vegetation is generally the abundance of
Mercurialis. Where its cover is high, as it tends to be in
long-established woods on better-aerated soils, the associated field-layer herbs and especially the ground-growing mosses are strongly depressed in variety and abundance. Few other species seem able to survive its dense
shade: the evergreen *Glechoma* and *Viola* spp. persist
patchily where the cover is a little more open but, of the
other associates, only *Circaea lutetiana* maintains its
normal pattern of occurrence, an interesting exception
since it puts up its leaves quite late. Poorer *Mercurialis*-
dominated floras of this kind are often associated in this
sub-community with dense coppice regrowth; sometimes there is a dominant canopy of *Fraxinus* or *Q. robur*
but the understorey is frequently very well developed
and often exceeds the trees in cover with *Corylus*,
Crataegus monogyna and *Prunus spinosa* well represented. However, *Fraxinus* saplings and, more especially,
seedlings are unusually sparse and regeneration of
woody species from seed is probably very rare (e.g.
Wardle 1959). Interestingly, *Rubus fruticosus* agg., which
often spreads in undisturbed woods, is also much
patchier where *Mercurialis* is overwhelmingly dominant.

Stands of this sub-community with low to moderate

(5–50%) *Mercurialis* cover show a much greater richness and diversity and three major trends can be recognised. First, where there is an increase in soil moisture, *Hyacinthoides* often takes over from *Mercurialis* as the dominant in vegetation which is otherwise little changed. *Hyacinthoides* is a common plant throughout this sub-community (indeed, the most frequent herb after *Mercurialis*) but it does not begin to become abundant until *Mercurialis* cover falls below about 30%. Often here this is where soil aeration is reduced in small hollows or on flatter plateaus where ground water cannot percolate away so readily through the usually heavy clays. A common pattern is thus for there to be small patches of dominant *Hyacinthoides* within a ground of dominant *Mercurialis* or, more strikingly, along gradients of increasing soil moisture, a belt of dominant *Mercurialis* giving way to a zone of dominant *Hyacinthoides*, a feature shown very well in Hayley Wood in Cambridgeshire (Abeywickrama 1949, Rackham 1975).

A second kind of variation is seen where *Mercurialis* declines in response to the soil compaction that results from coppicing. This automatically opens up the ground for colonisation by existing shade-tolerant species which can recover more rapidly. Among the most obvious of these here are *Glechoma* and *Lamiastrum*, both of which can spread quickly and extensively in recently-cut panels by putting out creeping shoots from surviving, untrampled plants, and, with *Glechoma*, from seed. *Hyacinthoides* seems to show a more complex reaction (e.g. Salisbury 1924) but, with a depression in *Mercurialis* cover, it can become the temporary dominant in this sub-community in the early years after cutting. Often, though, the most striking visual change in the vegetation is the great increase in flowering in the second, sometimes the third, spring with a subsequent decline again thereafter as the canopy closes once more. Both *Glechoma* and *Lamiastrum* can show this effect but it is the glorious show of *Hyacinthoides*, *Viola* spp. and *Primula vulgaris* that is the most renowned. On wetter soils occupied by this sub-community, which become very badly poached during coppicing, there is an increasing tendency for the post-coppice flora to show a more dramatic shift towards the *Deschampsia* sub-community but, even in drier conditions, there may be a marked and quite unpredictable variety among the adventive species which gain a temporary hold. Plants such as *Arctium minus* agg., *Epilobium angustifolium*, *Heracleum sphondylium* and *Anthriscus sylvestris* may become prominent with scattered tussocks of coarse grasses like *Dactylis glomerata* and *Holcus lanatus* or, where there is some slight surface depression of pH, *Silene dioica*, *Euphorbia amygdaloides* and *Rubus idaeus* can, with *Hyacinthoides*, *Rubus* and *Lonicera*, give a temporary feel of a *Quercus-Pteridium-Rubus* woodland. Each of these possible var-

iations may take some time to play itself out before the typical *Primula-Glechoma* flora is restored.

Finally, here, *Mercurialis* can show some depression in relation to *Hyacinthoides* where there is a reduction in the base status of the soils, as on less calcareous clays or where there is some eluviation of upper horizons with increased rainfall. Then, there tends to be a thinning of the more calcicolous element in this woodland leaving *Hyacinthoides*, *Rubus* and *Lonicera* prominent over a ground of clay-soil species like *Glechoma*, *Poa trivialis*, *Primula vulgaris*, *Viola* spp. (especially here *V. riviniana*) and *Ajuga reptans*, sometimes with a little *Oxalis acetosella* and *Dryopteris dilatata* and frequently with some *Atrichum undulatum* and *Mnium hornum*. These last four species are very characteristic of the upland *Fraxinus-Sorbus-Mercurialis* woodland but, in the south-east, this kind of *Primula-Glechoma* vegetation represents a clear transition to *Quercus-Pteridium-Rubus* woodland. It is especially characteristic of less base-rich clays in the wetter Weald where it often occurs under canopies with much *Fraxinus* and *Q. robur* with a little more *Betula* (almost always *B. pendula*) than is usual for this sub-community but it can be found, too, in very fragmentary form wherever there is local base-enrichment within tracts of *Quercus-Pteridium-Rubus* woodland, as on stream-side terraces, a feature noted by Salisbury (1916, 1918a) in his account of hornbeam-dominated stands of these kinds of woodland.

***Anemone nemorosa* sub-community:** Damp oakwood association Moss *et al.* 1910; *Quercetum roburis* Tansley 1911 *p.p.*; *Fraxinetum excelsioris*, moist sub-association Tansley 1911 *p.p.*; *Spiraea* society Adamson 1912 *p.p.*; *Quercus robur-Carpinus* woods, *Anemone* and *Ficaria* societies Salisbury 1916 *p.p.*; *Quercus sessiliflora-Carpinus* woods, *Anemone* and *Ficaria* societies Salisbury 1918a *p.p.*; *Quercetum roboris*, *Anemone* Society Tansley 1939 *p.p.*; Hayley Wood zones 2, 3 & 4 Abeywickrama 1949; *Querco-Fraxinetum filipendulietosum* Klötzli 1970 *p.p.*; Ash-maple-hazel woods Rackham 1980 *p.p.*; Lime-ash woods Rackham 1980 *p.p.*; Maple-hornbeam woods Rackham 1980 *p.p.*; Ash-wych elm stand type 1Bb Peterken 1981 *p.p.*; Ash-maple stand types 2A & 2B Peterken 1981 *p.p.*; Ash-lime stand type 4Ba Peterken 1981 *p.p.*; Hornbeam stand types 9Ab & 9Bb Peterken 1981 *p.p.*; *Querco-Ulmetum glabrae*, Typical subassociation Birse & Robertson 1976 *emend.* Birse 1984 *p.p.* In the south-east, the tree and shrub cover here often shows the same general composition as that of the *Primula-Glechoma* sub-community and it can encompass a similar diverse range of structures in relation to coppicing and its abandonment. Indeed, where a uniform sylvicultural treatment has been applied across boundaries between the two sub-communities, the difference between the vegetation types can lie entirely in the field layer.

Overall, however, some minor differences among the woody species can be seen with certain trees and shrubs being better represented here. The most obvious involves *Carpinus betulus* which, among its occurrences in more calcicolous woods, shows a marked preference for the spring-wet soils of the *Anemone* sub-community, stands of which often form the basis of hornbeam coppice. *Populus tremula* has a similar distribution. This may have something to do with exacting moisture requirements for seedling establishment because the soils here can provide the several weeks of surface moisture from May onwards that seem to be essential for successful germination and early growth (Watt in Rackham 1980), though many clones are single-sexed and the seed therefore seldom fertile, so, once established, the tree may often behave as a vegetatively persistent relic. It is not so prominent a tree as *Carpinus*, occurring, even here, as usually small patches of suckering individuals which can be shaded out of overgrown coppice or beneath invading elm (Rackham 1975). Although its bark is very palatable, its leaves are reputed not to be and this may give it some advantage where there is heavy browsing by stock or deer (Rackham 1975, 1980). *Alnus glutinosa*, generally a rare tree in *Fraxinus-Acer-Mercurialis* woodland, may also be encountered occasionally here, as isolated trees on plateaus or at the beginning of zonations to wetter woodland around flushes. Among the shrubs, *Sambucus nigra* is a little more frequent in this sub-community, though it never rivals *Corylus* and the hawthorns, again the major understorey components here.

In general, though, the distinctive features of this vegetation are in the field layer. Many of the community associates, like *Circaea lutetiana*, *Geum urbanum*, *Viola* spp., *Arum maculatum* and *Lamiastrum galeobdolon*, and typical south-eastern species, such as *Poa trivialis*, *Glechoma hederacea* and *Primula vulgaris* (or *P. elatior*), remain frequent here. But, in this sub-community, the usual dominance of *Mercurialis perennis* and, to a lesser extent, *Hyacinthoides non-scripta*, very commonly gives way to an abundance of *Anemone nemorosa* and *Ranunculus ficaria*. It should always be remembered, however, that the prominence of these two species is strictly a vernal event and, after June, when their last leaves have disappeared, this vegetation can be qualitatively very similar indeed to that of the *Primula-Glechoma* sub-community. *Anemone* is invariably present here, often at covers in excess of 10% and sometimes much greater; even where its abundance is low, it can be the most prominent spring plant in the often rather open field layer. *Ranunculus* is somewhat less common and generally less abundant, though it can be patchily prominent.

It also puts up its leaves, from late January, and begins to flower, from mid-February, a month or so before *Anemone*, so it may strike the early visitor as the more obvious. When both species are in bloom in April, they give this vegetation a delightful and unmistakeable aspect.

Other preferentials are few and weak. *Rumex sanguineus* occurs occasionally and there are sometimes scattered plants of *Carex pendula*, *C. remota*, *C. strigosa* and even *C. acutiformis* and these sedges may look prominent by virtue of their bulky habit.

As in the *Primula-Glechoma* sub-community, there is considerable quantitative variation here among the possible dominants. *Mercurialis* is less common than on better-aerated soils: it very often has less than 30% cover and frequently it is absent. When it is more abundant, however, *Hyacinthoides* and, more especially, *Anemone* fare badly because of the dense and early shade that it casts. As *Mercurialis* cover declines on the more anaerobic soils, *Hyacinthoides* tends to assume dominance: it is the more common of the two species overall and usually the more abundant here. Although *Anemone* becomes a little more prominent as *Hyacinthoides* takes over from *Mercurialis*, it is not until *Hyacinthoides* too has declined on the very spring-wet soils, that it really attains its greatest abundance: although *Hyacinthoides* flowers a good month after *Anemone*, it puts up its leaves a little earlier and they quickly bend over above, casting a shade (e.g. Pigott 1982). *Ranunculus ficaria*, though it is of patchier occurrence than *Anemone*, shows the same general response.

These kinds of interaction are especially well seen where there is a marked and even gradient of spring-waterlogging of the soils, as for example in moving up a gentle slope on to a plateau (as in Hayley Wood: Abeywickrama 1949, Martin 1968, Rackham 1975) where a single stand of this sub-community can show dominance by, first *Mercurialis*, then *Hyacinthoides*, then *Anemone*, with transitions between. And it is this kind of zonation which forms the framework for the increasing prominence of *Primula elatior* on spring-wet soils in the two parts of East Anglia to which it is confined. In these areas, oxlip shows a substantial geographical replacement of *P. vulgaris* in calcicolous woodlands (though not in less base-rich or more open habitats) (Christy 1897, 1922, 1924, Rackham 1975, 1980). It can occur in the *Primula-Glechoma* sub-community but it is sensitive to competition for light and rarely abundant where, as is usually the case there, higher covers of *Mercurialis* are maintained (Abeywickrama 1949, Martin 1968). Here, though, it follows *Anemone* in its behaviour, increasing, as *Mercurialis* and *Hyacinthoides* decline, to the spectacular abundance so characteristic of our wetter oxlip woods. *P. elatior* can set viable seed in Britain and seedlings can reach maturity after only two years (Valentine 1947) but, unlike *P. vulgaris*, it spreads only very slowly from existing colonies, expanding mostly by vegetative extension of its clones (Rackham 1975): like *Mercurialis* and *Anemone* and, to a lesser extent, *Hyacinthoides*, it is thus a good marker of longer-established stands. It hybridises freely with *P. vulgaris* where the two species occur together but there does not seem to be any evidence to support Christy's assertion (Christy 1922, repeated in Tansley 1939) that *P. elatior* is being hybridised out of existence (Valentine 1948, 1951, Woodell 1969). Oxlip certainly has shown a marked decline in Hayley Wood but Rackham (1975) adduces firm evidence that this is due to grazing by deer and small mammals. It also often appears less prominent because the decline of coppicing means that grand displays of flowering oxlips are rare.

Where coppicing of stands of this sub-community still occurs, as in Hayley (Rackham 1975) and the Norfolk wood described by Ash & Barkham (1976), the general response among the herbaceous plants is as described for the *Primula-Glechoma* sub-community with a decrease in such *Mercurialis* as is present, a vegetative expansion of certain existing shade-tolerant perennials and a great increase in flowering, especially noticeable here in *Anemone*, *Viola* spp., *Primula vulgaris* and within its range, *P. elatior*. There is also the same variety in the invasion of adventives, first annuals and biennials, then perennials. Here, though, on these generally wetter soils, there is a much more common tendency for the field layer to shift temporarily to the *Deschampsia* sub-community with its prominent Holco-Juncion and Filipendulion elements. This kind of post-coppice flora is described fully below.

The *Anemone* sub-community is a widespread kind of *Fraxinus-Acer-Mercurialis* woodland on the often heavy clay soils of the south-east but an essentially similar field layer occurs locally on wetter sites towards the north-west. Here *Anemone* and *Ranunculus* become prominent on more base-rich soils under somewhat different canopies. *Fraxinus* and *Corylus* and, to a lesser extent, *Acer campestre*, remain very frequent but *Quercus petraea* and hybrid oaks begin to replace *Q. robur* and *Ulmus glabra* and *Acer pseudoplatanus* become much more frequent than in the south-east. These woodlands sometimes show signs of past coppicing but often they have a high-forest structure and, though *Corylus* is very common and sometimes abundant, it usually forms, with hawthorn (always *C. monogyna* here), a rather open understorey.

As in the south-east, the field layer is dominated by various proportions of *Mercurialis*, *Hyacinthoides*, *Anemone* and *Ranunculus* with zonations often here running downslope from drier ground to wetter stream sides. Species such as *Poa trivialis*, *Glechoma hederacea* and *Primula vulgaris* are reduced somewhat in frequency and

others more characteristic of north-western *Fraxinus-Acer-Mercurialis* woods, like *Hedera helix*, *Urtica dioica* and *Galium aparine*, are a little more common. However, other north-western preferentials, like *Brachypodium sylvaticum* and *Geranium robertianum* and the distinctive bryophytes of the region, remain sparse and it seems better, on balance, to retain these stands with those of the south-east.

Deschampsia cespitosa **sub-community:** *Quercetum roburis* Tansley 1911 *p.p.*; *Spiraea-Deschampsia* society Adamson 1912; Oak-hornbeam woods, marginal and path floras Salisbury 1916, 1918; *Quercetum roboris, Filipendula ulmaria* society Tansley 1939; Hayley Wood variants C & D Rackham 1975; Foxley Wood coppice 0–16 years Ash & Barkham 1976; Ash-maple-hazel woodlands Rackham 1980 *p.p.*; Maple-hornbeam woods Rackham 1980 *p.p.*; Hatfield Forest oakwoods Rackham 1980 *p.p.*; Ash-wych elm stand type 1Ba Peterken 1981 *p.p.*; Ash-maple stand types 2A & 2B Peterken 1981 *p.p.*; Hornbeam stand type 9Ab Peterken 1981 *p.p.*; Woodland plot type 12 Bunce 1982. The *Deschampsia* sub-community does not show any major or consistent peculiarities in the species composition of its woody component and it can be found in the south-east under canopies of the same kind and variety as those typical of the *Primula-Glechoma* and *Anemone* sub-communities. Very often though, it is associated with a more open cover of trees and shrubs than is usual among *Fraxinus-Acer-Mercurialis* woodlands. In some cases, such openness is temporary, as where this vegetation occurs as a post-coppice flora, its herbs comprising a fairly short-lived replacement for other field layers (most frequently of the *Anemone* sub-community) in the early and middle years after cutting of the underwood. Then, the woody cover can show any of the numerous floristic permutations characteristic of south-eastern *Fraxinus-Acer-Mercurialis* woodlands but with the distinctive structure of young coppice regrowth with or without standards: a low profusion of shoots springing from the cut stools, fresh growth from suckering trees and shrubs and young maiden poles left uncut or newly developed from seedlings. In the samples available from this kind of woodland, *Quercus robur* is the commonest standard with occasional *Fraxinus* and scarce *Acer campestre* and the underwood is predominantly *Corylus* and *Acer* with some hawthorn (both *Crataegus monogyna* and *C. laevigata*) and a little *Salix caprea*. Recently-cut *Tilia cordata* and *Carpinus betulus* were not sampled but it is very likely that young regrowth of these species can also occur with a *Deschampsia* field layer.

Quite often in this kind of situation, the temporarily open conditions allow quick-growing woody occasionals of the community to gain a local prominence.

Suckering species such as *Populus tremula* and *Prunus spinosa* may become rapidly abundant on the moist soils here or, where the substrate is more calcareous, there may be a dense scrubby growth of *Cornus sanguinea, Viburnum opulus* and *Ligustrum vulgare*. Then, beneath the young spring, there often develops a thick under-scrub of *Rubus fruticosus* agg. (often *R. vestitus* on more calcareous clays), sometimes with *R. caesius* or *R. idaeus*, and much scrambling *Lonicera periclymenum*. Probably of greater long-term significance to the composition of the expanding canopy is the frequent abundance in more open areas of young *Fraxinus* saplings which, provided the canopy does not close over too quickly, can get away as maiden poles to augment the tree cover in future years. More locally, invasive elms can sucker into this vegetation. Often, however, there is a gradual re-establishment of something like the original underwood with the extinction of some smaller and more light-demanding woody species.

Very similar mixtures of trees and shrubs are found where the *Deschampsia* sub-community occurs as a more permanent vegetation cover around clearings and along rides. Here, the sub-community can form a fringe to intact stretches of *Fraxinus-Acer-Mercurialis* woodland, partly overhung by the marginal standards and underwood, and with an often untidy edge of *Prunus spinosa* and *Cornus sanguinea*, banks of trailing *Rubus* and climbing *Lonicera, Clematis vitalba* and *Tamus communis*.

As with the *Anemone* sub-community, the heavy, moist and often trampled soils characteristic of this vegetation can show a reduced cover of *Mercurialis* and, to a lesser extent, of *Hyacinthoides*. But, here, the most obvious feature of the field layer is the constancy, and often the great abundance, of *Deschampsia cespitosa*. In fragmentary stands, as in ill-drained hollows or along the margins of paths within tracts of the *Primula-Glechoma* or *Anemone* sub-communities, it occurs as scattered individuals, often not very vigorous and non-flowering. In the more open conditions of young coppice, clearings and rides, however, it can be truly a dominant in extensive stretches of bulky, floriferous tussocks. In coppiced woodland, such prominence is a cyclical event with the *Deschampsia* waxing and waning with the cutting rotation. Where stands of *Fraxinus-Acer-Mercurialis* woodland on moister ground are coppiced, the combination of heavy trampling and suddenly increased light seem to create ideal conditions for what can be an explosive spread of *Deschampsia* over the cleared ground (Davy & Taylor 1974*a*, Rackham 1975, 1980, Ash & Barkham 1976, Davy 1980). This expansion can overwhelm the slower increase in species such as *Poa trivialis, Glechoma hederacea* and *Lamiastrum galeobdolon* and mask the typical flush of flowers in

plants like *Primula vulgaris, P. veris, Anemone nemorosa* and *Ranunculus ficaria*. However, with the gradual closure of the canopy, the dominance of *Deschampsia* fades so that, when it is virtually extinguished, the original field layer is restored. Such a process can take some considerable time (more than 16 years in the woodland studied by Ash & Barkham 1976) and, where short-rotation cropping is maintained on wetter ground, the *Deschampsia* sub-community may become a permanent feature, though the ground cover is often attenuated with the shade. Around continuously-maintained clearings and along rides, *Deschampsia* retains its vigour so that, of the usual field-layer associates of the community, only taller species with stem leaves, like *Circaea lutetiana* and *Geum urbanum*, persist with their usual frequency. And where clearings are replanted, the *Deschampsia* can seriously compete with the young trees (Davy 1980).

Superimposed on to this general pattern, there is, in clearings and rides and especially in young coppice with its features of disturbance and change, a further level of floristic variation which makes the vegetation of the *Deschampsia* sub-community among the most diverse of woodland floras. Overall, the shift among these additional species is towards the Holco-Juncion and Filipendulion communities with a further weedy element, but individual stands often show marked peculiarities because of the varied interactions between the colonisers and the different situations available for their establishment and spread. The weeds are the more ephemeral, varied and patchily-represented component. They can show an early prominence in the developing ground vegetation of young coppice, first annuals, then biennials, though the smaller species tend to be quickly overwhelmed there and they survive in more permanent stands of this sub-community only where disturbance along paths and in ruts continues to provide small areas of open ground. Among such plants, species of *Rumex, Cirsium, Epilobium* and *Hypericum* are often conspicuous. Some of these are able to survive a little longer as the vegetation thickens up: *Rumex sanguineus* is one and, often forming persistent clumps, *Hypericum hirsutum, H. perforatum, Epilobium hirsutum* and *E. angustifolium*, this last often marking the sites of bonfires of coppiced brushwood. *Scrophularia nodosa* is also a scarce but quite characteristic plant in this kind of vegetation. But the most striking of the longer-lived companions of *Deschampsia* here are *Juncus effusus* and, less commonly, *J. conglomeratus*. These probably germinate from buried seed (e.g. Chippindale & Milton 1934, Milton 1936, Salisbury 1964) and they can be locally very abundant, sometimes rivalling *Deschampsia* itself in their cover. It is these bulky perennials, with occasional scattered plants of *Filipendula ulmaria, Geum rivale, Angelica sylvestris, Lysimachia nemorum, L. num-mularia* and clumps (now flowering in the open conditions) of *Carex pendula, C. remota, C. sylvatica* and *C. strigosa*, that give the vegetation of the middle years of the coppice rotation and of damper rides its distinctive stamp. Very wet areas, such as rain-filled ruts and hollows, can even show an abundance of such species as *Galium palustre* or *Callitriche stagnalis*. Varied mixtures of these plants and many other occasional adventives provide a great enrichment of the flora in woods which are still regularly coppiced (e.g. Rackham 1975, Ash & Barkham 1976).

As with the *Anemone* sub-community, stands of *Fraxinus-Acer-Mercurialis* woodland with field layers rich in *Deschampsia* can be found locally towards the northwest though, as usual in this part of the country, *Ulmus glabra* and *Acer pseudoplatanus* frequently join *Fraxinus* in the canopy and such oak as is present tends to be *Q. petraea*. Coppicing is less widespread here and the sub-community usually occurs as permanent vegetation over drift-derived soils and on damp north-facing slopes. Then, it can closely resemble the *Deschampsia* sub-community of the *Alnus-Fraxinus-Lysimachia* woodland (which it can adjoin around slope-flushes) but the increased representation of *Alnus glutinosa* there and the absence of many of the more calcicolous herbs of this community will usually serve to distinguish the two vegetation types.

***Hedera helix* sub-community:** *Quercetum roburis* Tansley 1911 *p.p.*; Broadbalk Wilderness oak-hazel wood Brenchley & Adam 1915; *Quercetum roboris* Tansley 1939 *p.p.*; Hayley Triangle woodland Rackham 1975; Ash-maple-hazel woodlands Rackham 1980 *p.p.*; Ash-maple stand type 2C Peterken 1981 *p.p.*; Hazel-ash stand type 3B Peterken 1981 *p.p.*; Woodland plot types 4 & 6 Bunce 1982. Unlike the *Deschampsia* sub-community, this kind of woodland is almost always found with a closed cover of woody plants: there is very often a continuous canopy of trees and the understorey, too, can be very extensive, so that stands frequently have a rather dense and gloomy appearance. However, in terms of its species composition, the woody component here is less rich and variable than in other south-eastern *Fraxinus-Acer-Mercurialis* woodlands. Quite commonly, various combinations of *Fraxinus, Corylus* and *Acer campestre* dominate and signs of past coppicing are frequent, but neglect is almost universal with the hazel and maple stools much overgrown and the (usually *Quercus robur*) standards often overtopped by emergent maidens of *Fraxinus*. Other stands have a much simpler structure resembling high forest: here *Fraxinus* is the usual dominant (less frequently *Q. robur*) and the understorey more sparse with hawthorn (almost always here

C. monogyna) often rivalling *Corylus*, very little or no *A. campestre* and patchily abundant *Sambucus nigra*. In this kind of woodland, the trees often appear to be more or less even-aged and the canopy can have a striking uniformity, structural features strongly suggestive of fairly recent spontaneous origin or of planting. Indeed, in some cases, as on the Broadbalk Wilderness at Rothamsted (Brenchley & Adam 1915, Tansley 1939) and in the Triangle in Hayley Wood (Rackham 1975), there is well-documented evidence of the development of this sub-community by invasion and young stands progressing from hawthorn scrub can commonly be found in abandoned quarries, marl pits and neglected field corners. Occasionally, this sub-community can be dominated by suckering elms and such woodland may sometimes have been produced by the invasion of previously unwooded ground. Although individual trees of *Tilia cordata* and *Carpinus betulus* can be found in stands, this kind of *Fraxinus-Acer-Mercurialis* woodland is only rarely encountered with canopies in which these species play a major role.

Whatever the particular composition and structure of the woody component, the field layer here is distinctive in the constancy, and often considerable abundance, of *Hedera helix* growing as a ground carpet. Where it is especially extensive, *Mercurialis* and, to a lesser extent, *Hyacinthoides* tend to be sparse and often limited to the margins of sites, adjacent to established populations in existing hedgerows or tracts of older woodland: these species spread only slowly (Rackham 1975, 1980, Pigott 1977, 1984) and their scarcity in some stands of the *Hedera* sub-community can be supportive evidence of a recent origin of the woodland. Through the sub-community as a whole, however, both *Mercurialis* and *Hyacinthoides* are common, often forming a patchy cover over the *Hedera*, and all being overlain by an underscrub of *Rubus fruticosus* agg. What does seem to be important for the spread of *Hedera* is not particularly a recent origin of the woodland but an undisturbed period of canopy closure (perhaps more than 20–50 years), whether this comes about through neglect of coppicing, or uninterrupted invasion of shrubs and trees, or planting.

That, and an influence of climate, because there is no doubt that growth of *Hedera* is favoured by more oceanic conditions (e.g. Matthews 1955); ivy-dominance in the field layer, both here and in the *Fagus* analogue of this community and in its counterpart on less base-rich soils, the *Quercus-Pteridium-Rubus* woodland, is very much a feature of the south-west. In this region, the general similarity of ivy-dominated field layers of each of these woodland types can give rise to some confusion, though there are usually sufficient other floristic features about the vegetation to effect a separation between them. *Fagus*-dominance is usually obvious enough to distinguish *Fagus-Mercurialis* woodlands with much *Hedera*, though younger stands where *Fagus* is only just beginning to take over from *Fraxinus* and *Quercus robur* may present more of a problem. Separating the *Hedera* sub-communities of *Fraxinus-Acer-Mercurialis* woodland and *Quercus-Pteridium-Rubus* woodland can be much more difficult, especially where the canopy consists of planted *Q. robur*, as it can in both these communities, and where there has been insufficient opportunity for *Mercurialis* to invade and give some indication of the affinities of the field layer. But this is a reflection of the very real fact that younger plantations of oak on both less and more base-rich soils tend to have a rather similar herbaceous vegetation.

With the abundance of the carpet of *Hedera* here, many of the typical associates of south-eastern *Fraxinus-Acer-Mercurialis* woodlands show a depressed frequency and abundance. *Geum urbanum* and *Circaea lutetiana*, with their taller leafy stems, hold up quite well throughout but *Poa trivialis*, *Glechoma hederacea*, *Primula vulgaris* and *Ajuga reptans* are generally much scarcer than in the *Primula-Glechoma* and *Anemone* sub-communities and very much restricted to areas where the ivy is not too thick on the ground and where it is not replaced by either *Mercurialis* or *Hyacinthoides* as the field-layer dominant. Bryophytes, too, are much sparser than usual with even the common species like *Eurhynchium praelongum*, *Brachythecium rutabulum* and *Plagiomnium undulatum* reduced in frequency and cover.

Other species of note are few but *Brachypodium sylvaticum* is much more generally frequent in this kind of woodland than in other south-eastern sub-communities and *Sanicula europaea*, though not common, can be locally very prominent. Both these species are animal-dispersed: *Brachypodium* has rough-awned fruits and those of *Sanicula* are densely clothed with hooks, features which aid their quick spread by transport on fur, feather (*Sanicula*) or trousers. Another bird-dispersed plant which can occasionally be found – its fruits are eaten and its seeds defaecated – is *Iris foetidissima*, but this is an Oceanic Southern species (Matthews 1955) and its local abundance here is strongly related to the more equable climate of the south-west. Like *Hedera*, it is a feature, too, of certain kinds of more base-rich *Fagus* woodlands in this region.

Geranium robertianum sub-community: *Fraxinetum excelsioris*, drier sub-association Moss 1911; *Fraxinetum excelsioris calcareum* Tansley & Rankin 1911; *Fraxinetum calcicolum*, dry soil society Tansley 1939; Ash-oak wood Tansley 1939 *p.p.*; *Dryopterido-Fraxinetum phyllitidetosum* Klötzli 1970 *p.p.*; Ash-wych elm

stand types 1Aa & 1D Peterken 1981 *p.p.*; Hazel-ash stand type 3C Peterken 1981 *p.p.*; Ash-lime stand types 4Bb & 4C Peterken 1981 *p.p.*; Woodland plot types 1, 2, 8 & 12 Bunce 1982; *Querco-Ulmetum glabrae*, Typical subassociation Birse & Robertson 1976 *emend.* Birse 1984 *p.p.* In this central type of north-western *Fraxinus-Acer-Mercurialis* woodland, both *Corylus* and, much more obviously in moving towards the limit of its British range, *Acer campestre* play a less prominent role than towards the south-east and these are not usually ash-maple-hazel woods in the sense of Rackham (1980) or Peterken (1981). The continuing importance of *Fraxinus* led early workers to class this vegetation as some kind of '*Fraxinetum*' but, though ash often predominates and is sometimes the only tree, the most obvious distinctive feature of the canopy in this sub-community is the great frequency, alongside *Fraxinus*, of *Acer pseudoplatanus* and *Ulmus glabra*. Occasionally, these species occur in various combinations as a large-coppice underwood, but coppicing is not as universal as among the south-eastern sub-communities and more often they are co-dominants in a high-forest canopy. The other striking difference here is the marked patchiness of oak and, when it does occur, it is usually *Quercus petraea* or hybrids rather than *Q. robur*. As noted earlier, this uneven representation of oak may be the result of extensive felling of these woodlands in the past, because in many areas (like the Yorkshire and Derbyshire Dales), stands of this sub-community have every appearance of being more or less even-aged spontaneous woodland of not very great antiquity (Pigott 1960, 1969, Merton 1970). Some tracts may have been planted.

No other tree here attains anything like the frequency of *Fraxinus*, *Ulmus* or *Acer pseudoplatanus* but a number of additional species occur, as in the south-east, as scattered occasionals. The most widespread of these is birch and, on the generally free-draining soils characteristically occupied by this sub-community, it is almost always *Betula pendula*. In certain situations, its abundance can show a local increase: it can be prominent, for example, in younger stands as a coloniser alongside *Fraxinus* and *Crataegus monogyna* (e.g. Merton 1970) and, in areas of complex geological and edaphic variation, it can pick out somewhat less base-rich substrates, a feature well shown over the Oolite sequence in the dales of the North York Moors where a belt of birch marks out a transition to more acidophilous woodland over calcareous grit interbedded with limestones.

Other species occurring less frequently are *Fagus sylvatica* (a tree beyond its natural limit over much of the range of this sub-community but occasionally seen in stands in south Wales and spreading in from planted trees further north), *Taxus baccata* (locally prominent in some areas, as around Morecambe Bay, and occurring as isolated specimens on cliffs elsewhere), *Prunus avium, Sorbus aria* and, in more southerly stands, *S. torminalis*. *Tilia cordata* and *Carpinus betulus* are sometimes found but usually as isolated trees, those of *Tilia* especially often being of very obvious antiquity: only rarely do these species show that marked local abundance in large-coppice underwood which is so characteristic of their occurrence in the south-east. Invasion by suckering *carpinifolia* or *procera* elms is rare. Species of *Larix* or *Pinus* have sometimes been planted into stands which are used for amenity purposes or have invaded from nearby conifer plantations.

Shrubs and smaller trees occur most commonly here as part of a distinct high-forest understorey and, though this can be quite dense in younger stands or in gaps, in mature woodland of this kind it is characteristically rather open. *Corylus* in particular, though it has sometimes obviously been coppiced, does not in general have that disproportionate abundance so typical of most south-eastern *Fraxinus-Acer-Mercurialis* woodlands. Indeed, it is often rivalled or even exceeded in cover by hawthorn which here is almost always *Crataegus monogyna*. This is especially abundant in younger stands but, even in mature woodland, individual hawthorns may be very conspicuous because they can grow to substantial size and contribute to the tree canopy. The same is true of *Acer campestre*, less frequent here than either *Corylus* or *Crataegus*, and also of *Ilex aquifolium* which, in response to the milder winter climate of this more oceanic region, shows a slight but distinct increase in frequency here compared with the south-eastern subcommunities. Usually, however, all these species occur in a well-defined lower tier beneath the tree canopy.

Two other features of the understorey are noticeable. The first is the high frequency of *Sambucus nigra* which is almost as common throughout as *Corylus*, a striking contrast to the south-east where it is very much an indicator of local disturbance and enrichment. Its abundance is still very variable, though in certain situations it becomes the most prominent shrub. Second, there is often a great abundance of saplings because, though oak regenerates as badly here as in the south-east, *Fraxinus, Ulmus* and *Acer pseudoplatanus* are all very well represented by young trees of all ages, except where shade from the canopy of field-layer plants like *Mercurialis, Urtica dioica, Phyllitis scolopendrium* or *Rubus* is too dense to allow seedlings to get away (e.g. Wardle 1959). Gaps quickly become filled by cores of young ash and sycamore (*Ulmus* seems less speedy a coloniser) and all three species can readily spread out around ungrazed margins of stands.

Finally, among the woody species, there are occasional records for *Salix caprea, S. cinerea, Prunus spinosa*,

Euonymus europaeus, Malus sylvestris and, less commonly, *Populus tremula*.

The field layer here shares with those of the southeastern sub-communities frequent records for *Mercurialis, Geum urbanum, Circaea lutetiana, Rubus fruticosus* agg. and, occurring less commonly, *Arum maculatum* and *Dryopteris filix-mas*. *Hyacinthoides* remains quite frequent but it is noticeably less common on the generally free-draining soils here and only rarely assumes dominance. Much more obviously reduced in frequency is that group of species so characteristic of the heavy clay soils of south-eastern *Fraxinus-Acer-Mercurialis* woodlands: *Poa trivialis, Glechoma hederacea, Primula vulgaris, Ajuga reptans, Viola* spp. and, to a lesser extent, *Carex sylvatica* and *Lamiastrum galeobdolon*. In the northwest, all these species tend to be restricted to locally-occurring stands of the *Anemone* and *Deschampsia* subcommunities or to flush surrounds carrying the *Alnus-Fraxinus-Lysimachia* woodland.

By contrast, there is a marked increase in the representation of species which, in the south-east, occur only locally, if at all. The commonest of these are *Urtica dioica* and *Galium aparine*, the high frequency of which, as with *Sambucus*, perhaps reflects the more rapid nutrient turnover that occurs in these better-aerated soils. *Geranium robertianum* and *Brachypodium sylvaticum*, two other readily-dispersed species, are also of widespread occurrence here. Then, with the milder climatic conditions, there is often an abundance of ground-growing *Hedera helix*. This feature can be seen too in the *Hedera* sub-community of the south-west but, here, *Hedera* is often accompanied by the oceanic fern *Phyllitis scolopendrium*, a very good marker of all the north-western kinds of *Fraxinus-Acer-Mercurialis* woodland. Although the early spring emergence of *Mercurialis* still gives this vegetation a distinct vernal aspect, it is often the winter-green character of the field layer that strikes the regular visitor most, with the carpet of ivy and the prominent sprays of *Phyllitis* fronds which persist until March or later, until the new growth emerges (Page 1982). Other ferns which can contribute to the lush appearance of the field layer, especially where ravines or large crevices further increase the shelter, are *Polystichum setiferum* (often strikingly gregarious), *P. aculeatum* (more local and typically solitary) and (much rarer but occurring under miniaturised versions of this woodland in Carboniferous Limestone grikes in the Yorkshire Dales), *P. lonchitis* and *Dryopteris villarii* (Gilbert 1970). Smaller crevices can provide a niche for *Asplenium trichomanes, A. viride, A. ruta-muraria* and *Cystopteris fragilis*. Note, however, that a high frequency of ferns like *Dryopteris filix-mas, D. dilatata* and *Athyrium filix-femina* is not characteristic here: this is one feature which helps separate the north-western

Fraxinus-Acer-Mercurialis woodlands from their upland counterpart, the *Fraxinus-Sorbus-Mercurialis* woodland.

The luxuriant quality of the vegetation is also frequently enhanced by an extensive and diverse bryophyte cover: bryophytes, mostly fairly bulky mosses, are more varied and abundant here than in the south-east, sometimes forming a virtually continuous vivid green carpet over the soil or rocky substrate. As well as the community species *Eurhynchium praelongum*, *Brachythecium rutabulum* and *Plagiomnium undulatum*, *Eurhynchium striatum* and *Thamnobryum alopecurum* become preferentially frequent in the north-west and there are occasional records, too, for *Ctenidium molluscum*, *Eurhynchium swartzii*, *Fissidens taxifolius*, *F. bryoides*, *Mnium hornum*, *Atrichum undulatum*, *Thuidium tamariscinum*, *Cirriphyllum piliferum*, *Rhytidiadelphus triquetrus*, *Plagiomnium rostratum* and *Rhizomnium punctatum*. In especially sheltered spots like ravines or rock clefts, even greater enrichment of the bryophyte flora can be found (e.g. Proctor 1960).

As in the south-eastern *Fraxinus-Acer-Mercurialis* woodlands there is considerable quantitative variation in the representation of these characteristic species. Once again, this is often most obviously seen in terms of the abundance of *Mercurialis*, though the resulting vegetation patterns are expressed in rather a different way over the commonly sharp valley-side topography typically occupied by this sub-community. Overall, *Mercurialis* is the commonest dominant here and it can maintain a high cover even over rocky and very sharply draining scree slopes from which *Hyacinthoides* is totally excluded and where such species as *Geum urbanum* and *Circaea lutetiana* are diminished not by direct competition with the *Mercurialis* but by the dry conditions. In this kind of woodland, *Brachypodium sylvaticum* is often the most frequent companion of *Mercurialis*; here and there can be scattered tussocks of *Arrhenatherum elatius*, especially where the cover of shrubs and trees thins out somewhat. Where *Mercurialis* maintains its cover in less extreme conditions, many of the herbaceous associates are diminished by the dense shade that it casts and seedlings and small saplings of trees become much less common. As in the south-east, the richest and most varied field layers tend to be associated with situations where, for one reason or another, *Mercurialis* itself is reduced in abundance. Most often, here, this occurs where the soil becomes naturally moister, rather than badly aerated because of the trampling that accompanies coppicing. Such conditions can be found where this sub-community extends on to less permeable calcareous bedrocks like shales or clays or occurs over heavy-textured superficials, or where there is flushing around springs, in slope-foot colluvium or on alluvial flats.

There, reduction in *Mercurialis* and an increase in other species may mark the beginning of a transition to the *Anemone* or *Deschampsia* sub-communities over ill-drained ground, or to the *Allium* sub-community over very moist but free-draining soils.

However, there is a further quite common kind of quantitative variation within this sub-community that seems to be related to the trophic state of the soils and which is marked most obviously by differences in the abundance of *Urtica dioica* and, to a lesser extent, *Galium aparine*. These species occur throughout this kind of vegetation, though usually at low cover, and Pigott & Taylor (1964) have shown that, though *Urtica* readily germinates on a wide range of soils occupied by this sub-community, it only makes good subsequent growth where supplies of all minerals, including phosphate, are high. Such conditions can be found here, as in the south-east, where there is local disturbance, dumping of waste or burning but they are quite often met in the habitat of this kind of woodland on the deep and largely stone-free colluvium which accumulates at the bottom of slopes. In such situations, dominance in the field layer frequently passes to *Urtica* with a consequent reduction in the abundance of *Mercurialis*, a feature well shown in Pigott & Taylor's (1964) transects from the Derbyshire Dales. *Geranium robertianum* tends to follow *Urtica* and *Galium*, perhaps because it, too, is more nutrient-demanding: on rocky soils, it often shows a marked reddening that betokens deficiency of phosphate or nitrogen. And, among the shrubs, *Sambucus nigra* often increases its cover in this kind of vegetation. However, other species in the field layer may become more abundant here because colluvial soils are also generally moister than those on the valley sides above.

***Allium ursinum* sub-community:** *Fraxinetum excelsioris*, moist sub-association Moss 1911 *p.p.*; *Fraxinetum calcicolum*, *Allium ursinum* society Tansley 1939; Ash-oak wood, *Allium ursinum* society Tansley 1939; Ashwych elm stand types 1Aa & 1Ab Peterken 1981; *Querco-Ulmetum glabrae*, *Allium ursinum* subassociation Birse & Robertson 1976 *emend.* Birse 1984 *p.p.* This sub-community maintains all the general features of the woody component of typical north-western *Fraxinus-Acer-Mercurialis* woodlands: a mixed canopy of *Fraxinus*, *Ulmus* and *Acer pseudoplatanus*, with or without oak, and an understorey with *Corylus*, *Crataegus monogyna*, *Sambucus nigra*, *Acer campestre* and abundant saplings of the leading trees. On the deeper, moister soils which this kind of woodland occupies, growth of these woody species is often vigorous and majestic trees here can contrast with poorer specimens in stands of the *Geranium* sub-community over more rocky substrates. Where these two sub-communities occur contiguously, there can be some sorting of trees between the two kinds of woodland with *Ulmus*, *Acer pseudoplatanus* and, if locally represented, oak being much more obvious and well grown in the *Allium* sub-community and *Fraxinus* leading in the *Geranium* sub-community on the drier slopes above. This kind of contrast was noted in early accounts of the Mendip woodlands (e.g. Moss 1907, Hope-Simpson & Willis 1955).

However, the really distinctive feature of this sub-community is the prominence of *Allium ursinum* in the field layer. This species can be found occasionally in pockets of moister soil within stands of the *Geranium* sub-community but here it is preferentially frequent and commonly the vernal dominant, often carpeting the ground from March to July with an even cover of its shiny leaves and providing a spectacular sight when flowering between April and June. *Mercurialis*, though, remains very frequent and *Hyacinthoides*, though less common, can be patchily abundant. On occasion, either or both of these can share dominance with *Allium* in the spring and, on these moister soils, there may also be occasional scattered plants of *Anemone nemorosa* or *Ranunculus ficaria*. When the leaves of *Allium* have died down in mid-summer, the more persistent shoots of *Mercurialis* can become for some time the most obvious feature of the field layer. Indeed, at this time of year, there may be little to separate this vegetation from more luxuriant stands of the *Geranium* sub-community: there is often a patchy carpet of *Hedera* and scattered individuals of *Urtica*, *Galium aparine*, *Geranium robertianum*, *Phyllitis scolopendrium*, *Geum urbanum*, *Circaea lutetiana* and *Arum maculatum* are common. In other stands, the flaccid, pale green remains of the *Allium* foliage finally disappear to leave only the fruiting stems and a remnant smell of garlic hanging over an expanse of bare ground or an extensive cover of bryophytes on the exposed soil.

***Teucrium scorodonia* sub-community:** *Fraxinetum excelsioris*, dry stony sub-association   Moss 1911 *p.p.*; *Fraxinetum calcicolum* Tansley 1939 *p.p.*; Uneven-aged limestone woodland   Pigott 1969; Ash-wych elm stand type 1D Peterken 1981 *p.p.*; Hazel-ash stand type 3C Peterken 1981 *p.p.*; Ash-lime stand type 4C Peterken 1981 *p.p.* This highly distinctive kind of woodland shares many floristic features with the *Geranium* and *Allium* sub-communities but it is richer and more varied in its composition and has a structural complexity to its woody component that is lacking from many north-western *Fraxinus-Acer-Mercurialis* woodlands. *Fraxinus*, *Ulmus* and *Acer pseudoplatanus* all remain frequent here, though the last two, and particularly *Acer*, play a less prominent role in the canopy than is usual. Furthermore, although *Fraxinus* is frequently

the most abundant tree, it is often accompanied here by a variety of other species that are generally of restricted occurrence in the region. First, oak (both *Quercus petraea* and *Q. robur* have been recorded) is a little commoner than is the rule; more accurately, in areas where oak is generally absent from *Fraxinus-Acer-Mercurialis* woodlands, such trees as do survive are more often in stands of this sub-community, rather than in (even adjacent) tracts of the *Geranium* or *Allium* sub-communities. Then, there may be some *Taxus baccata*, occurring as scattered trees or, in some stands, as in Matlock Dale in Derbyshire (Pigott 1969), forming a distinct lower tier to the canopy. *Fagus sylvatica* can occur too, though its representation differs according to the locality. To the north, as in the Derbyshire Dales, it is typically very rare, even though it has been widely planted throughout the region and often occurs close by; in the Wye valley, by contrast, this kind of woodland occurs within the supposed natural limit of *Fagus* and occasional trees, sometimes of grand size, can be found. Finally among these larger species, both our native limes, *Tilia cordata* and the much rarer *T. platyphyllos* can occur, together with apparently natural hybrids between them; indeed, in Derbyshire, trees of intermediate character are more common than the parents (Pigott 1969). Although trees of different ages occur in this locality, some individuals are clearly very old, with massive stools rooted into crevices or maintaining a hold, as bent and spreading specimens, on talus which has shifted beneath and around them (Pigott 1969, Merton 1970). *T. cordata*, as noted earlier, is not limited in Britain to *Fraxinus-Acer-Mercurialis* woodland, but *T. platyphyllos* almost always is and, within this community, is very characteristic of these rocky stands of the *Teucrium* sub-community. In some localities, where populations approach pure *cordata* or *platyphyllos*, the somewhat different ecological preferences of the two species can be seen, as in Kingley Wood at Ilam in Derbyshire, where the former occurs over less base-rich soils (Pigott 1969).

In some stands, all these species occur together as co-dominants in a mixed canopy; in others, there is a patchy mosaic with different trees (notably *Fraxinus*, *Taxus* and *Tilia*) showing local prominence here and there. Signs of coppicing can sometimes be seen (e.g. Merton 1970, Ratcliffe 1977) but typically the structure is that of high forest with an uneven and complicated age structure (Pigott 1960, 1969, Merton 1970). Together with the fact that this kind of woodland often occurs over the very uneven topography of talus slopes and rocky bluffs and cliffs, this means that the canopy is characteristically rather open and uneven-topped. Frequently, there is a complete gradation between the leading maidens, younger trees, smaller trees and shrubs and saplings, and the cover can be punctuated or fringed by stretches of more scrubby growth.

Among the smaller woody species, *Corylus*, *Crataegus* (always here *C. monogyna*) and *Acer campestre* are all well represented, though hawthorn does not usually have the prominence that it often shows in the *Geranium* and *Allium* sub-communities. *Sambucus nigra*, too, is noticeably less common and abundant here than in many north-western *Fraxinus-Acer-Mercurialis* woodlands. *Ilex aquifolium* occurs occasionally and there are sometimes scattered individuals of *Prunus spinosa*, *Malus sylvestris*, *Euonymus europaeus*, *Salix caprea* and *S. cinerea*. Again, however, what is most noticeable about the understorey is the additional occurrence of species which, when they do occur in the *Geranium* sub-community, are, like the preferential trees, usually found only as isolated individuals, often clinging to cliffs. Here, by contrast, there are frequent records for *Cornus sanguinea*, *Viburnum opulus*, *Rhamnus catharticus* and *Sorbus aucuparia*, with *Prunus padus* a little less common. All of these can occur scattered in an open understorey or in thicker scrubby patches beneath gaps and around the margins of stands.

A further feature of interest here is the occurrence of various of the rarer Sorbi. *Sorbus torminalis* can sometimes be found as a tall tree, but there are also records for the bushier *S. anglica* and a number of the disjunct polyploid apomicts of the *S. aria* group: *S. rupicola* and, in the Wye valley, *S. porrigentiformis* and *S. eminens* (Pigott 1969, Ratcliffe 1977).

The field layer of this sub-community also has its own peculiarities, though its general relationships to north-western *Fraxinus-Acer-Mercurialis* woodlands are fairly clear. Thus, there are frequent records for *Mercurialis perennis*, *Rubus fruticosus* agg., *Circaea lutetiana*, *Geum urbanum* and some of *Hyacinthoides*, *Urtica*, *Galium aparine*, *Geranium robertianum*, *Hedera helix* and *Phyllitis scolopendrium*, and a general absence of those plants more associated with the heavy clay soils of the southeast. Although substrate heterogeneity can allow species like *Deschampsia cespitosa*, *Filipendula ulmaria*, *Angelica sylvestris*, *Allium ursinum* and *Anemone nemorosa* to flourish locally in pockets of deeper soil or on damper ledges, the overall feel of the vegetation is of a drier kind of *Geranium* sub-community field layer, though with some notable additions. Typically, however, there is no consistent pattern of dominance because one important factor in the increased richness of the herbaceous component here is the patchy shade cast by the characteristically discontinuous canopy. The usual appearance of the field layer is thus of a diverse mosaic of vegetation distributed over less well lit areas interspersed with scrubby glades. Plants like *Brachypodium sylvaticum* and *Arrhenatherum elatius*, which become abundant in

drier open areas in the *Geranium* sub-community, are more common here and often joined by *Teucrium scorodonia*, *Melica uniflora*, the rarer *M. nutans*, *Campanula latifolia* and *Myosostis sylvatica*. Although not a constant feature, a further very distinctive associate in this kind of woodland is *Convallaria majalis*, a plant not confined in Britain to calcicolous woods but one which seems to perform better than *Mercurialis* on drier, more unstable limestone screes and which can be quite abundant here, even in areas of deeper shade (Pigott 1969, Merton 1970). Where the canopy thins out more markedly to a shrub-dominated cover, fragments of vegetation more characteristic of sunny, calcicolous scrub may be encountered with species like *Geranium sanguineum*, *Rosa villosa*, *R. pimpinellifolia* and national rarities such as *Polygonatum odoratum*, *Carex digitata* and *Epipactis atrorubens*. Other plants of restricted distribution which can be found here are *Daphne mezereum* (as well as the more widespread *D. laureola*) and, in deeper shade, *Cardamine impatiens*. Stands of this sub-community also include some of the southern-most stations of *Rubus saxatilis* (like *Melica nutans*, a Northern Montane species) and the oceanic fern *Polystichum aculeatum* is sometimes encountered in moister, shaded crevices.

On the drier, often well illuminated, substrates here, the typical bryophytes of the north-western *Fraxinus-Acer-Mercurialis* woodlands are rather poorly represented, apart from the community species *Eurhynchium praelongum* and *Plagiomnium undulatum*. However, in more open areas there is often a considerable enrichment of the ground cover by species characteristic of rocky, calcicolous grassland and there *Ctenidium molluscum*, *Homalothecium lutescens*, *Thuidium tamariscinum* and *Tortella tortuosa* can all be recorded.

Although stands of the *Geranium* sub-community sometimes approach the richness of this kind of woodland in one or another respect, the peculiar variety of this flora is seen in all its fullness in only a few localities. Stands can be extensive but it is a noticeable feature of this vegetation that regeneration appears very slow. Although the canopy typically comprises uneven-aged trees, younger saplings are very much less obvious than in other north-western *Fraxinus-Acer-Mercurialis* woodlands. For one thing, *Ulmus* and *Acer pseudoplatanus* are less common here and, though young *Fraxinus* often occur, oak regenerates badly and only in exceptionally warm summers does *Tilia cordata* set viable seed in these areas (Pigott 1969, Pigott & Huntley 1981), though *T. platyphyllos* fruits more regularly. *Cornus sanguinea*, too approaching the north-western limit of its range here, and *Prunus padus*, predominantly a northern species in which fruit development is favoured by cool, wet conditions (Jarvis 1960), also rarely reproduce sexually in Derbyshire, though both sucker freely (Pigott 1969). Some of the characteristic field-layer species, like *Convallaria, Melica nutans* and *Rubus saxatilis*, though they fruit, also seem to rely predominantly on vegetative spread (Pigott 1969). For all its variety and luxuriance therefore, this type of woodland appears to be remnant vegetation of quite a fragile kind and to have survived largely because of the intractable character of the topography.

Habitat

The *Fraxinus-Acer-Mercurialis* woodland is typically a community of calcareous mull soils in the relatively warm and dry lowlands of southern Britain. Over much of its range, it probably represents the climax forest type of more base-rich soils, though almost universally its structure and floristics have been affected by sylvicultural treatment. In some cases, stands occupy sites known to have been continuously wooded for very considerable periods of time but the community also includes vegetation of relatively recent spontaneous or planted origin.

This woodland occurs on soils derived from a wide variety of more calcareous parent materials, being most common over sedimentary limestones, shales and clays and superficial deposits like glacial drift. Over such substrates, it is confined to those parts of the country where the annual rainfall is less than 1000 mm (*Climatological Atlas* 1952, Chandler & Gregory 1976) and where there are fewer than 160 wet days yr^{-1} (Ratcliffe 1968). In such areas, these materials weather to produce soils in which leaching is generally limited to superficial depletion of calcium carbonate and mobilisation of clay minerals. Thus, though surface pH of the soils under the community can be reduced to quite low values (down to about 4.5 or even less), base-rich conditions are usually maintained below and there is often much exchangeable calcium in the lower horizons (e.g. Hodge & Seale 1966, Martin & Pigott 1975). Where the influence of the parent material is more dominant, the profiles can be base-rich and calcareous throughout, with surface pH between 6 and 7 or more. However, except in very rocky or heavily-waterlogged situations, these are usually mull soils with an active invertebrate fauna rich in lumbricid worms. Typically, the litter, made up predominantly here of the leaves of plants with softer and more palatable foliage, is incorporated very quickly; even by mid-winter, the ground between the evergreen herbs and bryophytes is often quite bare, apart from the more resistant leaves of oak and twigs. Sometimes, transitional moder conditions are encountered but surface mor is rare.

These general edaphic features are reflected in the vegetation by the strong representation here of woodland plants tolerant of more base-rich conditions. A more strictly calcicolous element is not always the most obvious feature though: its prominence is often muted by the abundance of plants characteristic of mull soils and variation in factors other than the base-status of the soil commonly produces floristic differences which mask its presence. Nonetheless, among British woodlands, this community (together with its analogues dominated by *Fagus* and *Taxus*) represents one extreme of variation in relation to pH. The frequent presence of *Fraxinus* and *Acer campestre* with plants like *Mercurialis, Geum urbanum, Circaea lutetiana, Viola reichenbachiana, Arum maculatum* and *Sanicula europaea* is unique to this kind of woodland and helps define edaphically-related boundaries between the community and other woodland types. Thus, though some of these plants continue northwards over more calcareous substrates which occur beyond the range of the *Fraxinus-Acer-Mercurialis* woodland, the higher rainfall there means that they are accompanied by plants indicative of more marked surface eluviation, such as *Dryopteris dilatata* and *Athyrium filix-femina*. These are species which, generally speaking, are rare here and which help separate this community from its northern analogue, the *Fraxinus-Sorbus-Mercurialis* woodland. Towards the drier south, too, the boundary between the *Fraxinus-Acer-Mercurialis* woodland and its closest relative, the *Quercus-Pteridium-Rubus* woodland, is essentially an edaphic one. Although this community also has frequent records for plants tolerant of a wide range of soil pH (e.g. *Quercus robur, Corylus, Crataegus* spp., *Hyacinthoides* and *Rubus*), more calcicolous plants fade out there with the shift to moder and mor soils of lower pH derived from non-calcareous bedrocks and superficials.

As well as helping define the general limits of the *Fraxinus-Acer-Mercurialis* woodland, edaphic variation plays a major role in influencing floristic differences between and within the two suites of sub-communities here and contributes to their striking geographical association with either the north-west or the south-east. In the latter part of the country, this kind of woodland is relatively rare on more free-draining soils developed over pervious calcareous deposits like the Chalk and Jurassic limestones. Here, *Fagus* and, more locally, *Taxus* are important competitors with *Quercus robur* and, to a lesser extent, *Fraxinus*, and usually come to dominate over rather similar field layers. Much more commonly in the south-east, the *Fraxinus-Acer-Mercurialis* woodland occurs over softer argillaceous rocks, such as the Jurassic, Cretaceous and Tertiary clays which underlie the basins and vales, and fine-textured superficials, like the Chalky Boulder Clay that is very widely distributed over the east Midlands and East Anglian plateaus. The soils derived from such materials typically have a clay or clay-loam texture and their chief characteristic is their general impermeability to excess rain which results in seasonal surface-water gleying (e.g. Hey & Perrin 1960, Hodge & Seale 1966, Martin & Pigott 1975, Curtiss *et al.* 1976, Fordham & Green 1980). Although the rainfall here is relatively low (less than 600 mm yr^{-1} over much of the area), the fine texture of the sub-surface horizons impairs the drainage so that, as winter advances, the upper layers of the soil become very wet, the clays swelling and closing the pores, and, in extreme cases, there can even be a thin patchy cover of standing water on the surface. In spring, the soils dry and, during early summer, shrink and crack above, sometimes to a distinctive coarse blocky structure, and let in air which re-oxidises iron and manganese compounds producing the characteristic red-brown/grey mottling of the B and C horizons (e.g. Hodge & Seale 1966, Martin & Pigott 1975). The degree of impermeability and the extent of gleying vary so that profiles can range from incipiently gleyed typical brown soils right through to stagnogleys (though some of these soils, like the Hanslope Series, which is very widespread in East Anglia, are now classed as pelosols because of the predominance of clay in the profile: Avery 1980, Soil Survey 1983). Although there is very little hard evidence, it is likely that such soils are rather poor in major nutrients (e.g. Martin & Pigott 1975), a condition accentuated in many cases by generations of cropping for underwood and timber, with no fertilisation apart from that derived from litter decay and the occasional bonfire. Both trophic state and pH are probably reduced as the tendency to gleying increases, perhaps because of poorer incorporation of litter by reduced earthworm populations (Martin & Pigott 1975, Rackham 1975).

It is where the *Fraxinus-Acer-Mercurialis* woodland occurs on soils such as these, that it is characterised by the frequent presence of *Quercus robur* and, in the field layer, of such plants as *Primula vulgaris* (or *P. elatior* in parts of East Anglia), *Poa trivialis*, *Glechoma hederacea*, *Ajuga reptans* and *Viola* spp. and by the preferentially common occurrence of *Hyacinthoides*, all of them plants tolerant, to varying degrees, of heavier and moister soils. Also here, more eutrophic species like *Urtica* and *Galium aparine* are generally rather local, occurring mostly where there is some obvious enrichment over old settlements and bonfires, along wood margins affected by fertiliser drift from adjacent farmland or beneath invasive elms with their calcium-rich litter (Martin & Pigott 1975).

Furthermore, among these south-eastern kinds of *Fraxinus-Acer-Mercurialis* woodland, there is every indication that many of the floristic differences between the *Primula-Glechoma* and *Anemone* sub-communities and, to a lesser extent, the *Deschampsia* sub-community, are related to the extent and duration of the soil waterlogging. The *Primula-Glechoma* sub-community is the central type and it is very widely distributed throughout the region on more base-rich brown soils with a less marked tendency to gleying. It even extends a little on to more permeable mull rendzinas over the southern Chalk and on to better-drained brown soils derived from generally less calcareous parent materials like Clay-with-Flints. For the most part, however, it occurs over deposits like the Weald, Oxford and London Clays, provided these do not remain waterlogged well into the spring. Often, this is where a natural slope (even a very slight slope) allows quicker run-off of surface water such that the soils are gleyed only in the lower part of the profile.

Where similarly-structured soils derived from the same parent materials remain wetter longer, as they do in surface hollows and on flatter plateaus, there is an increasing tendency for the *Anemone* sub-community to take over on soils that are gleyed right up to the base of the A horizon. Quite commonly, such patterns can be seen as zonations within the same wood, as at Hayley, where they have been very neatly related, by both observation and field laboratory experiments, to an increasing gradient of waterlogging (Abeywickrama 1949, Martin 1968, Martin & Pigott 1975, Rackham 1975). In fact, the switch between what are defined here as the *Primula-Glechoma* and *Anemone* sub-communities is often a rather gradual one and it seems to be based on the decreasing sensitivity of the major species involved to the ferrous ions which accumulate in the reduced atmosphere of gleyed soils (Martin 1968). Thus, as the most susceptible plants, first *Mercurialis*, then *Hyacinthoides*, lose their vigour over more markedly gleyed soils, others which are more tolerant, like *Anemone*, *Primula vulgaris* and, in East Anglian woods like Hayley, *P. elatior*, are able to assume dominance in striking patterns of zoned vegetation. In some places, the zonation continues to the *Deschampsia* sub-community over soils which are free from waterlogging for only a short period in the summer months. *D. cespitosa* was the species most tolerant of ferrous ions among those examined by Martin (1968), although its prominence in many stands of the *Deschampsia* sub-community is clearly related to the distinctive combination of reduced aeration due to trampling and the increased light that follows clearance or coppicing on heavy soils. In some of these species at least, tolerance of waterlogging appears to be due to an ability to export oxygen from aerenchymatous roots and so produce a protective oxidised rhizosphere (Martin 1968).

The increased soil moisture marked by these changes in the field layer also influences woody species in these sub-communities, and *Populus tremula* and, to a more obvious extent, *Carpinus betulus* both tend to increase over strongly-gleyed profiles. Seedlings of *Fraxinus* are also often very numerous here, though much of this response may be due to the extensive patchwork of more open ground characteristic of the wetter soils. *Alnus* can also become a little more frequent but, though the *Deschampsia* field layer is very similar to that found in some kinds of *Alnus-Fraxinus-Lysimachia* woodland, *Alnus* itself is probably limited here by the fact that, in summer, the soils can bake very hard.

Both the *Anemone* and *Deschampsia* sub-communities can occur to the north-west over pockets of less permeable parent materials like glacial drift or where there is some gleying by ground waters along stream sides and around slope flushes. For the most part, though, more base-rich soils occur in this region over pervious bedrocks like the Carboniferous Limestone (in the Mendips, Wales, Derbyshire, the Pennines and around Morecambe Bay), Jurassic limestones (in the North York Moors), Magnesian Limestone (in Yorkshire and Durham) and the Chalk (mainly in the Yorkshire Wolds). Even with the higher rainfall in these areas (750–1000 mm yr^{-1} with around 140–160 wet days yr^{-1}: *Climatological Atlas* 1952, Ratcliffe 1968), the soils derived from such deposits remain generally calcareous and of high base-status, particularly on the frequent steep slopes which are generally drift-free and often unstable and actively weathering. More importantly, though these soils can be quite moist, especially on north-facing slopes or where there is some accumulation of colluvium receiving downwash, they are, in contrast to those of the south-east, free-draining. The *Fraxinus-Acer-Mercurialis* woodland is thus most often found here over rendzinas and, in extreme cases, scrubby versions of the community can extend on to very fragmentary protorendzinas between talus fragments, in cliff crevices or, over Carboniferous Limestone, in pavement grikes. On such soils, the clay-soil species so characteristic of the south-eastern sub-communities occur very sparsely and *Hyacinthoides* becomes generally much less common. By contrast, plants like *Brachypodium sylvaticum* and *Geranium robertianum*, which readily colonise free-draining substrates, increase in frequency. Quite obvious here, too, is the much more widespread occurrence of *Sambucus nigra*, *Urtica dioica* and *Galium aparine*, perhaps indicating a faster nutrient turnover in these well-aerated calcareous soils. The frequent stoniness of the profiles often means that this more eutrophic character is not fully expressed because the amount of nutrient per unit area is reduced (Pigott & Taylor 1964) but it becomes very apparent in the *Geranium* sub-community where deeper stone-free colluvium has accumulated. Often, too, such slope-foot soils are moister and here the north-western kind of *Fraxinus-Acer-Mercurialis* woodland extends on to mull rendzinas and brown calcareous earths in the striking vegetation of the *Allium* sub-community.

Although much of the floristic character of this community and its various sub-divisions can be related to edaphic interactions between different parent materials and climates across the country, both rainfall and temperature also have some more direct effects on the vegetation. These influences are various, complex and often imperfectly understood but, in general, they tend to confirm the soil-related patterns of variation and the way in which the two suites of sub-communities are associated with distinct regions. First, the distribution of the community as a whole corresponds roughly with the limits of the more Continental influence in the British flora. By and large, the *Fraxinus-Acer-Mercurialis* woodland is restricted to those parts of the country with a mean annual maximum temperature in excess of 26 °C (Conolly & Dahl 1970) and the presence here of a number of species with Continental or Continental Southern distributions in Europe helps separate this community from the more northerly *Fraxinus-Sorbus-Mercurialis* woodland. Thus, plants such as *Acer campestre*, *Carpinus betulus*, *Cornus sanguinea*, *Euonymus europaeus*, *Rhamnus catharticus*, *Sorbus torminalis*, *Tilia cordata*, *T. platyphyllos*, *Viburnum lantana*, *Clematis vitalba*, *Tamus communis*, *Arum maculatum*, *Daphne laureola* and *Lamiastrum galeobdolon* are all of very sporadic occurrence beyond the northern bounds of the *Fraxinus-Acer-Mercurialis* woodland, though some of them, of course, occur widely in other kinds of woodland within their ranges. However, the extreme limits of these species are often of rather complicated configuration. *Acer campestre*, for example, and a number of the other trees and shrubs listed above, tend to have a gap in their distribution in the central Pennines (where this community is common) and to reappear with renewed frequency around Morecambe Bay. Then, some, like *Tilia cordata* and *Sorbus torminalis*, are much more frequent along the western side of the northern fringe of the *Fraxinus-Acer-Mercurialis* woodland than to the east, a slanted limit which runs in reverse direction to that of the community as a whole (Pigott & Huntley 1978). The reasons for these variations are probably varied and not necessarily of climatic origin: *A. campestre* is generally a hedgerow tree towards its north-west limit and many of the *Fraxinus-Acer-Mercurialis* woodlands in the central Pennines are of fairly recent spontaneous origin, so human interference may well have eaten away at more coherent past boundaries. Even where distributions can be generally related to climatic variation, the work of Pigott & Huntley (1978, 1980, 1981) on *T. cordata* has shown just how intricate the control can be. To set fertile seed, this tree requires three to four consecutive days during its flowering period when the mean daily maximum air temperature exceeds 19 °C so as to allow sufficient time for the pollen tubes to extend right to the ovular micropyle. At the present time, such conditions are only regularly met roughly south-east of the Humber–Severn line: in this area, seedlings can often be found and the tree maintains itself as widespread populations, often with great local abundance. Further north, however, where the climate is generally cooler and somewhat less predictable, fertilisation occurs only sporadically in exceptionally warm summers (like those of 1976 and 1984). Also, flowering is two weeks or so later than towards the south (late July to early August as opposed to early July) and the onset of cooler autumn temperatures allows less time for complete development of the seed, even when fertilisation has occurred. Added to this, some at least of the often very old and isolated trees in the north seem to show ovular sterility and self-incompatibility (Pigott & Huntley 1981). Beyond the area in which it is actively reproducing at the present time, *T. cordata* thus seems to persist largely because the great longevity of its individuals plus very occasional sexual regeneration has ensured its survival from the time in the Forest Maximum when conditions were much more widely suitable for the production of viable seed. The rough coincidence of its overall range with that of the community can thus be related to long-past climatic conditions; its preferential frequency in the south-eastern sub-communities reflects the present climatic regime. Other species (like *Cornus sanguinea*, for example, which does not readily set viable seed towards the north: Pigott 1969) may well show this kind of relationship.

A second more direct effect of climate on the floristics of this woodland can be seen in the somewhat more oceanic character of the north-western sub-communities. In fact, the influence of the more equable climate of the western parts of Britain can already be seen towards the south where the frequency of *Hedera helix* as a ground carpet increases greatly in moving from the heartland of the *Primula-Glechoma* sub-community towards the south-west. Here, the *Hedera* sub-community becomes very common, even in longer-established woodlands where the abundance of ivy is not due largely to the uninterrupted development of dense canopy shade as in young scrub and plantations and abandoned coppice. In this kind of *Fraxinus-Acer-Mercurialis* woodland, the south-eastern type of canopy, with frequent *Quercus robur* and little *Ulmus glabra* or *Acer pseudoplatanus*, is maintained, though most of the clay-soil herbs are uncommon: this may be because they are suppressed by the thick ivy cover. Another oceanic species, *Iris foetidissima*, is an uncommon but very characteristic associate of *Hedera* in some stands in the south-west.

Hedera continues as an important component of the field layer in most *Fraxinus-Acer-Mercurialis* woodlands in the north-west, though, throughout the sub-communities typical of this region, a much better preferential indicator of the oceanic conditions is *Phyllitis scolopendrium*. This fern is much commoner in Britain in those areas with a more continuously humid climate, freedom from frequent frost and a long growing season (Page 1982). It is generally much scarcer towards the south-east, though its local occurrences (as in cool and moist, north-facing coombes in the North Downs) help define far-flung stands of the largely north-western *Geranium* sub-community. *Polystichum setiferum*, though far less frequent than *Phyllitis*, has a similar distribution among these woodlands. Interestingly, all the forementioned plants are evergreen or winter-green and, together with the more abundant bryophytes of the north-western *Fraxinus-Acer-Mercurialis* woodlands, may carry out the bulk of their photosynthetic activity in the relatively mild winter months when the canopy is bare of leaves. Finally, among this group, there is *Ilex aquifolium*, also evergreen but probably slightly more frequent among the north-western sub-communities because of its sensitivity to low winter temperatures (Iversen 1944, Peterken & Lloyd 1967, Pigott 1970*b*).

Third, and last among these kinds of climatic influence, there is the marked association with the north-western sub-communities of *Ulmus glabra* and *Acer pseudoplatanus*, both as frequent and abundant canopy components and as often prolific saplings. Neither of these trees is confined to this community (though the former is a fairly good preferential) but, in all the woodland types in which they occur, they attain constancies in excess of III only where the annual rainfall is fairly high: the fit between the distribution of samples of such woodland types and areas with rainfall in excess of 762 mm (30″) yr^{-1} is especially good for *Acer* (Rodwell in Pigott 1984). Though the fact that this tree is essentially a species of damp ravine woodland is a commonplace in Europe, the reasons for its behaviour, and for the similar pattern in *U. glabra*, are unclear. However, the diagnostic value of these species here is very obvious and a recognition of the climatic response in *Acer* needs to temper statements about its value as an indicator of human interference: within the region where many ancient stands of this woodland survive, it may not be able to perform well for quite natural reasons. As with the geographical influence of a more oceanic climate, it should be noted that areas of higher rainfall extend across the extreme southern part of England to take in some of the western Chalklands and the downland of the Home Counties, so outlying stands of largely north-western sub-communities can be found here; indeed, in this region, the coincidental occurrence of abundant *Phyllitis* under sycamore-rich canopies is a very striking local feature.

Climate and soils, then, are of prime importance in

determining the general character of the *Fraxinus-Acer-Mercurialis* woodland and much of the floristic variation within it; and they play a major role in governing its overall distribution and the range of its two suites of sub-communities. Very often, though, their controlling influence is overlain by the effects of sylvicultural treatments. In general terms, treatments have operated within the constraints which the natural factors impose, though they have not always worked in the same direction. In some cases, as with selection for *Carpinus* as a large-coppice underwood crop on moister soils, treatment has accentuated variation related to, in this instance, edaphic conditions. In other cases, as with the reduction in the abundance of *Quercus robur* to occasional standards in many kinds of coppice in the southeast, treatment has worked against a natural trend, blurring floristic patterns. Also, very commonly, treatments have affected species of broad tolerance of different soils and climates (like *Corylus*) which have now come to vary quite independently of such factors or to show spurious associations with certain environmental conditions because of simple accidents of selection.

The net result of such effects is that variation in the woody component of the community has been dislocated to some extent from differences in the herbaceous element. This can be seen very clearly where the same field and ground layers occur under very different canopies derived by diverse manipulations of the same basic assortment of trees and shrubs. The *Primula-Glechoma* herbs and bryophytes, for example, can be found beneath coppiced underwood of *Corylus, Acer campestre* or *Fraxinus*, or various combinations of these, with or without an admixture of other shrubs and small-coppice trees and with or without varying numbers of standards, usually *Quercus robur*, occasionally *Fraxinus*, rarely *Acer campestre*; under large-coppice underwood of *Carpinus* or (provided it is not too dense) *Tilia cordata*, or various mixtures of these, again with or without standards; beneath invasive suckering elms; in semi-natural high forest with signs of only sporadic removal of timber or coppice-wood; and in older plantations, mainly of *Quercus robur* or *Fraxinus* (Figure 19). Conversely, where similar treatment has been applied to stands of different sub-communities, it has often evened out any natural variation among the woody species so that virtually identical canopies survive over different field and ground layers; this is a common occurrence with the *Primula-Glechoma* and *Anemone* sub-communities.

Of the various styles of treatment, it is coppicing that has left the most striking legacy in surviving stands of the *Fraxinus-Acer-Mercurialis* woodland, especially among the sub-communities of the south-east. Here, these kinds of woodland have been the major source of coppice-wood and timber over more base-rich soils, probably for many centuries. Here, too, the coppicing tradition survived longer, though only a small proportion of stands are now actively worked (Peterken 1972, 1981, Rackham 1976, 1980) so the effects of treatment are usually seen these days filtered through long periods of neglect. As in other woodland communities, the impact of coppicing in a given stand is a complex function of the nature of the existing vegetation, local traditions of management and market forces worked out through systems of underwood and timber extraction on a variety of rotations with additional manipulations like cleaning, planting, layering and promotion. The diligent work of Peterken (1974, 1977, 1981; see also Peterken & Harding 1975) and Rackham (1967, 1971, 1975, 1976, 1980) has shown just how multifarious and complex are the local manifestations of such practices on the surviving vegetation, but how powerful a tool an understanding of treatment can be in giving a historical perspective on this community, especially when it is combined with the use of archaeological and documentary evidence. Many older stands of the *Fraxinus-Acer-Mercurialis* woodland constitute an irreplaceable record of past land use and its effects on the vegetation.

Although coppicing has been responsible for the development of great floristic and physiognomic contrasts between the woody cover of different stands of this community, within stands it is, in the long run, a conservative kind of treatment. At any one site, its major impact is to be seen in response to the periodic perturbations that accompany each cut: a sudden increase and steady decline in the exposure of the ground to the unshaded climatic environment during each rotation (Salisbury 1924, Rackham 1975, 1980, Ash & Barkham 1976, Peterken 1981) together with the trampling and disturbance associated with the cropping and removal of underwood and occasional timber. Although the shrubs and trees may show some slight differences in the proportions of the species represented as regrowth occurs, it is the cyclical changes in the field layer that are more obvious. As in all coppiced woodland communities, these patterns can be very varied, but some general features can be recognised. First, there is often a quick decline and slow recovery of *Mercurialis*, in response to the consolidation of the ground with trampling and consequent development of adverse anaerobic conditions, especially marked when the winter and spring of the cut are wet and the soil becomes badly poached (e.g. Pigott 1977). *Mercurialis* can decline, too, when the increased exposure of the soil to higher surface temperatures induces droughting in the summer after the cut (Martin 1968, Ash & Barkham 1976). Either way, any reduction in this, the commonest field-layer dominant here and a strong competitor to other herbs, opens up more ground for subsequent prominence of the existing field-layer associates and the spread of any adventives.

Among the former, the characteristic post-coppice flowering can be seen in great splendour in this community with its various mixtures of *Hyacinthoides*, *Primula vulgaris*, *P. elatior*, *Anemone*, *Ranunculus ficaria*, *Viola* spp., *Glechoma* and *Lamiastrum*. Individual plants can increase greatly in vigour producing larger rosettes or spreading carpets and seedlings may appear in profusion. Other characteristic species in the community may be able to make more obvious temporary capital on areas of open ground. *Rubus* often increases with the prolonged summer light and, on lighter soils, *Sanicula*

europaea, *Brachypodium sylvaticum* and *Geranium robertianum* may appear in profusion; these latter species are often abundant in coppiced versions of the north-western sub-communities. To the south-east, it is often the spread of *Deschampsia cespitosa* which marks the combination of increased light and trampling associated with coppicing and, on moister ground, there may be sufficient shift in the vegetation as to mark the development of the *Deschampsia* sub-community. Such changes may outlast the repetition of the coppice cycle, especially where bulkier perennials of this woodland (like *Deschampsia* and *Filipendula ulmaria*) become well-established in between-times. Often, though, the characteristic herbs of the particular sub-community being coppiced settle down once more into an assortment which may show quantitative changes but which preserves the general qualitative features of the original

Figure 19. Variation in canopy and underwood in *Fraxinus-Acer-Mercurialis* woodland at House Copse, Rusper in Sussex.

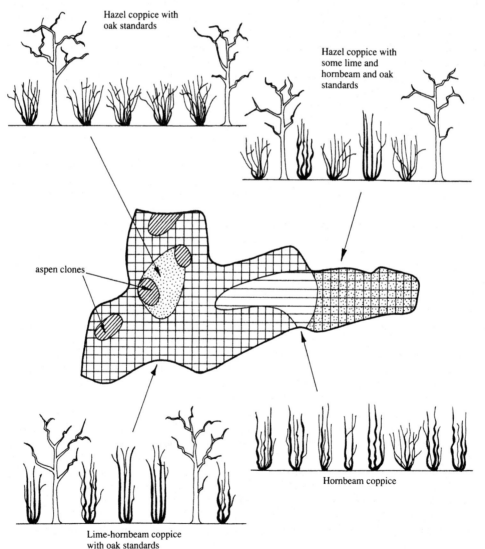

field layer. *Mercurialis* recovers and adds its own canopy to that of the regrowing shrubs and trees, the more shade-sensitive associates thin out and, if the next cut is long delayed, a ground carpet of ivy may spread and lead to the development of the *Hedera* sub-community.

Superimposed on these variations in the existing herbaceous flora of these woodlands, there is often a temporary abundance of adventive species in the more open conditions after the coppice cut. A clear sequence of life-forms is usually obvious, with annuals, then biennials, then perennials attaining prominence in turn, but the particular species represented vary greatly according to local edaphic and climatic conditions and the supply of propagules. Again, it is among the more frequently coppiced south-eastern sub-communities that this element has been best described (e.g. Salisbury 1916, 1918*a*, 1924, Rackham 1975, 1980, Ash & Barkham 1976, Peterken 1981). Some of these adventives, like the *Juncus* spp., appear to originate from buried seed (Rackham 1975, Peterken 1981, Brown & Oosterhuis 1981); many others colonise from permanently open rides and clearings, neighbouring hedgerows or field margins. Some favour moister soils, like the Junci, *Epilobium hirsutum* or *Cirsium palustre*; even semi-aquatic species can appear in patches of standing surface water. Others spread on drier ground and some species, like *Urtica dioica* and *Epilobium angustifolium*, mark out areas of local enrichment, like the sites of brushwood bonfires. Provided there is uninterrupted re-establishment of the canopy, few, if any, of these species last into the later stages of longer rotations, though clumped perennials can be quite persistent and may become a permanent feature where cutting is very frequent. Within whole woods, of course, where there is rotational cropping, medleys of these species represented in compartments at different stages of the cycle can become an enduring source of enrichment to the site.

However, even where extensive old coppiced stands of the *Fraxinus-Acer-Mercurialis* woodland survive, such temporal and spatial variety is usually a thing of the past. There are quite widespread signs of a last cut around the time of the Second World War or in the cold winter of 1946/7, a few woodlands are still cropped commercially or for research purposes and there is a renewed interest in the practice of woodmanship; in general, though, the picture now is one of long neglect.

Other stands give some clues as to how the community responds to different kinds of treatment. Some have obviously been clear-felled in the relatively recent past. This is a more drastic treatment than coppicing though, if there is opportunity for undisturbed regrowth and attention to the spread of woody adventives (notably birch), something like the original canopy composition (not structure) may re-establish itself over an essentially unchanged field layer. Monks' Wood in Cambridges-

hire, for example, was extensively cleared in the First World War but has substantially recovered (Hooper 1973, Ratcliffe 1977, Rackham 1980). Modest interplanting, too, provided the characteristic hardwoods are used, need not destroy the floristic richness and diversity of existing stands, though it alters the age structure and genetic diversity of the canopy.

The survival of the essential features of the community is much more problematical where clear-felling is combined with replanting. Some existing stands of *Fraxinus-Acer-Mercurialis* woodland are certainly plantations, derived either by re-afforestation or planting on soils which would naturally carry this community. In such woodland, the canopy is usually of *Quercus robur* and/or *Fraxinus* and it bears obvious signs of its origin in the ordered rows of even-aged trees of similar morphology. Younger stands of this kind are often of the *Hedera* sub-community with its rather dull field layer but, where there has been time for slow-spreading herbs to immigrate from the margins or spread from surviving patches, something like the richness of the field layers of the *Primula-Glechoma* or *Anemone* sub-communities (in the south-east) or the *Geranium* sub-community (to the north-west) can slowly be attained. We do not know how long such a process takes, though recorded rates of spread of *Mercurialis*, *Hyacinthoides*, *Anemone* and *Primula elatior* are of the order of only a few metres per century (Rackham 1975, Pigott 1984).

Where existing stands of the *Fraxinus-Acer-Mercurialis* woodland are felled and replanted, the crucial factors in maintaining field-layer diversity are probably an avoidance of damage to the existing herbs, the maintenance of or quick progression to a fairly light canopy and the continuance of a mull humus regime. Although ornamenting with occasional larch or pine is quite common, this community will stand only very dilatory coniferisation. Clear-felling, scrub-bashing, herbicide treatment and replanting with softwoods spell a quick and irreparable end. Sadly, this has been the fate of many stands.

Against this extensive loss must be placed an increase in the extent of *Fraxinus-Acer-Mercurialis* woodlands by relatively recent invasion of neglected farmland and disused chalk pits and quarries. As in plantation stands, the trees in such situations are often more or less even-aged (or grouped in pronounced age-classes where colonisation has been in waves), but the woody cover is often more varied and, even after many years, it may preserve elements of the early scrub (like large old hawthorns). Again, the *Hedera* sub-community is very characteristic of the dense woodland that often develops in such situations, though more open stands on heavier soils may show the *Deschampsia* sub-community or, on lighter soils, the *Geranium* sub-community. In the early stages, more slow-spreading herbs are usually limited to the margins. However, such younger *Fraxinus-Acer-*

Mercurialis woodlands are well worthy of study: sites such as the Hayley Wood Triangle (Rackham 1975) and Geescroft Wilderness at Rothamsted (Brenchley & Adam 1915, Tansley 1939, Pigott 1977) provide ready-made opportunities for monitoring the development of high-forest canopies and the establishment of the full richness of the field-layer flora here. In some areas, like the Yorkshire Dales, younger stands probably comprise the bulk of the woodland cover and, where old chalk pits and quarries lie close to urban areas, they offer vegetation robust enough for heavy amenity and educational use.

Zonation and succession

Most commonly, zonations between different kinds of *Fraxinus-Acer-Mercurialis* woodland and from this community to other woodland types are under the primary influence of soil variation. Treatment differences can accentuate or confuse such patterns and also produce direct effects of their own. Agricultural improvement has often truncated zonations and created a wide variety of artificially sharp transitions to herbaceous vegetation.

Over wooded tracts of suitably base-rich soils, the most widespread zonations between different sub-communities here are related to variations in soil moisture. On the heavy, impermeable clays which form the typical substrate for the community in much of the south and east, such variations are often directly dependent on the extent of surface-water gleying, itself a function of slope. Slope differences are often very slight over the clay superficials characteristic of many stands of *Fraxinus-Acer-Mercurialis* woodland in East Anglia and the east Midlands, but quite sufficient to induce a shift from the *Primula-Glechoma* sub-community to the *Anemone* sub-community wherever there are flatter plateaus on which water becomes perched or shallow hollows into which it runs. As in the patterns described above from Hayley Wood, there is often a superimposed zonation of dominance within these sub-communities, as the major species succumb in turn to the increased waterlogging (Abeywickrama 1949, Martin 1968, Martin & Pigott 1975, Rackham 1975). Quite frequently, too, the zonation continues to the *Deschampsia* sub-community on even more waterlogged soils. This kind of woodland can be found in wet hollows within stands of the *Anemone* sub-community but it also often forms a transition around flushes and springs to some type of alder woodland. Flushes in this kind of situation often carry the *Alnus-Fraxinus-Lysimachia* woodland but, where there is pronounced ground-water gleying and some accumulation of fen peat below spring-heads, the *Alnus-Carex* woodland can terminate the sequence. Repeated patterns of this kind are very characteristic of the more calcareous strata of the Wealden and Atherfield Clays in

Surrey and Kent, where series of springs emerge at the base of the overlying Lower Greensand. Where streams with some alluvial deposition flow through *Fraxinus-Acer-Mercurialis* woodlands, the *Deschampsia* sub-community can give way to the more eutrophic *Alnus-Urtica* woodland on the flats.

Flush-surround and stream-side zonations of this type are also quite frequent to the north and west, where *Fraxinus-Acer-Mercurialis* woodlands can sometimes be found on more calcareous shales of the Carboniferous and Oolite sequences or over superficials. Typically, though, the substrates of the community in this region are more free-draining limestones and the *Primula-Glechoma* sub-community is absent. More usually, here, there is a rather sharp transition from any stands of the *Anemone* or *Deschampsia* sub-communities to the *Geranium* sub-community on rendzina soils over steeply-sloping surrounding ground. On this more rugged topography of the north and west, however, aspect can confuse this basic pattern because, on cooler and wetter north-facing slopes, the *Anemone* and *Deschampsia* sub-communities can sometimes extend on to quite steep ground, forming patches within tracts of the *Geranium* sub-community (or the rarer *Teucrium* sub-community) over pockets of moister soil. Such patterns are well seen in some of the Derbyshire Dales and in the Mendips, where opposite faces of the valleys and gorges can present a sharp contrast in the extent of the different sub-communities. Finally, on the steeper slopes in this region, downwash is often important and a very typical pattern here is for the *Geranium* sub-community to give way to the *Allium* sub-community wherever there is an accumulation of deep, moist and free-draining colluvium.

Although all kinds of the *Fraxinus-Acer-Mercurialis* woodland are characteristic of more base-rich soils, individual woodland sites in which the community is represented often span geological transitions to less calcareous rocks or superficials. Then, zonations to less calcicolous woodland can be found, usually to the *Quercus-Pteridium-Rubus* woodland, which is the counterpart of this community throughout its range on brown earth soils of moderate to low base-status. Where there are sharp bedrock differences, as where limestones or calcareous shales or clays are interbedded with arenaceous deposits, a common feature of the Carboniferous, Jurassic and Cretaceous sequences, the corresponding vegetational boundaries can be quite clear. However, it is important to remember that some of the leading species in the *Fraxinus-Acer-Mercurialis* woodland continue to be prominent in the *Quercus-Pteridium-Rubus* woodland and can therefore run across such junctions with no change in their abundance. Among the woody species, for example, *Quercus robur*, *Carpinus betulus*, *Tilia cordata*, *Corylus* and *Crataegus monogyna* and, in the north-west, *Acer pseudoplatanus*, can all occur commonly in both communities; and, among the field-layer plants, *Hyacinthoides*, *Rubus fruticosus* agg. and, on moister soils, *Anemone* and *Ranunculus ficaria* can likewise be found in both. Usually, however, these transitions from one community to the other are marked by an obvious fading in the prominence of *Fraxinus*, *Acer campestre* (and, in the north-west, *Ulmus glabra*), *Mercurialis*, *Geum urbanum*, *Circaea lutetiana*, *Arum maculatum* and *Viola reichenbachiana* and by an increase, at least where the soils are not waterlogged, in *Pteridium*. In the south-east, the transitions commonly involve a switch from the *Primula-Glechoma*, *Anemone* or *Hedera* sub-communities of the *Fraxinus-Acer-Mercurialis* woodland to the Typical, *Anemone* or *Hedera* sub-communities of the *Quercus-Pteridium-Rubus* woodland respectively, depending on such factors as the amount of soil moisture and light. To the north-west, analogous transitions are usually from the *Geranium* sub-community of the *Fraxinus-Acer-Mercurialis* woodland to the *Holcus mollis* sub-community of the *Quercus-Pteridium-Rubus* woodland.

Over superficial deposits where there can be very diffuse or disorderly variations in calcium carbonate content and base-status, transitions may be much harder to discern or understand. Sometimes, sandier patches are marked by obvious islands of the *Quercus-Pteridium-Rubus* woodland within tracts of the *Fraxinus-Acer-Mercurialis* woodland, as at Gamlingay in Cambridgeshire (Adamson 1912) and many other East Anglian woods (Rackham 1980). In other cases, there may be no more than the haziest impression that one kind of woodland is trying to 'break through' the other with scattered *Mercurialis* and *Pteridium* intermixed over soils of intermediate quality.

There are situations, too, where variations in the base-status of the soils within woodlands are related, not to geological differences, but to flushing with calcium-rich waters. Where streams flow through stands of the *Quercus-Pteridium-Rubus* woodland, for example, the better drained of the alluvial flats often carry fragments of the *Fraxinus-Acer-Mercurialis* woodland, sometimes little more than small patches of *Mercurialis* inside every loop of the stream with an occasional scattered *Fraxinus*. These were a marked feature of the *Carpinus*-dominated woodlands described from Hertfordshire by Salisbury (1916, 1918a) but the phenomenon is very widespread.

Very commonly, and especially in the south-east, these kinds of soil-related zonations within stands of the community and to other types of woodland are overlain by treatment-derived patterns of variation. Sometimes, treatments have reinforced elements of natural zonations by selecting for trees or shrubs favouring particular soil conditions and accentuating their association with a certain field layer (as with *Carpinus* in the

Anemone sub-community). Usually, however, treatments have been applied with no reference to the disposition of soils of different moisture content or base-status, so that coppice compartments or patterns derived from timber extraction or replanting are superimposed independently over the natural transitions. This means that woods often show fairly clear soil-related variations within the field layer and equally clear but artificial and quite differently-disposed differences in the cover of underwood and timber.

As described above, different styles of treatment can affect the field-layer composition and physiognomy in the various sub-communities in different ways, but much of this variation can be seen as temporary (often cyclical) disruptions which do not, in the long term, modify the soil-related patterns. By and large, then, zonations persist while treatment-derived mosaics come and go within their components and across their boundaries. But there are some exceptions to this general rule. First, where trampling and increased light are combined, as they are when tracts are coppiced or clear-felled and allowed to regenerate (or planted with the usual hardwoods of the community), the parcel or compartment is often clearly marked out by a stretch of the *Deschampsia* sub-community which interrupts the natural zonation. Where short-cycle rotations are maintained on soils approaching the wet state of those normally occupied by this vegetation, the stand may persist. And, where trampling and high light levels are maintained permanently along paths and rides, there can be similar sudden switches to the *Deschampsia* sub-community, winding in narrow strips through the wood or marking out regular compartments in a grid.

Second, there can be abrupt transitions to the rather species-poor and gloomy vegetation of the *Hedera* sub-community at the boundaries of tracts where regrowth has been left unhindered for more than 20–50 years or so as in long-neglected coppice parcels or in planted compartments at the pole stage. Often, too, newer additions to older woods, originating from planting or by spontaneous invasion, show a sharp switch to the *Hedera* sub-community at the original boundary, or a little way beyond it if there has been time for invasion of the more slow-spreading herbs (as in the Hayley Triangle: Rackham 1975).

Third, at those few localities to the north and west where the *Teucrium* sub-community survives, the boundaries between it and the more widespread *Geranium* sub-community seem to mark a treatment-related distinction between older woodland that has escaped gross interference and younger woodland derived by a mixture of planting and natural regeneration on open ground over the past few centuries (Pigott 1960, 1969, Merton 1970). As noted earlier, such boundaries are often compounded with a topographic and edaphic transition to less intractable ground where treatment has been easier to execute and colonisation speedier.

Finally, here, topographic alterations to the woodland environment associated with sylvicultural treatments have sometimes modified soil conditions themselves and produced artificial edaphic zonations. Woodbanks, for example, are often better-drained than their surrounds and can provide, in the south-east, a belt of the *Primula-Glechoma* sub-community terminating (at the wood boundary) or interrupting (within woods) a stretch of the *Anemone* or *Deschampsia* sub-community. Where older stands of the *Fraxinus-Acer-Mercurialis* woodland have been cleared and replanted, ancient banks can provide narrow strips of richer field-layer vegetation with *Mercurialis*, *Hyacinthoides* and *Primula vulgaris*, often with interesting relic boundary specimens of *Tilia cordata* or *Carpinus*, surviving within tracts of the *Hedera* sub-community and providing nuclei from which the herbs can subsequently spread. Wood-banks are often too substantial to have been totally destroyed when compartments have been turned over to agricultural use in the past, but ploughing has often evened out natural surface undulations, so any new woodland developing on such sites shows much less variation in the ground topography and soil water content. Quite frequently, though, ploughing has created patterns of its own, notably ridge-and-furrow and, where woodland has been long established over abandoned farmland, the field-layer herbs can sort themselves over the drier ridges and wetter furrows such that stripes of the *Primula-Glechoma* and *Anemone* sub-communities alternate one with the other.

Almost always now, stands of the *Fraxinus-Acer-Mercurialis* woodland in the south-east, whether they comprise whole woods or the marginal parts of more varied sites, have artificially sharp boundaries. Where older woodlands survive intact, the boundaries may be very old, but, even in such cases, stands are invariably very closely hemmed in by intensive agricultural land. Many stands, of course, have been reduced in extent: here the boundaries are younger but generally equally abrupt. The usual picture now is of isolated fragments, often fringed by a narrow band of the shrubs and climbers of the community or by a belt of stabilised *Crataegus-Hedera* scrub with *Rubus-Holcus* underscrub below and a basal zone of the *Arrhenatheretum*, then a sharp transition to an arable crop (the usual pattern over much of East Anglia) or intensive pasture (as in wetter areas like parts of the Midlands). Where zonations are very tight, elements of all these vegetation types may be compressed into a confused mixture that can defy classification.

To the north and west, the pattern is a little different. The losses here have also been extensive though, in many places, they have been to commercial forestry, not to agriculture. But, where stands do persist or have developed more recently within stretches of agricultural land, their boundaries are frequently not so sharply defined from the surrounding vegetation. Very often this is pasture of a less improved kind and under less intensive management than in the south-east. Where grazing (it is usually by sheep or mixtures of sheep and cattle) is lighter or more sporadic, stands can develop quite an extensive fringe of *Crataegus-Hedera* scrub or mixtures of young *Fraxinus*, *Betula* spp., *Acer pseudoplatanus* and *Ulmus glabra*. A typical picture in many parts of the Yorkshire and Derbyshire Dales is for patchworks of stands of the *Geranium* sub-community of different ages to be disposed over Carboniferous Limestone scree slopes with intervening stretches of open calcicolous grassland, either the more northerly types of the *Festuca-Avenula* grassland or *Sesleria*-dominated swards (e.g. Merton 1970).

Where complex surface topography hinders grazing, rather specialised mosaics with herbaceous vegetation can be found. Over Carboniferous Limestone pavements, for example, fragmentary miniaturised stands of the *Fraxinus-Acer-Mercurialis* woodland are frequently found in the deeper, shaded grikes, the canopies of the trees and shrubs browsed and wind-pruned at the level of the pavement surface and the herbs of the community disposed over grike-bottom soil and on small ledges and in crevices. Fern-dominated vegetation often occurs intermixed with the woodland and, where the clefts give way to shallower solution hollows in the more accessible and exposed clint surfaces, the mosaics include developing stands of *Sesleria-Galium* grassland.

The quite common existence of younger stands of the community to the north-west has meant that it has been possible there to monitor its seral development or make deductions about succession from spatial patterns (e.g. Scurfield 1959, Okali 1966, Merton 1970). In the south-east, evidence is more fragmentary: opportunities provided by some early studies, mostly on abandoned farmland, (e.g. Adamson 1912, 1921, Brenchley & Adam 1915, Salisbury 1918*b*, Tansley & Adamson 1925) have largely been lost and neglected open ground is now very rare. The most that can be done here is to collate existing results (e.g. Tansley 1939), add such continuing observations on these sites as there have been (Pigott 1977), set these in the context of the ecology of the mature stands of the community (e.g. Rackham 1980, Peterken 1981) and make some educated guesses. And, also, consider the conclusions in the light of more detailed work on successional development of related communities like the *Fagus-Mercurialis* and *Taxus* woodlands (e.g. Watt 1923, 1925, 1926, 1934*a, b*).

This latter is of some considerable importance because it seems clear that, within the natural British range of *Fagus*, the starting point for the development of climax *Fraxinus-Acer-Mercurialis* woodland is not usually bare limestone or shallow rendzina soils with calcicolous grasslands. *Quercus robur* is, in any case, at some disadvantage against *Fraxinus* in the more excessively draining of these situations but, more importantly, neither of these trees can ultimately compete with *Fagus*, especially in areas like the Chilterns where the climate is drier. *Fraxinus-Acer-Mercurialis* woodland certainly can develop over deeper rendzinas carrying swards like those of the *Festuca-Avenula* grassland (or its less heavily grazed derivatives dominated by *Bromus erectus* or *Brachypodium pinnatum*) and perhaps even from more fragmentary soils over Chalk spoil in quarries and pits (e.g. Tansley 1922, Tansley & Adamson 1925, Hope-Simpson 1940*b*, 1941*b*, Wells 1969, 1973). But, even where rainfall is higher, as in the North and South Downs, it appears to be often a seral stage in the eventual development of beech forest (Watt 1925, 1934*a, b*). Much more locally, *Taxus* can overtake young *Fraxinus*-dominated stands of the community and come to dominate in yew forest (Watt 1926).

Much more likely precursors of the *Fraxinus-Acer-Mercurialis* woodland in the south-east are the more calcicolous forms of mesotrophic grasslands developed over deeper, moister base-rich soils derived from argillaceous bedrocks, like calcareous clays and shales, and calcareous superficials. Here, *Quercus robur* is very much at home and can maintain its important role amongst the mixtures of more calcicolous trees and shrubs that eventually come to dominate. However, because the vast majority of these soils have been under cultivation (or existing woodland) for very long periods of time, we do not know what the natural course of succession might be. Where the community has arisen secondarily on abandoned farmland, quite diverse lines of development have been seen to converge into young stands of the community. On old arable land with loamy brown earths, early mixtures of weeds have been seen to progress to some kind of *Arrhenatheretum*, then scrub, then woodland (as on Broadbalk Wilderness: Brenchley & Adam 1915, Tansley 1939); on heavier clay soils (as in Geescroft Wilderness: Brenchley & Adam 1915, Tansley 1939, Pigott 1977), *Holcus-Deschampsia* grassland has gained a dominant hold and a slower invasion of shrubs and trees has been observed; within abandoned pasture, the more vigorous of the existing grasses have grown rank as woody species have colonised to form 'tumble-down' scrub and then woodland (e.g. Adamson 1912).

From what we know of the ecology of existing stands of these precursors and of the *Fraxinus-Acer-Mercurialis* woodland itself, the limits defining the possible development of the community are almost certainly edaphic and, though they are quite wide, these seem to be fairly well defined. On better-drained soils, base-status is likely to be the controlling factor, so the most

obvious precursors will be the *Centaurea* and *Pastinaca* sub-communities of the *Arrhenatheretum*, the *Avenula pubescens* grassland or, where old pasture runs down, the more calcicolous forms of the *Centaureo-Cynosuretum* and the *Lolio-Cynosuretum*. Less base-rich soils with other sub-communities of these mesotrophic grasslands are likely to develop eventually into the *Quercus-Pteridium-Rubus* woodland. With increasing soil moisture, the *Holcus-Deschampsia* grassland, and perhaps also certain kinds of *Holco-Juncetum*, mark the probable limits of invasion: where waterlogging becomes more extreme, *Salix cinerea* and *Alnus* begin to assume importance as invaders and succession moves to wetter kinds of woodland.

The major trees of the south-eastern types of *Fraxinus-Acer-Mercurialis* woodland can invade these communities directly, provided there is no undue restriction on colonisation in general by the growth of very rank grasses (like *Deschampsia*): even *Quercus robur* which now regenerates badly within existing stands of woodland has no difficulty in appearing within old pasture. However, very frequently, there is a preliminary stage of scrub development before the woodland proper, especially where existing grassland is being colonised. In this scrub, *Crataegus monogyna* is almost invariably a major component and early stages in succession usually converge into the *Crataegus-Hedera* scrub, through which the trees of the community eventually emerge.

The appearance of a stand of the south-eastern kind of *Fraxinus-Acer-Mercurialis* woodland with all its diagnostic woody species (*Fraxinus*, *Q. robur*, *Acer campestre*, *Corylus* and *Crataegus*) does not necessarily take very long: the early studies were dealing with tracts which had already attained this degree of maturity after only 50 years or so. However, it probably takes very much longer for the other woody associates to appear in the kinds of mixtures that are widespread in the south-east with *Tilia cordata*, *Carpinus*, *Sorbus torminalis* and *Crataegus laevigata*. And, some of the most characteristic herbs of the community are very slow to colonise: young stands are often clearly of the *Hedera* or, over moister soils, the *Deschampsia* sub-community. Even after 90 years, the Geescroft woodland had acquired little more than a marginal fringe of *Mercurialis* (Pigott 1977) and Rackham (1975) estimated that it would take at least another 150 years for the Hayley Triangle (already 50 years old) to acquire something like the herbaceous flora of the older parts of the wood.

To the north-west, the *Fraxinus-Acer-Mercurialis* woodland can also probably develop from mesotrophic swards over more heavy-textured calcareous soils, though, with the higher rainfall here, there is an increased tendency for succession in moister situations to move towards its upland counterpart, the *Fraxinus-Sorbus-Mercurialis* woodland (as in higher-altitude dale heads) or the *Alnus-Fraxinus-Lysimachia* woodland (where there is pronounced local flushing). More frequently in this region, which is largely outside the natural range of *Fagus*, the *Fraxinus-Acer-Mercurialis* woodland is the culmination of the invasion of bare limestones and shallow, free-draining rendzinas. Its usual precursors are therefore calcicolous grasslands, either the *Dicranum* sub-community of the *Festuca-Avenula* grassland (in the Mendips, Derbyshire and Durham), the *Sesleria-Scabiosa* grassland (in Durham) or the *Sesleria-Galium* grassland (in the Yorkshire Dales and around Morecambe Bay).

Usually, it is grazing which mediates the succession on more accessible slopes here. These often have a somewhat more intact and deeper soil cover, derived by long weathering or from colluvium, glacial drift or loess and, in such situations, *Crataegus* is again an important early invader, sometimes thickening up to form dense *Crataegus-Hedera* scrub before being overtaken by colonising trees (Scurfield 1959, Pigott 1969, Merton 1970). Over more broken slopes with scattered outcrops and fine talus intermixed with downwash, *Corylus* can colonise quickly and, on warmer slopes, it is often accompanied by *Cornus sanguinea*, *Rhamnus*, *Euonymus* and *Rosa* spp. (in the so-called 'retrogressive scrub' of Moss 1913: see Merton 1970). It is this kind of scrub which seems to persist as the understorey in the more open stands of the *Teucrium* sub-community (Pigott 1969).

Very often, however, and especially on coarser, and even mobile, talus, *Fraxinus* invades very open herbaceous vegetation in the early stages and, with little or no grazing, it can quickly overtake *Crataegus* or *Corylus* and establish the basis of a canopy of *Fraxinus-Acer-Mercurialis* woodland. Very commonly, too, it is accompanied or sometimes largely replaced by *Acer pseudoplatanus*: although this tree is often found associated with disturbed sites (like mine spoil or failed plantations: Merton 1970), its prominence in successions in the region is closely related to the wetness of climate and soil. How well it competes with *Fraxinus* is unclear: Okali (1966) suggested that it was less tolerant than *Fraxinus* of sub-optimal conditions, though availability of seed-parents may have been an important factor in the development of the local variation in the relative abundance of these trees that can be seen now (Merton 1970). Clearly, though, both species and, more locally, birch (usually *Betula pendula* on the drier soils), have behaved almost like woody weeds in the speedy and diverse successions that have given rise to many stands of the *Geranium* sub-community.

Although Merton (1970) reported the presence in scrub of certain herbaceous species usually associated with older stands of north-western *Fraxinus-Acer-Mercurialis* woodlands (*Convallaria majalis* for example), there is again little doubt that it takes some time for

developing stands to acquire the full range of canopy and field-layer species. *Ulmus glabra* is usually slower to colonise than *Fraxinus* and *Acer pseudoplatanus* and, towards the north-west, *A. campestre* is approaching the limit of its range. *Quercus*, too, has been notably unsuccessful in gaining a place in younger stands. The herbaceous component in the early stages also often has a strong representation of the grassland or open scree species with *Arrhenatherum elatius*, *Brachypodium sylvaticum* and *Geranium robertianum* prominent. The complexity found in the *Teucrium* sub-community may take many generations to develop and some of its typical species (like *Tilia cordata*) now regenerate very infrequently.

Distribution

The *Fraxinus-Acer-Mercurialis* woodland is widespread over more base-rich soils in lowland Britain, becoming progressively sparser in moving to the smaller exposures of calcareous rocks and superficials in the cooler and wetter climate of the north-western uplands where it is replaced by the *Fraxinus-Sorbus-Mercurialis* woodland. The sub-communities show a well-defined geographical division. To the south-east, the central type is the *Primula-Glechoma* sub-community which is especially well represented in the east Midlands, East Anglia and the Weald. On moving to the more oceanic south-west, there is an increasing tendency for this to be replaced by the *Hedera* sub-community, though this kind of woodland can also be found throughout the region in younger stands. The *Anemone* and *Deschampsia* sub-communities are also commoner on the generally heavy soils of the south-east but they can occur locally, where edaphic conditions permit, to the north-west. Usually, however, in this region, the community is represented by the *Geranium* sub-community which is especially characteristic of the Yorkshire and Derbyshire Dales, the Welsh Marches and scattered sites further south wherever there is locally high rainfall. The *Allium* sub-community follows essentially the same pattern. The *Teucrium* sub-community is much rarer, having been recorded only in parts of the Wye valley and Derbyshire with more fragmentary stands in the Yorkshire Dales.

Affinities

The *Fraxinus-Acer-Mercurialis* woodland brings together vegetation types which, in early descriptive accounts and in Klötzli's (1970) phytosociological treatment, were separated into 'damp oakwood' (e.g. Moss *et al.* 1910, Tansley 1911, 1939), the major woodland of heavier base-rich soils in the drier south-east and included within Klötzli's *Querco-Fraxinetum*, and 'ashwood' (e.g. Moss *et al.* 1910, Moss 1911, Tansley & Rankin 1911, Tansley 1939), its counterpart on the more free-draining calcareous soils in the wetter north-west, constituting the core of Klötzli's *Dryopterido-Fraxinetum*. This floristic distinction is still visible here in the recognition of two suites of sub-communities but the general similarities between their constituents are a powerful argument for retaining them within a single community characteristic of our relatively warm and dry lowlands. In this scheme, then, the major distinction among British calcicolous mixed deciduous woodlands is between this community as a whole and the more obviously sub-montane *Fraxinus-Sorbus-Mercurialis* woodland, a vegetation type not prominent in early accounts and one subsumed by Klötzli (1970) in his *Dryopterido-Fraxinetum* as the *Blechnum* sub-community. In geographical terms, the dividing line has thus been pushed somewhat further to the north and west, corresponding here with a rough lowland/upland distinction along the 1000 mm isohyet and at the extreme limit of continental influence in our woodland flora.

The community includes only more calcicolous woody vegetation and does not subsume, under a general dominance of *Quercus robur* or mixtures of this species with *Carpinus* or *Tilia cordata*, field layers in which *Mercurialis* and its associates give way to *Rubus* and *Pteridium* with a vernal dominance of *Hyacinthoides*. Here, such vegetation is considered as a different kind of woodland altogether, the *Quercus-Pteridium-Rubus* woodland. In this respect, the treatment returns to the earliest British accounts (Moss *et al.* 1910, Tansley 1911) and does not follow the later tradition, enshrined in Tansley (1939), of recognising a single, compendious 'oakwood' with a very wide range of field-layer societies. There is a very clear floristic basis for this kind of separation between 'mercury' and 'bluebell' woodlands in Britain: the distinction is visible among some important woody species as well as herbs and it corresponds to a major edaphic contrast which is repeated in both lowlands and uplands. Klötzli (1970), at least in the lowlands, draws the edaphic bounds of his *Querco-Fraxinetum* somewhat more broadly than here, taking in vegetation which is considered as part of the *Quercus-Pteridium-Rubus* woodland. Rackham (1980) and Peterken (1981) too, though they have a narrower view of 'oakwood' than Tansley (1939), take in both 'mercury' and 'bluebell' woodlands within their communities dominated by *Carpinus* and *Tilia*.

With such schemes as those of Rackham (1980) and Peterken (1981), there are other difficulties of comparison because of the weight they give to treatment-derived variations among the trees and shrubs. Here, such differences have not been used as a basis for making major sub-divisions within this community (or any other), but regarded as constituting a finer tier of variation within a framework related primarily to climate and soils. The sub-communities of the *Fraxinus-Acer-Mercurialis* woodland thus cross-cut distinctions

made in these other classifications on the basis of dominance in canopy or underwood by such species as *Fraxinus*, *Acer campestre*, *Corylus*, *Tilia cordata*, *Carpinus* and *Ulmus glabra*. But the two approaches are not irreconcilable: groups of stands within sub-communities could, for example, be recognised as variants with different woody dominants. The reverse procedure, using sub-community species to characterise variants within Rackham woodland communities or Peterken stand types, produces a much more cumbersome result and obscures what are here taken to be the major lines of natural variation. Correspondence between the sub-communities recognised here and the woodland types diagnosed in the other recent classification, by Bunce (1982), is unfortunately patchy. Roughly speaking, the *Fraxinus-Acer-Mercurialis* woodland is equivalent to the first quarter of Bunce's hierarchy but half of the sub-communities have no clear counterpart.

Although the kinds of woodland included in this community have clear equivalents on the Continent, it is not easy to set them in a European context because many of them lie beyond the natural limits of *Fagus* and *Carpinus*, the two trees which have tended to control the phytosociological perspective on north temperate forests on neutral to base-rich soils. In the south-east, where both these species can occur, relationships are clearer. The consensus would be that, here, the community is obviously part of the alliance Carpinion which comprises oak-hornbeam forests typical of those parts of Europe with only relatively low annual rainfall and moderately high summer temperatures (Neuhäusl 1977, Ellenberg 1978). In north-west Europe, in a zone which Noirfalise (1968) saw as encompassing northern France, Belgium, the southern Netherlands and south-eastern England, such oak-hornbeam woods occur widely on soils too wet for *Fagus* to thrive or where beech-dominance is prevented by treatment, exactly the behaviour described by Watt (1923, 1924, 1925, 1934*a*, *b*). Much of the *Fraxinus-Acer-Mercurialis* woodland could thus be taken in as a calcicolous portion of a community like the *Endymio-Carpinetum* (Noirfalise 1968, 1969; see also Noirfalise & Sougnez 1963) or, in the older terminology, *Querceto-Carpinetum* (e.g. Lemée 1937, Dethioux 1955) or *Quercetum atlanticum* (LeBrun *et al.* 1949). However, as argued above, there is a strong case, at least in Britain, for maintaining a sharp distinction between more and less calcicolous woodlands, so a better solution would be to regard the *Fraxinus-Acer-Mercurialis* woodland as a quite separate Carpinion community from such as these, which are more closely equivalent to the *Quercus-Pteridium-Rubus* woodland. One would then have a pair of British Carpinion communities, exactly parallel to those dominated by *Fagus*. The *Fraxinus-Acer-Mercurialis* woodland would thus be the British equivalent of Belgian and

Dutch communities like the *Fraxino-Ulmetum* (LeBrun *et al.* 1955) or the *Fraxineto-Ulmetum* (Westhoff & den Held 1969).

Whether the phytosociological affiliations of the north-western sub-communities are as clear is a moot point. Here, we are well beyond the natural limits of *Fagus* and *Carpinus*, though both trees grow well when planted within *Fraxinus-Acer-Mercurialis* woodland in this region. Oak is of low frequency but this is probably treatment-related and, though *Quercus petraea* is the commoner species, this tree is well represented in continental Carpinion woodlands on more free-draining substrates. Furthermore, *Acer pseudoplatanus* and *Ulmus glabra*, the best woody preferentials in the north-west, are common in mainstream Carpinion woods in wetter parts of north-west Europe: indeed, Noirfalise (1968) regarded the abundance of the former as a good distinguishing feature of his *Endymio-Carpinetum*. Dis-

tinct phytosociological equivalents of these north-western sub-communities have been described from damp ravines in Europe (e.g. the *Scolopendrieto-Fraxinetum* of Schwickerath 1944, Vanden Berghen 1953, Durin *et al.* 1968, and the *Acereto-Fraxinetum* of Le Brun *et. al.* 1949) and their relationships with the Carpinion stressed. Birse (1984) also placed his Scottish *Querco-Ulmetum* in the Carpinion. Klötzli (1970), on the other hand, saw the distinctive character of these woodlands, with their abundance of *Acer pseudoplatanus*, occasional presence of *Tilia platyphyllos* and abundance of evergreen ferns, as arguing for a place among the montane ravine woodlands of the Tilio-Acerion. Shimwell (1968*b*) recommended the erection of a new alliance within the Fagetalia, the Fraxino-Brachypodion, to contain the bulk of the north-western *Fraxinus-Acer-Mercurialis* woodlands he described from Derbyshire.

Floristic table W8

	a	b	c	d	e	f	g	8
Fraxinus excelsior	IV (1–10)	IV (1–7)	IV (1–9)	V (2–10)	V (1–10)	III (2–8)	V (2–10)	IV (1–10)
Acer campestre	II (1–7)	I (2–5)	II (1–4)	II (1–8)	I (3–6)	II (4–6)	I (2–6)	II (1–8)
Salix caprea	I (2–7)	I (1–3)	I (2)	I (1–4)	I (1–5)	I (7)	I (1–3)	I (1–7)
Fagus sylvatica	I (1–3)	I (4–7)		I (3–6)	I (1–8)	I (6–7)	I (2–4)	I (1–8)
Taxus baccata	I (4)	I (1–3)		I (1–7)	I (3–4)		I (4)	I (1–7)
Larix spp.	I (3–8)	I (4)	I (3)	I (2–4)	I (1–4)			I (1–8)
Quercus hybrids	I (2–9)	I (4–6)	I (6–7)		I (2–6)	I (4–6)		I (2–9)
Ilex aquifolium	I (1–2)			I (2–3)	I (1–4)		I (2–8)	I (1–8)
Salix cinerea	I (3–5)			I (3)	I (1–3)		I (1–3)	I (1–5)
Sorbus aria				I (1–4)	I (1)	I (4)		I (1–4)
Betula pubescens	I (3–6)		I (1–4)	I (3–4)	I (1)			I (1–6)
Malus sylvestris	I (1–3)				I (1)		I (1–3)	I (1–3)
Prunus avium	I (1)	I (1–5)				I (4)		I (1–5)
Sorbus aucuparia					I (2–5)			I (2–5)
Alnus glutinosa	I (3)	I (1–4)	I (2–3)					I (1–4)
Quercus robur	IV (1–10)	III (1–8)	III (1–7)	IV (2–10)	I (1–7)	I (5–8)	I (1–5)	III (1–10)
Carpinus betulus	III (1–10)	II (5–7)		I (2–8)	I (1)			I (1–10)
Betula pendula	II (1–10)	I (1–5)	I (3)	I (1–4)	I (1–6)			I (1–10)
Tilia cordata	II (1–10)	I (1)		I (1–2)	I (3)		I (4–6)	I (1–10)
Ulmus carpinifolia	II (3–10)	I (8–9)		I (4)				I (3–10)
Populus tremula	I (3–8)	I (5)	I (3–7)	I (4)				I (3–8)
Ulmus procera	I (2–6)			I (7)				I (2–7)
Ulmus spp.	I (4–10)		I (5)					I (4–10)
Castanea sativa	I (3–4)	I (6)						I (3–6)
Sorbus torminalis	I (1–3)							I (1–3)
Acer pseudoplatanus	I (2–5)	II (1–10)	II (1–6)	I (4–8)	IV (2–10)	III (4–10)	III (1–8)	II (1–10)
Ulmus glabra		II (4–7)	II (1–5)	I (4)	IV (1–10)	III (5–7)	III (1–7)	II (1–10)
Quercus petraea	I (1–8)	I (2–9)	I (1–5)	I (2–9)	II (2–8)	I (4)	I (1–6)	I (1–9)
Tilia platyphyllos							II (4–8)	I (4–8)
Corylus avellana	V (2–10)	IV (1–9)	V (4–9)	V (2–10)	V (1–10)	IV (2–9)	V (2–8)	V (1–10)
Crataegus monogyna	III (1–7)	IV (1–5)	III (4–7)	V (2–7)	IV (1–7)	III (1–4)	IV (1–6)	III (1–7)

Floristic table W8 *(cont.)*

	a	b	c	d	e	f	g	8
Acer campestre	II (1–6)	I (3)	III (2–5)	III (1–7)	III (1–6)	II (2–4)	III (1–5)	III (1–7)
Fraxinus excelsior sapling	II (1–8)	II (1–3)	IV (2–6)	II (2–6)	III (1–5)	I (3)	III (1–5)	III (1–8)
Sambucus nigra	I (1–5)	II (1–7)		II (1–7)	III (1–6)	II (1–5)	II (1–6)	II (1–7)
Cornus sanguinea	II (2–8)	I (3)	II (2–4)	I (2–4)	I (2–3)	I (3)	III (1–6)	II (1–8)
Prunus spinosa	I (1–8)	I (1–4)	II (2–5)	I (2–5)	I (1–6)	I (1)	I (2–3)	I (1–8)
Euonymus europaeus	I (2–3)		I (2)	I (1–5)	I (1–5)	I (3)	I (1–3)	I (1–5)
Fagus sylvatica sapling	I (1–10)	I (1–4)		I (3)	I (1–4)	I (1)		I (1–10)
Malus sylvestris	I (1–2)			I (1–3)	I (1)			I (1–3)
Taxus baccata sapling			I (1–2)	I (1)	I (1)		I (1–4)	I (1–4)
Crataegus laevigata	I (3–6)	I (3)	I (3–4)	I (5)				I (3–6)
Quercus robur sapling	I (2–3)		I (2–3)	I (3)				I (2–3)
Viburnum lantana	I (4)	I (3)		I (1–8)	I (1–4)			I (1–8)
Crataegus hybrids	I (3–5)	I (2)		I (3)				I (2–5)
Carpinus betulus	I (2–10)	I (3–10)		I (3)				I (2–10)
Betula pendula sapling	I (2–6)	I (2)						I (2–6)
Castanea sativa sapling	I (3)	I (3–4)						I (3–4)
Ulmus carpinifolia suckers	I (3–7)							I (3–7)
Ulmus spp. suckers	I (3–9)							I (3–9)
Acer pseudoplatanus sapling	I (1–4)	II (1–5)	II (1–5)	I (2–6)	III (1–5)	II (2–4)	I (1–2)	II (1–6)
Ilex aquifolium	I (3–4)	I (6)	I (4)	I (1–3)	II (1–4)	II (1–4)	II (1–8)	II (1–8)
Ulmus glabra sapling		II (1–5)	II (1–4)	I (3–4)	II (1–6)	II (1–4)	I (1–6)	I (1–6)
Viburnum opulus	I (2–4)	I (1–3)	I (2–4)	I (2–5)	I (1–2)		III (1–4)	I (1–5)
Sorbus aucuparia		I (1–2)			I (1–3)		III (1–6)	I (1–6)
Rhamnus catharticus							III (1–6)	I (1–6)
Prunus padus		I (1)					II (1–5)	I (1–5)
Mercurialis perennis	IV (1–10)	III (1–10)	II (2–6)	V (2–10)	IV (1–10)	V (4–9)	V (4–10)	V (1–10)
Eurhynchium praelongum	IV (1–9)	IV (1–7)	III (3–6)	II (1–8)	IV (1–8)	V (4–7)	V (1–6)	IV (1–9)
Rubus fruticosus agg.	IV (1–10)	III (2–8)	V (3–8)	IV (2–9)	III (1–9)	III (1–6)	II (2–4)	IV (1–10)
Poa trivialis	III (1–9)	II (1–8)	III (3–5)	I (1–7)	I (1–5)		I (1–3)	II (1–9)
Glechoma hederacea	III (2–8)	II (1–4)	I (3–4)	I (2–6)	I (1–5)		I (1–4)	II (1–8)

Primula vulgaris	III (1–4)	II (3–5)	I (3)	I (2–6)	I (1–4)	I (1)	I (1–2)	II (1–6)
Viola riviniana/reichenbachiana	II (2–6)	II (2–5)	II (4–6)	II (1–7)	I (1–5)		II (1–4)	II (1–7)
Ajuga reptans	II (1–6)	II (2–3)	II (2–3)	I (1–6)	I (3)			I (1–6)
Primula elatior	I (2–7)	I (4–5)	I (3)					I (2–7)
Primula vulgaris × elatior	I (5)	I (5)						I (5)
Anemone nemorosa	I (2–6)	V (1–9)	I (4)	I (1–5)	I (1–8)	II (1–4)	I (1–3)	I (1–8)
Ranunculus ficaria	I (1–5)	IV (1–7)		I (2–5)	I (1–6)	II (2–4)	I (1–3)	I (1–6)
Lamiastrum galeobdolon	I (1–6)	II (1–4)	I (1–6)	I (1–6)	I (1–6)	I (2–4)	I (1–5)	I (1–6)
Rumex sanguineus	I (1–4)	II (2–3)		I (1–3)	I (1–4)			I (1–4)
Deschampsia cespitosa	I (1–4)	I (1–4)	V (4–9)	I (1–4)	I (2–7)	I (4)	II (2–7)	I (1–9)
Filipendula ulmaria	I (1–4)	I (5)	II (3–7)	I (4)	I (2–5)	I (3)	II (1–3)	I (1–7)
Potentilla sterilis	I (1–4)	I (1–2)	II (3–4)	I (3)	I (1–3)		I (1–4)	I (1–4)
Lysimachia nemorum	I (1–3)	I (2–3)	II (1–2)					I (1–3)
Juncus effusus			II (1–5)					I (1–5)
Hedera helix	II (2–9)	II (1–7)	II (4–6)	IV (2–10)	III (1–10)	III (2–8)	II (2–5)	III (1–10)
Urtica dioica	II (1–8)	II (1–7)	II (1–7)	I (1–5)	III (1–9)	III (1–4)	III (1–6)	II (1–9)
Galium aparine	I (1–7)	II (1–7)	I (4)	I (2–5)	III (1–6)	III (2–5)	II (1–4)	II (1–7)
Geranium robertianum	I (1–4)	I (3)		I (1–7)	III (1–7)	II (2–3)	II (1–3)	II (1–7)
Eurhynchium striatum	I (1–6)	I (1–6)		I (3–7)	III (1–7)	II (3–5)	II (2–5)	II (1–7)
Thamnobryum alopecurum	I (1–8)	I (3–4)	I (5–6)	I (1–6)	II (1–7)	II (4–6)	I (1–4)	I (1–8)
Phyllitis scolopendrium				I (1–8)	II (1–5)	II (1–4)	I (3)	I (1–8)
Ctenidium molluscum					I (1–7)	I (3)	I (1–5)	I (1–7)
Allium ursinum	I (3–4)	I (2–5)	I (4)	I (3)	II (1–4)	V (6–10)	I (1–2)	I (1–10)
Brachypodium sylvaticum	II (2–8)	I (4–5)	II (3–4)	III (2–7)	II (1–8)	I (1)	IV (1–6)	II (1–8)
Teucrium scorodonia		I (3)		I (4)	I (2–4)		IV (1–4)	I (1–4)
Melica uniflora	I (2–6)	I (2–4)	I (5)	I (2–5)	I (2–7)		III (2–4)	I (2–6)
Arrhenatherum elatius					I (1–5)	I (3)	III (1–4)	I (1–5)
Campanula latifolia		I (2)			I (1–2)		II (1–4)	I (1–4)
Polystichum aculeatum	I (2–4)				I (1–6)		II (1–4)	I (1–6)
Myosotis sylvatica					I (1–3)		II (1–3)	I (1–3)
Plagiothecium denticulatum					I (5)		II (1–5)	I (1–5)
Convallaria majalis					I (1–3)		II (1–7)	I (1–7)

Floristic table W8 *(cont.)*

	a	b	c	d	e	f	g	8
Melica nutans							II (1–7)	I (1–7)
Rubus saxatilis							II (2–3)	I (2–3)
Rosa villosa							I (1–2)	I (1–2)
Hyacinthoides non-scripta	III (2–9)	IV (1–9)	III (2–4)	III (3–10)	II (1–9)	II (2–6)	II (1–3)	III (1–10)
Brachythecium rutabulum	III (1–8)	II (1–8)	III (3–6)	II (2–5)	III (1–9)	V (2–8)	I (2)	III (1–9)
Plagiomnium undulatum	III (2–7)	III (1–6)	II (2–4)	I (1–4)	III (1–5)	II (2–4)	III (1–5)	III (1–7)
Circaea lutetiana	III (2–5)	I (1–5)	III (2–4)	II (2–4)	I (1–7)	II (3–6)	III (1–6)	III (1–7)
Geum urbanum	III (1–6)	I (1–4)	II (3–5)	II (1–4)	II (1–7)	I (2)	III (2–5)	III (1–7)
Fissidens taxifolius	II (1–4)	I (1–3)	I (3)	I (2–5)	II (1–4)	II (1–3)		II (1–5)
Arum maculatum	II (1–6)	II (1–4)	I (3)	II (1–7)	II (1–5)	II (2–4)	I (1–3)	II (1–7)
Atrichum undulatum	II (2–6)	I (1–4)	II (3–4)	I (4)	I (1–4)	I (1–3)	II (2–4)	II (1–6)
Mnium hornum	II (1–7)	II (1–6)	I (2–3)	I (4)	I (1–5)		II (1–5)	II (1–7)
Fraxinus excelsior seedling	II (1–3)	I (1–3)	II (2–4)	II (1–4)	II (1–4)	I (1–2)	I (2)	II (1–4)
Dryopteris filix-mas	II (1–4)	II (1–4)		I (1–4)	II (1–5)	I (5)	I (1–4)	II (1–5)
Rosa canina agg.	II (2–6)	I (1–4)	II (3–6)	I (1–6)	I (1–4)		I (1–4)	II (1–6)
Lonicera periclymenum	II (1–6)	II (1–6)	II (2–6)	I (1–5)	I (1–4)	I (2)		II (1–6)
Thuidium tamariscinum	II (2–7)	I (1)	III (3–7)	I (2–4)	I (1–7)	I (3)		II (1–7)
Carex sylvatica	II (1–4)	I (4)	I (1–3)	I (2–5)	I (1–3)		I (2–3)	I (1–5)
Tamus communis	I (2–4)	I (3)	II (2–4)	I (1–4)	I (1–3)		II (1–4)	I (1–4)
Eurhynchium swartzii	I (2–3)	II (1–5)		I (4–7)	II (1–5)	II (1–4)	II (1–4)	I (1–7)
Silene dioica	I (1–5)	I (1–4)	I (3)	I (1–5)	II (1–7)	I (1)	I (1–2)	I (1–7)
Lophocolea bidentata s.l.	I (1–4)	I (1)	I (4)		I (1–4)		II (1–4)	I (1–4)
Ligustrum vulgare	I (1–8)	I (2)	I (1–3)		I (2–4)		I (1–6)	I (1–8)
Sanicula europaea	I (1–3)	I (3–4)		I (1–4)	I (1–4)		I (1–4)	I (1–5)
Angelica sylvestris	I (1–3)	I (2)	I (1–3)	I (2–5)	I (1)		I (1–2)	I (1–4)
Stachys sylvatica	I (2–5)	I (1–3)	I (3)	I (4)	I (1–5)	I (1)	I (1–2)	I (1–5)
Dryopteris borreri	I (4)	I (1)	I (1)	I (1–4)	I (1–4)	I (2)	I (2–3)	I (1–4)
Poa nemoralis	I (1–6)	I (3)	I (3–4)	I (1–4)	I (1–4)	I (1)	I (1–4)	I (1–6)
Festuca gigantea	I (1–4)	I (1–3)	I (2–4)	I (2–4)	I (1–4)		I (1)	I (1–6)
Bromus ramosus	I (1–3)	I (4)	I (4)	I (1–6)	I (1–4)		I (1–3)	I (1–4)
Veronica chamaedrys	I (1–3)	I (2–3)	I (3)	I (1–3)	I (1–3)		I (3)	I (1–3)
Euphorbia amygdaloides	I (1–4)	I (3)	I (3)	I (1–5)	I (1–2)	I (1)		I (1–5)

Species	1	2	3	4	5	6	W8
Amblystegium serpens	I(1-3)	I(3)	I(2)	I(1-5)	I(3)		I(1-5)
Hypnum cupressiforme	I(1-4)	I(3)	I(1)	I(1-3)	I(1-3)		I(1-4)
Taraxacum officinale agg.	I(1)	I(1-4)	I(1-4)	I(1-3)		I(1-4)	I(1-4)
Pteridium aquilinum	I(1-5)	I(3)	I(3-5)	I(1-4)			I(1-5)
Ranunculus repens	I(1-3)	I(2-3)	I(3)	I(1)			I(1-3)
Holcus lanatus	I(3-4)	I(3-4)	I(2)			I(1-6)	I(1-6)
Oxalis acetosella	I(1-7)	I(2)		I(1-3)	I(2)		I(1-7)
Athyrium filix-femina	I(4-5)		I(5)	I(1-5)		I(2-3)	I(1-5)
Dactylis glomerata	I(2-3)		I(1)	I(1-5)		I(1-3)	I(1-5)
Galium odoratum	I(2-4)		I(6-7)	I(1-6)		I(2-4)	I(1-7)
Adoxa moschatellina	I(3)		I(4-5)	I(1-5)		I(3)	I(1-5)
Fragaria vesca	I(2-4)	I(3)	I(2-4)	I(1-2)		I(2-4)	I(1-4)
Conopodium majus	I(1-6)		I(1-5)	I(1-3)	I(1-3)	I(1-3)	I(1-5)
Acer pseudoplatanus seedling	I(1-4)		I(1-4)	I(1-3)	I(2)	I(1-2)	I(1-4)
Dryopteris dilatata	I(1-4)		I(1-3)	I(1-4)	I(1-3)		I(1-4)
Heracleum sphondylium	I(2-3)		I(1-3)	I(1-6)	I(1-2)		I(1-6)
Stellaria holostea	I(2-3)		I(3-4)	I(1-5)	I(3)		I(1-5)
Isothecium myosuroides	I(2)			I(1-4)	I(1-4)		I(1-4)
Eurhynchium confertum	I(2-6)			I(1-4)	I(2-4)		I(1-6)
Moehringia trinervia	I(1-4)	I(4)		I(1-4)		I(2-4)	I(1-4)
Plagiochila asplenoides major	I(2)	I(4)	I(1-8)	I(1-4)		I(1-4)	I(1-4)
Clematis vitalba	I(1-4)			I(1-4)			I(1-8)
Listera ovata	I(1-3)		I(2)	I(2-5)			I(1-5)
Anthriscus sylvestris	I(2-3)		I(3-4)	I(3-4)			I(2-4)
Polystichum setiferum	I(4)		I(3)	I(1-8)			I(1-8)
Crataegus monogyna seedling	I(1-3)	I(1-2)	I(1-4)	I(1-3)			I(1-4)
Luzula sylvatica	I(1-4)	I(2-4)		I(5)			I(1-4)
Prunella vulgaris	I(2-5)	I(3)		I(1-2)		I(1-3)	I(1-5)
Vicia sepium	I(2-3)	I(3)		I(1-2)		I(1)	I(1-3)
Arctium minus agg.	I(2-3)	I(3)		I(1-2)		I(1-3)	I(1-3)
Rubus idaeus	I(3-6)			I(1-4)		I(1-2)	I(1-6)
Cardamine flexuosa	I(2-5)			I(1)		I(1-2)	I(1-5)
Holcus mollis	I(3-6)			I(1-4)		I(2-6)	I(1-6)
Campanula trachelium	I(1)			I(1-2)		I(1-3)	I(1-3)
Milium effusum	I(3-6)			I(1-5)		I(1-3)	I(1-6)
Veronica montana	I(3-4)			I(1-4)	I(3)		I(1-6)

Floristic table W8 (*cont.*)

	a	b	c	d	e	f	g	8
Plagiothecium sylvaticum	I(1–4)	I(1)			I(1–3)	I(1)		I(1–4)
Plagiomnium affine	I(2–3)				I(2)	I(2)	I(2–3)	I(2–3)
Ribes uva-crispa				I(1–2)	I(1–3)	I(1)	I(1)	I(1–3)
Cirriphyllum piliferum		I(1)			I(1–6)	I(1)	I(2)	I(1–6)
Rhytidiadelphus loreus		I(2)			I(1–4)	I(1–5)	I(1–2)	I(1–5)
Ranunculus auricomus	I(3)			I(2–3)		I(3)		I(2–3)
Alliaria petiolata	I(1–3)			I(1–2)		I(1)		I(1–3)
Iris foetidissima	I(1–3)			I(4–6)	I(1–4)			I(1–6)
Orchis mascula	I(2–3)	I(1)			I(2–3)			I(1–3)
Polygonatum multiflorum	I(1–3)	I(4)			I(1–4)			I(1–4)
Digitalis purpurea	I(2–4)		I(4)		I(1–5)			I(1–5)
Fagus sylvatica seedling		I(1–3)		I(1)	I(1–3)			I(1–3)
Chrysosplenium oppositifolium		I(1)		I(4)	I(1–5)			I(1–5)
Acer campestre seedling			I(3)	I(1–4)	I(1–2)			I(1–4)
Daphne laureola				I(2–4)	I(1–2)		I(1)	I(1–4)
Epilobium angustifolium	I(1–4)				I(2–7)		I(3)	I(1–7)
Plagiomnium rostratum	I(3)				I(1–2)		I(1–2)	I(1–3)
Climacium dendroides	I(4–5)				I(3)		I(1–4)	I(1–5)
Epilobium montanum		I(1)			I(1–3)		I(1–2)	I(1–3)
Lapsana communis		I(1)			I(1–3)		I(1–2)	I(1–3)
Carex remota	I(1–4)		I(4)					I(1–4)
Platanthera chlorantha	I(1)		I(1)					I(1)
Rumex crispus	I(3–4)			I(3)				I(3–4)
Plagiochila asplenoides	I(3–4)				I(1–3)			I(1–4)
Lophocolea heterophylla	I(1–4)				I(1)			I(1–4)
Dicranella heteromalla	I(2–3)							I(1–3)
Hypericum hirsutum			I(3)				I(1–2)	I(1–3)
Rubus caesius			I(1–3)				I(1–2)	I(1–3)
Ilex aquifolium seedling				I(1–3)	I(1–2)		I(1–3)	I(1–3)
Corylus avellana seedling				I(1)	I(1)			I(1)
Polypodium vulgare				I(1–2)	I(1)			I(1–2)
Viola odorata				I(1–4)	I(1–4)			I(1–4)
Ribes rubrum					I(1–3)	I(2)		I(1–3)

	a	b	c	d	e	f	g	8
Rhizomnium punctatum					I (1-3)		I (1-2)	I (1-3)
Ranunculus acris					I (1-3)		I (1)	I (1-3)
Rhytidiadelphus triquetrus					I (1-4)		I (1-4)	I (1-4)
Fissidens bryoides					I (1)		I (1)	I (1)
Mycelis muralis					I (1-2)		I (1-2)	I (1-2)
Pellia epiphylla					I (1-4)		I (2-3)	I (1-4)
Daphne mezereum					I (1)		I (1)	I (1)
Narcissus pseudonarcissus						I (4)	I (3)	I (3-4)
Number of samples	128	79	12	67	81	27	35	429
Number of species/sample	24 (5-48)	26 (12-35)	24 (17-43)	20 (7-53)	27 (6-53)	27 (16-37)	29 (19-64)	25 (5-64)
Tree height (m)	15 (8-26)	19 (9-35)	16 (12-20)	19 (8-32)	18 (8-32)	19 (15-20)	13 (8-15)	17 (8-35)
Tree cover (%)	79 (10-100)	85 (10-100)	53 (10-90)	89 (30-100)	83 (15-100)	84 (70-100)	92 (40-100)	83 (10-100)
Shrub height (m)	5 (1-12)	3 (1-7)	5 (2-8)	5 (1-10)	3 (1-12)	3 (1-4)	3 (3-4)	4 (1-12)
Shrub cover (%)	63 (1-100)	37 (1-100)	65 (10-100)	46 (1-100)	33 (0-90)	23 (0-100)	54 (10-100)	48 (0-100)
Herb height (cm)	36 (3-140)	32 (15-70)	44 (15-80)	33 (10-100)	42 (10-150)	29 (10-45)	no data	36 (3-150)
Herb cover (%)	81 (1-100)	92 (5-100)	90 (50-100)	87 (5-100)	96 (30-100)	94 (65-100)	93 (70-100)	87 (1-100)
Ground height (mm)	18 (10-50)	19 (5-50)	15 (5-40)	14 (10-30)	19 (5-80)	16 (10-30)	no data	17 (5-80)
Ground cover (%)	29 (0-100)	21 (0-100)	24 (0-95)	13 (0-100)	34 (1-95)	35 (5-85)	38 (10-80)	27 (0-100)
Altitude (m)	91 (15-203)	85 (10-250)	71 (30-120)	98 (5-230)	115 (8-240)	100 (40-235)	224 (60-290)	105 (5-290)
Slope (°)	2 (0-45)	10 (0-30)	0 (0-4)	4 (0-30)	25 (0-85)	18 (5-50)	40 (10-45)	11 (0-85)

a *Primula vulgaris-Glechoma hederacea* sub-community

b *Anemone nemorosa* sub-community

c *Deschampsia cespitosa* sub-community

d *Hedera helix* sub-community

e *Geranium robertianum* sub-community

f *Allium ursinum* sub-community

g *Teucrium scorodonia* sub-community

8 *Fraxinus excelsior-Acer campestre-Mercurialis perennis* woodland (total)

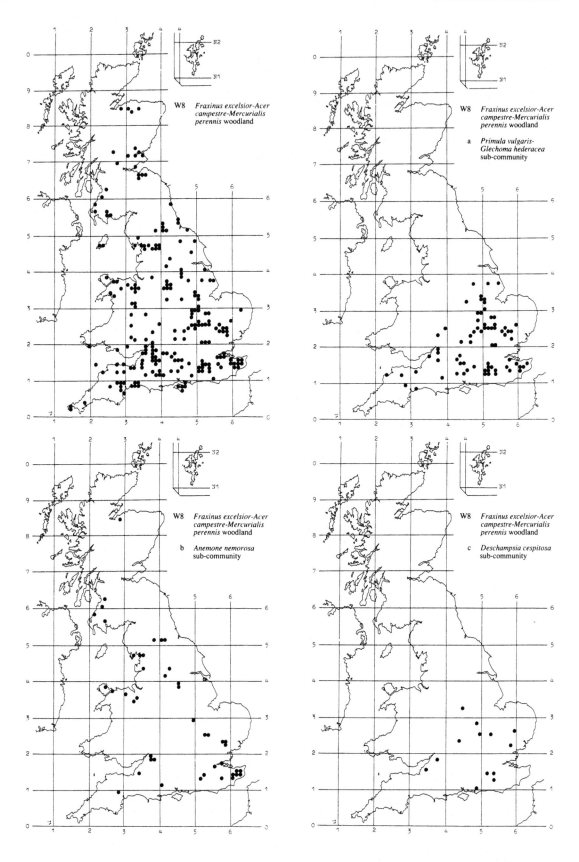

W8 *Fraxinus excelsior-Acer campestre-Mercurialis perennis* woodland

W8 *Fraxinus excelsior-Acer campestre-Mercurialis perennis* woodland

a *Primula vulgaris-Glechoma hederacea* sub-community

W8 *Fraxinus excelsior-Acer campestre-Mercurialis perennis* woodland

b *Anemone nemorosa* sub-community

W8 *Fraxinus excelsior-Acer campestre-Mercurialis perennis* woodland

c *Deschampsia cespitosa* sub-community

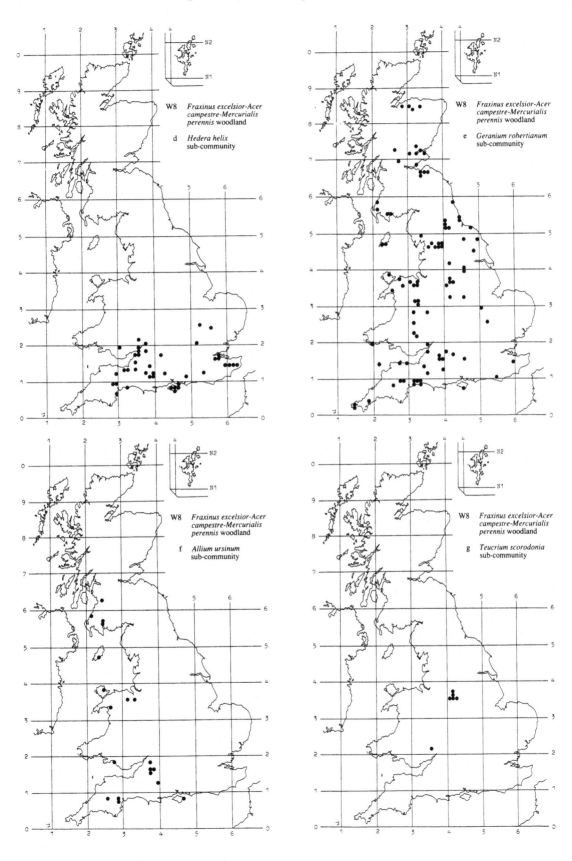

W8 *Fraxinus excelsior-Acer campestre-Mercurialis perennis* woodland

d *Hedera helix* sub-community

W8 *Fraxinus excelsior-Acer campestre-Mercurialis perennis* woodland

e *Geranium robertianum* sub-community

W8 *Fraxinus excelsior-Acer campestre-Mercurialis perennis* woodland

f *Allium ursinum* sub-community

W8 *Fraxinus excelsior-Acer campestre-Mercurialis perennis* woodland

g *Teucrium scorodonia* sub-community

W9

Fraxinus excelsior-Sorbus aucuparia-Mercurialis perennis woodland

Synonymy

Upland ashwood Tansley 1939; *Corylo-Fraxinetum* Br.-Bl. & Tx 1952 *p.p.*; Herb-rich birchwood Pigott 1956a *p.p.*; *Fraxinus-Brachypodium sylvaticum* nodum McVean & Ratcliffe 1962; *Corylus* scrub McVean & Ratcliffe 1962; Mixed deciduous woodland McVean & Ratcliffe 1962; *Betula*-herb nodum, basiphilous facies McVean & Ratcliffe 1962; Ashwood McVean 1964; *Dryopterido-Fraxinetum blechnetosum* Klötzli 1970; *Sorbo-Brachypodietum* Graham 1971; *Fraxinus excelsior-Brachypodium sylvaticum* Association Birks 1973; *Corylus avellana-Oxalis acetosella* Association Birks 1973 *p.p.*; *Betula pubescens-Cirsium heterophyllum* Association Birks 1973; Mixed Deciduous Woodland Ferreira 1978 *p.p.*; Ashwych elm stand types 1Ab & 1D Peterken 1981 *p.p.*; Hazel-ash stand type 3C Peterken 1981 *p.p.*; Alder stand type 7D Peterken 1981 *p.p.*; Birch stand type 12B Peterken 1981 *p.p.*; *Primulo-Quercetum* J. Tüxen *apud* Birse 1982 *p.p.*; *Querco-Ulmetum glabrae* Birse & Robertson 1976 *emend.* Birse 1984 *p.p.*

Constant species

Fraxinus excelsior, Corylus avellana, Dryopteris filix-mas, Mercurialis perennis, Oxalis acetosella, Viola riviniana, Eurhynchium praelongum, E. striatum, Plagiomnium undulatum, Thuidium tamariscinum.

Rare species

Actaea spicata, Bromus benekenii, Crepis mollis, Gagea lutea, Polygonatum verticillatum.

Physiognomy

In the *Fraxinus-Sorbus-Mercurialis* woodland, as in its southern lowland counterpart, the *Fraxinus-Acer-Mercurialis* woodland, *Fraxinus excelsior* and *Corylus avellana* play a major role in the definition of the canopy. Both species are constant here and frequently abundant, often dominating the woody cover in various proportions. Typically, though, trees and shrubs with more continental affinities are very scarce. *Tilia cordata* survives on some Lake District crags in association with the community (Pigott & Huntley 1978), some Cumbrian, Pennine and southern Scottish stands provide northerly localities for *Acer campestre, Rhamnus catharticus* or *Euonymus europaeus* (Ratcliffe 1977) and *Viburnum opulus* extends further north in the more oceanic parts of north-west Scotland (Birks 1973, Ratcliffe 1977) but, for the most part, this kind of woodland lies beyond the north-western limit of these species.

By contrast, birch, almost always *Betula pubescens* on the moist soils characteristic of this woodland and increasingly to the north ssp. *carpatica*, and *Sorbus aucuparia* are much more frequent than in most *Fraxinus-Acer-Mercurialis* woodlands. *B. pubescens* is the more common and generally the more abundant but *S. aucuparia* can be locally prominent, especially in ungrazed stands (McVean 1964a) and quite commonly the two co-dominate with *Fraxinus* and *Corylus*. *Alnus glutinosa* occurs very occasionally where there is local flushing but it is rarely more than as scattered individuals, a good distinction between this community and the *Alnus-Fraxinus-Lysimachia* woodland, a common kind of flush alderwood in the sub-montane parts of northern and western Britain.

There is a trend within the community from well-developed high-forest canopies to scrubby woodland with only occasional trees. In the former, *Fraxinus* can be accompanied by some *Ulmus glabra, Acer pseudoplatanus* and oak, usually here *Quercus petraea* or hybrids, in canopies which, apart from the frequent presence of scattered *B. pubescens* and *S. aucuparia*, can closely resemble those of north-western *Fraxinus-Acer-Mercurialis* woodlands. In such stands, *Corylus* usually dominates in a distinct understorey, often with some other small trees and shrubs, notably hawthorn, here *Crataegus monogyna* with the single striking exception of a far-flung Pennine location for *C. laevigata* (Ratcliffe 1977). At the other extreme, scattered trees, often just *Fraxinus*, emerge from scrubby mixtures of *Corylus, B. pubescens* and *S. aucuparia*. On the typically irregular topography occupied by this kind of woodland, even the high-forest canopies tend to be uneven-topped and of somewhat patchy cover, but there is a distinct association between the more scrubby stands and more exposed situations like ravine tops and sides in the windy far north-west of Scotland. In the latter area, though, there may have been a long history of timber removal from these woods (McVean & Ratcliffe 1962, McVean 1964a, Birks 1973, Birse 1982, 1984). In other stands, a dense shrub layer, and especially an abundance of *Corylus*, reflects past coppicing but this treatment is nothing like as systematic or widespread as in the *Fraxinus-Acer-Mercurialis* woodland. Surviving stands are often remote from existing settlements and ground is frequently intractable and the usual pattern here is of occasional, ill-defined parcels in those woods which are nearer to villages or upland farmsteads.

The irregular topography also has a strong influence on the appearance of the field layer. In marked contrast to many *Fraxinus-Acer-Mercurialis* woodlands, where the vegetation is disposed over undulating plateaus and graded slopes, there is here no consistent pattern of dominance among the herbs. Very commonly, the plants form complex mosaics over the highly uneven, often steep and sometimes unstable deposits of drift or head which can choke upland ravines, or over the intricately-weathered surfaces of rock exposures in pavements or tumbles of boulders. Often, too, local flushing adds further variety. There are also considerable differences in total herb cover. Slumping can create temporarily bare areas and where denser canopy shade is combined with grazing, as in the Typical sub-community, rather open field layers are commonplace, a feature which accentuates the appearance of a diverse patchwork of plants. Even under the lighter canopies of the *Crepis* sub-community, however, where there is no grazing, there is no consistent dominant, but varied mixtures of clumps and tussocks of many different species.

In floristic terms, though, there are some important similarities between the field layers of the *Fraxinus-Sorbus-Mercurialis* and *Fraxinus-Acer-Mercurialis* woodlands. *Mercurialis perennis* itself and *Hyacinthoides non-scripta* are both very frequent and, in many stands, they are patchily dominant in spring and early summer. But they are not so consistently abundant as further south: *Mercurialis* especially tends to do badly on the moist soils here and it thins out markedly towards the north of the range of the community. Both species also tend to suffer from competition from taller herbs in the ungrazed *Crepis* sub-community. The *Fraxinus-Sorbus-Mercurialis* woodland also provides the major northern locus for *Circaea lutetiana* and *Geum urbanum*, and *Geranium robertianum* and *Brachypodium sylvati-*

cum are very common, too. The last is especially prominent on steeper, somewhat drier banks and overall is as abundant in the community as *Mercurialis*: it has been used in some earlier accounts to define this kind of woodland (e.g. McVean & Ratcliffe 1962, Birks 1973) or sub-divisions of it (e.g. Birse 1984).

But it is often among plants characteristic of heavier, moister soils that the most obvious similarities between the two kinds of woodland are to be seen. Thus *Primula vulgaris, Poa trivialis* and *Deschampsia cespitosa*, all of them species which become very local where the *Fraxinus-Acer-Mercurialis* woodland extends on to the permeable limestones of north-west England, reappear here with renewed frequency. *Viola riviniana* is very common, too, though its frequent companion in the south, *V. reichenbachiana*, only just reaches into the range of the *Fraxinus-Sorbus-Mercurialis* woodland. Low-growing ground-cover plants are also often prominent especially in the more open field layer of the Typical sub-community, though *Lamiastrum galeobdolon* is recorded only in the more southerly stands (probably because low summer temperatures further north inhibit sexual reproduction: Packham 1983) and *Glechoma hederacea* is likewise scarce. But here this niche is well filled by *Potentilla sterilis, Veronica chamaedrys* and *V. montana. Anemone nemorosa* and *Ranunculus ficaria* can be found occasionally too, though they do not occur as vernal dominants in distinctive zones but with patchy local prominence or scattered throughout. Finally, by contrast with *Fraxinus-Acer-Mercurialis* woodlands on the better-aerated soils of the limestones of north-western England, *Urtica dioica* and *Galium aparine* are of only occasional and local occurrence. In all these respects, the *Fraxinus-Sorbus-Mercurialis* woodland resembles not so much the geographically close north-western *Fraxinus-Acer-Mercurialis* woodlands but the edaphically similar south-eastern types, especially those of wetter areas like the western Weald.

Other floristic features, though, are very distinctive and help separate the two communities. First, in the much more humid climate of this sub-montane woodland, ferns are very prominent. *Dryopteris filix-mas* becomes constant here and *D. borreri* also occurs occasionally but it is the high frequency of *Athyrium filix-femina* that is an especially good indicator of soils that are kept moist throughout the year and which show a marked tendency towards superficial eluviation. Where the soils are somewhat drier, as in the Typical sub-community, *Dryopteris dilatata* provides a further enrichment of the fern flora; and locally pronounced surface depression of pH, as occurs very obviously over nidus-capped boulders in the extremely wet climate of north-west Scotland, can allow *Blechnum spicant* to flourish. It may be this tendency towards surface acidity that is responsible for the markedly low frequency of

Phyllitis scolopendrium here, because climatic conditions could scarcely be better for this strongly oceanic fern. Many stands lack the exposures of fissured limestone that it favours, though where the community does extend over scree or pavement (as on the Pennine Carboniferous, the Cleveland Oolite or the Durness Limestone of Skye) it can appear in some local abundance together with smaller calcicolous ferns such as *Asplenium trichomanes*, *A. ruta-muraria*, *A. adiantum-nigrum*, the rarer *A. viride* and *Cystopteris fragilis*. Other ferns recorded in the community at low frequency are *Polystichum aculeatum* and *P. setiferum*. *Pteridium aquilinum* is characteristically uncommon here and hardly ever abundant; with *Thelypteris limbosperma* and increased frequencies of *Blechnum spicant*, it is very much more typical of the analogue of this community on more base-poor sub-montane soils, the *Quercus-Betula-Oxalis* woodland.

The wetness of the soils throughout the year may also play a part in the second distinctive features of the field layer here, the high frequency and abundance of *Oxalis acetosella*. This is not a common species in the *Fraxinus-Acer-Mercurialis* woodland, though it can become locally abundant there, as for example when stands are coppiced, and it seems to be limited by dryness of the soils towards the southern limit of its range (Packham 1978). Here, though, it thrives, whether under the denser tree and shrub canopies of the Typical sub-community or among the luxuriant cover of tall herbs in the *Crepis* sub-community. Interestingly, both *Oxalis* and *Viola riviniana* show the same pattern of a rise to constancy in the sub-montane communities of both mercury and bluebell woods in Britain. And, in geographically intermediate regions, like Derbyshire, they both tend to be more abundant on north-facing slopes (e.g. Grime & Lloyd 1973).

A third characteristic is that the field layer is often quite grassy in appearance. Quite apart from species such as *Brachypodium sylvaticum*, *Deschampsia cespitosa* and *Poa trivialis* and more occasional records for *Bromus ramosus*, *Festuca gigantea*, *Milium effusum* and *Melica uniflora*, there can be prominent *Arrhenatherum elatius* and an abundance of species whose establishment is probably related to grazing by stock and deer, but which flourish best in the more open and less heavily grazed vegetation of the *Crepis* sub-community, e.g. *Dactylis glomerata*, *Holcus lanatus*, *H. mollis*, *Agrostis capillaris*, *Anthoxanthum odoratum*.

Finally, among the herbs, there are some plants which give a distinctly northern feel to this vegetation. The Northern Montane *Rubus saxatilis* and *Actaea spicata* occur very occasionally and, among the preferentials of the *Crepis* sub-community, *Crepis paludosa* itself, *Cirsium helenioides* (both Continental Northern) and *Trollius europaeus* (Northern Montane).

The bryophytes of the community are also distinctive in their abundance and variety, commonly forming a patchy mat, sometimes a virtually continuous carpet, over the surface of the soil and exposed rock. *Eurhynchium praelongum*, *E. striatum*, *Plagiomnium undulatum* and *Thuidium tamariscinum* are all constant, *Mnium hornum* and *Atrichum undulatum* frequent and *Cirriphyllum piliferum*, *Rhytidiadelphus triquetrus*, *Hypnum cupressiforme*, *Plagiochila asplenoides* and *Lophocolea bidentata s.l.* occasional. Various kinds of enrichment of this basic flora are visible in the community. Where the soil is more calcareous, stony or firm or where outcrops of limestone occur, species such as *Fissidens cristatus*, *F. bryoides*, *Thamnobryum alopecurum*, *Tortella tortuosa*, *Homalothecium sericeum* and *Neckera crispa* can appear. Then, with the increasingly wet climate of the extreme north-west, *Hylocomium splendens*, *H. brevirostre*, *Isothecium myosuroides*, *I. myurum* and *Thuidium delicatulum* become more common, and more strictly Atlantic species such as *Metzgeria leptoneura*, *Saccogyna viticulosa*, *Plagiochila spinulosa* and *Scapania gracilis* can be recorded. Many of these are of quite wide edaphic tolerance (Ratcliffe 1968) and some help accumulate distinctly acid mats of humus over the limestone boulders such that quite calcifugous bryophytes make a prominent contribution. Also in this region, the very humid climate encourages a lush growth of species characteristic of rotting wood, on which *Riccardia palmata* and *Scapania umbrosa* are sometimes found, and a very striking epiphytic flora of *Corylus* and *Fraxinus* twigs with festoons of *Ulota calvescens*, *U. phyllantha*, *U. crispa*, *Frullania teneriffae* and *F. dilatata*. Tree boles can support a distinctive clothing of *Hypnum mammillatum* and *Isothecium mysuroides* with *Metzgeria furcata* and *Radula complanata* and ground-growing species can extend up the tree bases (Ratcliffe 1968, 1977, Birks 1973). Fragments of all these components can be found in the community with decreasing frequency and richness in moving away from the far north-west as the climate becomes less humid and locally polluted (e.g. Graham 1971).

Sub-communities

Typical sub-community: *Fraxinus-Brachypodium sylvaticum* nodum McVean & Ratcliffe 1962 *p.p.*; Betula-herb nodum, basiphilous facies McVean & Ratcliffe 1962; Ashwood McVean 1964a *p.p.*; *Dryopterido-Fraxinetum blechnetosum*, *Ulmus* variant Klötzli 1970; *Sorbo-Brachypodietum* Graham 1971 *p.p.*; *Fraxinus excelsior-Brachypodium sylvaticum* Association Birks 1973; *Corylus avellana-Oxalis acetosella* Association Birks 1973 *p.p.*; Mixed Deciduous Woodland Ferreira 1978 *p.p.*; *Primulo-Quercetum*, *Deschampsia cespitosa* subassociation J. Tüxen *apud* Birse 1982 *p.p.*;

Querco-Ulmetum glabrae, Typical subassociation Birse & Robertson 1976 *emend.* Birse 1984 *p.p. Fraxinus* and *Corylus* are the most frequent and often the most abundant woody species here with *Betula pubescens* and *Sorbus aucuparia* less common and generally in smaller amounts. There is also a considerable diversity among the associated trees and shrubs. Where the vegetation occurs as high forest, a frequent occurrence here, *Fraxinus* generally dominates in a distinct, though often quite low (usually 12–20 m) and uneven-topped canopy, but there is commonly some *Ulmus glabra* and *Acer pseudoplatanus* and occasionally a little oak, usually *Quercus petraea* or the hybrids, much more rarely *Q. robur*. *Alnus glutinosa* occurs very occasionally and there is sometimes an occasional pine or larch. *Fagus sylvatica* is found very rarely, though some of the planted 'herbaceous beechwoods' described from the Aberdeen area by Watt (1931*a*) are perhaps best considered in relation to this community.

In such stands as these, *B. pubescens* (very occasionally with *B. pendula*) and *S. aucuparia* usually form a slightly lower tier, though well-grown specimens can break the main canopy. Beneath them, there is generally a well-defined understorey in which *Corylus* (sometimes increased in cover by coppicing) almost always predominates but *Crataegus monogyna* is very common in this sub-community and it can occasionally rival *Corylus*. *Ilex aquifolium*, *Sambucus nigra* and *Prunus padus* occur sparsely and usually at low cover though *Ilex* is sometimes found in local abundance and with individuals of grand stature. Sapling trees are quite common with young *Fraxinus*, *U. glabra*, *A. pseudoplatanus*, *B. pubescens* and *S. aucuparia*. On more unstable slopes, on the exposed margins of stands and on moving towards the north-west seaboard of Scotland, much scrubbier woodlands of this kind are found with a dominant layer of *Corylus*, *B. pubescens* (often obviously ssp. *carpatica*) and *S. aucuparia*, at around 10–12 m, and with scattered emergent trees, usually *Fraxinus*, much less frequently *U. glabra*, *Q. petraea* or *A. pseudoplatanus*.

The floristics of the field layer are well defined in qualitative terms but the quantitative composition is very variable and quite often the total cover is rather open, especially in denser shade or where the substrate is unstable: extensive stretches of bare soil are common here. Structurally, the most prominent plants are often ferns with large spreading crowns of *Athyrium filix-femina* and dryopteroids, including here *Dryopteris dilatata*. Between these, *Mercurialis perennis*, *Hyacinthoides non-scripta*, *Brachypodium sylvaticum*, *Oxalis acetosella* and *Primula vulgaris* all occur frequently and with patchy local abundance. *Viola riviniana*, *Geranium robertianum*, *Veronica chamaedrys*, *Lysimachia nemorum*, *Poa trivialis*, *Epilobium montanum* and *Sanicula europaea* are quite common too and, again, these can be prominent in small patchy mosaics. More preferential for this sub-community are *Geum urbanum*, *Circaea lutetiana*, *Potentilla sterilis* and, less frequently *Urtica dioica*, *Galium aparine*, *Rubus fruticosus* agg., *Hedera helix*, *Silene dioica* and *Veronica montana*. Seedlings of *Fraxinus* are also very much commoner here than in the *Crepis* sub-community. *Lamiastrum galeobdolon* and, somewhat more extensively, *Arum maculatum* occur in more southerly stands and *Circaea × intermedia*, though rather infrequent, is a good marker of this kind of woodland. *Galium odoratum* likewise occurs rather sparsely but it characteristically picks out little cascades of talus down which water periodically runs. Where larger boulders or outcrops of limestone provide crevices, *Phyllitis scolopendrium* and other calcicolous ferns appear.

Although these woodlands seem to be quite commonly grazed, grasses, apart from *B. sylvaticum*, *P. trivialis* and occasional *Deschampsia cespitosa*, are generally of low abundance here. *Dactylis glomerata* occurs quite frequently and there is sometimes a little *Arrhenatherum elatius* and *Agrostis capillaris*, but they do not attain the lush growth that helps characterise the *Crepis* sub-community. Grazing also restricts the occurrence of plants such as *Cirsium helenioides*, *Crepis paludosa*, *Geranium sylvaticum*, *Filipendula ulmaria* and *Luzula sylvatica* to very occasional specimens in more inaccessible places.

On the more stable areas of bare soil, bryophytes are often very abundant, frequently accounting for up to 50% cover of the ground surface between and among the herbs. All the community constants are well represented and, on patches of compact soil, these together with *Mnium hornum*, *Atrichum undulatum*, *Fissidens cristatus*, *F. bryoides* and *Pellia epiphylla* form a very distinctive suite. Rockier stands and those in more oceanic regions exhibit the characteristic kinds of enrichment described earlier.

***Crepis paludosa* sub-community:** Herb-rich birchwood Pigott 1956*a p.p.*; Hazel scrub McVean & Ratcliffe 1962; Mixed deciduous woodland McVean & Ratcliffe 1962; *Dryopterido-Fraxinetum blechnetosum* Klötzli 1970 *p.p.*; *Sorbo-Brachypodietum* Graham 1971 *p.p.*; *Betula pubescens-Cirsium heterophyllum* Association Birks 1973; Mixed Deciduous Woodland Ferreira 1978 *p.p.*; *Primulo-Quercetum, Deschampsia cespitosa* subassociation J. Tüxen *apud* Birse 1982 *p.p.* The woody cover here is less diverse than in the Typical sub-community with *U. glabra*, *A. pseudoplatanus*, oaks and *C. monogyna* occurring only rarely. Quite commonly, too, the canopy is rather open with scattered *Fraxinus*, larger specimens of *B. pubescens* and *S. aucuparia* and very occasional *Alnus*, emerging from a patchy shrub layer with *Corylus* and smaller *B. pubescens* and *S.*

aucuparia. Prunus padus and *Salix cinerea* occur occasionally and *Populus tremula* rarely. Apart from young *B. pubescens* and *S. aucuparia*, saplings are uncommon.

Most of the characteristic community herbs maintain high frequency, though some show a depressed abundance in this vegetation and *Mercurialis* especially is less common and prominent on the sometimes moister soils here. *Dryopteris dilatata* is also very scarce and there is some reduction in ground cover plants like *Potentilla sterilis*; *Fraxinus* seedlings are extremely rare. By contrast, there is a marked increase in a variety of different plants, most colourfully among taller herbs which, where the woody cover is more open, grow luxuriant and flower profusely. *Filipendula ulmaria, Conopodium majus, Geum rivale* (largely replacing *G. urbanum* in this sub-community), *Rumex acetosa, Succisa pratensis, Senecio jacobaea, Stachys sylvatica, Cruciata laevipes, Alchemilla glabra* and trailing *Vicia sepium* are all preferential here as well as some distinctly northern species such as *Crepis paludosa, Cirsium helenioides* and *Trollius europaeus*. The nationally rare *Crepis mollis* is also found in some stands, providing a rather different phytogeographic link with more continental mountain vegetation of this kind (Pigott 1956a, Ratcliffe 1977).

Then, grasses are numerous and patchily abundant with frequent records for *Deschampsia cespitosa, Arrhenatherum elatius, Agrostis capillaris, Anthoxanthum odoratum, Holcus lanatus* and *H. mollis*, as well as the community species *Dactylis glomerata* and *Poa trivialis. Luzula sylvatica* is also sometimes prominent. *Carex pallescens* is a rather infrequent but very characteristic plant in this vegetation and *Rubus idaeus* can produce a patchy underscrub. Where the substrate is more unstable, as, for example, where local flushing causes a slumping of soil, a more open and lower cover of herbs can develop with mats of *Agrostis stolonifera* and *Ranunculus repens* and scattered *Tussilago farfara. Ranunculus ficaria* and *Allium ursinum* may also occur in moister places and *Paris quadrifolia* is scarce but distinctive. Where this sub-community has developed in bouldery woods in the extreme north-west, such grassy tall-herb cover may form a mosaic with fragments of more calcifugous vegetation with *Blechnum spicant* and *Potentilla erecta* on boulder tops but species such as *Vaccinium myrtillus* and *Deschampsia flexuosa* are typically absent.

As in the Typical sub-community, there is generally a diverse and extensive bryophyte flora here, though some bare soil acrocarps such as *Mnium hornum, Atrichum undulatum* and *Fissidens* spp. are rather less common. Larger pleurocarps, however, thrive among the damp bases of the herbs and over the extensive litter and some species, like *Cirriphyllum piliferum, Rhytidiadelphus triquetrus* and the liverwort *Plagiochila asplenioides* (including var. *major*) increase in frequency. There are

some stronger preferentials here, too, with *Rhytidiadelphus squarrosus, R. loreus, Plagiothecium denticulatum, Rhizomnium punctatum* and *Plagiomnium rostratum* enriching the flora. In more oceanic regions, there is the further variety of species able to take advantage of enhanced superficial eluviation on the tops of boulders or the increasingly humid atmosphere.

Habitat

The *Fraxinus-Sorbus-Mercurialis* woodland is characteristic of permanently moist brown soils derived from calcareous bedrocks and superficials in the sub-montane climate of north-west Britain. Over much of its range, it is typically a community of valley heads in the upland fringes but, in the cool oceanic parts of north-west Scotland, it descends almost to sea-level. Timber and underwood removal have affected the floristics and physiognomy and grazing is important in mediating the differences between the two sub-communities.

The climate over the range of the community is cool, wet, windy and cloudy (e.g. Manley 1936, *Climatological Atlas* 1952, Chandler & Gregory 1976) and it affects the vegetation directly, through temperature, humidity and perhaps also through exposure to wind, and indirectly, through the influence of rainfall on soil development. Mean annual maximum temperatures throughout the range are always less than 26 °C, mostly less than 25 °C and substantially lower over much of north-west Scotland (Conolly & Dahl 1970). It is this summer coolness, combined with the high frequency of daytime cloudiness, that seems to affect pollination, fertilisation and seed-ripening in the more continental species characteristic of our mixed deciduous woodlands. Thus, the northern and western limits of woody plants like *Tilia cordata, Acer campestre, Cornus sanguinea, Euonymus europaeus* and *Rhamnus catharticus* (e.g. Pigott & Huntley 1978, 1981) and herbs such as *Lamiastrum galeobdolon* (Packham 1983) and *Arum maculatum*, correspond crudely with the southern and eastern limit of this community and help mark it off from the *Fraxinus-Acer-Mercurialis* woodland. On the positive side, there is an appearance here, especially well seen in less heavily grazed stands, of plants with Northern Montane affinities (e.g. *Prunus padus, Rubus saxatilis, Actaea spicata, Trollius europaeus*) or with Continental Northern distributions (*Crepis paludosa, Cirsium helenioides*) which serve to emphasise the links between this vegetation and woodlands in the cooler parts of Northern Europe (e.g. Pigott 1956a, 1958, Bradshaw 1962, McVean & Ratcliffe 1962, Birks 1973).

Winter climate, however, is not always severe. Towards the western fringes of the range of the community, February minima are usually above freezing (Chandler & Gregory 1976) and annual accumulated temperatures are, for the most part, above 800 °C (Page 1982). In many areas, morning snow-lie is nil or negligible (Ratcliffe 1968, Page 1982). Even where the community extends into the upland fringes, where the winter climate becomes much more bitter (e.g. Manley 1936, 1945), deeply-incised valleys can afford some protection from extreme cold. This moderate to pronounced mildness, combined with the extreme humidity of many sites (with 160 to more than 220 wet days yr^{-1}: Ratcliffe 1968), helps give the woodland, especially its field and ground layers which have the additional protection of the woody cover, its markedly oceanic and winter-green character, with an abundance of ferns and bryophytes. Although this feature becomes much more obvious on the north-west seaboard of Scotland (e.g. McVean & Ratcliffe 1962, Ratcliffe 1968, Birks 1973, Birse 1982, 1984), such that the community takes on much of the character of Irish woodlands of this kind (e.g. Braun-Blanquet & Tüxen 1952), the presence of these elements throughout helps distinguish the community from the *Fraxinus-Acer-Mercurialis* woodland. Ferns and bryophytes do increase somewhat in that vegetation towards the north and west of England and Wales but sharpness of drainage over the permeable limestones there often offsets the effect of increased rainfall. Although the more Atlantic bryophytes disappear fairly rapidly in moving away from the extreme north-west, it is possible that atmospheric pollution plays some part in restricting their occurrence in Pennine stands of the community (e.g. Graham 1971).

In the less continental and more sub-montane and cool oceanic parts of Britain, the *Fraxinus-Sorbus-Mercurialis* woodland is probably the climax community of more base-rich soils. However, certain extremes of climate may play a part in maintaining the vegetation as permanent scrub, even though its essential floristic character is preserved. In the upland fringes, for example, as in valley heads in the Yorkshire Dales and along the Tees and Wear in Durham, stands sometimes extend up over the top of the dale sides and here tree cover thins out with increased exposure to wind and cold. In such situations, the influence of near-freezing ground water in spring may further inhibit the development of an extensive woody cover. Even in the much more equable climate of the far north-west, the very frequent gales may be sufficient to maintain the woody cover as low, wind-pruned shrubs and trees, though here it needs to be remembered that there has probably been extensive removal of any larger timber trees, perhaps over long periods of time (e.g. McVean & Ratcliffe 1962, McVean 1964a, Birks 1973, Birse 1982, 1984).

Within the north-western parts of Britain, exposures of calcareous bedrocks are generally small and scattered, with the notable exception of the Carboniferous Limestone which underlies the quite numerous stands of the community clustered in the Pennines. The *Fraxinus-*

Sorbus-Mercurialis woodland also occurs over this rock at a few localities in south Wales. Other more extensive calcareous sedimentary deposits on which the community is found are the Jurassic limestones of the northern part of the North York Moors and Skye and the Cambrian/Ordovician Durness Limestone, also on Skye (Birks 1973) and along the Moine thrust on the Scottish mainland (McVean & Ratcliffe 1962). Elsewhere on sedimentary rocks, this woodland can be found picking out more calcareous strata occurring within a mass of more acidic rocks, as over Ordovician and Silurian shales exposed in deeply-cut valleys in parts of Wales and south-east Scotland (Ratcliffe 1977, Ferreira 1978) and, more rarely, on parts of the Devonian Old Red Sandstone in south Wales and Cumbria. Some more basic igneous and metamorphic deposits also provide a suitable substrate, for example calcite beds within the Borrowdale Volcanics in the Lake District, Tertiary basalts on Skye and Mull, and Moine and Dalradian schists in the Scottish Highlands (McVean & Ratcliffe 1962, Birks 1973).

The pronounced wetness of the climate in this region, with annual rainfall always in excess of 1200 mm and, in many areas, more than 1600 mm (*Climatological Atlas* 1952, Chandler & Gregory 1976), means that, even where the soils are derived entirely from native parent materials, there is a strong tendency for them to be continually moist and superficially eluviated of free calcium carbonate. Thus, even where such soils are shallow, they are rarely of the typical dry and calcareous rendzina form so characteristic of *Fraxinus-Acer-Mercurialis* woodlands over the permeable limestones of north-western England. Indeed, where the rainfall is very heavy, quite acid rankers can develop directly over the surface of exposed limestones within stands of this woodland and encourage a sporadic representation of calcifuges, as on Skye (Birks 1973).

Very commonly, however, the direct influence of the calcareous bedrocks on pedogenesis is further reduced by the presence of superficial deposits, especially till or head or more recently redistributed material that has slumped down the often steep slopes here. Even where the rainfall is somewhat less heavy, soils therefore tend strongly towards the brown earth type. Where the superficials have a substantial fine fraction, a frequent occurrence, this further increases the tendency to permanent wetness, often making drainage poor or strongly impeded with marked gleying below (Pigott 1956a, 1978b). Local flushing with ground water can accentuate this character. It is this heavy and moist nature of many of the soils here that accounts for the general similarity between the field layer of the *Fraxinus-Sorbus-Mercurialis* woodland and that of the south-eastern types of *Fraxinus-Acer-Mercurialis* woodland, with the resurgence of species such as *Primula vulgaris, Poa trivialis, Deschampsia cespitosa, Filipendula ulmaria, Anemone nemorosa*, the ground-cover replacements for *Lamiastrum galeobdolon* and *Glechoma hederacea* and a relative scarcity of *Urtica dioica* and *Galium aparine*, typical of better-aerated and more eutrophic mull-rendzinas. And, since the soils do not experience summer droughting, *Oxalis acetosella* is able to establish itself as a constant element (Packham 1978).

Where the superficials are more free-draining, plants typical of drier and more calcareous conditions may be able to attain local prominence on soils which tend towards the brown calcareous earth type, but the susceptibility to leaching inhibits this. Moreover, the coarser and more permeable elements in the till are often not of a calcareous nature, but derived rather from siliceous rocks, so that pedogenesis often moves towards the development of surface-acid brown earths.

Such superficial deposits have often been laid down or redistributed in very irregular fashion over the rugged topography of valley sides and ravines or in intricate patterns in weathered crevices on pavements and between talus fragments. Flushed areas, too, can have a very complex disposition as ground water meets impervious interbedded deposits in the bedrock sequence or encounters patches of heavier-textured till. Although the *Fraxinus-Sorbus-Mercurialis* woodland is thus generally associated with brown earths, individual stands characteristically have a fine, often complex and sometimes chaotic mosaic of soils which helps account for the diversity of the field layer and the absence of consistent dominants. Commonly, fairly moist and somewhat surface-leached brown earths form the matrix, but steeper, drier banks can have brown calcareous earths and with fragmentary rendzinas over outcrops of harder limestone; here and there, more acid brown earths or rankers can pick out coarse sandy drift or elevated bluffs or boulder tops; and scattered throughout, in hollows, along the foot of slopes or around flushes or below small seepage lines, the soils tend towards surface-water gleys.

It is probable that flushing plays some part in the distinctive character of the *Crepis* sub-community. Increased soil moisture and substrate instability could help retard the development of a dense tree cover and so maintain the patchy shade in which the tall herbs and grasses thrive (e.g. Pigott 1956a). Several of the preferentials (e.g. *Filipendula ulmaria, Deschampsia cespitosa, Geum rivale, Crepis paludosa*) are also species which show a clear predilection for moister soils. But grazing is undoubtedly a major factor in limiting the growth of many of its characteristic plants in the Typical sub-community. Stock, usually sheep but sometimes cattle, and both roe (*Capreolus capreolus*) and red deer (*Cervus elaphus*) commonly graze and browse in these woodlands and it is only where they are excluded, either by difficult terrain or by fencing (as in the Rassal Ashwood

enclosure: Ratcliffe 1977) that the *Crepis* sub-community develops its full richness and luxuriance.

In contrast to many *Fraxinus-Acer-Mercurialis* woodlands, we know very little of the treatment history of stands of this community. Timber removal has probably contributed to the scrubby physiognomy of many Scottish woods of this kind (e.g. McVean & Ratcliffe 1962, McVean 1964a, Birks 1973, Birse 1982, 1984) and confused the limiting effect of climatic exposure. Such coppices as there are seem to be universally neglected (Peterken 1981). It is possible that renewed cutting in wetter sites could favour a temporary development of the *Crepis* sub-community.

Zonation and succession

Transitions between the two sub-communities of the *Fraxinus-Sorbus-Mercurialis* woodland are primarily related to grazing pressure. The *Crepis* sub-community tends to replace the Typical sub-community in areas inaccessible to large herbivores, like ledges and steeper ravine slopes or pavements with numerous deep grikes (e.g. McVean & Ratcliffe 1962, Birks 1973, Ratcliffe 1977) and, because of the topographic irregularity of many sites, such zonations are usually complex mosaics with small fragments of the former scattered through the latter. However, there is no doubt that the *Crepis* sub-community attains its most luxuriant development where lack of grazing is combined with abundant soil moisture, so stands also often pick out dripping rock faces or flushed areas.

Zonations to other kinds of woodland are usually related more directly and exclusively to changes in parent materials and soil conditions. Where the soils are derived entirely from native bedrocks, a common pattern is for the *Fraxinus-Sorbus-Mercurialis* woodland to give way to the *Quercus-Betula-Oxalis* woodland as calcareous rocks with more base-rich mull brown earths pass to more acid deposits with leached brown earths with moder or mor humus. *Betula pubescens, Sorbus aucuparia* and *Corylus* may run through both communities, together with *Hyacinthoides, Oxalis, Viola riviniana* and many of the grasses, but *Fraxinus* (and *Ulmus glabra* if it is present) fade, and herbs such as *Mercurialis, Geum urbanum* and *Circaea lutetiana* are replaced by *Potentilla erecta, Galium saxatile* and *Pteridium*.

In regular sequences of limestones or calcareous shales and arenaceous strata, a frequent feature of Carboniferous deposits, this kind of zonation can be well ordered and obvious: in parts of the Yorkshire Dales, for example, where deep valleys have been cut through more or less level-bedded rocks, a typical pattern is for the *Fraxinus-Sorbus-Mercurialis* woodland to occupy the valley-side slopes over the Great Scar limestone with the *Quercus-Betula-Oxalis* woodland on the brow tops, over Yoredale deposits (Figure 20). Where sequences are less ordered, deposits thinner or

Figure 20. Mixed deciduous and oak-birch woodlands at Kisdon Force, North Yorkshire. Topography (*a*) with (*b*) the disposition of W9 *Fraxinus-Sorbus-Mercurialis* and W11 *Quercus-Betula-Oxalis* woodlands and (*c–e*) sections through and across the hollow and spur.

steeply inclined and incised by actively-eroding streams into younger irregular ravines, the vegetation patterns can be more complex with small fragments of both kinds of woodland succeeding one another up and across the valley sides. This is a feature of sites on Ordovician and Silurian deposits in Wales and south-east Scotland (Ratcliffe 1977, Ferreira 1978). On many igneous and

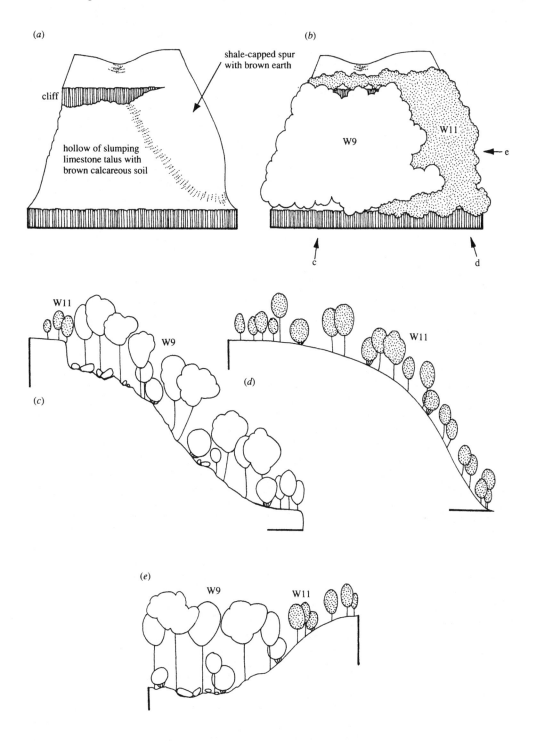

metamorphic rocks, more calcareous stretches are exceptional and here very small stands of the *Fraxinus-Sorbus-Mercurialis* woodland can be found embedded within extensive tracts of *Quercus-Betula-Oxalis* woodland or, where there is a more marked shift in rock and soil acidity, the *Quercus-Betula-Dicranum* woodland, as in north Wales, the Lake district and north-west Scotland (Ratcliffe 1977).

The presence of superficial deposits frequently confuses such zonations. Over more calcareous till or head, the *Fraxinus-Sorbus-Mercurialis* woodland can still be extensive but more arenaceous drift may restrict its occurrence over suitable bedrocks and tends to shift the whole sequence of woodlands towards the more calcifugous extreme. Very heavy rainfall may have the same effect so that, in the far north-west of Scotland, the *Fraxinus-Sorbus-Mercurialis* woodland occupies only the most lithomorphic soils and, even then, can have fragments of much more calcifugous herbaceous vegetation within it on nidus-capped boulders (e.g. McVean & Ratcliffe 1962, Birks 1973).

Zonations to woodland on strongly-flushed soils are also quite common and they almost always involve the *Alnus-Fraxinus-Lysimachia* woodland, the typical alderwood of more base-rich stagnogleys. Species such as *Filipendula ulmaria, Deschampsia cespitosa, Athyrium filix-femina* and *Lysimachia nemorum* remain prominent in this community and *Viola riviniana, Oxalis acetosella* and *Dryopteris dilatata* can run some way into the surrounds but increasing *Alnus* dominance and the appearance of plants like *Chrysosplenium oppositifolium, Carex remota, C. pendula* and *Juncus effusus* usually serve as a good distinction. Where flushing is regular, as along an impervious parting, the pattern of repeated slumped triangles of this woodland within stretches of *Fraxinus-Sorbus-Mercurialis* woodland can be very striking. Elsewhere the emergence of ground water is much more disordered and the mosaics consequently more complex. However, the frequent association of the *Fraxinus-Sorbus-Mercurialis, Quercus-Betula-Oxalis* and *Alnus-Fraxinus-Lysimachia* woodlands on sequences of base-rich, base-poor and gleyed brown earths in the north-west is exactly analogous to zonations of the *Fraxinus-Acer-Mercurialis, Quercus-Pteridium-Rubus* and *Alnus-Fraxinus-Lysimachia* woodlands on comparable soils in the south-east, as for example in the Weald.

The landscape context of the *Fraxinus-Sorbus-Mercurialis* woodland is usually that of pastoral agriculture along the upland fringes, where stands of the community represent some of the last vestiges of a woodland cover now confined to the more intractable slopes, crags and pavements. Grazing by stock as well as by deer is frequent in the Typical sub-community and this vegetation sometimes gives way, with varying degrees of sharpness, to semi-natural calcicolous pasture. Usually,

this is *Festuca-Agrostis-Thymus* grassland or, in the Pennines, the more surface-leached kinds of *Sesleria-Galium* grassland or, in northern Scotland, the *Dryas-Carex* heath (e.g. Birks 1973). Where limestone crops out as pavement, as on the Yorkshire Dales Carboniferous or the Skye Durness Limestone, fragmentary stands of the community can be virtually confined to drift-filled grikes, forming complex mosaics with the grasslands on the clint surfaces (e.g. Birks 1973). Where grazed stretches are flushed, base-rich small-sedge mires, like the *Carex-Pinguicula* flush, may occur in close juxtaposition with *Fraxinus-Sorbus-Mercurialis* woodland.

Although successions to this community have never been studied, such grasslands as those mentioned above would probably revert back to *Fraxinus-Sorbus-Mercurialis* woodland if neglected: this kind of woodland probably represents the climax vegetation over more calcareous soils in sub-montane and cool oceanic regions. However, on a very local scale, one other kind of agricultural treatment has been important in deflecting progressions to the community. Where tree and shrub cover thins out very substantially in the less-grazed *Crepis* sub-community, the field layer becomes very similar to the *Anthoxanthum-Geranium* grassland. This is the vegetation of the unimproved hay-meadows of sub-montane parts of the Pennines and it seems highly likely that such swards have been derived from this woodland by repeatedly setting back invasion of woody plants by an annual mowing for hay (Pigott 1956a, Bradshaw 1962). Zonations between the two communities are rare, but sometimes contiguous stands occur on identical soils, separated only by a wall. Open woodlands very similar to the *Crepis* sub-community have traditionally been mown in other parts of northern Europe (e.g. Nordhagen 1928, 1943, Sjörs 1954, Böcher 1954).

Distribution

The *Fraxinus-Sorbus-Mercurialis* woodland replaces the *Fraxinus-Acer-Mercurialis* woodland in the cooler and wetter north-western parts of Britain. In the Pennines, where Carboniferous Limestone provides more extensive potential sites, the switch between the two communities is especially well seen. Here the *Fraxinus-Acer-Mercurialis* woodland is the more common community over talus slopes in the drier and warmer lower reaches of the dales; the *Fraxinus-Sorbus-Mercurialis* woodland replaces it in damp ravines (like Ling Ghyll), on higher-altitude pavements (like Colt Park) and, more consistently, in the drift-choked tributaries of wetter and cooler Swaledale, Teesdale and Weardale. Further west and north, in Wales, southern and central Scotland, suitable substrates tend to be scarce and stands small and scattered, though the community shows a resurgence in the far north-west where limestones and calcareous igneous and metamorphic rocks are again more abundant. Increasingly here, too, the community shows an enrichment of its bryophyte flora which brings it close to equivalent Irish vegetation (e.g. Braun-Blanquet & Tüxen 1952). The Typical sub-community occurs throughout the range with the *Crepis* sub-community more restricted, though good stands of both types can be found in the Pennines and north-west Scotland.

Affinities

This community brings together a variety of more calcicolous sub-montane woodlands that have often been recognised as distinctive but usually split on the basis of canopy or field layer differences. More orthodox mixed deciduous stands have sometimes been fused with lowland woodlands, as in Klötzli's (1970) *Dryopterido-Fraxinetum*, Peterken's (1981) ash-wych elm stand type 1D or the *Querco-Ulmetum* of Birse (1982, 1984) while types dominated by *Betula pubescens* and *Corylus* have been put alongside more calcifugous woodlands, as in the *Betula*-herb nodum of McVean & Ratcliffe (1962), the *Corylus-Oxalis* Association of Birks (1973) and the 12B hazel-birchwood of Peterken (1981).

As defined here, the community holds together quite well, though its phytosociological affinities are fairly diverse because it characterises sites that, while generally cool and wet, range from mild and oceanic through to harshly sub-montane. The soils too, though comparatively base-rich, often show a combination of wetness with surface eluviation. In one direction, then, there are similarities with the extreme oceanic scrub of western Ireland which Braun-Blanquet & Tüxen (1952) described as *Corylo-Fraxinetum* and, on the other, with 'Park Meadow' communities described from Norway (e.g. the *Geranium sylvaticum*-reicher Birchenwald of Nordhagen (1928) or the *Betuletum geraniosum subalpinum* of Nordhagen (1943)), Sweden (Sjörs 1954) and Greenland (Böcher 1954). These latter have usually been placed in the Cicerbition alpini where Birks (1973) located the ungrazed stands of his *Betula-Cirsium* Assocation.

Such diagnoses stress the more peculiar features of the vegetation included here. Its more general affinities place it somewhere in the Fagetalia, though exactly where is more of a problem. It is the north-western counterpart of the calcicolous Carpinion woods of the south-east and Birse (1980, 1982, 1984) favoured locating his communities in this alliance. Both Graham (1971) and Birks (1973), on the other hand, noted the suggestion of Shimwell (1968b, 1971c) that a new alliance, the Fraxino-Brachypodion, might be needed to contain British woodlands beyond the geographical limit of the Fagion. A more satisfactory position might be within the Alno-Ulmion (*sensu* Ellenberg 1978 and equivalent to the Alno-Padion Knapp 1942): this would recognise the general similarities of the vegetation with more calcicolous mixed deciduous woodland while acknowledging the links with woodland of more strongly-gleyed mineral soils like the *Alnus-Fraxinus-Lysimachia* woodland.

Floristic table W9

	a	b	9
Fraxinus excelsior	IV (1–9)	III (3–5)	IV (1–9)
Sorbus aucuparia	III (1–4)	V (3–5)	III (1–5)
Betula pubescens	III (1–7)	III (1–9)	III (1–9)
Alnus glutinosa	I (1–7)	I (4–9)	I (1–9)
Ulmus glabra	III (1–10)	I (1–6)	II (1–10)
Acer pseudoplatanus	III (1–9)	I (5–9)	II (1–9)
Quercus petraea	II (1–9)		I (1–9)
Quercus robur	I (1–8)		I (1–8)
Betula pendula	I (1–5)		I (1–5)
Fagus sylvatica	I (1–10)		I (1–10)
Pinus sylvestris	I (1–7)		I (1–7)
Quercus hybrids	I (1–7)		I (1–7)
Corylus avellana	IV (1–9)	V (4–9)	IV (1–9)
Crataegus monogyna	III (1–7)	I (5)	II (1–7)
Fraxinus excelsior sapling	II (1–7)	I (3)	I (1–7)
Ulmus glabra sapling	II (1–4)		I (1–4)
Acer pseudoplatanus sapling	I (1–5)		I (1–5)
Sambucus nigra	I (1–4)		I (1–4)

Floristic table W9 *(cont.)*

	a	b	9
Ilex aquifolium	I (1–4)		I (1–4)
Betula pubescens sapling	I (2–4)	II (1–9)	I (1–9)
Sorbus aucuparia sapling	I (1–4)	II (1–5)	I (1–5)
Prunus padus	I (1–4)	II (3–4)	I (1–4)
Salix cinerea		I (3–4)	I (3–4)
Populus tremula		I (4–6)	I (4–6)
Oxalis acetosella	IV (1–7)	V (3–5)	IV (1–7)
Thuidium tamariscinum	IV (1–6)	IV (2–5)	IV (1–6)
Viola riviniana	IV (1–4)	IV (1–4)	IV (1–4)
Plagiomnium undulatum	IV (1–6)	IV (1–2)	IV (1–6)
Mercurialis perennis	IV (1–10)	III (1–4)	IV (1–10)
Dryopteris filix-mas	IV (1–9)	III (2–5)	IV (1–9)
Eurhynchium striatum	IV (1–8)	III (2–4)	IV (1–8)
Eurhynchium praelongum	V (1–7)	III (2–3)	IV (1–7)
Circaea lutetiana	III (1–7)	II (1–3)	III (1–7)
Geum urbanum	III (1–7)	II (1–3)	III (1–7)
Potentilla sterilis	III (1–4)	II (1–4)	III (1–4)
Dryopteris dilatata	III (1–6)	I (1–3)	II (1–6)
Fraxinus excelsior seedling	III (1–5)	I (1)	II (1–5)
Fissidens taxifolius	II (1–5)	I (1)	I (1–5)
Pellia epiphylla	II (1–4)	I (2–4)	I (1–4)
Galium aparine	II (1–4)		I (1–4)
Urtica dioica	II (1–6)		I (1–6)
Rubus fruticosus agg.	II (1–10)		I (1–10)
Silene dioica	II (1–7)		I (1–7)
Hedera helix	II (1–8)		I (1–8)
Brachythecium rutabulum	II (1–4)		I (1–4)
Veronica montana	II (1–4)		I (1–4)
Milium effusum	I (1–2)		I (1–2)
Eurhynchium swartzii	I (1–6)		I (1–6)
Arum maculatum	I (1–2)		I (1–2)
Adoxa moschatellina	I (1–4)		I (1–4)
Anthriscus sylvestris	I (1–2)		I (1–2)
Lamiastrum galeobdolon	I (1–5)		I (1–5)
Circaea × intermedia	I (2–6)		I (2–6)
Festuca gigantea	I (1–2)		I (1–2)
Phyllitis scolopendrium	I (1–6)		I (1–6)
Filipendula ulmaria	II (1–6)	IV (1–6)	III (1–6)
Brachypodium sylvaticum	II (1–7)	IV (4–7)	III (1–7)
Conopodium majus	II (1–5)	IV (1–4)	III (1–5)
Crepis paludosa	I (1)	IV (1–4)	II (1–4)
Cirriphyllum piliferum	II (1–8)	III (1–4)	II (1–8)
Deschampsia cespitosa	II (1–7)	III (3–7)	II (1–7)
Rhytidiadelphus triquetrus	II (1–9)	III (1–6)	II (1–9)
Plagiochila asplenoides	II (1–6)	III (1–4)	II (1–4)

Arrhenatherum elatius	II (1–4)	III (1–5)	I (1–5)
Agrostis capillaris	II (1–5)	III (1–6)	I (1–6)
Geum rivale	I (1–2)	III (2–5)	I (1–5)
Cirsium helenioides	I (1)	III (4–6)	I (1–6)
Vicia sepium	I (1–2)	III (1–2)	I (1–2)
Geranium sylvaticum	I (1)	III (1–5)	I (1–5)
Anthoxanthum odoratum	I (1)	III (3–6)	I (1–6)
Holcus lanatus	I (1–6)	III (1–4)	I (1–6)
Rhytidiadelphus squarrosus	I (2)	III (1–5)	I (1–5)
Ranunculus ficaria	I (1)	III (1–3)	I (1–3)
Ranunculus repens	I (1–2)	III (1–4)	I (1–4)
Holcus mollis	I (1–7)	III (1–5)	I (1–5)
Rumex acetosa		III (1–3)	I (1–3)
Succisa pratensis	I (1)	II (1–4)	I (1–4)
Senecio jacobaea	I (1)	II (1–2)	I (1–2)
Agrostis stolonifera	I (2–4)	II (2–5)	I (2–5)
Tussilago farfara	I (2–5)	II (3–5)	I (2–5)
Stachys sylvatica	I (1–4)	II (1–3)	I (1–4)
Plagiothecium denticulatum	I (1–2)	II (1–2)	I (1–2)
Stellaria holostea	I (1–5)	II (1–3)	I (1–3)
Blechnum spicant	I (1)	II (1–4)	I (1–4)
Melica uniflora	I (1–4)	II (1–4)	I (1–4)
Rubus idaeus	I (1–6)	II (1–5)	I (1–6)
Ranunculus acris	I (1–2)	II (1–3)	I (1–3)
Potentilla erecta	I (1)	II (1–4)	I (1–4)
Rhytidiadelphus loreus	I (2–4)	II (2–6)	I (2–6)
Allium ursinum	I (1–5)	II (1–2)	I (1–2)
Rhizomnium punctatum	I (1–5)	II (1–4)	I (1–5)
Carex pallescens		II (1–2)	I (1–2)
Cruciata laevipes		II (1–4)	I (1–4)
Alchemilla glabra		II (1–2)	I (1–2)
Plagiomnium rostratum		II (1–3)	I (1–3)
Trollius europaeus		I (3–4)	I (3–4)
Mnium hornum	III (1–8)	III (1–7)	III (1–8)
Hyacinthoides non-scripta	III (1–9)	III (1–4)	III (1–9)
Geranium robertianum	III (1–5)	III (1–5)	III (1–5)
Athyrium filix-femina	III (1–8)	III (2–5)	III (1–8)
Atrichum undulatum	III (1–7)	III (3–5)	III (1–7)
Poa trivialis	III (1–6)	III (1–3)	III (1–6)
Dactylis glomerata	II (1–4)	III (1–5)	II (1–5)
Dryopteris borreri	II (1–6)	II (3–6)	II (1–6)
Primula vulgaris	II (1–5)	II (3–5)	II (1–5)
Veronica chamaedrys	II (1–5)	II (2–4)	II (1–5)
Lophocolea bidentata s.l.	II (1–6)	II (1–3)	II (1–6)
Lysimachia nemorum	II (1–4)	II (1–4)	II (1–4)
Epilobium montanum	II (1–5)	II (1–2)	II (1–5)
Sanicula europaea	II (1–8)	II (1–3)	II (1–3)
Hypnum cupressiforme	II (1–4)	II (1–6)	II (1–6)
Prunella vulgaris	I (1–3)	I (1–3)	I (1–3)
Galium odoratum	I (1–6)	I (1–3)	I (1–6)
Bromus ramosus	I (1–3)	I (1)	I (1–3)

Floristic table W9 *(cont.)*

	a	b	9
Valeriana officinalis	I (1–4)	I (1–4)	I (1–4)
Pteridium aquilinum	I (1–6)	I (1–9)	I (1–9)
Galium saxatile	I (1–2)	I (2–4)	I (1–4)
Isothecium myurum	I (1)	I (1–3)	I (1–3)
Ctenidium molluscum	I (1–6)	I (1–5)	I (1–6)
Digitalis purpurea	I (1–3)	I (1–3)	I (1–3)
Acer pseudoplatanus seedling	I (1–6)	I (1)	I (1–6)
Plagiothecium succulentum	I (1–4)	I (1–3)	I (1–4)
Heracleum sphondylium	I (1–4)	I (1–3)	I (1–4)
Rosa canina agg.	I (1–4)	I (2)	I (1–4)
Fissidens bryoides	I (1–5)	I (1–4)	I (1–5)
Hylocomium brevirostre	I (1–2)	I (3–5)	I (1–5)
Polytrichum formosum	I (1–2)	I (2–3)	I (1–3)
Cardamine flexuosa	I (1–4)	I (1–4)	I (1–4)
Ajuga reptans	I (1–5)	I (1–2)	I (1–5)
Anemone nemorosa	I (1–5)	I (1–3)	I (1–5)
Luzula sylvatica	I (1–8)	I (4–7)	I (1–8)
Thuidium delicatulum	I (1–2)	I (3–6)	I (1–6)
Hylocomium splendens	I (1–2)	I (3–4)	I (1–4)
Isothecium myosuroides	I (1–5)	I (1–4)	I (1–5)
Fragaria vesca	I (1–3)	I (1–2)	I (1–3)
Angelica sylvestris	I (1–3)	I (5)	I (1–5)
Carex sylvatica	I (1–2)	I (1–4)	I (1–4)
Chrysosplenium oppositifolium	I (1–5)	I (3)	I (1–5)
Dicranum scoparium	I (1–3)	I (1–2)	I (1–3)
Cirsium palustre	I (1)	I (2)	I (1–2)
Alchemilla xanthochlora	I (1)	I (1–2)	I (1–2)
Rubus saxatilis	I (1)	I (3)	I (1–3)
Pellia endiviifolia	I (1–4)	I (1)	I (1–4)
Number of samples	93	24	117
Number of species/sample	32 (18–74)	47 (35–73)	36 (18–74)
Tree height (m)	17 (7–25)	12 (10–16)	16 (7–25)
Tree cover (%)	77 (10–100)	41 (0–100)	63 (0–100)
Shrub height (m)	5 (1–8)	5 (4–7)	5 (1–8)
Shrub cover (%)	38 (0–100)	28 (0–90)	32 (0–100)
Herb height (cm)	44 (10–122)	65 (15–80)	48 (10–122)
Herb cover (%)	76 (10–100)	68 (55–90)	68 (10–100)
Ground height (mm)	28 (10–80)		28 (10–80)
Ground cover (%)	45 (0–100)	24 (5–70)	37 (0–100)
Altitude (m)	113 (6–335)	209 (16–305)	139 (6–335)
Slope (°)	21 (0–70)	25 (2–50)	21 (0–70)

a Typical sub-community

b *Crepis paludosa* sub-community

9 *Fraxinus excelsior-Sorbus aucuparia-Mercurialis perennis* woodland (total)

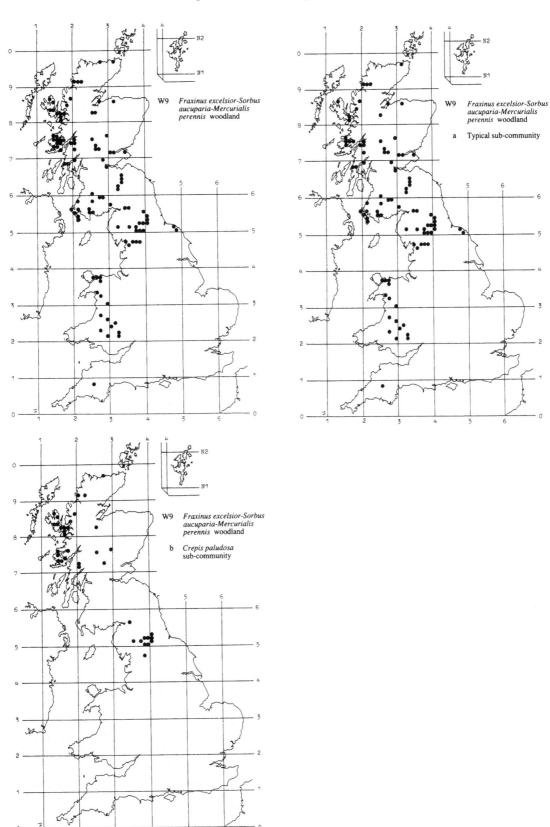

W9 *Fraxinus excelsior-Sorbus aucuparia-Mercurialis perennis* woodland

W9 *Fraxinus excelsior-Sorbus aucuparia-Mercurialis perennis* woodland

a   Typical sub-community

W9 *Fraxinus excelsior-Sorbus aucuparia-Mercurialis perennis* woodland

b   *Crepis paludosa* sub-community

W10
Quercus robur-Pteridium aquilinum-Rubus fruticosus woodland

Synonymy

Damp oakwood association Moss *et al.* 1910 *p.p.*; *Quercetum roburis* Tansley 1911 *p.p.*; *Quercetum arenosum roburis et sessiliflorae* Tansley 1911 *p.p.*; *Quercetum sessiliflorae* Moss 1911, 1913, *p.p.*; *Quercus robur-Carpinus* woodland Salisbury 1916 *p.p.*; *Quercus sessiliflorae-Carpinus* woodland Salisbury 1918*a p.p.*; *Quercetum roboris* Tansley 1939 *p.p.*; *Quercetum petraeae sessiliflorae* Tansley 1939 *p.p.*; Ash-maple-hazel woods Rackham 1980 *p.p.*; Hornbeam-woods Rackham 1980 *p.p.*; Limewoods Rackham 1980 *p.p.*; Oakwoods Rackham 1980 *p.p.*; Birchwoods Rackham 1980 *p.p.*; Chestnut-woods Rackham 1980 *p.p.*; Hazel-ash woodland Peterken 1981 *p.p.*; Ash-lime woodland Peterken 1981 *p.p.*; Oak-lime woodland Peterken 1981 *p.p.*; Birch-oak woodland Peterken 1981 *p.p.*; Hornbeam woodland Peterken 1981 *p.p.*; Woodland plot types 9, 17, 19, 20 & 24 Bunce 1982; *Lonicero-Quercetum* (Birse & Robertson 1976) Birse 1984 *p.p.*; *Querco-Betuletum* Klötzli 1970 *p.p.*; *Blechno-Quercetum fraxinetosum* Klötzli 1970 *p.p.*

Constant species

Quercus robur, Lonicera periclymenum, Pteridium aquilinum, Rubus fruticosus agg.

Physiognomy

The *Quercus robur-Pteridium aquilinum-Rubus fruticosus* woodland, like its counterpart on more base-rich soils, the *Fraxinus-Acer-Mercurialis* woodland, is a variable community in which floristic differences related to climate and soil are overlain by treatment-derived variation in the canopy and underwood. Here, though, the resultant combinations of these patterns are not so numerous or complex and the overall definition of the community is a little more straightforward.

Essentially, these are oakwoods, though in a narrower sense in which that term was applied in early British studies (e.g. Moss *et al.* 1910, Tansley 1911, 1939). Oak is the most characteristic tree here and, though its cover has been very widely modified by treatment, it remains far and away the commonest tree throughout and is the only woody constant. Quite often, it is very abundant, dominating in semi-natural high-forest canopies and, where the community has been worked as coppice (a frequent, though now largely discontinued, practice), it is invariably the standard. It can also sometimes be found as a component of the underwood itself and the community includes, too, oak plantations whose general floristic character clearly places them here.

Of the two species of oak, *Quercus robur* is very much the commoner throughout, being especially characteristic of this kind of woodland over most of the lowland south-eastern part of the range of the community. Towards the west and north, it is partly replaced by *Q. petraea* and a perplexing range of hybrids between the two, which add to the already considerable amount of variation within each species (Jones 1959, 1968, Cousens 1965, Wigston 1974). But this general geographic distinction, which reflects a shift of the community on to more free-draining soils to the north-west, is not a hard and fast one and there are some prominent enclaves in the lowland zone where *Q. petraea* is the usual oak in these woodlands, most notably in south-east Essex, parts of Hertfordshire and Sherwood in Nottinghamshire: this replacement does not always show a clear edaphic correlation (Jones 1959, Gardiner 1974). The picture is further complicated by the bias, of long standing until recent times, towards *Q. robur* as the preferred oak for planting (Gardiner 1974, Penistan 1974) which might explain some of the more unexpected occurrences of this tree in northerly stands. Whatever the origin of this geographical patterning among the oaks, it is not accompanied here by any other absolutely exclusive floristic features. The prominence of *Q. petraea* to the north-west helps define the *Acer-Oxalis* sub-community and local abundance to the south-east is sometimes associated with the presence of other trees such as *Carpinus betulus* (in Essex and Hertfordshire) or

Castanea sativa (in Essex) but there is no sound basis for using the species of oak to make sub-divisions within the community (cf. Peterken 1981).

The next most common tree throughout is birch, almost always here *Betula pendula*. This is often a very prominent species in younger stands of the community developing by colonisation of open ground on neglected commons and in field corners, but it is also the most frequent and initially successful invader of gaps, recently-cut coppice and clear-felled areas in established tracts of this woodland and has attained increased patchy abundance with the decline in traditional kinds of management. It often marks, for example, areas where post-coppice cleaning was neglected after the cuts of the 1940s. Where oak is dominant in the surrounding woodland in such situations, its own poor regeneration may leave successive generations of birch in occupation as locally-established groves. However, though this tree has undoubtedly spread in recent years, it also occurs very often here as an integral low-cover component of mature high-forest canopies, turning over at a more rapid rate than the other tree species, but apparently maintaining its contribution continually and attaining a grand stature that can belie its reputation as simply a forest weed. In some stands of the community, birch has also been included within a coppice crop (Rackham 1980). The prominence of birch here was recognised by Klötzli (1970) in his definition of a lowland *Querco-Betuletum* and by Peterken's (1981) diagnosis of many of these woodlands as belonging to his birch-oak stand types (parts of 6C and 6D). There is also some small overlap between this community and the birchwoods of Rackham (1980).

The shift in emphasis towards the pre-eminence of oak and birch here is also marked by the great scarcity in the community of *Acer campestre* and the restricted occurrence of *Fraxinus excelsior*. *Fraxinus* is only occasional in this kind of woodland as a whole and distinctly patchy towards the south-east. Here, it can attain local prominence, sometimes on quite acidic, though probably fairly fertile, soils, as in Rackham's (1980) 'pure ashwoods' in East Anglia and some of Peterken's (1981) 3A oak-hazel-ashwoods and it figures, too, in some *Castanea sativa*-dominated woodlands of this community and in plantations. But only in the *Acer-Oxalis* sub-community of the wetter north-west does it attain the kind of frequency that is commonplace through the *Fraxinus-Acer-Mercurialis* woodland and, even then, it is usually of low abundance. *Acer pseudoplatanus* shows the same general pattern of distribution, effectively marking out areas of higher rainfall and being prominent in the drier south-east more as a local invader of gaps and cleared areas (though often a very successful invader) than as a consistent component of this woodland. To the north-west, by contrast, it is very much at

home here, occurring along with *Fraxinus* and the oaks in well-established high-forest canopies and with a full complement of the community herbs. It has also been treated in this region as a coppice crop, supplying wood for bobbins for the textile industry. A third tree, *Ulmus glabra*, follows *Fraxinus* and *A. pseudoplatanus* geographically, though it is less frequent here, even in the north-west, than it is in the *Fraxinus-Acer-Mercurialis* woodland and decidedly rare in south-eastern stands of the community.

Three other trees are of major importance in certain kinds of *Quercus-Pteridium-Rubus* woodland, occurring quite widely in the warmer and drier south-east as infrequent companions to oak and birch but also being found in this region with very marked local prominence. *Tilia cordata* and *Carpinus betulus* are not exclusive to this community: they occur in other kinds of woodland, notably the *Fraxinus-Acer-Mercurialis* woodland, though the stands in which they dominate here tend to be larger. As elsewhere, the naturally gregarious tendency of these trees has very frequently been accentuated by generations of cropping for large-coppice underwood and now they can be found as dominants, alone or in mixtures of the two, in extensive tracts of woodlands in which oaks have been reduced to occasional standards and from which birch has traditionally been eliminated. The impoverishment of the field layer in dense stands of these species, especially striking under *Tilia*, means that it may sometimes be difficult to separate lime- or hornbeam-dominated stretches of this community from areas of *Fraxinus-Acer-Mercurialis* woodland in which they are abundant and some schemes (e.g. Rackham 1980, Peterken 1981) unite them in 'lime-' and 'hornbeam-woodlands'. When the canopy is opened up, however, in coppicing or clearance, the flush of field-layer growth is rather different in each case.

Much more strictly confined to the *Quercus-Pteridium-Rubus* woodland and prominent on a more restricted geographical scale than *Tilia* or *Carpinus*, though with equally distinctive effect when it does occur, is *Castanea sativa*. This tree, which has a Sub-Mediterranean distribution in Europe, is almost certainly an introduction to Britain (Godwin 1975, Rackham 1980), but it is eminently successful on the moister soils over which this community occurs and has been strongly selected as a prolific coppice crop to supply poles and stakes. It is especially associated here with the *Anemone* sub-community though, again, there is little justification for separating off this kind of woodland from the community as a whole and erecting a separate 'chestnut-woodland' (cf. Rackham 1980).

Quite a wide variety of other tree species occurs at low frequency throughout the community and only very exceptionally at high cover. First, larger specimens of *Ilex aquifolium* and *Sorbus aucuparia* sometimes make a contribution to the canopy. Both of these, and especially *Ilex*, are rather more common here than in the *Fraxinus-Acer-Mercurialis* woodland, though generally they form part of an understorey, occurring as scattered individuals or with patchy local prominence. But, in some situations, they can be the major woody associates of oak and birch in this community and then it may be difficult to separate samples from the *Quercus-Betula-Deschampsia* woodland on the basis of the trees and shrubs alone. Field-layer differences (see below) will usually effect a separation in such cases, though there are some quite widespread situations where these two communities grade one into another with their herbs distributed in complex mosaics.

The occasional presence of *Fagus sylvatica* here may also create problems of definition. Woodlands in which *Fagus* dominates over similar, though often rather impoverished, field layers to those found in this community, are treated separately in this scheme as the *Fagus-Rubus* woodland but, in certain circumstances, transitional stands can be found. *Fagus* is generally a sparse canopy component in *Quercus-Pteridium-Rubus* woodland and *Q. robur* similarly infrequent within *Fagus-Rubus* woodland but, where oak succeeds in regenerating in larger gaps within the latter community or where beech has been introduced into stands of the former, more balanced mixtures of the two trees can occur over the typical herbs. Often, this is in areas like the Chiltern plateau and the New Forest where established tracts of both communities occur in close proximity, though interplanting with *Fagus* has been widely pursued well beyond its natural British limit. *Taxus baccata*, a locally prominent associate of beech in the *Fagus-Rubus* woodland, is sometimes conspicuous here, too, as in some New Forest stands where it occurs in a patchy lower tier to a canopy of oak. Overall, however, it remains an infrequent tree in the community.

A further scarce associate is *Alnus glutinosa*. Generally speaking, this also occurs only as scattered individuals, in this case picking out areas of poorer drainage, and any increase in its cover usually marks out a transition to flushes (where it is a major canopy component in the *Alnus-Fraxinus-Lysimachia* woodland) or to alluvial flats (where it figures in the *Alnus-Urtica* woodland). But, since springs and streams are a common feature of *Quercus-Pteridium-Rubus* woodlands, some more gradual zonations between these communities should be expected.

As in the *Fraxinus-Acer-Mercurialis* woodland, wetter soils here can also have some *Populus tremula*, characteristically forming clones of small suckering trees. Another suckering tree which can also attain local prominence in the community is *Prunus avium*: very occasionally, it can dominate in this kind of woodland over considerable areas, as in some East Anglian sites

described by Rackham (1980), but frequently it is found as a solitary tree, sometimes of magnificent size and, with its peculiar angular branches, looking very distinctive. Much more strictly non-gregarious is *Sorbus torminalis*, which often needs an extensive search to find it, and *Malus sylvestris*.

Finally, among the canopy components, mention must be made of planted softwoods because this community, unlike the *Fraxinus-Acer-Mercurialis* woodland, can stand fairly intensive coniferisation without a complete destruction of its overall floristic integrity, though not, of course, without often irremediable damage to its historic continuity. Interplanting or replacement of the canopy with softwoods is quite widespread here and stands of such species as *Pinus sylvestris*, *P. nigra* var. *maritima*, *Pseudotsuga menziesii* and *Larix* spp. over field layers with *Pteridium aquilinum* and *Rubus fruticosus* agg. are best incorporated, on general ecological grounds, within this community.

Semi-natural high-forest stands of the *Quercus-Pteridium-Rubus* woodland generally have a distinct understorey of shrubs and small saplings, often quite low (less than half the height of the trees) though sometimes fairly dense, at least in patches. But, very commonly, and especially towards the south-east, this community has been treated as various kinds of coppice, sometimes with trees such as *Tilia*, *Carpinus* or *Castanea* forming the crop, but on a much more widespread scale with shrubs and smaller trees being cut. With the general neglect of coppicing, a complicated range of physiognomies, between well-defined coppice and high forest, can now be encountered, though traces of the treatment can still often be detected in stands of this woodland by careful observation.

As in the *Fraxinus-Acer-Mercurialis* woodland, *Corylus avellana* is the commonest shrub here and its high frequency helps separate this community from oak-birch woodlands of more acid soils where, in the southern lowlands at least, *Corylus* is scarce. This difference in representation was recognised by Peterken (1981) in his diagnosis of birch and hazel sub-types among his lowland birch-oak woodlands. Even in uncoppiced high forest, *Corylus* is usually the most abundant element in an understorey but, in coppiced stands, it has been very widely favoured, encouraged and perhaps planted so it is often here an effective dominant, with or without a tier of standards, almost always oak, above. With similar stands in the *Fraxinus-Acer-Mercurialis* woodland, these kinds of *Quercus-Pteridium-Rubus* woodlands constitute the vast bulk of hazel coppice in lowland Britain and, where uniform treatment has been imposed over natural boundaries between these two communities, an identical physiognomy can extend deceptively over different field layers. Usually, sufficient herbs occur to discern a switch from the one to the other but, where there is especially dense shade, as in stands which are reaching the end of their rotation of growth or which have been long neglected, it may be difficult to draw a boundary; and, on soils which show a gradual change in their calcium content or base-status, the zonation may be continuous for quite natural, edaphic reasons.

As noted above, *Acer campestre* is rare in this community and the most frequent woody companions of *Corylus* are generally the hawthorns. *Crataegus monogyna* is very much the more common species overall and, though it rarely rivals *Corylus* in its abundance, it is a characteristic feature of the understorey in high-forest stands, sometimes growing very tall and breaking the canopy, and an occasional in coppice, where it has sometimes been included in the crop. To the north-west, *C. monogyna* is the usual woodland hawthorn here, but in the warmer and drier south-east, it is sometimes replaced, especially within longer-established stands of the community, by *C. laevigata* which is a naturally bushier plant and more readily coppiced. Hybrids between the two can also be found (Bradshaw 1953).

Smaller trees of *Ilex aquifolium* are also a quite frequent feature of the understorey or coppice-underwood here and there is occasionally some *Sorbus aucuparia*. *Viburnum lantana* and *V. opulus* occur sparsely and, in younger scrubbier stands or along margins and rides, *Prunus spinosa* may join *Crataegus monogyna* in a dense spiny cover. *Sambucus nigra* can be patchily prominent, too, especially on more disturbed and enriched areas and, though it is not so frequent overall as in the *Fraxinus-Acer-Mercurialis* woodland, it shows the same tendency as there, in that it increases its representation towards the north-west where the soils are perhaps more consistently aerated and have a faster nutrient turnover as well as having the benefit of the rich foliage of *Fraxinus* and *Ulmus glabra*. In contrast to the *Fraxinus-Acer-Mercurialis* woodland, however, shrubs such as *Cornus sanguinea*, *Euonymus europaeus* and *Ligustrum vulgare* are noticeably rare on the typically less base-rich soils here. Indeed, where stands of this community are coppiced, the place of these species as invaders is sometimes taken by *Cytisus scoparius* or *Ulex europaeus*. The soils here are also sometimes sufficiently acid to support *Rhododendron ponticum* and, where this shrub has been planted into these woodlands, a widespread practice since its original introduction in the eighteenth century, as game cover or as ornamentation of what is sometimes visually dull vegetation, it can become a serious weed (Cross 1975).

Except in young, unthinned plantations and in older and neglected coppice, the cover of trees and shrubs in the *Quercus-Pteridium-Rubus* woodland is often not complete. Some stands occur more as wood-pasture than as woodland proper, being found in association with open 'plains' (or 'lawns' or 'launds') in ancient parkland or forests in the old legal sense. Even in high forest, the shade cast is frequently quite light and, provided there is no intense competition from herbs or consistent predation by small herbivores, deer or stock, tree saplings can be quite common, thickening up in gaps to form dense regeneration cores. As noted earlier, young birch are often a conspicuous feature here but *A. pseudoplatanus* saplings can also be very prominent, more consistently in the wetter north-west, where, along with *Fraxinus*, this tree is usually the most prolific regenerator. After good mast years, *Fagus* saplings can also occur frequently, near to their parents. In marked contrast, the most characteristic woody species of the community, oak, is often the most poorly represented among younger trees, a feature which was noted early by Watt (1919), which may be a relatively recent development (Rackham 1980) and which probably has a complex number of causes (see below). Certainly, the general scarcity of oak saplings within established stands of *Quercus-Pteridium-Rubus* woodland is very different from the good representation of young trees where the community is developing by invasion of open ground.

The most obvious general feature of the field layer here, especially when it is compared with that of the *Fraxinus-Acer-Mercurialis* woodland, its counterpart on more base-rich soils, is its general species-poverty. In essence, the herbaceous component of this community as a whole comprises variations on the representation of only three constants and a very small number of occasional to frequent species. But the proportions of these plants are quite varied and the general impression created by the field layer can be very different from one stand to another. In common parlance, these are 'bluebell woods' and a visit to a woodland of this kind when *Hyacinthoides non-scripta* is flowering in abundance from April to June can be an unforgettable experience. *Hyacinthoides* is the commonest vernal dominant here, indeed the only frequent vernal dominant apart from *Anemone nemorosa*, which tends to replace it on the moister soils occupied by the *Anemone* sub-community. But, though its leaves emerge in early spring, they have almost totally decayed by mid-summer (Blackman & Rutter 1954), so apart from a competitive relationship with *Anemone* (Pigott 1982), this species does not exert the same kind of controlling influence on its associates here as does *Mercurialis perennis* in the *Fraxinus-Acer-Mercurialis* woodland. *Hyacinthoides* is, of course, a frequent and often abundant plant in that community, too, but *Mercurialis* is characteristically very sparse here, even on the better-aerated soils where it can perform well. Indeed, if anything, its very occasional appearances in the *Quercus-Pteridium-Rubus* woodland tend to be on fairly moist soils because these often indicate some slight base-enrichment from streams or flushes: typically, here, a few plants mark out little alluvial flats in stream-loops or transitions to *Alnus-Fraxinus-Lysimachia* flushes. Generally speaking, however, the soils here are too base-poor to support any abundant growth of *Mercurialis* or frequent representation of its typical associates in the *Fraxinus-Acer-Mercurialis* woodland: plants like *Geum urbanum, Circaea lutetiana, Arum maculatum, Viola reichenbachiana, Geranium robertianum* and *Brachypodium sylvaticum* are all markedly uncommon in this community. It is in the *Quercus-Pteridium-Rubus* woodland that *Hyacinthoides* extends on to the most acid of the woodland soils on which it is represented in Britain and the more species-poor 'bluebell woods' which this community comprises are actually of rather restricted distribution in Europe, being largely confined to the north-west of France, Belgium and the lowlands of this country (Noirfalise 1968, Neuhäusl 1977).

Despite its characteristic pre-eminence here, *Hyacinthoides* is not a constant of the *Quercus-Pteridium-Rubus* woodland, though it probably would be were not so many stands long-overgrown and gloomy coppices or modified plantations. More frequent overall and, even where *Hyacinthoides* is well represented, more obvious by mid-summer, are *Pteridium aquilinum, Rubus fruticosus* agg. and *Lonicera periclymenum*, and various combinations and amounts of these species typically form a ferny underscrub that can attain a metre or more in height. One further component of the pattern, of patchy frequency to the south-east, more consistently common to the north-west, is *Holcus mollis*, lush swards of which can become prominent as *Hyacinthoides* is fading and before the *Pteridium* fronds have attained their full height and the *Rubus* its summer leaves. Spatial variations on this basic theme are very diverse: some stands are intimate mixtures or patchy mosaics of just *Hyacinthoides* and *Holcus*, in others there are intermixed sparse small fronds of *Pteridium* and a few strands of *Rubus*; or *Rubus* and *Lonicera* can form a virtually impenetrable underscrub with very little else; and *Pteridium* can increase its height and cover to present a virtual jungle of fronds, man-high, and with such a dense shade and thick litter that no associates, not even the other community constants, can gain a hold. All these variations, which are a complex function of canopy shade, soil moisture, grazing and competition between the plants themselves, can be found within each of the sub-communities here and, apart from the general feature of impoverishment with thicker cover, they are often unaccompanied by any consistent floristic differences among other species. These variations are therefore not employed here for distinguishing the sub-communities of the *Quercus-Pteridium-Rubus* woodland, though it should always be remembered what a distinct visual impression each of them can make and it may be useful to recognise variants according to the particular dominant.

The other component of the field layer that is sometimes physiognomically prominent consists of dryopteroid ferns. *Dryopteris filix-mas* occurs occasionally throughout and *D. dilatata* is quite common overall, too, though increased frequency of this species is a fairly good diagnostic feature of the north-western *Acer-Oxalis* sub-community. *D. borreri* occurs much less often than these two and there is sometimes, again more so to the north-west, a little *Athyrium filix-femina*. All these ferns tend to be more obvious when *Pteridium* is less abundant, and scattered crowns, emerging through a low patchy underscrub of *Rubus* and sprawling *Lonicera*, are a very characteristic feature of many stands.

Against this general background, a variety of other species make a contribution to the field layer which is not more than occasional overall but which can, in certain circumstances, become prominent, helping to define some of the sub-communities of the *Quercus-Pteridium-Rubus* woodland. First, there is *Hedera helix* which, as well as occurring as a sometimes conspicuous climber, quite often forms a ground carpet here, except where *Pteridium* is very abundant. As in the *Fraxinus-Acer-Mercurialis* woodland, *Hedera* shows a distinct tendency to be a more frequent and extensive component here in more oceanic regions but it is also, as in that community, a marker of stands of *Quercus-Pteridium-Rubus* woodland where there has been a substantial period of uninterrupted canopy or underwood growth. The *Hedera* sub-community thus contains many neglected coppices, younger secondary woods and plantations approaching a first thinning.

Second, there are grasses. Many stands here have a grassy appearance, especially in the period between the spring dominance of *Hyacinthoides* and the summer dominance of *Pteridium*. But this cover is generally rather species-poor with *Holcus mollis* as the major component. *Deschampsia cespitosa* may make a locally conspicuous contribution on moister soils, as around transitions to flushes, and along the edges of paths and in rides, and there is sometimes a little *Poa trivialis, Milium effusum* or *Melica uniflora*. And on more free-draining soils, as on drier banks or on transitions to more leached brown earths, *Deschampsia flexuosa* can become more frequent in mosaics with *H. mollis*, occasional *Blechnum spicant* and patchily prominent *Luzula sylvatica*. In sharp contrast to the sub-montane *Quercus-Betula-Oxalis* woodland, however, *D. flexuosa* remains generally scarce and is not accompanied here by grasses like *Anthoxanthum odoratum* and *Agrostis capillaris* which, in that community, are such an obvious indicator of grazing on more surface-leached soils. Here, increased representation of grasses generally involves more mesophytic species, notably *Holcus lanatus*, less so *Dactylis glomerata*, and these are very much a marker of the kind of disturbance associated with replacement of the canopy with oak or conifers. Such vegetation is here placed in a *Holcus lanatus* sub-community.

Then there is a wide range of low-frequency associates. Some of these, though not very common and not exclusive, are quite characteristic of this kind of woodland: *Stellaria holostea*, *Silene dioica*, *Luzula pilosa*, *Digitalis purpurea* and *Solidago virgaurea*. Often these occur as scattered individuals, though *S. holostea* is sometimes patchily prominent within swards of *Holcus mollis*, especially in the north-western *Acer-Oxalis* sub-community. But *S. dioica* and *D. purpurea* can increase very markedly where this community is opened up and disturbed and they are a frequent component here of post-coppice vegetation sometimes with *Corydalis claviculata*. Also, in such situations, *Epilobium angustifolium*, *Rumex acetosa*, *R. sanguineus*, *Urtica dioica*, *Teucrium scorodonia* and *Euphorbia amygdaloides* may increase their representation and shrubs like *Cytisus scoparius* and *Ulex europaeus* appear. Occasional plants of these species may survive for some time as the canopy closes.

Finally, among the herbs, there are ground-cover species which can attain local prominence over more open areas of bare soil, especially where this is heavy and moist and where the cover of *Pteridium* in particular is reduced. This is the kind of situation where some species very characteristic of south-eastern *Fraxinus-Acer-Mercurialis* woodlands may make an appearance. As well as *Poa trivialis* already noted, there can be some *Glechoma hederacea*, *Ajuga reptans* and *Lamiastrum galeobdolon*, sometimes with *Veronica chamaedrys*, *Lysimachia nemorum* and *L. nummularia*. These plants, too, may show some vegetative spread here after coppicing.

It is in the more open areas of herb cover that tree seedlings are especially numerous. *Acer pseudoplatanus* is generally the best represented species with *Fraxinus*, *Crataegus monogyna* and, in drier situations, *Betula pendula*. Oak seedlings occur in larger gaps and in young stands where the canopy has not yet closed.

Bryophyte cover in the *Quercus-Pteridium-Rubus* woodland is usually low, especially where, as amongst lush *Holcus mollis* or beneath *Pteridium*, litter is extensive and thick. The commonest species throughout are *Eurhynchium praelongum*, on soil and less acidic litter, and *Mnium hornum* on fallen twigs and bark, around tree bases and on soil. Other less frequent species are *Brachythecium rutabulum*, *Thuidium tamariscinum*, *Pseudoscleropodium purum*, *Isopterygium elegans* and *Plagiothecium undulatum*. Strongly calcicole or calcifuge bryophytes are characteristically sparse but they may be represented in transitional stands.

Sub-communities

Typical sub-community: *Quercus robur* wood, *Pteris-Holcus* and *Holcus* societies Adamson 1912; *Quercus robur-Carpinus* woodland, *Pteris* and *Scilla* societies Salisbury 1916; *Quercus sessiliflora-Carpinus* woodland,

Pteris, *Rubus* and *Scilla* societies Salisbury 1918*a*; *Quercetum roboris* and *Quercetum petraeae*, *Pteridium*, *Rubus-Pteridium*, *Scilla* and post-coppice societies Tansley 1939; *Querco-Betuletum* Klötzli 1970 *p.p.*; Ash-maple-hazel woodland Rackham 1980 *p.p.*; Hornbeam, hornbeam-ash and oak-hornbeam woods Rackham 1980 *p.p.*; Lime and lime-oak woods Rackham 1980 *p.p.*, Oakwoods Rackham 1980 *p.p.*; Lowland birchwoods Rackham 1980 *p.p.*; Hazel-ash stand types 3Aa & 3Ab Peterken 1981 *p.p.*; Ash-lime stand type 4A Peterken 1981 *p.p.*; Oak-lime stand type 5A Peterken 1981 *p.p.*; Birch-oak stand types 6Cc, 6Db & 6Dc Peterken 1981 *p.p.*; Hornbeam stand types 9Aa & 9Ba Peterken 1981 *p.p.*; Woodland plot types 17 & 24 *p.p.* Bunce 1982. This is the central type of *Quercus-Pteridium-Rubus* woodland over most of the south-eastern lowlands, forming the core of what has traditionally been described as 'dry oakwood' in Britain (Moss *et al.* 1910, Tansley 1911, 1939). Certainly, oak is the most frequent tree throughout, usually *Q. robur*, though in its distinctive enclaves and more generally on more free-draining soils, *Q. petraea*. In some stands, oak dominates in a high-forest canopy with frequent, though usually less abundant, *Betula pendula* and sometimes a little *Fagus*, *Acer pseudoplatanus* or *Pinus sylvestris*, seeding in from nearby or deliberately introduced for ornament. Oak is predominant, too, usually as *Q. robur*, in some long-established plantations here. Very often, however, oak is not of great abundance in stands of this sub-community, having been reduced to occasional standards in, or altogether eliminated from, coppice of hazel, lime, hornbeam or sometimes chestnut. *Tilia* and *Carpinus*, in fact, occur as low-frequency canopy trees in this kind of *Quercus-Pteridium-Rubus* woodland, typical as it is of the warmer and drier parts of the country and, when they do occur, the woody component of the vegetation can closely resemble that of continental Carpinion high forest. Very often, though, they have been coppiced and locally selected for as the basis of an underwood crop, so stands of this sub-community dominated by overgrown lime or hornbeam, or diverse mixtures of the two, are widespread. Those kinds of lime and hornbeam woodland characterised by Rackham (1980) and Peterken (1981) from more base-poor and free-draining soils therefore belong here in the scheme, together with the drier types of oak-hornbeam woodlands described in the early account of Salisbury (1916, 1918*a*). *Castanea* is occasionally found here in association with *Carpinus* and rarely it is locally dominant, but most chestnut coppice is best included in the *Anemone* sub-community.

Much more widespread than these kinds of coppiced versions of this sub-community are those dominated by *Corylus*, with or without a cover of standards, almost invariably oak, very occasionally *Fraxinus*. Even in high-forest stands, *Corylus* is generally the most prominent shrub but here it is often extremely abundant in neglected coppice, the bushes growing dense and tall where the tree canopy remains more open but eventually thinning out with increasing age and canopy extension. Under such gloomy conditions, transitions to the *Hedera* sub-community are common.

The most frequent companions of *Corylus* in this sub-community are the hawthorns, usually *C. monogyna*, but quite commonly in older stands of these south-eastern woods, *C. laevigata*, with the former then characteristic of younger tracts and scrubby margins, where *Prunus spinosa* can also increase in abundance. *Ilex* occurs occasionally, together with less frequent *Sorbus aucuparia* and, on more base-poor, free-draining soils, these two species tend to become proportionately more prominent. Other associates are *Sambucus nigra*, often clearly associated with disturbed areas, *Viburnum opulus*, *Malus sylvestris* and introduced *Rhododendron ponticum*. In coppiced stands, *Crataegus laevigata*, *Ilex* and *Sorbus* would often have been included with *Corylus* in a mixed underwood crop. In high forest here, all these smaller trees and shrubs occur as a generally open and low understorey with, in gaps, saplings of *Betula pendula* (especially on the drier soils), *Acer pseudoplatanus* (more on the moister soils), *Fagus* and occasional *Fraxinus*.

In the field layer of this sub-community, *Hyacinthoides* is a very common and often extremely abundant vernal dominant. Completing the bulk of its growth before the canopy comes into leaf, it grows quite freely in stands with a fairly dense woody cover though, like most herbs here, it declines under thicker stands of *Carpinus* and especially of *Tilia* and beneath neglected *Corylus* underwood. With regular coppicing, it shows the characteristic flush of flowering in the second or third spring after the cut (Salisbury 1924), though this is apparently not so marked on the more acid soils on to which this kind of woodland extends (Rackham 1980). *Hyacinthoides* can also thrive amongst moderately dense *Pteridium* here, the fronds of which are still expanding as the former flowers, but it is much reduced where very vigorous bracken growth leads to an accumulation of thick litter.

By mid-summer, *Pteridium* is often the most prominent feature of the field layer, especially on the more free-draining soils which here tend to be the more acidic (though bracken cannot be seen as a simple calcifuge). Its abundance is also very dependent on light penetration, so its vigour and cover are at their maximum here under more open canopies such as are found in recently-cut coppices and clear-felled areas, in gaps, around glades and where young birch and oak are establishing themselves on previously open ground (Adamson 1912, Salisbury 1916, 1918*a*, Tansley 1939). In such situations, the fronds, which emerge in late May and are fully unfurled by early July, can be very densely

packed and robust, commonly reaching well over a metre high and producing a dense shade in which there is little other summer growth. *Pteridium* litter can then be very bulky, the dead fronds decaying only slowly into a layer 10 cm or more thick. This may prevent establishment of seedling trees or even regrowth of coppice shoots, especially if the additional hindrance of browsing is severe (Tansley 1939, Rackham 1980). Beneath denser canopies of trees or shrubs, the abundance of *Pteridium* is much reduced with sparser, shorter and less vigorous fronds occurring scattered through a richer summer field layer.

In stands of this kind, it is often *Rubus* and *Lonicera* which become most obvious as *Hyacinthoides*-dominance fades in early summer. Brambles, or some of them at least, have some advantage over *Pteridium* in denser woodlands, in that they have shoots which overwinter in full leaf (Taylor 1980). But maximum growth coincides with canopy closure, summer leaves unfolding on new lateral shoots which can form a dense underscrub with its own shading canopy of foliage beneath which other species are again quite sparse. Through this tangle of arching shoots, *Lonicera* sprawls, producing knee-high vegetation which is virtually impenetrable. As with *Pteridium*, however, *Rubus* grows best under more open canopies and it shows a similar response to coppicing. It is also sensitive to browsing and increases its cover dramatically if access to deer and stock is prevented.

Holcus mollis is not so consistent a feature of this kind of *Quercus-Pteridium-Rubus* woodland as it is to the north-west, but on moderately moist soils and where the canopy shade is not too dense, it can be prominent as a patchy sward in the short period between the waning of *Hyacinthoides* and the closure of a bramble or bracken cover. In this period, too, scattered crowns of *Dryopteris filix-mas* and *D. dilatata* become fully unfurled and occasional tussocks of *Milium effusum* and *Melica uniflora* flower. In more densely shaded stands, where sparse bluebells are replaced by a thin summer cover of bracken or brambles, ground-cover plants such as *Poa trivialis*, *Glechoma hederacea* and *Veronica chamaedrys* may attain prominence with small patches of *Eurhynchium praelongum*, *Mnium hornum*, *Isopterygium elegans* and *Atrichum undulatum*. This kind of very open, impoverished field layer is very characteristic of dense *Tilia* coppice here.

Vegetative expansion of more shade-tolerant ground-cover plants is often the first response when this kind of *Quercus-Pteridium-Rubus* woodland is coppiced, followed by increased flowering in *Hyacinthoides* and a marked spread of *Silene dioica*, *Euphorbia amygdaloides* and *Digitalis purpurea* which are normally only present as scattered individuals here. *S. dioica* may mark out areas of local nutrient enrichment but often much more obvious around brushwood bonfires are patches of

Epilobium angustifolium and *Urtica dioica*. Larger Rumices may also appear with occasional *Cytisus scoparius*, *Ulex europaeus* and *Rubus idaeus*. *Juncus effusus* and *Deschampsia cespitosa*, though they can become locally prominent on wetter soils, are not a consistent feature here and the more frequent gross physiognomic change in such post-coppice floras is the spread of *Pteridium* (especially on the free-draining substrates) and/or *Rubus* (extending on to moister soils). Increased cover of these species may initially hinder coppice regrowth and remain a prominent feature of the field layer until well into the cutting cycle. Bracken-dominated glades may develop within stands of this sub-community where there has been coppicing or clear-felling and intense browsing.

Anemone nemorosa **sub-community:** Provisional hornbeam association Tansley 1911; *Quercetum sessiliflorae*, damp sub-association Moss 1911, 1913; *Quercus robur-Carpinus* woods, *Anemone* society Salisbury 1916; *Quercetum roboris*, *Anemone* society Tansley 1939; Sweet chestnut coppice Ford & Newbould 1970; Hornbeam & oak-hornbeam woods Rackham 1980 *p.p.*; Lime and lime-oak woods Rackham 1980 *p.p.*; Chestnut-woods Rackham 1980 *p.p.*; Hazel-ash stand type 3Ab Peterken 1981 *p.p.*; Ash-lime stand type 4A Peterken 1981 *p.p.*; Oak-hornbeam stand type 9Aa Peterken 1981 *p.p. Q. robur* is very much the characteristic oak of the *Anemone* sub-community and it can be found here as a dominant, along with a little *Betula pendula*, *Fraxinus*, *Alnus* and *Acer pseudoplatanus* and clones of *Populus tremula*, in high-forest stands, over a thin understorey of *Corylus* and usually much less hawthorn. *Tilia* and *Carpinus* occur as sparse associates in such woodland but, again, they can be locally dominant here in large-coppice underwood with a much-reduced canopy of oak standards, now commonly over-topped, and few associated shrubs. Hazel coppice is also a widespread form of this sub-community.

But the most striking peculiarity is the strong association between the *Anemone* field layer and the local prominence of *Castanea*. This long-standing introduction has been widely planted throughout the British lowlands (Perring & Walters 1962) but, in certain well-defined areas, notably Kent and eastern Essex, which seem to be the ancient centre of its distribution (Godwin 1975, Rackham 1980), it is a common dominant in coppiced versions of *Quercus-Pteridium-Rubus* woodland and these stands are very often of this sub-community. Some show obvious signs of having been planted, sometimes quite recently, chestnut having continued in favour as a source of hop-poles and cleft pales for fencing, but others have a much more natural appearance with ill-defined boundaries and stools of different sizes and ages in irregular arrangements (Rackham 1980). *Castanea*, though it can be a long-lived tree, is a fast and prolific producer of underwood: stools put up a very large number of shoots which quickly self-thin to leave a few substantial poles. In Ford & Newbould's (1970) study of a Kent chestnut coppice, only about 6% of the flush of sprouts survived an 18-year growth cycle, though their winter biomass at the close was 88 tonnes ha^{-1}. Further work (Ford & Newbould 1971) showed that the leaf canopy was fully closed after eight years and quantified the very poor light transmission at less than 4% of visible radiation for a single leaf thickness, a formidable shading effect, not only on its own coppice shoots but on any herbs beneath. Denser stands of *Castanea*, like those of *Tilia*, can thus have a very sparse field layer beyond the middle years of their coppice cycle.

When there is a reasonable amount of light, however, the three community constants, *Pteridium*, *Rubus* and *Lonicera* are all frequent here, though the cover of *Pteridium* is often fairly low: it can become more prominent in the first year after coppicing but typically does not form the dense, tall stands found in other sub-communities. Before these plants have reached their summer maximum, *Hyacinthoides* can be prominent but the more characteristic vernal dominant here is *Anemone nemorosa*. This species does occur elsewhere in the community but at low frequency and cover: though it flowers before *Hyacinthoides*, its leaves emerge a little later and are shaded out by those of bluebell (Pigott 1982). On these typically moister soils, *Hyacinthoides* is, for reasons which are unclear, not so successful so *Anemone* is able to attain abundance before the canopy of trees and shrubs closes and the *Rubus* attains its summer leaves. In contrast to the *Fraxinus-Acer-Mercurialis* woodland, such a switch is not accompanied here by a rise to prominence of *Primula elatior* in those parts of East Anglia where this plant is such a feature of wetter woodland soils.

Other distinctive characteristics are weak. *Holcus mollis* tends to be less frequent here than in most other *Quercus-Pteridium-Rubus* woodlands: it can tolerate fairly moist soils in the community but is probably reduced here by the marked inclination to gleying. *H. lanatus*, on the other hand, makes an occasional appearance, though it is not so prominent here as in plantation stands. *Lamiastrum galeobdolon* is sometimes present and it can become locally abundant after coppicing. The high frequency of dense stands of heavily-shading trees like *Tilia*, *Castanea* and *Carpinus* also produces a slight rise in the occurrence of *Atrichum undulatum* which, with *Mnium hornum*, scattered plants of *Anemone*, *Hyacinthoides*, *Lamiastrum* and *Ajuga reptans* and a few trailing shoots of *Hedera* and *Rubus*, comprise the very sparse ground cover in many older coppices included here.

Hedera helix **sub-community:** Hazel-ash stand type 3Aa, *Hedera-Ilex* sub-type Peterken 1981. This kind of *Quercus-Pteridium-Rubus* woodland has rarely been recognised as a distinct vegetation type, but it has some quite clear floristic and structural characteristics. Again oak is the most common tree, almost always *Q. robur*, and, with a little *Fagus* and *Fraxinus*, it can dominate here in high forest, forming a tall, often closed, canopy that sometimes shows clear signs of planting. In such stands, shrubs are often rather sparse, the understorey comprising some scattered *Corylus* and a little *Crataegus monogyna*, *Ilex* and *Sambucus*. In other cases, *Corylus* is extremely abundant in dense, long-abandoned coppice with few oak standards or none. Some old *Tilia*, *Carpinus* and *Castanea* coppices occur here too. Yet other stands look to be fairly young, but undisturbed, woodland derived by recent invasion: here *Crataegus monogyna* and *Prunus spinosa* can be abundant among scattered oaks. The striking scarcity of birch in all these kinds of woodland is a good indication of their closed and often gloomy character.

Beneath these canopies, *Pteridium* remains very frequent though its cover is usually low. *Rubus*, on the other hand, is often abundant, especially where it is trees, rather than shrubs, which constitute the bulk of the woody cover, and, with *Lonicera*, it can form a thick underscrub. *Holcus mollis* occurs fairly frequently and it is sometimes prominent in early summer but *Hyacinthoides* is uncommon and does not usually provide the striking springtime displays of flowers characteristic of the Typical and *Acer-Oxalis* sub-communities.

Much more obvious throughout the winter and spring is a ground cover of *Hedera helix* which can occur here as a virtually continuous carpet, thinning out only beneath the very densest bramble. *Dryopteris filix-mas* and *D. dilatata* occur occasionally and there can be patches of *Galium odoratum* and grasses such as *Milium effusum*, *Melica uniflora* and *Brachypodium sylvaticum*. In high-forest stands with these species, this kind of vegetation closely approaches the *Fagus-Rubus* woodland and transitional stands can be found.

Holcus lanatus **sub-community:** Woodland plot types 19 & 20 Bunce 1982. In this again little-described but very widespread and distinctive woodland type, *Q. robur* is the most frequent tree and quite often it forms the bulk of a high-forest canopy, commonly with some *Betula pendula*. In some cases, such stands are young woodlands where oak and birch are invading together but have not yet formed a dense woody cover (in contrast to the closed sub-spontaneous canopies of the *Hedera* sub-community). But often here, there are clear structural indications that the oak has been planted and this sub-community incorporates many younger plantations or thinned stands of oak in their middle years. The other

distinctive feature here is the frequency of conifers, either naturally invading trees which have needed in from nearby (into gaps within established stands or amongst newly-planted oaks) or planted trees, introduced for occasional ornament or as a more substantial replacement of oak or as an oak nurse. The characteristic field layer of this kind of *Quercus-Pteridium-Rubus* woodland is thus commonly encountered under young or thinned stands of such softwoods as *Pinus sylvestris*, *P. nigra* var. *maritima*, *Pseudotsuga menziesii* and *Larix* spp., forming the characteristic herbaceous component of many conifer plantations on base-poor but less free-draining soils throughout the south-eastern lowlands.

In these woodlands, smaller trees and shrubs are very sparse and often there is no understorey at all. *Corylus*, especially, is strikingly infrequent and the usual picture is of scattered *Crataegus monogyna*, *Sambucus* and *Prunus spinosa*, very occasional *Ilex* or *Sorbus aucuparia*, birch saplings in more open areas and, locally prominent, *Rhododendron*.

In the field layer, all the community constants are very common and often extremely abundant. In sub-spontaneous stands and many younger plantations, *Pteridium* in particular can show high cover and vigorous growth, over a very impoverished associated flora. In other cases, the exclusion of grazing animals from plantations is accompanied by a marked spread of *Rubus* and *Lonicera*, early on among the newly-planted trees and again after thinning. Such field layers are not quite so species-poor as those under dense bracken, but this kind of bramble underscrub beneath an even-aged canopy of virtually identical oaks or conifers presents one of the dreariest scenes among British woodlands.

The most obvious loss in these kinds of *Quercus-Pteridium-Rubus* woodlands is *Hyacinthoides* which is extremely infrequent. It can sometimes be found here migrating in from marginal colonies around longer-established plantations or sub-spontaneous woods but it spreads only slowly (6–10 m per century in a Surrey stand: Pigott 1984). And it is eliminated under conifers by the time these reach an age of 15–20 years, so slow re-invasion under a subsequent rotation of oak will not allow it to regain dominance (Pigott 1984). *Anemone* likewise is absent from the moister soils here.

Most of the other characteristic species of the community are uncommon. *Holcus mollis* is occasionally prominent and there are sometimes scattered dryopteroid ferns and sparse plants of *Stellaria holostea*, *Silene dioica*, *Teucrium scorodonia* and, on heavier soils, some *Deschampsia cespitosa*.

Among this generally impoverished picture, the one striking preferential is *Holcus lanatus*, scattered tussocks of which are a typical feature here. And, accompanying this in more open stands there are occasional records for a number of species which increase the untidy appearance of the vegetation. *Epilobium angustifolium* occurs with patchy prominence (as in the stands of Bunce 1982) and there can be some *Urtica dioica, Senecio jacobaea, Arrhenatherum elatius, Heracleum sphondylium, Rubus idaeus, Cytisus scoparius* and *Ulex europaeus*. Young plantations here thus have much in common with the post-coppice vegetation of the *Quercus-Pteridium-Rubus* woodland.

Acer pseudoplatanus-Oxalis acetosella sub-community:
Quercetum sessiliflorae, Pteris-Holcus-Scilla complementary society Woodhead 1906; *Quercetum sessiliflorae*, drier sub-association Moss 1911, 1913; *Quercetum petraeae, Pteridium-Holcus-Scilla* society Tansley 1939; *Blechno-Quercetum fraxinetosum* Klötzli 1970 *p.p.*; Hazel-ash stand type 3D Peterken 1981 *p.p.*; Birch-oak stand types 6Ac & 6Bc Peterken 1981 *p.p.*; *Lonicero-Quercetum, Endymion* subassociation (Birse & Robertson 1976) Birse 1984 *p.p.*. Oak is still a frequent and usually the most abundant tree in this sub-community but there is a marked shift here from *Q. robur* to *Q. petraea* and hybrids. And, very commonly, *Acer pseudoplatanus* and *Fraxinus* make some contribution to the woody cover with, less often, *Ulmus glabra*. Birch is surprisingly sparse, *Fagus* is rare and *Tilia* and *Carpinus* have not been recorded, though in some Lake District ravines, *Tilia* survives in close association with this kind of woodland (Pigott & Huntley 1978). A high-forest structure predominates, though some stands are obviously of fairly recent natural origin and others are clearly plantations or much-modified woods. There are also some coppices, though this treatment is not so universally widespread among these north-western *Quercus-Pteridium-Rubus* woodlands as it is to the south-east: the usual underwood crop is hazel, though oak itself has sometimes been cut and there are some sycamore coppices.

Corylus is the most frequent shrub here, even in high-forest stands where it forms the bulk of a generally open low understorey. Hawthorn is common, though it is often rivalled in frequency and cover by *Sambucus* and *Ilex*: the latter is sometimes locally very abundant, as around the Pennine fringes where it seems to have figured as a winter foliage crop for stock. *Sorbus aucuparia* also occurs occasionally and *Rhododendron* can be a prominent introduction or invader. Saplings are more numerous here than is usual in the community and there are commonly some young *Fraxinus* and *A. pseudoplatanus* with less frequent birch.

The basic components of the field layer are as in the Typical sub-community with *Hyacinthoides* a common vernal dominant, succeeded by various mixtures of *Pteridium* and *Rubus* and *Lonicera*. But here *Holcus mollis* is much more generally frequent, its lush cover developing as the bluebell flowers fade, and dryopteroid

ferns, especially *D. dilatata*, tend to be more prominent too, along with occasional *Athyrium filix-femina*. *Stellaria holostea* and *Deschampsia cespitosa* are also slightly preferential to this sub-community.

But the most distinctive herb in this woodland of the wetter north-west is *Oxalis acetosella* which attains constancy here and is one of the very few plants that will tolerate a dense cover of *Pteridium* (Packham 1978). As in the *Fraxinus-Acer-Mercurialis* woodland, this rise is accompanied by a less marked increase in the frequency of *Viola riviniana* and by a greater representation of bryophytes, though on these more base-poor soils, more calcicolous species are not involved. Most prominent are *Eurhynchium praelongum, Mnium hornum, Brachythecium rutabulum, Thuidium tamariscinum* and on more acidic litter and soils, *Plagiothecium undulatum* and *Isopterygium elegans*. Among the patches of mosses and on areas of open ground, seedlings of *A. pseudoplatanus* can be extraordinarily abundant in spring, though few get away except in gaps.

The other characteristic feature of the field layer here is that its herbs often occur in intimate mosaics with species of more free-draining and surface-leached soils, especially on steeper slopes where there can be frequent outcrops of rocks or terracing, as in some of the sites described by Woodhead (1906) and Scurfield (1953) from the Pennines. It is in this kind of situation that plants such as *Deschampsia flexuosa, Blechnum spicant, Vaccinium myrtillus* and, where there is grazing, *Anthoxanthum odoratum* and *Agrostis capillaris*, can appear. Strictly, patches with such species should be regarded as fragments of more calcifugous oak-birch woods, but, in practice, it is often impossible to sample separately every terrace riser and nidus-capped rock.

Habitat
The *Quercus-Pteridium-Rubus* woodland is characteristic of base-poor brown soils throughout the temperate lowlands of southern Britain. Here, its distinctive assemblage of species probably approximates to the natural climax forest on such soils, but the floristics and physiognomy of the community have been very widely affected by sylvicultural treatments. Some stands are of known antiquity, but many have been derived by relatively recent invasion or planting.

Edaphically, this kind of woodland occupies a broad middle ground between rendzinas and brown calcareous earths on the one hand, and brown podzolic soils and true podzols on the other. It is thus absent from limestones unless they are overlain by non-calcareous superficials, even towards the north-west of its range where higher rainfall induces extensive surface leaching in calcareous lithomorphic soils. It tends to be scarce, too, over calcareous superficial deposits like the Chalky Boulder Clay which typically weathers to produce profiles that can be quite acid above but which usually have much exchangeable calcium below. Over such rocks and soils as these, the community is characteristically replaced by the *Fraxinus-Acer-Mercurialis* woodland, the more calcicolous species of which make an appearance here only where there is some local base-enrichment, as along stream sides or around flushes.

Towards the other extreme, the *Quercus-Pteridium-Rubus* woodland does not extend far on to soils derived from more pervious, acidic deposits. It can be found in the south-east over some of the sandstones of the Cretaceous sequence in the Weald, on Eocene sands and gravels in Essex and Kent and around the Southampton Basin and over more free-draining acid superficials but, in general, it gives way in such situations to the *Quercus-Betula-Deschampsia* woodland where species such as *Deschampsia flexuosa*, *Vaccinium myrtillus* and calcifugous bryophytes become very important. Towards the north-west, these plants show an increasing tendency to move into all drier woodlands apart from those on freely-weathering limestones and they can make an occasional appearance in the *Acer-Oxalis* sub-community. Beyond the 1000 mm isohyet (*Climatological Atlas* 1952), which forms an effective limit to the *Quercus-Pteridium-Rubus* woodland, they become an important component of the *Quercus-Betula-Oxalis* woodland, its sub-montane analogue on moderately acid rocks and superficials.

Within these fairly wide limits, the *Quercus-Pteridium-Rubus* woodland is characteristic of soils with a superficial pH that is generally between 4 and 5.5 but which show a great variety of textures and of water and humus regimes. In the British lowlands, such profiles are widespread over many different deposits but this community is especially associated with argillaceous rocks and superficials. As mentioned above, it can be found on some sandstones and sands, but it is especially frequent on the Eocene London Clay north and south of the Thames and in Hampshire, on the Triassic Keuper Marl which crops out in Devon and forms very extensive parts of the Midlands and the Cheshire Vale, on Carboniferous Culm and Coal Measure shales, again in Devon, but particularly important in this community in the Pennine foothills, on Welsh Silurian shales and on long-weathered Granite in the South-West Peninsula and in south-west Scotland. Among superficial deposits, Clay-with-Flints forms a characteristic substrate over some of the southern Chalk (though here *Fagus-Rubus* woodland often replaces this community) and, in valleys on the upland fringes, this kind of woodland often marks out patches of glacial drift or head. It can occasionally be found on the drier alluvium of river terraces.

Among the more free-draining profiles derived from such parent materials, many approximate to the classic brown earth (Avery 1980: brown forest soil in the older

terminology). In some of these soils, mor accumulation can be quite marked and there may be some slight bleaching of sand grains in the A horizon, with a depression of surface pH down to 3.5, but podzolisation is typically no more than incipient. It is when the *Quercus-Pteridium-Rubus* woodland occurs in such situations that it makes its closest approach to the *Quercus-Betula-Deschampsia* woodland with an increase in *Betula pendula* and *Pteridium*. Stands of this kind occur quite commonly in the *Acer-Oxalis* sub-community to the north-west and, more locally, in the Typical sub-community in East Anglia and the Weald. Some locally dominant trees also seem to encourage the development of mor humus: *Castanea* is one, even oak and, of course, the coniferous replacements of the *Holcus* sub-community.

Generally, however, the most marked feature of these more free-draining soils is the mobilisation of clay minerals and very commonly here the B horizon shows a clearly argillic character. Where the clay fraction is not too dominant, the soil structure can be well developed with good incorporation of mull humus by an active soil fauna and brown earths of this type are quite frequent beneath the Typical, *Hedera* and *Holcus* sub-communities (except where conifers have been long established) and also in the *Acer-Oxalis* sub-community where, as is sometimes the case, it occurs on slopes with fairly free drainage.

However, very often here, there is so much clay in the profile that the soils are rendered impermeable to excess rain and seasonal surface-water gleying is very widespread beneath stands of the community, with the frequent characteristic signs of mottling above and sometimes manganese concretions below (e.g. Curtiss *et al.* 1976). This is true even to the south-east because, though rainfall there is considerably lower than to the north-west (less than 600 mm yr^{-1} for the most part), the topography is typically flat or only gently undulating with little surface drainage. Frequently, then, the soils are stagnogleyic brown earths or true stagnogleys (though some clay-rich profiles would now be classed as pelosols: Avery 1980, Soil Survey 1983) and, on a national scale, there is a rather striking coincidence between the distribution of these profiles and that of the *Quercus-Pteridium-Rubus* woodland (e.g. Soil Survey 1974).

This tendency to surface-water gleying can be seen throughout the community and it probably plays a part in influencing the balance between some of the major species here. It is likely to be involved to some extent in controlling the proportions of the two oaks (Jones 1959), with *Q. robur* being more characteristic of the heavier soils, *Q. petraea* rising to prominence on the lighter, more freely drained (though not always drier) soils. And it is probably partly responsible for the variation in the amounts of *Hyacinthoides*, *Holcus mollis* and *Pteridium* that is so characteristic a feature of the field layer in this kind of woodland. Most strikingly among the herbs, however, a pronounced tendency to winter and spring waterlogging is marked here, as in the *Fraxinus-Acer-Mercurialis* woodland, by a switch from *Hyacinthoides* to *Anemone* as the vernal dominant. This floristic transition, which seems to be based on a decreasing ability on the part of *Hyacinthoides* to maintain its competitive advantage against *Anemone* on the wettest soils, is a continuous one but the *Anemone* sub-community is very obviously associated with waterlogged plateaus and hollows over undulating topographies and becomes much commoner on the very heaviest substrates on which this community is found. It is particularly frequent, for example, on the sticky Eocene clays in Essex and Kent where many stands have been treated as hornbeam or chestnut coppice. It can also be found picking out areas of ground-water gleying as where this kind of woodland extends down on to small alluvial flats.

As in the *Fraxinus-Acer-Mercurialis* woodland, climate also plays a direct role in influencing the distribution and floristics of this community, as well as being involved in these edaphic relationships, and its effects here are similar, though not so sharply-defined. First, like its more calcicolous counterpart, the *Quercus-Pteridium-Rubus* woodland is confined to the warmer parts of Britain, where mean annual maximum temperatures are generally in excess of 26 °C (Conolly & Dahl 1970). In this community, though, a Continental/Continental Southern element is not very prominent and only *Tilia cordata, Carpinus betulus, Sorbus torminalis, Crataegus laevigata, Euphorbia amygdaloides* and *Lamiastrum galeobdolon*, none of which is more than occasional throughout, serve to mark off the community from its analogue in the cooler uplands, the *Quercus-Betula-Oxalis* woodland. *Tilia*, though it tends to be a more extensive dominant in more southerly stands of the *Quercus-Pteridium-Rubus* woodland than in the *Fraxinus-Acer-Mercurialis* woodland, shows exactly the same complex pattern of reproductive behaviour here in relation to the progressively cooler summers as one moves north into the range of the *Acer-Oxalis* sub-community (Pigott & Huntley 1978, 1980, 1981).

The second more direct impact of climate is seen in the increasingly oceanic character of the community towards the west and north. Indeed, in comparison with similar woodlands throughout Europe (e.g. Noirfalise 1968, Neuhäusl 1977), the most distinctive feature of this woodland as a whole is the Atlantic aspect of its flora, with *Hyacinthoides* playing such a prominent role. But, in moving away from the south-east, other floristic elements reflect more clearly the shift to milder winters and a generally wetter climate. Increasingly to the south-west, for example, the *Hedera* sub-community replaces the Typical sub-community, even in longer-established

stands, where the abundance of an ivy carpet is not related to uninterrupted growth of underwood or canopy. *Ilex*, too, shows a slight increase throughout in more westerly stands of the community.

Much more obvious are the various characteristic features of the *Acer-Oxalis* sub-community where the *Quercus-Pteridium-Rubus* woodland extends on to base-poor soils in those parts of Britain with an annual rainfall approaching 800 mm or more (about 140 wet days yr^{-1} or more: Ratcliffe 1968). Among the trees, *Acer pseudoplatanus* itself is the best marker of this shift, as in the *Fraxinus-Acer-Mercurialis* woodland, though its abundance here is not so consistently high, perhaps because of the tendency towards gleying in so many of the soils. But, where the profiles are free-draining and moist, it does supremely well. *Fraxinus* follows it in increasing its frequency in a marked transgression into less calcicolous woodlands noted by Tansley (1939), and *Ulmus glabra* shows a similar but smaller response.

Among the field layer species of the *Acer-Oxalis* sub-community, the marked increase in the frequency of *Oxalis acetosella* and, to a lesser extent, in *Viola riviniana*, are also probably a reflection of a more consistently moist soil surface than is usual among south-eastern *Quercus-Pteridium-Rubus* woodlands where even less free-draining soils can become severely droughted in summer (e.g. Packham 1975, 1978). Bryophytes, too, tend to be more extensive and lusher to the north-west, and, marking the combination of fairly high soil moisture with surface acidity, there is a rise in the representation of *Dryopteris dilatata* and *Athyrium filix-femina*.

Variations in climate and soils set the basic framework of floristic variation within the community and influence the distribution of certain of the sub-communities but, as in the *Fraxinus-Acer-Mercurialis* woodland, sylvicultural treatment has commonly influenced the physiognomy and composition of the vegetation here. Sometimes, such treatment has accentuated what are probably natural associations between certain trees and particular field layers, as with the selection for *Carpinus* and especially for *Castanea* on the moister soils of the *Anemone* sub-community. But often it has worked against natural patterns of variation, as with the reduction of *Q. robur* to occasional standards in many coppices or the favouring of this species as against *Q. petraea* in plantations on some lighter soils, or the introduction of exotic canopy dominants such as various conifers. In general terms, treatment has produced variation within the trees and shrubs that is not always consonant with differences in the field layer. Individual sub-communities can thus have a wide variety of underwood and canopy covers and different sub-communities exactly the same woody component.

The traditional treatment here has been coppicing with or without standards and, over more base-poor

lowland soils, the *Quercus-Pteridium-Rubus* woodland
has been the major source of underwood and timber
through historic times. Stands with obvious signs of
coppicing are much more widespread and frequent to
the south-east and here the Typical and *Anemone* sub-
communities have been extensively cropped for hazel
and mixed small underwood, for lime and hornbeam,
occasionally for oak and, mostly in the *Anemone* sub-
community, for chestnut. In many (though now fast-
decreasing) woodlands in this region, tracts of these sub-
communities survive as an irreplaceable record of past
land use, often with archaeological features, like wood-
banks and rides, marking out the coppice compartments
(e.g. Rackham 1980, Peterken 1981). Almost universally
now, such woodlands are not used in the traditional
fashion. Many show a spread of birch with the neglect of
cleaning after the last cuts and, where the underwood
has remained undisturbed for a number of decades,
there has often been a progression to the gloomy conditions that favour the development of the *Hedera* sub-
community. To the north-west, evidence of coppicing is
generally rarer but stands of the *Acer-Oxalis* sub-
community can be seen converted to hazel or mixed
coppice and also to sycamore coppice; there are even a
few far-flung stands of lime coppice in this region.

With the neglect of coppicing, the distinctive post-
coppice floras associated with this kind of woodland are
now rarely seen but, where cutting continues, the general
pattern of development is as in the *Fraxinus-Acer-
Mercurialis* woodland (e.g. Salisbury 1916, 1918*a*, 1924,
Rackham 1980). First, there is a spread of the more
shade-tolerant herbs of the community, notably here
Silene dioica and *Euphorbia amygdaloides*, and, in the
second or sometimes third spring, a dramatic flush of
flowering from *Hyacinthoides* and/or *Anemone*. Second,
Rubus and *Pteridium* can show a marked increase,
sometimes very early on and overwhelming these species. Then, more weedy plants can spread or appear
anew, most commonly here *Epilobium angustifolium* and
Urtica, Rumices and more calcifugous shrubs like *Cytisus scoparius* and *Ulex europaeus*, or even *Calluna
vulgaris* on the most acid soils. Mixtures of these species
present a rather different kind of flora to that characteristic of *Fraxinus-Acer-Mercurialis* coppices in their middle years though, as there, many of the adventives are
eliminated as the new underwood growth closes over.
But *Rubus* and, especially, *Pteridium* can compete with
the springing shoots, especially where these are subject
to the additional hindrance of heavy browsing and some
bracken glades within stretches of these woodlands may
have resulted from *Pteridium* gaining a firm hold after
coppicing.

Many of the older coppiced stands of the *Quercus-
Pteridium-Rubus* woodland have now been felled with a
permanent loss in the extent of the community. However, plantation stands of this kind of woodland are
quite widespread and, where these have a long-established canopy of hardwoods, usually here oak, sometimes with a little *Fraxinus* or *Fagus*, the full complement
of the characteristic field-layer species can be present.
Some older plantations of this kind are accommodated
in the Typical and *Acer-Oxalis* sub-communities. Quite
commonly, though, even hardwood plantations lack a
good representation of *Hyacinthoides* and have the
typically impoverished and untidy field layer of the
Holcus sub-community. Many of these, of course, may
be woodlands planted on previously open ground where
there were no existing patches of *Hyacinthoides* to
provide centres for subsequent migration. But, even
where such plantations have been established on the site
of clear-felled stands of the *Quercus-Pteridium-Rubus*
woodland, *Hyacinthoides* may be eliminated or severely
reduced by a combination of physical disturbance and a
dense growth of, first, *Rubus* or *Pteridium* on the cleared
ground or, second, the young trees. This certainly
happens under conifers, such that *Hyacinthoides* can be
completely suppressed by the second decade of their
growth. Ornamentation with occasional softwoods is
quite widespread in the community and, where the trees
are not too densely spaced, the vegetation can retain the
greater richness of the Typical or *Acer-Oxalis* sub-
communities. But a dense canopy of conifers, with its
heavy shade and accumulation of mor, excludes *Hyacinthoides* and alternations with oak may not allow time for
re-invasion (Pigott 1984). Older softwood plantations,
where there is more light, often have the *Holcus* sub-
community and the development of this kind of *Quercus-Pteridium-Rubus* woodland may be very dependent
on the opening up and disturbance that occurs with
thinning. Fairly young and undisturbed woody covers
here frequently show the *Hedera* sub-community and
this may be the usual precursor to the *Holcus* sub-
community under both softwood and hardwood
replacement canopies.

The *Hedera* sub-community is characteristic, too, of
young stands of this kind of woodland which have
sprung up by natural invasion of open ground. With
intensive agricultural pressure on many of the lowland
clay areas, such sub-spontaneous tracts are not very
common or extensive but they can be found quite widely
in neglected field corners, old clay pits, on ungrazed
commons and on waste ground. In such situations, the
uniform or grouped age-classes of the trees and a
prominent element of spinose shrubs can give a persistent clue as to the origin of the woodland.

Zonation and succession

Zonations between different kinds of *Quercus-Pteridium-Rubus* woodland and transitions to other woodland types are primarily related to edaphic variation.

Treatment differences produce effects of their own but they can also emphasise or confuse soil-related patterns. Marginal zonations sometimes represent successional changes but very often now agricultural use of the surrounding land produces artificially abrupt transitions to herbaceous vegetation.

By far the commonest kinds of edaphic zonations within stands of the community are related to differences in soil moisture, most frequently changes in the extent and frequency of surface-water gleying. Often these are related directly to variations in slope with the Typical or *Acer-Oxalis* sub-communities picking out areas with better surface drainage, often here on ground with only a very gentle slope, and the *Anemone* sub-community marking hollows and plateau tops. Similar patterns can be seen where ground-water gleying produces local areas of wetter soil along the sides of streams flowing through *Quercus-Pteridium-Rubus* woodland (as in the transect in Pigott 1982). Quite frequently, such zonations continue into some kind of alder woodland. Around slope flushes, for example, there is often a transition to the *Alnus-Fraxinus-Lysimachia* woodland. This may be quite gradual, with a progressive increase in herbs such as *Deschampsia cespitosa* or *Juncus effusus* and a slow switch to *Alnus*-dominance. In other cases, there is an abrupt appearance of *Carex remota* or *C. pendula* in sharply-defined patches. Both these kinds of zonation are very well seen along the base of the Lower Greensand in the Weald and on slopes cut into the Coal Measure shales of the Pennines. Where there is local peat accumulation in small river valleys, the *Quercus-Pteridium-Rubus* woodland may give way to small stands of the *Alnus-Carex* woodland, quite a common pattern in the Weald again and in some Essex woods (e.g. Rackham 1980, Wheeler 1980c). In other cases, it is the *Alnus-Urtica* woodland which terminates the sequence on alluvial flats, the appearance of *Urtica dioica* and *Galium aparine* marking the transition to more eutrophic soils.

The other very common kind of zonation to other woodland types is related to differences in the base status of the soils. Very often, sites in which the *Quercus-Pteridium-Rubus* woodland is represented cover geological transitions to bedrocks or superficials which are either more or less calcareous. Where there is a switch to limestones, lime-rich shales or clays or to drift with more calcium carbonate, the community characteristically passes to the *Fraxinus-Acer-Mercurialis* woodland on more base-rich soils, usually brown earths, calcareous stagnogleys or pelosols, more rarely free-draining rendzinas. Sharp alternations of different bedrocks, especially typical here of the Cretaceous and Eocene sequences of the south-east, or abrupt differences in the nature of superficials (as at Gamlingay: Adamson 1912), can be marked by equally well defined floristic zona-

tions, with an abrupt switch in dominance from *Pteridium* or *Holcus mollis* to *Mercurialis* and its calcicolous associates. In the south-east, such transitions generally involve a move from the Typical, *Anemone* or *Hedera* sub-community of the *Quercus-Pteridium-Rubus* woodland to the *Primula-Glechoma*, *Anemone* or *Hedera* sub-community of the *Fraxinus-Acer-Mercurialis* woodland respectively. To the north-west, the zonation is typically from the *Acer-Oxalis* sub-community of the former woodland to the *Geranium* sub-community of the latter. It should be remembered, though, that a number of important species occur in both these kinds of woodland and, where there is a continuation of a *Hyacinthoides* carpet, say, under a canopy of *Q. robur*, *Tilia*, *Carpinus* and *Corylus*, the boundary between the communities may be blurred. Such gradual transitions are especially characteristic of heavy superficial deposits in the south-east, where there may be diffuse variations in the calcium carbonate content of the parent material and a complex mosaic of brown earths or stagnogleys with small but critical differences in base-status.

Where the characteristic soils of the *Quercus-Pteridium-Rubus* woodland give way to strongly-leached brown earths, brown podzolic soils or podzols, the community is replaced by the *Quercus-Betula-Deschampsia* woodland. Typically, here, this kind of transition is related to a geological switch from shales or clays to some kind of arenaceous bedrock or sandy superficial deposit. Such patterns can be seen in the south-east in the Cretaceous sequences of the Weald and they are very common to the north and west where, in Carboniferous deposits, alternations of shales and grits form the basis of much of the scenery along the fringes of the Pennine uplands in Derbyshire, South and West Yorkshire and Durham. Here, the *Acer-Oxalis* sub-community characteristically occupies the much-weathered dips of cuestas and the lower slopes of the valley sides, with the *Quercus-Betula-Deschampsia* woodland clothing the more resistant grit scarps. Sharp transitions between these kinds of woodland are quite common and, though *Pteridium* continues through both with *Q. petraea* and *B. pendula* figuring in the canopy of each, the springtime boundary of *Hyacinthoides* and the restriction of *Fraxinus*, *A. pseudoplatanus* and *Corylus* to the *Quercus-Pteridium-Rubus* woodland generally serve to delimit the communities.

In some situations, though, the zonation is not so clear. On steeper slopes, slipping is quite common, with grit blocks tumbling down over the incompetent shales and then there can be a patchy inter-digitation of the two kinds of woodland along the slope foot. Even on more stable slopes, terracing is frequent and then every tread may be clothed with vegetation resembling the *Quercus-Pteridium-Rubus* field layer, every riser with *Quercus-Betula-Deschampsia* herbs (e.g. Woodhead 1906, Scurfield 1953). When *Hyacinthoides* has faded from the patches of the former, such mosaics are typically marked by a patterning of *Holcus mollis* and *Deschampsia flexuosa* (Jowett & Scurfield 1949, 1952). *Holcus* seems to hold its own on the less sharply draining areas and also tends to hold the litter, the decay of which helps maintain the more mull-like qualities of the soils. Indeed, there is evidence that *Holcus* can invade the *Deschampsia* patches in these kinds of mosaic (Ovington 1953).

Particularly in the south-east, where the effects of traditional sylviculture have persisted more extensively, coppicing treatments can confuse these soil-related transitions. Most obviously, where a uniform treatment has been applied across edaphic zonations, identical covers of underwood can continue over what are really the field layers of different communities. This is very common here across junctions between the *Quercus-Pteridium-Rubus* woodland and the *Fraxinus-Acer-Mercurialis* woodland, both of which have been widely cropped for hazel, lime and hornbeam. Under dense underwood, especially of the heavily-shading *Tilia*, herbs become very sparse so there may be only the very slightest indication of where one community ends and the other begins.

Even within sites made up wholly of *Quercus-Pteridium-Rubus* woodland, coppicing produces its own patterns of variation which are superimposed, often quite independently, over floristic differences related to, say, soil moisture. In actively-coppiced woodland, compartments marked out by varied and striking post-coppice herbaceous vegetation at different stages of development can thus provide an interruption to zonations between the Typical and *Anemone* sub-communities, or a cyclically changing patchwork of parcels within tracts of each. Treating a whole woodland in this way results in great spatial variation at any one time, though in any one section of a stand, the differences are essentially a temporal perturbation of an underlying floristic pattern related to natural variation in the soils. With neglect of coppicing, such edaphic transitions can become more obvious again, though after several decades of uninterrupted underwood growth the shape of old compartments can become marked on the ground once more by abrupt transitions to the *Hedera* sub-community with a dense woody cover. And many woodlands subject to intensive use in the past still preserve complex patterns of zonations between the *Quercus-Pteridium-Rubus* woodland and other vegetation types which reflect the structural organisation of the site. Commonest among these are the transitions in larger rides and glades which are here often picked out by stands of *Pteridium* or *Rubus-Holcus* underscrub, frequently now invaded by birch, *Crataegus monogyna* and *Prunus spinosa*.

External boundaries to stands of the community can show similar zonations, though where the woodlands survive within intensive agricultural landscapes, there is generally only a very compressed fringe of *Crataegus* or *Prunus* scrub with *Rubus-Holcus* underscrub or, on more free-draining soils, a narrow strip of *Betula* and *Pteridium*, with an abrupt transition to pasture or arable.

In other situations, more extensive marginal transitions to herbaceous vegetation can be seen and these, together with young stands developing anew on railway verges (e.g. Sargent 1984) and on commons where grazing rights have fallen into disuse, can give some clues as to the seral development of the community. However, systematic studies of successions to the *Quercus-Pteridium-Rubus* woodland are very few (Salisbury 1918*b*, Adamson 1921, Tansley 1939), usually concerned with stands developing on abandoned arable land and set within a very broad understanding of what 'oak woodland' is or ought to be. The following observations are therefore very brief and generalised.

Within the lowlands, the natural limits to the development of the *Quercus-Pteridium-Rubus* woodland are set, for the most part, by the character of the soils such that the community is excluded from situations where the substrate is markedly base-rich, base-poor or very wet. Within this fairly broad compass, it probably approximates to the climax forest type, except within the natural range of *Fagus* which competes successfully with *Q. robur* for dominance over similar field layers to those characteristic here on all but the moister brown soils on which the *Quercus-Pteridium-Rubus* woodland can occur (e.g. Watt 1923, 1924, 1925, 1934*a, b*).

The natural herbaceous precursors to the community are probably very varied. Over the central type of brown earth, with free but not excessive drainage, less calcicolous types of the *Arrhenatheretum* are likely to figure prominently with, on more waterlogged mesotrophic soils, communities like the *Holco-Juncetum* and the *Holcus-Deschampsia* grassland. Towards the extreme of more strongly leached soils, this kind of woodland can probably supersede the less calcifuge types of *Festuca-Agrostis* grassland, though it is not usually a natural successor to ericaceous heath.

Where these types of vegetation are not regularly grazed, mown or burned, invasion by some of the characteristic shrubs and trees of the community can be very rapid. On moister soils, *Crataegus monogyna* and *Prunus spinosa*, *Rubus fruticosus* and *Rosa* spp. are frequent early colonisers, quickly thickening up to form a patchwork of scrub. In drier situations, *Betula pendula* becomes increasingly prominent in the early stages and it too may form dense thickets. In certain cases, this kind of young woodland can persist for some time: birch-dominated stands in particular seem to be more resistant to further development than might be expected from the relatively short life of the individual trees (e.g. Rackham 1980).

But it is a very common feature of this type of succession that oak, especially *Q. robur*, invades very early. The present poor regeneration of this tree within closed woodlands is well known (e.g. Watt 1919, Jones 1959, Rackham 1980) and probably a fairly natural consequence of a number of factors, chiefly the neglect of coppicing which created a regular sequence of large well-lit gaps, and loss of seedlings to moth caterpillars and oak mildew (e.g. Jones 1959, Shaw 1974, Rackham 1980). In the open, however, young oaks can appear quickly and in profusion. Acorns are distributed much further from the parent tree than, for example, beech mast, being carried hundreds of metres by pigeons, corvids and squirrels (Jones 1959, Mellanby 1968) and they are readily able to germinate in closed swards, even if the shade cast by tall grasses, open bracken or nearby young shrubs is moderately dense (Jones 1959). Early growth may be slow but young oaks are very resilient and there seems little doubt that in many younger stands, *Q. robur* has played an important role from the outset.

We do not know how long it takes for *Quercus-Pteridium-Rubus* woodlands to acquire the richer field layers typical of the community. Young dense stands are often of the *Hedera* type, ivy carpets spreading even when the vegetation is in the scrub stage. And older stands derived from invasion may perhaps develop the character of the *Holcus* sub-community as the canopy opens and the woodland becomes more accessible to disturbance. But complete invasion by the more slow-spreading herbs probably takes centuries and, though *Carpinus* and *Castanea* may find a place in the developing woody cover in the south-east, the more diverse and complex mixtures of trees and shrubs characteristic of ancient stands are not found.

Distribution

The *Quercus-Pteridium-Rubus* woodland is widely distributed and common over the lowlands of England and Wales. It does extend into southern and eastern Scotland but, in the cooler and wetter upland fringes of the west and north, it is replaced by the *Quercus-Betula-Oxalis* woodland. The central types over the more south-easterly parts of the range are the Typical and *Anemone* sub-communities, the former more widespread, the latter more local, though particularly prominent on the heavy clays of the Weald and Essex. Towards the more oceanic south-west and especially in Devon and Cornwall, the *Hedera* sub-community becomes increasingly common, though this kind of woodland is quite widespread elsewhere in neglected

coppices and younger stands. The *Holcus* sub-community occurs throughout most of the range but has been most frequently recorded in plantations in the Weald, East Anglia and the Midlands. The *Acer-Oxalis* sub-community is confined to the upland margins of Wales, northern England and Scotland and represents a clear transition to the *Quercus-Betula-Oxalis* woodland.

Affinities

This community has long been recognised as one of the major kinds of woodland in lowland Britain but comparisons with early definitions are difficult because of the particular conceptions of 'oak woodland' which have long been current in the English descriptive tradition. In the first place, it has generally been the practice to make some kind of separation between different kinds of oak woodland according to whether *Q. robur* or *Q. petraea* was the dominant tree. Some schemes explicitly recognised that either or both could form a canopy over what was essentially the same kind of field layer, like Tansley's (1911) proposal for a *Quercetum arenosum roburis et sessiliflorae*. And, indeed, this fact was acknowledged in many accounts; but, generally, classifications have characterised a pair of communities, a *Quercetum roboris* and a *Quercetum sessiliflorae/petraeae* (e.g. Moss 1911, 1913; Tansley 1939), or equivalents in which other trees figured prominently, like the pair of oak-hornbeam woodlands described by Salisbury (1916, 1918a). Such communities were then diagnosed as having overlapping series of field-layer societies, many of which were shared, though some more consistently associated with the dominance of one or the other oak. A similar, though less exclusive division on the basis of *Quercus* spp. characterises the much more recent scheme of Peterken (1981) within his birch-oak woodlands proper and in his lime and hornbeam woods.

Among British woodlands, there certainly is a pattern in the distribution of the two oaks in relation to climate and soil but it is quite a complex pattern, even when the more obvious effects of sylvicultural preferences are excluded, and it is not entirely consistent as one moves from calcicolous woodlands, through those of more neutral soils, to the most calcifuge. In this community, typical of soils of moderate to low pH and often with impeded drainage, the predominant oak is *Q. robur*, but *Q. petraea* is locally important to the south-east and becomes much more prominent to the wetter and cooler north-west. Overall, there is no justification for making a precise split within the community according to which oak is the dominant. Thus, although the *Quercus-Pteridium-Rubus* woodland approximates to the older *Quercetum roboris*, and to the pedunculate oak stand types in Peterken's (1981) scheme, the correspondence is not exact.

The second difficulty is that, in early accounts, oak-wood of the *Quercetum roboris* type was a very broadly defined amalgam of virtually all woodlands in which *Q. robur* was a frequent canopy component. It therefore often took in more calcicolous vegetation which is here considered part of the *Fraxinus-Acer-Mercurialis* wood-land, where *Q. robur* is certainly common (*Q. petraea*, too, to the north-west) but which can hardly be des-cribed as oak woodlands. The *Quercus-Pteridium-Rubus* woodland does grade floristically into the *Fraxinus-Acer-Mercurialis* woodland, especially in its woody component, but the two communities are really quite distinct. In vernacular terms, the former comprise 'blue-bell woods', the latter 'mercury woods'.

The schemes of Rackham (1980) and Peterken (1981) also split off from the older *Quercetum roboris* more calcicolous woodlands like those separated here into the *Fraxinus-Acer-Mercurialis* woodland. But comparisons between what remains and the *Quercus-Pteridium-Rubus* woodland are difficult because of the concentra-tion by these two authors on the woody component of the vegetation. Peterken (1981) does have a birch-oak woodland which corresponds in part with this commun-ity; Rackham (1980) does not and most of what he calls oak woodland would fit in this scheme into the *Quercus-Betula-Deschampsia* community. But in the classifica-tions of both these workers, many of the kinds of woodland included here are placed in separate commu-nities defined by the dominance of lime, hornbeam, chestnut or hazel. The approach in this scheme is to regard such stands as treatment-derived variants of sub-communities related primarily to differences in climate and soils.

Retaining such stands within the general ambit of the *Quercus-Pteridium-Rubus* woodland helps to integrate the community within a European framework. The presence of *Tilia* and *Carpinus* is especially important in this respect because both these trees are prominent components of this kind of woodland right across Northern Europe into the USSR (Neuhäusl 1977). What is especially distinctive about the British stands, like those of north-west France and southern Belgium is the presence of *Hyacinthoides* as the characteristic vernal dominant (Noirfalise 1968, Neuhäusl 1977). This community is thus a clear equivalent of woodlands like the *Querceto-Carpinetum* (Lemée 1937, Dethioux 1955), the *Quercetum atlanticum* (LeBrun *et al.* 1949) and the *Endymio-Carpinetum* (Noirfalise & Sougnez 1963, Noir-falise 1968, 1969) described from these areas and includes all our Carpinion woodlands on less calcareous soils. The north-western limit of this kind of woodland in Britain is mapped by Noirfalise (1969) and Neuhäusl (1977) at the supposed natural limit of *Carpinus* itself, but there is no strong reason to exclude lowland 'blue-bell woods' which lie beyond this line.

Floristic table W10

	a	b	c	d	e	10
Quercus robur	III (2–10)	IV (3–10)	IV (2–10)	IV (1–10)	II (1–8)	IV (1–10)
Betula pendula	III (2–9)	III (2–8)	I (1–8)	III (1–10)	I (1–8)	II (1–10)
Fagus sylvatica	I (1–10)	I (3)	II (1–10)	I (3–5)	I (1–6)	I (1–10)
Sorbus aucuparia	I (1–5)	I (3)	I (1–4)	I (3–5)	I (1–5)	I (1–5)
Ilex aquifolium	I (1–5)	I (2–7)	I (1–4)	I (2–6)	I (1–4)	I (1–7)
Alnus glutinosa	I (1–5)	I (4)	I (4–7)	I (9)	I (3–6)	I (1–9)
Prunus avium	I (3)	I (3–5)	I (2–3)	I (1–4)	I (2–4)	I (1–5)
Betula pubescens	I (4–7)	I (4–7)	I (2–9)	I (4–8)	I (1–7)	I (1–9)
Taxus baccata	I (1–5)	I (1–3)	I (2–6)	I (9)		I (1–9)
Tilia vulgaris	I (3–7)			I (3–4)	I (4)	I (3–7)
Carpinus betulus	I (1–9)	I (4–9)	I (3)			I (1–9)
Tilia cordata	I (2–5)	I (2–5)	I (1–4)			I (1–5)
Populus tremula	I (1–4)	I (1–4)		I (4)		I (1–4)
Quercus petraea	III (3–10)	III (3–10)	I (3)	I (3–9)	II (3–9)	II (3–10)
Castanea sativa	I (1–5)	I (4)	I (3–7)	I (3–5)	I (2–4)	I (1–10)
Pinus sylvestris	II (3–4)		I (3–4)	II (2–10)	I (1–10)	I (1–10)
Pinus nigra var. *maritima*				II (6–10)		I (6–10)
Pseudotsuga menziesii				II (6–10)		I (6–10)
Larix spp.				I (6–10)	I (1–8)	I (1–10)
Acer pseudoplatanus	II (1–9)	I (5)	I (5–8)	I (3–9)	III (1–7)	II (1–9)
Fraxinus excelsior	I (1–6)	II (2–7)	II (1–7)	I (1–6)	III (1–8)	II (1–8)
Quercus hybrids					II (1–10)	II (1–10)
Ulmus glabra	I (1–8)				II (1–7)	I (1–7)
Corylus avellana	III (1–9)	III (2–9)	IV (1–10)	I (3–7)	III (1–9)	III (1–10)
Crataegus monogyna	II (1–6)	I (3–7)	II (1–5)	I (4–6)	II (1–5)	II (1–7)
Ilex aquifolium	II (1–6)	I (2)	II (2–9)	I (3–7)	II (1–6)	II (1–9)
Viburnum lantana	I (2)	I (2–3)	I (3–4)	I (4)		I (2–4)
Carpinus betulus sapling	I (8)	I (3–5)	I (2–4)			I (2–8)
Viburnum opulus	I (1–4)	I (2–3)	I (1–3)			I (1–4)
Crataegus laevigata	I (3)	I (4)	I (2–3)			I (2–4)

Floristic table W10 (*cont.*)

	a	b	c	d	e	10
Fagus sylvatica sapling	II (1–5)		I (2–4)	I (1–8)	I (1–8)	I (1–8)
Rhododendron ponticum	I (1–8)		I (1–5)	I (2–5)	I (1–5)	I (1–8)
Sorbus aucuparia	I (1–4)		I (3–5)	I (3)	I (1–5)	I (1–5)
Betula pendula sapling	I (2–3)		I (3–4)	I (3–7)	I (1–5)	I (1–7)
Betula pubescens sapling	I (2–4)		I (2–6)	I (6)	I (1–4)	I (1–6)
Malus sylvestris	I (1–2)		I (2)		I (1)	I (1–2)
Prunus spinosa	I (1–3)		I (1–7)	I (4–5)		I (1–7)
Quercus robur sapling	I (2–3)		I (1–5)	I (1–4)		I (1–5)
Acer campestre	I (1–3)		I (2–4)			I (1–4)
Quercus petraea sapling	I (2–4)			I (4)	I (1)	I (1–4)
Quercus hybrids sapling	I (1–3)				I (1–4)	I (1–4)
Castanea sativa	I (1–3)	II (3–9)	I (1–3)	I (2–5)		I (1–9)
Acer pseudoplatanus sapling	II (1–7)	I (2–4)	I (1–5)	I (3)	II (2–5)	I (1–7)
Fraxinus excelsior sapling	I (1–5)	I (3)	I (1–4)	I (4)	II (1–6)	I (1–6)
Sambucus nigra	I (2–3)	I (2–3)	I (1–7)	I (3)	II (1–5)	I (1–7)
Ulmus glabra sapling					I (1–6)	I (1–6)
Rubus fruticosus agg.	V (3–10)	IV (2–9)	V (1–10)	IV (1–10)	III (1–8)	IV (1–10)
Pteridium aquilinum	IV (1–9)	III (2–7)	IV (1–10)	V (1–10)	III (1–8)	IV (1–10)
Lonicera periclymenum	III (2–8)	IV (3–7)	V (1–8)	III (1–7)	II (1–8)	IV (1–8)
Anemone nemorosa	I (1–2)	IV (3–8)	I (3–8)		I (1–6)	I (1–8)
Atrichum undulatum	I (1–4)	II (2–7)	I (2–4)	I (2)	I (1–4)	I (1–7)
Lamiastrum galeobdolon	I (1–5)	II (2–5)	I (1–5)		I (1–5)	I (1–5)
Hedera helix	II (2)	II (2–8)	IV (2–10)	I (2–4)	I (2–7)	II (2–10)
Galium odoratum			I (2–3)		I (1)	I (1–3)
Geranium robertianum			I (2–4)		I (1–5)	I (1–5)
Holcus lanatus	I (1–6)	II (3–9)	I (1–7)	IV (1–8)	I (1)	I (1–9)
Dactylis glomerata				I (3–4)	I (1–3)	I (1–4)
Senecio jacobaea				I (1–3)		I (1–3)

Oxalis acetosella	I (1–4)	I (2–3)	I (2)	IV (1–9)	II (1–9)
Holcus mollis	II (1–10)	I (2–8)	II (2–10)	IV (1–9)	II (1–10)
Dryopteris dilatata	II (1–7)	I (2–5)	II (1–6)	III (1–8)	II (1–8)
Eurhynchium praelongum	II (1–7)	II (3–5)	I (1–6)	III (1–8)	III (1–8)
Mnium hornum	II (1–5)	II (1–6)	I (1–5)	III (1–9)	II (1–9)
Viola riviniana	I (1–3)	I (2)	I (1–3)	II (1–4)	I (1–4)
Thuidium tamariscinum	I (1–8)	I (5)	I (2–7)	II (1–5)	I (1–8)
Stellaria holostea	I (1–5)	I (2–4)	I (1–5)	II (1–6)	I (1–6)
Deschampsia cespitosa	I (1–4)	I (2–4)	I (1–6)	II (1–9)	I (1–9)
Brachythecium rutabulum	I (1–3)	I (3–5)	I (2–5)	II (1–3)	I (1–5)
Plagiothecium undulatum	I (2)	I (3)	I (2–4)	II (1–6)	I (1–6)
Isopterygium elegans	I (2–4)		I (4)	II (1–4)	I (1–4)
Pseudoscleropodium purum	I (1)		I (1–8)	II (1–5)	I (1–8)
Athyrium filix-femina	I (1–4)		I (2–4)	II (1–7)	I (1–7)
Eurhynchium striatum				I (1–5)	I (1–5)
Thelypteris limbosperma				I (1–5)	I (1–5)
Hyacinthoides non-scripta	III (3–9)	IV (4–10)	II (1–10)	III (1–9)	III (1–10)
Acer pseudoplatanus seedling	II (1–9)	I (3)	I (1–2)	II (1–2)	II (1–9)
Dryopteris filix-mas	II (1–5)	I (2–6)	II (1–7)	II (1–6)	II (1–8)
Epilobium angustifolium	I (1–6)	I (2–4)	I (1–4)	I (1–3)	I (1–6)
Conopodium majus	I (1–4)	I (2–3)	I (2–4)	I (1–3)	I (1–5)
Poa trivialis	I (1–5)	I (2–3)	I (1–4)	I (1–7)	I (1–7)
Luzula pilosa	I (2–3)	I (4)	I (1–5)	I (1–4)	I (1–5)
Luzula sylvatica	I (1–7)	I (2–4)	I (2–8)	I (1–9)	I (1–9)
Rumex acetosa	I (4)	I (3)	I (3)	I (1–4)	I (1–5)
Silene dioica	I (3)	I (3)	I (2–3)	I (1–5)	I (1–6)
Melica uniflora	I (2–5)	I (2–5)	I (3–6)	I (1–5)	I (1–6)
Fraxinus excelsior seedling	I (1–3)	I (3)	I (1–3)	I (1–2)	I (1–3)
Stellaria media	I (3–4)	I (3)	I (2–3)	I (3)	I (2–4)
Teucrium scorodonia	I (1–5)	I (2)	I (1–6)	I (1–3)	I (1–6)
Urtica dioica	I (1–2)	I (4)	I (2–5)	I (1–9)	I (1–9)
Crataegus monogyna seedling	I (1–3)	I (2)	I (1–3)	I (1)	I (1–3)
Dicranella heteromalla	I (1–4)	I (2–3)	I (1–4)	I (1–4)	I (1–4)

Floristic table W10 *(cont.)*

	a	b	c	d	e	10
Hypnum cupressiforme	I (1–4)	I (2–3)	I (1–3)	I (3–4)	I (1–3)	I (1–4)
Carex sylvatica	I (1–2)	I (2)	I (1–3)		I (1)	I (1–3)
Euphorbia amygdaloides	I (3)	I (2–4)	I (3)	I (3)		I (2–4)
Heracleum sphondylium	I (2–3)	I (2)	I (1–3)	I (1–3)		I (1–3)
Glechoma hederacea	I (2–3)	I (2–3)	I (4)	I (3–6)		I (2–6)
Melampyrum pratense	I (2–4)	I (4)	I (3–5)	I (3–4)		I (2–5)
Blechnum spicant	I (5)	I (4)	I (2–7)	I (2–3)		I (2–7)
Rumex sanguineus	I (3)	I (2–4)	I (2–3)	I (2–3)		I (2–4)
Solidago virgaurea	I (2–5)	I (2)	I (1–3)	I (2–3)		I (1–5)
Quercus robur seedling	I (3–4)	I (3)	I (1–3)	I (1–3)		I (1–4)
Sanicula europaea	I (1–4)	I (2–3)	I (3)		I (2–3)	I (1–4)
Poa nemoralis	I (2)	I (2)	I (2–7)		I (1–3)	I (1–7)
Milium effusum	I (1–4)	I (3–8)	I (3)		I (1)	I (1–8)
Ligustrum vulgare	I (2)	I (3)	I (2–4)		I (3)	I (2–4)
Circaea lutetiana	I (1–4)	I (2)	I (2–4)		I (1–2)	I (1–4)
Ajuga reptans	I (1–3)	I (2)	I (1–2)		I (1–2)	I (1–3)
Stachys sylvatica		I (3–4)	I (1–5)	I (2)	I (1)	I (1–5)
Veronica chamaedrys	I (2–4)	I (2)		I (3)	I (1)	I (1–4)
Lysimachia nemorum	I (1–3)	I (3)		I (2)	I (1–3)	I (1–3)
Amblystegium serpens	I (3)	I (3–5)		I (3)		I (1–5)
Agrostis capillaris	I (1–9)		I (1)	I (3–8)	I (1–7)	I (1–9)
Anthoxanthum odoratum	I (1–2)		I (3)	I (2–4)	I (1–5)	I (1–5)
Brachypodium sylvaticum	I (2–4)		I (2–4)	I (2–3)	I (1–4)	I (1–4)
Deschampsia flexuosa	I (1–5)		I (3–6)	I (2–6)	I (1–9)	I (1–9)
Digitalis purpurea	I (1–3)		I (1–3)	I (1–4)	I (1–6)	I (1–6)
Galium saxatile	I (2–3)		I (3)	I (2–5)	I (1–7)	I (1–7)
Juncus effusus	I (2)		I (3)	I (3–4)	I (2–4)	I (2–4)
Betula pendula seedling	I (2)		I (1)	I (2)	I (1–3)	I (1–3)
Vaccinium myrtillus	I (2–5)		I (5)	I (3–4)	I (1–2)	I (1–5)
Rubus idaeus	I (3)		I (2–3)	I (4–5)	I (1–4)	I (1–5)
Rosa canina agg.	I (2–3)		I (1–6)	I (3–4)	I (1–2)	I (1–6)
Sorbus aucuparia seedling	I (1–2)		I (2–3)	I (1–3)	I (1–2)	I (1–3)
Lophocolea bidentata s.l.	I (1–3)		I (2–3)	I (1–4)	I (1–4)	I (1–4)

Plagiothecium denticulatum	I(1-3)		I(3)	I(3)	I(1-3)	I(1-3)
Fagus sylvatica seedling	I(1-4)		I(1)	I(1-2)	I(1)	I(1-4)
Isothecium myosuroides	I(1-3)		I(1-3)	I(1)	I(1-2)	I(1-3)
Ilex aquifolium seedling	I(1-4)		I(2-6)	I(1-2)	I(1-4)	I(1-6)
Polytrichum formosum	I(1-4)		I(2-5)	I(5)	I(1-4)	I(1-5)
Cytisus scoparius	I(1)		I(4)	I(4-6)	I(1)	I(1-6)
Ranunculus ficaria	I(3)	I(3)			I(3-5)	I(3-5)
Pellia epiphylla	I(2)	I(4)			I(1-5)	I(1-5)
Plagiomnium undulatum	I(3)	I(5)			I(1-4)	I(1-5)
Plagiomnium rostratum	I(1)	I(3)			I(1)	I(1-3)
Bromus ramosus	I(3)	I(3)	I(2-4)			I(2-4)
Fragaria vesca		I(3)	I(2)		I(2)	I(2-3)
Potentilla sterilis		I(2-3)	I(3)		I(1-2)	I(1-3)
Ranunculus repens		I(3)		I(2)	I(1-4)	I(1-4)
Mercurialis perennis	I(2-4)		I(4)		I(1-4)	I(1-4)
Primula vulgaris	I(3-4)		I(1-3)		I(1)	I(1-4)
Dryopteris borreri	I(3)		I(4)		I(1-5)	I(1-5)
Galium aparine	I(2)		I(3-4)		I(1-4)	I(1-4)
Dicranum scoparium	I(3)			I(2-3)	I(1-4)	I(1-4)
Ulex europaeus	I(3)		I(2-6)	I(3-7)		I(2-7)
Luzula multiflora			I(4)	I(2-6)	I(1-3)	I(1-6)
Lysimachia nummularia	I(2-3)	I(2)				I(2-3)
Dicranoweissia cirrata	I(2)	I(2-3)				I(2-3)
Ceratodon purpureus	I(4)	I(3)				I(3-4)
Prunella vulgaris	I(3)	I(3)				I(3)
Aegopodium podagraria		I(3)	I(2)			I(2-3)
Anthriscus sylvestris		I(2)	I(3-4)			I(2-4)
Acer campestre seedling		I(3)	I(1-3)			I(1-3)
Narcissus pseudonarcissus		I(3)	I(3)			I(3)
Arrhenatherum elatius			I(2-4)	I(2-4)		I(2-4)
Corydalis claviculata			I(5)	I(1-5)		I(1-5)
Festuca ovina			I(3)	I(2-5)		I(2-5)
Number of samples	51	22	77	150	79	379
Number of species/sample	18 (7-35)	13 (1-31)	16 (7-23)	10 (1-27)	24 (11-39)	15 (1-39)

Floristic table W10 (*cont.*)

	a	b	c	d	e	10
Tree height (m)	17 (5–25)	12 (8–20)	18 (8–32)	13 (2–30)	18 (6–30)	16 (2–32)
Tree cover (%)	89 (40–100)	88 (2–100)	89 (60–100)	84 (0–100)	77 (20–100)	84 (0–100)
Shrub height (m)	4 (1–8)	5 (1–10)	4 (1–8)	3 (1–8)	4 (1–10)	4 (1–10)
Shrub cover (%)	24 (0–90)	19 (0–95)	39 (0–100)	12 (0–100)	17 (0–85)	21 (0–100)
Herb height (cm)	51 (15–100)	33 (15–60)	57 (1–130)	66 (10–200)	46 (5–150)	56 (1–200)
Herb cover (%)	81 (20–100)	69 (1–100)	88 (20–100)	77 (1–100)	83 (1–100)	81 (1–100)
Ground height (mm)	15 (10–30)	13 (10–20)	11 (10–20)	21 (10–50)	18 (5–30)	17 (5–50)
Ground cover (%)	6 (0–60)	6 (0–40)	9 (0–100)	7 (0–80)	14 (0–90)	9 (0–100)
Altitude (m)	120 (10–250)	89 (16–190)	102 (8–270)	95 (4–250)	110 (9–282)	102 (4–282)
Slope (°)	7 (0–60)	1 (0–15)	2 (0–30)	1 (0–30)	17 (0–45)	5 (0–60)

a   Typical sub-community

b   *Anemone nemorosa* sub-community

c   *Hedera helix* sub-community

d   *Holcus lanatus* sub-community

e   *Acer pseudoplatanus-Oxalis acetosella* sub-community

10 *Quercus robur-Pteridium aquilinum-Rubus fruticosus* woodland (total)

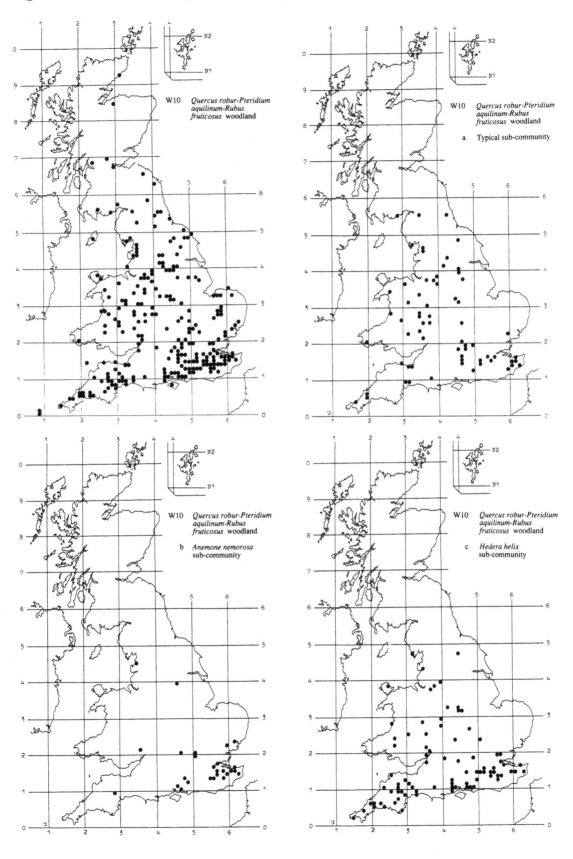

W10   *Quercus robur-Pteridium aquilinum-Rubus fruticosus* woodland

W10   *Quercus robur-Pteridium aquilinum-Rubus fruticosus* woodland

a   Typical sub-community

W10   *Quercus robur-Pteridium aquilinum-Rubus fruticosus* woodland

b   *Anemone nemorosa* sub-community

W10   *Quercus robur-Pteridium aquilinum-Rubus fruticosus* woodland

c   *Hedera helix* sub-community

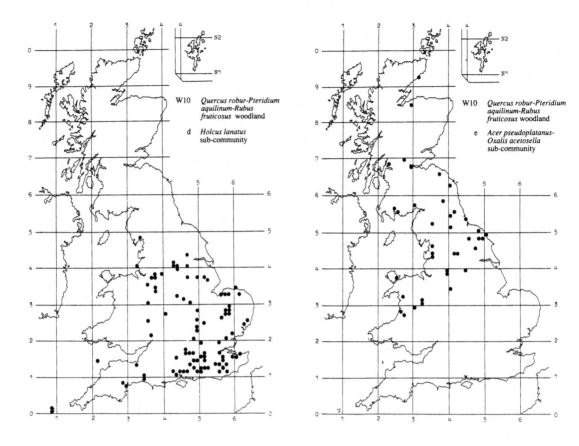

W10   *Quercus robur-Pteridium aquilinum-Rubus fruticosus* woodland

d   *Holcus lanatus* sub-community

W10   *Quercus robur-Pteridium aquilinum-Rubus fruticosus* woodland

e   *Acer pseudoplatanus-Oxalis acetosella* sub-community

W11
Quercus petraea-Betula pubescens-Oxalis acetosella woodland

Synonymy
Caithness birchwoods   Crampton 1911 *p.p.*; *Betuletum tomentosae* Moss 1911 *p.p.*; Scottish beechwoods Watt 1931 *p.p.*; Highland birchwoods Tansley 1939 *p.p.*; Highland oakwoods Tansley 1939 *p.p.*; Heathy birchwood Pigott 1956; *Betuletum Oxaleto-Vaccinietum* McVean & Ratcliffe 1962 *p.p.*; *Betula*-herb nodum   McVean & Ratcliffe 1962 *p.p.*; *Vaccinium*-rich birchwood association   McVean 1964a *p.p.*; Herb-rich birch and oakwood association   McVean 1964a *p.p.*; *Blechno-Quercetum* (Br.-Bl. & Tx. (1950) 1952) Klötzli 1970 *p.p.*; *Oxalido-Betuletum*   Graham 1971 *p.p.*; Loch Lomond oakwoods Tittensor & Steele 1971 *p.p.*; *Betula pubescens-Vaccinium myrtillus* Association Birks 1973 *p.p.*; *Corylus avellana-Oxalis acetosella* Association   Birks 1973 *p.p.*; Hazel-ash woodland Peterken 1981 *p.p.*; Oak-lime woodland   Peterken 1981 *p.p.*; Birch-oak woodland   Peterken 1981 *p.p.*; Birch woodland   Peterken 1981 *p.p.*; Woodland plot types 21, 22, 23, 26 & 29   Bunce 1982; *Trientali-Betuletum pendulae* Birse 1982 *p.p.*; *Luzulo-Betuletum odoratae*   Birse 1984 *p.p.*; *Blechno-Quercetum* (Br.-Bl. & Tx. (1950) 1952) Birse 1984 *p.p.*; *Lonicero-Quercetum* (Birse & Robertson 1976) Birse 1984 *p.p.*

Constant species
Betula pendula/pubescens, Agrostis capillaris, Anthoxanthum odoratum, Deschampsia flexuosa, Galium saxatile, Holcus mollis, Oxalis acetosella, Potentilla erecta, Pteridium aquilinum, Viola riviniana, Hylocomium splendens, Pseudoscleropodium purum, Rhytidiadelphus squarrosus, Thuidium tamariscinum.

Physiognomy
The *Quercus petraea-Betula pubescens-Oxalis acetosella* woodland is almost invariably dominated by either oak or birch or various mixtures of the two. Of the oaks, *Quercus petraea* is very much the more frequent and characteristic species and it can be an abundant component of this community in south-west England, Wales,

the Lake District and into southern and central Scotland. Sometimes, it forms a high forest canopy of tall, well-grown trees; in other stands, it dominates in oak coppice with a fairly low cover of thin or multi-stemmed and often crookedly-growing individuals.

Moving north-westwards, *Q. petraea* becomes rarer and increasingly confined to lower altitudes: in the southern and east-central Highlands, it can dominate in this community up to about 275 m, but in the far west, its general limit is around 150 m (McVean 1964a). However, there is every indication that oak has been extensively removed from woodlands towards the north-western limit of this community, so its absence there is probably partly artificial (McVean & Ratcliffe 1962, McVean 1964a). Sylvicultural activity may also be involved in the low frequency of *Q. petraea* in stands in the eastern lowlands of Scotland. Here, oak is quite common in high-forest and coppice stands of this community, but it is often *Q. robur* or, more frequently, hybrid (Birse 1982, 1984) and this may reflect the preference for *Q. robur* as the plantation oak that grew up in the past two centuries (Anderson 1967, Gardiner 1974).

In oak-dominated stands such as these, birch is often only a low-cover canopy component or a gap-coloniser; and, in coppice, it was often weeded out. Sometimes, though, it was itself cropped to provide bobbin-wood or brushwood for besoms and, with the felling of oak and the abandonment of oak-coppicing, it has spread considerably to become the dominant in neglected and run-down stands. And, at high altitudes and to the far northwest, it is usually the predominant tree. Overall, it is now the most frequent woody species of the community. *Betula pubescens* is the typical birch here and in locally exposed situations towards the lowlands and more consistently at higher latitudes (north-west from Invernessshire), the trees have the distinctly shorter and bushier stature of ssp. *carpatica* (Walters 1964, Forbes & Kenworthy 1973). But *B. pendula* also occurs, though its predominance is very much concentrated geographi-

cally in eastern Scotland. Here, it can be as common as
B. pubescens and shows a fairly precise altitudinal switch
to ssp. *carpatica* (as in the Dee valley woods described by
Forbes & Kenworthy 1973 where one taxon gave way to
the other at about 350 m: see also Hulten 1950, Birse &
Dry 1970). Throughout the community, birches intermediate in character between *B. pubescens* and *B. pendula* can be found and these seem to be genuine hybrids
(Kennedy & Brown 1983).

When birch dominates here, it can form quite dense
thickets but usually the cover is rather open, sometimes
very much so, with widely-spaced, rather moribund
trees showing signs of early damage by fire or from
browsing. The height of the canopy is very variable,
though rarely tall: *B. pubescens* ssp. *pubescens* and *B.
pendula* commonly attain 15 m but, where *B. pubescens*
ssp. *carpatica* predominates, there is generally a bushy
cover about 10 m high, quite often lower. Frequently,
the birch canopy looks uniform with no great mixture of
age-classes: the usual picture is of a single generation of
taller individuals and a very few young saplings or
seedlings; more rarely, older pioneer birch can be surrounded by a second generation of trees (McVean &
Ratcliffe 1962).

Apart from oak and birch, other trees are scarce in the
community as a whole. *Fraxinus excelsior*, which
becomes a distinctive component of the *Quercus-Pteridium-Rubus* woodland towards the north-western limit
of its range, does not extend far into this community. It
is preferential for more southerly stands on deeper and
less-leached soils and, even then, is no more than
occasional. And *Acer pseudoplatanus*, despite the high
rainfall and free-draining character of many of the soils
here, is very rare. *Tilia cordata* survives on some cliffs
and in gorges in close association with the community
towards the very limit of its distribution in Cumbria
(Pigott & Huntley 1978): such vegetation would fall
within Peterken's (1981) oak-lime stand types 5B. Then,
there are sparse records for *Fagus sylvatica* and some
conifers, planted in as occasionals or seeding in from
nearby plantations: some planted stands of *Fagus* (e.g.
Watt 1931*a*) and of *Larix* spp. can be accommodated
within the community.

More typical than any of these species, are larger
specimens of two of the understorey elements which
occasionally grow tall enough to find a place in the
canopy. *Sorbus aucuparia* is occasional throughout and
distinctly preferential for one sub-community but generally it is of low cover in high forest, occurring as
scattered trees. In ungrazed stands, it can thicken up
considerably and, being shade-tolerant, can gain abundance where birch cannot (Anderson 1950, McVean
1958). Towards the far north-west, it can be co-dominant with birch in low scrubby canopies. Grazing and
browsing, which are very prevalent in this community,

probably also account for the great scarcity of *Ilex
aquifolium*, a tree which one would expect to be prominent here in the rather equable climate experienced by
more westerly and low-altitude stands.

The only other common smaller woody species
throughout the community is *Corylus avellana* and even
this is no more than occasional, but it can become locally
prominent in ungrazed tracts over deeper soils and has
sometimes been selected as a coppice crop. More often,
it occurs as scattered bushes beneath an oak canopy or,
to the far north-west, amongst a low cover of birch and
rowan (as on Skye: Birks 1973). Hawthorn (always
Crataegus monogyna) is scarce and very much confined
to deeper and less strongly leached soils. Particularly in
eastern Scotland, but also locally elsewhere (as in Teesdale: Graham 1971), *Juniperus communis* ssp. *communis*
can be found in more open places, its increasing abundance often marking transitions to stands of the *Juniperus-Oxalis* woodland. *Rhododendron ponticum* has sometimes been planted into the *Quercus-Betula-Oxalis*
woodland and it can spread to become locally prominent. Finally, there can be a few birch saplings within
gaps.

Generally, however, the combined cover of all these
species is low and the striking feature on entering stands
of this community is their openness. Quite apart from
this typical scarcity of smaller trees and shrubs and the
absence of anything more than patchy regeneration, the
field layer is often short. Grasses frequently make a
major contribution. As in the *Quercus-Pteridium-Rubus*
woodland, *Holcus mollis* is characteristic of this
community, though it is much more consistently frequent and abundant here and is accompanied by constant *Deschampsia flexuosa*, *Anthoxanthum odoratum*
and *Agrostis capillaris*, frequent *A. canina* ssp. *montana*
and occasional *Festuca ovina*, *F. rubra* and *Holcus
lanatus*. These grasses become especially prominent with
their flush of new growth in early summer and they can
form an extensive sward, trimmed short where grazing is
heavy, so that the vegetation looks more like a *Festuca-
Agrostis* grassland with trees than a woodland proper.
In other stands, their cover is broken by boulders, when
they extend between the rocks or occur more patchily on
top of them; or, where *Pteridium* is very vigorous, they
can thin out to scattered, rather puny plants.

This predominance of grasses (and the abundance of
bryophytes: see below) can give this vegetation a rather
uniform appearance throughout the year but, where the
soil cover is extensive, the field layer often shows a clear
pattern of phenological change. In more westerly
stands, *Hyacinthoides non-scripta* is the typical vernal
dominant, extending northwards the very characteristic
role it plays in more Atlantic woodlands in Britain.
Where the soils are moister, with some measure of spring
waterlogging, it may be joined or locally replaced by

Anemone nemorosa, but this plant becomes most important here where the community extends into the more continental regions of eastern Scotland. There, *Hyacinthoides* is rare and, in long-established woodlands at least, *Anemone* is the typical vernal dominant, with the Northern Montane *Trientalis europaea* continuing the picture of scattered white flowers into early summer.

Another typical early spring flowerer common throughout the community, though usually at low cover, is *Viola riviniana.* With *Oxalis acetosella,* another constant here, this species already shows a rise in frequency in the more north-westerly stands of the *Quercus-Pteridium-Rubus* woodland but their increased prominence in the *Quercus-Betula-Oxalis* woodland is a very good marker of the consistently moist character of the soils in this wetter part of the country. *Oxalis* is often most conspicuous here in winter, its pale green foliage set off against the darker shades of the grass and bryophyte carpet and, in shadier stands, it may carry out the bulk of its photosynthetic activity in this season, becoming hidden later by herb growth, though often persisting even under a dense canopy of *Pteridium* (Packham & Willis 1977, Packham 1978).

A very distinctive feature of the *Quercus-Betula-Oxalis* woodland is that such species as *Oxalis* and *Viola,* indicative of the moistness of the soils, occur together with Quercion herbs which favour surface-leached profiles. Foremost among these are *Galium saxatile* and *Potentilla erecta* which, in the north-west of Britain, transgress far into more mesophytic woodlands. This feature makes the floristic boundary between this community and the *Quercus-Betula-Dicranum* woodland a rather indistinct one, especially where the soils are transitional or disposed in complex mosaics over heterogenous parent materials and where very heavy rainfall induces strong leaching overall. By and large, the best diagnostic feature is the scarcity in the *Quercus-Betula-Oxalis* woodland of ericoid sub-shrubs but even this criterion may be unreliable where stands of the community include siliceous boulders or drier banks. In such situations, *Vaccinium myrtillus* may make an occasional appearance (as in some of the stands in McVean & Ratcliffe's (1962) *Betuletum Oxaleto-Vaccinietum* or Birks' (1973) *Betula-Vaccinium* Association in the far north-west of Scotland), though *Calluna vulgaris* and *Erica cinerea* are typically absent, even under more open canopies. The fact that grazing tends to produce a floristic convergence of the *Quercus-Betula-Oxalis* and *Quercus-Betula-Dicranum* woodlands (see below) further confounds this boundary.

Other characteristic herbs of the *Quercus-Betula-Oxalis* woodland include *Teucrium scorodonia, Stellaria holostea, Luzula pilosa,* (with, more occasionally, *L. multiflora*), *Conopodium majus, Veronica chamaedrys,*

V. officinalis, Hypericum pulchrum and *Succisa pratensis.* Scattered plants of all these species, flowering under more open canopies, typically occur through the grassy carpet, but they become especially prominent in areas where there is some protection from grazing, as amongst boulders. Here, too, where there is some slight flushing, *Primula vulgaris* and *Lysimachia nemorum* can occur as in some stands distinguished by Tittensor & Steele (1971) and Birse (1984). Where flushing brings a more pronounced measure of base-enrichment, the field layer may acquire some of the characteristics of more calcicolous communities like the *Fraxinus-Sorbus-Mercurialis* or *Alnus-Fraxinus-Lysimachia* woodlands, but more calcicolous herbs are generally absent from the *Quercus-Betula-Oxalis* woodland.

By mid-summer, many stands of the community have a cover of *Pteridium* fronds, though in bouldery woods bracken is restricted by and large to deeper soils in crevices. Quite a common pattern is for bracken to become much more vigorous and extensive where the community extends down from more rocky slopes with very thin soils to deeper creep soils at the slope foot. Being shade-sensitive, it is also more prominent in open areas of the canopy and it can become abundant, too, in stands where there has been some relaxation of grazing or a neglect of coppicing (e.g. Tittensor & Steele 1971). The same is probably true of *Lonicera periclymenum*: this is generally only occasional here with sparse, trailing stems but it can thicken up locally to form a dense tangle. The third member of this trio, *Rubus fruticosus* agg., so conspicuous in sub-scrub throughout the *Quercus-Pteridium-Rubus* woodland, is here very much confined to less heavily grazed stands on less markedly leached soils. However, when all three of these species are present in such situations, succeeding a bluebell carpet and with abundant *Dryopteris dilatata, D. filix-mas* and *D. borreri* (as in the *Dryopteris* sub-community), the similarity between the two woodland types is very obvious.

Ferns can be a conspicuous element in other kinds of *Quercus-Betula-Oxalis* woodland, too. *Blechnum spicant* is the most frequent and characteristic of these overall, though it is strongly preferential for one sub-community. *Thelypteris limbosperma* is generally less common, though it is a very good marker of certain sub-communities and often conspicuous around the margins of stands and in transitions to closely-related tall-herb vegetation (as in McVean & Ratcliffe's (1962) 'fern-dominated treeless facies' of their *Betula*-herb nodum). *T. phegopteris* and *Gymnocarpium dryopteris* are somewhat less frequent, though both can become prominent in zonations to the *Juniperus-Oxalis* woodland. *Athyrium filix-femina* is occasional on moister soils and *Polypodium vulgare* agg. (probably *sensu stricto*) can sometimes be found, often epiphytic on oak. Sheltered sites with some slight base-enrichment can provide localities for *Polystichum aculeatum* or *P. setiferum* but the more calcicolous *Phyllitis scolopendrium* is absent.

One other species that can attain great abundance in ungrazed stands is *Luzula sylvatica*, a prominent feature of this kind of woodland on inaccessible slopes and on islands, as in Loch Lomond (Tittensor & Steele 1971) and Windermere, where it can crowd out most of the other herbs (usually not *Oxalis*, however) and bryophytes. More extensive sampling of this kind of *Quercus-Betula-Oxalis* woodland would probably permit the diagnosis of a distinct sub-community, though it seems best to retain such vegetation within the parent community (cf. Birse (1984) who characterised a distinct *Luzulo-Betuletum*).

Other herbs of note which occur very occasionally here are *Rubus saxatilis*, like *Trientalis* a species with northern continental affinities, and, at some of its most far-flung stations, *Convallaria majalis*.

Throughout the *Quercus-Betula-Oxalis* woodland, bryophytes make a consistent and important contribution, thriving among close-cropped grassy swards where competition is reduced and over boulders blown free of litter and only fading in variety and abundance under denser covers of tall herbs and bracken. They are especially extensive in sites which have a degree of shelter additional to that provided by the trees themselves, as on north-facing slopes or where the community extends into ravines, and in the cool and humid climate of the far north-west where they frequently cover more than half of the ground in a thick and luxuriant carpet that is enriched by a variety of more strictly Atlantic species. In more oceanic areas, there can also be specialised epiphytic floras on the trunks, branches and twigs of the trees and shrubs.

Further details of such richer suites are given below but, among the more generally characteristic bryophytes here are *Rhytidiadelphus squarrosus*, *Pseudoscleropodium purum*, *Thuidium tamariscinum*, *Hylocomium splendens* (all community constants), *Pleurozium schreberi*, *Dicranum majus*, *Polytrichum formosum*, *Rhytidiadelphus triquetrus* (all frequent but unevenly represented in different sub-communities) and *Dicranum scoparium*, *Mnium hornum*, *Plagiothecium undulatum*, *Atrichum undulatum* and *Eurhynchium praelongum* (occasional to frequent throughout). Except in more north-westerly stands, hepatics are usually few in number, but *Lophocolea bidentata s.l.* occurs quite commonly and *Plagiochila asplenoides* is occasional. As among the herbs, it may again be noted that the balance among the bryophyte element is towards the calcifugous, with species such as *Thamnium alopecurum* and *Eurhynchium striatum* of restricted occurrence, one further reflection of the prevalence of surface leaching in the soils here.

Sub-communities

***Dryopteris dilatata* sub-community:** *Blechno-Quercetum fraxinetosum* Klötzli 1970 *p.p.*; *Oxalido-Betuletum typicum* Graham 1971; Loch Lomond Community types 4, 8, 9 & 10 Tittensor & Steele 1971; Hazel-ash stand types 3C & 3D Peterken 1981 *p.p.*; Oak-lime stand type 5B Peterken 1981 *p.p.*; Birch-oak stand types 6Ac & 6Bc Peterken 1981 *p.p.*; *Lonicero-Quercetum*, *Endymion* & Typical subassocations (Birse & Robertson 1976) Birse 1984 *p.p.*; *Luzulo-Betuletum*, *Rubus saxatilis* subassociation Birse 1984. Oak, almost invariably *Q. petraea*, is more frequent than birch in this sub-community and quite commonly it dominates in high forest with a tall (sometimes more than 20 m) canopy that is almost closed. In such situations, *B. pubescens* (only rarely *B. pendula* or intermediates) may be relegated to being a coloniser of gaps: a frequent picture is for there to be occasional birch and some large *Sorbus aucuparia* forming a sparse second tier to the canopy below the oak. In disused oak coppices, too, which are quite widespread here, birch may be sparse, though where it does spread in, it sometimes overtops the often quite low and poorly-growing oak. In other stands, birch may be proportionately much more abundant, forming, with occasional *Sorbus*, a patchily dense cover with just a few oaks. *Fraxinus* is occasional in the canopy of this kind of *Quercus-Betula-Oxalis* woodland though its saplings are rare.

Corylus is also preferentially frequent here and it may be plentiful enough to form a distinct understorey in oak high forest, though the bushes are usually fairly small and their total cover low. In some sites, it has clearly been selected for in what is now disused hazel coppice with oak reduced to occasional standards. Elsewhere, *Corylus* joins *B. pubescens* and *Sorbus* as a co-dominant in open scrubby thickets (as in some of the Teesdale stands: Graham 1971). *Crataegus monogyna*, though not nearly so common here as in the *Quercus-Pteridium-Rubus* woodland, is preferential for this sub-community. The most frequent saplings are those of birch.

In the field layer here, at least where the soil cover is moderately deep and not too much broken by boulders, *Hyacinthoides* is a common vernal dominant. As it fades, ferns generally become the most prominent element and by late July, *Pteridium* fronds are fully expanded and the crowns of *Dryopteris dilatata*, *D. filix-mas* and *D. borreri*, all preferential here, have their new foliage. *Pteridium* tends to follow *Hyacinthoides* in favouring more extensive soil covers and then it can produce a dense canopy; the dryopteroids, on the other hand, can extend amongst or on top of moss-covered boulders or grow around the tree bases, a feature well seen in Graham's (1971) *Oxalido-Betuletum*, where distinct variants are recognised on this basis. *Gymnocarpium dryopteris*, *Thelypteris phegopteris* and *T. limbosperma* can all be locally prominent here, though the last is not so common or characteristic as in the *Blechnum* sub-community. *Blechnum* itself is also rather infrequent.

Lonicera periclymenum and, especially, *Rubus fruticosus* agg. are both more frequent in this sub-community than in other kinds of *Quercus-Betula-Oxalis* woodland, though their abundance is distinctly patchy. When plentiful, they can provide an important structural element by mid-summer, but quite often they form a rather sparse cover which does not completely mask the herbs that grow between the ferns. Among these, there is a slight shift away from species indicative of more markedly leached soils and a somewhat more luxuriant physiognomy. All the characteristic grasses of the community are well represented with, in addition, *Deschampsia cespitosa* becoming occasional in moister areas but *Galium saxatile* and, more noticeably, *Potentilla erecta* are reduced in frequency. The shade-tolerant *Oxalis* can be very abundant and taller herbs like *Teucrium scorodonia*, *Digitalis purpurea* and the sprawling *Stellaria holostea* are frequent. The small rosettes of *Viola riviniana*, on the other hand, are rather uncommon.

The same trends can be seen among the bryophytes which have consistently lower cover beneath this extensive layer of bulkier herbs. *Hylocomium splendens*, *Pleurozium schreberi*, *Thuidium tamariscinum*, *Dicranum majus* and *Polytrichum formosum*, which are such a prominent component of bryophyte mats in the *Blechnum* sub-community, are all rather infrequent here and sometimes this element of the vegetation is reduced to *Rhytidiadelphus squarrosus*, *Pseudoscleropodium purum*, *Hypnum cupressiforme* and *Eurhynchium praelongum* growing among the herbs and over their litter. More open areas of soil fern stools and tree bases may have *Atrichum undulatum*, *Mnium hornum* and *Dicranum scoparium* and wherever boulders break up the surface, there can be a greater richness and cover among the bryophytes.

***Blechnum spicant* sub-community:** Heathy birchwood Pigott 1956; *Betuletum Oxaleto-Vaccinietum* McVean & Ratcliffe 1962 *p.p.*; *Betula* herb nodum McVean & Ratcliffe 1962 *p.p.*; *Blechno-Quercetum typicum* and *coryletosum* Klötzli 1970 *p.p.*; *Oxalido-Betuletum vaccinietosum* Graham 1971 *p.p.*; Loch Lomond Community types 2, 3 & 4 *p.p.* Tittensor & Steele 1971; *Betula pubescens-Vaccinium myrtillus* Association Birks 1973 *p.p.*; *Corylus avellana-Oxalis acetosella* Association Birks 1973 *p.p.*; Birch-oak stand types

6Ab & 6Bb Peterken 1981 *p.p.*; Birch stand type 12A Peterken 1981 *p.p.*; *Blechno-Quercetum*, Typical subassociation Birse 1984 *p.p.* *B. pubescens*, often ssp. *carpatica*, is the usual dominant here, forming, with frequent *Sorbus* and occasional *Corylus*, a low and often rather open canopy. Commonly, there is no understorey, though smaller birch and rowan can occur in gaps and may form a scrubby fringe to stands where grazing is less intense. Oak is scarce but the occasional trees that are encountered are almost always *Q. petraea*. *Fraxinus* can be found very rarely.

As in the *Dryopteris* sub-community, *Hyacinthoides* occurs frequently as a vernal plant and quite commonly here it is preceded by the flowering of *Primula vulgaris* or, in very moist places, *Anemone nemorosa*. But the cover of these species is very much restricted in more bouldery woods to patches of deeper soil and it is a marked feature of many of the more rocky or heavily-grazed stands that there is relatively little seasonal change in the appearance of the vegetation.

Where the soil mantle is more extensive, grasses typically make up the bulk of the field layer, with mixtures of all the characteristic community species forming a varied and fairly open-textured sward. In late spring, there is a flush of new growth but heavy grazing often keeps the cover fairly short and smaller herbs like *Oxalis*, *Galium saxatile* and *Potentilla erecta* occur frequently and with patchy abundance. Less commonly, there can be some *Luzula pilosa*, *Veronica chamaedrys*, *V. officinalis*, *Teucrium scorodonia*, *Conopodium majus*, *Succisa pratensis*, *Melampyrum pratense* and *Stellaria holostea* and, in inaccessible spots, as in boulder crevices, these may thicken up to form a more luxuriant cover.

In many stands, too, ferns add further variety by midsummer. *Pteridium* is common and, on deeper soils, it can be abundant, taking over dominance from the grasses and herbs by August or permanently extinguishing most of them with its thick accumulations of litter and heavy shade. But more characteristic here is *Blechnum spicant* which is very frequent both in the grassy swards and amongst boulders, and often very luxuriant in the characteritically moist climate in which these woods grow. *Thelypteris limbosperma* is frequent too and *Athyrium filix-femina* occasional and there can be locally abundant *Thelypteris phegopteris*, *Gymnocarpium dryopteris* and *Dryopteris aemula*. Where this subcommunity extends over the very broken ground of boulder-choked ravines, all these can occur together creating the feel of an extraordinarily lush hanging garden.

The other component which contributes very greatly to this impression is the bryophytes. Even amongst the grasses, these can be abundant and varied provided, as is usually the case, grazing keeps the height of the vascular plants low. Then, *Pleurozium schreberi*, *Rhytidiadelphus loreus*, *Dicranum majus* and *Polytrichum formosum* frequently join the community constants in a diverse patchwork over the ground. But the bryophytes become especially prominent here where the soil cover thins out over stable rock surfaces and they can provide a virtually continuous cover over nidus-capped boulders. On the flat and gently-sloping tops of such rocks, the vegetation is essentially a bryophyte-rich and fragmented version of the field layer, though with a shift towards a more calcifugous composition with the increased acidity of the very shallow and strongly-leached soils. In effect, such a flora represents a local occurrence of one of the suite of species that is fully developed in the *Quercus-Betula-Dicranum* woodland. Most of the grasses, *Deschampsia flexuosa* being the notable exception, fall in frequency and many of the mesophytic herbs disappear, though *Oxalis*, *Potentilla erecta* and *Galium saxatile* can remain common. *Blechnum* occurs occasionally, too, and it is here that *Vaccinium myrtillus* can make an appearance. The bulk of the cover, though, is provided by such mosses as *Hylocomium splendens*, *Rhytidiadelphus loreus*, *Dicranum majus*, *Pleurozium schreberi*, *Polytrichum formosum*, *Thuidium tamariscinum* and sometimes its more Atlantic counterpart *T. delicatulum*. Less frequently, there may be some *Hylocomium brevirostre*, *H. umbratum*, *Isothecium myosuroides*, *Rhytidiadelphus triquetrus*, *Ptilium crista-castrensis* and the hepatics *Diplophyllum albicans* and *Plagiochila spinulosa* (McVean & Ratcliffe 1962, Ratcliffe 1968, Birks 1973).

An increasing prominence of siliceous boulders in this kind of situation, with the progressive extinction of the more mesophytic flora on the drift or downwash soils between them and the appearance of a distinctive suite of bryophytes on the boulder sides, marks a transition to the *Quercus-Betula-Dicranum* woodland. Quite often, though, such zonations are gradual and complex mosaics can occur, a feature well seen in the patterns over drift and Torridonian sandstone in Skye (Birks 1973). In such regions, too, enrichment with more Atlantic bryophytes can occur throughout both communities, exaggerating the convergence between the vegetation types. Full details of the extraordinary abundance and diversity of this flora is given under the *Quercus-Betula-Dicranum* woodland where there is also an account of the specialised suite found on rotting logs and the species epiphytic on birch. On hazel, rather more characteristic of the *Quercus-Betula-Oxalis* woodland, epiphytes include *Ulota crispa* (subsuming *U. bruchii*), *U. phyllantha*, *U. calvescens*, *Frullania dilatata* and *F. teneriffae* on twigs (the *Uloteto-Frullanietum* of Barkman 1958), *Dicranum scoparium*, *Isothecium mysuroides*, *Hypnum mammilatum*, *Neckera complanata*,

Radula complanata and *Metzgeria furcata* on trunks (the *Scoparieto-Hypnetum* of Barkman 1958) and *Eurhynchium striatum*, *Hylocomium brevirostre* and *Mnium hornum* on the tree bases (the *Eurhynchietum striatae* of Barkman 1958) (Birks 1973). On hazel bark, too, and on the occasional oaks, larger foliose lichens may occur in Lobarion pulmonariae communities (e.g. Rose 1974, James *et al.* 1977).

***Anemone nemorosa* sub-community:** Caithness birchwoods Crampton 1911 *p.p.*; Scottish beechwoods, *Holcus* type Watt 1931*a*; Birch stand type 12A Peterken 1981 *p.p.*; *Trientali-Betuletum*, typical subassociation Birse 1982 *p.p.* Either oak or birch, or mixtures of the two, can dominate here but, in this eastern Scottish sub-community, there is a clear shift in the taxa represented. In contrast to more western stands of the *Quercus-Betula-Oxalis* woodland, when oak is present here it generally shows some obvious *robur* characteristics though pure *Q. robur*, like pure *Q. petraea*, is rare. And, though *B. pubescens* remains quite frequent, it is not as common, especially in the lower-altitude stands, as *B. pendula*. The structure of the woodland varies: sometimes oak is predominant as a fairly tall and quite closed canopy with birch occasional in gaps; in other cases birch is much the more abundant, forming a cover that can be quite open. Other trees are rare, though *Fagus* can sometimes be found and certain of the beech plantations described by Watt (1931) are essentially this kind of *Quercus-Betula-Oxalis* woodland in which there has been a canopy replacement.

Typically, shrubs and saplings are very sparse and rarely is there a true understorey. *Corylus* occurs very occasionally and sometimes there is a little *Sorbus aucuparia* and some young birch.

Although the field layer here preserves the general characteristics of the community as a whole, it has a number of distinctive features. First, by contrast with western stands, *Hyacinthoides* is rare and the most prominent plant in early spring, even on drier ground, is usually *Anemone nemorosa*. Then, as this finishes flowering, another good preferential, *Trientalis europaeus*, often becomes obvious. Among the carpet of grasses and herbs that gives the vegetation its typical early summer appearance, there are some further species which show a slight rise in frequency here. *Luzula pilosa* is especially common and there is occasionally some *Melampyrum pratense*, *Lathyrus montanus* and *Rubus idaeus*. *Veronica chamaedrys*, *V. officinalis*, *Hypericum pulchrum* and *Ajuga reptans* can also be found, though they are more strongly preferential for the *Stellaria-Hypericum* sub-community. Some stands have *Rubus saxatilis*. A further striking feature is that, apart from *Pteridium* which quite commonly assumes dominance by August, ferns are scarce: *Blechnum spicant* is found only occasionally, *Thelypteris limbosperma* is only very locally prominent and dryopteroids are generally very infrequent.

Bryophytes in this sub-community can be quite extensive, though in the markedly drier climate of the region, they do not show the variety or luxuriance characteristic of western stands. But all the community species remain frequent, *Pleurozium schreberi* and *Dicranum majus* are also common and, in addition, there is often an abundance of *Rhytidiadelphus triquetrus*. As usual in the *Quercus-Betula-Oxalis* woodland, when the flush of spring and summer growth among the herbs is over, it is the bryophyte carpet which continues to give this vegetation its verdure throughout the winter.

Stellaria holostea-Hypericum pulchrum sub-community:

Trientali-Betuletum, typical subassociation Birse 1982 *p.p.*; *Lonicero-Quercetum*, *Endymion* and typical subassociations (Birse & Robertson 1976) Birse 1984 *p.p.* The general character of the woody cover here is very similar to that of the *Anemone* sub-community, though the shift among the oak and birch taxa is a little more pronounced. *Quercus* hybrids and pure *Q. robur* tend greatly to outnumber *Q. petraea* and the birch is generally *B. pendula* with *B. pubescens* only occasional. Again, the structure is variable, with either oak or birch predominating but typically the canopy is open and quite low. Some stands are abandoned oak coppice but, even among those which have a high forest physiognomy, shrubs are sparse with only occasional *Corylus* and very scarce *Crataegus monogyna*.

In the field layer, there is generally no marked vernal aspect to the vegetation, though *Anemone* does occur occasionally and locally it can be abundant. *Trientalis*, too, shows a depressed frequency compared with the previous sub-community. More obviously distinctive here is a shift in the character of the grass and herb sward towards a somewhat more mesophytic composition. Species such as *Anthoxanthum odoratum*, *Agrostis tenuis* and, more especially, *A. canina* ssp. *montana* and *Deschampsia flexuosa*, are less generally prominent here; even where they remain frequent, their cover is patchy or consistently reduced. By contrast, *Holcus mollis* is often abundant and *Festuca rubra* and *Holcus lanatus* become preferentially frequent. Among the dicotyledons, *Potentilla erecta* is somewhat less common than in the *Blechnum* and *Anemone* sub-communities and there is a pronounced rise in *Veronica chamaedrys*, *V. officinalis*, *Ajuga reptans*, *Cerastium fontanum* and, especially characteristic here, *Stellaria holostea* and *Hypericum pulchrum*. *Luzula multiflora* tends to replace *L. pilosa* and there is occasionally some *Rumex acetosa* and *Angelica sylvestris*.

The bryophyte carpet too, though it can retain the high cover typical of the community, reflects this trend.

Hylocomium splendens, though frequent, is distinctly patchy in its abundance and species such as *Pleurozium schreberi*, *Dicranum majus*, *Polytrichum formosum* and *Rhytidiadelphus loreus* are very rare or totally absent. *R. triquetrus*, as in the *Anemone* sub-community, is constant but here its most consistent companions are *R. squarrosus*, *Pseudoscleropodium purum*, *Thuidium tamariscinum*, *Eurhynchium praelongum* and, good preferentials here, *Plagiomnium undulatum* and *Lophocolea bidentata s.l.*

Habitat

The *Quercus-Betula-Oxalis* woodland is typically a community of moist but free-draining and quite base-poor soils in the cooler and wetter north-west of Britain. Grazing by stock and deer contributes greatly to the character of the field layer and affects the physiognomy of the woody cover of the community by hindering regeneration. Many stands have been treated as coppice and there is evidence of widespread timber removal and some planting.

The *Quercus-Betula-Oxalis* woodland is confined to those parts of Britain where the annual rainfall exceeds 1000 mm (*Climatological Atlas* 1952), and where there are over 160 wet days yr^{-1} (Ratcliffe 1968). Indeed, except in eastern Scotland, precipitation is generally more than 1200 mm yr^{-1} (with about 180 wet days) and, in the far west, often approaches 3000 mm (with more than 220 wet days). Mean annual maximum temperatures over most of the region are less than 25 °C, in much of Scotland less than 23 °C (Conolly & Dahl 1970). Within the areas characterised by this kind of climate, the *Quercus-Betula-Oxalis* woodland is typically found on the slopes of the upland fringes, extending up to more than 450 m in some places, but generally below 180 m.

Within this zone, it is characteristic of substrates that are neither markedly calcareous nor strongly acidic: it is absent from limestones, even in the very wet far northwest, and does not extend far on to sandstones and grits or pervious and acidic igneous or metamorphic rocks unless there is a mask of superficial deposits or some moderate degree of base-enrichment from downwash waters or by diffuse flushing along seepage lines. It occurs widely on argillaceous rocks like Ordovician and Silurian shales in south-west Scotland, the Lake District and north Wales and, more locally, on Upper Carboniferous shales in the northern Pennines. It also picks out interbedded or intruded rocks of less extreme character occurring within masses of predominantly acidic deposits like the Lewisian gneiss along the north-west seaboard of Scotland, Palaeozoic plutonic rocks in Aberdeenshire, the Borrowdale Volcanics of the Lake District and the Devonian Old Red Sandstone around the Moray Firth and, more locally, in southern Perthshire. Colluvium, head, till and, especially in eastern

Scotland, fluvioglacial deposits, also provide substrates.

Typically, the soils developed from such deposits show no more than incipient podzolisation under this deciduous woodland. The surface pH is usually in the range 3.5–5.0 but the humus is very often of the transitional moder type and the profile generally a brown earth or brown podzolic soil (Avery 1980). There is typically a weathered B horizon with some accumulation of iron, manganese and organic matter but usually no distinct iron pan or humus-rich B_h (e.g. McVean & Ratcliffe 1962, Tittensor & Steele 1971, Birks 1973, Birse 1982, 1984). In soils derived from shales or colluvium or superficials that are neither too heavy-textured nor excessively free-draining, there is often a somewhat better structure with the profile approaching a mull brown earth, conditions often associated here with the *Dryopteris* sub-community. To the other extreme, as often in the *Blechnum* sub-community, this woodland can extend over areas where there are accumulations of siliceous boulders which carry thin and fragmentary rankers.

The general confinement of the community to base-poor but not excessively-leached brown soils gives the *Quercus-Betula-Oxalis* woodland much of its floristic character and helps to define its limits against its north-western counterparts on more calcareous or more strongly-podzolised profiles, distinctions which are very well seen in the range of communities defined from Loch Lomond by Tittensor & Steele (1971) and, on a finer scale, in the mosaics described from Skye by Birks (1973). On the one hand, calcicoles are fairly rigidly excluded from this community and the nearest approach to the *Fraxinus-Sorbus-Mercurialis* woodland is seen where species characteristic of more enriched mull soils appear in the *Dryopteris* sub-community or, less consistently, in other sub-communities where there is some local flushing.

On the other hand, edaphically-related distinctions between the *Quercus-Betula-Oxalis* woodland and the *Quercus-Betula-Dicranum* woodland, which replaces it on rankers and podzols, are much less sharp. The balance in the floristic composition here, among the grasses, dicotyledons, ferns and bryophytes, is clearly towards the calcifugous (much more so than in its southern analogue, the *Quercus-Pteridium-Rubus* woodland), and towards the very wet far north-west, it becomes increasingly difficult to define the limits of the community, a very real reflection of the overwhelming tendency towards leaching of all but the most calcareous lithomorphic soils in moving towards this region. In such extreme situations, the distinction between the *Quercus-Betula-Oxalis* and the *Quercus-Betula-Dicranum* woodlands is the scarcity here of *Vaccinium myrtillus* and *Calluna vulgaris*, the persistence of less markedly calcifuge grasses like *Anthoxanthum odoratum* and

Agrostis capillaris and, more especially, of *Hyacinthoides*. It should be noted, though, that grazing, which is very widespread in both communities, can reduce *Hyacinthoides* here and eliminate *Vaccinium* from the *Quercus-Betula-Dicranum* woodland, so tending to make the two vegetation types converge.

The other floristic boundary directly affected by soil conditions is that with wetter woodlands. Typically, the profiles here are kept moist throughout the year, a feature which is important to the maintenance of high frequencies of *Oxalis acetosella* and *Viola riviniana* and to the abundance of bryophyte mats. But they are characteristically free-draining, sometimes excessively so where the community runs some way on to pervious rocks over steeper slopes. This probably plays some part in the general scarcity here of *Q. robur*, at least in the west (e.g. Jones 1959), and the confinement of more mesophytic ferns and dicotyledons to the *Dryopteris* sub-community. Occasionally, flushing can be quite pronounced, especially where soil water encounters shale bands or heavy drift, and then the vegetation may take on some of the character of the *Alnus-Fraxinus-Lysimachia* woodland (well seen in the Community type 5 of Tittensor & Steele 1971), but gleying is rare.

As well as affecting the vegetation through variation in the soils, climate is directly important to the composition of the *Quercus-Betula-Oxalis* woodland in a number of ways. In the first place, the generally low summer temperatures virtually exclude any Continental or Continental Southern elements from the flora. Such species are not very numerous even in the southern analogue to this community, the *Quercus-Pteridium-Rubus* woodland, but they hardly penetrate into the range of the *Quercus-Betula-Oxalis* woodland. *Tilia cordata* persists as a relic on some cliffs and ravines within stands of the community in the Lake District (Pigott & Huntley 1978) and such woodland would therefore fall into the oak-lime stand type 5B of Peterken (1981) but species such as *Carpinus betulus*, *Crataegus laevigata*, *Euphorbia amygdaloides* and *Lamiastrum galeobdolon* are generally absent.

On the other hand, the difference between the mean temperatures of the warmest and coldest months of the year within the range of the *Quercus-Betula-Oxalis* woodland is, for the most part, significantly smaller than that of the lowland south and east and this is very much a community of the more equable parts of Britain. This, and the markedly higher rainfall to the north-west, gives the climate a distinctly oceanic character. The impact of this, as might be expected, can be best seen in the two more westerly sub-communities. These can be understood as continuing a floristic trend within British 'bluebell woods', first visible in the *Acer-Oxalis* sub-community of the *Quercus-Pteridium-Rubus* woodland, continuing through the *Dryopteris* sub-community of the *Quercus-Betula-Oxalis* woodland and terminating in the *Blechnum* sub-community. In this sequence of woodlands, the maintenance of vernal dominance by *Hyacinthoides* (itself a markedly Atlantic species in Europe: Noirfalise 1968, Roisin 1969), is accompanied by a switch among the oaks and birches to *Q. petraea* and *B. pubescens*, a rise in the prominence of ferns and a lush bryophyte component and, more particularly, an increasing representation of Widespread Atlantic (*sensu* Ratcliffe 1968) mosses and hepatics. Elements of this bryophyte flora can be seen sporadically in *Quercus-Betula-Oxalis* woodlands of north Wales and northwest England but, along the north-west seaboard of Scotland, with its relatively mild, extremely wet and very cloudy climate, these species become a very prominent feature, especially where local topographic variation (like north-facing slopes or deep ravines) accentuates the shelter even further. Exactly what it is among the complex of factors involved in oceanicity that favours these plants is probably quite varied and still uncertain, but the constancy of the humid conditions is probably of major importance, ameliorating desiccation by wind and sun and muting the effects of frosts (Ratcliffe 1968). Most of the species are not strictly associated with this community, or even with woodland in general, but the fact that the *Quercus-Betula-Oxalis* woodland provides an insulating canopy over diverse topography in the far north-west is probably of considerable importance in maintaining the conditions that encourage a rich and diverse flora of this kind.

On moving to the east of Scotland, climatic differences produce some obvious effects in the community. Compared with much of the lowland south-east, the climate here is still very wet but much less so than at identical latitudes on the west coast, with annual rainfall and the number of wet days more like the levels reached along the Lakeland fringes and the foothills of Snowdonia. Furthermore, though the summers are warmer than to the west, the winters are considerably colder, so the regional climate has an element of continentality. This may play a part in the rise to prominence of *B. pendula* in the *Anemone* and *Stellaria-Hypericum* sub-communities here and is certainly involved in the scarcity of *Hyacinthoides*, here replaced by *Anemone*, not in the sort of edaphic shift common in the south-east, but in some kind of response to climate. The appearance of the Northern Montane *Trientalis* is probably also related to the cooler conditions. Although the longer snow cover in this region (Manley 1940) may help some more drought-sensitive bryophytes to persist (Ratcliffe 1968), most of the Atlantic species of the *Blechnum* sub-community are absent. What factor lies behind the striking rise in *Rhytidiadelphus triquetrus* in these two sub-communities (and in eastern Scottish types of *Quercus-Betula-Dicranum*, *Pinus-Hylocomium* and *Juniper-*

us-Oxalis woodlands) is unclear.

Despite these internal variations within the *Quercus-Betula-Oxalis* woodland in relation to climate and soils, it is of general significance that it replaces the *Quercus-Pteridium-Rubus* woodland in the north-west, where predominantly non-calcareous parent materials have weathered to produce base-poor soils in a cool and wet climate. But there is a further factor of great importance to the floristic distinctions between these two kinds of woodland, the intensity of which roughly corresponds to this major climatic and edaphic divide: it is grazing. Very often, stands of the *Quercus-Betula-Oxalis* woodland are open to stock: many occur on the slopes which lie towards the upper limit of enclosure on upland farms and they provide grazing, mostly for sheep but also sometimes for cattle, especially in winter, when they also give shelter in the frequently inclement weather. They are also extensively grazed by deer, mainly red and roe deer in northern England and Scotland, with naturalised fallow deer (*Cervus dama*) in north Wales and locally elsewhere (around Loch Lomond, for example, where feral goats are also present: Tittensor & Steele 1971). The effects of grazing and browsing can be seen throughout the community, though they are particularly obvious in the *Blechnum*, *Anemone* and *Stellaria-Hypericum* sub-communities: it is likely that the distinctive features of the *Dryopteris* sub-community can sometimes reflect lighter or less consistent grazing or, at least, that relaxation of grazing favours its development on less strongly leached soils (e.g. Tittensor & Steele 1971).

The activities of these larger herbivores are probably partly responsible, first, for the general scarcity of saplings and shrubs in most *Quercus-Betula-Oxalis* woodlands (cf. Yapp 1953). Even given the fact that many stands have been kept clean of shrubs for oak-coppicing and remembering that birch regenerates with difficulty even under its own canopy, the typically sparse understorey here is likely to depend to some extent on the continual devouring of seedlings and the browsing of such saplings as do get away. Young trees of *Q. petraea*, for example, which seem to be better able to regenerate within closed woodland than *Q. robur* (e.g. Jones 1959) are very sparse; saplings of birch, *Sorbus* and, when it is present, *Fraxinus* are often very slow to appear in gaps or around the fringes of stands; and *Corylus* thickens up to form a distinct understorey only in the *Dryopteris* sub-community. Furthermore, when access to stock and deer is restricted, as where stands occur on islands or are fenced, there is often a clear response among the woody species, notably in the shade-tolerant *Sorbus* within the woods and in birch in the adjacent unwooded areas (e.g. McVean 1958).

Second, grazing is certainly of prime importance in maintaining the high frequency and abundance of such grasses as *Anthoxanthum odoratum*, *Deschampsia flex-*

uosa, *Agrostis* spp. and *Festuca* spp. in the field layer and in restricting the prominence of tall luxuriant herbs and underscrub. These latter components attain consistently high frequency here only in the *Dryopteris* sub-community, though lusher herbs can be found elsewhere where there is a measure of local protection. *Luzula sylvatica*, for example, can become very prominent in island stands (Tittensor & Steele 1971) and on precipitous slopes (Birse 1984) and, where the community extends over acid boulders, *Vaccinium myrtillus* may attain local abundance (as in some of the stands of McVean & Ratcliffe 1962 and Birks 1973). Sets of woodlands with different intensities of grazing, like those described from Loch Lomond (Tittensor & Steele 1971), may show a continuous gradation in the prominence of these grazing-sensitive elements in the vegetation, but, as a rule, they are not very obvious.

Third, since grazing generally keeps the field layer short, it encourages an abundance of bryophytes, especially those bulky pleurocarpous mosses which can grow amongst grass tillers and expand to fill any gaps developing in the cover. In very heavily grazed stands, as in those where there is much talus, bryophytes may account for the bulk of the ground cover.

Very often now, grazing in these woods is uncontrolled, representing the renewal of a treatment which has operated for many centuries and which has probably been the major factor in reducing the extent of the community (e.g. McVean & Ratcliffe 1962, McVean 1964a, Pearsall 1968, Pearsall & Pennington 1973). But there is also abundant evidence in many surviving stands of the past use of the *Quercus-Betula-Oxalis* woodland as coppice, perhaps managed in conjunction with controlled grazing, but carefully treated to yield a continual supply of underwood and timber. Some tracts have been used as hazel coppice, sometimes with oak standards (falling within the hazel-ash stand types 3C and 3D or hazel-rich birch-oak woods of Peterken 1981) but the most widespread crop was oak itself. In the north-west of Britain, this community, together with the more calcifugous *Quercus-Betula-Dicranum* woodland, has been the major source of oak for charcoal (particularly important for iron-smelting, especially from the sixteenth to the eighteenth centuries), tan-bark, pit-timber and constructional timber (e.g. Tittensor 1970a, b; Tittensor & Steele 1971, Pearsall & Pennington 1973, Linnard 1982). Birch and hazel also provided wood for cotton-bobbins in Lake District stands (Pearsall & Pennington 1973). For the most part, such treatments have become defunct but they have left a lasting legacy in the physiognomy and floristics of the woody component in many areas. Quite commonly, oak dominates as regularly-disposed old coppice stools and scattered through the woods there are often flat, circular platforms (pit-steads) where the charcoal heaps were con-

structed and, between them, interconnecting paths used repeatedly by the burners (e.g. Pigott & Huntley 1978, Barker 1985). In other cases, especially in stands towards the north-west of Scotland, the scarcity of oak is probably partly due to extensive extraction of the best timber over many generations. And, in eastern Scotland, too, repeated cutting of oak has led to a general run-down of many woods with an increasing prominence of birch.

There is evidence from some parts of the country that some oak coppices belonging to the *Quercus-Betula-Oxalis* woodland were planted to supplement the relatively small amount of woodland surviving from previous times. Planting in may explain the prominence of *Q. robur* or hybrid derivatives in eastern Scottish stands of the *Anemone* and *Stellaria-Hypericum* sub-communities (Jones 1959, Gardiner 1974). The causes of the floristic differences between these two kinds of *Quercus-Betula-Oxalis* woodland are uncertain, but it is possible that the *Stellaria-Hypericum* type includes younger or more modified stands. *Q. robur* elements are a little more prominent here, the slow-spreading *Anemone* is scarce and herbs such as *Veronica chamaedrys*, *V. officinalis*, *Hypericum pulchrum*, *Festuca rubra* and *Holcus lanatus* create the same effect as can be seen in the *Holcus* sub-community of the *Quercus-Pteridium-Rubus* woodland, often a planted replacement for southern 'bluebell woodland'. However, it is likely that, in time, hardwood plantations can acquire the essential features of the *Quercus-Betula-Oxalis* woodland, as seen in some of Watt's (1931) Scottish beechwoods which were established in the mid-nineteenth century.

Zonation and succession

Zonations within stands of the community and between the *Quercus-Betula-Oxalis* woodland and other vegetation types are most frequently related to edaphic differences and variation in grazing intensity. Often, the effects of these factors can be seen in isolation but, since lighter grazing and a tendency towards mull development can produce similar results in the field layer, it is sometimes difficult to disentangle their influence. Transitions from the *Blechnum* to the *Dryopteris* sub-community, for example, may result from either of these causes; and, in some cases, they may be operating together (e.g. Tittensor & Steele 1971).

Soil-related zonations here directly parallel the patterns found among analogous woodlands in the lowland south, with transitions to more calcicolous, more calcifugous and wetter woodlands occurring commonly. On more base-rich soils, the *Quercus-Betula-Oxalis* woodland typically gives way to the *Fraxinus-Sorbus-Mercurialis* woodland, provided the soils do not, at the same time, become very wet. Such zonations are frequent but local in the west and they often reflect a change in the character of the underlying rock, as where limy partings occur interbedded with shales in Silurian deposits in Wales, the Lake District and southern Scotland or in Carboniferous sequences in the Pennines, where andesite occurs within the Borrowdale Volcanics of the Lake District (Pearsall 1968, Pearsall & Pennington 1973), where Carboniferous Limestone replaces the intruded dolerite in Teesdale (Pigott 1956a, Graham 1971) and where Durness Limestone and Torridonian deposits are juxtaposed on Skye (Birks 1973). In such situations, there may be a fairly sharp switch from the *Blechnum* sub-community here to the Typical sub-community of the *Fraxinus-Sorbus-Mercurialis* woodland or, on less sharply contrasted soils and where grazing is less intense, from the *Dryopteris* sub-community to the *Crepis* sub-community of the *Fraxinus-Sorbus-Mercurialis* woodland.

Quite often, however, changes in soil conditions are much more gradual or uneven than this, being related to quite slight differences in the calcareous character of a drift cover or bedrock or to diffuse flushing by base-rich waters. Then, particularly if the woodlands are grazed, the characteristic grasses of the *Quercus-Betula-Oxalis* woodland may continue some way into the *Fraxinus-Sorbus-Mercurialis* woodland and more calcicolous herbs like *Mercurialis perennis*, *Geum urbanum* and *Circaea lutetiana* make but a sporadic appearance at first. Over more complex topographies, as in ravines, there may be a very disorderly inter-digitation of the two communities over the surface of tumbled boulders, slumped masses of soil, rock faces and fragmentary patches of alluvium along the stream-side. Changes in the canopy, too, can be slight, especially in the far north-west, where *Q. petraea* and *Fraxinus* are scarce and where both these communities have a scrubby cover in which *B. pubescens* and *Sorbus* figure prominently. In such situations, a rise in the prominence of *Corylus* may mark the transition but, further south, it is *Fraxinus*, less commonly *Ulmus glabra* and *Acer pseudoplatanus*, that increase. Where the *Quercus-Betula-Oxalis* woodland gives way to the *Fraxinus-Sorbus-Mercurialis* type along streams, these can often be picked out from afar as very distinct streaks of brighter green, a feature well caught in Pearsall (1968). In the Lake District, where such situations are quite frequent, relic stools of *Tilia cordata*, sometimes of immense size, can be found growing on the ravine edge which marks the boundary between the two communities (Pigott & Huntley 1978). Here, too, treatment has sometimes sharpened up the distinction between the communities, the *Quercus-Betula-Oxalis* woodland of the interfluves having been used as oak coppice, the transitional zone and *Fraxinus-Sorbus-Mercurialis* woodland proper of the ravine slopes having been inaccessible (Pigott & Huntley 1978).

The fact that flushing is often the factor responsible for base-enrichment within sites where the *Quercus-Betula-Oxalis* woodland is represented, means that transitions of this kind often involve elements of the *Alnus-Fraxinus-Lysimachia* woodland disposed over the very wettest ground with local patches of plants such as *Lysimachia nemorum*, *Deschampsia cespitosa*, *Chrysosplenium oppositifolium* and large sedges such as *Carex remota*, *C. pendula* or *C. laevigata*. On gentler slopes, this community may form fairly well defined flushes within the *Quercus-Betula-Oxalis* woodland: these are very characteristic of Carboniferous and Silurian sequences where ground water emerges when it hits impervious shales. In analogous situations where less base-rich ground water emerges, the *Alnus-Fraxinus-Lysimachia* woodland is replaced by soligenous *Betula-Molinia* woodland.

Zonations to woodland of freely-drained, acidic soils with pronounced mor accumulation and/or podzolisation typically involve replacement by the *Quercus-Betula-Dicranum* woodland and they can be seen throughout the range of the community wherever the influence of pervious and acidic parent materials increases and leaching can have its full effect. Sometimes, this involves a switch in the nature of the bedrock, as in moving on to Torridonian sandstones or Lewisian gneiss in north-west Scotland (e.g. Birks 1973, Ratcliffe 1977), or to more acidic strata in the Devonian Old Red Sandstone in eastern Scotland and in Silurian sequences in the Lake District, southern Scotland and Wales, to sandstones and grits in the Pennine Carboniferous or to acidic igneous rocks in the Lake District. In other cases, it is related to the thinning of a cover of colluvium or drift over such deposits or to the lessening of a flushing effect.

The floristic transitions involved here can already be seen to some extent within the *Quercus-Betula-Oxalis* woodland itself in the way in which the more strictly calcifuge element in the *Blechnum* sub-community rises to prominence as the soil cover thins to a humic cap over acidic boulders. But often the zonation continues over rankers and podzols to the *Isothecium-Diplophyllum* or Typical sub-communities of the *Quercus-Betula-Dicranum* woodland in the west (Figure 21). In eastern Scotland, there are analogous transitions from the *Anemone* and *Stellaria-Hypericum* sub-communities to the *Rhytidiadelphus* sub-community of the *Quercus-Betula-Dicranum* woodland. Locally, in this region, more calcifugous kinds of *Juniperus-Oxalis* woodland can replace the *Quercus-Betula-Dicranum* woodland in zonations of this kind and, in both east and west Scotland, the *Pinus-Hylocomium* woodland is locally associated with the *Quercus-Betula-Oxalis* woodland on more podzolised soils.

As noted earlier, the effect of grazing on the ericoid sub-shrubs which help distinguish the field layers of these communities, can blur the boundaries. But grazing also mediates zonations of its own, most notably to different kinds of the *Festuca-Agrostis-Galium* grassland, which is composed of virtually the same species as the field-layer constants of the *Quercus-Betula-Oxalis* woodland and which, in wetter regions, can have a number of its characteristic bryophytes. Very often, on the lower slopes of the uplands, the *Festuca-Agrostis* grassland runs straight into unenclosed *Quercus-Betula-Oxalis* woodland, the only changes across the junction being the appearance of trees, an increase in the cover and variety of bryophytes with the greater shelter and a patchy dominance of *Hyacinthoides* or *Anemone* in the spring. In more inaccessible spots within such woodlands, there may be some patchy regeneration of the canopy but there is no doubt that such zonations often represent the near-final stages in the conversion of the woodland to the grassland and that the completion of the process is then attendant only on the final destruction of the trees, an event which has been only too frequent in the long history of upland grazing.

Continued grazing maintains the *Festuca-Agrostis-Galium* grassland as a plagioclimax and recolonisation is rarely seen but it can be observed in some extensive fenced reserves or where tracts of upland have been enclosed for commercial forestry or around lake and reservoir catchments. Around Thirlmere, in the Lake District, for example, recolonisation of open grassland

has been speedy: *B. pubescens*, *Sorbus* and *Corylus* have established themselves in local profusion and young *Q. petraea* can be found growing up through the thickets (Pearsall & Pennington 1973). Such enclosures could also provide opportunities for studying uninterrupted primary invasion of bare talus, free of the disturbance which the roaming of sheep and walkers bring. We do not know what the natural precursors of the *Quercus-Betula-Oxalis* woodland in such situations might be. The most likely candidates would seem to be open communities of bryophytes and ferns such as *Thelypteris limbosperma* and *Blechnum spicant* and, where there is a more extensive soil cover, vegetation like that of the *Luzula sylvatica-Vaccinium myrtillus* community. Fragments of such vegetation can often be found in more rocky situations close to the *Quercus-Betula-Oxalis* woodland; and, on inaccessible ledges with seepage of fairly base-poor waters within tracts of *Festuca-Agrostis-Galium* grassland, they can reproduce fairly faithfully the field layer of ungrazed *Quercus-Betula-Oxalis* woodland without a cover of trees. They also extend the distribution of this kind of herbaceous vegetation above the present tree-line in the Scottish Highlands and were considered by McVean & Ratcliffe (1962) as a tree-less facies of the community.

Some of the characteristic herbs of the *Quercus-Betula-Oxalis* woodland can also be found under a cover of bracken in the *Pteridium aquilinum-Galium saxatile*

Figure 21. Patterns among oak-birch woodlands in relation to soils and treatment.
At a grit-shale junction (*a*), a common pattern on ungrazed slopes is for W17b Typical *Quercus-Betula-Dicranum* woodland to give way below to
W11a   *Dryopteris* sub-community of *Quercus-Betula-*

Oxalis woodland (*b*). With grazing, there is often a gradual transition between these two,
W17c   *Anthoxanthum-Agrostis* sub-community of the former passing imperceptibly to W11b   *Blechnum* sub-community of the latter (*c*).

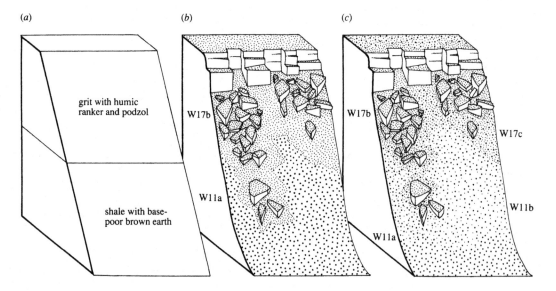

(*a*)   (*b*)   (*c*)

grit with humic ranker and podzol

shale with base-poor brown earth

W17b

W11a

W17b

W11a

W17c

W11b

community and stands of this vegetation type are commonly seen in close association with the woodland and *Festuca-Agrostis-Galium* grassland. A characteristic pattern is for *Pteridium* to be very dense and vigorous over the colluvium on the lower slopes adjacent to the wood, thinning out to a sparse cover in the grassland above. This is just the kind of distribution that bracken shows within stands of the *Quercus-Betula-Oxalis* woodland but there is no doubt that, with the destruction of the woodland cover and freedom from shade, it can increase greatly in abundance. Stands of the *Pteridium-Galium* community sometimes preserve elements of the richness of the woodland flora, with a vernal carpet of *Hyacinthoides* or *Anemone*, but very often they are poor in associates and the great extension of bracken over the deeper soils of the uplands is an unfortunate side effect of the widespread destruction of the community, and a change that is difficult and costly to reverse.

There have been other permanent losses to coniferous forestry. The *Quercus-Betula-Oxalis* woodland will stand canopy replacement by *Fagus* (e.g. Watt 1931) and a moderate amount of coniferisation, but extensive planting of softwoods (except perhaps *Larix* spp.) greatly increases shade and speeds mor accumulation, thus accentuating the tendency to podzolisation in the wet climate, a process which is ameliorated by the natural cover of oak and birch.

Although the history of the loss of the *Quercus-Betula-Oxalis* woodland by these treatments has been long and complex, with local or temporary halts and reversals related to shifts in the balance of the upland economy, its decline has been progressive and very widespread. It seems to represent the climax type of oak-birch forest on less markedly leached soils in the north-west but now survives as fragments within its potential range, usually much affected by grazing and in landscapes devoted primarily to pastoral agriculture and commercial forestry. In one area, however, it is proportionately more prominent than might be expected and that is in east-central Scotland. Here, McVean & Ratcliffe (1962) proposed that pine-dominance was the more natural development in post-Glacial forests, a suggestion largely borne out by subsequent palynological studies (e.g. Birks 1970, Gunson 1975, O'Sullivan 1977). In fact, the tree cover of the original pine forest in this region seems to have been more diverse and intimately mixed than in surviving stands of the *Pinus-Hylocomium* woodland, such that both pine and birch probably occurred together over more acidic soils. With clearance, burning and grazing, birch has increased its relative prominence in this region, now holding sway in both the *Quercus-Betula-Dicranum* woodland on podzols and in the *Quercus-Betula-Oxalis* woodland on less strongly leached profiles in sites where *Pinus sylvestris* was once much better represented.

Distribution

The *Quercus-Betula-Oxalis* woodland is widely distributed through the upland fringes of Wales, north-west England and Scotland with a few stands in south-west England. To the west, the community is represented by the *Dryopteris* and *Blechnum* sub-communities, the former predominating in England, Wales and south-west Scotland, the latter present locally there but becoming much more prominent along the north-western seaboard of Scotland. Occasionally in southern Scotland but much more exclusively in north-east Scotland, these kinds of *Quercus-Betula-Oxalis* woodland are replaced by the *Anemone* and *Stellaria-Hypericum* sub-communities.

Affinities

The *Quercus-Betula-Oxalis* woodland as defined here unites some of the less heathy oak-dominated communities described in early studies (e.g. Crampton 1911, Tansley 1911, 1939) as falling within the general ambit of the *Quercetum petraeae*, and certain kinds of birch woodland characterised from higher altitudes and latitudes (e.g. Pigott 1956a, McVean & Ratcliffe 1962, Graham 1971, Birks 1973, Birse 1982, 1984). While recognising that, in more extreme habitats, the absence of oak may have some climatic basis, replacement by birch is often here a treatment-related phenomenon and, whichever tree is the dominant, the essential character of the field layer is preserved throughout (e.g. McVean 1964a, Tittensor & Steele 1971, Birks 1973). The diagnosis thus unites vegetation which has figured in some schemes under two different heads (e.g. Tansley 1939), or more, where weight has also been given to treatment-related prominence of *Corylus* or the presence of relic *Tilia* (e.g. Peterken 1981).

The distinctive features of the field layer here, though often much modified by grazing, help to place the *Quercus-Betula-Oxalis* woodland in a central position between communities of more calcareous soils on the one hand and more markedly leached soils on the other. Such a trio of woodland types has not generally been recognised in the north-west, though the need for the distinction is hinted at in McVean & Ratcliffe (1962) and given expression at sub-community level within associations described by Klötzli (1970) and Birse (1982, 1984). The difficulty has been that the most detailed studies have concentrated on areas where soil-related differences within the suite of woodlands tend to be masked by striking climatically-influenced similarities. In north-west Scotland, for example, all three communities occur under a scrubby cover in which *B. pubescens* predominates and have a very prominent bryophyte element in which many species are shared. In this region, the *Quercus-Betula-Oxalis* woodland as defined here includes only the central core of Klötzli's (1970) modification of Braun-Blanquet & Tüxen's (1952) *Blechno-Quercetum*, and occupies a middle ground between the *Betuletum Oxalido-Vaccinietum* and *Betula*-herb nodum of McVean & Ratcliffe (1962) and the *Betula-Vaccinium* and *Corylus-Oxalis* Associations of Birks (1973), taking some of the less extreme samples from each. Likewise, it includes only the less calcifugous vegetation within Birse's (1984) version of the *Blechno-Quercetum*.

In eastern Scotland, the woodlands have been less extensively described but the prominence of such species as *Anemone nemorosa*, *Trientalis europaea* and *Rhytidiadelphus triquetrus* within communities of less- and more-leached soils led Birse (1982, 1984) to group together vegetation which is here divided between the *Quercus-Betula-Oxalis* and *Quercus-Betula-Dicranum* woodlands. The occurrence of *Trientalis* in these woodlands is not given as much diagnostic weight as in Birse (1982, 1984), so the *Quercus-Betula-Oxalis* woodland also subsumes most of his *Lonicero-Quercetum* which is virtually identical to less calcifugous stands of his *Trientali-Betuletum* apart from the absence of this Northern Montane plant.

Although Scottish stands of the *Quercus-Betula-Oxalis* woodland in both the east and west closely approach in their floristics the composition of more heathy woodlands dominated by *B. pubescens*, *Juniperus* and *Pinus*, the affinities of the community as a whole lie not with the Vaccinio-Picetea but with the Quercetea robori-petraeae, and the alliance Quercion robori-petraeae. This includes West European deciduous woodlands and their modified derivatives on acidic to only moderately base-rich soils, characterised by the prominence of oak and birch (and often in mainland Europe, beech) and the presence of such herbs as *Lathyrus montanus*, *Hypericum pulchrum*, *Teucrium scorodonia*, *Melampyrum pratense* and Hieracia of the Sabauda and Umbellata sections. The field layers are often grassy, with *Holcus mollis* and *Deschampsia flexuosa* well represented, frequently have some *Lonicera* and *Pteridium* and varying amounts of *Vaccinium myrtillus* and *Calluna*. The *Quercus-Betula-Oxalis* woodland is thus the British equivalent of such associations as the Quercetum medioeuropaeum. Br.-Bl. 1932 or the *Querceto-Betuletum* Tx. 1937 or their subsequent modifications described from France (Issler 1926, Malcuit 1929, Dethioux 1955), Switzerland (Braun-Blanquet 1932), Germany (Tüxen 1937, 1955, Oberdorfer 1957, Hartmann & Jahn 1967), the Netherlands (Westhoff & den Held 1969) and Belgium (LeBrun *et al.* 1955). The distinctive feature of these woodlands in north-west Europe, as in their counterparts in the Carpinion, is the vernal dominance of *Hyacinthoides* and those *Quercus-Betula-Oxalis* woodlands where this is combined with an abundance of bryophytes present a unique spectacle.

Floristic table W11

	a	b	c	d	11
Betula pubescens	III (1–9)	IV (1–9)	III (1–9)	II (1–7)	III (1–9)
Quercus petraea	V (5–10)	II (2–9)	I (4–8)	II (2–9)	II (2–10)
Betula pendula	I (1–10)	I (6)	III (1–8)	III (1–7)	II (1–10)
Quercus robur	I (1–6)	I (2–5)	I (7–9)	II (1–8)	I (1–9)
Betula hybrids	I (6–7)	I (6)	I (7)	I (2)	I (2–7)
Larix spp.	I (1–5)		I (1)	I (5–7)	I (1–7)
Fagus sylvatica	I (1–5)		I (1–10)	I (1–10)	I (1–10)
Fraxinus excelsior	II (1–7)	I (3–6)			I (1–7)
Sorbus aucuparia	II (1–3)	IV (1–7)	I (1)	I (5)	II (1–7)
Quercus hybrids	I (6)	I (2–4)	III (5–10)	II (3–9)	I (2–10)
Corylus avellana	III (1–7)	II (1–7)	II (1–5)	II (1–8)	II (1–8)
Betula pubescens sapling	II (1–6)	I (4–5)	I (1–7)		I (1–7)
Crataegus monogyna	II (1–3)	I (1–2)		I (1–4)	I (1–4)
Juniperus communis communis	I (1–2)		I (1)	I (6)	I (1–6)
Betula pendula sapling	I (1–2)		I (2–4)		I (1–4)
Quercus robur sapling		I (1)	I (3)		I (1–3)
Anthoxanthum odoratum	IV (1–7)	V (3–8)	V (1–8)	V (2–7)	V (1–8)
Oxalis acetosella	IV (1–9)	V (2–7)	IV (1–8)	V (1–7)	V (1–9)
Agrostis capillaris	IV (1–9)	IV (1–6)	IV (1–9)	V (1–7)	IV (1–9)
Deschampsia flexuosa	IV (1–7)	IV (2–8)	V (1–8)	III (1–6)	IV (1–8)
Holcus mollis	III (1–8)	III (1–7)	IV (1–8)	V (1–7)	IV (1–8)
Rhytidiadelphus squarrosus	III (1–5)	IV (1–4)	III (1–5)	V (1–8)	IV (1–8)
Pteridium aquilinum	III (1–9)	IV (1–5)	IV (1–9)	IV (1–8)	IV (1–9)
Galium saxatile	III (1–6)	IV (1–4)	V (1–5)	IV (1–6)	IV (1–6)
Pseudoscleropodium purum	II (1–5)	III (1–6)	IV (1–8)	V (1–7)	IV (1–8)
Viola riviniana	II (1–4)	IV (1–4)	V (1–6)	V (1–5)	IV (1–6)
Thuidium tamariscinum	II (1–9)	V (2–7)	V (1–8)	IV (1–8)	IV (1–9)
Potentilla erecta	I (2–5)	V (2–7)	IV (1–6)	III (1–4)	IV (1–7)
Hylocomium splendens	I (4)	IV (1–7)	V (1–8)	IV (1–6)	IV (1–8)
Rubus fruticosus agg.	III (1–8)	I (1)	I (1–2)		I (1–8)
Dryopteris dilatata	III (1–9)	I (1)	I (1–5)		I (1–9)
Dryopteris borreri	II (1–6)	I (1–3)	I (1–4)	I (1–4)	I (1–6)

Floristic table W11 (*cont.*)

	a	b	c	d	11
Digitalis purpurea	II (1–7)		I (1–7)	I (1)	I (1–7)
Dryopteris filix-mas	II (1–5)	I (3)	I (1)	I (4)	I (1–5)
Deschampsia cespitosa	II (1–3)	I (1–5)	I (1–6)	I (1–4)	I (1–6)
Pleurozium schreberi	II (1–5)	IV (2–6)	III (1–7)		III (1–7)
Dicranum majus	II (1–5)	V (1–6)	III (1–4)		III (1–6)
Hyacinthoides non-scripta	III (1–10)	IV (2–5)	I (1–4)	II (1–2)	III (1–10)
Polytrichum formosum	III (1–5)	IV (1–4)	I (1–5)	I (1–3)	II (1–5)
Blechnum spicant	I (1–5)	V (1–7)	II (1–8)	I (1–2)	II (1–8)
Hypnum cupressiforme	III (1–4)	III (1–5)	I (1–4)	I (1–3)	II (1–5)
Primula vulgaris	II (2–3)	III (1–4)	I (1–3)	II (1–3)	II (1–4)
Isothecium myosuroides	II (2–4)	III (1–6)	I (1–2)	I (1–3)	II (1–6)
Rhytidiadelphus loreus	I (1–7)	III (1–5)	I (1–5)	I (2)	I (1–7)
Thelypteris limbosperma	I (1–5)	III (2–5)	I (1–5)	I (4)	I (1–5)
Athyrium filix-femina	I (1–2)	II (1–3)	I (1)	I (2)	I (1–3)
Plagiothecium denticulatum	I (1–3)	II (1–3)	I (1–3)	I (1–5)	I (1–5)
Corylus avellana seedling	I (1–3)	II (1)			I (1–3)
Diplophyllum albicans	I (2–4)	II (1–3)			I (1–4)
Hylocomium brevirostre		II (1–5)	I (4)		I (1–5)
Sphagnum quinquefarium		I (1–4)			I (1–4)
Plagiochila spinulosa		I (1–2)			I (1–2)
Rhytidiadelphus triquetrus	I (1–3)	II (2–6)	IV (1–9)	IV (3–6)	III (1–9)
Luzula pilosa	II (1–4)	II (2–3)	IV (1–4)	I (1–3)	III (1–4)
Anemone nemorosa	I (1–4)	II (1–4)	IV (1–7)	II (1–7)	III (1–7)
Trientalis europaea		I (2)	III (1–5)	I (1–4)	I (1–5)
Lathyrus montanus		I (1–3)	II (1–4)	I (1–4)	I (1–4)
Melampyrum pratense	I (3–9)	I (4–5)	II (1–8)	I (5)	I (1–9)
Rubus idaeus	I (1–2)	I (2)	II (1–3)	I (1)	I (1–3)
Plagiomnium affine			I (2–5)		I (2–5)
Vaccinium vitis-idaea			I (1–3)		I (1–3)
Convallaria majalis			I (6)		I (6)
Pyrola minor			I (1)		I (1)
Brachypodium sylvaticum			I (1)		I (1)

Species				
Veronica chamaedrys	I (1)	II (1-3)	V (1-3)	III (1-5)
Lophocolea bidentata s.l.	III (1-5)	I (1-3)	IV (1-5)	III (1-5)
Plagiomnium undulatum	I (1-5)	I (2-3)	IV (1-3)	II (1-5)
Hypericum pulchrum		II (1-3)	III (1-2)	II (1-3)
Veronica officinalis		II (1-4)	III (1-2)	II (1-4)
Stellaria holostea	II (2-6)	I (3-4)	III (1-3)	I (1-6)
Luzula multiflora	I (1-2)	I (1)	III (1)	I (1-3)
Ajuga reptans	I (1-2)		III (1-4)	I (1-4)
Festuca rubra		I (4)	III (2-5)	I (1-5)
Cerastium fontanum			III (1)	I (1-2)
Holcus lanatus	I (4-7)	I (3-4)	II (1-4)	I (1-7)
Rumex acetosa	I (1-2)		II (1-2)	I (1-2)
Fraxinus excelsior seedling		I (1)	II (1)	I (1-3)
Angelica sylvestris		I (1)	II (1-3)	I (1-3)
Lonicera periclymenum	III (1-6)	II (1-4)	II (1-6)	II (1-6)
Teucrium scorodonia	III (1-7)	II (1-4)	II (1-7)	II (1-7)
Agrostis canina montana	II (1-7)	III (2-5)	I (2-3)	III (1-7)
Dicranum scoparium	II (1-3)	II (2-4)	II (1-2)	II (1-4)
Mnium hornum	III (1-4)	III (1-6)	II (1-3)	II (1-6)
Conopodium majus	I (1-3)	II (1-4)	II (1-3)	II (1-4)
Eurhynchium praelongum	III (1-5)	I (1-2)	III (1-4)	II (1-5)
Plagiochila asplenoides	I (1-3)	II (1-3)	II (1-4)	I (1-4)
Vaccinium myrtillus	I (1-5)	II (1-4)	II (1)	I (1-9)
Plagiothecium undulatum	II (1-4)	II (1-3)	I (1-4)	I (1-4)
Atrichum undulatum	II (1-3)		II (1-3)	I (1-3)
Poa pratensis		I (1)	II (2-5)	I (1-5)
Succisa pratensis		II (3-6)	I (1)	I (1-6)
Festuca ovina		II (1-5)	I (1-4)	I (1-6)
Luzula sylvatica	I (1-9)	I (1-4)	I (1)	I (1-9)
Betula pubescens seedling	I (1-3)	I (1)	I (1-2)	I (1-3)
Poa trivialis	I (1-2)	I (1)	I (2-5)	I (1-5)
Lysimachia nemorum	I (1-3)	I (1-3)	I (1-2)	I (1-4)
Eurhynchium striatum	I (1)	I (2)	I (2-4)	I (1-4)
Polytrichum commune	I (1)	I (1)		I (1-2)
Quercus hybrids seedling	I (2)		I (1)	I (1-3)
Crataegus monogyna seedling	I (1-3)		I (1)	I (1-3)
Carex pilulifera	I (1-2)		I (1)	I (1-2)

Floristic table W11 *(cont.)*

	a	b	c	d	11
Ilex aquifolium seedling	I (1–3)		I (1–2)	I (1)	I (1–3)
Galium aparine	I (2)		I (1)	I (2)	I (1–2)
Cirriphyllum piliferum	I (1)		I (1–3)	I (1–3)	I (1–3)
Geranium robertianum	I (2–3)		I (2)	I (2)	I (2–3)
Ranunculus repens	I (1)		I (2)	I (1)	I (1–2)
Betula hybrids seedling		I (2)	I (1–2)	I (2)	I (1–2)
Rubus saxatilis		I (1)	I (2–6)	I (1)	I (1–6)
Prunella vulgaris		I (1–2)	I (1)	I (1–3)	I (1–3)
Campanula rotundifolia		I (1)	I (1–2)	I (1–2)	I (1–2)
Luzula campestris		I (1–3)	I (1–2)	I (1–3)	I (1–3)
Ranunculus acris		I (1–2)	I (1)	I (5)	I (1–5)
Quercus robur seedling		I (3)	I (1)		I (1–3)
Calluna vulgaris		I (1–2)	I (1–4)		I (1–4)
Erica cinerea		I (1–3)	I (1–2)		I (1–3)
Polytrichum longisetum			I (1–5)	I (1)	I (1–5)
Cytisus scoparius			I (1)	I (1)	I (1)
Number of samples	40	18	61	20	139
Number of species/sample	27 (11–65)	34 (25–52)	29 (19–42)	31 (20–35)	29 (11–65)
Tree height (m)	20 (5–35)	10 (8–15)	15 (8–22)	15 (6–22)	16 (5–35)
Tree cover (%)	80 (30–95)	56 (10–100)	59 (10–80)	46 (5–75)	63 (5–100)
Shrub height (m)	3 (1–8)	5 (4–6)	4 (1–9)	4 (4–5)	4 (1–9)
Shrub cover (%)	23 (0–95)	10 (8–15)	1 (0–40)	12 (0–80)	10 (0–95)
Herb height (cm)	37 (10–100)	27 (15–47)	49 (13–120)	49 (10–137)	43 (10–137)
Herb cover (%)	82 (10–100)	72 (50–100)	77 (45–100)	81 (55–95)	78 (10–100)
Ground height (mm)	24 (10–40)	45 (30–80)	40	20 (10–30)	25 (10–40)
Ground cover (%)	13 (2–70)	63 (10–85)	47 (1–90)	44 (4–85)	39 (1–90)
Altitude (m)	132 (15–280)	131 (40–458)	156 (30–314)	124 (50–366)	141 (15–458)
Slope (°)	19 (0–45)	31 (5–70)	14 (0–50)	16 (0–37)	18 (0–70)

a   *Dryopteris dilatata* sub-community
b   *Blechnum spicant* sub-community
c   *Anemone nemorosa* sub-community
d   *Stellaria holostea-Hypericum pulchrum* sub-community
11   *Quercus petraea-Betula pubescens-Oxalis acetosella* woodland (total)

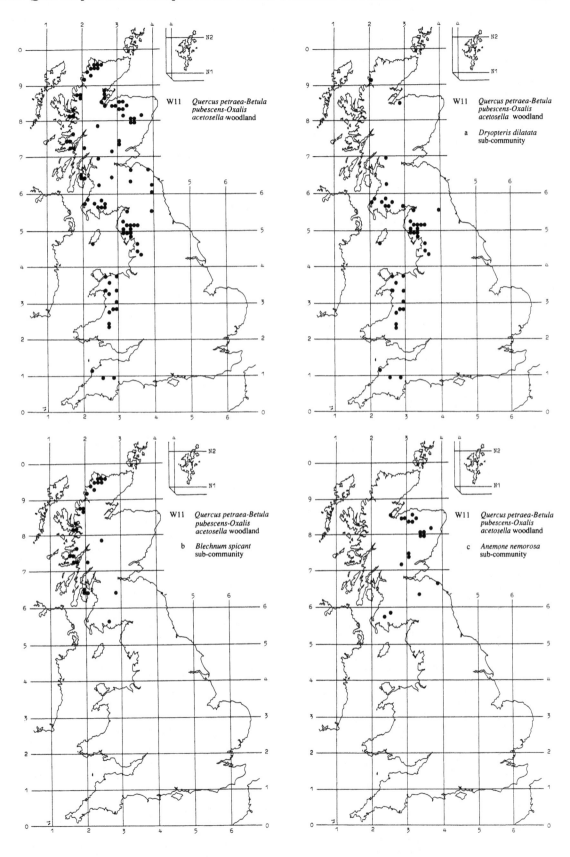

W11 *Quercus petraea-Betula pubescens-Oxalis acetosella* woodland

W11 *Quercus petraea-Betula pubescens-Oxalis acetosella* woodland

a *Dryopteris dilatata* sub-community

W11 *Quercus petraea-Betula pubescens-Oxalis acetosella* woodland

b *Blechnum spicant* sub-community

W11 *Quercus petraea-Betula pubescens-Oxalis acetosella* woodland

c *Anemone nemorosa* sub-community

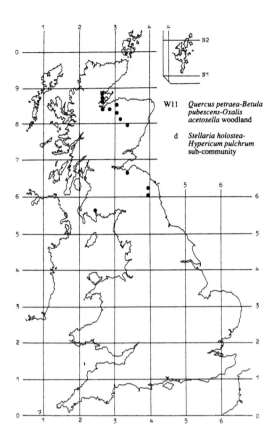

W11   *Quercus petraea-Betula pubescens-Oxalis acetosella* woodland

d   *Stellaria holostea-Hypericum pulchrum* sub-community

W12
Fagus sylvatica-Mercurialis perennis woodland

Synonymy

Beechwood association   Moss *et al.* 1910 *p.p.*; *Fagetum sylvaticae calcareum*   Tansley & Rankin 1911; *Fagetum sylvaticae* beechwoods on Chalk   Adamson 1921; Beech associes, seres 3 & 4   Watt 1924; Beech consociation, Juniper and Hawthorn seres   Watt 1934a; Beech consociation, seres 3 & 4   Watt 1934b; *Fagetum sylvaticae calcicolum*   Tansley 1939; *Fagetum rubosum* Tansley 1939 *p.p.*; Beech-Ash-Yew Association   McNeil 1961; Beech-Oak-Ash Association   McNeill 1961 *p.p.*; Cotswold beechwoods   Barkham & Norris 1967 *p.p.*; Beechwood   Rackham 1980 *p.p.*; Beech stand types 8Cb & 8Cc   Peterken 1981; Woodland plot types 1 & 8   Bunce 1982 *p.p.*

Constant species

Fagus sylvatica, Mercurialis perennis.

Rare species

Buxus sempervirens, Cephalanthera longifolia, C. rubra, Cynoglossum germanicum, Epipactis leptochila, Epipogium aphyllum, Hordelymus europaeus, Orchis purpurea.

Physiognomy

The *Fagus sylvatica-Mercurialis perennis* woodland is one of three woodland communities in Britain characterised by the great pre-eminence of *Fagus sylvatica*. In this community, *Fagus* is the only woody species that is constant throughout and is always the most abundant tree: in mature stands, it is an overwhelming dominant in a usually quite distinct topmost tier to the canopy. In general, though, and especially on the more shallow soils of the *Sanicula* and *Taxus* sub-communities, the individual trees do not attain the majestic stature so typical of the *Fagus sylvatica-Rubus fruticosus* woodland. There can also be considerable variation in the overall canopy height and in the distribution and physiognomy of the trees according to the age of the woodland and younger sub-spontaneous stands can continue to betray the pattern of invasion by *Fagus* for some time. In some of

the woodlands which Watt (1924, 1934a) examined, for example, he detected varied mixtures of pioneers, families and groups of different sizes and degrees of maturity and noted a contrast between the stocky, richly-branched early invaders with their spreading crowns and individuals of subsequent generations with taller, thinner boles, often obliquely set and with lop-sided crowns crowded by the existing trees. With increasing age, however, the *Fagus* cover takes on a more even and regular look and younger trees become progressively confined to gaps with slow and sometimes sporadic regeneration.

As defined here, this community also includes semi-natural or planted stands in which gap creation and regeneration have been controlled by a selection system of timber extraction (Troup 1966), an uncommon form of forest management in Britain but one especially associated with Chiltern stands of *Fagus-Mercurialis* woodland (Watt 1934b, Peterken 1981), and also stands treated by clear-felling and replanting or under a shelter-wood regime (Troup 1966, Peterken 1981). Coppiced stands, however, are rare: *Fagus* pollards well but it produces rather weak coppice shoots or is perhaps more prone to disease when coppiced, and is only very occasionally seen as the dominant in coppice underwood forms of the community (Rackham 1980, Peterken 1981). But, quite commonly, sites which would be expected naturally to carry the *Fagus-Mercurialis* woodland seem to have been converted to *Corylus* or mixed small coppice by the removal of *Fagus* (Watt 1934b, Tansley 1939), usually the total removal since *Fagus* standards have a strong shading effect on the underwood (Rackham 1980). In this scheme, such vegetation would fall within the *Fraxinus-Acer-Mercurialis* woodland, though neglect and *Fagus*-invasion means that many stands are reverting to this community.

Apart from *Fagus*, no other tree is consistently frequent throughout the community or dominant over more than very local areas. Overall, the commonest companion is *Fraxinus excelsior*, especially in the *Mer-*

curialis sub-community, where older trees can be survivors from the *Fraxinus-Acer-Mercurialis* woodland that typically precedes this community in successions over somewhat deeper soils in less exposed situations (Watt 1924, 1934*a*). Occasional individuals can persist in mature *Fagus* canopies here, though they very often have a drawn-up and rather spindly appearance. *Fraxinus* is also the most frequent coloniser of gaps here and, when *Fagus* mast is in short supply, a leading ash maiden may grow up and fill the space in the canopy (Watt 1923, 1925). *Quercus robur* can also remain as a scarce associate though, in the end, it does not persist as well here as on the substantially deeper and moister soils that come to support the *Fagus-Rubus* woodland. But it is more shade-tolerant than *Fraxinus* and can be proportionately more abundant in the early stages of *Fagus* invasion, to be eventually overtaken again by *Fraxinus* because of the more prolific seeding-in of the latter (Watt 1924): it can, however, regenerate in larger gaps. Another very successful gap-coloniser in areas of higher rainfall like the western end of the South Downs is *Acer pseudoplatanus* and this likewise can be found as an occasional in developing and mature canopies (Watt 1924, 1934*a*). *Ulmus glabra*, which often accompanies *A. pseudoplatanus* in calcicolous woodlands in the wetter north and west, is noticeably rare in this community.

Figure 22. Canopy and understorey in *Fagus-Mercurialis* woodland at White Hill, Surrey. Section shows a canopy of tall beech with occasional horse-chestnut (A) and an understorey of yew (T), box (B) and ash saplings (F) in a regeneration core beneath a gap.

Two other characteristic trees here are *Sorbus aria* and *Taxus baccata*. Both can be relics of the early stages of invasion and, on shallower soils where *Fagus* does not grow so tall, they can survive as occasional components of the main canopy tier. This is the usual role here of *S. aria*, but the more slow-growing *T. baccata* is often overtaken to remain as a distinct lower contributor to the woody cover and, being very shade-tolerant, it can persist and regenerate to attain some local abundance (Figure 22). Then it is often possible to distinguish the older early colonisers branching from the ground and the later more diffusely branched invaders (Watt 1924). Both these trees can occur occasionally in each of the sub-communities, though increased frequencies of both, especially of *T. baccata*, help define the *Taxus* sub-community.

Other trees are generally very sparse. Birch, almost always *Betula pendula*, occurs occasionally throughout, though the dense shade here limits it to gaps. A somewhat better and rather surprising low-frequency survivor is the seventeenth-century Balkan introduction *Aesculus hippocastanum*: though widely planted in Britain, this tree seems to have seeded into this community naturally in some places and it performs quite well. *Prunus avium* can also be found, though it is more characteristic of the *Fagus-Rubus* woodland. *Carpinus betulus* is very infrequent and, though *Fagus* and *Tilia cordata* occur together in certain kinds of *Fraxinus-Acer-Mercurialis* woodland (the *Teucrium* sub-community in the Wye valley, for example), this association is not characteristic here. *Tilia platyphyllos* and *T. vulgaris*, however, both occur in close proximity to

stands of the *Taxus* sub-community at Box Hill in Surrey.

The main feature of the shrub layer of the *Fagus-Mercurialis* woodland is that it is very poorly developed, especially in mature stands. A thicker understorey may persist around the margins of stands or within enclaves where invasion of the preceding woodland is not so far advanced; and there may be a resurgence of shrubs within gaps. Generally speaking, however, shrub cover becomes increasingly patchy as *Fagus* asserts its dominance and the shade increases and the feeble growth of most of the survivors is very marked.

Overall, the most common elements of this lowest tier of woody vegetation are *Corylus avellana*, *Crataegus monogyna*, *Acer campestre*, *Sambucus nigra* and *Ilex aquifolium* with *Euonymus europaeus*, *Cornus sanguinea*, *Ligustrum vulgare*, *Viburnum lantana* and *V. opulus* rather less frequent; in other words, very much the same suite as is characteristic of the south-eastern kinds of *Fraxinus-Acer-Mercurialis* woodland. Here, however, none of these species survives to constancy through the community as a whole; indeed, none persists to constancy in any of the sub-communities, though some are preferentially frequent in different kinds of *Fagus-Mercurialis* woodland and can have locally high cover there. *Cornus*, *Ligustrum* and *V. lantana* tend to be more characteristic of the *Sanicula* sub-community and, though the last two are quickly extinguished as *Fagus* invades, they can show a marked resurgence in gaps over shallow, stony soils. *Juniperus communis*, which is a locally prominent species with these shrubs in the scrub which precedes this kind of *Fagus-Mercurialis* woodland, only very rarely persists long enough to qualify as a living member of this woodland, but its woody skeletons quite commonly testify to its previous abundance (Watt 1934a). *Crataegus* and *Corylus*, by contrast, are a little more frequent in the *Mercurialis* sub-community and the former especially can remain as drawn-up individuals in more mature woodlands. Overall, however, the best survivor among all these species seems to be the evergreen *Ilex* and this small tree can persist in some abundance, taller individuals sometimes growing up to contribute to a lower canopy tier with *Taxus*. Where *Taxus* itself is very prominent, as in the *Taxus* sub-community, the shade is doubly dense and very few of the smaller woody associates of the community can survive, though *Buxus sempervirens* appears as a distinctive companion at some sites.

Tree saplings, like the shrubs and smaller trees, are also noticeably patchy. Generally, the most frequently encountered are those of *Fraxinus* and *A. pseudoplatanus* but these are very much more common on the deeper soils of the *Mercurialis* sub-community and, even there, are largely confined to gaps. Here, though, they can be extremely abundant, forming distinctly domed regene-

ration cores (Watt 1924). Young *Fagus* are most frequent in these situations, too, but their abundance is much more sporadic through time because of the pronounced mast-year pattern of fruiting (Watt 1923, 1925). *Taxus* and *Sorbus aria* saplings can also sometimes be found. *Quercus robur* generally regenerates poorly in this community.

By and large, climbers and lianes are scarce, apart from the evergreen *Hedera helix* which, as well as forming a sometimes abundant ground carpet here, is sometimes found up the trunks and among the branches of the trees. In more open stands of the *Sanicula* sub-community, *Clematis vitalba*, *Tamus communis* and *Bryonia dioica* sometimes occur.

In qualitative terms, the field layer of the *Fagus-Mercurialis* woodland shows a general similarity to that of the *Fraxinus-Acer-Mercurialis* woodland with plants characteristic of more base-rich soils providing the core of the herbaceous component of the vegetation. Thus, among the most frequent species here are *Mercurialis perennis*, *Sanicula europaea*, *Geum urbanum*, *Circaea lutetiana*, *Arum maculatum*, *Brachypodium sylvaticum*, *Geranium robertianum* and *Viola riviniana/reichenbachiana* (incompletely separated in the available data but probably mostly *V. reichenbachiana*, e.g. Watt 1934a). However, by contrast with the geographically close south-eastern kinds of *Fraxinus-Acer-Mercurialis* woodland, plants typical of moister calcareous soils are relatively uncommon in this community. *Primula vulgaris*, *Poa trivialis*, *Ajuga reptans*, *Lamiastrum galeobdolon*, *Carex sylvatica*, *Deschampsia cespitosa*, *Anemone nemorosa* and *Ranunculus ficaria* can all attain a measure of local prominence in particular stands but overall they are no more than occasionals here. *Hyacinthoides nonscripta* is also much less common in this community and even *Mercurialis perennis*, though it is the most frequent species throughout and the only herbaceous constant, is a common vernal dominant only on the deeper soils of the *Mercurialis* sub-community. This means that the field layer of this woodland, especially in the *Sanicula* sub-community, shows a much closer general kinship to that of the north-western kinds of *Fraxinus-Acer-Mercurialis* woodland typical of more free-draining substrates.

Two other frequent species can attain some measure of prominence in the field layer. *Hedera helix* can form a ground carpet of varying density and *Rubus fruticosus* agg. (often *R. vestitus* or members of the Triviales section) is sometimes sufficiently thick as to constitute a local underscrub, though it is not so consistent here as in the *Fagus-Rubus* woodland and, even in older stands, is hardly ever accompanied by *Oxalis acetosella*, so characteristic a herald of *Rubus*-dominance in well-established woodlands of that community (Watt 1934b). Less common than these two but sometimes attaining patchy abundance are ground-cover herbs such as *Fragaria vesca* and *Veronica chamaedrys* and, more distinctive here, *Galium odoratum*, *Mycelis muralis* and *Melica uniflora*, the last two clearly preferential to the *Sanicula* sub-community. *Urtica dioica* and *Galium aparine* can be found occasionally, too, though they are not so frequent as in north-western *Fraxinus-Acer-Mercurialis* woodlands, and ferns, likewise, are rarely a prominent feature here: *Dryopteris filix-mas* occurs in some stands and, less commonly, *D. dilatata*. Other occasional species include *Rumex sanguineus*, *Campanula trachelium*, *Teucrium scorodonia*, *Milium effusum*, *Bromus ramosus*, *Euphorbia amygdaloides* and, especially associated with gaps, *Verbascum thapsus*, *Atropa belladonna* and *Epilobium angustifolium*. More locally, there are records for *Daphne laureola*, *Aquilegia vulgaris*, *Polygonatum multiflorum*, *Helleborus viridis*, *Iris foetidissima* and *Hordelymus europaeus*. There are quite commonly some seedlings of *Fraxinus*, *A. pseudoplatanus*, *Taxus* and, after good mast years, of *Fagus*.

Rich and fairly luxuriant field layers are quite widespread in this community but they are by no means as consistently common as in the *Fraxinus-Acer-Mercurialis* woodland. As with the understorey, much of this effect can be due to the increasingly dense shade that the developing *Fagus* canopy casts: in stands of thickly-set pioneers and in more mature woodland that has not yet opened up (especially where *Fagus* and *Taxus* are both abundant, as in the *Taxus* sub-community), the herb cover may be negligible or nil. Where the field layer is very sparse, very large samples are necessary to gain an adequate impression of its constitution and, where there are next to no herbs at all, only an examination of gaps will give a sure diagnosis of the community. When a reasonably regular turnover of canopy trees has been established, a much more balanced and complete field layer can become prominent. In general, the herbs which withstand the heavy shade best are those which put up their leaves before the canopy foliage emerges, like *Mercurialis* and *Sanicula*, or which are evergreen like *Hedera*; *Rubus* may also carry out much of its vegetative growth during the winter. These four are the commonest field-layer species of the community and, since they all retain their leaves throughout the growing season, they can themselves exert some controlling influence over the prominence of their herbaceous associates where they are themselves dense; and they may also influence the success of tree seedlings in getting away (e.g. Watt 1923, 1925, Wardle 1959).

But other kinds of quantitative variation can also be encountered. Stands on shallower and drier soils, for example, tend to have more open and less luxuriant field layers, and here a general shortage of moisture is probably exacerbated by the very thorough ramification of the soils by the tree roots which are all concentrated in

sometimes but a few centimetres of mantle above the bedrock (e.g. Adamson 1921, Watt 1934*a*, Tansley 1939). In stands on steeper slopes, there can also be exposures of bare bedrock or unstable tumbles of talus which help to keep the field layer open. Such conditions are especially associated here with the *Sanicula* and *Taxus* sub-communities. Even in well-illuminated stands of the *Mercurialis* sub-community, however, where deeper and somewhat moister soils can theoretically support a better cover of herbs, root competition may be important and, on the exposed margins of woodlands, wind too may play a part in keeping herb cover down. Where there is no protective marginal fringe or little understorey to break their force, winds can blow about the substantial quantities of *Fagus* litter, blanketing smaller species and exposing bare soil. Then the most prominent feature of the woodland floor may be the snaking pattern of tree roots and the intervening stretches of decaying litter.

All these factors also have an effect on the bryophytes of the community which are few in number, infrequent and often sparse. The most widespread species on exposed soil are *Brachythecium rutabulum* and *Eurhynchium praelongum* with, less commonly, *Fissidens taxifolius*, *Plagiomnium undulatum*, *Mnium hornum*, *Thamnobryum alopecurum* and *Ctenidium molluscum*. A clothing of *Hypnum mammilatum* is often a conspicuous feature of the bases of the tree boles and the exposed surfaces of roots and their immediate surrounds.

Sub-communities

Mercurialis perennis **sub-community:** Beech associes, seres 3 & 4 Watt 1924; Mercury beechwoods Watt & Tansley 1930; Beech consociation, Hawthorn sere Watt 1934*a*; Beech consociation, seres 3 & 4 Watt 1934*b*; *Fagetum sylvaticae calcicolum*, Hawthorn-Mercury sere Tansley 1939; *Fagetum rubosum* Tansley 1939 *p.p.*; Beech-Ash-Yew Association type 6 McNeill 1961; Beech-Ash-Oak Association type 7 McNeill 1961; Cotswold beechwoods, mercury type Barkham & Norris 1967; Cotswold beechwoods, ivy type Barkham & Norris 1967 *p.p.* *Fagus* is very much the dominant tree here and, on the somewhat deeper and moister soils characteristic of this sub-community, it makes manifestly better growth than in the other kinds of *Fagus-Mercurialis* woodland, forming the bulk of a canopy that is usually 15–25 m tall (mean 19 m). But other trees can be quite common, either as survivors of a preceding woodland cover that has been invaded by *Fagus*, or within gaps. *Fraxinus* is the most frequent of the associates with *A. pseudoplatanus* a little less so and more restricted to areas with higher rainfall; occasional *Q. robur* can be found too. Each or all of these trees can attain a measure of local abundance as canopy subdominants and transitions to *Fraxinus-Acer-Mercurialis* woodland are quite widespread. Less common woody companions are *Taxus* (often forming a distinct lower tier), *Sorbus aria*, *Betula pendula* (with *B. pubescens* also making a very occasional appearance here), *Aesculus*, *Prunus avium* and *Carpinus*. Conifers such as larches and pines occasionally seed into gaps from nearby plantations.

The understorey, though it is rarely dense, often patchy and quite frequently very sparse, is generally better developed than in the other sub-communities. Both *Corylus* and *Crataegus* are more frequent here and there is occasionally some *Acer campestre*, *Sambucus* and *Ilex. Cornus sanguinea*, *Viburnum lantana*, *V. opulus*, *Euonymus europaeus* and *Ligustrum vulgare* occur at lower frequencies but are very much confined to margins and gaps. In gaps, too, saplings of *Fraxinus*, *A. pseudoplatanus* and, following good mast years, *Fagus*, can be very prominent.

Provided that shade, root competition and exposure to winds blowing between the trees are not too severe, the somewhat deeper and moister soils typical of this sub-community favour the development of the kind of extensive and quite tall field layer characteristic of many *Fraxinus-Acer-Mercurialis* woodlands. *Mercurialis* is very common here, can make good growth and is the usual vernal dominant but, as always where it attains this kind of prominence, it exerts a strong control over the cover and variety of its associates because of the dense shade that its own leaves cast from early spring right through to late summer (Watt 1934*a*, Wardle 1959, Wilson 1968, Martin 1968). Overall, then, its most frequent companions are hemicryptophytes with taller leafy stems, most notably here *Circaea lutetiana*, *Brachypodium sylvaticum* and *Galium odoratum*, which protrude here and there through the *Mercurialis* carpet, and the shade-bearing evergreen *Hedera* which very typically forms a ground cover beneath it. Indeed, *Hedera* may replace *Mercurialis* as the field-layer dominant in the areas of deepest shade (Barkham & Norris 1967), though available samples of this kind of vegetation are not sufficiently distinct to warrant the erection of a *Hedera* sub-community. *Rubus* is also much commoner here than on the drier, shallower soils of the *Sanicula* and *Taxus* sub-communities: usually it occurs as rather sparse spindly shoots sprawling over the other herbs but it can thicken up locally to form a patchy underscrub which can itself limit the dominance of *Mercurialis*, though not usually the extent of *Hedera*. In contrast to these species, *Sanicula*, a rosette hemicryptophyte, is notably infrequent here.

Other herbs which occur a little more frequently than in the other two sub-communities are *Arum maculatum*, *Dryopteris filix-mas* and, sometimes with local abundance, *Lamiastrum galeobdolon* and *Hyacinthoides non-*

scripta, all of them perhaps indicative of the somewhat moister soils here. *Allium ursinum*, *Anemone nemorosa*, *Ranunculus ficaria* and *Deschampsia cespitosa* can also be found with patchy prominence, though they are generally infrequent and the striking patterns of zoned dominance involving these species so typical of the *Fraxinus-Acer-Mercurialis* woodland are not a regular feature here. There are certainly insufficient samples to erect parallel sub-communities in which these plants are preferential, though Rackham (1980) distinguishes an *Allium* field layer in the beechwoods of the Cotswolds.

Community occasionals found in this kind of woodland include *Viola* spp., *Geranium robertianum*, *Urtica dioica*, *Galium aparine*, *Fragaria vesca* and *Primula vulgaris* but seedlings of the woody species are usually infrequent. Under the denser covers of *Mercurialis*, bryophytes too are sparse but *Brachythecium rutabulum* and *Eurhynchium praelongum* can sometimes be found and *Fissidens taxifolius* is quite common on litter-free patches of soil.

***Sanicula europaea* sub-community:** Sanicle beechwoods Watt & Tansley 1930; Beech consociation, Juniper sere Watt 1934*a*; *Fagetum sylvaticae calcicolum*, Juniper-Sanicle sere Tansley 1939; Beech-Ash-Yew Association type 5b McNeill 1961; Cotswold beechwoods, hawkweed type Barkham & Norris 1967. The tree canopy is more overwhelmingly dominated by *Fagus* here with only very occasional records for *Fraxinus*, *A. pseudoplatanus* or *Q. robur*. However, the individual trees are usually obviously less well grown than in the *Mercurialis* sub-community, forming a less tall canopy that is generally only 10–15 m high (mean 14 m). Occasional *Taxus*, *S. aria* or large *Ilex* can be represented in the tree layer.

The shrub cover, too, is typically less extensive than in the *Mercurialis* sub-community with *Corylus* and *Crataegus* noticeably less common, though the latter persists longer in younger sub-spontaneous stands than does its local co-dominant in the preceding scrub, *Juniperus*. *Sambucus* and *Ilex* occur occasionally and, where invading *Fagus* makes slow progress, *Cornus sanguinea* and *Viburnum lantana* may survive for some time. In older stands, though, these more light-demanding species are confined to margins or gaps where, with *Ligustrum vulgare*, *Clematis vitalba* and *Tamus communis*, they can form a dense tangle of woody growth. Saplings of all the major trees of the community are infrequent.

Although quite extensive and luxuriant field layers can be found in woodlands of this type, the herb cover is often a little sparser than in the *Mercurialis* sub-community and, even in areas of good illumination, intense root competition in the shallow soils and instability of slope helps keep the cover open. *Mercurialis* in particular, though still common here, is generally less abundant and vigorous (e.g. Watt 1934a) and, though it can attain a patchy local prominence, it is not usually the vernal dominant. Here, this role is assumed by *Sanicula europaea* which, though often very abundant, is a rosette hemicryptophyte casting much less shade than *Mercurialis* and thus not exerting such a controlling influence on the distribution and variety of the associated flora. The herbaceous component here is thus a little richer and somewhat more diverse.

Among the associates are some strong preferentials, most notably *Mycelis muralis*, here attaining a tall, delicate, almost stately form far removed from its sturdy physiognomy on sunlit walls, and *Melica uniflora* and *Poa nemoralis* which, together with occasional *Brachypodium sylvaticum*, help give the field layer an open grassy appearance. Then, there are occasional scattered plants of such species as *Rumex sanguineus, Heracleum sphondylium, Anthriscus sylvestris, Arctium minus* agg., *Bromus sterilis, Cynoglossum officinale* (and in some sites, its rare relative *C. germanicum*) and, especially in more open sunny situations, Glandulosa Hieracia such as *H. exotericum* and *H. pellucidum*, all of which can lend a distinctly weedy feel to the vegetation.

The other striking component of the field layer comprises geophytes, more particularly a variety of orchids which, when flowering in early summer, can present a splendid sight. Most characteristic among these is *Cephalanthera damasonium*, a Continental Southern helleborine whose British distribution almost exactly matches the range of the *Fagus-Mercurialis* woodland but which is clearly preferential to this sub-community, quite common within it, locally very consistent and sometimes abundant. The more showy but rarer *C. longifolia* can also sometimes be found, particularly in Hampshire, but this species is not so confined, geographically and ecologically, to this kind of woodland. The now very rare *C. rubra* survives in this sub-community in the Chilterns and the Cotswolds and seems to have some preference for scrubbier stands, or at least to flower better under a more open cover (Tansley 1939, Summerhayes 1968). *Epipactis helleborine*, which is widely distributed in Britain, also occurs here, again in the more open conditions of rides and margins. Much more characteristic of a denser tree cover and sometimes appearing locally in great numbers is the saprophytic *Neottia nidus-avis*, which seems to favour the moist accumulations of rotting leaves over a permeable substratum that shadier stands of this vegetation provide. Our other saprophytic bird's nest, the non-orchidaceous *Monotropa hypopitys s.l.* (both sub-species: Perring 1968), is sometimes also found and similar situations have provided some of the few and very sporadic British records for *Epipogium aphyllum*. Other orchids recorded rarely here are *Epipactis leptochila, Orchis purpurea,*

Ophrys apifera, O. insectifera and *Listera ovata*.

In addition to these more distinctive elements, there is usually some *Hedera* in this vegetation, again often forming a ground carpet, very occasional *Rubus* and *Rosa canina* agg. and scattered plants of *Geum urbanum, Viola* spp., *Veronica chamaedrys, Geranium robertianum, Urtica dioica, Galium aparine, Primula vulgaris* and, preferential here at low frequency, *Carex flacca*, usually not flowering in the shade. Tree seedlings are a little commoner here in the more open herb cover and bryophytes, too, can be more extensive with some enrichment by species typical of stony base-rich soils such as *Ctenidium molluscum, Homalothecium sericeum* and *Encalypta streptocarpa*.

Taxus baccata sub-community: Beech consocation, Juniper sere   Watt 1934a *p.p.*; *Fagetum sylvaticae calcicolum*, Juniper-Sanicle sere   Tansley 1939 *p.p.*; Beech-Ash-Yew Association type 5a   McNeill 1961. As in the *Sanicula* sub-community, the growth of *Fagus* here is generally poor: indeed, on average, the canopy is even lower than there with a mean height of only 12 m. But *Taxus* is a constant and can be present in abundance in a lower tier beneath the *Fagus*. *Sorbus aria* is somewhat more frequent in this kind of woodland too, and there is occasionally some *Fraxinus*. When all these trees are in leaf, the scarps on which this sub-community typically occurs present an unmistakeable patchwork of colour with the predominant mid- and dark green of the *Fagus* and *Taxus* and the scattered pale green and dusty white crowns of the *Fraxinus* and *Sorbus*.

In the very deep shade which this canopy casts, shrubs and small trees are very sparse indeed. The typical understorey species of the community are all poorly represented and such individuals as do persist present a very spindly appearance. The usual picture is of infrequent and much drawn-up *Sambucus* and *Crataegus* with, in somewhat more open places, a little *Ligustrum* and scrambling *Clematis*. At some sites, however, *Buxus sempervirens* is a distinctive (and probably native: Pigott & Walters 1953) associate in this sub-community, tolerating the gloom to form a patchy understorey or growing up among the trees and adding further to the shade with its dense evergreen foliage.

Very few herbs are able to survive the combination of this intense shade and the usually very inhospitable soil conditions, with intense root competition from both *Fagus* and *Taxus*, and the most obvious feature of the woodland floor is often the great extent of bare, rubbly substrate with litter and loose rock tumbling down between the tree boles. *Sanicula* and *Mercurialis* occur only occasionally and are generally very sparse, thickening up only locally where there is a little more light or a somewhat more extensive and stable soil cover. *Rubus* is likewise very infrequent and even *Hedera* largely fades out here. A few widely-scattered plants of *Mycelis muralis*, *Melica uniflora*, *Arum maculatum*, *Geum urbanum*, *Circaea lutetiana*, *Viola* spp., very sparse tree seedlings and pale wisps of *Eurhynchium praelongum* and *Brachythecium rutabulum* complete the impoverished scene.

Habitat

The *Fagus-Mercurialis* woodland is the kind of beech forest which has developed where *Fagus* has attained dominance on free-draining, base-rich and calcareous soils in the south-eastern lowlands of Britain. It is essentially a community of limestone scarplands in this region and, at the present time, seems to represent a stable end-point of successions in such situations. However, although the *Fagus-Mercurialis* woodland has some claim to be a natural climax, individual stands of the community are not necessarily very old. Some are certainly relatively recent plantations and human agency may have been very widely important for the great pre-eminence of *Fagus* in this kind of woodland.

Despite early doubts to the contrary, we now know that *Fagus* is a certain native in Britain but, though the limits of its natural distribution are generally agreed, they are not very simply explained. It gained ascendancy as a forest dominant before the great era of tree-planting in the warmer and drier south-eastern corner of the country, within a line from The Wash, across to south Wales and down to Dorset (e.g. Tansley 1939, Noirfalise 1968, Godwin 1975, Rackham 1980): all the three kinds of *Fagus* woodland recognised in this scheme are centred in this region. It is possible that, in the more continental parts of East Anglia, seedling establishment and survival are limited by summer drought (e.g. Watt 1923, 1925, Rackham 1980), especially over more permeable substrates like the Chalk, a characteristic bedrock of the *Fagus-Mercurialis* woodland, the range of which stops short there in the Chilterns. Towards the north and west, late frosts, low summer temperatures and heavier rainfall may hinder *Fagus*-dominance by their effects on masting and regeneration (e.g. Watt 1923, 1925, 1931, Watt & Tansley 1930, Godwin 1975): in this direction, it is the Oolite scarp of the Cotswolds and Lincolnshire that forms the general bounds of this community and the north-western limit of widespread *Fagus*-dominance in our woodlands. But correlations between this apparently natural range and present climatic conditions are not entirely convincing. *Fagus* is widely planted further north, it can grow very well there and can actively regenerate (even in Scotland: see Watt 1931): plantation stands of the *Mercurialis* sub-community of the *Fagus-Mercurialis* woodland, for example, occur in South Yorkshire and South Humberside. It is possible, then, that *Fagus* never reached its northern limit in Britain and that, in the way the three

beech woodlands are concentrated in the south-east, we see the results of some hindering of its spread here.

The post-Glacial advance of *Fagus* in Britain, as elsewhere in north-west Europe, was certainly late: it seems to have had a quite widespread distribution, though no more than local prominence, until the Sub-Atlantic (e.g. Firbas 1949, van Zeist 1959, 1964, Godwin 1975, Huntley & Birks 1983). Climatic conditions during this period would hardly seem to have favoured an extensive natural migration and the gathering consensus is that it expanded from existing small, scattered populations in response to some anthropogenic change, such as more systematic clearance of the natural woodland cover and the associated abandonment of traditional farmlands. The assumption here is that it had been previously largely excluded from the mixed deciduous forest but was then able to compete supremely well against other tree species in the invasion of newly-opened ground. As far as the *Fagus-Mercurialis* woodland is concerned, it has been suggested that the community now occupies sites which perhaps first became widely available for *Fagus*-dominance with the shift in population from the southern limestones to the heavier claylands in Roman and Anglo-Saxon times (Godwin 1975). Certainly, the studies of Watt (1924, 1934a) have shown just how successful *Fagus* can be as an invader of abandoned agricultural land over the Chalk and documentary and archaeological evidence confirms the general south-eastern bias in the distribution of *Fagus* within historic time (Rackham 1980). But there is room for much further work here on the exact status of *Fagus-Mercurialis* woodland within this region and on its absence further north: for example, why, when *Fagus* seems to have been quite well represented in Derbyshire in the Sub-Atlantic, did it not take off there to become a major dominant in woodlands over limestone slopes which now experience a climate not far removed from that characteristic of the western end of the South Downs?

Whatever the climatic and historical reasons for the geographical distribution of the *Fagus-Mercurialis* woodland, its edaphic relationships within its range are quite clear. *Fagus* is a fairly catholic tree as far as soil conditions are concerned: it can become dominant in soils that range from the very base-rich and highly calcareous through to markedly surface-acid podzols; and, though it avoids profiles with strongly-impeded drainage, it can extend some way on to moister or more heavy-textured soils (e.g. Watt & Tansley 1930, Watt 1934b, Tansley 1939, Rackham 1980). Within this broad spectrum, the *Fagus-Mercurialis* woodland is associated with the more base-rich extreme: typically, it is found over rendzinas which are shallow, free-draining, rich in free calcium carbonate and with a surface pH generally within the range of 7–8 (e.g. Adamson 1921, Watt 1924,

1934*a*; Avery 1958, 1964). In the south-east, such profiles are generally limited to the steeper drift-free faces of
the escarpments of the Chalk and the Oolite and this is
very much the type site of the community, along the
North and South Downs, westwards into Hampshire,
where Gilbert White long ago extolled the beauties of
the 'beech hangers' (White 1788), on into parts of
Wiltshire and up the cuestas of the Chilterns and the
Cotswolds. These soils can also be found, more patchily,
over the steeper slopes of the gaps cut through the Chalk
and in some dry valleys.

In this region, these profiles are now mapped as
Upton grey rendzinas (Soil Survey 1983), though they
were long known as the Icknield series for which Avery
(1958, 1964) provided classic descriptions from the
Chilterns (see also Curtiss *et al.* 1976, Smith 1980).
Typically, there is a shallow, quite well structured A
horizon, within which the tree roots are concentrated
and where, provided drainage is not too excessive, a
numerous and varied soil fauna, including earthworms,
incorporates the beech litter to produce a mull humus
well-bound with the clays into crumby aggregates.
Where the soils are shallower, clay-deficient and more
sharply draining, structural integration can be much
poorer and, as we shall see, this is of major importance in
influencing floristic variation within the community.
But, throughout, the A horizon is never very deep
(usually less than 20 cm), is rich in calcium carbonate (up
to 60% or more) and quickly passes through a rubbly,
humus-stained A/C to the bedrock proper at 40–60 cm
or so below the surface.

These general soil conditions are reflected in the
community as a whole by the predominance of more
calcicolous associates among the woody and herbaceous
components: *Fraxinus, Taxus, Acer campestre, Cornus
sanguinea, Rhamnus catharticus, Euonymus europaeus,
Mercurialis, Circaea lutetiana, Geum urbanum, Viola
reichenbachiana* and *Arum maculatum.* Also, they help
define the *Fagus-Mercurialis* woodland against its
counterpart on deeper, moister and less base-rich brown
earths, the *Fagus-Rubus* woodland where these species
fade and *Rubus* and *Pteridium* exert a major influence on
the structure and composition of the field layer. *Rubus* is
quite common in the *Fagus-Mercurialis* woodland but
only occasionally and locally abundant and often chlorotic on the shallower soils; *Pteridium* is hardly ever
present. It is not until the strong influence of the
underlying permeable limestone is masked by a cover of
superficials that these species become really abundant
and sometimes a very sharp switch between these communities can be seen in beech-dominated woodlands
that run up a scarp and on to the dip slope (see, for
example, the striking Figure 8.3 in Smith 1980). In other
situations, the distinction is not so clear and, as defined
here, the *Fagus-Mercurialis* woodland takes in a little of

what early workers would have included within *Fagetum
rubosum* (like the sere 3 woodlands of Watt (1924) where
Fraxinus and *Mercurialis* were frequently recorded).
Such more marginal stands can be found where the
Fagus-Mercurialis woodland extends on to brown rendzinas (like the Andover series) or brown calcareous
earths (like the Coombe series) on the somewhat gentler
slopes on scarp crests and the upper parts of Chalk dry
valleys (Avery 1958, 1964, Hodgson 1967, Jarvis 1973).
Here the soils may be partly derived from a thin downwash of superficials like Clay-with-Flints and loess, can
have a fragmentary (B) horizon and show some superficial decalcification with a surface pH as low as 6 (Soil
Survey 1983). However, on more base-rich brown soils
where the drainage becomes impeded, as where heavy-
textured superficials have contributed to the parent
materials, the *Fagus-Mercurialis* woodland is replaced
by the *Fraxinus-Acer-Mercurialis* woodland. Such soils
are rarely contiguous with the types characteristic here,
but they replace them in the clay vales that alternate with
the limestone scarps and dips in the south-east. The
edaphically-related switch between these two kinds of
calcicolous woodland is a very distinctive feature of this
region and of the geologically similar parts of northern
France and Belgium (e.g. Noirfalise 1968).

Within the fairly narrow range of soils characterised
by the *Fagus-Mercurialis* woodland, the importance of
edaphic differences for the structure and composition of
the community was hinted at by Adamson (1921) on the
South Downs but first given unequivocal expression by
Watt (1934*a*) in his study of the Chiltern beechwoods. It
has since been confirmed by Barkham & Norris (1967,
1970) in the Cotswolds, related more thoroughly to
profile structure by Avery (1958, 1964) and to the
performance of *Fagus* as a commercial timber tree by
Brown (1953, 1964) and Wood & Nimmo (1962). All
these studies indicate that the major axis of floristic
variation within the *Fagus-Mercurialis* woodland can be
related to a complex of edaphic differences of which soil
moisture or soil depth can serve as a convenient summary. Thus, the *Mercurialis* sub-community is characteristically found on deeper grey rendzinas (and some
brown rendzinas and brown calcareous earths) which
are more moisture-retentive, the *Sanicula* sub-community on shallower, more fragmentary grey rendzinas
where drainage tends towards excessive. In the latter,
though organic matter can be quite high, there is usually
not a well-developed and stable aggregation into a good
mull. Earthworms are noticeably sparse, insect larvae,
millipedes and ants comprise the bulk of soil fauna and
the A horizon is often of the type which Kubiena (1953)
termed a mull-like rendzina moder with recognisable
litter fragments and invertebrate droppings (Avery
1958, 1964). The *Taxus* sub-community, though it has
not traditionally been distinguished from the 'mercury'

and 'sanicle' types, can be regarded in part as a continuation of this edaphic trend: typically it occurs on the most fragmentary of soil covers where much of the ground consists of talus and exposed bedrock.

Within individual stands of the community or in stands which occur in close proximity in the same region, this kind of edaphic contrast is often expressed in terms of slope: commonly, the *Mercurialis* sub-community occupies the more gently sloping ground, the *Sanicula* and *Taxus* sub-communities the steeper. Or, it may be related to the degree of exposure to drying winds (not always a simple function of aspect on markedly embayed scarps like the Chilterns) or to sun: the *Sanicula* sub-community tends to be associated with more exposed sites and the majority of the stands of the *Taxus* sub-community face south. Or, again, it may reflect differing degrees of susceptibility to weathering in the underlying bedrock: the *Mercurialis* sub-community can thus pick out softer strata, the *Sanicula* sub-community the harder, like Chalk Rock. Often, more than one of these kinds of physiographic variation can be invoked to understand sub-community distribution as in Watt's (1934*a*) Chiltern stands or in the pair of woodlands which Barkham & Norris (1970) contrasted. It should be remembered, too, that, in moving from one district to another, differences in regional climate can result in a shift in the absolute definition of sites occupied by the different sub-communities. As Watt (1934*a*) noted, on the moister South Downs the *Mercurialis* sub-community described by Adamson (1921) occupied some of the ground that, in the drier Chilterns, would have carried the *Sanicula* sub-community. This is why, when samples are pooled from throughout the range of the *Fagus-Mercurialis* woodland, sharp correlations between the different sub-communities and variables like slope and aspect do not emerge.

Although the general effect of edaphic variation is clear, its influence on the vegetation is quite complex and partly indirect. First, there is the control it exerts on the growth of the trees, especially *Fagus*. It was a major part of Watt's (1934*a*) seral scheme for this community that this effect began very early by its influence on the kind of scrub and woodland that could precede *Fagus* and affect its pattern of invasion, but it is also clearly visible in mature stands of the community whether these are of sub-spontaneous or planted origin. Among more natural populations of beech, there is often quite a wide variation in tree physiognomy which perhaps suggests some genetic differences (Watt & Tansley 1930) but, even here, it is usually possible to see that *Fagus* makes distinctly better growth on deeper, moister soils. In the available data, there is a small but obvious increase in the mean height and height range of the trees in moving from the *Taxus*, through the *Sanicula* and to the *Mercurialis* sub-community. Girth was not measured but early

surveys (Adamson 1921, Watt 1924, 1934*a*) showed that this, too, is greater in the *Mercurialis* than in the *Sanicula* type and commercial foresters have observed that there is often a whole quality class difference between the beech timber from the two sub-communities (Brown 1953, McNeill 1961; see also Wood & Nimmo 1962). The exact reasons for this are uncertain: mycorrhizal growth does not seem to be poorer on the shallower soils (Harley 1937, 1949) and variation in water content itself and in the amounts of iron, nitrogen and phosphorus have all been suggested as of importance (Day 1946, Fourt in Smith 1980). Again, it needs to be remembered that such differences may be visible in comparative rather than absolute terms: on the South Downs, for example, *Fagus* generally makes better growth overall than in the drier Chilterns (Watt & Tansley 1930, Brown 1964).

Some other trees of the community, such as *Fraxinus*, *A. pseudoplatanus* and *Q. robur*, also tend to grow better in the *Mercurialis* sub-community. They are more frequent there since they can survive from the mixed woodland that can precede this sub-community on deeper, moister soils and the first two are common gap-colonisers. But, since *Fagus* also does so well, its shading effect on these woody associates is that much greater and they maintain or gain their position in the canopy only with some difficulty. On the other hand, although these species are much scarcer in the *Sanicula* and *Taxus* sub-communities, *Fraxinus* especially can sometimes keep pace with *Fagus* there. In the *Taxus* sub-community, too, *Sorbus aria* is not so readily overtopped and the slow-growing *Taxus* can sometimes break the lower *Fagus* canopy.

Another effect of better *Fagus* growth is probably felt through root competition for available water and nutrients, the former likely to be in generally short supply here where the A horizon is so shallow, the substrates so permeable and the ground water-table so low. Apart from the early observations of Watt & Fraser (1933), we know next to nothing about how this factor operates, though a visit to a stand of *Fagus-Mercurialis* woodland where trees have been wind-thrown will readily demonstrate the great extent of the horizontal beech root system through the thin soil mantle. It might be expected that, since *Fagus* makes better growth on soils which are somewhat deeper and more retentive, the greater extent of its root system in such situations would help even out natural variations in moisture content. As Barkham & Norris (1967) observed with respect to *Fagus* density, a *Fagus-Mercurialis* woodland on sheltered, gentle slopes with closely-packed trees can thus have a similar field layer to one on exposed, steep slopes where the *Fagus* density is much lower. Their study of Cotswold beechwoods certainly showed a strong negative correlation between *Fagus* density and soil moisture, though, in managed woodlands, tree density may not be a reliable indicator of the natural capacity of the soils: thinning may be delayed over more unfavourable soils because the trees make slower growth there. Clearly, there is room here for much research.

Whatever the interactions between the beech cover and the natural water content of the soils here, the amount of available soil moisture certainly has a strong influence on the field layer, an effect seen most clearly in the simple contrast between the *Sanicula* and *Mercurialis* sub-communities (the *Taxus* sub-community being a little more complex: see below). On drier, shallower soils, there is probably an edaphic limitation on the representation of the more mesophytic species of the *Mercurialis* sub-community within the *Sanicula* sub-community, most importantly a reduction in the vigour and abundance of *Mercurialis* itself. On deeper, moister soils where this species can produce an extensive canopy of foliage, there is probably a competitive exclusion of species typical of the *Sanicula* sub-community within the *Mercurialis* sub-community (Watt 1934a). To the dense and long-lasting shade of *Mercurialis*, Watt (1934a) attributed the general scarcity of rosette hemicryptophytes, notably *Sanicula*, in the 'mercury' field layer, the virtual exclusion there of *Mycelis muralis* and most of the grasses, which perhaps require high light intensities for seed germination, and a general impoverishment of the bryophyte flora, floristic distinctions which were, in general, confirmed here in the separation of the two sub-communities. The predominant association of orchids with the 'sanicle' field layer, again confirmed here, Watt (1934a) found more puzzling: we still do not have an explanation for this.

In one way or another, then, soil differences can account for a considerable amount of the floristic and physiognomic variation within the *Fagus-Mercurialis* woodland. The other important factor is canopy shade, already touched upon in noting how the trees of the community compete for a place in the topmost structural tier but very obvious, too, in its influence on the understorey and field and ground layers. *Fagus*, of course, can cast a very deep shade: it is one of the few British deciduous trees that can eliminate everything beneath it, a feature which causes problems in classifying the darkest stands and which means that the field layers here are seen in their characteristic form only where there has been some lessening of competition among the components of the canopy and a measure of opening up (Watt 1924, 1934a). Even then, the cover and composition can be decidedly patchy, varying from an extensive and varied carpet of herbs beneath the places where adjacent crowns join, through a sparser clothing in which a mantle of *Hedera* is often the most conspicuous feature, to very bare areas around the bases of the trees, where occasional bryophytes and the smears of

algae down the trunks and over the roots provide the only splash of green. In general, however, this kind of variation does not work in line with the floristic trends noted above in the *Sanicula* and *Mercurialis* sub-communities: rather, it cuts across it to produce an additional range of variation (see, for example, the ordinations in Barkham & Norris 1967, 1970). The *Taxus* sub-community is perhaps an exception to this. Although it can be seen as a continuation of the edaphic trend visible in the other kinds of *Fagus-Mercurialis* woodland, the combination of *Fagus, Taxus* and occasional *Buxus*, makes for a more consistently dense shade here than elsewhere. For the most part, though, shade differences are best considered as producing variants.

The appearance of gaps creates a further range of variation in response to the local, sudden and temporary increase in light. In unmanaged stands, this happens naturally as individual trees age and succumb to disease. In practice, since so many beechwoods are under commercial forestry, such degeneration of the canopy is often pre-empted, though the selection system was essentially a way of creating artificial gaps in which natural *Fagus* regeneration was supposed to occur. But, even in managed stands, *Fagus* can fall early because of beech-snap following infestation with beech-bark disease (Brown 1953, Parker 1974) or be wind-thrown where exposed trees are rooted in thin soils. Conditions in gaps are very variable, much probably depending on their size, but a fairly natural pattern of recolonisation can often be observed, with a flush of growth among the herbaceous survivors and stunted shrubs and saplings of the preceding shade flora and a rapid appearance of young trees from seed. In more disturbed areas, adventive herbs can mask for a time the more usual kind of field layer with plants such as *Verbascum thapsus*, *Epilobium angustifolium*, *Arctium minus* agg., *Atropa belladonna* and *Scrophularia nodosa* becoming very prominent; on somewhat moister soils, *Deschampsia cespitosa*, generally speaking an uncommon grass here, may spread and, in sunnier spots over more stony soils, *Hieracium* spp. and *Inula conyza* can appear in abundance. Again, such vegetation has not been recognised here as constituting a distinct sub-community but is regarded as small-scale variation within the overall framework.

The commonest woody invaders in this kind of situation here are *Fraxinus* and, more locally in areas of higher rainfall, *Acer pseudoplatanus*, and, in smaller gaps, dense thickets of their young saplings commonly form the distinctive dome-shaped regeneration cores described so graphically by Watt (1924, 1925). Single survivors of the intense competition can eventually take their place in the canopy unless the neighbouring *Fagus* crowns expand to fill the space above, but they are often very drawn: perversely, on the deeper and moister soils where these species grow best, *Fagus* itself tends to have a taller canopy. *A. pseudoplatanus*, being more shade-tolerant than *Fraxinus*, may be a better survivor in this respect (e.g. Watt 1925). Theoretically, *Fagus* itself should be able to emerge from enough gaps sufficiently frequently to maintain its own position in the canopy, but as Watt (1923, 1925) demonstrated, and as many foresters have found to their dismay, its natural regeneration is very unreliable. In essence, only where nearby trees produce sufficient fruit is there any great likelihood of *Fagus* getting away in gaps and, even then, intense shade and invertebrate and vertebrate pests can severely hinder its chances of reaching maturity. Grey squirrel (*Sciurus carolinensis*) damage at the pole stage is now a particularly widespread cause of failure. Where the appearance of small gaps within *Fagus* stands coincides with a mast year, conditions can therefore approach the optimal but, where the seed source is more remote, as in the centre of large gaps, and when fruiting is poor, canopies dominated by *Fraxinus* and *A. pseudoplatanus* may persist for some considerable time, perhaps decades (Rackham 1980, Smith 1980). Such stands are best classified as *Fraxinus-Acer-Mercurialis* woodland which, on free-draining but moister, base-rich soils within the natural range of *Fagus*, is regarded in this scheme as a seral precursor to *Fagus-Mercurialis* woodland.

Quite apart from the fact that this community may have originally attained prominence only as a biotically-assisted climax over the south-eastern limestones, many stands bear obvious signs of more recent treatment. In a very few places, there is evidence of the apparently extensive local use of *Fagus* as an underwood crop, as for example in the Chilterns from medieval times until the eighteenth century, where it was grown with oak or ash standards and cut mainly for fuel (Mansfield 1952, Roden 1968). For the most part, though, over the past 300 years, *Fagus* has been valued as a timber tree, extracted from and encouraged in existing stands of all three kinds of *Fagus* woodland and extensively planted. The present disposition of tracts of the *Fagus-Mercurialis* woodland and the canopy structure of individual stands can thus reflect a complex pattern of the waxing and waning interest in ornamental and commerical forestry in recent centuries. In the Chilterns, for example, evidence can be seen of the abandonment of the older coppices, the natural invasion of neglected land after enclosure, early ornamental and shelterbelt planting, the treatment of existing stands and plantations on a selection system to supply local industry, the shift to shelterwood or clear-felling regimes with the rise of modern forestry and the depradations of the wartime fellings (e.g. Brown 1964).

In this region and elsewhere, since *Fagus-Mercurialis* woodlands typically occupy the relatively narrow scarp-land portion of the landscape, they have contributed much less to the beech economy than the more extensive *Fagus-Rubus* woodlands of the dip slopes. Also, growing on poorer soils, they yield lower-quality timber (commonly class III or worse: Brown 1953, 1964, McNeill 1961), so promise a lower return on investment. But, in former days, when foresters had to be more content with what there was and labour was cheaper, this community seems to have been widely exploited under the selection system on the Chilterns. One lasting heritage of this seems to be a common preponderance there of smaller, less well formed and poorly-fruiting trees as a result of the progressive removal of the larger and better timber (Brown 1953, 1964, Peterken 1981). In recent years, *Fagus* has been widely planted over former Chalk downs or on cleared land in Sussex, Hampshire, Wiltshire and Dorset, often with a short-lived conifer nurse crop, usually *Pinus sylvestris* or *P. nigra* var. *nigra* or var. *maritima*, and with a battery of techniques to encourage good establishment, such as land preparation, application of herbicides and fertilisers, and prevention of predation by rabbits and deer. Such stands have not been included in this survey and it remains to be seen whether they will, in time, progress to *Fagus-Mercurialis* woodland.

Zonation and succession

Zonation within individual stands of *Fagus-Mercurialis* woodland are generally related to edaphic differences, upon which are superimposed variations related to canopy shade, regeneration and treatment, a pattern very well illustrated in the Cotswold woodlands studied by Barkham & Norris (1967, 1970). In fact, though the sub-communities can be clearly defined in relation to soil differences, transitional stands are very common because of the fairly gentle character of many slopes over the southern limestones and the frequently contorted fronts of their scarps. Generally speaking, however, within individual woodlands which cover a fair range of topographic variation, the trend is for the *Mercurialis* sub-community to give way to the *Sanicula* sub-community over steeper, more exposed slopes with shallower, drier soils, with the *Taxus* sub-community making a more local appearance over the steepest, rockiest ground with high exposure to wind and sun. Very locally, as on Box Hill, in Surrey, the transition continues into *Taxus* woodland proper with a thick undercanopy of *Buxus*.

Many stands of the community now remain as remnant strips of woodland isolated on the more intractable ground of scarps by the conversion of the dip slope above and the vale beneath to intensive agriculture. But in some places, the *Fagus-Mercurialis* woodland may run on some way over the dip slope as the *Mercurialis* sub-community on brown rendzinas or brown calcareous earths derived from but thin covers of superficials: this pattern seems most characteristic of the South Downs where fairly base-rich soils are quite extensive on the dip. Often, though, where there is continuity of woodland cover, there is an edaphically-related transition to the *Fagus-Rubus* woodland on deeper, more base-poor and patchily-gleyed brown earths formed from substantial thicknesses of superficials like Clay-with-Flints. Such zonations can be gradual but, where they are sharp, they can be very striking, with an abrupt termination of *Mercurialis*-dominance and a switch to thick tangles of *Rubus*, and a replacement of the poor *Fagus* of the scarp by magnificent trees on the dip.

Intensive farming and forestry of limestone slopes in the south-east, with conversion of the vegetation to improved pasture, arable or plantations, has severely restricted gradual transitions from the *Fagus-Mercurialis* woodland to more natural kinds of herbaceous communities, so that stands are now abruptly defined by artificial boundaries or have, at most, a very narrow fringe of marginal scrub and rank grassland. But, in larger gaps within existing woodlands, and where there is more long-lasting neglect of the surrounding land, fragmentary zonations can be seen which give some clue as to the seral development of this kind of woodland: these were the sorts of situations from which Watt (1924, 1934*a*) accumulated the data for his hypothesis on succession to beech-domination over more base-rich soils. Although Watt's proposals were based on observation of spatial patterns rather than monitoring through time and undertaken when a combination of severe agricultural depression and good masting in *Fagus* (in 1922) tended to maximise colonisation, his theory still provides a convincing account of how the *Fagus-Mercurialis* woodland can develop and how the major floristic differences within it are prefigured in the scrub, and perhaps also in the grassland, that precede it.

Although Watt recognised a continuous range of variation between both the two major types of mature *Fagus-Mercurialis* woodland, the *Sanicula* and *Mercurialis* sub-communities, and their seral precursors, he defined two extreme lines of succession, the 'Juniper sere' and the 'Hawthorn sere' (termed the 'Juniper-Sanicle sere' and the 'Hawthorn-Mercury sere' by Tansley 1939). In both, a calcicolous scrub was the first major development in pasture where grazing was neglected. Qualitatively, little difference was to be perceived within this vegetation but there was some quantitative polarisation in the frequency and abundance of certain shrubs and trees according to soil conditions. On the more drought-prone, shallower soils developed on steeper slopes in exposed situations, *Juniperus communis* was especially frequent, often abundant and quite vigorous, with a slightly better representation, too, of *Taxus*, *Ilex* and lianes. On deeper, moister soils over gentler slopes

with some measure of shelter, *Crataegus* was usually the dominant and, provided rabbit-grazing did not give it an advantage over *Crataegus*, *Juniperus* was not so prominent, being susceptible to shading. There was thus, early on in the succession, the same exclusion, in the one direction, by edaphic limitation and, in the other, by competition, as could be seen in the mature woodlands in the relationship between *Mercurialis* and *Sanicula*.

Fagus was found to invade both types of scrub though, even with abundant mast, colonisation was slow because of the frequent remoteness of seed-parents along the upper scarp slopes and the relative scarcity of denser patches of scrub to afford protection against rabbit-grazing. However, although the greater resistance of *Juniperus* to rabbit attack might have been expected to allow *Fagus* to thrive better there, it in fact colonised more extensively on the moister soils favoured by *Crataegus*: among the *Juniperus* scrub, there was a characteristically local pattern of dense young *Fagus* clumps; among the *Crataegus*, a more diffuse invasion over more extended areas of ground. More importantly in the short term, the moister soils also favoured a prominence of *Fraxinus* which, though eventually outgrown by the *Fagus*, could come to dominate in a temporary 'ash consocies' interposed between scrub and the mature 'beech consociation', and survive more frequently in the woodland derived from the 'Hawthorn sere' than from the 'Juniper sere'. In the end, however, both lines of succession terminated with the inexorable dominance of *Fagus*, though the different kinds of 'mercury' and 'sanicle' field layer preserved the important edaphic contrast that had been of major importance throughout.

In fact, in the relatively dry climate of the Chilterns, Watt's (1934*a*) 'ash consocies' was relatively rare but, on the moister South Downs, his earlier work (Watt 1924) and the initial survey by Adamson (1921) showed that woodlands dominated by *Fraxinus* or mixtures of *Fraxinus* and *Quercus robur* (termed 'ash-oak associes'), were of extensive and perhaps quite long-lasting occurrence, on ground which they presumed would eventually carry *Fagus-Mercurialis* woodland (Watt 1934*b*). As described earlier, large gaps or clear-felled areas within stands of the community can also carry this kind of vegetation, classified in this scheme as *Fraxinus-Acer-Mercurialis* woodland, for long periods before *Fagus* is able to reassert its dominance. Watt (1924) did demonstrate that *Fagus* was able to invade mature stands of this community (an interesting observation in the light of the supposed confinement of the tree in prehistoric forests) but this may be a very haphazard process. There may be some quite subtle edaphic limitations to its expansion, especially on transitional soils which tend to favour either *Fraxinus* or *Q. robur* (e.g. Watt 1934*b*) and some general climatic restrictions on its spread but, as

Rackham (1980) succinctly puts it, *Fagus* is much more sensitive to weather than to climate and we may have to concede that its local, as well as its regional, prominence and thus the balance between *Fagus-Mercurialis* and *Fraxinus-Acer-Mercurialis* woodlands in any particular area, are partly the result of coincidences of good fruiting and human disturbance, the details of which are frequently irrecoverable.

On the most extreme kinds of rendzina, where *Fraxinus* is very unlikely to gain any prominence in the developing woodland, it seems as if the main rival to *Fagus* is *Taxus*. No evidence is available to reveal the successional origin of the *Taxus* sub-community here but, edaphically, it can be seen as an extreme form of the *Sanicula* sub-community and, in the light of Watt's (1926) study of *Taxus* woodlands proper, it seems possible that it develops where *Taxus* gains early prominence in *Juniperus*-dominated scrub and where invasion by *Fagus* is not entirely excluded but where its growth is locally very restricted on extremely shallow and impoverished soils on steep, sun-baked slopes.

Distribution

The *Fagus-Mercurialis* woodland has a wide distribution within the natural range of *Fagus* in Britain, being especially characteristic of the scarps of the North and South Downs, the Chilterns and the western end of the Cotswolds. The *Sanicula* sub-community has been encountered somewhat less frequently than the *Mercurialis* sub-community which has also been recorded in *Fagus* plantations in North Humberside and West Yorkshire. The *Taxus* sub-community is much more local but good examples are to be seen on the south-facing scarp of the North Downs and within some gaps in the Chilterns.

Affinities

In general terms, there is a very good correspondence between this community and its major sub-types and the *Fagetum calcicolum* described in the classic studies of Watt (1924, 1934a, b) and summarised in Tansley (1939) and Rackham (1980), although, as defined here, the *Fagus-Mercurialis* woodland takes in a little of the early workers' *Fagetum rubosum* and some stands transitional to yew woods. In this respect, the treatment is similar to that adopted by Peterken (1981) in the recognition of his stand type 8C of calcareous pedunculate oak-ash-beech-woods, though the internal division of his community is based on the representation of different woody species and does not therefore correspond to the sub-communities defined here. He also includes within this type certain woodlands with *Fagus* from south Wales which are considered in this scheme to fall within the *Teuc-*

rium sub-community of the *Fraxinus-Acer-Mercurialis* woodland.

The widespread occurrence of calcicolous beech-woods throughout Europe was first revealed in the exchanges of the Fifth International Botanical Congress (Watt & Tansley 1930 and accompanying papers) and has since been amplified in a variety of phytosociological investigations. These reveal a close general similarity between the *Fagus-Mercurialis* woodland and a range of associations now usually grouped in the Galio odorati-Fagion or Cephalanthero-Fagion sub-alliances (*sensu* Ellenberg 1978) within the Fagion: for example, the *Fagetum boreoatlanticum* from north-west Germany (Tüxen 1937) and Belgium (LeBrun *et al.* 1949), the *Elymo europaei-Fagetum* from the Schwäbischer Alps (Kühn 1937) and the Mittelgebirge (Hartmann & Jahn 1967), the *Endymio-Fagetum mercurialetosum* from north-west France (Noirfalise & Sougnez 1963), the *Carici albae-Fagetum* from the Swiss Jura (Moor 1952), the *Violo reichenbachianae-Fagetum* from Czechoslovakia (Moravec 1979) and the *Cephalanthero-Fagetum* in which Oberdorfer (1957) subsumed a number of communities previously described from southern Germany and France. Within and between some of these woodland types, it is possible to see a reflection of the differences which characterise the different sub-communities here: in the contrasts between the *Cephalanthero-Fagetum typicum* and *caricetosum digitatae* and the closely-related *Taxo-Fagetum* which Oberdorfer (1957) describes, for example, and in the woodlands compared in Moravec's (1979) synopsis. But these Continental Fagion woodlands are often much richer in species which are only occasional or often rare in British stands (e.g. *Cephalanthera damasonium, Neottia nidus-avis, Polygonatum multiflorum, Convallaria majalis, Daphne mezereum, Hordelymus europaeus, Cardamine bulbifera*) and they span a much greater altitudinal range, since calcareous bedrocks occur commonly in more mountainous areas within the range of *Fagus*-prominence, a combination of circumstances not met with in this country (see, for example, Ellenberg 1978).

As in other parts of north-west Europe, the occurrence of more calcicolous *Fagus* woodlands in Britain is strongly restricted by edaphic conditions since many of the more base-rich soils are of too impeded drainage for *Fagus* to thrive. Our Cephalanthero-Fagion community (*Fagus-Mercurialis* woodland) and our major calcicolous Carpinion community (*Fraxinus-Acer-Mercurialis* woodland) thus show the same general edaphic replacement but close floristic relationships that their counterparts exhibit in northern France, Belgium and The Netherlands (e.g. Noirfalise 1968).

Floristic table W12

	a	b	c	12
Fagus sylvatica	V (4–10)	V (3–10)	V (5–10)	V (3–10)
Aesculus hippocastanum	I (2–6)	I (2)	I (2–5)	I (2–6)
Betula pendula	I (1–3)	I (4)	I (2)	I (1–4)
Fraxinus excelsior	IV (1–7)	I (2–5)	II (3–5)	III (1–7)
Acer pseudoplatanus	III (1–6)	I (2–6)	I (3–5)	II (1–6)
Quercus robur	II (1–6)	I (2)	I (3)	I (1–6)
Prunus avium	I (2)		I (3)	I (2–3)
Carpinus betulus	I (2–4)			I (2–4)
Betula pubescens	I (3)			I (3)
Taxus baccata	I (1–5)	I (3)	V (3–8)	II (1–8)
Sorbus aria	I (3)	I (2)	II (3–5)	I (3–5)
Corylus avellana	III (2–7)	II (2–6)	I (3)	II (2–7)
Crataegus monogyna	II (1–7)	I (2–7)	I (3–5)	II (1–7)
Acer pseudoplatanus sapling	III (2–6)	I (3–5)	I (3–5)	II (2–6)
Fagus sylvatica sapling	III (1–7)	I (1–4)		II (1–7)
Acer campestre	II (2–4)	I (2–7)	I (4)	I (2–7)
Sambucus nigra	II (2–4)	I (2–5)	I (3–4)	I (2–5)
Fraxinus excelsior sapling	II (2–6)			I (2–6)
Euonymus europaeus	I (1–3)	I (2)	I (5)	I (1–5)
Viburnum opulus	I (2–4)	I (4)		I (2–4)
Betula pubescens sapling	I (1–4)			I (1–4)
Ulmus glabra sapling	I (3–4)			I (3–4)
Taxus baccata sapling	I (3–4)			I (3–4)
Sorbus aria sapling	I (3)			I (3)
Quercus robur sapling	I (3)			I (3)
Ligustrum vulgare	I (3–5)	II (2–4)	I (3–6)	I (2–6)
Viburnum lantana	I (3–4)	II (1–5)	I (2)	I (1–5)
Ilex aquifolium	I (2–7)	II (2–4)	I (5)	I (2–7)
Cornus sanguinea	I (2–3)	II (3–5)		I (2–5)
Buxus sempervirens			II (4–6)	I (4–6)
Mercurialis perennis	IV (2–10)	IV (2–7)	II (2–7)	IV (2–10)
Rubus fruticosus agg.	IV (1–10)	II (2–7)	II (2–4)	III (1–10)
Circaea lutetiana	II (2–8)	I (3–4)	I (3)	I (2–8)
Hyacinthoides non-scripta	II (2–10)	I (3)	I (3)	I (2–10)
Arum maculatum	II (2–4)	I (2–4)	I (2)	I (2–4)
Fissidens taxifolius	II (2–5)	I (3)	I (4)	I (2–5)
Galium odoratum	II (1–5)	I (2–4)		I (1–5)
Dryopteris filix-max	II (1–3)	I (3)		I (1–3)
Lamiastrum galeobdolon	II (3–7)	I (3)		I (3–7)
Allium ursinum	I (3–7)			I (3–7)
Anemone nemorosa	I (3–7)			I (3–7)
Dryopteris dilatata	I (1–4)			I (1–4)
Ranunculus ficaria	I (3–6)			I (3–6)

Epilobium montanum	I (1–5)			I (1–5)
Phyllitis scolopendrium	I (1–6)			I (1–6)
Thamnium alopecurum	I (3–8)			I (3–8)
Solidago virgaurea	I (1–3)			I (1–3)
Sanicula europaea	I (3–4)	III (3–8)	II (2–3)	II (2–8)
Mycelis muralis	I (1–3)	III (2–4)	I (2–3)	II (1–4)
Melica uniflora	I (3–6)	III (3–6)	I (3)	II (3–6)
Poa nemoralis	I (3–5)	III (3–6)		II (3–6)
Tamus communis	I (1–3)	III (2–6)	I (1)	II (1–6)
Clematis vitalba	I (1–4)	II (2–5)	I (1–8)	I (1–8)
Bromus ramosus	I (1–4)	II (2–7)	I (3)	I (1–7)
Rumex sanguineus	I (1–2)	II (3–4)		I (1–4)
Cephalanthera damasonium		II (3–5)	I (2–3)	I (2–5)
Heracleum sphondylium		I (2–4)		I (2–4)
Ranunculus bulbosus		I (2–3)		I (2–3)
Galium mollugo		I (2–5)		I (2–5)
Anthriscus sylvestris		I (3–4)		I (3–4)
Arctium minus agg.		I (2–3)		I (2–3)
Arrhenatherum elatius		I (2–4)		I (2–4)
Carex flacca		I (1–2)		I (1–2)
Ajuga reptans		I (2–3)		I (2–3)
Neottia nidus-avis		I (2)		I (2)
Cynoglossum officinale		I (2)		I (2)
Bromus sterilis		I (3)		I (3)
Hypericum hirsutum		I (2)		I (2)
Hedera helix	IV (3–10)	IV (2–10)	I (6)	III (2–10)
Brachypodium sylvaticum	III (2–5)	II (2–5)		II (2–5)
Viola riviniana/reichenbachiana	II (2–6)	II (2–4)	I (2–3)	II (2–6)
Fraxinus excelsior seedling	II (1–5)	II (2–3)	I (3–4)	II (1–5)
Fagus sylvatica seedling	I (1–3)	III (2–8)	II (2–4)	II (1–8)
Brachythecium rutabulum	I (2–4)	II (3–4)	II (3–4)	II (2–4)
Eurhynchium praelongum	II (2–5)	I (3)	II (2–4)	II (2–5)
Urtica dioica	II (2–4)	II (1–4)		I (1–4)
Geranium robertianum	II (1–5)	II (2–7)		I (1–7)
Galium aparine	I (2–3)	II (2–4)	I (3)	I (2–4)
Veronica chamaedrys	I (4)	II (2–3)	I (2)	I (2–4)
Acer pseudoplatanus seedling	I (1–3)	II (1–3)	I (2)	I (1–3)
Taxus baccata seedling	I (1–3)	II (2)	I (1–2)	I (1–3)
Fragaria vesca	I (3–5)	I (2–3)	I (2)	I (2–5)
Geum urbanum	I (1–4)	I (2–5)	I (2)	I (1–5)
Iris foetidissima	I (6)	I (2)	I (3)	I (2–6)
Carex sylvatica	I (2–5)	I (2–3)		I (2–5)
Rosa canina agg.	I (1–3)	I (2–3)		I (1–3)
Crataegus monogyna seedling	I (1–4)	I (2–3)		I (1–4)
Epilobium angustifolium	I (2–3)	I (2–3)		I (2–3)
Stachys sylvatica	I (2–4)	I (2–5)		I (2–5)
Primula vulgaris	I (3–4)	I (3–4)		I (3–4)
Campanula trachelium	I (2–3)	I (5)		I (2–5)
Mnium hornum	I (3–4)	I (3)		I (3–4)

Floristic table W12 *(cont.)*

	a	b	c	12
Teucrium scorodonia	I (3–6)	I (3)		I (3–6)
Dactylis glomerata	I (3)	I (2–3)		I (2–3)
Milium effusum	I (3–4)	I (3)		I (3–4)
Daphne laureola	I (1–3)	I (1–2)		I (1–3)
Alliaria petiolata	I (2–4)	I (3)		I (2–4)
Ilex aquifolium seedling	I (1–5)	I (3)		I (1–5)
Plagiomnium undulatum	I (3–4)	I (5)		I (3–5)
Euphorbia amygdaloides	I (1–3)	I (3)		I (1–3)
Listera ovata	I (2–3)	I (2–4)		I (2–4)
Silene dioica	I (3–4)	I (2–3)		I (2–4)
Deschampsia cespitosa	I (1–7)	I (4)		I (1–7)
Lonicera periclymenum	I (2–4)	I (3)		I (2–4)
Glechoma hederacea	I (3)	I (4)		I (3–4)
Poa trivialis	I (2–5)	I (3)		I (2–5)
Number of samples	57	30	22	109
Number of species/sample	21 (9–43)	24 (16–36)	9 (3–16)	19 (3–43)
Tree height (m)	19 (10–35)	14 (8–25)	12 (8–15)	16 (8–35)
Tree cover (%)	92 (50–100)	90 (50–100)	97 (90–100)	93 (50–100)
Shrub height (m)	4 (2–10)	2 (1–5)	3 (1–7)	3 (1–10)
Shrub cover (%)	18 (0–60)	10 (0–50)	8 (0–40)	14 (0–60)
Herb height (cm)	37 (15–100)	31 (10–75)	14 (2–40)	31 (2–100)
Herb cover (%)	87 (25–100)	81 (10–100)	19 (0–100)	72 (0–100)
Ground height (mm)	13 (0–50)	10 (10–20)	10	12 (10–50)
Ground cover (%)	8 (0–90)	9 (0–100)	1 (0–10)	7 (0–100)
Altitude (m)	129 (30–260)	112 (56–200)	143 (60–210)	127 (30–260)
Slope (°)	10 (0–85)	8 (2–15)	9 (0–30)	9 (0–85)

a   *Mercurialis perennis* sub-community
b   *Sanicula europaea* sub-community
c   *Taxus baccata* sub-community
12 *Fagus sylvatica-Mercurialis perennis* woodland (total)

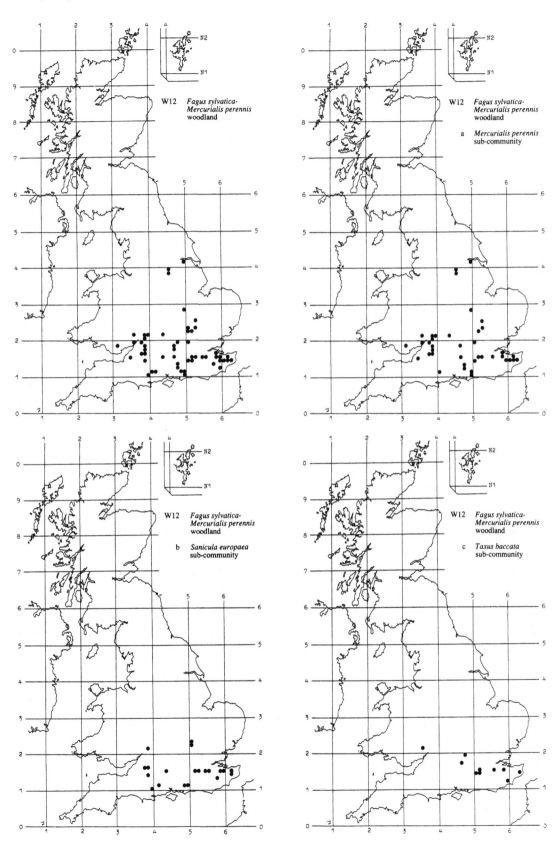

W12 *Fagus sylvatica-Mercurialis perennis* woodland

W12 *Fagus sylvatica-Mercurialis perennis* woodland

a *Mercurialis perennis* sub-community

W12 *Fagus sylvatica-Mercurialis perennis* woodland

b *Sanicula europaea* sub-community

W12 *Fagus sylvatica-Mercurialis perennis* woodland

c *Taxus baccata* sub-community

W13
Taxus baccata woodland

Synonymy
Yew-woods Tansley & Rankin 1911, Watt 1926, Tansley 1939, Ratcliffe 1977.

Constant species
Taxus baccata.

Rare species
Buxus sempervirens.

Physiognomy
Only rarely can such species-poor vegetation as mature *Taxus baccata* woodland present such a memorable spectacle. *Taxus* is the only constant woody species here, indeed the only constant, being an uncompromising dominant in a canopy that is rarely higher than 10 m but typically closed and very dense. Beyond the outer margins of stands, which can be wind-pruned and surrounded by a fringe of shrubs and climbers of the neighbouring scrub, *Taxus* reigns supreme in sometimes quite extensive stretches of striking floristic poverty and uniformity. In such a scene, the different character of the individual trees is very impressive, with pioneers of venerable appearance, richly branched from the ground and often with their trunks fused into weird shapes, and, between them, younger trees, less branched but frequently grown up lop-sided in the shade of the earlier invaders (Watt 1926, Williamson 1978).

No other tree is more than occasional throughout the community, though *Sorbus aria* is a very characteristic associate and, in one sub-community, it becomes a little more frequent, its crowns characteristically taller than the *Taxus* canopy and, with their dusty white foliage, forming a sharp contrast to the sea of dark green around. Quite commonly, it carries *Viscum album*. Emergent *Fraxinus excelsior* can also sometimes be seen and, rarely, there can be widely-scattered *Fagus sylvatica*, *Acer pseudoplatanus* or *Quercus robur*.

The picture beneath the *Taxus* is characteristically gloomy and bare. In the first place, there is no true understorey here. Associated shrubs are rare, except around the margins of stands and in gaps, and usually there is nothing more than a few sparse and spindly specimens of *Sambucus nigra* and a very occasional drawn-up *Ilex aquifolium* or *Crataegus monogyna*. Although seedlings of *Taxus* appear in large numbers in most years, they disappear by autumn and saplings occur only rarely. Where the canopy is a little thinner there may be some young *Fraxinus* or *A. pseudoplatanus*. In a few stands, *Buxus sempervirens* is a distinctive associate, and it can grow up as a very local canopy dominant.

One other frequently noticeable feature beneath the trees is the amount of dead woody remains of plants from the preceding scrub. *Juniperus communis*, a locally important precursor to *Taxus*, can be especially distinctive with its very persistent contorted skeletons, the wood smelling strongly of cigar-boxes, marking the base of each yew as a long-redundant nurse (Figure 23).

The field layer is typically very sparse indeed: at most there is but a thin and patchy cover of herbs and quite often just a bare expanse of soil and rock with brown carpets of slowly-rotting yew needles. The most frequently encountered species is *Mercurialis perennis* with very occasional *Urtica dioica*, *Hedera helix*, *Brachypodium sylvaticum*, *Arum maculatum*, *Rubus fruticosus* agg., *Viola* spp., *Glechoma hederacea* and *Fragaria vesca*, all of them plants found in other of our more calcicolous woodlands, though often with an abundance and luxuriance that is quite unknown here.

Bryophytes, too, are generally extremely poorly developed. *Eurhynchium praelongum* is, at most, occasional and, even then, its cover is low and other species occur only as infrequent small patches: *Fissidens viridulus*, *Ctenidium molluscum*, *Eurhynchium murale*, *Thuidium tamariscinum* and *Brachythecium velutinum* on soil and stones, *Mnium hornum* on twiggy litter and *Isopterygium elegans* on humus.

Sub-communities

***Sorbus aria* sub-community.** *S. aria* is a fairly frequent woody associate of *Taxus* here, its scattered crowns breaking the yew canopy at about 2 or 3 trees ha⁻¹, and, more rarely, there can be some *Fraxinus, Fagus, Q. robur* or *A. pseudoplatanus.* In most stands, *Sambucus* is the only shrub, but it is in this kind of *Taxus* woodland that *Buxus* can become locally important, forming a patchy slightly lower tier beneath the yew or growing up amongst it, its sinuous trunks and dense terminal branching producing a very distinctive effect.

Only exceptionally is there anything that could remotely be termed a field layer here. *Mercurialis* is totally absent and quite often there are no herbs at all: sometimes a diligent search may reveal a single individual of plants like *Arum maculatum, Viola* spp. or *Fragaria vesca* and very rarely there can be some frail-looking *Hedera.* More often, the ground is completely bare apart from carpets of litter over exposed flinty soil and tumbles of talus. On more stable patches of soil or on rock fragments, some bryophytes can be present.

***Mercurialis perennis* sub-community.** In this sub-community, *Taxus* is almost always the sole canopy tree with only very infrequent *S. aria* or *Fraxinus.* But the cover can be a little (only a little) more open and this can permit a more frequent and somewhat denser growth of *Sambucus* beneath, sometimes with sparse *Ligustrum*

vulgare, Euonymus europaeus and *Cornus sanguinea* and scrambling *Clematis vitalba* or *Tamus communis.*

Herbs, too, are a little more numerous and of somewhat greater cover here. *Mercurialis* is constant, though it is rarely more than patchily abundant and sometimes wilted by August, and there can be some *Urtica dioica* and scattered *Brachypodium sylvaticum, Rubus, Viola* spp. and *Fragaria. Iris foetidissima* is a striking associate at some localities and the characteristic Fagion gap species, *Atropa belladonna* and *Inula conyza*, can occur. Sparse seedlings of *Taxus* and *Fraxinus* may also be found. But such abundance is very much a comparative thing and, again, the bulk of the ground is often quite bare apart from occasional bryophytes.

Habitat

Taxus is an occasional and sometimes prominent associate in various kinds of British woodlands, most notably the *Fraxinus-Acer-Mercurialis* and *Fagus-Mercurialis* woodlands, but only very locally does it attain dominance in the type of extensive and uniform canopy characteristic of this community. Existing stands of the *Taxus* woodland are typically associated with moderate to very steep limestone slopes carrying shallow, dry rendzinas. With a single striking exception on the Durham Magnesian Limestone, the community is confined to the Chalk of south-east England where it characteristically occurs along the sides and bottom of dry valleys and on scarps. Here, the soils are generally grey rendzinas of the Upton series (Soil Survey 1983), thin, often rich in downwashed flints and poor in earthworms. Along valley bottoms, the *Taxus* woodland may extend on to somewhat deeper and moister soils derived from head. Many sites face south with slopes experiencing increased insolation and wind-exposure.

Figure 23. Canopy physiognomy in *Taxus* woodland at Juniper Bottom, Surrey.
The section shows first- and second-generation yews with occasional whitebeam (Sa) and a cemetery of junipers (J).

In such situations, the ultimate success of *Taxus* seems to depend on its ability to capitalise, slowly but inexorably, on a more prolific colonisation than other potential invaders, such as *Fagus* and *Fraxinus*, are able to achieve. Both these species probably have difficulty in sustaining growth on the drier and more exposed sites here. In addition, *Taxus* may gain some initial advantage, at least against *Fagus*, by its prolific and regular fruiting and by having seeds that are bird-distributed: yew fruits are eaten in formidable quantities by certain birds, notably by winter-flocking members of the thrush family which disgorge or excrete the seeds (Williamson 1978, Fuller 1982). The seeds are also hoarded and eaten by rodents.

Taxus can invade ungrazed grasslands directly (Williamson 1978), but its progress is very much assisted by the presence of scrub and the existence of some younger stands of the community can be traced to particular episodes of grazing relaxation on the south-eastern Chalk, as in the period of agricultural neglect in the 1920s (Watt 1926) or after myxomatosis (Williamson 1978). Scattered bushes may be important in providing breaks in the intact sward with patches of shaded and sheltered bare soil beneath them but, more importantly, they offer protection from remaining browsers. For, contrary to some early views (e.g. Tansley & Rankin 1911) and to popular opinion, *Taxus* foliage is not deadly to all larger herbivores, at least not invariably (see Lowe 1897), and is quite regularly browsed by rabbits, hares, deer and sheep, though not usually by cattle and horses which can suffer quick and fatal effects (Watt 1926, Williamson 1978); rabbits also bark stems and branches (Elwes & Henry 1906). Initial invasion is thus very much controlled by the distribution of resistant shrubs, most notably by the unpalatable *Juniperus*, itself a plant of local distribution and one tolerant of exposed sites with dry soils where *Taxus* can do well. In many of the areas examined by Watt (1926), every young juniper had its protected cluster of yew seedlings and every maturing yew its dying or dead juniper. Other shrubs too, he found, such as *Crataegus* and *Prunus spinosa*, offered some defence against browsers, though they did not provide such complete or consistent protection to the young *Taxus*. Moreover, on the deeper, moister soils where these shrubs tended to replace *Juniperus* as the scrub dominant, other trees, notably *Fraxinus*, increased in importance as invaders.

Once established within scrub, a further feature aids the survival of *Taxus* against other tree species, probably even against *Fagus*, and that is its extreme tolerance of shade. Saplings can persist, growing very slowly, for very many years before breaking through and, though estimates and borings usually give *Taxus* 4–12 annual rings cm^{-1}, shaded trees can show as many as 20 (Edlin 1958, Newbould 1960, Williamson 1978). Moreover,

Taxus can survive its own formidable shade and family groups of trees are a very characteristic feature of young yew woods, especially where invasion has occurred within more open scrub, and they can sometimes still be discerned within mature stands (Watt 1926).

Without exception, all other woody and herbaceous associates of the scrub suffer from the severely-reduced light as the *Taxus* canopy closes. Well-grown *Fraxinus* or *Sorbus aria*, which have outstripped *Taxus* in the early years, may persist as emergents in mature stands or reappear in gaps, but the bulk of these are not replaced as they die and the groups of the longer-lived *Taxus* fuse into a virtually continuous cover. Likewise, it is only in gaps that there is any substantial development of shrubs, herbs or bryophytes characteristic of the base-rich soils here and, even then, plants such as *Sambucus* and *Urtica*, reflecting the enrichment provided by past rabbit-infestation, may be more prominent.

The striking impoverishment of the flora under *Taxus* is probably also a reflection of intense root competition in soils which are often already dry. Like *Fagus*, *Taxus* has a very extensive horizontal root system with a thick felted mat ramifying the soil near the surface. And, in extreme situations, it can survive rooted into virtually bare rock from the surface of which almost all else has disappeared. This is well seen on the slopes above the river Mole at Box Hill in Surrey where *Taxus* and *Buxus* persist on very steep ground which is being actively undercut, their roots exposed on the down-slope side, their trunks piled with tumbled rubble above, and the surface of the bedrock washed clean by sheet-rain erosion.

Apart from a romanticised, though real, function as a source of wood for longbows and as a provider of material much prized for turnery, carving and veneering, it seems doubtful whether our *Taxus* woodlands have ever been actively treated for timber production. Stands may have been extensively felled to fuel the iron-smelting industry of the south-east and, more recently, in the renowned woods of Kingley Vale in Sussex, yews were used for wartime target practice (Williamson 1978). *Taxus* has been very widely planted throughout the country, though usually as hedging or for solitary ornament and groups seldom comprise more than a few trees. These eventually develop the characteristically bare floor but are scarcely a substitute for natural stands with their long and complex topographic associations.

It is possible that some occurrences of *Buxus* in British woodlands originate from planting but, as Pigott & Walters (1953) demonstrated, there is no convincing evidence that this species is not a native in this country (see also Godwin 1975). Its consistent association with *Taxus*, both in this community and in related kinds of *Fagus-Mercurialis* woodland, in sites which are characterised by steep, unstable slopes and shallow soils (matched in many of its French localities), argue strongly for a natural origin and its patchy occurrence may relate to sporadic survival right through the Forest Maximum (Pigott & Walters 1953). Stands of pure *Buxus* scrub can also be found very locally in Britain, notably at Ellesborough Warren in Buckinghamshire and at Boxwell in Gloucestershire (Ratcliffe 1977). At the latter site, which has been known since the time of Domesday, the trees have been coppiced: box-wood is prized for turnery and carving and for providing end-grain blocks for engraving (Clapham & Nicholson 1975). These stands have not been sampled but their typically sparse associated flora includes *Sambucus* and the herbs typical of the *Taxus* woodland.

Zonation and succession

Typically, the *Taxus* woodland is found in close association with grasslands and scrub which reflect grazing-relaxation over the south-eastern Chalk. Thus, stands can be found in mosaics and zonations with a complete sequence of vegetation types from close-cropped turf of the *Festuca-Avenula* grassland, rank swards dominated to varying degrees by coarse grasses such as *Bromus erectus*, *Brachypodium pinnatum*, *Avenula pubescens*, *Festuca rubra* and *Arrhenatherum elatius*, and various kinds of *Crataegus-Hedera* scrub. Where rabbit-infestation has become important again, patches of *Festuca-Hieracium-Thymus* grassland or stands of *Urtica dioica* may also occur. The exact disposition of these communities, their proportions and the age structure of the *Taxus* woodland can reflect long and complex histories of grazing by stock and wild herbivores. Much of this information is now irrecoverable though, as Williamson (1978) has shown at Kingley Vale, the effects of recent particular events can often be revealed by careful observation.

It was from a consideration of spatial patterns in a number of sites in the South Downs that Watt (1926) proposed his scheme for the general seral development of the *Taxus* woodland. From the varied examples of *Taxus* invasion, Watt characterised two main trends in succession associated with the proportional importance in the protective scrub of *Juniperus* and *Crataegus*. Where the former predominated, as on shallower soils in more exposed sites, its greater resistance to browsing allowed *Taxus* to gain a more extensive and rapid hold; among *Crataegus*, invasion was slower and more diffuse. On the deeper and somewhat moister soils typical of *Crataegus* scrub, *Fraxinus* also became increasingly important so that, in the more gradual progress to a closed *Taxus* cover, an intermediate stage of 'yew-ash wood' supervened. Eventually, however, *Taxus* exerted its overwhelming dominance throughout, often leaving just either dead juniper nurses beneath or occasional emergent ash to give any clue as to the particular origin of a stand.

As Watt went on to demonstrate (Watt 1934*a, b*), exactly the same kinds of seral development are characteristic of succession to the *Fagus-Mercurialis* woodland, though in that community the generally richer survival of different field layers in mature stands provides a lasting indication of the scrubby precursors and the edaphic conditions which they favour. But the great similarity of the successions does raise the question of why *Fagus* should invade and triumph in some situations and *Taxus* in others. The answer probably lies in the greater tolerance of *Taxus* of drier, more exposed conditions but, in many marginal cases, the success of the latter may be due to its more reliable fruiting and better seed-dispersal. Even so, *Taxus*-dominance of the kind seen here is a very local phenomenon. Like *Fagus*, it seems to have shown a late post-Glacial spread in Britain (Firbas 1949, Godwin 1975) but it has been nowhere near as successful in establishing itself as a widespread canopy component.

Once developed, however, the *Taxus* woodland is a climax community which cannot naturally be replaced by *Fagus*, or any other canopy dominant tolerant of base-rich soils, without death of the yews or felling. Some existing stands contain very old trees, though probably not as old as Watt (1962) thought: 500 years, rather than 1000, seems to be a more accurate estimate for the most ancient individuals at Kingley Vale (Newbould 1960, Williamson 1978). Stands of the community may, of course, have been in occupation in such sites for much longer, though, as Watt (1926) showed, the *Taxus* woodland tends to show small-scale migrations around Chalk coombes, colonising first in the head, then along the flanks and eventually, after many generations, dying from behind and invading again. In many sites, afforestation of surrounding plateau and slopes with mixtures of *Fagus* and conifers has isolated stands of the community, reduced their room for manoeuvre and destroyed their grand landscape context.

Distribution

The *Taxus* woodland is almost wholly confined to the Chalk of the North and especially the South Downs with some stands in gaps on the Chilterns. A stand of *Taxus*-dominated woodland on the Magnesian Limestone at Castle Eden Dene in Durham is included here but local prominence of *Taxus* on the Carboniferous Limestone around Morecambe Bay is best considered as variation within the north-western *Fraxinus-Acer-Mercurialis* woodland.

Affinities

British stands of *Taxus* woodland have long attracted the admiration of Continental ecologists but it is not easy to accommodate them in phytosociological schemes. Tüxen (1952) and Delelis-Dusollier & Géhu (1972) allocated samples of fairly species-rich yew scrub from south-east England to a *Roso-Sorbetum ariae* which they saw as the British equivalent to the *Taxo-Prunetum mahalebis*, a community of Chalk cliffs along the Seine (Delelis-Dusollier & Géhu 1972). These woodlands did not accommodate species-poor mature stands

and their equivalent in this scheme is part of the *Crataegus-Prunus* scrub. The *Taxus* woodland as understood here is probably best seen as part of the Fagion alliance in the Querco-Fagetea, alongside the *Fagus-Mercurialis* woodland and European communities like the *Taxo-Fagetum* Elter 1947.

Floristic table W13

	a	b	13
Taxus baccata	V (9–10)	V (6–10)	V (6–10)
Fraxinus excelsior	I (3)	I (4)	I (3–4)
Sorbus aria	III (3–5)	I (4–6)	II (3–6)
Fagus sylvatica	I (3)		I (3)
Acer pseudoplatanus	I (4)		I (4)
Quercus robur	I (3)		I (3)
Sambucus nigra	II (3–4)	III (2–5)	III (2–5)
Crataegus monogyna	I (1–2)	I (1–2)	I (1–2)
Ilex aquifolium	I (1–2)	I (1–2)	I (1–2)
Buxus sempervirens	II (4–8)		I (4–8)
Fraxinus excelsior sapling	I (2)		I (2)
Acer pseudoplatanus sapling	I (3)		I (3)
Taxus baccata sapling		I (4)	I (4)
Euonymus europaeus		I (3)	I (3)
Ligustrum vulgare		I (3)	I (3)
Arum maculatum	I (2)		I (2)
Ctenidium molluscum	I (6)		I (6)
Rhynchostegium murale	I (2)		I (2)
Brachythecium velutinum	I (3)		I (3)
Mercurialis perennis		V (1–9)	III (1–9)
Eurhynchium praelongum	I (2)	II (1–3)	I (1–3)
Clematis vitalba	I (3)	II (1–4)	I (1–4)
Fissidens viridulus	I (3)	II (1–4)	I (1–4)
Urtica dioica		II (1–2)	I (1–2)
Taxus baccata seedling		II (1–2)	I (1–2)
Atropa belladonna		I (2–3)	I (2–3)
Inula conyza		I (1–2)	I (1–2)
Glechoma hederacea		I (3)	I (3)
Fraxinus excelsior seedling		I (2)	I (2)
Iris foetidissima		I (7)	I (7)
Thuidium tamariscinum		I (4)	I (4)
Isopterygium elegans		I (3)	I (3)
Brachypodium sylvaticum		I (2)	I (2)
Tamus communis		I (3)	I (3)
Eupatorium cannabinum		I (3)	I (3)

Viola reichenbachiana/riviniana	I (2–3)	I (1–2)	I (1–3)
Rubus fruticosus agg.	I (1)	I (1–3)	I (1–3)
Mnium hornum	I (1)	I (2–3)	I (1–3)
Hedera helix	I (8)	I (2)	I (2–8)
Fragaria vesca	I (2–3)	I (3)	I (2–3)
Rhytidiadelphus loreus	I (2)	I (1)	I (1–2)
Number of samples	11	11	22
Number of species/sample	4 (1–8)	7 (4–10)	6 (1–10)
Tree height (m)	9 (6–10)	10 (8–20)	10 (6–20)
Tree cover (%)	100	97 (90–100)	98 (90–100)
Shrub height (m)	1 (1–2)	3 (1–4)	2 (1–4)
Shrub cover (%)	1 (0–5)	6 (0–15)	4 (0–15)
Herb height (cm)	8 (5–10)	40 (20–60)	24 (5–60)
Herb cover (%)	6 (0–60)	27 (1–90)	15 (0–90)
Ground height (mm)	20 (10–30)	10	15 (10–30)
Ground cover (%)	5 (0–30)	1 (0–3)	3 (0–30)
Altitude (m)	174 (76–205)	161 (76–205)	167 (76–205)
Slope (°)	15 (5–30)	15 (0–30)	15 (0–30)

a   *Sorbus aria* sub-community

b   *Mercurialis perennis* sub-community

13   *Taxus baccata* woodland (total)

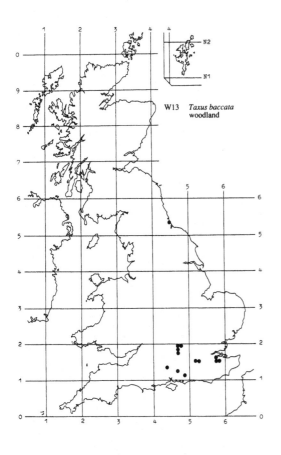

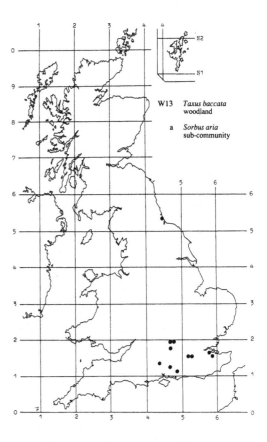

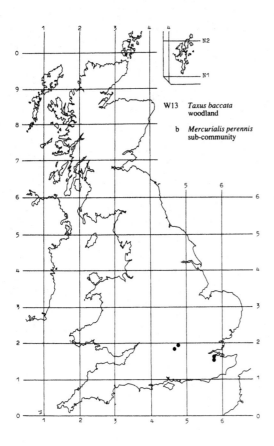

W13   *Taxus baccata*
       woodland

b   *Mercurialis perennis*
       sub-community

W14
Fagus sylvatica-Rubus fruticosus woodland

Synonymy

Beechwood association   Moss *et al.* 1910 *p.p.*; *Fagetum sylvaticae* plateau beechwoods   Adamson 1922 *p.p.*; Beech associes, seres 1 & 2   Watt 1924; Chalk plateau beechwoods (b)   Watt & Tansley 1930; Beech associes, seres A *p.p.* & B   Watt 1934*b*; Beech consociation, seres 1 & 2   Watt 1934*b*; Beech consociation, seres A *p.p.* & B   Watt 1934*b*; *Fagetum rubosum*   Tansley 1939 *p.p.*; Beech-Oak-Ash Association, types 3 & 7   McNeill 1961; Beech-oak-holly woods   Peterken & Tubbs 1965 *p.p.*; Cotswold beechwoods   Barkham & Norris 1967 *p.p.*; Beechwood   Rackham 1980 *p.p.*; Beech stand type 8D   Peterken 1981; Woodland plot types 17 & 20   Bunce 1982.

Constant species

Fagus sylvatica, Ilex aquifolium, Rubus fruticosus agg.

Rare species

Epipactis purpurata.

Physiognomy

The *Fagus sylvatica-Rubus fruticosus* woodland is a floristically simple community organised on a very large scale. Typically, in mature stands, the canopy is over-whelmingly dominated by *Fagus sylvatica* which forms a closed, even-topped cover of trees that can attain a magnificent stature. It is in this community that beech makes its best general height growth in Britain with individuals commonly attaining 30 m or more and yielding timber of the top-quality classes (e.g. Watt 1934*b*, Brown 1953, McNeill 1961). Where the trees are well spaced, tall and unbranched below, they give the canopy an architectural quality, creating the impression of a vast, spacious vault. Even here, however, growth is frequently not so good as this, varying considerably, even on the most favourable soils, with exposure, and becoming distinctly poorer where edaphic conditions are not so congenial (Watt 1934*b*, Brown 1953, 1964). Moreover, stands often have a greater measure of structural complexity related to patterns of natural invasion or treatment or both. In younger, sub-sponta-neous woodlands of this kind, the height and density of the beech can be much less uniform, and it is sometimes possible to distinguish older, more richly branched pioneers that led the colonisation of open ground from subsequent generations of straighter, more crowded trees that followed (Watt 1924, 1934*b*). Even in more long-established stands, grouped age-classes and phy-siognomic variation may give some clues to the history of the vegetation, as Peterken & Tubbs (1965) demon-strated in the New Forest (see also Tubbs 1968). Many tracts of the community show signs of treatment. Some have obviously been planted, with regularly-disposed, morphologically-similar and even-aged trees giving the canopy great structural uniformity. Others, especially in the Chilterns where the *Fagus-Rubus* woodland is common, have been managed on a selection system which produced no clear segregation of age-classes but which has often left a legacy of smaller, less well grown trees with removal of the best timber (Brown 1953, 1964, Peterken 1981). Coppiced *Fagus-Rubus* woodland is very rare, though beech seems to have been extensively cut as underwood in medieval and later times (Mansfield 1952, Roden 1968), and some stands may have deve-loped by overgrowth of coppiced *Quercus-Pteridium-Rubus* woodland (e.g. Adamson 1921). Pollarding of beech, on the other hand, is quite commonly seen here: it is a striking feature of older stands in the New Forest, where the practice was forbidden by statute in 1698 (Peterken & Tubbs 1965, Tubbs 1968), and can be seen in Burnham Beeches in Buckinghamshire.

The most characteristic associate of *Fagus* in this community is oak, almost invariably *Q. robur*, very locally on lighter soils *Q. petraea*. *Q. robur* grows well here, maintaining itself at the same height as *Fagus* even in the tallest canopies and sometimes exceeding it where soil conditions are somewhat less favourable for *Fagus* (Watt 1934*b*). Generally speaking, it is only occasional in mature canopies but it can become more frequent and locally co-dominant in transitions to the *Quercus-Pteridium-Rubus* woodland. In some places, as in the New Forest, mosaics of these two communities are common (Peterken & Tubbs 1965, Tubbs 1968): their edaphic requirements overlap extensively and, in younger woodlands, oak may have an invasive advantage and, in larger gaps, a regenerative edge, over *Fagus* so that stands resembling *Quercus-Pteridium-Rubus* woodland become prominent for some time before beech exerts its final dominance (e.g. Watt 1924, 1934*b*). Transitional stands with a greater proportion of *Q. robur* are also characteristic of zonations to soils which show more consistent gleying, where this oak is better able than *Fagus* to maintain itself as a permanent dominant: this is well seen on the sequence of soils on the Chiltern plateau (Avery 1958, 1964). In some older stands of the *Fagus-Rubus* woodland in the New Forest, oak, like the beech, has been pollarded (Peterken & Tubbs 1965, Tubbs 1968).

Other tree species of the main canopy tier are scarce. Birch (usually *Betula pendula* though sometimes *B. pubescens*) is generally the first and most common coloniser of gaps and individuals can persist for 70–80 years in intact canopies though they are rare. *Fraxinus excelsior* and *Acer pseudoplatanus* can also be found very occasionally maintaining themselves at the same height as *Fagus*, sometimes as survivors of a pre-existing woodland cover that beech has invaded (e.g. Watt 1924, 1934*a*), more often as remnants of regeneration cores in gaps, which they are both eminently successful at colonising if seed-parents are close. But neither of these is as common here as in the more calcicolous *Fagus-Mercurialis* woodland or the oak-dominated *Quercus-Pteridium-Rubus* woodland and, in some New Forest stands, their presence may originate from trees planted as ornamentals (Peterken & Tubbs 1965, Tubbs 1968).

Prunus avium can sometimes be found: it was locally abundant in Watt's (1934*b*) Chiltern stands and can be a conspicuous feature when in full flower before the beech comes into leaf. *Sorbus aria* can also occur but it is not so characteristic here as in the *Fagus-Mercurialis* woodland. *Pinus sylvestris* may be locally prominent: it is quite common in younger stands in the New Forest having seeded in from extensive nearby plantations, and sometimes colonises gaps.

Very characteristically here, there is no understorey of shrubs but a second tier of trees beneath the beech canopy, the most frequent and distinctive species being *Ilex aquifolium*. Generally speaking, this is also the most abundant woody associate of beech and oak in this community, though its cover is very variable. In more oceanic areas, like Sussex and the New Forest, it is especially prominent, sometimes forming a dense layer of trees and occasionally being a local dominant in scrubby enclaves or isolated stands which may or may not have emergent trees, as in the holms (or 'hats') of the New Forest (Peterken & Tubbs 1965, Peterken & Newbould 1966, Tubbs 1968). In Chiltern stands, by contrast, *Ilex* is much sparser, though it usually remains the numerically most frequent of the smaller woody species of the community (Watt 1934*b*, Tansley 1939). Although well able to tolerate the dense shade of the *Fagus* canopy (Peterken 1965, 1966, Peterken & Lloyd 1967), *Ilex* is, however, very susceptible to browsing and this often affects its cover and the height of the trees in open woodlands and is probably a major factor in hindering regeneration, especially where light approaches critically low levels (Peterken 1966). Hollies here can be reduced to a low, procumbent underscrub or occur as conical bushes, sometimes with an apical tuft of inaccessible growth; and though larger trees can reach 5–8 m, their lower branches may be nibbled away to form a clean, horizontal browse line. Where grazing and shade are both heavy, *Ilex* can be largely limited to gaps and margins. Older hollies in the New Forest have often been pollarded, apparently repeatedly in some cases (Peterken & Tubbs 1965, Tubbs 1968).

Occasionally in the New Forest and sometimes in Surrey and Sussex, much more rarely in the Chilterns, there is also some *Taxus baccata*, scattered among the *Ilex* and usually taller, except in dense young holms. *Sorbus aucuparia* occurs occasionally, too, and there may be a very little *Crataegus monogyna*, *Sambucus nigra*, *Corylus avellana*, *Ligustrum vulgare* or *Salix caprea*, and spindly, overtopped specimens of *Fraxinus* or *Acer pseudoplatanus*. Younger saplings are never abundant except in gaps. Those of *Fagus* are the most frequent throughout and some can usually be found even under intact canopies, as scattered individuals making slow growth, but their prominence varies ultimately with masting and is much affected by the predations of grey squirrel, mice and woodpigeons on the seed and whether the young saplings have been able to penetrate the underscrub of *Rubus* (Watt 1923, 1925, 1934*b*). Where creation of gaps coincides with good mast years, *Fagus* saplings can be very prominent beneath the parent trees around the edge of gaps (Watt 1924), though the cores are generally filled by young *B. pendula*, *Fraxinus* and *A. pseudoplatanus*, with occasional *Q. robur* in larger openings, forming a dense thicket which can retain its distinctive composition for many years and from which any of these associated trees may promote themselves to the canopy.

In younger stands of the *Fagus-Rubus* woodland (which, in Sussex, Watt (1924) considered could include tracts up to 80 years or so), the ground beneath the densely-crowded beech can be totally bare and, even where the canopy has begun to open up somewhat, the field layer is often very sparse with extensive areas of beech litter and nothing else. But the most characteristic plant in well-established *Fagus-Rubus* woodland is *Rubus fruticosus* agg. Where the shade is less intense, brambles can come to form a virtually continuous cover up to 1 m or more in height, impassable to the human visitor and, having some evergreen shoots, of very great significance to the prominence of potential herbaceous companions and the regeneration of the trees, a role which Watt (1934*b*) characterised as mischievous in its frustration of the expression of the associated flora here.

Essentially, the *Fagus-Rubus* woodland is the beech-dominated counterpart of the south-eastern *Quercus-Pteridium-Rubus* 'bluebell woodland' but because *Rubus* deepens the shade of the beech canopy even further, many of the typical species of that community are very poorly represented here. *Hyacinthoides non-scripta* itself, for example, is very infrequent, even in longer-established stands where it has had ample time to migrate in, often being confined to more open margins: typically, therefore, this community has no marked vernal aspect. The other two components of the under-scrub trio, *Pteridium aquilinum* and *Lonicera periclymenum* are quite frequent, but generally not abundant, though they can increase locally and temporarily in gaps; but, in so far as they gain higher covers, they further reduce the light reaching the ground by mid-summer. Against this kind of competition, even *Hedera helix* seems to suffer: it is not common here and, even in more oceanic areas, does not often form an extensive carpet so typical of more south-westerly *Quercus-Pteridium-Rubus* woodlands, though deer-grazing may also play some part in reducing its cover in unenclosed stands.

The general effect of this intense shade on the remaining species of the community is that it so reduces their cover that they typically occur as widely-scattered individuals, whose frequency is very low if small field-layer samples are taken. Many of the plants are very typical of this kind of woodland but they are not found very often unless extensive areas of ground are covered. Among the most characteristic of these associates is *Oxalis acetosella*: Watt (1924, 1934*b*) found this species to be especially prominent in the decade or so before the *Rubus* underscrub attained dominance but later to be largely overwhelmed, despite its great shade tolerance (e.g. Packham & Willis 1977, Packham 1978). Then, there are typically some scattered tussocks of grasses, notably *Holcus mollis*, *Milium effusum*, *Melica uniflora* and, on heavier soils, *Deschampsia cespitosa* with, less frequently, *Dactylis glomerata*, *Poa trivialis*, *Agrostis capillaris*, *Bromus ramosus*, *Festuca gigantea*, *Brachypodium sylvaticum* and *Holcus lanatus*. *Luzula pilosa* is quite common, too, and there can be scattered crowns of *Dryopteris filix-mas* and *D. dilatata*.

More noteworthy is the presence in some stands of *Galium odoratum*, sometimes in local abundance, a feature which suggests that, though the soils here are usually quite acidic and have no free calcium carbonate, the exchange capacity in the better-structured mulls may be quite high (e.g. Avery 1958). In general, however, species characteristic of more calcicolous woodlands, like *Mercurialis perennis*, *Circaea lutetiana*, *Geum urbanum*, *Sanicula europaea* and *Arum maculatum*, are rare here. They may make an appearance on profiles transitional to brown calcareous earths but Watt's beech-woods of sere 3 in Sussex (Watt 1924) and seres A_0 and part of A in the Chilterns (Watt 1934*b*), where such plants are common, are considered to fall within the *Fagus-Mercurialis* woodland. More exacting calcifuges such as *Deschampsia flexuosa* and certainly ericoid sub-shrubs are likewise absent. On moister soils, *Deschampsia cespitosa* is occasionally accompanied by *Anemone nemorosa*, *Lamiastrum galeobdolon*, *Carex sylvatica*, *C. remota* or even *Juncus effusus*, but such plants are not widely distributed throughout nor ever more than locally abundant in contrast to some wetter *Quercus-Pteridium-Rubus* woodlands. The Carices and *J. effusus*, for example, often mark tractor and cart ruts where water lies in winter.

In more open areas, as in gaps and around the margins of stands, other species may become prominent. *Rubus idaeus* can be very plentiful along with brambles where trees fall and there is often some *Digitalis purpurea*, *Euphorbia amygdaloides* or *Arctium minus* agg. in such places. *Epipactis helleborine* is sometimes found on the more open edges of the *Fagus-Rubus* woodland and the rarer *E. purpurata* can occur too, its distinctive clusters of pale flower spikes appearing late in the summer and sometimes in deep shade. One other occasional plant worthy of note is *Ruscus aculeatus*.

Finally, growing among the herbs, there are usually some tree seedlings. Those of *Fagus* itself are the most frequently encountered, though their abundance is very dependent on the irregular masting. Following good years, they can be extraordinarily plentiful, though the vast majority succumb to the predations of small mammals and invertebrate damage or to the dense shade cast by the *Rubus* underscrub (Watt 1925, 1934*b*). *Fraxinus* seedlings can also be quite common and can perhaps more readily penetrate a bramble cover; maybe *Q. robur* too, though its seedlings are quite rare, except in larger gaps. But this greater tolerance may give both these species an edge over *Fagus* in short-term regeneration, beech bringing up a slow rear but eventually benefiting by the shading out of bramble, tolerating the canopy shade of ash and oak and finally overtopping them (Watt 1934*b*).

Ilex seedlings can also be abundant, but they are of patchy occurrence both through time and spatially, according to the pattern of fruiting, which is somewhat irregular, and the activities of the birds which eat the berries and excrete the seeds (Peterken 1965, 1966). The seedlings are shade-tolerant but very susceptible to being nibbled by larger herbivores and, though they show remarkable persistence in the face of such damage (Peterken 1966), it may be critical in denser shade. *Sorbus aucuparia* seedlings, like those of *Ilex*, can often be found beneath bird-perches and there may also be some *Taxus* and *A. pseudoplatanus* seedlings.

Bryophytes are not a consistently prominent feature of this kind of woodland and their total cover is generally low but, again, certain species are very characteristic and, though occurring as widely-scattered patches, they can provide virtually the only splash of green in shadier stands. The most frequent species are *Mnium hornum* (especially on twiggy litter and humus), *Isopterygium elegans* (on humus), *Hypnum cupressiforme s.l.* (often *H. mammilatum* over exposed beech roots), *Atrichum undulatum* (on exposed mineral soil), *Polytrichum formosum* and *Dicranella heteromalla*. Less common are *Thuidium tamariscinum*, *Eurhynchium praelongum* and *Isothecium myosuroides*. *Dicranum scoparium* and *Leucobryum glaucum* are rare here and much more characteristic of the *Fagus-Deschampsia* woodland.

Habitat

The *Fagus-Rubus* woodland is confined to brown earths of low base-status and with moderate to slightly impeded drainage in southern England. It probably represents the climax forest in such situations within the natural British range of beech but, although some stands are undoubtedly old, very many have been modified by sylvicultural treatments and grazing and some are relatively young plantations.

Within the broad spectrum of soil types on which *Fagus* can become dominant, the *Fagus-Rubus* woodland occupies a middle ground between rendzinas and brown calcareous earths on the one hand and podzolised soils on the other. It occurs widely over the southern Chalk but only where the influence of the underlying limestone is masked by a cover of superficials: on its very typical Chalk plateau sites, it is characteristic of Clay-with-Flints and Plateau Drift, probably with some local admixture of loess (Watt 1924, 1934b, Brown 1953, Avery 1958, 1964, Loveday 1962). But, on such deposits, it does not extend far on to more free-draining materials where the tendency for surface-leaching is very pronounced. Likewise, where it is found on soils derived directly from the underlying bedrock, acidic and pervious arenaceous substrates are usually avoided: in the Weald, for example, the community occurs on certain of the Cretaceous Lower Greensand beds and, in the New Forest, Oligocene Headon Beds and Eocene Barton Clays provide important parent materials (Tubbs 1968).

Typically, the brown earths derived from such

deposits under the *Fagus-Rubus* woodland are of the classic *sol brun lessivé* type (Duchaufour 1956). Superficial pH is low, generally between 4 and 5, but leaching is usually limited to eluviation of any free calcium carbonate and the mobilisation of clay minerals with accumulation below in an often distinctly argillic B horizon. Some of these soils have a good mull structure: there is a surface accummulation of spongy beech and bramble litter, but steady integration by an active soil fauna in which earthworms figure quite prominently. Fairly often, however, there is a tendency towards the development of mor, with coated or laminated humus building up beneath the leaf litter and a much more compact subsurface horizon, a textural difference often detectable to the foot (Watt 1934b, Avery 1958, 1964). There may even be some mobilisation of humus and sesquioxides such that a discontinuous bleached layer forms in the top few centimetres of the A horizon, constituting a micropodzol (Brown 1953, Avery 1958). Such trends may be influenced by *Fagus* itself, more especially by certain treatments of beech (see below), but they are also affected by topography (mor formation being more pronounced on exposed brows: Avery 1958) as well as by variation in the parent material itself. Quite commonly, the *Fagus-Rubus* woodland is disposed over mosaics of brown earths, some areas tending to mull soils, some to mor.

The general association with leached brown earths sets much of the floristic character of the field and ground layers here and helps define the limits of the community against the two other kinds of beech woodland in Britain. This is especially well seen in the Chilterns where all three communities occur in close proximity over a sequence of profiles running from rendzina to podzolised soils, disposed over the free Chalk scarp and the dip slope with its mantle of superficials (Watt 1934b, Avery 1958, 1964). Here, the *Fagus-Rubus* woodland is confined to more base-poor brown earths in the Batcombe, Winchester and Charity series (Avery 1958, 1964), largely on the plateau, sometimes running over its edge where there is appreciable downwash of superficials or deposition of decalcified head in shallower valleys. Wherever profiles of the mull type become richer in calcium carbonate, as where the superficials are not so deep or are themselves more calcareous, there is an edaphic switch to the *Fagus-Mercurialis* woodland. Within the *Fagus-Rubus* woodland, the beginnings of this transition are probably marked by the appearance of *Galium odoratum*, a slight increase in *Fraxinus* and often extremely good growth of beech, but it is the appearance of such plants as *Mercurialis perennis*, *Geum urbanum*, *Circaea lutetiana* and *Viola reichenbachiana* and the increasingly patchy chlorotic cover of *Rubus* which provide the best separation between these communities. In the Chilterns, the changeover is well under way on the Coombe brown calcareous earths and completed on the Upton (one-time Icknield) rendzinas on the scarp. On the South Downs (Adamson 1921, Watt 1924) and in some Cotswold woodlands (Barkham & Norris 1967, 1970), the contrast is not always so clearly associated with a topographic difference, because brown rendzinas are quite widespread there on the limestone dip slopes.

Towards the opposite extreme on the Chilterns, fully-developed podzols are rare, but any strong tendency towards mor accumulation and strong eluviation in the Batcombe soils is characterised by the appearance beneath beech of *Deschampsia flexuosa*, *Calluna vulgaris* (in more open places) and markedly calcifuge bryophytes. Such species are typically absent from the *Fagus-Rubus* woodland and mark the change to the *Fagus-Deschampsia* woodland (Watt 1934b, Avery 1958). An identical soil-related contrast between these two kinds of beechwood can be seen in moving from brown earths to more strongly developed podzols on sands in the Weald and on sands and Pleistocene Plateau Gravels in the New Forest.

With some change in definition of one of its floristic boundaries (to exclude the more calcicolous plateau woods of Watt 1924, 1934b), the *Fagus-Rubus* woodland thus takes its place as the central example of three kinds of beechwood developed over soils of differing acidity and calcareousness (Watt 1934b, Tansley 1939, Rackham 1980). But a further important edaphic limit is set by drainage impedence. Typically, here, though the soils can be distinctly argillic, there is no more than slight gleying and then only in the B horizon, a feature well seen in the Batcombe brown earths on the Chilterns (Avery 1958, 1964). Any increase in this tendency is marked by a rising prominence of *Quercus robur* on these brown earths and a switch to the oak-dominated analogue to this community, the *Quercus-Pteridium-Rubus* woodland, the characteristic forest type of more base-poor stagnogleys on argillaceous bedrocks and heavier-textured superficials throughout the south-east of Britain. Transitional stands with more *Q. robur* and an increase in such herbs as *Deschampsia cespitosa*, *Lamiastrum galeobdolon* and *Anemone nemorosa*, are quite common and intimate mosaics of the two communities can be seen, for example, in the New Forest where they are disposed over more and less impeded profiles of the Wickham stagnogleys and Bursledon stagnogleyic brown earths (Soil Survey 1983).

Within the bounds set by these edaphic conditions, beech has attained natural dominance in the *Fagus-Rubus* woodland by late (and perhaps man-assisted) expansion in the post-Glacial within those parts of southern Britain that are relatively warm but not too drought-prone (Watt 1923, 1925, Watt & Tansley 1930, Godwin 1975, Rackham 1980). As in the other beech woodlands, *Fagus* tends to show better growth here in areas with a moister, milder climate, like Sussex and the New Forest (Watt 1934b, Brown 1964) and the greater prominence of *Ilex* in these localities is a further testimony to their more oceanic character (Tansley 1939, Iversen 1944). But of much more obvious importance to the floristics of the community is the microclimate which beech itself, and the evergreen *Ilex* when it is present, create, casting a very dense shade which is unrelieved until the beech canopy begins to open up. The field layer is thus slow to develop here and in its early stages consists largely of a carpet of the shade-bearing *Oxalis* with only scattered grasses and ferns. Moreover, any further development of the herbs in this community is very severely constrained by the slow but often extensive establishment of the *Rubus* underscrub, which increases the shade still further and, being partly evergreen, precludes any abundance of vernal plants: *Hyacinthoides* would be the expected spring dominant here, but it is actually infrequent and often very sparse (Watt 1924, 1934b).

Root competition with *Fagus* may also play some part in the contrast between the very open herb cover in this community and the more extensive field layer typical of the *Quercus-Pteridium-Rubus* woodland because beech, unlike oak and the other common dominants of that community, is a surface-rooter. This may not be so widely important here as in the *Fagus-Mercurialis* woodland where the substrates are pervious and the soils often shallow and excessively-draining, but it may be more critical in drier areas like the Chilterns: Watt (1934b) noted that *Oxalis*, for example, which has a high demand for surface moisture (Packham 1978), was less abundant there than in Sussex stands.

It has also been suggested that beech dominance in the *Fagus-Rubus* woodland modifies the environment by influencing pedogenesis. Especially on inherently poorly-draining soils, its shallow rooting may reduce aeration and nutrient turnover, leading to a deterioration in sub-soil structure (Brown 1953, Manil 1956) and this, together with accumulation of beech mor, could favour clay deflocculation and an increased tendency to podzolisation. More likely, according to Avery (1958), is that it is certain treatments of *Fagus* that enhance such trends, mor being especially associated with dense stands of undersized, slow-grown trees, such as develop with repeated application of the selection system (Brown 1953, 1964). In curtailing this development, *Rubus* may play an important role: it can spread rapidly where gaps form, adding its own less acid litter to the humus and increasing nutrient turnover. Moreover, as Watt (1923, 1925, 1934b) noted, and as practitioners of the selection system discovered, a bramble cover can suppress beech regeneration but allow *Fraxinus* and, in larger gaps, *Q. robur*, to develop and maintain their occasional presence in the canopy and, beneath these trees, there is a less dense shade, a more luxuriant herb cover and a local development of mull. In such situations, *Fagus* should theoretically be able to come through given time, regaining its position as the canopy dominant, shading out the field layer and shifting the humus regime back towards mor: a cyclical pattern of mor and mull formation would thus be a natural feature under stands of this community (Avery 1958). In practice, masting in *Fagus* is very erratic, its fruits are not transported far and, even if the seedlings get away, predation by a variety of invertebrate and vertebrate pests (especially, now, grey squirrel attacking at the pole stage) severely reduce the chances of the trees reaching maturity (Watt 1923, 1925, Brown 1953). The cycle of regeneration may thus be very slow and unreliable and, in larger gaps, vegetation approximating to the *Quercus-Pteridium-Rubus* woodland may be in occupation for some time (Rackham 1980, Smith 1980).

Hopeful attendance on natural regeneration in gaps created by timber removal was the essential feature of the selection system, applied extensively to beech woodlands in the Chilterns in the nineteenth and early twentieth centuries (Brown 1953, 1964, Peterken 1980). Here, the *Fagus-Rubus* woodlands of the Chalk plateau have been a major contributor to the local beech-based economy for many centuries, supplying coppice underwood for fuel from medieval times (e.g. Mansfield 1952, Roden 1968) and later, with the rise of the furniture industry, timber, small material being worked in the woods by itinerant 'chair-bodgers', larger material going to factories. Applied with skill, the selection system, drawing timber of varying sizes from coupes at intervals of 8–15 years, seems to have been adequate to ensure regeneration. But increased mechanisation and the use of imported timber meant that cuts became less regular and concentrated on larger timber, such that stands developed an over-abundance of densely-placed small trees beneath which regeneration, never very reliable, became extremely difficult. Some tracts of the community still show signs of this neglect; others have a structure related to the subsequent change to shelterwood and clear-felling treatments with planting of beech from outside the Chilterns; many are derived from restocking after the heavy fellings of the Second World War, sometimes with intermixtures of conifers (Brown 1953, 1964). Beech from the *Fagus-Rubus* woodland still finds a ready market: although quality varies greatly with topographic situation, the community can yield timber of class I or II on the better mull soils, especially in less drought-prone areas like Sussex (Brown 1953, 1964, McNeill 1961).

Planting or sowing of beech has also played a part in the origin of some stands of the community in the New Forest (Tubbs 1968), as in the statutory enclosures of the eighteenth and mid-nineteenth centuries and in the more recent Forestry Commission plantations, within which there is an undertaking to maintain a proportion of hardwoods. In this region, however, the *Fagus-Rubus* woodland also constitutes a large proportion of the 'Ancient and Ornamental Woodlands' which are an unenclosed element of a landscape derived from the medieval Royal Forest with common grazing rights. In fact, the bulk of the woodlands in which beech predominates are not very ancient (Tubbs 1964, 1968, Peterken & Tubbs 1965) and its prominence may be partly due to the preferential extraction of oak for ship timber in the eighteenth and nineteenth centuries. But it does seem to have been a very successful invader of open ground or derelict or cleared woodland in the first wave of woodland development, which Peterken & Tubbs (1965) dated to the period from the mid-seventeenth to the mid-eighteenth centuries. The cause of this phase of regeneration, which seems to have been matched in other Royal Forests, is unknown: deliberate enclosure against grazing and browsing animals may have played some part but does not seem a totally adequate explanation.

Canopy closure in these older stands, which now comprise a mosaic of *Fagus-Rubus* and *Quercus-Pteridium-Rubus* woodlands, together with their calcifuge counterparts, the *Fagus-Deschampsia* and *Quercus-Betula-Deschampsia* woodlands, seems to have been sufficiently sustained in some stands to prevent any extensive internal regeneration. Of much more general importance here, though, was the long-continued intensive grazing and browsing of these woods by deer (mostly fallow, with some roe and red), ponies and cattle, and the turning out of pigs in the pannage season. Not until this pressure fell with the Deer Removal Act of 1851 was there any marked regeneration. From this time onward dates the second age-class group of stands, apparently mostly *Quercus-Pteridium-Rubus* woodlands, with a wide variety of canopy species, including sometimes *Ilex* as a scrubby dominant, and among which *Fagus* found only an occasional place. Such stands occupy gaps within the *Fagus-Rubus* woodlands and represent some marginal expansion. Regeneration in this phase seems to have been checked by a variety of factors: dense shading by holly, an increased incidence of heath-burning and a renewed burst of pony-browsing towards the end of the First World War. However, a further decline in pony-browsing and cattle-grazing in the Second World War permitted another phase of regeneration, often represented by woodlands with much *Ilex, Fraxinus* and *Acer pseudoplatanus*. With deer and stock numbers again reaching very high levels, these woodlands provide one of the best examples of wood-pasture in Britain and an evocative reminder of an earlier landscape, whose openness is much valued by huge numbers of visitors. But continued browsing makes further regeneration of the *Fagus-Rubus* wood-lands unlikely and precludes the possibility of seeing whether any of the younger *Quercus-Pteridium-Rubus* woodland will succumb to eventual beech-dominance.

Zonation and succession

Zonations from the *Fagus-Rubus* woodland to other vegetation types most commonly reflect edaphic differences or represent stages in seral successions, though both of these kinds of transition are much affected by sylvicultural treatments and grazing or by the isolation of stands within predominantly agricultural landscapes.

Some of the most intact edaphic zonations are to be seen on the Chilterns. Here, though the stands of beech are of diverse origin and have canopy structures derived by manipulations pursued, in large measure, independently of soil variation, the *Fagus-Rubus* woodland can still be seen in direct continuity with *Fagus-Mercurialis* woodland on the one hand and *Fagus-Deschampsia* woodland on the other, disposed in the soil-related sequence given classic expression by Watt (1934*b*: see also Brown 1953 and Avery 1958). Incomplete sequences, lacking the *Fagus-Deschampsia* woodland, occur on the South Downs (e.g. Adamson 1921, Watt 1924), and the more calcifuge end of the series can be seen in the Weald and the New Forest.

The sharpness of the transitions between the different kinds of beech-dominated woodlands is very variable. Zonations from the *Fagus-Rubus* woodland to the *Fagus-Mercurialis* woodland can be very striking, with an abrupt switch from a vigorous bramble underscrub on the plateau brown earth to a luxuriant carpet of *Mercurialis* on the scarp rendzina. But, where there is some downwash of Clay-with-Flints over the scarp brow or deposition of decalcified head in shallow valleys, the junction between the communities may be ill defined and not so clearly coincident with topography. And plateau superficials may themselves be more calcareous so that the *Fagus-Mercurialis* woodland extends some way on to the typical *Fagus-Rubus* site-type, a feature especially widespread on the South Downs.

Transitions from the *Fagus-Rubus* to the *Fagus-Deschampsia* woodland are often less sharp, partly because the plateau superficials from which their soils are often derived are very variable in composition. In some localities on the Chilterns (e.g. Watt 1934*b*), the zonation relates clearly to increasing depth of superficials with increasing distance from the scarp edge, *Rubus* underscrub and the more mesophytic grasses passing in well-ordered fashion to *Deschampsia flexuosa*, *Calluna* and calcifuge bryophytes. But local sandy patches and exposed brows swept clear of loose litter by wind (e.g. Avery 1958) may show unexpected transitions to the *Fagus-Deschampsia* woodland and there may be local development of mor beneath dense plateau beech so that

mosaics of the two communities develop.

Throughout these zonations, long and intensive exploitation for beech often makes canopy distinctions between the different communities very much less obvious than those of the field layer. Even in more natural circumstances, of course, *Fagus* is an overwhelming dominant in each but changes in woody associates can sometimes also be seen. In mature stands, *Fraxinus*, *Acer pseudoplatanus* and *Sorbus aria* appear in the canopy more frequently with the move to the *Fagus-Mercurialis* woodland and *Ilex* is often replaced by *Corylus* as the leading smaller companion. In the *Fagus-Deschampsia* woodland, the differences are much less: both *Q. robur* and *Ilex* remain common but there is sometimes more *Betula pendula* and, less often, a little *Q. petraea*. More recently, treatments have fragmented larger stretches of these beech-dominated sequences, interposing stands with some pine and larch or plantations composed entirely of conifers. The more recent kinds of landscape with beech are well illustrated for the Chilterns by Figure 68 in Tansley (1939) and for the New Forest by the map in Tubbs (1968).

The other common kind of woodland zonation involving the community, that to the *Quercus-Pteridium-Rubus* woodland, is more complex. Sometimes this, too, is related to soil differences: there seems little doubt that, within the natural range of beech, that kind of woodland is an edaphic replacement for *Fagus-Rubus* woodland on strongly-gleyed soils (e.g. Watt 1934*b*, Noirfalise 1968), and such transitions form part of the variation in the proportions of *Fagus* and *Q. robur* in the Weald, on the Chiltern plateau and in the New Forest. But complex mosaics are frequent because of fine differences in drainage impedence on gently-undulating surfaces (especially where the soils derive from superficials), and the situation is complicated by the fact that the *Quercus-Pteridium-Rubus* woodland seems to be a seral precursor to the *Fagus-Rubus* woodland. Thus, where neglected farmland or heathland on moderately base-poor soils is being colonised, or where *Fagus-Rubus* woodland has been felled and the site abandoned, or where large gaps develop within it, there is an initial, sometimes lengthy, period of oak-dominance before beech wins through (Watt 1924, 1934*b*, Tansley 1939, Peterken & Tubbs 1965). Similarly, where *Fagus-Rubus* woodlands have been coppiced, they may have an underwood and sapling population more like that of the *Quercus-Pteridium-Rubus* woodland (Adamson 1921). That *Fagus* can effect the conversion seems undeniable from the careful studies of Watt (1924, 1934*b*); that it is a precarious and sometimes very protracted process is also clear (e.g. Watt 1923, 1925, 1934*b*, Peterken & Tubbs 1965, Rackham 1980).

The ultimate precursors to the eventual development of the *Fagus-Rubus* woodland are probably the less calcicolous kinds of mesotrophic sward (various sub-communities of the *Arrhenatheretum* and neglected meadows and pastures of the *Centaureo-Cynosuretum* and the *Lolio-Cynosuretum*) and the less extreme kinds of calcifuge sward in the *Festuca-Agrostis* grasslands and perhaps also, around the New Forest, the grassier types of *Ulex minor-Agrostis curtisii* heath. More open stands of *Pteridium* may also be colonised. Spinose shrubs are important invaders of all these communities, typically forming patchy *Crataegus-Hedera* scrub with a marginal tangle of *Rubus-Holcus* underscrub (e.g. Tansley 1922, Tansley & Adamson 1925, Watt 1924, 1934*b*). On more free-draining soils, *Ulex europaeus* may figure prominently in these early stages. Quite commonly, however, *Q. robur* and *Ilex* invade early and, if there is a seed source close by, *Fagus* itself. How long it takes for open herbaceous vegetation to progress fully to *Fagus-Rubus* woodland is unknown, but beech high forest of this type has clearly grown up in the New Forest within the last 300 years or so (Peterken & Tubbs 1965) and on the Chilterns in perhaps 200 years (Watt 1934*b*). Active development is now rare: stands are often embedded within commercial forests with much coniferisation or are sharply marked off from intensive arable land; where they remain unenclosed, as in the New Forest, heavy grazing prevents expansion.

Finally, there remains the possibility that the three kinds of beech forest in Britain are themselves related in a seral sequence, the *Fagus-Rubus* woodland developing from the *Fagus-Mercurialis* woodland and progressing to the *Fagus-Deschampsia* woodland, the process being assisted by the dominance of beech itself. Duchaufour (1950) has proposed that the establishment of a beech cover plays some part in the evolution of rendzinas to brown earths on the Lorraine Jurassic limestones but, in the Chilterns, Avery (1958) could adduce little evidence for progression beyond calcareous lithomorphic soils on drift-free Chalk, more acidic profiles being primarily related to the presence of superficials and largely independent of whether there was a cover of beech woodland. Over calcareous drift, matters are not quite so clear. Here, brown calcareous soils (which typically carry the *Fagus-Mercurialis* woodland) and brown earths (usually with the *Fagus-Rubus* woodland) may well represent successive stages of leaching and suggest a succession of one community by the other. Certainly, the floristic boundary between these two woodlands can be very indistinct but this is much affected by differences in the superficials and topography (Avery 1958): again,

such pedogenic variation also occurs outside beech forests, notably between calcicolous and calcifuge grass-lands on limestone brows.

Development of the *Fagus-Rubus* woodland to the *Fagus-Deschampsia* woodland is likewise unattested. Mor accummulation and the development of micropodzols certainly occur under beech in the *Fagus-Rubus* woodland, both in the Chilterns (Watt 1934*b*, Brown 1953, Avery 1958) and the New Forest (Dimbleby & Gill 1955, Tubbs 1968) but, once more, variation in parent materials and topography play a part and a beech canopy is not a necessary pre-requisite: even when *Fagus* is present, how it is treated may be very important (Avery 1958). On balance, podzolisation under the *Fagus-Rubus* woodland does not seem to be widespread or progressive.

Distribution

The natural range of the *Fagus-Rubus* woodland is confined to southern England with especially good representation on the dip-slope plateaus of the Chilterns and, less commonly, the North and South Downs, and in the New Forest. Where beech has been planted as a replacement for oak on moderately base-poor soils in lowland England and Wales beyond its natural limit, such stands could be included here.

Affinities

Apart from the exclusion of more calcicolous woodlands on brown calcareous earths, a change also advocated in Peterken (1981), the *Fagus-Rubus* woodland corresponds with the *Fagetum rubosum* of Watt's classic beechwood series (Watt 1924, 1934*b*, Watt & Tansley 1930, Tansley 1939, Rackham 1980). Similar woodlands are quite widespread on more base-poor, but not excessively leached, soils in Europe and are now usually placed in the Galio odorati-Fagion sub-alliance, separated off from the more calcicolous forests of the Cephalanthero-Fagion. The community is thus the British equivalent of associations like the *Melico-Fagetum* Knapp 1942 described from Germany (Oberdorfer 1957) and Poland (Matuszkiewicz 1981) and the less calcifuge forms of the *Fagetum arduennense* (LeBrun *et al.* 1949) from Belgium. The more Atlantic character of this kind of beech woodland in north-west Europe is acknowledged in such designations as the *Endymio-Fagetum* (Noirfalise & Sougnez 1963, Klötzli 1970) and *Ilici-Fagetum* (Durin *et al.* 1968, Noirfalise 1968, Klötzli 1970).

Floristic table W14

Fagus sylvatica	V (6–10)		*Carex sylvatica*	I (1–3)
Quercus robur	II (2–7)		*Circaea lutetiana*	I (1–3)
Betula pendula	I (2–3)		*Digitalis purpurea*	I (1–3)
Betula pubescens	I (3–4)		*Fraxinus excelsior* seedling	I (1–3)
Fraxinus excelsior	I (1–4)		*Galium verum*	I (1–2)
Acer pseudoplatanus	I (1–4)		*Euphorbia amygdaloides*	I (1–3)
Prunus avium	I (2–6)		*Thuidium tamariscinum*	I (2–4)
Sorbus aria	I (1–4)		*Acer pseudoplatanus* seedling	I (2–4)
Ulmus glabra	I (2–4)		*Holcus lanatus*	I (2–8)
Pinus sylvestris	I (6)		*Urtica dioica*	I (3–4)
			Bromus ramosus	I (3–5)
Ilex aquifolium	IV (1–8)		*Atrichum undulatum*	I (3–4)
Fagus sylvatica sapling	IV (1–5)		*Dicranum scoparium*	I (3–4)
Taxus baccata	I (5)		*Sorbus aucuparia* seedling	I (1–2)
Corylus avellana	I (1–4)		*Ranunculus repens*	I (2–3)
Sambucus nigra	I (1–3)		*Polytrichum commune*	I (3–5)
Fraxinus excelsior sapling	I (1–4)		*Isothecium myosuroides*	I (3–5)
Acer pseudoplatanus sapling	I (3–4)		*Eurhynchium praelongum*	I (2–4)
Sorbus aucuparia	I (2–3)		*Geum urbanum*	I (2–3)
Betula pendula sapling	I (3)		*Agrostis stolonifera*	I (3)
Salix caprea	I (3)		*Carex remota*	I (3)
Ligustrum vulgare	I (3)		*Deschampsia flexuosa*	I (1–3)
			Hyacinthoides non-scripta	I (5)
Rubus fruticosus agg.	V (1–7)		*Taxus baccata* seedling	I (3)
Fagus sylvatica seedling	IV (1–7)		*Plagiothecium undulatum*	I (2)
			Anemone nemorosa	I (1)
Pteridium aquilinum	III (1–6)		*Arum maculatum*	I (1)
Mnium hornum	III (1–8)		*Fissidens taxifolius*	I (3)
Milium effusum	II (3–6)		*Viola riviniana*	I (2)
Oxalis acetosella	II (1–4)		*Brachythecium rutabulum*	I (4)
Deschampsia cespitosa	II (2–7)		*Stellaria media*	I (4)
Luzula pilosa	II (2–4)		*Poa nemoralis*	I (3)
Melica uniflora	II (3–7)		*Moehringia trinervia*	I (3)
Holcus mollis	II (2–5)		*Mercurialis perennis*	I (3)
Lonicera periclymenum	II (1–4)		*Hypericum pulchrum*	I (2)
Dryopteris filix-mas	II (1–4)		*Festuca gigantea*	I (4)
Hedera helix	II (2–4)		*Epipactis helleborine*	I (2)
Ilex aquifolium seedling	I (1–3)			
Poa trivialis	I (1–8)		Number of samples	49
Isopterygium elegans	I (2–3)		Number of species/sample	12 (6–27)
Hypnum cupressiforme	I (2–3)			
Galium odoratum	I (4–5)		Tree height (m)	20 (10–30)
Polytrichum formosum	I (4–6)		Tree cover (%)	96 (60–100)
Dryopteris dilatata	I (2–4)		Shrub height (m)	4 (1–9)
Rubus idaeus	I (2–7)		Shrub cover (%)	15 (0–70)
Ruscus aculeatus	I (1–4)		Herb height (cm)	23 (1–80)
Dicranella heteromalla	I (2–5)		Herb cover (%)	41 (1–100)
Plagiothecium denticulatum	I (1–4)		Ground height (mm)	12 (10–20)
Quercus robur seedling	I (1–3)		Ground cover (%)	6 (0–20)
Brachypodium sylvaticum	I (2–7)			
Dactylis glomerata	I (3–4)		Altitude (m)	155 (20–226)
Lamiastrum galeobdolon	I (2–4)		Slope (°)	2 (0–15)
Agrostis capillaris	I (3–4)			

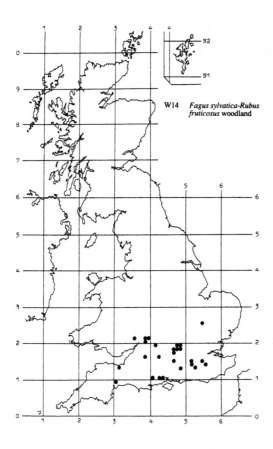

W14   *Fagus sylvatica-Rubus*
      fruticosus woodland

W15
Fagus sylvatica-Deschampsia flexuosa woodland

Synonymy

Chalk plateau beechwoods (c) Watt & Tansley 1930; Beechwoods on heath Watt & Tansley 1930; Beech associes, sere C Watt 1934*b*; Beech consociation, sere C Watt 1934*b*; *Fagetum arenicolum/ericetosum* Tansley 1939; Beech-Oak-Birch Association, types 1 & 2 *p.p.* McNeill 1961; Beech-oak-holly woods Peterken & Tubbs 1965 *p.p.*; Beechwood Rackham 1980 *p.p.*; Beech stand types 8A & 8B Peterken 1981; Woodland plot type 17 Bunce 1982 *p.p.*

Constant species

Fagus sylvatica (Deschampsia flexuosa, Pteridium aquilinum, Dicranella heteromalla, Mnium hornum).

Physiognomy

The *Fagus sylvatica-Deschampsia flexuosa* woodland is the third of the forest types in which beech attains great pre-eminence in Britain but, as in its counterpart at the opposite edaphic extreme, the *Fagus-Mercurialis* woodland, beech does not make such generally good growth here, nor is it so overwhelmingly dominant, as in the *Fagus-Rubus* woodland. Thus, though beech is the most frequent tree in this community as a whole and often the most abundant component of individual stands, the mean canopy height is usually less than 20 m and the trees are quite commonly of manifestly poor quality, sometimes growing crookedly and generally attaining no more than quality class III (Watt 1934*b*, Brown 1953, McNeill 1961, Rackham 1980). Coppiced stands are very rare (though they may have been more abundant in the past) and the usual structure is high forest, though signs of a variety of different treatments are widespread. Some tracts are obviously of planted origin, with morphologically-similar trees disposed in more or less even spacings (Watt 1934*b*, Brown 1953, 1964) and, in the Chilterns, where this community is represented among the plateau woodlands, a preponderance of smaller, poorly-grown individuals sometimes indicates long treatment under the selection system (Brown 1953, 1964,

Peterken 1981). Other stands have been treated as wood-pasture, as in Burnham Beeches (Tansley 1939), in Epping Forest (Paulson 1926, Tansley 1939) and parts of the New Forest (Peterken & Tubbs 1965, Tubbs 1968) and, in some of these places, magnificent pollard beeches survive. Even in stands which have been more actively managed, it is sometimes still possible to discern variation between individual trees and their groupings which can give some clue of the original development of the woodland by sub-spontaneous invasion with much-branched pioneers and unbranched followers (Watt 1934*b*) or clumped age-classes related to waves of colonisation (Peterken & Tubbs 1965, Tubbs 1968).

The most frequent canopy associate of beech in this community is again, as in the *Fagus-Rubus* woodland, oak. *Quercus robur* is much the more common of the two species within the natural range of beech, though *Q. petraea* is locally prominent in parts of the south (as in the New Forest, for example: Géhu 1975*b*) and it increases its representation towards south Wales (Peterken 1981); and, to the north of England, where planted stands occur widely, it is the more usual species. Where both oaks occur, hybrids can be frequent. Oak grows well here, often maintaining itself at constant, or near to constant, frequencies and commonly attaining the same height as *Fagus* or exceeding it: where it is less common, there is often the suspicion that it has been selected against. Generally, it has lower cover than beech, though it can be locally abundant and quite often the canopy presents the appearance of a mosaic of the two trees, a feature especially characteristic of the 'Ancient and Ornamental Woodlands' of the New Forest (Peterken & Tubbs 1965, Géhu 1975*b*), though also found on parts of the Chiltern plateau (Watt 1934*b*) and in Burnham Beeches (Tansley 1939). This means that spatial boundaries between this community and the *Quercus-Betula-Deschampsia* woodland, which grows in similar situations, are often somewhat unclear, a fact of some importance when considering the stability and successional status of both these communities.

Birch, too, is quite often a conspicuous feature in gaps and clearings and it can remain abundant for some time in newly-developing stands, with individuals surviving as a fairly frequent but low cover component of mature canopies. *Betula pendula* is by far the more frequent of the two species, with *B. pubescens* making only a very occasional contribution. *Acer pseudoplatanus* is much scarcer than birch, though it increases in frequency a little towards the wetter north-west, and *Fraxinus excelsior* is typically absent: even in regeneration cores, these species do not provide the early challenge to *Fagus* growth that is a feature of other kinds of beechwood. *Sorbus aria*, which has its best representation in the *Fagus-Mercurialis* woodland, and *Prunus avium*, a good occasional of the *Fagus-Rubus* woodland, are likewise very scarce here. *Pinus sylvestris*, on the other hand, can be locally prominent: it is sometimes planted as an ornamental here and can seed into young stands and gaps from nearby plantations.

The characteristically dense shade cast by the canopy of this kind of woodland is inimical to the development of rich understoreys, and in the very darkest stands there may be no smaller woody plants at all. The shade-tolerant *Ilex aquifolium* is, however, very frequent overall and, as in the *Fagus-Rubus* woodland it can thicken up to form an extensive second tier of trees beneath the beech, larger individuals, occasionally accompanied by *Taxus baccata*, almost breaking the canopy. *Ilex* becomes more frequent and abundant here outside drier areas like the Chilterns (Watt 1934b, Tansley 1939), though its prominence is very much related to the incidence of grazing and browsing, which hinder regeneration (Peterken 1965, 1966, Peterken & Lloyd 1967) and affect the physiognomy of any surviving trees. Holly is sometimes abundant in younger stands of the community that have sprung up in response to grazing relaxation (e.g. Peterken & Tubbs 1965, Tubbs 1968; also Géhu 1975b where such vegetation was included in a *Frangulo-Ilicetum*) and, being itself densely shading, can retain a local dominance, as in some parts of the New Forest (Peterken & Tubbs 1965, Peterken & New-bould 1966, Tubbs 1968).

Corylus avellana and *Crataegus monogyna* are excluded here, not by the shade but by the extreme edaphic conditions, and other associates of *Ilex* are few. Apart from occasional *Sorbus aucuparia* (and sometimes, in younger stands, *Frangula alnus*: e.g. Géhu 1975b), the most characteristic companions are saplings, especially of *Fagus* but also of *Betula pendula* and *Quercus robur*. Beech saplings can be found even under closed canopies, though their frequency and abundance depend ultimately on masting and predation of seed and seedlings and they are often checked under denser tree covers: a characteristic feature of many sites is an abundance of young beech, several years old but only

25–40 cm tall. Even around gaps, early growth may be hindered by the presence of dense bracken or prior occupation by thickly-set holly or birch (Watt 1923, 1925, 1934b). Oak saplings are generally confined to larger openings or newly-colonised ground (Jones 1959).

As in all beech-dominated woodlands, the field layer here is often reduced to very low cover by the dense shade of the *Fagus* and, if it is present, *Ilex*, younger stands which have not yet opened up by the natural death of the trees being especially impoverished. Then, the bulk of the ground may be bare of herbs, with great expanses of beech litter or, where this has been blown away, exposed patches of raw mor humus, among which the shallow roots of the beech can be seen winding. Even where the field layer is more extensive, its cover is often patchy and only continuous in areas of greater light penetration, as around the margins of stands or in gaps or, less luxuriantly, beneath the scattered oaks and birches, which cast a less dense shade than beech. A further feature of importance on the typically very free-draining soils here is that beech, being shallow-rooted, probably exerts considerable competition for water, conditions which are ameliorated under oak, the roots of which grow obliquely downwards. It should also be noted that, since the soils are also consistently base-poor, the potential vascular flora is considerably less rich than in other kinds of beechwood so, even in situations where shade and water shortage present less severe problems, cover and diversity are provided by variations in but a few species.

None of these attain constancy, though some would, were the shade not so habitually deep. The most frequent vascular plants are *Pteridium aquilinum* and *Deschampsia flexuosa*, the former very obviously limited by light penetration, often being represented by very sparse fronds and attaining real abundance and luxuriance only under more open canopies. In such places, its annual pattern of growth represents the only really obvious phenological change in the herbaceous component here. *D. flexuosa* is more shade-tolerant, though it, too, is rarely abundant and never forms the kind of continuous sward that can occur under the oak/birch-dominated canopies of the *Quercus-Betula-Deschampsia* woodland: quite frequently, it is reduced to widely-scattered plants with but a few lax leaves.

Other field-layer species are considerably less common overall. *Rubus fruticosus* agg. occurs occasionally throughout but it is often present as very weak sprawling shoots and, though it may present some check on beech regeneration in gaps (Watt 1934b), it does not here have the important role in suppressing the associated flora that it assumes in the *Fagus-Rubus* woodland. *Lonicera periclymenum*, which typically accompanies *Rubus* on more mesotrophic soils, is likewise only occasional in this community and usually not abundant.

The grass component of the field layer is sometimes enriched by a variety of species, particularly in transitions to heaths or where grazing animals have access. *Agrostis capillaris* is the most consistent companion to *D. flexuosa* in such situations but *Holcus mollis* sometimes occurs and, less frequently, *Poa nemoralis*. In stands towards the south-west, most notably in the New Forest, *Molinia caerulea* may make a small contribution. *Melica uniflora* and *Milium effusum*, which are a very characteristic feature of beechwoods on less base-poor soils, are here rare. Among other monocotyledons, *Luzula pilosa* is quite a good indicator of this kind of beech woodland, though it is not very common. *L. sylvatica* is very much confined to ungrazed situations and, though scarce overall, it can attain local abundance. *Carex pilulifera* can also occur, again in ungrazed stands and transitions to open heath. That most characteristic species of British woodlands on more base-rich and moister soils, *Hyacinthoides non-scripta*, is hardly ever found, even where the shade is not too dense.

Other field-layer plants occurring with varying frequency, though never very commonly, include *Melampyrum pratense*, *Oxalis acetosella* and, particularly distinctive in the New Forest and locally elsewhere, *Ruscus aculeatus*. Ferns are typically sparse, though *Blechnum spicant* is sometimes found and, to the north, *Dryopteris dilatata*. Tree seedlings occur rather patchily, the exposed patches of the bare mor mat presenting an especially uncongenial substrate on which to gain a hold and herbivores often devouring the bulk of the more palatable plants. However, after good mast years, *Fagus* seedlings can be plentiful and less frequently, there can be some *S. aucuparia*, *Ilex* and, in gaps, *Q. robur* and birch. In western and northern stands, *A. pseudoplatanus* seedlings can also appear in considerable numbers but, even where shade or grazing are not a threat to survival, they rarely get away and often show severe signs of nutrient deficiency, probably of nitrogen.

Grazing and browsing also reduce the prominence of *Vaccinium myrtillus* which is strongly preferential to this kind of beech woodland, but very variable in its frequency of occurrence and its abundance. And, as in the *Quercus-Betula-Deschampsia* woodland, it tends to have a better potential representation outside drier areas, being very characteristic of ungrazed stands in the western Weald and the New Forest, where it occurs in both more open transitions to heath and under closed canopies with fairly dense shade. Much more strictly confined to the former situations and to gaps and rides is *Calluna vulgaris*. When either or both of these ericoids are present, they give this vegetation a heathy aspect never found in the *Fagus-Rubus* woodland.

Also very distinctive here is the bryophyte flora which, though not providing consistently high cover,

often gives the only touch of green to the ground beneath the densest stands of *Fagus* and includes a number of good preferentials against other kinds of beechwoods. The most frequent species overall are *Dicranella heteromalla* and *Mnium hornum* with *Hypnum cupressiforme s.l.* (often *H. mammilatum* over the beech roots), *Polytrichum formosum*, *Dicranum scoparium* and *Isopterygium elegans* somewhat less common and consistent; but the most striking species is *Leucobryum glaucum*, especially when it is present in abundance as pale green cushions set amongst a dark expanse of beech litter. Other species found more occasionally are *Lepidozia reptans*, *Calypogeia fissa/muellerana*, *Plagiothecium undulatum*, *Diplophyllum albicans*, *Rhytidiadelphus loreus*, *Thuidium tamariscinum*, *Hylocomium splendens* or, in heavily-polluted stands in the Pennines, *Orthodontium lineare* and *Gymnocolea inflata*. There can also be some sparse patches of *Cladonia* squamules and, of course, this kind of woodland is renowned for its autumn fungi among which various species of *Boletus*, *Russula*, *Amanita* and chantarelles are especially distinctive.

Sub-communities

***Fagus sylvatica* sub-community.** Here, the vegetation is overwhelmingly dominated by beech in a canopy that is consistently closed and densely shading, with almost total exclusion of all other plants. Woody associates are very scarce with even *Q. robur* reduced to an occasional, generally low-cover, contribution; smaller trees and shrubs are virtually absent with only very sparse *Ilex* or a few *Fagus* saplings; the field layer is of negligible cover with even *D. flexuosa* eliminated and just beech seedlings occurring with any frequency or abundance (and, even then, varying very much with masting); and an impoverished ground layer comprises a few scattered tufts of *Mnium hornum*, *Dicranella heteromalla*, *Eurhynchium praelongum*, *Isopterygium elegans* and *Hypnum cupressiforme*.

***Deschampsia flexuosa* sub-community:** *Rusco-Fagetum* Géhu 1975*b p.p.* In this sub-community, there is a rather more varied canopy than in the above and all the other layers show at least some increase in cover and diversity. *Fagus* is still the most frequent tree and it is typically still the overwhelming dominant but oak (usually *Q. robur* but sometimes *Q. petraea* and hybrids) is quite frequent and it can be locally abundant; *A. pseudoplatanus* also makes an occasional contribution in stands to the west and north. Beneath, the prominence of smaller woody species is patchy and still generally of a low order but both *Ilex* and beech saplings are more common and abundant, *S. aucuparia* is sometimes found and, in more open places, there can be some young birch. Much more obviously, there is a richer and more extensive herb cover, especially beneath the scattered oaks. *D. flexuosa* becomes constant and *Pteridium* occurs occasionally and there can be scattered plants of *Agrostis capillaris*, *Holcus mollis* and *Luzula pilosa*, sparse trailing shoots of *Rubus* or *Lonicera* and sometimes a little *Dryopteris dilatata*. *Holcus lanatus* and *Epilobium angustifolium* can occur in disturbed places, as in plantation stands. Bryophytes are also noticeably more varied and extensive: *Mnium hornum* and *Dicranella heteromalla* become frequent and *Polytrichum commune*, *Dicranum scoparium*, *Isopterygium elegans*, *Hypnum cupressiforme* and *Leucobryum glaucum* are occasional.

Vaccinium myrtillus sub-community: *Rusco-Fagetum*

Géhu 1975*b p.p.* Oak further increases its representation here, becoming as frequent as *Fagus* and quite often comprising 10–25% of the canopy. *B. pendula* is frequent, too, though its cover is usually low, and there can be occasional large *Ilex* or *Taxus*, but the main contribution of holly is to a second tier of trees which is here quite extensive and sometimes very dense. Beech saplings are frequent and *S. aucuparia* occurs occasionally.

The field layer is usually no more extensive than in the *Deschampsia* sub-community but it is a little richer in species and quite distinctive. Both *D. flexuosa* and *Pteridium* are constant but *Vaccinium myrtillus* is the most striking feature, sometimes growing up to half a metre tall and, in areas of less deep shade, attaining high cover values. Some herbs, such as *Carex pilulifera* and *Melampyrum pratense*, are preferentially frequent too and there can be a patchy ground cover of *Hedera*. The most numerous plants, however, are often tree seedlings with, as well as those of *Fagus*, many *Ilex* and *Q. robur* and occasionally some *S. aucuparia*. Bryophyte cover can be quite extensive though, apart from a slightly increased frequency for *Leucobryum* and an occasional preferential occurrence of *Calypogeia fissa/muellerana*, the species represented are very much as in the *Deschampsia* sub-community.

Calluna vulgaris sub-community: *Frangulo-Ilicetum*

Géhu 1975*b p.p.* The woody cover in this sub-community is not quite so continuous as elsewhere and, though *Fagus* remains constant and sometimes provides the bulk of the canopy, both *Q. robur* and *B. pendula* are common and patchily abundant and, quite often, mixtures of *Ilex* and young birch make up some of the cover. Beneath this more open canopy, the field layer is more extensive than usual in this kind of woodland, though it is largely made up of areas of *Pteridium*, occasional patches of *Vaccinium* and, especially distinctive here, *Calluna vulgaris*. Under the often quite tall growth of these species, casting a shade of their own and, in the

case of bracken, accumulating bulky litter, herbs, tree seedlings and bryophytes are all reduced. *D. flexuosa* and *Agrostis capillaris* can occasionally be found and there may be a few individuals of *Rubus*, *Lonicera* or *Luzula pilosa* but otherwise the ground is largely bare apart from sparse plants of *Dicranella heteromalla*, *Mnium hornum*, *Hypnum cupressiforme* and *Leucobryum*.

Habitat

The *Fagus-Deschampsia* woodland is confined to very base-poor, infertile soils in the southern lowlands of Britain. Within the native range of beech in this country, the community has some claim to be the climax forest type but it has developed widely in plantations, even in this region, and many stands bear signs of sylvicultural treatment, being managed now as high forest for timber. In some places, it forms part of a wood-pasture landscape and grazing and browsing continue to be important in the floristics and regeneration of the community.

The *Fagus-Deschampsia* woodland is found over one extreme kind of afforested lowland soils, being limited to lime-free profiles with a superficial pH usually less than 4 and with mor humus (Watt 1934*b*, Rackham 1980, Peterken 1981). Among such soils, shallow humic rankers are quite rare here: they are not common anyway within the natural range of *Fagus* and where they are more abundant beyond this, as on Pennine sandstones and grits, they have been avoided in the planting of beech. Most often, then, the profiles are deeper and more mature, being brown earths with incipient podzolisation (frequently having discontinuous micropodzols in the top few centimetres) or podzols proper with prominent bleaching and humus iron pans. Typically, drainage is free to excessive, though some deep profiles are influenced by ground water below, when they are classified as stagno-podzols or stagnogley-podzols (Brown 1953, Avery 1958, 1964, 1980, Tubbs 1968).

Such soils are found under beech over a variety of parent materials. Arenaceous bedrocks are important substrates in the Weald where the community occurs on Cretaceous Ashdown and Tunbridge Wells beds with Poundgate stagnogley-podzols in the High Weald and on Folkestone and Hythe Beds with Shirrell Heath humo-ferric podzols around the western edge (Wooldridge & Goldring 1953, Soil Survey 1983). Further north, plantations occur over podzolised soils derived from Triassic Keuper and Bunter sandstones and Carboniferous sandstones and grits, notably the Belmont stagno-podzols along the Pennine fringes (Carroll *et al.* 1979, Jarvis *et al.* 1984). Quite commonly, however, and especially in the south, the profiles are developed in part from coarse-textured superficials. In the New Forest, for example, Eocene Bagshot and Barton sands and gravels overlain with Plateau Gravels carry extensive podzolised soils and similar mixtures of parent materials underlie stands in parts of Surrey and Essex, as in Epping Forest (Paulson 1926, Tansley 1939). Northwest of here, running up the dip slope of the Chilterns, sandy remnants of Eocene Reading Beds, intermixed with Clay-with-Flints and thick Plateau Drift, have given rise to podzolised Batcombe and Berkhamsted soils, supporting this acidic extreme of the beechwood series described by Watt (1934*b*) and Avery (1958, 1964).

The general edaphic conditions here help define this community against its counterpart on somewhat less base-poor brown earths, the *Fagus-Rubus* woodland, though the very dense shade of beech often reduces the associates to such low levels of occurrence that visual differences between the two communities are not always very striking; and some species, such as *Ilex* and *Pteridium*, are common throughout both. Nonetheless, the lower frequency of *Rubus* here and its much less consistent abundance provide one good marker of the move towards more sharply draining and acidic conditions; the replacement of grasses like *Milium effusum*, *Melica uniflora*, *Poa trivialis* and *Deschampsia cespitosa* by *D. flexuosa* and *Agrostis capillaris* is another; then, there is the patchy prominence, where grazing and light permit, of the ericoid sub-shrubs and the preponderance, among the bryophytes, of markedly calcifuge species. And, overall, though *Fagus* remains generally dominant, there is an obvious reduction in the quality of the trees on these extreme soils (Brown 1953, 1964, McNeill 1961).

The general importance of this edaphic boundary can sometimes be clearly seen where both communities occur in close proximity in the kind of soil-vegetation series described from the Chiltern plateau (Watt 1934*b*, Avery 1958, 1964), the switch from the spongy mull of the *Fagus-Rubus* woodland to the firm mor mat of the *Fagus-Deschampsia* woodland often being felt underfoot. But small-scale variation within the kinds of soils that support both these communities is quite common, even over relatively short distances, being related in part to differences in superficials and in part to topography (exposed slopes blown free of litter being particularly likely to develop mor: Avery 1958, 1964). More accurately, then, the *Fagus-Deschampsia* woodland occupies soil mosaics in which podzolised profiles with mor predominate, the *Fagus-Rubus* woodland mosaics in which less strongly leached profiles with mull cover most of the ground. And it seems likely that temporal changes from one soil type to the other can also occur, partly as a consequence of natural replacement of beech by oak or *vice versa* at particular points in space, partly under the influence of certain kinds of treatment (see below). So small-scale edaphic and vegetational instability may be

a frequent, and in some respects quite normal, feature here.

Climate varies considerably across the range of the *Fagus-Deschampsia* woodland and plays some part in the exclusion of natural stands of the community from the drier parts of East Anglia and the cooler and wetter north and west (Godwin 1975, Rackham 1980), although planted beech grows well in this kind of woodland way beyond the limit which the tree was able to attain by natural spread. The effects of climatic differences on the associated flora are, however, not very obvious because of the striking ability of beech to create its own climate beneath the canopy, impoverishing the vegetation with its deep shade and through root competition, probably very severe again here, as on the rendzinas of the *Fagus-Mercurialis* woodland which are likewise very free-draining. Shade-tolerant species show the clearest response with *Ilex* and, to a lesser extent, *Hedera*, increasing in areas of more equable climate like the New Forest, as against the Chilterns. *Vaccinium* also is largely restricted to stands receiving more than 800 mm rainfall yr^{-1} and *Dryopteris dilatata* and *Acer pseudoplatanus* increase somewhat as annual precipitation approaches 1000 mm, though the former is not as common here as in north-western stands of the *Quercus-Betula-Deschampsia* woodland and both may be hindered by the way in which beech accentuates the natural dryness of the soils. The bryophyte component shows no marked enrichment in stands towards the more humid north-west, though atmospheric pollution may restrict their abundance in the southern Pennines: badly-affected *Fagus-Deschampsia* woods in this area can have very few species.

Such floristic differences of this kind as there are, are not a sufficient basis in this community for erecting subdivisions, but the influence of the local light climate is. At one extreme, the *Fagus* sub-community includes stands with the very deepest canopy shade where beech reigns supreme in a gap-free cover, typically associated with younger woodlands of natural origin where the trees have not yet begun to die or with unthinned plantations. In fact, so many of these beechwoods are under commercial forestry that natural gap formation can be quite rare, but where it does occur or where trees are extracted, the *Calluna* sub-community typically develops, with its abundance of heather and bracken and profusion of birch, holly and oak. Similar vegetation is characteristic of those places where beech is establishing itself on previously open heathland (Watt 1934*b*, Peterken & Tubbs 1965) and of stands of the community which occur in more heathy wood-pasture landscapes of some Forests and parks, where grazing helps maintain a mosaic of closed and open areas (Rackham 1980, Peterken 1981). In terms of their relationship with the degree of light penetration, the *Des-*

champsia and *Vaccinium* sub-communities lie between the two extremes, being associated with more or less intact canopies, though ones in which high frequencies and local abundance of oak and, to a lesser extent, birch, can provide some relief from the overwhelming shade of beech. In these kinds of *Fagus-Deschampsia* woodland, the field layer is often distributed rather patchily, thickening up most obviously beneath these other trees, which often occupy old gaps and which perhaps represent a stage in a slow cyclical replacement of the local dominants, one by the other.

The floristic differences between the *Deschampsia* and *Vaccinium* sub-communities illustrate very clearly the continuing importance of a further factor in these woods and that is grazing and browsing by stock and deer which influence not only the composition of the field layer but also the regeneration of the trees. *Vaccinium*, though its occurrence is limited to some extent by climate (it is sparse in drier regions and under very dense shade), only attains any prominence in the absence of herbivores which can reduce it to sparse and leafless shoots. Freedom from grazing also allows some herbs to increase their frequency and cover, but the most obvious response is often among tree seedlings. Regeneration of *Ilex* is under the close control of herbivores here (Peterken 1965, 1966, Peterken & Tubbs 1965, Peterken & Lloyd 1967) and its seedlings can become very common and plentiful in the *Vaccinium* sub-community. *S. aucuparia* is likewise very palatable and, though seed-parents are infrequent in the community, its seedlings too show an increase in this kind of *Fagus-Deschampsia* woodland. *Q. robur* is also well represented though it is less shade-tolerant than either holly or rowan and will only get away in gaps. And finally, seedlings of *Fagus* itself are more frequent here than in any other sub-community, subject, of course, to the vagaries of masting. No details of grazing history were available for the samples but the fact that well-grown holly and beech saplings are also abundant in the understorey of the *Vaccinium* sub-community here suggests that there has been some continuity in the exclusion of herbivores from these particular stands. Clearly, this need not always be the case because, where grazing has recently declined, it will take a considerable time for seedlings to progress and contribute to the taller tiers of the vegetation. Conversely, stands of the *Deschampsia* sub-community may be denuded of seedlings as each season progresses and yet have a dense understorey related to past periods of relative freedom from herbivores, as in the New Forest where waves of regeneration in woodlands to which this community contributes have been clearly related to grazing history (Peterken & Tubbs 1965, Tubbs 1968).

The regeneration of beech and oak here, though affected by grazing, is, however, rather more complex than that of holly. *Fagus* maintains its general dominance, in the end, by the very considerable shade-tolerance of its own seedlings and saplings but it has difficulty in getting a hold if, either all litter is blown away leaving a bare mat of mor which sucks up water like a sponge, or the field layer is very densely-shading. It also fruits very erratically and the mast is not transported far, often just dropping straight down from the tree (Watt 1923, 1924, 1925). Conditions seem to be optimal where, after good mast years, fruit falls towards the edge of gaps which have not yet become clothed with dense *Pteridium* or *Calluna* or acquired a core of thickly-set *Ilex*, birch or oak, requirements whose importance is well illustrated here by the scarcity of beech seedlings in the *Calluna* sub-community. Oak is rather different: it, too, favours the presence of some loose litter to prevent desiccation but it can grow up through quite dense herbage (Jones 1959) and, more importantly, acorns can be transported considerable distances (Mellanby 1968) so, free of any challenge other than a previously-developed thick canopy of *Ilex*, it can gain the advantage over *Fagus* in the middle of gaps here and on open ground where new stands are developing. Where such an advantage prevails over substantial areas and/or for considerable lengths of time, the woodland cover can be seen as stabilising into the *Quercus-Betula-Deschampsia* woodland, the oak- (or birch-) dominated analogue of this community, but local and relatively short-term alternations of beech and oak are probably an integral part of canopy variation here and, as explained above, they can themselves affect the patterning of the field layer and perhaps alter the balance between mor and mull development in the soils.

The relative prominence of beech and oak in particular stretches of mature high forest has, however, been much affected by treatment. Even in the New Forest, where intimate mosaics are a prominent feature, the balance has been swung towards beech by preferential extraction of oak, often, in this case, for providing ship timber (Tubbs 1968). Elsewhere, *Fagus* has been favoured where the community has contributed to a local beech economy, either as semi-natural stands or, often, as plantations. Coppicing of beech is now defunct, though it was important in the Chilterns in medieval and later times (Mansfield 1952, Roden 1968), and stands are usually now treated on shelterwood or clear-felling systems for timber, restocking sometimes involving conifers (Brown 1953, 1964). However, there is still evidence in some stands in the Chilterns of the results of treatment on the selection system which depended on natural regeneration in gaps created artificially by extraction of individual trees. The unreliability of such regeneration, coupled with shifts in demand for beech, led eventually to the abandonment of this style of treatment but stretches of *Fagus-Deschampsia* woodland with a preponderance of densely-set, poorly-grown trees bear witness to continued removal of better-quality timber. Such woodland seems especially likely to maintain mor soils and is perhaps more resistant to the kinds of vegetational and edaphic changes seen elsewhere in the community (Avery 1958).

Zonation and succession

Two kinds of zonation are commonly found within stands of the *Fagus-Deschampsia* woodland. The first relates to differences in canopy shade and it can be a direct reflection of the maturation of the vegetation, with the *Calluna* sub-community occupying recently-colonised ground, the *Fagus* sub-community forming dense woodland in the establishment phase, the *Deschampsia* or *Vaccinium* sub-community occurring where the canopy has begun to open up a little and the *Calluna* sub-community marking gaps. In fact, actively-colonising *Fagus-Deschampsia* woodlands are of rather restricted occurrence and, in commercial forests, natural gap formation is pre-empted, re-establishment is from planted stock and the resultant stands have trees of uniform age with a more homogeneous field layer. Here, then, the patterning of the different sub-communities is much more regular in both space and time with sharply-defined compartments of the *Fagus* sub-community before thinning, the *Deschampsia* or *Vaccinium* sub-community after, and the *Calluna* sub-community making only a brief appearance between rotations apart from along the edges of rides where it may mark a transition to a narrow strip of heath or bracken without trees. Characteristically, the outer margins of such treated stands are sharply defined with an abrupt switch to other types of forest (often conifer-dominated stands of the *Quercus-Betula-Deschampsia* woodland) or to surrounding agricultural land.

The other kind of zonation relates to the intensity of grazing or browsing. Sometimes differences in the present pattern of access by herbivores can be seen in a change from the *Deschampsia* to the *Vaccinium* sub-community at an artificial stock- and deer-proof boundary; in other cases, the structure of the vegetation itself may hinder access, creating protected enclaves (as in some stands surrounded by a ring of holly, a feature in parts of the New Forest: Tubbs 1968) or a diffuse mosaic of more- and less-grazed areas with these two sub-communities. Such patterns can be found in some ancient Forests and parks and they often also involve more open glades with the *Calluna* sub-community and stretches of heath where the establishment of a woody cover is held in check by the herbivores. Withdrawal of grazing in such situations (and the cessation of burning or the cutting of heather and bracken, once important on commons in the south) may allow more gradual zonations to develop in which light penetration through the canopy becomes the primary factor in determining

the distribution of the sub-communities. In fact, wood-pasture landscapes usually present complex patchworks of the various kinds of *Fagus-Deschampsia* woodland related to the interplay of grazing and shading, both at the present time and in the past, as Peterken & Tubbs (1965) showed in the New Forest.

Mosaics of this kind, developed over fairly uniform podzolised soils, also frequently involve tracts of woodland in which beech is so poorly represented in proportion to oak and birch that the vegetation is best considered part of the *Quercus-Betula-Deschampsia* woodland. The relative ease with which these trees outperform beech in the early colonisation of open ground means that this community often functions as a seral precursor to the *Fagus-Deschampsia* woodland, interposed between calcifugous grassland, heath and bracken on the one hand and mature beech forest on the other, both in time and in spatial zonations. The *Quercus-Betula-Deschampsia* woodland can also be a temporary replacement for the *Fagus-Deschampsia* woodland in larger gaps and clearings, before beech re-establishes by seeding in around the margins or is planted in; and, where *Quercus-Betula-Deschampsia* woodland seems more permanently ensconced, there is sometimes an assumption that it occupies ground that, had it not suffered gross disturbance, would naturally carry *Fagus-Deschampsia* woodland (e.g. Géhu 1975a). In fact, even within the native range of beech in Britain, this tree is rather slow to assert or reassert its dominance in these situations: on newly-colonised ground, the *Quercus-Betula-Deschampsia* woodland often seems to attain a stability of its own; and, even when beech is abundant in the immediate neighbourhood, it has some difficulty in getting away in larger gaps. It may therefore be too simplistic to regard the *Fagus-Deschampsia* woodland as the natural climax forest on more surface-leached soils in southern Britain: if anything, it is beech, with its unreliable fruiting and poor seed-dispersal, which requires a measure of assistance to succeed, a fact borne in on practitioners of the selection system of treatment (Brown 1953, 1964), and one invoked to explain the tardy and limited migration of this tree in the post-Glacial period in Britain (e.g. Godwin 1975). And, even where beech has become well established on such profiles, it may be more accurate to see its dominance as under a measure of threat from oak and birch, such that slowly-shifting patchworks of *Fagus-Deschampsia* and *Quercus-Betula-Deschampsia* woodlands are a quite natural feature, moved towards the former community by selection for beech, in favour of the latter wherever there is opportunity for expansion of the forest cover.

There may also be room for some reassessment of the status of the *Fagus-Deschampsia* woodland in relation to its counterpart on less base-poor brown earths, the *Fagus-Rubus* woodland. In general, the floristic distinctions between these two communities are better defined than those between the *Fagus-Deschampsia* woodland and its oak-birch analogue, having a firm foundation in edaphic differences. Transitions from the one community to the other can sometimes be observed within individual stands, beech-dominance being maintained throughout, but the field and ground layers showing a marked change with the switch from leached brown earths with mull to podzols with mor, a zonation forming part of the soil-vegetation series described from the Chilterns (Watt 1934b, Avery 1958) but also seen in more complex mosaics in the New Forest (Tubbs 1968). As noted earlier, however, small-scale variation in topography and parent materials may make this kind of transition quite ill defined, so that again patchworks of the two communities are a fairly normal feature. It is also possible that, in more marginal situations, there is some measure of cyclical alternation in time between the *Fagus-Deschampsia* and *Fagus-Rubus* woodlands at any one place. Despite the fact that beech can encourage the formation of mor and thus favour the development of soil conditions inimical to the survival of some of its characteristic associates in the *Fagus-Rubus* woodland, this is not necessarily a progressive and irreversible process, indeed it seems to be best favoured by certain treatments of beech, rather than being a wholly natural development (Avery 1958). And, once again, it may be that, in less strictly-treated high forest, oak and birch play an important part here because with *Rubus* underscrub, they can pre-empt beech in gaps, allow the spread of a more extensive field layer in their less dense shade and perhaps shift the edaphic and vegetational balance back to less extreme conditions. Along both its boundaries with other woodland types, the soil-related junction with the *Fagus-Rubus* woodland and the competitive/successional one with the *Quercus-Pteridium-Rubus*

woodland, the relationships of the *Fagus-Deschampsia* woodland are thus more complex than they seem at first sight.

Distribution
Within the natural British range of beech, this community is best represented in the Weald, the New Forest and on the Chiltern plateau with isolated sites elsewhere in the south. Planted stands, indistinguishable in their floristics, occur well beyond this limit, especially around the Pennine fringes.

Affinities
As defined here, the *Fagus-Deschampsia* woodland is the more or less exact equivalent of the most calcifugous beechwood in Watt's (1934b) Chiltern series, described by Tansley (1939) and referred to by Rackham (1980) as *Fagetum ericetosum* (or *arenicolum*). It includes the bulk of Peterken's (1981) 'acid oak-beechwoods', split in his scheme on the basis of whether the associated oak is *Q. petraea* (stand type 8A) or *Q. robur* (8B).

Apart from Watt's (1934b) original account and a phytosociological description provided from the New Forest by Géhu (1975b, where immature stands were treated as part of a separate community), systematic and complete treatments of this kind of woodland have not been published. Nonetheless, the parallels with similar communities in other parts of north-west Europe are very clear and the *Fagus-Deschampsia* woodland has obvious equivalents in such associations as the *Ilici-Fagetum*, the *Luzulo-Fagetum*, the *Deschampsio-Fagetum* and the *Fago-Quercetum* described from France, Belgium and the Netherlands (e.g. Durin *et al.* 1968, Westhoff & den Held 1969, Géhu 1975a). In phytosociological terms, then, this community represents the Luzulo-Fagion sub-alliance in Britain.

Floristic table W15

	a	b	c	d	15
Fagus sylvatica	V (7–10)	V (6–10)	V (6–9)	V (6–9)	V (6–10)
Quercus robur	II (2–7)	III (1–4)	V (4–5)	III (3–4)	III (1–7)
Betula pendula	I (3)	I (1)	III (1–4)	III (4–5)	II (1–5)
Ilex aquifolium	I (3)	I (2–4)	II (2–7)	I (3)	I (2–7)
Quercus petraea	I (4)	II (1–6)	II (1–4)		I (1–6)
Taxus baccata	I (3–5)	I (4)	I (4)		I (3–5)
Acer pseudoplatanus	I (2–4)	II (1–8)			I (1–8)
Pinus sylvestris	I (2–4)	I (4)			I (2–4)
Quercus hybrids		I (1–4)	I (1–4)		I (1–4)
Betula pubescens		I (3–4)			I (3–4)
Ilex aquifolium	I (3)	III (1–2)	IV (3–7)	IV (2–4)	III (1–7)
Fagus sylvatica sapling	I (1–2)	II (3–4)	III (3–4)	I (4)	II (1–4)
Sorbus aucuparia		I (2–3)	I (1–2)		I (1–3)
Quercus robur sapling		I (1)	I (7)	I (2–3)	I (1–7)
Betula pendula sapling		II (3–6)		III (3–6)	I (3–6)
Rhododendron ponticum		I (1)		II (1–6)	I (1–6)
Fagus sylvatica seedling	III (1–6)	III (1–4)	IV (1–3)	I (1–3)	III (1–6)
Dicranella heteromalla	II (1–3)	III (1–4)	III (2–4)	III (2–4)	III (1–4)
Pteridium aquilinum	I (1–2)	II (2–7)	V (3–8)	V (3–9)	III (1–9)
Mnium hornum	II (1–5)	IV (1–8)	V (1–6)	III (2–3)	III (1–8)
Deschampsia flexuosa		V (2–8)	V (1–5)	II (3–5)	III (1–8)
Eurhynchium praelongum	II (1–4)	III (1–5)	III (2–6)		I (1–4)
Polytrichum formosum		III (1–5)	II (5–6)		II (1–6)
Dicranum scoparium		II (2–5)	I (3)	I (3)	I (2–6)
Luzula pilosa		II (1–3)		I (4)	I (1–3)
Lepidozia reptans		II (2–3)	I (4)		I (2–4)
Isopterygium elegans	I (2)	II (2–5)		I (3)	I (2–5)
Cladonia fimbriata		II (2–4)	I (3)		I (2–4)
Dryopteris dilatata		II (1–3)			I (1–3)
Holcus lanatus		II (1–4)			I (1–4)
Epilobium angustifolium		I (3–4)			I (3–4)

Floristic table W15 *(cont.)*

	a	b	c	d	15
Orthodontium lineare		I (3–4)			I (3–4)
Gymnocolea inflata		I (2–4)			I (2–4)
Quercus petraea seedling		I (2–3)			I (2–3)
Hylocomium splendens		I (5–6)			I (5–6)
Vaccinium myrtillus			V (4–8)	II (4–7)	II (4–8)
Ilex aquifolium seedling		I (2)	IV (2–5)	I (3)	II (2–5)
Leucobryum glaucum		II (3–6)	III (4–6)	II (3–4)	II (3–6)
Quercus robur seedling		I (2)	III (1–3)		I (1–3)
Carex pilulifera		I (3)	III (2–4)	I (3)	I (2–4)
Sorbus aucuparia seedling		I (1–2)	II (2)		I (1–2)
Calypogeia fissa/muellerana		I (1–2)	II (2)		I (1–2)
Melampyrum pratense		I (1)	II (3)		I (1–3)
Hedera helix		I (3)	II (3–4)		I (3–4)
Calluna vulgaris				V (2–7)	I (2–7)
Hypnum cupressiforme	I (3–4)	II (1–4)	II (1–4)	II (2–3)	II (1–4)
Rubus fruticosus agg.	I (3–5)	II (3–6)	II (3–4)	II (3–4)	II (3–6)
Agrostis capillaris		II (2–5)	II (1–4)	II (2–3)	II (1–5)
Lonicera periclymenum	I (2)	II (1–4)	II (2–3)	I (2)	I (1–4)
Holcus mollis	I (3)	I (3–8)	I (3–5)		I (3–8)
Plagiothecium undulatum	I (3–4)	I (1–3)	I (1–3)		I (1–4)
Acer pseudoplatanus seedling	I (2–3)	I (1–3)			I (1–3)
Poa nemoralis	I (3–4)	I (1–3)			I (1–4)
Hyacinthoides non-scripta	I (4)	I (1–2)			I (1–4)
Luzula sylvatica	I (3)	I (1–7)			I (1–7)
Blechnum spicant	I (3)		I (4)		I (3–4)
Oxalis acetosella		I (1–3)	I (1–3)		I (1–3)
Tetraphis pellucida		I (2)	I (2)		I (2)
Cladonia squamules		I (1)	I (3)		I (1–3)
Thuidium tamariscinum		I (1)	I (3)		I (1–3)
Rhytidiadelphus loreus		I (1)	I (2)		I (1–2)
Diplophyllum albicans		I (3)	I (2)		I (2–3)

	a	b	c	d	15
Ruscus aculeatus		I (1–3)	I (1–3)		I (1–3)
Molinia caerulea		I (1–3)	I (1–3)	I (5)	I (1–5)
Number of samples	19	25	8	7	59
Number of species/sample	5 (1–15)	15 (8–23)	19 (10–32)	12 (9–14)	12 (1–32)
Tree height (m)	18 (15–25)	21 (15–25)	26 (10–35)	16 (9–30)	20 (9–35)
Tree cover (%)	98 (90–100)	92 (60–100)	100	85 (50–100)	94 (50–100)
Shrub height (m)	3 (1–6)	3 (1–5)	4 (2–5)	3 (1–4)	3 (1–6)
Shrub cover (%)	1 (0–5)	5 (0–50)	23 (1–60)	14 (0–40)	7 (0–60)
Herb height (cm)	11 (2–50)	19 (8–40)	29 (15–50)	81 (20–120)	25 (2–120)
Herb cover (%)	6 (0–30)	55 (10–90)	49 (10–100)	77 (40–100)	41 (0–100)
Ground height (mm)	10	17 (5–40)	10	15 (10–30)	14 (5–40)
Ground cover (%)	6 (0–30)	29 (0–90)	38 (1–90)	7 (1–15)	22 (0–90)
Altitude (m)	138 (61–270)	146 (75–380)	83 (25–122)	94 (20–152)	129 (20–380)
Slope (°)	11 (0–30)	11 (0–85)	7 (0–30)	5 (0–10)	10 (0–85)

a   *Fagus sylvatica* sub-community
b   *Deschampsia flexuosa* sub-community
c   *Vaccinium myrtillus* sub-community
d   *Calluna vulgaris* sub-community
15 *Fagus sylvatica-Deschampsia flexuosa* woodland (total)

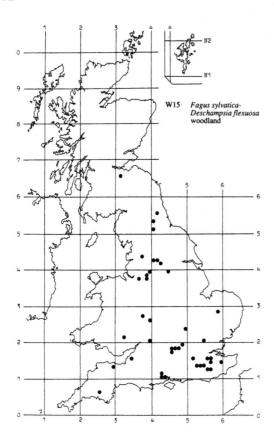

W15 *Fagus sylvatica-*
     Deschampsia flexuosa
     woodland

W16
Quercus spp.-*Betula* spp.-*Deschampsia flexuosa* woodland

Synonymy

Oak-birch-heath association   Moss *et al.* 1910, Tansley 1911; *Quercetum arenosum roburis et sessiliflorae* Tansley 1911 *p.p.*; *Quercetum ericetosum* Tansley 1911, 1939; *Quercetum sessiliflorae* Moss 1911, 1913 *p.p.*; Dry oak-birchwood   Hopkinson 1927 *p.p.*; Southern Pennine oakwoods Scurfield 1953 *p.p.*; Oakwoods Pigott 1955 *p.p.*; *Querco-Betuletum* Klötzli 1970 *p.p.*; Oakwoods Rackham 1980 *p.p.*; Birchwoods Rackham 1980 *p.p.*; Birch-oak woodland Peterken 1981 *p.p.*; Birch woodland Peterken 1981 *p.p.*; Woodland plot type 18 Bunce 1982 *p.p.*

Constant species

Betula pendula/pubescens, Quercus petraea/robur, Deschampsia flexuosa, Pteridium aquilinum.

Physiognomy

The *Quercus* spp.-*Betula* spp.-*Deschampsia flexuosa* woodland is a much less species-rich and less variable community than its mixed deciduous counterparts and much less complicated by the effects of treatment. In essence, these are oak-birch woodlands: oak and birch are very much the most frequent trees throughout and the usual dominants, alone or in mixtures. Both species of oak are well represented, their distribution showing the same general pattern as in the *Quercus-Pteridium-Rubus* woodland. *Quercus robur* predominates over most of the southerly parts of the range, apart from those striking enclaves, such as parts of the Weald, south-east Essex and Sherwood Forest in Nottinghamshire, where it is partly replaced by *Q. petraea* and hybrids (Jones 1959, Gardiner 1974). In these areas, the favouring of *Q. robur* as the plantation oak and the greater facility with which it colonises open ground, tend to sharpen up the difference between older stands, many of which are coppiced, with *Q. petraea*, and younger stands with *Q. robur*, a feature reflected in the views of Rackham (1980) on East Anglian examples of this kind of woodland. Towards the north-western fringes, *Q. petraea* prevails almost exclusively, even in younger tracts.

Oak can dominate here in high-forest canopies which are virtually closed, though the trees are not always tall, especially in more exposed sites at higher altitudes along the Pennine fringes (e.g. Moss 1911, 1913, Scurfield 1953, Pigott 1955). A complete cover of oak is also characteristic of some plantation stands included here. Quite commonly, however, the oak canopy is somewhat open, not only in younger woodlands where colonisation is not far advanced, but also in wood-pasture stands, in which form this community makes an important contribution to the landscape of some of our older Forests like Sherwood (Hopkinson 1927), Epping and parts of the New Forest (Tubbs 1968, Rackham 1980, Peterken 1981) and smaller parks like Staverton (Peterken 1969). Here, magnificent venerable oaks can survive with obstinate individuality among extensive lawns of the field layer of the community, providing historical continuity with an earlier kind of land use and invaluable habitats for invertebrates and lichens (e.g. Welch 1972, Rose 1974). Many of these ancient oaks were pollarded but coppiced oak underwood is quite common in older stands of the community too, with or without oak standards, and long-neglected moots can attain great size (Rackham 1980, Peterken 1981).

Birch can occur as an occasional in the canopy here but, throughout the community as a whole, it easily rivals oak in frequency and is quite commonly the more abundant tree, especially with the neglect of traditional treatments of coppice and wood-pasture and the abandonment of the grazing, cutting and burning of heathlands, over which the community has greatly increased its spread over the past century or so. The *Quercus-Betula-Deschampsia* woodland therefore includes many stands in which oak is sparse or absent, and where birch dominates as an open or patchy canopy or as dense thickets. In some existing schemes (e.g. Tansley 1911, Rackham 1980, Peterken 1981), such woodlands have been treated separately but, apart from the variation in the proportions of oak and birch, there is no floristic justification for this. Both species of birch (and intermediates) occur but their regional separation is less well marked than that of the oaks: *Betula pubescens* is quite common here in the south-east among constant *B. pendula*; to the north-west, the latter species maintains itself at similar frequencies to *B. pubescens*.

No other larger woody species attains anything like the predominance of oak and birch here but, quite commonly, there is some *Pinus sylvestris*. This can be an abundant invader of heathlands, especially where, as is often the case, there are some plantations nearby, and it can persist as a prominent canopy component in sub-spontaneous stands, regenerating well in gaps, even where the woodlands are browsed and grazed (Tansley 1911, 1939, Wooldridge & Goldring 1953; Tubbs 1968). Plantations of pine themselves, where they occur on soils that would normally support this community, can be included within it. Denser stands are virtually bereft of any associates but margins, rides and clearings and older, more open tracts, which are very common in long-established plantations in the Weald and Breckland, have the characteristic field layer. Less commonly, *P. nigra* var. *maritima*, *Larix* spp., *Picea sitchensis*, *P. abies*, *Tsuga heterophylla* and *Pseudotsuga menziesii* can be found as canopy replacements.

One other tree which finds an occasional place in the canopy here is *Fagus sylvatica* but typically it is not abundant, and stands where more balanced mixtures of oak and beech occur over this kind of calcifugous field layer found in this community, a common feature in the New Forest and parts of the Chilterns, fall within the *Fagus-Deschampsia* woodland. *Castanea sativa* can sometimes be found (the community perhaps includes some of the more extreme 'chestnut-oak' woods of Rackham 1980) and, very occasionally, *Sorbus aria* and *Populus tremula*. *Acer pseudoplatanus* is typically scarce, even to the wetter north-west, and *Fraxinus excelsior* and *Ulmus* spp. are hardly ever found.

In some stands, the most characteristic of the smaller woody species of the community, *Sorbus aucuparia* and *Ilex aquifolium*, can also contribute to the main tier of trees. Neither of these is very frequent here, though *S. aucuparia* is markedly more common to the north-west, and, throughout, their occurrence is probably much affected by grazing and browsing, to which both are very sensitive. On ungrazed pasture and heath, both can be abundant invaders, *Ilex* especially having the facility to persist as a local dominant in this kind of woodland, as in some of the New Forest holms (Peterken & Tubbs 1965, Tubbs 1968, Géhu 1975b) and the remarkable Thicks at Staverton Park (Peterken 1969, Rackham 1980) where huge hollies, with a continuous range of descendants down to seedlings, prevail over many hectares. Both *S. aucuparia* and *Ilex* are shade-tolerant and will persist if overtopped by oak: in high-forest stands here, they usually comprise the bulk of the understorey, though generally as an open cover of scattered bushes.

Small woody associates are rare, with *Crataegus* spp. and *Corylus avellana* both being strikingly uncommon. The scarcity of the latter is especially noteworthy: in the southern lowlands of Britain, the frequency of *Corylus* provides a good separation among oak woodlands between this community and the *Quercus-Pteridium-Rubus* woodland, where it is very common, often abundant and has been widely favoured as a coppice crop. In the sub-montane analogues of these two communities, the same distinction does not hold true, *Corylus* transgressing a considerable way on to more surface-leached soils (cf. Peterken 1981).

Other species encountered occasionally are *Frangula alnus*, sometimes prominent too in young stands among invading trees (e.g. Géhu 1975b), *Sambucus nigra*, particularly in disturbed situations like plantations, and *Rhododendron ponticum*, widely planted as an ornamental in this kind of woodland and sometimes spreading to become an abundant and virtually ineradicable weed (Cross 1975).

Where there has been some relaxation of grazing or burning, saplings can be abundant in gaps and, in younger tracts of the community, can provide the bulk of the woody cover around pioneer trees and on expanding margins of more well-established stands. Along with smaller specimens of *Ilex* and *S. aucuparia*, young birch and oak can be plentiful, though the representation of these too is very dependent on the treatment history (e.g. Jones 1959, Pigott 1983). Saplings of *A. pseudoplatanus* can also sometimes be found, more often than canopy trees of this species, though much less commonly than seedlings.

The field layer here is characteristically species-poor but variations on its simple composition are quite wide. The two most consistent elements are *Deschampsia flexuosa* and *Pteridium aquilinum*. The high frequency of the former provides a good separation between this community and the *Quercus-Pteridium-Rubus* woodland, where *D. flexuosa* is uncommon and sparse, especially in the drier south-east. Here, it occurs very frequently and often in considerable abundance, particularly on the shallower soils where the community occurs over more massively bedded sandstones (a typical situation to the north-west) and where the vegetation is open to grazing. Then it can form a virtually continuous sward, close-cropped but rather hummocky, non-flowering in deep shade, and giving the field layer an unvarying dark green colour throughout the year, a very distinctive feature of Pennine woodlands of this kind that are open to sheep and deer (e.g. Moss 1911, 1913, Scurfield 1953, Pigott 1983). Where grazing is relaxed, *D. flexuosa* grows taller and flowers more prolifically but it also begins to trap tree leaf litter which was previously blown off its own short, fine foliage. This is of profound significance not only for its own survival (it tends to be smothered and to die), but also for the subsequent establishment of birch (Pigott 1983: see below). In ungrazed stands, then, the cover of *D. flexuosa* becomes more patchy and confined to knolls and the surfaces of buried boulders standing proud of the slope.

Often, in stands with an uneven occurrence of soils of different depths, such patterning is accentuated by the inverse distribution of *Pteridium* but, where the soils are uniformly of moderate depth and the shade cast by the trees not too intense, bracken may be overwhelmingly dominant by mid-summer, forming a vigorous and dense canopy of fronds often more than a metre high and excluding virtually all other herbs with its own deep shade and thick, slowly-rotting litter. Under more densely shading covers of trees, whether in long-established oak high forest or coppice or beneath thick stands of birch, *Pteridium* may be more restricted, occurring in gaps and rides and along the margins, and continuing its cover into larger clearings, the lawns of wood-pasture areas or on to stretches of as yet uncolonised heathland. It can also come and go somewhat as trees fall or are cropped in coppices and then replaced by regeneration or regrowth, but rapid and extensive expansion of bracken can severely hinder the re-establishment of a woody cover. In stands with heavy amenity use (as on the commons of south-eastern England), the patterning in the *Pteridium* canopy, with its colour change through the year from fresh green to russet (virtually the only phenological variation shown by the field layer), provide a much-valued delight.

In contrast to the *Quercus-Pteridium-Rubus* woodland, *Pteridium* is not consistently accompanied here by an underscrub of *Rubus fruticosus* agg. and *Lonicera periclymenum*. *Rubus* occurs occasionally, *Lonicera* more rarely and, though they may thicken up somewhat when grazing is relaxed or after coppicing, they are typically present only as sparse, trailing shoots, providing little cover for small mammals, a feature of significance to the survival of tree seedlings (Pigott 1984). In some localities, as in parts of Surrey and Kent, *Gaultheria shallon*, an exotic undershrub used as pheasant cover, has become naturalised in this kind of woodland, spreading by seed and stolons.

Much more characteristic here, as bulkier components of the field layer, are ericoid sub-shrubs, particularly *Vaccinium myrtillus* and *Calluna vulgaris*, less commonly *Erica cinerea* and rarely, to the north-west, *V. vitis-idaea*. *V. myrtillus* is the most typical species of closed woodlands here, except where the shade is very dense, though it is more frequent in areas of higher rainfall (like the Weald, the north-west Midlands and more so the Pennines) and its prominence is very dependent on the intensity of grazing (Moss 1913, Scurfield 1953, Pigott 1983). Where the field layer is grazed, it is typically nibbled down to sparse shoots in a *Deschamp-*

sia sward, or to closely-browsed mats distributed over the surfaces of boulders, but it expands its cover and height substantially with enclosure, to the eventual detriment of *V. vitis-idaea* where both occur together, as in rockier Pennine woods (Pigott 1983). *Calluna* also occurs quite commonly, but it is very much a plant of more open places, surviving in colonised stretches of heathland until the canopy closes, after which it becomes leggy and eventually disappears, occurring thereafter largely in gaps or after coppicing (even in stands well away from heaths and then presumably surviving as buried seed: Rackham 1980). *E. cinerea* is likewise a species of more open mosaics of the community among heathland.

By and large, it is differences in the proportions of these elements, a *D. flexuosa* sward, a canopy of *Pteridium* and clumps of ericoids, which give the field layer of this kind of woodland its distinctive stamp and variety, whether on the wooded commons of south-eastern England, in oak coppices and the wood-pasture of ancient Forest or in the typical Pennine oakwood. But other species are quite characteristic at low frequencies and provide local enrichment and, when they are present in some abundance, a small measure of physiognomic variety. Quite commonly there are some other grasses along with *D. flexuosa*. *Agrostis capillaris* and *Anthoxanthum odoratum* occur occasionally and may attain prominence in more closely grazed field layers. *Holcus mollis* can also be found. This is more obviously a plant of the *Quercus-Pteridium-Rubus* woodland in southern Britain and its edaphic preferences as against those of *D. flexuosa*, which prefers and helps to maintain mor, are quite well marked (Jowett & Scurfield 1949, 1952). *H. mollis* can occur here wherever there is a slight shift towards mull conditions on moister and somewhat less acid soils, as where less freely draining drift occurs among sandy superficials or where head or downwashed soil has accumulated on gentler slopes or among boulders (e.g. Woodhead 1906, Moss 1913, Scurfield 1953). *H. mollis* also traps tree leaf litter better than does *D. flexuosa* even in grazed swards, and survives its blanketing effect more successfully, being more extensively rhizomatous, so it may encourage the local maintenance of mull conditions and, where it occurs in mosaics with *D. flexuosa*, expand its cover against it (Ovington 1953).

Other grasses occurring occasionally are *Holcus lanatus*, which is often associated with disturbed situations, as after coppicing or in plantations, and *Deschampsia cespitosa* and *Molinia caerulea*, both of which are generally scarce here but which sometimes mark areas with moister soils or transitions to flushes or wetter stretches of heath. Sedges are characteristically rare, but *Luzula pilosa* sometimes occurs and *L. sylvatica*, though uncommon, can show striking local abundance in ungrazed stands, spreading over more inaccessible steeper slopes in valley-side woodlands, often to the exclusion of all other herbs.

One very marked and valuable distinction between this community and the *Quercus-Pteridium-Rubus* woodland is the great scarcity here of *Hyacinthoides non-scripta*: at most, this is of very local occurrence, usually marking out patches of mull, with *Holcus mollis*. In the absence of this plant and other spring-flowerers, the community typically has no vernal aspect to its field layer, though *Convallaria majalis* occurs locally, providing an attractive show of flowers in May and June. Other vascular associates are often limited to scattered individuals in areas free of substantial bracken or ericoid cover, nibbled down in grazed stands and often not flowering under more complete tree canopies. Among the more characteristic of these plants are *Galium saxatile*, *Potentilla erecta*, *Teucrium scorodonia*, *Corydalis claviculata*, *Digitalis purpurea*, *Solidago virgaurea* and *Rumex acetosella*. All of these may become more prominent in gaps and clearings or after coppicing, together with *Epilobium angustifolium*, persistent clumps of which may mark quite localised areas of disturbance or fire, a not uncommon occurrence in stands on heaths (e.g. Hopkinson 1927, Rackham 1980, Pigott 1983). *Ulex europaeus* and *Cytisus scoparius* can also figure in such situations.

One other occasional component of the field layer is *Hedera helix* but this is nothing like so common here as in the *Quercus-Pteridium-Rubus* woodland, in the range of which moister soils occur widely in areas of more equable climate. Edaphic and atmospheric dryness also restrict the occurrence of ferns in the community: *Blechnum spicant* occurs very sparsely throughout but *Dryopteris dilatata* is confined to north-western stands and more mesophytic dryopteroids and *Athyrium filix-femina* are typically absent.

Seedlings of all the major woody species of the community can also be found in the field layer, though their frequency and abundance, as well as their ultimate fate, are strongly influenced by the past and present treatment, the subtle effects of which on regeneration are discussed below. Seedlings of *A. pseudoplatanus* can also be found in small quantities though, even when there is no grazing, these do not get away here with anything like the facility this species shows on moister and more base-rich soils.

The dry character of the soils here and the low atmospheric humidity also play a part in limiting the contribution which bryophytes make to this vegetation over the more easterly parts of its range. And, though there is a clear increase in their variety and abundance to the wetter north-west, where they help define the *Vaccinium-Dryopteris* sub-community, many stands there are heavily affected by atmospheric pollution which res-

tricts the expression of this element. However, there are usually some small patches of a variety of species and cover may thicken up under the shelter of more open bushes of *Vaccinium* or *Calluna*, over tree bases and on the more open risers of terracettes which commonly develop on steeper slopes. Among the characteristic species are *Dicranum scoparium* and *Hypnum cupressiforme s.l.* (both of which can persist sparsely under dense bracken or in conifer plantation replacements), *Dicranella heteromalla*, *Isopterygium elegans*, *Mnium hornum*, *Campylopus paradoxus*, *Dicranoweissia cirrata*, *Pleurozium schreberi*, *Eurhynchium praelongum*, *Lophocolea bidentata s.l.* and the more pollution-tolerant *Orthodontium lineare* and *Pohlia nutans*. *Leucobryum glaucum* sometimes occurs but not so often as in the beech-dominated analogue of this community, the *Fagus-Deschampsia* woodland.

Sub-communities

***Quercus robur* sub-community:** Oak-birch-heath association Moss *et al.* 1910, Tansley 1911; *Quercetum arenosum roburis et sessiliflorae* Tansley 1911 *p.p.*; *Quercetum ericetosum* Tansley 1911, 1939; Dry oak-birchwood Hopkinson 1927 *p.p.*; *Querco-Betuletum* Klötzli 1970 *p.p.*; Oakwoods Rackham 1980 *p.p.*; Birchwoods Rackham 1980 *p.p.*; Birch-oak stand types 6Cb & 6Db Peterken 1981 *p.p.*; Woodland plot type 18 Bunce 1982 *p.p.* When oak occurs in the woodlands of this sub-community, which is more often than not, it is typically *Q. robur*, except in the *petraea* enclaves. Quite commonly, oak is the woody dominant, in high forest or neglected coppice, where it also provides such standards as are present, or is a long-established tree in wood-pasture where the individuals can be more widely spaced. In other stands, oak is frequent but of low abundance, as where younger woodlands are developing by invasion of open heath or where, with abandonment of traditional treatments, birch (or pine or, more locally, holly and rowan) have increased their proportion of the cover. In fact, stands dominated wholly by birch, usually *B. pendula* here with occasional *B. pubescens*, or by admixtures of birch and pine, to the exclusion of oak, are quite widespread. They may have been a long-standing feature in some places, like the striking landscape of Birklands in Sherwood Forest (Hopkinson 1927, Rackham 1980), but elsewhere are clearly related to a relaxation of the agricultural use of heaths, over which there characteristically develops a mosaic of stands of differing age and composition with intervening stretches of open ericoid vegetation, as on the commons of Surrey and the western Weald, where this patchwork of communities was early described as 'oak-birch-heath' (e.g. Moss *et al.* 1910, Tansley 1911, 1939, Wooldridge & Goldring 1953) and in the New

Forest (Peterken & Tubbs 1965, Tubbs 1968).

Apart from *Pinus sylvestris*, which is quite common here not only as an invader but as a canopy replacement in older plantations (it has fallen from favour somewhat in the south-east in more recent times) and *Fagus sylvatica*, which can become common in areas where beech figures prominently in semi-natural woodlands or plantations, other larger canopy trees are scarce. Smaller trees and shrubs are also infrequent here in well-established stands and, in high forest, the understorey is typically sparse with occasional *Ilex*, a little *Sorbus* and, in ungrazed situations, a few saplings of oak and birch in gaps.

Generally, however, the view beneath the trees is dominated by an abundance of *Pteridium* or, in areas of deeper shade, a sparser bracken cover over a *D. flexuosa* sward. In gaps, around margins and amongst newly-colonising stands on heaths, *Calluna* becomes common and there may also be some *E. cinerea* and *Ulex europaeus* and, in southerly stands, *U. gallii* or *U. minor*, sometimes with a little *Molinia*, the sequence of purple and golden flowers of these shrubs giving the patchwork of vegetation a delightful aspect. *Vaccinium* occurs here, too, surviving a quite dense shade, though it is restricted to stands in areas of higher rainfall, like the western Weald, the New Forest and the north-west Midlands, and its abundance is much affected by grazing.

The frequency of disturbance in many of these woodlands, whether through occasional fires on heaths or planting and felling, makes for a slightly preferential occurrence of species such as *Holcus lanatus*, *Festuca rubra*, *Agrostis stolonifera* and *Pseudoscleropodium purum* but, apart from these plants, the field and ground layers show no distinctive features.

Vaccinium myrtillus-Dryopteris dilatata sub-community: *Quercetum sessiliflorae*, heathy sub-association Moss 1911, 1913; Southern Pennine oakwoods, *Deschampsia-Vaccinium* type Scurfield 1953; Oakwoods Pigott 1956*b* *p.p.*; Birch-oak stand type 6Ab Peterken 1981 *p.p.*; Woodland plot type 18 Bunce 1982 *p.p.* In contrast to the former sub-community, the oak here is almost invariably *Q. petraea* and commonly it is an overwhelming dominant in high forest with a well-defined, though not always very tall, canopy in which birch (both species and intermediates) is limited to more occasional occurrences in gaps or to margins (e.g. Moss 1911, 1913, Scurfield 1953, Pigott 1956*b*, 1983). Some stands are obviously of planted origin and may show ornamentation with *Pinus, Larix* spp. or *Fagus*, well beyond its natural range here. Other tracts show signs of coppicing for oak.

Many of these woodlands have a poorly-developed understorey, a reflection of the fact that they are often open to sheep and deer. However, *S. aucuparia* is

preferentially frequent in this sub-community and *Ilex* occurs quite commonly too and these are the typical components of any second tier of trees, sometimes growing large enough to break the main canopy. In those situations where younger woodlands are developing, quite a rare occurrence in the region but sometimes to be seen on heaths which are no longer burned or grazed or in abandoned quarries, *S. aucuparia* can figure prominently alongside birch (and pine if seed-parents are close), with scattered oak, in low and patchy thickets.

The field layer here preserves the general picture in the community of mixtures of *D. flexuosa*, *Pteridium* and ericoids but often here these are disposed over the varied topography of valley sides with rock exposures and talus above, head and colluvium below, the distribution of which can produce distinctive patterns among these dominants. *Pteridium*, for example, is often limited largely to the deeper soils of the lower slopes, becoming sparse above and increasingly restricted to pockets between the scree fragments. *Vaccinium*, by contrast, though preferentially frequent in this sub-community is often much reduced by grazing on gentler slopes, where *D. flexuosa* forms a sward in unenclosed woods, and attains high and luxuriant cover only over more blocky talus, where it thrives on shallow rankers. Here, too, where the tree cover often thins out to smaller and more widely scattered individuals, *Calluna* can make a contribution.

One other species which frequently figures in the field layer is *Dryopteris dilatata*, a better preferential than *Vaccinium*, and a good indicator of the fact that the soils, though strongly leached, are kept moderately moist by the higher rainfall in the region. It can be very abundant, creating a rather stately impression where its crowns occur thickly scattered among a *D. flexuosa* sward. However, the soils are not so moist as to allow *Oxalis acetosella* to make more than a very occasional appearance here and never have that combination of moisture and somewhat greater fertility so favoured by *Viola riviniana*, its frequent associate in woods towards the north-west of Britain.

A further response to the moister climate, though one often offset in the southern Pennines by atmospheric pollution from the nearby industrial conurbations, is seen in the increased richness and abundance of the bryophyte flora here. *Hypnum cupressiforme s.l.*, *Mnium hornum*, *Isopterygium elegans*, *Dicranella heteromalla* and *Lepidozia reptans* are all preferential, and in more unpolluted and sheltered sites, especially where sandstone blocks provide a variety of niches, there can be records for a wide variety of other species including *Plagiothecium undulatum*, *Isothecium myosuroides*, *Dicranum majus*, *D. fuscescens*, *Polytrichum formosum* and *Rhytidiadelphus loreus* (e.g. Evans 1954, Pigott 1956*b*).

These, together with bulkier lichens, such as *Cladonia squamosa* and *C. polydactyla*, give a foretaste in such sites of the much greater diversity and importance of the cryptogam flora of the sub-montane analogue of this community, the *Quercus-Betula-Dicranum* woodland.

Habitat

The *Quercus-Betula-Deschampsia* woodland is confined to very acid and oligotrophic soils in the southern lowlands of Britain and the upland fringes of the Pennines. In some areas, long-established stands occur as high forest or old oak coppice or as components of ancient wood-pasture landscapes in Forests and parks, but to the south especially many stands are a relatively recent development on heathy commons whose traditional exploitation has fallen into disuse. Grazing continues to be important in some areas and afforestation with hardwoods and, more importantly, softwoods, has had a widespread influence on the composition and structure of the vegetation.

Edaphically, this community occupies one extreme among wooded lowland soils, being limited to lime-free profiles with a superficial pH that is rarely above 4 and with mor humus (Moss 1911, 1913, Tansley 1939, Rackham 1980, Peterken 1981). More lithomorphic soils are shallow humic rankers, sometimes amounting to little more than accumulations of organic remains and rock particles disposed fragmentarily over and among boulders, but deeper profiles are widespread, especially over superficials. Typically, however, the soils are free-draining, sometimes excessively so, being very commonly of a sandy texture and, when more mature, they characteristically show signs of strong eluviation, being of the brown podzolic type (sometimes with micropodzols) or humo-ferric podzols with marked bleaching and pan formation or, where there is some influence of ground water deep in the profile, stagnopodzols or stagnogley-podzols (Avery 1980: see also Tubbs 1968, Corbett 1973, Furness 1978, Carroll et al. 1979, Pigott 1983).

Within its range, the *Quercus-Betula-Deschampsia* woodland effectively serves to mark out the major areas of such soil types. In some places, their occurrence is related to the distribution of pervious arenaceous sedimentaries. In the Weald, for example, sands and sandstones of the Cretaceous sequence are important substrates, Ashdown and Tunbridge Wells Sands with Poundgate stagnogley-podzols occurring in Ashdown Forest on the High Weald, and Folkestone Sands and Hythe Beds carrying Shirrell Heath humo-ferric podzols running along the western rim (Wooldridge & Goldring 1953, Soil Survey 1983). The great frequency of the community in this area is carried over on to the Surrey commons where Eocene Bagshot sands and gravels have Holidays Hill stagnogley-podzols and a variety of profiles of this kind underlie stands on Bagshot and Barton beds in the New Forest (Tubbs 1968). In a great arc around the Midlands, running from Cheshire to Nottinghamshire, Triassic Keuper and Bunter sandstones provide localities with podzolised soils (Hopkinson 1927, Mackney 1961, Furness 1978) and, all around the Pennine fringes, a very characteristic sequence of Anglezarke humo-ferric podzols, Revidge humic rankers and Belmont stagno-podzols is disposed over the tops, freely-weathering faces and colluvial slopes of Carboniferous sandstones and grits carrying the *Vaccinium-Dryopteris* sub-community (Furness 1978, Carroll et al. 1979, Pigott 1983, Jarvis et al. 1984).

In other places, the *Quercus-Betula-Deschampsia* woodland occurs over more free-draining superficials, as in the New Forest where Pleistocene Plateau Gravels carry stagnogley-podzols (Tubbs 1968) and on the Cheshire Plain and in Breckland, down through Suffolk and into Essex, where extensive tracts of brown sands derived from fluvioglacial or aeolian material show patchy podzolisation (Corbett 1973, Furness 1978, Rackham 1980).

Variation in these kinds of profiles under individual stands of the community is often quite marked, even within very short distances (e.g. Mackney 1961, Pigott 1983), being related to differences in topography and in the character of the parent materials and perhaps also to developmental interactions between the vegetation and the soil (see below). As noted earlier, this can have some marked effects on the distribution and abundance of some of the important species here, notably *Pteridium*, *Vaccinium* and the bryophytes. But the general nature of the edaphic conditions is well marked in the overall floristic character of the community in both a negative and positive fashion. More calcicolous plants never find a place here and even more tolerant species like *Corylus, Crataegus* spp., *Rubus, Holcus mollis* and *Hyacinthoides*, which are so important on brown earths of low base-status, are largely excluded. By contrast, calcifuges, notably *D. flexuosa*, the ericoids, a variety of other low-frequency herbs and bryophytes, which are only very sparsely represented on such profiles, are strongly preferential. By and large, then, the soil-related boundary between this community and its counterpart on moister and less markedly leached profiles, the *Quercus-Pteridium-Rubus* woodland, is quite well defined, though transitional stands do occur, especially where the character of superficial deposits changes gradually; where the two communities are found contiguously, mosaics of their smaller herbs, sometimes overlain by an unbroken canopy of *Pteridium*, are fairly frequent. Coppicing for oak, extended over such transitions, may further blur their sharpness.

The floristic boundary with calcifuge woodlands of wetter, acid soils is generally a clear one. The free-draining nature of the profiles here prevents any tendency to surface-water gleying and, though deeper soils are intermittently affected by the ground water-table (in the Weald, for example, and the New Forest), this never reaches high in the profile. Thus, though *Molinia caerulea* can occasionally be found in the community, this is more a reflection of the increasing prominence of this grass in fairly dry situations towards the south-west, rather than any obvious confusion of the junction between the *Quercus-Betula-Deschampsia* woodland and its counterpart on acid soils which show accumulation of surface peat, the *Betula-Molinia* woodland.

However, though the soils here are always free-draining, they are not necessarily kept exceedingly dry; and neither is atmospheric humidity consistently low. Indeed, there is quite a sharp difference in the value of these variables between stands in, say, East Anglia and the Pennines and this probably plays a considerable part in determining the floristic differences between the sub-communities. The effect of increasing rainfall and humidity can already be seen within the *Quercus* sub-community in the way in which *Vaccinium* appears in areas with a mean annual rainfall in excess of 800 mm but, with the further shift to precipitation levels approaching 1000 mm (*Climatological Atlas* 1952) and 140 wet days yr^{-1} (Ratcliffe 1968), the *Vaccinium-Dryopteris* sub-community becomes the characteristic type of *Quercus-Betula-Deschampsia* woodland with strongly preferential *D. dilatata* and a greater representation of bryophytes. Increasing richness among the latter component means that the ground flora of stands in this region begins to take on some of the character of the sub-montane *Quercus-Betula-Dicranum* woodland. Although, in their extreme forms, this community and the *Quercus-Betula-Deschampsia* woodland look very different, much of the contrast is provided by the bryophytes and, as rainfall and humidity rise progressively in moving to the north-western uplands, there is a fairly gradual switch from the one community to the other. Indeed, it is possible that, without such high levels of atmospheric pollution towards the northern fringes of the *Quercus-Betula-Deschampsia* woodland, some Pennine stands here would move over to the *Quercus-Betula-Dicranum* woodland, shifting the geographical boundary between the two communities. The demise of industrial activity in some Pennine valleys and an increasing attention to air-cleanliness could perhaps produce some change in this direction.

A further environmental variable affecting the *Quercus-Betula-Deschampsia* woodland is grazing and browsing by stock and deer which influence both the physiognomy and quantitative composition of the field layer (and of the bryophyte component in so far as it makes a contribution here) and the regenerative capa-

city of the major woody species. Even in such a floristically simple community as this, the ways in which larger herbivores, or their exclusion, operate are quite subtle though, in the long term, very striking. Thus, long-continued grazing and browsing play a major part in the appearance of this kind of woodland where it contributes, usually as the *Quercus* sub-community, to the wood-pasture landscape of some ancient Forests and parks, with a preponderance of older trees set amongst extensive grassy lawns and bracken (e.g. Hopkinson 1927, Rackham 1980, Peterken 1981). And in other situations where stock and deer have access, including many stands of the *Vaccinium-Dryopteris* sub-community which, in Pennine valleys, are often contiguous with open moorland, the striking scarcity of smaller trees and saplings can be directly attributed in many cases to the predation of herbivores (e.g. Pigott 1983). Conversely, where grazing and browsing have been reduced, there is often a clear response here among the more sensitive members of the field layer which increase in height and cover, in the bryophytes, which become more confined to rock outcrops and tree bases, and among the woody species which are able to regenerate more effectively. Such curtailment of grazing and browsing has occurred within stands of the community as a result of reduction in the numbers of particular herbivores (as in the New Forest: Peterken & Tubbs 1965, Tubbs 1968) or because of enclosure (as at Staverton Park: Peterken 1969, Rackham 1980) or as part of more widespread cultural changes (as on the Surrey and Wealden commons where grazing, and the burning and cutting of bracken and heather have all declined: Wooldridge & Goldring 1953, Hoskins & Stamp 1963). In many areas, there has been extensive renewed regeneration of this kind of woodland and a spread on to previously open ground in larger clearings within existing woods and on adjacent heath, and in some cases it is possible to trace the origin of individual stands to particular local events in the grazing history.

The intricacies involved in such events, and especially the different responses of oak and birch to the direct and indirect effects of grazing reduction, are well shown in the enclosure experiment which Pigott (1983) conducted in a Pennine stand of the *Vaccinium-Dryopteris* sub-community. Among all the woody species represented here, birch is the most consistent and prolific fruiter and wind-dispersal is very effective in producing a potentially widespread distribution of offspring. But birch is a great light demander, even at the germination stage, the largest numbers of seedlings being found in short turf or on patches of bare ground. Moreover, having small food reserves in the seed, it cannot get a hold on thicker deposits of loose litter. Grazing (or burning), which creates such close-cropped, litter-free conditions, thus tends to favour seedling establishment, though it can be quite sufficient, even where canopy shade is not a problem, to eliminate each generation of offspring completely. When grazing was withdrawn in Pigott's experiment, some of the existing seedlings got away to form saplings in gaps but, as the *D. flexuosa* sward grew taller and began to trap litter, and *Vaccinium myrtillus* extended its cover, conditions became uncongenial for any further seedling establishment. The developing prominence of birch was thus related to a quite narrowly defined period of transition from the grazed to the ungrazed state. This kind of response, together with the fact that birch cannot regenerate under its own canopy, is probably responsible for the fact that, even where this species has become very abundant in stands on previously unwooded ground, it often occurs in patchy mosaics of more or less even-aged groups of trees.

Oak is rather different. It fruits less regularly and abundantly than birch but, after good summers, acorns can be plentiful. However, though it is one of the trees here that are less favoured by herbivores, and can be quite resistant to repeated damage even when very young (Jones 1959, Rackham 1980), seedlings are again often totally devoured where grazing continues. In Pigott's experiment, oak (*Q. petraea* in this case) got away only in the enclosure. In contrast to birch, however, seedlings continued to establish intermittently even in the ungrazed sward: in oak, germination is actually favoured by burial among taller herbage and litter, which affords protection against desiccation and predation, and, having bulky food reserves, fairly thick accumulations of unconsolidated material present no great barrier to radical growth down to the mineral horizons (Jones 1959). In the particular wood in which Pigott erected his enclosure, he considered that *Q. petraea* would come to outstrip birch as the eventual gap-filler. In *Quercus-Betula-Deschampsia* woods where *Q. robur* is the usual oak, as in most stands of the *Quercus* sub-community, oak regeneration under established canopies is more problematical. *Q. petraea* is about as shade-tolerant as *Vaccinium* but *Q. robur* is not and its great prominence in the *Quercus* sub-community (on soils which are not those most favoured by this species) may result from its greater ability to colonise new ground. Many stands in this sub-community are fairly young and tracts with *Q. petraea* are often long-established high forest or coppice. Though acorns are heavy, they can be transported considerable distances by small mammals and birds (Jones 1959, Mellanby 1968) and young saplings of *Q. robur* are a common feature among advancing birch here.

Grazing and browsing also have a marked effect on the smaller woody companions of oak and birch, *Ilex* and *S. aucuparia*. In Pigott's enclosure, the latter became quite plentiful, especially over block scree, and being shade-tolerant, grew well, even under more intact areas

of the canopy. *Ilex*, though rare in this particular wood, responds in the same way and is likewise shade-tolerant, even more so than *S. aucuparia* and, itself casting dense evergreen shade, can have a much more obvious effect on the accompanying flora where it becomes prominent (Peterken 1965, 1966). In stands in the New Forest (Peterken & Tubbs 1965) and in Staverton Park (Peterken 1969, Rackham 1980), local holly dominance in this community has been related to specific events in grazing reduction.

It is possible that grazing also has some effects on the soils under this community by helping maintain compact mor. Enclosure in Pigott's experiment did not disrupt the stability of the podzolised profiles but it did permit the development of more friable humus beneath the accumulations of decaying leaves and, within this, fine tree roots and fungal mycorrhizae became more profuse. Enhanced root growth may have been a response to decreased trampling by herbivores and the moister conditions developing beneath the leaf litter; mycorrhizal growth may have benefited by the decrease in *D. flexuosa* which seems to exert an inhibiting effect (Jarvis 1964). Whatever the particular causes, the scene seemed set for better tree growth within the enclosure and perhaps even reduced mortality (Pigott 1983). Local and temporary small-scale shifts in the character of the soils here, in response to grazing differences, may be quite widespread. They certainly seem to operate in edaphically transitional situations, as in mosaics with the *Quercus-Pteridium-Rubus* woodland, where litter accumulation, as a result of reduced grazing, can alter the balance in the field layer between *D. flexuosa* and *Holcus mollis* over soils that vary from brown earths with mull to podzols with mor (e.g. Ovington 1953).

Treatment of the *Quercus-Betula-Deschampsia* woodland as wood-pasture has sometimes been combined with pollarding of oak, and also in some places, of holly, in attempts to maintain a continuing balance of herbage and canopy. Other stands have been used as oak coppice and, locally, holly seems to have been cut as a supplement to winter fodder. Oak timber has also been widely extracted from this kind of woodland and some stands represent oak plantations. Where long established, these do not present a markedly different assemblage of species from semi-natural stands of the community, though they often have a sparse understorey and the trees lack individuality and great age. Older plantations of pine with some thinning can likewise conform in general terms to tracts with an oak and birch canopy, though recently-disturbed stands (and younger deciduous plantations) may show an abundance of herbs such as *Epilobium angustifolium*, *Holcus lanatus*, *Digitalis purpurea* and sub-shrubs like *Calluna*, *E. cinerea* and *Ulex* spp., an assemblage which probably approximates to the post-coppice flora here. Younger conifer plantations on soils which could carry the *Quercus-Betula-Deschampsia* woodland look very different: here, in the very dark conditions beneath the canopy, the ground cover may be reduced to but a few sparse wefts of mosses over a carpet of needles.

Zonation and succession

Zonations from this community to other kinds of woodland are most frequently related to differences in soils. Quite often, these are complicated by the effects of sylvicultural treatments and the impact of larger herbivores, but grazing and browsing also mediate other kinds of transitions to grassland and heath which are a spatial expression of succession or its reversal. Burning is also a factor in these latter kinds of zonations. Although neglect of heathland exploitation has allowed extensive development of progressive successional change in some areas, many stands have artificially sharp boundaries with adjacent land converted to agricultural use.

Edaphically-related transitions usually reflect changes to soil parent materials that are not so markedly acidic or pervious (commonly both) as those which typically weather to the humic rankers and podzolised profiles characteristic here. Such changes are very common within whole woods in which this community is represented and they most often result in zonations to the *Quercus-Pteridium-Rubus* woodland, the forest type of base-poor brown earths, frequently with some drainage-impedence, in the British lowlands. Boundaries between the two communities are not always very sharp, especially where superficials provide the underlying substrate. Fluvioglacial deposits, for example, and aeolian material are commonly very heterogeneous, even over short distances, and, in parts of the Midlands, in Suffolk and Essex and in some areas of the New Forest, small-scale variation in the proportions of sand, silt and clay may be sufficient to produce mosaics of brown earths or brown sands and podzols which bear patchworks of these two kinds of woodlands, in these regions usually the Typical sub-community of the *Quercus-Pteridium-Rubus* woodland and the *Quercus* sub-community of the *Quercus-Betula-Deschampsia* woodland. The distribution of more mesophytic species characteristic of the former and the calcifuge herbs of the latter often provides a firm diagnosis of the components but, by mid-summer, when the vernal dominance of *Hyacinthoides* has faded from the *Quercus-Pteridium-Rubus* woodland and a dense canopy of bracken extends throughout, the distinctions may be masked. Uniform treatment as wood-pasture, which can eliminate most of the understorey, and pollarding throughout, can cause further confusion. Oak coppice sometimes extends over such boundaries, though the post-coppice floras in actively-treated stands of each of the two communities

can indicate the underlying soil differences afresh. Coppicing for hazel, largely confined to the *Quercus-Pteridium-Rubus* woodland, may actually highlight the transition.

Where these kinds of soil differences are related to changes in sedimentary rocks, the zonation between the two kinds of woodland can be a much sharper one. Interbedded clays are a characteristic feature among the Cretaceous sands and sandstones of the Weald and they occur, too, within the Eocene sequences of Essex and the New Forest. Similar alternations are found further north in the Triassic deposits of the Midlands. All of these can carry bands of the *Quercus* sub-community of the *Quercus-Betula-Deschampsia* woodland and the Typical sub-community of the *Quercus-Pteridium-Rubus* woodland, the former usually above the latter on the more resistant topography of the arenaceous rocks, giving way below to the more weathered argillaceous slopes. On the Pennine fringes this kind of geological and topographic contrast, reflected in the soils and woodland cover, is even more striking with Carboniferous sandstones and grits forming cliffs and screes with the *Vaccinium-Dryopteris* sub-community, giving way below to the *Acer-Oxalis* sub-community of the *Quercus-Pteridium-Rubus* woodland on shales. Edaphic variation over the extensive and varied grit exposures may also here introduce much more field-layer variation into the *Quercus-Betula-Deschampsia* woodland than is usual over the gentler topography of the southern lowlands. Quite a common complication, too, is for landslips to occur where incompetent shales collapse and spill down great blocks of the overlying grits on to the lower slopes: then perched fragments of the calcifuge flora of the community may occur among a ground of the *Quercus-Pteridium-Rubus* woodland, a striking sight in spring when a carpet of bluebells runs up between them.

Very often, on shedding slopes, these junctions between arenaceous and argillaceous sedimentaries are marked by flushing, as ground waters percolating through the sands, sandstones or grits hit the impervious clays or shales and emerge in springs or seepage lines. Then the transition from the *Quercus-Betula-Deschampsia* woodland to the *Quercus-Pteridium-Rubus* woodland can have some kind of interposed flush woodland, frequently of small extent, though often occurring repeatedly along the geological boundary. Where the flush waters are not too base-poor, the *Carex-Cirsium* sub-community of the *Alnus-Fraxinus-Lysimachia* woodland is typical in such situations, being a common feature in the western Weald and along Pennine valley sides; where they are more acid, frequently the case to the north where they drain from expanses of moorland, the *Juncus* sub-community of the *Betula-Molinia* woodland replaces it in analogous topographic situations.

Where flush waters emerge at such geological junctions within more level-bedded rocks and maintain the soils over the clays and shales in a more permanently waterlogged state, some surface peat may accumulate. Then, these kinds of flush woodland are replaced by their topogenous equivalents, the *Carex-Alnus* woodland in more base-rich sites (usually the *Chrysosplenium* sub-community), the *Sphagnum* sub-community of the *Betula-Molinia* woodland in the more base-poor. Often, the stands of these kinds of wet woodland are more extensive than their flush counterparts, sometimes wholly replacing the *Quercus-Pteridium-Rubus* woodland in its position below the *Quercus-Betula-Deschampsia* woodland. Again, the western Weald and the Pennine fringes have good examples of this kind of pattern.

The other very common kind of zonation in which this woodland can be found involves transitions to heaths, or to complexes of heath with calcifugous grasslands and stands of bracken. The type of heath occurring in association with the community varies considerably according to the region: over the range of the *Quercus-Betula-Deschampsia* woodland, heaths show wide phytogeographic differences. In East Anglia and down into parts of the Weald, the *Calluna-Festuca* heath is typical, then from there westwards the *Calluna-Ulex minor* and *Ulex minor-Agrostis curtisii* heaths and, in the north Midlands, the *Calluna-Ulex gallii* heath. In the Pennines, the *Calluna-Deschampsia* and *Calluna-Vaccinium* heaths are the usual companions. Sometimes the zonations from woodland to heath are well ordered and have some clear basis in topographic and edaphic variation, as where tree cover becomes very sparse over fragmentary humic rankers on grit boulders or cliffs in the Pennines (e.g. Moss 1913, Pigott 1956*b*), but very often such patterns are more complex and independent of substrates and soils. Frequently, here, fragments of heath occur within stretches of the woodland, confined to gaps or forming larger areas in the kind of landscape seen in some ancient wood-pastures (e.g. Hopkinson 1927, Mackney 1961, Rackham 1980, Peterken 1981), or stands of the woodland are distributed among extensive tracts of heath, as on the Wealden and Surrey commons and in parts of the New Forest (e.g. Tansley 1939, Wooldridge & Goldring 1953). And in such situations, it is usually abundantly clear that the zonations are an expression of seral progression to woodland or its reconversion to heath as a result of a variety of biotic activities.

Thus, heath sub-shrubs and *Pteridium* form the characteristic gap or coppice flora here, developing where trees fall or are extracted or cut as underwood. And these plants can spread at the expense of a tree cover in the immediate aftermath of destruction of the community by fire. Grazing and browsing may work more slowly, though to the same ultimate end: a very open canopy of trees can persist in wood-pasture stands of the *Quercus-Betula-Deschampsia* woodland for centuries, but with no hope of unaided regeneration if access to herbivores is maintained, a situation seen only too well in some old Forests and parks (Peterken & Tubbs 1965, Rackham 1980, Peterken 1981). Where the community has declined in extent or vigour, then, or where its expansion has been checked, it is these factors, working separately or often in complex combinations (e.g. Peterken & Tubbs 1965, Peterken 1969, 1981, Rackham 1980), that seem generally to blame, rather than any natural demise in the face of extreme edaphic conditions (cf. Tansley 1911, 1939) or a harsh climate (Pigott 1983). Moreover, the ability of the community to regenerate and colonise new ground has been observed with striking clarity wherever and whenever the incidence of these factors has declined, the neglect of the exploitation of heathy commons being especially important in the recent extensive spread of this kind of woodland in the south-east.

However, though the community seems to be a quite natural development from heaths at the present time, two questions arise over its status. The first is of long-term historical interest and concerns the possibility that, in many areas, heath vegetation has prepared the ground for the spread of the *Quercus-Betula-Deschampsia* woodland by enhancing podzolisation on land that was previously of higher fertility and perhaps bore more mesophytic forest but which was progressively exhausted by early agricultural activity (e.g. Dimbleby & Gill 1955, Dimbleby 1962, Gimingham 1972). This may well have been the case in some places but whether it has everywhere been a necessary precondition is another matter. The community can certainly maintain podzols in a stable state (Mackney 1961, Pigott 1983) and oak-birch forest of this kind may itself be able to promote podzolisation (Scheys *et al.* 1954, Munault 1959).

The second issue concerns the stability of the community in relation to its beech-dominated analogue, the *Fagus-Deschampsia* woodland, in particular the competitive ability of oak and beech on the markedly dry, acid soils characteristic of both communities. On moister soils of low base-status, *Fagus* is at some disadvantage against *Q. robur* (Watt 1924, 1934*b*), so the distinction between the *Quercus-Pteridium-Rubus* and the *Fagus-Rubus* woodlands often has a clear edaphic basis in semi-natural woods within the natural range of beech. Here, oak and beech are more closely matched in their potential ability to dominate, each is found in stands where the other is pre-eminent and, in those areas where both communities are well represented, as in the New Forest (Peterken & Tubbs 1965, Tubbs 1968) and parts of the Chilterns (Watt 1934*b*, Avery 1958), intimate mosaics of the two are common. A number of factors are probably important in determining which species will come to prevail. First, though *Fagus* grows well way beyond its natural limit to the north, there may be some climatic restriction on its vigour towards the drier parts of East Anglia (Rackham 1980). Second, outside this area, *Fagus* is still a very erratic fruiter and oak, certainly *Q. robur*, though itself not very consistent in acorn production, has the advantage that its fruits can be transported over considerable distances, whereas beech mast generally drops straight down off the parent (Watt 1924, Jones 1959, Mellanby 1968). In larger gaps, therefore, and on previously open ground, colonisation by oak, as against beech, is more widespread and more regular, though grazing and browsing, which favour the latter, can offset this (Dengler 1930, Jones 1959). Third, under closed canopies, the advantage moves in the other direction (provided adult beech is present and has produced mast) because *Q. robur* regenerates badly under established woodland and beech seedlings and saplings are more shade-tolerant than those of *Q. petraea*. Here, then, beech may triumph eventually (Watt 1924, 1934*b*), even well outside its range (Pigott 1983).

The outcome of such conflicting factors may, in many places and at any given time, be a fairly even match so that the dominants, and their communities, form a mosaic, each coming and going, here and there, in a complex pattern of regenerative replacement. On the Continent, the consensus of opinion is that the equivalents of the *Fagus-Deschampsia* woodland represent the climax on the kinds of soil characteristic here, the counterparts of the *Quercus-Betula-Deschampsia* woodland, their seral precursors or degraded derivatives (e.g. Géhu 1975*a*) and, given ideal conditions for beech, this may well be the situation here. Often, however, conditions are not ideal so that the *Quercus-Betula-Deschampsia* woodland, though theoretically a sub-climax, attains a considerable measure of stability.

The other tree species which frequently complicates successions to this community is *Pinus sylvestris* which is widely distributed as an ornamental in this kind of woodland and extensively planted on lowland podzols. It fruits heavily, has light wind-dispersed seeds and does especially well, with birch, as a coloniser of open ground after fires have cleared a cover of bracken and heath (e.g. Tubbs 1968). Where its canopy remains open or where it is present in mixed covers of trees, the vegetation can retain the general character of the community but densely-shaded stands of pine quickly eliminate virtually all the field-layer species and bryophytes. The same poverty is characteristic of plantations of *P. sylvestris* and other conifers on the soils typical here, though thinning may restore some of the richness of the vegetation (such as it is). Often, now, in intensively afforested

areas, the *Quercus-Betula-Deschampsia* woodland is reduced to a fringe of birch, bracken and ericoids along the edges of rides and on recently-cleared compartments.

The generally poor quality of the soils under the community means that conversion of this kind of woodland to intensive agricultural land is exceptional. Quite commonly, stands have been reduced to isolated remnants but they usually persist, outside the heathland or wood-pasture context, in pastoral landscapes, marked off sharply from some kind of improved grassland, either various types of *Festuca-Agrostis* sward or more calcifugous sub-communities of mesotrophic grassland. Narrow strips of bracken, ericoids or *Ulex* spp. may occur as a compressed fringe around the woodland.

Distribution

The *Quercus-Betula-Deschampsia* woodland has a widespread but patchy distribution throughout the southern part of Britain, occurring in concentrations on areas of marginal land, notably in south-east England, the north-west Midlands and on the Pennine fringes and occasionally in other places where private parkland or common rights have afforded protection against clearance or favoured its advance.

Affinities

This kind of woodland has long been recognised as constituting a calcifugous extreme in the more southerly parts of Britain, though early descriptions tended to make distinctions within it according to whether it occurred as young stands in complexes with heath, where birch was often pre-eminent (e.g. Moss *et al.* 1910, Tansley 1911, 1939) or as more mature high forest where oak was the usual dominant (e.g. Moss 1911, 1913, Tansley 1939). Among the latter types of woodland, there has also been some hesitation as to whether stands should be separated on the basis of which oak species was the dominant. A distinction between birch- and oak-dominated woodlands persists in the scheme of Rackham (1980), where true acidophilous oak-woodland is consequently considered very rare in his lowland zone; separation on the basis of the oaks remains in Peterken's (1981) classification. Both these authors also include within their oak- and birchwoods vegetation which is, in this scheme, placed in the *Quercus-Pteridium-Rubus* woodland.

In phytosociological terms, the *Quercus-Betula-Deschampsia* woodland represents the Quercion robori-petraeae in the southern lowlands of Britain, the alliance which (with its beech analogue, the Luzulo-Fagion) replaces Carpinion woodlands on markedly-leached soils here and throughout the more temperate parts of Europe. It has clear counterparts in associations like the *Quercetum sessiliflorae* Lemée 1937, the *Betulo-Quer-*

cetum roboris Tüxen 1937, the *Querco petraeae-Betu-letum*, the *Luzulo luzuloidis-Quercetum* Noirfalise & Sougnez 1956 and the *Vaccinio-Quercetum sessiliflorae* Clement *et al.* 1975 described from France, Belgium and Germany (see also Dumont 1975, Duvigneaud 1975, Sougnez 1975, Frileux 1975, Tombal 1975). Some of these communities are considerably richer than the *Quercus-Betula-Deschampsia* woodland (being more akin to our sub-montane *Quercus-Betula-Oxalis* woodland) and they sometimes include wetter forms which are here placed in the *Betula-Molinia* woodland. But, in general floristics and in their edaphic and biotic relationships, they show a close similarity to their British counterpart.

Floristic table W16

	a	b	16
Betula pendula	V (4–10)	III (1–7)	IV (1–10)
Quercus robur	III (1–8)	I (1–8)	II (1–8)
Quercus petraea	II (1–9)	V (4–10)	II (1–10)
Betula pubescens	II (2–8)	III (1–7)	II (1–8)
Acer pseudoplatanus	I (1–4)	I (1–4)	I (1–4)
Castanea sativa	I (3–9)	I (6)	I (3–9)
Fagus sylvatica	I (3–7)	I (1–5)	I (1–7)
Ilex aquifolium	I (2–4)	I (1–4)	I (1–4)
Populus tremula	I (3–4)	I (3)	I (3–4)
Pinus sylvestris	II (3–8)	I (1–5)	I (1–8)
Frangula alnus	I (3–4)		I (3–4)
Sorbus aria	I (3)		I (3)
Ilex aquifolium	II (3–4)	II (2–4)	II (2–4)
Corylus avellana	I (5)	I (3)	I (3–5)
Crataegus monogyna	I (2)	I (1)	I (1–2)
Acer pseudoplatanus sapling	I (3–4)	I (1–2)	I (1–4)
Rhododendron ponticum	I (2–4)	I (1)	I (1–4)
Fagus sylvatica sapling	I (3)	I (1–3)	I (1–3)
Betula pendula sapling	I (4–5)	I (1–5)	I (1–5)
Quercus hybrids sapling	I (1–5)	I (1)	I (1–5)
Quercus robur sapling	I (1–4)		I (1–4)
Sorbus aucuparia	I (2–3)	III (2–4)	II (2–4)
Betula pubescens sapling	I (3–7)	II (1–5)	I (1–7)
Quercus petraea sapling	I (2–3)	II (2–6)	I (2–6)
Deschampsia flexuosa	IV (3–9)	V (1–9)	V (1–9)
Pteridium aquilinum	IV (1–10)	III (2–9)	IV (1–10)
Holcus lanatus	II (1–7)	I (4)	I (1–7)
Ulex europaeus	I (1–10)	I (2)	I (1–10)
Ulex gallii	I (6–10)		I (6–10)
Agrostis stolonifera	I (2–3)		I (2–3)
Festuca ovina	I (2–4)		I (2–4)
Festuca rubra	I (3–5)		I (3–5)
Pseudoscleropodium purum	I (3–5)		I (3–5)
Vaccinium myrtillus	II (2–7)	III (3–10)	II (2–10)
Dryopteris dilatata	I (1–6)	III (1–8)	II (1–8)
Dicranella heteromalla	I (2–5)	II (1–3)	I (1–5)

Hypnum cupressiforme	I (2–6)	II (1–4)	I (1–6)
Isopterygium elegans	I (2)	II (1–3)	I (1–3)
Mnium hornum	I (2–3)	II (1–5)	I (1–5)
Lepidozia reptans		II (1–5)	I (1–5)
Oxalis acetosella		I (2–3)	I (2–3)
Plagiothecium undulatum		I (1–4)	I (1–4)
Plagiothecium denticulatum		I (1–3)	I (1–3)
Tetraphis pellucida		I (1–3)	I (1–3)
Lophocolea heterophylla		I (1–2)	I (1–2)
Campylopus pyriformis		I (2)	I (2)
Dicranum fuscescens		I (6)	I (6)
Isothecium myosuroides		I (3)	I (3)
Plagiothecium succulentum		I (3)	I (3)
Plagiothecium sylvaticum		I (1)	I (1)
Rhytidiadelphus loreus		I (2)	I (2)
Rhytidiadelphus triquetrus		I (3)	I (3)
Thuidium tamariscinum		I (3)	I (3)
Barbilophozia attenuata		I (2)	I (2)
Calypogeia fissa		I (2)	I (2)
Cephalozia bicuspidata		I (1)	I (1)
Gymnocolea inflata		I (2)	I (2)
Cladonia coccifera		I (3)	I (3)
Cladonia coniocraea		I (1)	I (1)
Calypogeia trichomanis		I (3)	I (3)
Dicranodontium denudatum		I (3)	I (3)
Barbilophozia floerkei		I (5)	I (5)

Calluna vulgaris	II (2–6)	II (1–9)	II (1–9)
Rubus fruticosus agg.	II (1–7)	II (1–6)	II (1–7)
Agrostis capillaris	I (3–7)	I (4–5)	I (3–7)
Galium saxatile	I (1–5)	I (1–4)	I (1–5)
Epilobium angustifolium	I (1–3)	I (1–3)	I (1–3)
Hedera helix	I (3–7)	I (3–4)	I (3–7)
Holcus mollis	I (1–8)	I (4–8)	I (1–8)
Lonicera periclymenum	I (1–6)	I (3–4)	I (1–6)
Dicranum scoparium	I (1–4)	I (1–2)	I (1–4)
Potentilla erecta	I (1–4)	I (1–3)	I (1–4)
Teucrium scorodonia	I (1–5)	I (1–3)	I (1–5)
Corydalis claviculata	I (1–3)	I (1–3)	I (1–3)
Leucobryum glaucum	I (1–3)	I (3)	I (1–3)
Eurhynchium praelongum	I (1–3)	I (1–2)	I (1–3)
Hypnum jutlandicum	I (2–5)	I (1–4)	I (1–5)
Orthodontium lineare	I (1–2)	I (1–4)	I (1–4)
Lophocolea bidentata s.l.	I (1–3)	I (1–2)	I (1–3)
Sorbus aucuparia seedling	I (1–5)	I (1–3)	I (1–5)
Quercus petraea seedling	I (1–2)	I (1–2)	I (1–2)
Betula pubescens seedling	I (1–2)	I (1–2)	I (1–2)
Ilex aquifolium seedling	I (1–3)	I (1–2)	I (1–3)
Anthoxanthum odoratum	I (1–2)	I (4)	I (1–4)
Blechnum spicant	I (1–3)	I (1)	I (1–3)
Deschampsia cespitosa	I (1–3)	I (2)	I (1–3)

Floristic table W16 *(cont.)*

	a	b	16
Digitalis purpurea	I (2–3)	I (3)	I (2–3)
Hyacinthoides non-scripta	I (2–4)	I (4)	I (2–4)
Erica cinerea	I (2–7)	I (3)	I (2–7)
Luzula sylvatica	I (1–5)	I (4)	I (1–5)
Molinia caerulea	I (2–8)	I (1)	I (1–8)
Rumex acetosella	I (3–4)	I (3)	I (3–4)
Solidago virgaurea	I (2–6)	I (1)	I (1–6)
Campylopus paradoxus	I (3)	I (1–3)	I (1–3)
Dicranoweissia cirrata	I (1)	I (1–3)	I (1–3)
Pleurozium schreberi	I (2–5)	I (2)	I (2–5)
Pohlia nutans	I (2–3)	I (1)	I (1–3)
Cladonia fimbriata	I (2)	I (2)	I (2)
Cladonia polydactyla	I (1)	I (1–2)	I (1–2)
Cladonia squamosa	I (1–2)	I (1–3)	I (1–3)
Quercus robur seedling	I (2–3)	I (2)	I (2–3)
Fagus sylvatica seedling	I (2)	I (1)	I (1–2)
Acer pseudoplatanus seedling	I (1)	I (1–3)	I (1–3)
Betula pendula seedling	I (1)	I (2)	I (1–2)
Quercus hybrids seedling	I (1–5)	I (1)	I (1–5)
Number of samples	118	31	149
Number of species/sample	9 (3–26)	15 (8–29)	10 (3–29)
Tree height (m)	11 (3–17)	13 (3–23)	11 (3–23)
Tree cover (%)	77 (10–100)	81 (35–100)	78 (10–100)
Shrub height (m)	3 (1–6)	3 (2–8)	3 (1–8)
Shrub cover (%)	8 (0–80)	9 (0–60)	8 (0–80)
Herb height (cm)	71 (5–180)	40 (5–110)	65 (5–180)
Herb cover (%)	81 (1–100)	77 (10–100)	80 (1–100)
Ground height (mm)	2 (1–5)	11 (2–30)	4 (1–30)
Ground cover (%)	4 (0–75)	9 (1–40)	5 (0–75)
Altitude (m)	120 (5–335)	161 (30–230)	129 (5–335)
Slope (°)	13 (0–60)	22 (2–60)	15 (0–60)

a   *Quercus robur* sub-community

b   *Vaccinium myrtillus-Dryopteris dilatata* sub-community

16 *Quercus* spp.-*Betula* spp.-*Deschampsia flexuosa* woodland (total)

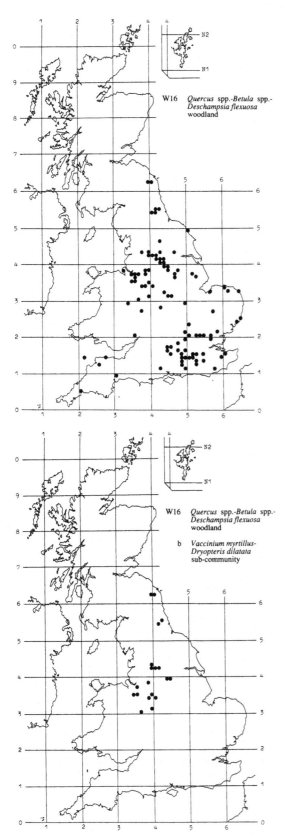

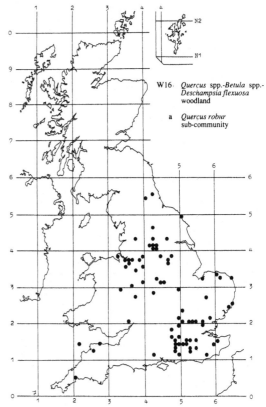

W16 *Quercus* spp.-*Betula* spp.-
Deschampsia flexuosa
woodland

W16 *Quercus* spp.-*Betula* spp.-
Deschampsia flexuosa
woodland

a *Quercus robur*
sub-community

W16 *Quercus* spp.-*Betula* spp.-
Deschampsia flexuosa
woodland

b *Vaccinium myrtillus-
Dryopteris dilatata*
sub-community

W17

Quercus petraea-Betula pubescens-Dicranum majus woodland

Synonymy

Caithness birchwoods   Crampton 1911 *p.p.*; *Betuletum tomentosae* Moss 1911 *p.p.*; *Quercetum roboris* Tansley 1939 *p.p.*; *Quercetum petraeae/sessiliflorae* Tansley 1939 *p.p.*; *Betuletum Oxaleto-Vaccinetum* McVean & Ratcliffe 1962 *p.p.*; *Vaccinium*-rich birchwood association McVean 1964*a p.p.*; *Blechno-Quercetum* (Br.-Bl. & Tx. (1950) 1952) Klötzli 1970 *p.p.*; Loch Lomond oakwoods Tittensor & Steele 1971 *p.p.*; *Betula pubescens-Vaccinium myrtillus* Association Birks 1973 *p.p.*; Birch-oak woodland Peterken 1981 *p.p.*; Birch woodland Peterken 1981 *p.p.*; *Trientali-Betuletum pendulae* Birse 1982 *p.p.*; Woodland plot type 18 Bunce 1982 *p.p.*; *Blechno-Quercetum* (Br.-Bl. & Tx. (1950) 1952) Birse 1984 *p.p.*

Constant species

Betula pubescens, Quercus petraea, Deschampsia flexuosa, Vaccinium myrtillus, Dicranum majus, Hylocomium splendens, Plagiothecium undulatum, Pleurozium schreberi, Polytrichum formosum, Rhytidiadelphus loreus.

Rare species

Goodyera repens, Adelanthus decipiens, Sematophyllum demissum, S. micans.

Physiognomy

The *Quercus petraea-Betula pubescens-Dicranum majus* woodland, like its counterpart on less base-poor soils, the *Quercus-Betula-Oxalis* woodland, is almost always dominated by either oak or birch or various mixtures of the two trees, though in this community there is a more pronounced shift towards *Quercus petraea* and *Betula pubescens* as the characteristic species. *Q. petraea* is very common here and, in western Britain, where the bulk of the stands of this community are located, *Q. robur* makes only a very local, though sometimes conspicuous, contribution: it predominates, for example, in certain localities on Dartmoor, in Devon, notably at Wistman's Wood (Harris 1921, Christy & Worth 1922, Tansley

1939, Wigston 1974), and *robur* elements can also be detected elsewhere, as among the oaks at Keskadale in Cumbria and in some stands in Ross & Cromarty (Jones 1959, Gardiner 1974, Wigston 1974). However, it is only in eastern Scotland that there is any more widespread switch to *Q. robur* and hybrids, and here there seems a stronger likelihood of the oaks having originated from planted stock (Jones 1959, Gardiner 1974, Birse 1982, 1984).

Oak is often abundant in the canopy of this community though, even where it dominates, the tree cover is typically rather open and low, only rarely exceeding 20 m and generally being less than 15 m. In extreme situations, oak forms a very dwarfed canopy, as at Wistman's Wood (Harris 1921, Christy & Worth 1922) and Keskadale (Leach 1925), where many trees are less than 4 m and the most exposed individuals even shorter, their crowns wind-shaped into the contours of a rocky landscape. Elsewhere, in many stands in Wales, up through Cumbria and into south-west Scotland, the canopy presents an appearance of more orthodox high forest with well-grown oaks predominating though, moving further to the north-west of Scotland, oak becomes increasingly rare in the community and restricted to progressively lower altitudes: towards the limit of the range of this kind of woodland, it dominates up to only about 150 m (McVean 1964*a*) and is totally absent from many stands. It is likely that past treatment plays some part in the scarcity of oak in this region: there are strong suspicions that large timber has been widely removed from all kinds of woodland here (McVean & Ratcliffe 1962, McVean 1964*a*). Even where oak is dominant in this community, there is abundant evidence from throughout its range, that many stands have been treated as oak coppice. In western Britain, the *Quercus-Betula-Dicranum* woodland, together with the *Quercus-Betula-Oxalis* woodland, has been the major source of oak for charcoal and tan-bark, and tracts of multi-stemmed trees forming a fairly pure and structurally-even cover are widespread.

It is in such stands that birch is generally most poorly represented, though it has often spread with the abandonment of coppicing and it was by no means always cleaned out as a weed, being itself cut to provide wood for bobbins or brushwood for besoms. Through the community as a whole, birch is very frequent, with *B. pubescens* being by far the more common species; *B. pendula* occurs only very occasionally overall though, like *Q. robur*, it tends to increase its representation in eastern Scotland. In high-forest stands, birch varies in abundance from occasional scattered trees growing amongst the oak to a much more prominent contribution, when the two trees form mosaics often, where the woodland is actively expanding, with a birch-dominated fringe around the margins. As oak decreases in frequency towards the north-west, birch becomes proportionately more obvious, just as it does in the *Quercus-Betula-Oxalis* woodland, until it provides the bulk of the woody cover. Towards higher latitudes, too, there is an increasing tendency for *B. pubescens* to be very obviously of ssp. *carpatica* with its distinct short and bushy habit (Walters 1964); elsewhere in this community, this taxon can be found in locally-exposed situations and at high altitudes.

Like oak, birch varies considerably in its physiognomy. In taller high forest, *B. pubescens* ssp. *pubescens* and *B. pendula* can readily keep pace with oak but birch-dominated stands often have a shorter canopy, especially where *B. pubescens* ssp. *carpatica* provides the cover. Dense thickets of birch can occur here, though the canopy is frequently more open with widely-spaced, and often rather sickly, individuals. Typically, stretches of birch have a rather uniform appearance: sometimes scattered pioneers are surrounded by younger trees but, more commonly, there is a single generation which shades out any offspring (McVean & Ratcliffe 1962).

As in the *Quercus-Betula-Oxalis* woodland, the most consistent woody associate of oak and birch is *Sorbus aucuparia*, though it is somewhat more frequent in this community than there. In high forest, it usually occurs as scattered trees, often of quite considerable size and breaking the canopy, though sometimes a little shorter than the oak and contributing to an ill-defined second tier of trees. However, it can be locally abundant (especially where there has been some relief from grazing: Anderson 1950, McVean 1964a) and, where oak becomes scarce towards the north-west it may share dominance with birch in a low, scrubby cover. The frequency and abundance of *Ilex aquifolium* are also affected by grazing and browsing: throughout this community, holly is rather uncommon, despite the favourable character of the climate.

Other tree species are scarce. *Fraxinus excelsior* and *Acer pseudoplatanus* occur very occasionally but they are rather strictly confined to pockets of more enriched soil. And, though *Tilia cordata* can sometimes be found growing on cliffs within Cumbrian stands of the community, it is very much a relic tree here, surviving at the extreme north-western limit of its British range (Pigott & Huntley 1978): such stands probably fall within Peterken's (1981) oak-lime stand type 5B, though the association of this species with the community is now essentially a topographic one, rather than of a vegetational kind. *Fagus sylvatica* occurs at very low frequencies, presumably having seeded in from introduced stock, and there are occasional records for conifers, extensive stands of which have been planted on soils that could support this community throughout the upland fringes of the north and west.

The contribution of smaller woody plants to the *Quercus-Betula-Dicranum* woodland is rather variable and only a few species, apart from saplings of those already mentioned, occur. In high-forest stands, there can be a discrete understorey, though its cover is somewhat patchy and usually low, sometimes negligible in old coppice or in woodlands which have been consistently grazed. In shorter canopies, stratification is indistinct and mature trees, shrubs and saplings merge into a single layer.

The most surprising feature among this element of the vegetation here, in sharp contrast to the lowland analogue of the community, the *Quercus-Betula-Deschampsia* woodland, is the continuing importance of *Corylus avellana*. In southern Britain, the frequency and abundance of this shrub provides a good floristic and ecological separation between the Carpinion 'bluebell woodland', where it is very common, and the Quercion 'bilberry woodland', from which it is virtually absent. The same is not true of the sub-montane north and west where, in the moister climate, it continues to be quite well represented even in those stands of the *Quercus-Betula-Dicranum* woodland which present a very calcifugous appearance (cf. Peterken 1981). However, there is no doubt that it does better on deeper and more fertile profiles, being confined to pockets of flushed soil in more bouldery woods and attaining its greatest cover where there is some accumulation of colluvium. And, although it has sometimes been selected for as a coppice crop here, it rarely forms a naturally dense understorey, occurring in high forest usually as scattered bushes. In the far north-west, it can contribute, with *B. pubescens* and *S. aucuparia*, to a scrubby cover over this community, though it is not so prominent in such mixtures here as it is in the canopy of *Quercus-Betula-Oxalis* woodlands of that region (e.g. Birks 1973).

Apart from *Corylus*, it is generally young trees and saplings of birch and rowan, together with occasional *Ilex*, that make up such understorey as there is in this kind of woodland. But the regeneration of all these is much affected by grazing, which is widespread in these woods, and also, in the case of birch, by canopy shade. Saplings of birch, then, are generally confined to gaps and margins and the usual picture beneath intact canopies where grazing and browsing are not severe is one of scattered and patchily dense rowan and holly. Young *Q. petraea* can also sometimes be found in ungrazed stands, even where the shade is quite deep (Jones 1959), and more occasionally there may be saplings of *Fraxinus* or *A. pseudoplatanus*. Hawthorn (always *Crataegus monogyna*) occurs very sparsely, even more confined to deeper, more fertile soils than *Corylus*, and there is sometimes a little *Salix caprea*. *Rhododendron ponticum* can occur, sometimes with local prominence: it has occasionally been planted into these woodlands and thrives on the acid soils in the moist climate (Cross 1975). In eastern Scotland, *Juniperus communis* may be found: in that region, the *Quercus-Betula-Dicranum* woodland can form mosaics with the more calcifugous stands of the *Juniperus-Oxalis* woodland, patchworks of birch and juniper extending over both.

Three components are especially important in giving the field layer of the *Quercus-Betula-Dicranum* woodland its distinctive stamp: grasses, bracken and ericoid sub-shrubs. However, the relative proportions of these elements varies considerably with a number of factors (notably topography and soils, shade and grazing) so that the gross physiognomy of the field layer is quite diverse; and bracken also provides a measure of phenological change. As in the *Quercus-Betula-Oxalis* woodland, grasses provide the background and they can be particularly prominent in grazed woods over more even topography when they show their flush of renewed growth in early summer. But, in contrast to that community, there is a shift among the species represented towards the more calcifugous. *Deschampsia flexuosa* is now the most consistently frequent grass throughout; it is quite often abundant and can maintain its cover even over very shallow soils extending over the tops of boulders. *Holcus mollis*, on the other hand, becomes less common and is more restricted to pockets of deeper moister soil, though it tends to increase its representation in consistently-grazed stands. *Anthoxanthum odoratum* and *Agrostis capillaris*, though somewhat more frequent overall than *H. mollis*, also become more common and abundant under grazing. Other grasses occurring occasionally throughout are *Agrostis canina* ssp. *montana* and *Festuca ovina* and, more obviously towards the far west, *Molinia caerulea*. Mesophytic species such as *Holcus lanatus* and *Deschampsia cespitosa* are, however, distinctly scarce.

The presence of *Pteridium aquilinum* in this community presents a further floristic similarity to the *Quercus-Betula-Oxalis* woodland, though its prominence here is rather more restricted by the frequent occurrence of rock exposures and boulders with, at most, a very thin capping of soil. In such situations, bracken is confined to deeper accumulations in larger crevices between the talus fragments, though it can be much more extensive wherever there is some downwash or creep and, over slope bases with much colluvium, it can provide a dense canopy by the time its fronds are fully unfolded in midsummer, severely restricting the richness of the associated flora with its heavy shade and thick litter. It is also itself influenced by the shade of the canopy, so that it tends to thin out under closely-set trees, thickening up in gaps and clearings; and it tends to become more prominent, too, where there has been some relief from grazing (e.g. Tittensor & Steele 1971). The two characteristic companions of bracken in woodlands of moister and less base-poor soils, *Rubus fruticosus* agg. and *Lonicera periclymenum*, are here not very frequent, though they can make some contribution to a patchy underscrub where edaphic conditions are less extreme than usual and they, too, can increase a little where grazing is relaxed. A very characteristic bramble in this community is *R. sprengelii*.

The plants which provide the major distinction between the field layer of this community and that of the *Quercus-Betula-Oxalis* woodland are, however, the ericoids which are generally of rather restricted occurrence there but which here attain quite high frequencies and often great abundance. The most common among them is *Vaccinium myrtillus*, which can tolerate quite dense shade and therefore continue to make a fairly prominent contribution here even under intact canopies. It is, though, very sensitive to grazing and browsing. The rougher, more inaccessible topography provided by large boulders may offer some protection against the attention of sheep and deer and, in such situations, *V. myrtillus* can grow vigorously, even on quite shallow soils, forming a patchy cover half a metre or more high. But, with continued grazing, it can be quickly reduced to sparse, leafless shoots or even eliminated altogether. Under such conditions, and with the attendant increase in such grasses as *Agrostis capillaris, Anthoxanthum odoratum* and *Holcus mollis*, it may be very difficult to separate the vegetation of the *Anthoxanthum-Agrostis* sub-community here from that of the *Blechnum* subcommunity of the *Quercus-Betula-Oxalis* woodland: this is a reflection of the very real fact that, over less extreme soils in the north-western uplands, grazing produces a floristic convergence in these two kinds of woodland.

Calluna vulgaris, which is a little less common overall than *V. myrtillus* but equally characteristic, is also reduced by the presence of herbivores, though it is much more sensitive to shade than bilberry, and only attains abundance here in more open areas, as in gaps and around the margins of stands, where it can figure prominently in transitions to heaths. *Erica cinerea* also occurs occasionally, again mostly in places with less dense canopy shade and sometimes *V. vitis-idaea* is encountered, though where this grows among *V. myrtillus*, it can be shaded out by the more vigorous growth of the latter when grazing is relaxed (e.g. Pigott 1983). In this community, *V. vitis-idaea* is especially associated with transitions to the *Juniperus-Oxalis* woodland in which the *Calluna* sub-community is sometimes found in eastern Scotland (e.g. Birse 1982, 1984). *Erica tetralix* is rare, though it can be found on wetter, acid soils.

Apart from these three elements, two further groups of herbs can make a contribution to the field layer of the *Quercus-Betula-Dicranum* woodland. Smaller angiosperms are often not very prominent in terms of their cover but a number of species are characteristic. Among the most frequent are *Galium saxatile* (often increasing in abundance in grazed woods), *Potentilla erecta*, *Melampyrum pratense*, *Teucrium scorodonia*, *Succisa pratensis*, *Solidago virgaurea*, *Carex pilulifera*, *Luzula pilosa*, *L. multiflora* and *L. sylvatica* (the last sometimes becoming extremely abundant when grazing is withdrawn and over inaccessible ledges and cliff faces). In contrast to the southern *Quercus-Betula-Deschampsia* woodland, *Oxalis acetosella* is also very frequent here, a good indicator of the continual moistness of the soils (Packham 1978) and one of the plants best able to survive a thick cover of bracken, bilberry or *Luzula sylvatica* (Packham & Willis 1977). Its typical companion on moister and less base-poor profiles, *Viola riviniana*, is sometimes found too where there is gentle seepage. It is only in such situations that *Hyacinthoides non-scripta* makes an appearance here: the general scarcity of this species, and its replacement on very moist soils, *Anemone nemorosa*, provides a good separation between this community and the *Quercus-Betula-Oxalis* woodland (though, once again, grazing may blur the distinction by reducing the cover of *Hyacinthoides* in the latter). Flushed areas may also have a little *Lysimachia nemorum* but species characteristic of marked base-enrichment by flushing are typically absent.

Ferns (other than bracken) constitute a final element in the field layer. *Blechnum spicant*, though it is still somewhat preferential for one particular sub-community, is more frequent throughout this community than in the *Quercus-Betula-Oxalis* woodland and it can attain great abundance and luxuriance in the mild climate of the far west (where this kind of vegetation forms part of the *Blechno-Quercetum* Br.-Bl. & Tx. (1950) 1952 redefined by Klötzli (1970) and Birse (1982, 1984)). *Thelypteris limbosperma* also occurs, but it is much less frequent than *Blechnum* and, though it can attain local abundance, is less characteristic here than it is in the *Quercus-Betula-Oxalis* woodland. *Dryopteris dilatata* is found occasionally and, rather less commonly, *D. filix-mas* and *D. borreri*, *Athyrium filix-femina* and *Gymno-*

carpium dryopteris. *Polypodium vulgare* (probably *s.s.*) is quite common as an epiphyte on oak. Then, among dripping mats of bryophytes, *Hymenophyllum tunbrigense* and *H. wilsonii* can sometimes be seen and, on the tops and shoulders of boulders or on ledges, *Dryopteris aemula* can occur. As in other sub-montane woodlands, this component of the field layer tends to attain maximum diversity and luxuriance where this community forms part of the vegetation mosaic of ravines: in such situations, a profusion of different niches in a very cool, humid environment can carry a spectacular fern-dominated cover beneath the overhanging trees.

Underlying this considerable variation in the part played by the different components of the vascular flora, there is a consistent and rich contribution from bryophytes. Six out of the eight field- and ground-layer constants of this kind of woodland are mosses; bryophytes often exceed vascular plants in their diversity and abundance in particular stands; and very frequently, they make an immediate visual impression, forming a mottled carpet which keeps its colours throughout the year and which, in rocky woods in the very wet far west, extends virtually everywhere, from more grassy swards, up over boulders, on to the faces and ledges of rock exposures and over the roots and bole bases of the trees. Different species tend to be prominent in different microhabitats over more diverse topographies but among the many species represented throughout the community are *Dicranum majus*, *Rhytidiadelphus loreus*, *Polytrichum formosum*, *Pleurozium schreberi*, *Plagiothecium undulatum* (the constancy of all of which helps separate the bryophyte element here from that in the *Quercus-Betula-Oxalis* woodland), *Hylocomium splendens*, *Dicranum scoparium*, *Mnium hornum*, *Thuidium tamariscinum*, *Isothecium myosuroides*, *Hypnum cupressiforme* (including *H. mammilatum*), *H. jutlandicum*, *Rhytidiadelphus squarrosus*, *R. triquetrus*, *Sphagnum quinquefarium*, *Pseudoscleropodium purum* and *Lophocolea bidentata s.l.* Less commonly, there are records among the community companions for *Dicranella heteromalla*, *Dicranum fuscescens*, *Eurhynchium praelongum*, *Rhizomnium punctatum*, *Polytrichum commune*, *Tetraphis pellucida*, *Sphagnum palustre*, *Ptilium crista-castrensis*, *Plagiochila asplenoides*, *Lophocolea cuspidata*, *Calypogeia fissa*, *C. muellerana*, *Cephalozia media*, *C. bicuspidata*, *Barbilophozia floerkii*, *Lepidozia pearsonii* and *Frullania tamarisci*. Against this general background, each of the sub-communities shows some peculiarities in its bryophyte flora. This is most marked in the *Isothecium-Diplophyllum* sub-community, where there is a further consistent enrichment and an occasional occurrence of a very large number of other species, including many of restricted distribution in Britain and some national rarities, which give this kind of *Quercus-Betula-Dicranum* woodland a unique character.

Apart from these ground-growing bryophytes, there are often here specialised epiphytic floras of tree bases, branches and twigs, differing somewhat according to the woody species being colonised. Epiphytic lichens can also be very prominent. Each of these assemblages is mentioned below but, among the more widespread lichens of soil surfaces, rotting logs and rocks are *Cladonia impexa*, *C. squamosa*, *C. polydactyla*, *C. arbuscula*, *C. furcata*, *C. chlorophaea*, *Sphaerophorus globosus*, *S. fragilis* and *Parmelia saxatilis*.

Sub-communities

***Isothecium myosuroides-Diplophyllum albicans* sub-community:** *Betuletum Oxaleto-Vaccinietum* McVean & Ratcliffe 1962 *p.p.*; *Vaccinium*-rich birchwood association McVean 1964a *p.p.*; *Blechno-Quercetum typicum* Klötzli 1970 *p.p.*; *Betula pubescens-Vaccinium myrtillus* Association and associated *Hymenophyllum wilsonii-Isothecium myosuroides* Association Birks 1973; Birch-oak stand types 6Ab & 6Ac Peterken 1981 *p.p.*; Birch stand type 12A Peterken 1981 *p.p.*; *Blechno-Quercetum, Saccogyna viticulosa* Subassociation Birse 1984. This kind of *Quercus-Betula-Dicranum* woodland exhibits almost the full range of floristic and physiognomic variation found among the woody element of the community. In more southerly stands, in areas like Dartmoor, North Wales and the Lake District, oak-dominated high forest is the norm though, over the more rugged topographies which are characteristic here, the cover and stature of the trees are often reduced, sometimes to extremes. In more closed canopies, *B. pubescens* generally plays a minor role, occurring as occasional scattered trees with some *S. aucuparia* and a little *Ilex*, and only thickening up in gaps and around margins but, in ungrazed stands where the oak cover is less extensive, it can be co-dominant in mosaics of the two trees. And, increasingly to the far north-west of Scotland, oak becomes much scarcer so that, at the limits of this sub-community's range, lower, scrubby canopies of *B. pubescens* (usually now ssp. *carpatica*) and *S. aucuparia* usually provide the woody cover (e.g. McVean & Ratcliffe 1962, Birks 1973). Although this sub-community can be found within *Quercus-Betula-Dicranum* woodlands which have been coppiced, the often poor growth of the oak and the rough character of the ground have frequently limited this kind of treatment.

Corylus is rather infrequent here, probably because of the scarcity of pockets of deeper, moister soil, and it is usually found only as sparse, scattered bushes; to the north-west, it does not make the prominent contribution to the canopy that is typical of scrubby versions of the *Quercus-Betula-Oxalis* woodland. Smaller woody plants as there are, then, are usually saplings of *B. pubescens* and *S. aucuparia*, sometimes with a little *Q.*

petraea.

The field layer here preserves the general features of the community as a whole and few vascular plants are preferential. The most prominent species are generally *Vaccinium myrtillus* and *Pteridium*, though the occurrence of both can be patchy. *Vaccinium* is restricted to some extent by the grazing of more accessible ground though it often gains some protection from herbivores over larger boulders and then can grow tall and bushy; over such areas, though, *Pteridium* can be sparse and confined to crevices with some downwashed soil, only growing really dense in less rocky ground. *Calluna* is found only occasionally in more open places and around the edges of stands.

Apart from *Deschampsia flexuosa*, which can extend its cover over the tops of more well-colonised boulders, the frequency and abundance of grasses in this sub-community is often restricted by the rough topography. On patches of soil, however, *Festuca ovina* (including *F. vivipara*), *Agrostis capillaris*, *A. canina* ssp. *montana* and *Anthoxanthum odoratum* can all occur, forming the basis of an open-textured sward. And, preferentially here, there may be some *Molinia caerulea*, occasionally in abundance. *Luzula sylvatica* can be prominent, too, sometimes becoming a local field-layer dominant in ungrazed stands. Then, there are scattered dicotyledons such as *Galium saxatile*, *Potentilla erecta*, *Melampyrum pratense* and, often most conspicuous of all, *Oxalis acetosella*. *Blechnum spicant* is rather more frequent here than in other sub-communities, often attaining majestic size, and there can be very occasional *Athyrium filix-femina*, *Thelypteris limbosperma*, *Dryopteris dilatata*, *D. filix-mas*, *D. borreri* and, in some localities, *D. aemula*. In sheltered ravines, this fern element can be very extensive and luxuriant.

Much more consistently obvious, however, and very distinctive of this kind of *Quercus-Betula-Dicranum* woodland, are the bryophytes. All the characteristic community species are very frequent and, as always, grazing may favour an abundance of more robust mosses among areas with a grassy turf. But, very commonly here, the bryophyte cover extends as a thick mat over the surfaces of boulders. Over flat or gently-sloping surfaces with a fairly substantial and more stable nidus, the community constants often provide the bulk of the cover, though even in such situations there may be some enrichment from mosses such as *Hylocomium brevirostre*, *Thuidium delicatulum* and *Isothecium myosuroides*. Some vascular species continue to be represented here, notably *D. flexuosa*, *Oxalis* and *P. erecta* and there may be a patchy cover of *Vaccinium* forming the kind of suite which Birks (1973) called the *Oxalis acetosella-Rhytidiadelphus loreus* Association.

More strikingly preferential in this sub-community, however, is a group of bryophytes characteristic of thinner and less stable humus and mineral mats which can extend down over the shoulders of boulders and hang over their steeper sides. Making a frequent appearance here are such species as *Isothecium myosuroides*, *Thuidium delicatulum*, *Diplophyllum albicans*, *Leucobryum glaucum*, *Campylopus paradoxus*, *Hylocomium umbratum*, *Lepidozia reptans*, *Plagiochila spinulosa*, *Scapania gracilis*, *Bazzania trilobata*, *Dicranodontium denudatum* and *Saccogyna viticulosa*. Where such mats are kept very moist, by the general shelter provided between the boulders and by dripping of condensation and rain, the filmy ferns *Hymenophyllum wilsonii* and *H. tunbrigense* can also be found constituting the vegetation which Birks (1973) characterised as the *Hymenophyllum wilsonii-Isothecium myosuroides* Association. Some of these species, notably *I. myosuroides*, *D. albicans* and *C. paradoxus*, can extend on to barer rock surfaces on the very steeply inclined sides of boulders and drier areas of this kind can provide a location for the Western Atlantic *Plagiochila punctata*, *P. tridenticulata* and *Dicranum scottianum*. Where there is periodic wetting in such situations, other western British species such as *Harpanthus scutatus*, *Scapania umbrosa*, *Jamesoniella autumnalis* and *Tritomaria exsecta* may be found and, in areas south of the north-west Highlands, the rare South Atlantic *Sematophyllum micans* and *S. demissum* (Ratcliffe 1968). More regular splashing with water, as by waterfalls and along stream sides, is often marked by an abundance of *Heterocladium heteropterum*, *Marsupella emarginata*, *Hyocomium flagellare* with some *Racomitrium aquaticum*, *R. fasciculare* and *Hypnum callichroum* (Birks' (1973) Open Boulder Association). Other interesting bryophytes recorded occasionally in this sub-community are the South Atlantic *Adelanthus decipiens*, the Western British *Bazzania tricrenata*, *Plagiochila killarniensis*, *P. atlantica*, *Anastrepta orcadensis* and *Jungermannia gracillima*.

Elements of the boulder bryophyte cover may run some considerable way up the trunks and branches of trees in this kind of woodland, with *Isothecium myosuroides*, *Dicranum scoparium*, *Hypnum mammilatum*, *Scapania gracilis*, *Plagiochila spinulosa*, *P. punctata* and *P. tridenticulata* often figuring in such assemblages, sometimes with filmy ferns (equivalent to the *Hymenophylleto-Isothecium myosuroidis* of Barkman 1958). Then, smaller branches of birch can have a rich epiphytic flora in which *Ulota crispa* (including var. *norvegica*) and *U. phyllantha* play a prominent part, together with *Lejeunea ulicina* and *Frullania tamarisci* (Ulotion crispae alliance: Barkman 1958). Rotting logs provide a further habitat: here *Nowellia curvifolia*, *Lophozia ventricosa*, *L. incisa*, *Cephalozia bicuspidata*, *C. connivens*, *C. media* and *Tetraphis pellucida* are best represented, together with *Tritomaria exsecta*, *Scapania gracilis*, *S. umbrosa*, *Plagiochila spinulosa* and some of the common

community species such as *Dicranum scoparium*, *Hypnum cupressiforme s.l.* and *Isothecium myosuroides* (alliances Blepharostomion Barkman 1958 or Nowellion Philippi 1965).

Typical sub-community: *Betuletum Oxaleto-Vaccinietum* McVean & Ratcliffe 1962 *p.p.*; *Vaccinium*-rich birchwood association McVean 1964a *p.p.*; *Blechno-Quercetum typicum* Klötzli 1970 *p.p.*; Loch Lomond Community Types 1, 6 & 7 Tittensor & Steele 1971; *Betula pubescens-Vaccinium myrtillus* Association Birks 1973 *p.p.*; Birch-oak stand types 6Ab & 6Ac Peterken 1981 *p.p.*; Birch stand type 12A Peterken 1981 *p.p.*; *Blechno-Quercetum*, Typical Subassociation Birse 1984 *p.p.* Oak-dominated high forest and coppice are very common here and, in many such stands, *B. pubescens* is reduced to occasional occurrences in the canopy with patchy abundance in gaps and around margins. However, more balanced mosaics of oak and birch sometimes provide the woody cover and, though this sub-community does not extend to the far north-west of Britain, some stands have only occasional oaks in a scrubby canopy dominated by mixtures of *B. pubescens* (often ssp. *carpatica* in exposed situations) and *S. aucuparia*. *Corylus* is somewhat more frequent here than in the *Isothecium-Diplophyllum* sub-community, which is partly a reflection of the less bouldery character of this kind of *Quercus-Betula-Dicranum* woodland. It has sometimes clearly been coppiced in the past though, generally, scattered hazel bushes form part of an understorey, along with *S. aucuparia* and *Ilex*, and the cover of these smaller woody species can be quite extensive here because grazing and browsing are not consistent.

The field layer preserves the general character typical of the community as a whole with very few preferential species. However, apart from *D. flexuosa*, grasses are rather poorly represented here in terms of frequency and cover, whereas *Vaccinium*, and, in more open places, *Calluna* and occasionally *Erica cinerea*, can be abundant and vigorous. *Pteridium* is common, too, thickening up to a dense cover on deeper soils where the shade is not too deep, and *Dryopteris dilatata* is a little more frequent here than in other sub-communities. With the tall growth of these species, smaller dicotyledons tend to be rather uncommon and even the shade-tolerant *Oxalis* is scarce. With the lack of heavy grazing, seedlings of some trees increase a little: *Q. petraea* is especially noticeable in this respect and there can be some *S. aucuparia* and *Ilex*.

But it is among the bryophytes that the most obvious distinctions from the *Isothecium-Diplophyllum* sub-community are to be seen. All the general community species remain common here and they can be quite abundant in places where the cover of vascular plants is not so dense but, with the scarcity of more open bouldery areas, the numerous colonisers of thinner humus mats become markedly infrequent: even *Isothecium* and *Diplophyllum* themselves are recorded only occasionally and the rest are rare, just making an appearance in twos or threes where there are scattered outcrops. But these sites still experience very high rainfall and humidity, so the potential for enrichment with more Atlantic species is still present and the epiphytic floras of trunks and twigs can be profuse and varied.

Anthoxanthum odoratum-Agrostis capillaris subcommunity: *Betuletum Oxaleto-Vaccinietum* McVean & Ratcliffe 1962 *p.p.*; *Vaccinium*-rich birchwood association McVean 1964*a p.p.*; *Blechno-Quercetum coryletosum* Klötzli 1970 *p.p.*; Loch Lomond Community Type 2 Tittensor & Steele 1971 *p.p.*; Birch-oak stand types 6Ab & 6Ac Peterken 1981 *p.p.*; Birch stand type 12A Peterken 1981 *p.p.* Again, a variety of canopy compositions and structures can be found in this subcommunity. Oak figures prominently and it often dominates in high forest or coppice but stands with much birch also occur. *S. aucuparia* and *Ilex*, however, are somewhat less frequent than in the Typical subcommunity with saplings of these species being especially sparse. *Corylus* remains frequent and is often the most abundant member of an open or patchy understorey, occasionally accompanied by a little *Crataegus monogyna*.

In this sub-community, it is the field layer that provides most of the distinctive character of the vegetation. Ericoids are noticeably uncommon and of low cover, even in more open areas of canopy, with *Vaccinium* showing an especially striking reduction from the generally high levels that it attains in the *Quercus-Betula-Dicranum* woodland. And there is a corresponding increase in the number and abundance of grasses with, in addition to *D. flexuosa*, frequent records for *Anthoxanthum odoratum*, *Agrostis capillaris* and *Holcus mollis* and more occasional occurrences of *Agrostis canina* ssp. *montana*, *Festuca ovina*, *Holcus lanatus*, *Dactylis glomerata*, *Poa nemoralis* and *P. pratensis*. Typically, these species form considerable expanses of grassy turf which can be kept quite closely cropped where grazing is heavy.

Pteridium remains frequent, occurring as sparse fronds in more grassy areas and on shallower soils in denser shade, thickening up to a dense cover in glades and on deeper colluvium. And there is usually a fairly obvious contribution from dicotyledonous herbs with *Galium saxatile* and *Oxalis acetosella* both being especially common and often abundant, some *Digitalis purpurea* and *Rumex acetosa* (both weakly preferential here) and scattered plants of *Potentilla erecta*, *Melampyrum pratense*, *Teucrium scorodonia*, *Luzula pilosa* and

Viola riviniana. Very occasionally, there may be some *Rubus fruticosus* agg. or *R. idaeus* and, in such stands, the vegetation approaches very closely to the more calcifugous forms of the *Quercus-Betula-Oxalis* woodland. In the north-western uplands, continuous variation in soil characteristics means that these two communities grade almost imperceptibly one into the other, and grazing often accentuates the floristic convergence over transitional profiles.

One element of the vegetation that can provide some guidance in separating the two kinds of woodland is the bryophytes. This sub-community has the less rich suite of mosses and liverworts characteristic of the Typical sub-community (bouldery ground again being rather scarce here) but more calcifugous mosses remain very frequent and can be abundant in the short-cropped sward, with bulky wefts of *Rhytidiadelphus loreus*, *Hylocomium splendens*, *Plagiothecium undulatum* and, a little less commonly here, *Pleurozium schreberi*, and tufts of *Dicranum majus*, *D. scoparium* and *Polytrichum formosum*.

Rhytidiadelphus triquetrus sub-community: Birch-oak stand type 6Bb Peterken 1981 *p.p.*; Birch stand type 12A Peterken 1981 *p.p.*; *Trientali-Betuletum pendulae*, *Vaccinium vitis-idaea* Subassociation Birse 1984 *p.p.* In terms of its canopy composition, this is the most peculiar kind of *Quercus-Betula-Dicranum* woodland. Oak and birch remain the most frequent and characteristic trees but *Q. petraea* is now more consistently replaced by *Q. robur* and hybrids than in the other sub-communities; and, though *B. pubescens* is still very common, *B. pendula* increases to become occasional. The cover of the canopy is also frequently quite open with one or the other birch often predominating over oak. *S. aucuparia* is frequent, too, and, with smaller birch, but only very sparse *Corylus*, it usually forms the bulk of whatever understorey is present. Often here, though, stratification is indistinct and the canopy presents the appearance of a thicket-like cover. Very occasionally, *Juniperus communis* is present in more open places and this sub-community can occur in mosaics with the *Juniperus-Oxalis* woodland.

Ericoids are quite well represented here and, though their cover is patchy, they give an obvious heathy character to many stands. *Vaccinium myrtillus* is common, though more distinctive in the frequent open areas are *Calluna*, which is preferential to this sub-community, and, more occasionally, *Erica cinerea* and, especially in transitions to *Juniperus-Oxalis* woodland, *V. vitis-idaea*. *Pteridium* remains frequent and sometimes forms dense stands in glades.

Stretches of ground which do not have a thick cover of these taller species can be quite rich in herbs. *Oxalis*, *Melampyrum pratense*, *Potentilla erecta*, *Viola riviniana* and, especially common, *Galium saxatile*, all occur and, with slightly increased frequency here, *Luzula pilosa*. Also very distinctive, though no more than occasional, is *Trientalis europaea* which, as in other woodlands of eastern Scotland, from where most of the stands here originate, provides a Northern Montane feel to the vegetation. *Goodyera repens* has also been recorded in this sub-community and the Arctic-Subarctic *Cornus suecica*.

Bryophytes remain an important component here, though western or more strictly Atlantic species are very rare. Usually the bulk of the cover is provided by the common community species but a strong preferential in this kind of *Quercus-Betula-Dicranum* woodland, as again in other eastern Scottish woodland types, is *Rhytidiadelphus triquetrus*.

Habitat

The *Quercus-Betula-Dicranum* woodland is a community of very acid and often shallow and fragmentary soils in the cooler and wetter north-west of Britain. Local differences in climate and topography have an important influence on the vegetation and frequently interact with grazing to determine the distinctive floristics of the sub-communities and affect the pattern of regeneration of the woody species. Many stands have been treated as coppice and there is evidence of timber removal and planting.

This kind of woodland is largely confined to those parts of Britain where the annual rainfall exceeds 1600 mm (*Climatological Atlas* 1952) and where there are usually more than 180 wet days yr^{-1} (Ratcliffe 1968). And, typically, within this zone, it occurs only on the most acid soils, where the influence of underlying arenaceous bedrocks is not masked by drift or counteracted by flushing with base-rich waters, where there is a strong tendency to mor accumulation and where the high rainfall induces strong leaching in more mature profiles. The soils are thus typically lime-free with a surface pH generally below 4.

Such soils are widely distributed throughout north-western Britain, though they are strongly associated with the more rugged topographies produced by the weathering of harder rocks. Important substrates here are Pre-Cambrian Torridonian sandstones along the north-west seaboard of Scotland, Cambrian, Silurian and Ordovician rocks in Wales and deposits of the last two periods in the Lake District, Devonian Old Red Sandstone around the Moray Firth and in the south-central Highlands of Scotland, Carboniferous sandstones in the Pennines and around Dartmoor, intrusive igneous rocks, especially granitic rocks on Dartmoor and in various parts of Scotland, lavas like the Borrowdale Volcanics of the Lake District and also quartzites and gneisses in various localities. Stands on superficials

are rare, though some occur over free-draining, coarse fluvioglacial deposits in Aberdeenshire. Pedogenesis over all these materials is typically slow and often repeatedly set back in more extreme situations by gross active weathering of the rocks, but the general trend is from very shallow accumulations of organic detritus and mineral fragments, which constitute humic rankers, through brown podzolic profiles to podzols proper with fully-developed humus and iron pans. The community occurs over the full range of these soils and, as detailed below, the degree of maturity of the profile is of considerable importance in determining the floristics of the different kinds of *Quercus-Betula-Dicranum* woodland.

At a more general level, however, the prevailing acid environment of these soils is reflected throughout the community and helps to define it floristically against its counterpart on less base-poor profiles in the north-west, the *Quercus-Betula-Oxalis* woodland. Transgression of less-demanding calcifuges into that kind of woodland is a major feature in the strongly-leaching environment of the sub-montane zone in Britain but, here, that element becomes much more obviously defined, with a marked increase in *Deschampsia flexuosa*, the ericoids and bryophytes like *Dicranum majus*, *Pleurozium schreberi*, *Polytrichum formosum*, *Rhytidiadelphus loreus* and *Plagiothecium undulatum*. Conversely, more mesophytic species, like *Rubus*, *Holcus mollis* and, especially, *Hyacinthoides non-scripta* (and its replacement in less oceanic areas, *Anemone nemorosa*) are much less important here. There is also a shift among the fern flora with dryopteroids, even *D. dilatata*, giving way to *Blechnum spicant* and, in some regions, *D. aemula*. However, it must be realised that such transitions among the various elements of the vegetation in these two communities are continuous, a reflection of gradual changes in the edaphic conditions and, frequently, of the impact of grazing and browsing which eliminate sensitive differentials from each and favour a spread of grasses through both.

Prominent among the shared species are plants which it would be very surprising to see in south-eastern calcifugous woodlands. *Corylus*, for example, though clearly associated with less extreme soils here, extends some way into the community and cannot readily be used to separate woodlands of more and less base-poor soils in the north-west (cf. Peterken 1981). *Oxalis* also remains very frequent (e.g. Packham 1978) and even *Viola riviniana* can be found very occasionally. All of these reflect the more consistently moist character of the soil surface here, in contrast to the typically parched upper horizons of rankers and podzols in the south-eastern lowlands. It should be noted, though, that this moisture is often provided by rain and condensation and that, where flushing occurs, it does not involve any enrichment in bases. Such a development would tend to favour the occurrence of either the *Quercus-Betula-*

Oxalis woodland or, where there was more extreme enrichment, the *Fraxinus-Sorbus-Mercurialis* or *Alnus-Fraxinus-Lysimachia* woodlands.

A much more obvious effect of the higher rainfall and humidity here, though, is to be seen in the very striking increase in the bryophyte component of this kind of woodland, in both the greater abundance of the carpet of mosses and liverworts and in the higher frequency of a large number of species. By and large, it is this element which provides the best distinction between the *Quercus-Betula-Dicranum* and *Quercus-Betula-Deschampsia* woodlands, with the 1600 mm isohyet or 180 wet days yr^{-1} line showing a close correlation with the geographical boundary between the two communities. In only one region is this correlation broken and that is in the southern Pennines where the latter community occurs in places where the former might be expected on climatic grounds. This anomalous zone corresponds very well with one area of high atmospheric pollution in Britain (as measured by, for example, the SO_2 concentration: Seaward & Hitch 1982) and it is a noticeable feature of calcifugous woodlands in the area that they most closely approach the *Quercus-Betula-Dicranum* type in deeper valleys more remote from the great industrial conurbations. The boundary between the communities here may thus be artificially maintained.

Within the community, there is a distinct association between the prominence and richness of the bryophyte element and the wetness of climate, as measured by the number of wet days yr^{-1} (a better index of the consistency of humidity than simple rainfall totals: Ratcliffe 1968). Lower levels of cover and smaller numbers of species characterise the *Rhytidiadelphus* and *Anthoxanthum-Agrostis* sub-communities where the number of wet days yr^{-1} either only just reaches 180 or, in parts of eastern Scotland, falls below this figure. In the area bounded by the 180–200 wet days yr^{-1} contour, the Typical sub-community is the usual type of *Quercus-Betula-Dicranum* woodland where abundance and variety of bryophytes are greater, but the most striking enrichment is seen in the *Isothecium-Diplophyllum* sub-community, most of the stands of which fall within the 200+ wet days yr^{-1} line, the richest of all occurring in the far north-west of Scotland where the figure exceeds 220 (Ratcliffe 1968). A considerable proportion of the wealth of bryophytes involved in this gradient consists of quite common and widespread species: here the difference is one of increased frequency of occurrence and abundance, such that these plants come to form the basis of an extensive and luxuriant mat. But superimposed on this, there is also a rise to prominence of mosses and liverworts with a more obvious western bias to their British distribution, either fairly widespread European sub-montane plants whose range in Britain reflects the predominance of higher ground towards the

west (Ratcliffe's (1968) Western British group including *Bazzania trilobata*, *Dicranodontium denudatum*, *Hypnum callichroum*, *Scapania umbrosa*, *Tritomaria exsecta*, *Harpanthus scutatus*, *Sphagnum quinquefarium*) or more strictly Atlantic species (including the Widespread Atlantic *Saccogyna viticulosa*, *Plagiochila punctata*, *P. tridenticulata* and *Dicranum scottianum* and the Southern Atlantic *Sematophyllum micans*, *S. demissum* and *Adelanthus decipiens*, which do not penetrate quite so far to the north). Exactly the same pattern of a general increase in abundance and a rising prominence of a western component is seen among the epiphytic species, with trees in the Typical and especially the *Isothecium-Diplophyllum* sub-communities having their trunks and branches festooned with rich and luxuriant assemblages. And, among the vascular plants, there is a preferential occurrence of ferns like *Hymenophyllum tunbrigense*, *H. wilsonii* and *Dryopteris aemula*, the first two typically in close spatial association with bryophyte mats.

Atmospheric humidity, as measured by the annual number of rain days provides only a crude index of the increasing oceanicity of the climate towards the west and other climatic factors are undoubtedly important in encouraging the bryophyte richness found in the *Isothecium-Diplophyllum* sub-community, such as a fairly equable range of temperature variation, with moderately cool summers and largely frost-free winters (Ratcliffe 1968). The region where the sharpest gradients in all these variables can be seen is in northern Scotland where, in running from west to east, from say Skye to Aberdeenshire, the climate becomes markedly more continental. And it is here that the most striking contrasts in the bryophyte component of the *Quercus-Betula-Dicranum* woodland can be seen, with an obvious loss of more western species and an increasing prominence of *Rhytidiadelphus triquetrus* (a difference seen also in the *Pinus-Hylocomium* woodland which spans the same geographical range). The appearance of *Trientalis* marks the same climatic shift and, perhaps also, the partial switch from *Betula pubescens* to *B. pendula* (Figure 24).

Topography plays an important part in the climatic features characteristic of the area where the *Quercus-Betula-Dicranum* woodland occurs. This can be seen on a large scale in the orographic influence of the high ranges of hills of more resistant rocks that typify the western parts of Britain, which force the wet westerly winds upwards with a release of rain. But there are other small-scale effects which produce local modifications to the general distribution pattern of the sub-communities in relation to annual patterns of rainfall. This can sometimes be seen as an effect of aspect, north-facing slopes preserving more consistently the cool, humid conditions favouring the bryophyte richness of the *Isothecium-Diplophyllum* sub-community and it is visible, too, in the preferential occurrence of this kind of *Quercus-Betula-Dicranum* woodland in damp, shady ravines. Even on the level of individual large boulders, there can be a marked difference in the bryophyte flora

Figure 24. Distribution of sub-communities of W17 *Quercus-Betula-Dicranum* woodland through Lochaber and Strathspey (after NCC reports).

of more sheltered and more exposed faces.

Various physical characteristics of the rocks over which this community is typically found are important in this respect (e.g. Ratcliffe 1968). Not only are they generally hard and resistant but, when they do weather and are eroded, many of them produce very angular landscapes, at all scales. Glacial erosion and freeze-thaw have resulted in rugged cliffs and crags, and streams have cut steep-sided ravines; great tumbles of often huge

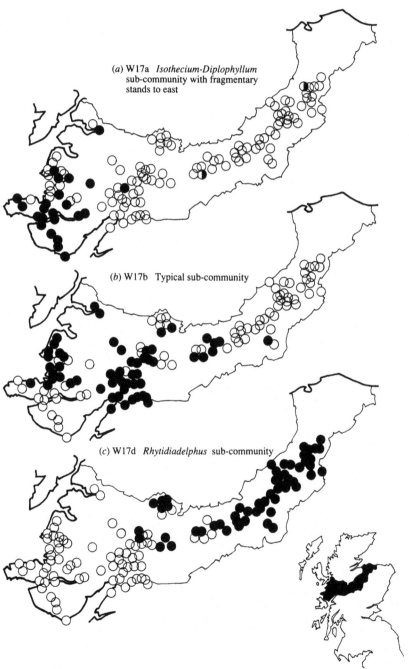

(*a*) W17a *Isothecium-Diplophyllum* sub-community with fragmentary stands to east

(*b*) W17b Typical sub-community

(*c*) W17d *Rhytidiadelphus* sub-community

boulders present a great diversity of surfaces with complex patterns of crevices between. All of these features enhance local interactions with climate of the kind noted above. And they have a direct effect, too, in presenting a large and often continually renewed rock surface for colonisation by the bryophyte flora. The difference between the *Isothecium-Diplophyllum* sub-community and the other kinds of *Quercus-Betula-Dicranum* woodland is thus partly an index of rockiness of the environment (Kelly & Moore 1975). Here, immature rock-dominant soils are characteristic and vascular plants are often largely confined to crevices and more stable boulder tops and ledges. In particular, the habitat of this sub-community provides an abundance of gently-sloping to steep faces on the sides of large boulders and over crags where the distinctive suite of bryophytes which Birks (1973) termed the *Isothecium-Hymenophyllum* Association can gain a hold on fragmentary accumulations of humus and mineral particles (see also Ratcliffe 1968). It is probable that, even within stands of this woodland which are stable on a large scale, there is a small-scale successional turnover among the bryophyte element, species of more open boulders giving way to mosses and liverworts of thin rankers (the *Isothecium-Diplophyllum* suite) and finally to the more stable mats of bulky bryophytes characteristic of intact swards over more mature profiles (the community constants) (Richards 1938, Evans 1954). In fact, on steeper faces, the effect of gravity often slows or repeatedly sets back colonisation, as fragments of the hanging bryophyte mat drop off (Lye 1967, Ratcliffe 1968, Birks 1973). Certain lithological features of the rocks may also encourage colonisation by bryophytes: Ratcliffe (1968) noted that the best substrates were those which presented a coarse-grained and somewhat porous surface, like granites and grits. Such deposits figure prominently throughout the community and they may be more frequent in the *Isothecium-Diplophyllum* sub-community.

The rockiness of the environment interacts with a further factor of importance in the *Quercus-Betula-Dicranum* woodland, grazing and browsing, the influence of which is seen not only among the bryophytes but in all other layers of the vegetation. Very often, stands of this community are open to stock, usually sheep, and deer, frequently backing on to open moorland which forms part of the grazing land of upland farms, and providing a valuable bite and shelter, especially in winter. Where there are large boulders, though, or generally precipitous and rocky ground, access is more limited, so even where there is a potentially large influence of herbivores, the actual effects may be much reduced. The impact of this, and of variation in actual herbivore numbers in more accessible stands, can be seen in all kinds of *Quercus-Betula-Dicranum* woodland but are of particular importance in helping define some of the sub-communities.

The most obvious visual effect of grazing and browsing is usually to be seen in the proportions of ericoids to grasses. The former, especially *Vaccinium myrtillus*, the most generally common sub-shrub of this kind of woodland, are much reduced, indeed can be totally eliminated, by the predations of sheep and deer; whereas the latter, particularly species like *Anthoxanthum odoratum* and *Agrostis capillaris* and, to a lesser extent on these very acid soils, *Holcus mollis*, thrive under continuous cropping. Variations in other herbs also reflect the impact of herbivores: the very palatable *Luzula sylvatica*, for example, is only prominent where there is no grazing (as on cliff ledges or on islands: well seen in Tittensor & Steele's (1971) Loch Lomond stands), whereas other low-growing and less palatable plants like *Oxalis acetosella* and *Galium saxatile* can thrive in a close-grazed sward. These contrasts are very well seen in the differences between the *Anthoxanthum-Agrostis* sub-community, which is more consistently grazed, and the other kinds of *Quercus-Betula-Dicranum* woodland, especially the Typical sub-community, which is characteristic of similar, slightly rocky ground.

Grazing also favours the abundance of bryophytes by keeping down the cover of larger vascular plants and, even where there are no boulders, the cover of bulkier mosses is usually high, though the species involved are generally common ones. Increased rockiness restricts the impact of grazing, though the species which respond are only those which can tolerate the shallower soils. *Vaccinium* typically increases, though its cover can be patchy and restricted to more stable accumulations of soil; *Luzula sylvatica*, too, can form great hanging sheets in quite precarious situations. But, in the *Isothecium-Diplophyllum* sub-community, the great preponderance of boulders restricts the lushness of the vascular flora that might be expected with the limitation of grazing. Response from this element is better seen, though even there not very clearly, in the Typical and *Rhytidiadelphus* sub-communities.

The other important influence of herbivores is seen in the regeneration pattern of the woody species. In the *Quercus-Betula-Dicranum* woodland, younger individuals of the trees are often sparse and very much restricted to less heavily grazed or ungrazed stands. All the major species of the community are affected but the interactions between factors like frequency of fruiting, dispersal, food reserves in the seed and germination patterns on the one hand, and release from or imposition of grazing and browsing on the other, are complex; they sometimes work indirectly through features like the increased frequency with which ungrazed swards trap and hold litter; and they are compounded by the character of the canopy shade (e.g. Jones 1959, Peterken 1966, Pigott 1983). In any particular stand, then, the prominence of saplings and seedlings often reflects rather particular combinations of past and present conditions. Among the available data, the best illustration of the general affects of herbivore predation on regeneration is seen in the contrasts between the Typical and *Anthoxanthum-Agrostis* sub-communities. In the former, which is less consistently grazed, the understorey is not generally denser, but saplings, especially of *S. aucuparia* and *Ilex*, the two most palatable trees here, are much commoner than in the latter: a rise in the abundance of these species, especially of the less cold-sensitive *S. aucuparia*, is often a marked feature of island stands of this kind of woodland (e.g. Anderson 1950, McVean 1964a, Tittensor & Steele 1971). And there is a slight increase in the occurrence of seedlings, including those of *Q. petraea*, which is able to regenerate under its own canopy given freedom from grazing (Jones 1959). Birch is more complex in its response: it is more light-demanding than any of the above and has smaller seed reserves so its progression from the seedling stage may depend on a more precise combination of freedom from canopy shade, from a blanket of leaf litter and from grazing. As Pigott (1983) demonstrated in the closely-related *Quercus-Betula-Deschampsia* woodland, such a coincidence of favourable conditions may be very short-lived which is perhaps one reason why more extensive stands of birch here (as elsewhere) are often more or less even-aged.

Again, it should be noted that, though the rocky terrain of the *Isothecium-Diplophyllum* sub-community offers the kinds of protection from herbivore predation that favours the active regeneration of trees, actual growth of the plants beyond the seedling stage is often prevented by the thin and sponge-like character of the bryophyte mat and the absence of any substantial amounts of mineral material. Soil can accumulate in crevices though, in deeper clefts, heavy shade becomes a problem for seedling survival.

Usually, now, if stands of the *Quercus-Betula-Dicranum* woodland are grazed, there is little interest in exercising any control over access to stock and, in many places, the community is regenerating only slowly or sporadically, so that we can see the kind of agriculturally-related decline that has played such a large role in the disappearance of this kind of woodland over many centuries (e.g. McVean & Ratcliffe 1962, McVean 1964a, Pearsall 1968, Pearsall & Pennington 1973). In certain periods in the past, however, control of grazing has been important because many stands of the community bear testimony to use as coppice. Sometimes, hazel has been the crop, or birch (both trees yielding wood for bobbins: e.g. Pearsall & Pennington 1973) but much more widespread has been coppicing for oak. Towards the north-west of Britain, this community and its less calcifugous counterpart, the *Quercus-Betula-Oxalis* woodland, have been the mainstay of woodland industries yielding charcoal, tan-bark and pit- and constructional timber (e.g. Tittensor 1970a, b, Tittensor &

Steele 1971, Pearsall & Pennington 1973, Linnard 1982). Remains of such activities can often be seen here in stands with regularly-disposed oak moots among which other tree species still play only a minor role; and there are frequently physical features such as disused charcoal pit-steads and interconnecting burners' pathways over the slopes (e.g. Pigott & Huntley 1978, Barker 1985).

Although oak shows some climatically related reduction in its growth in this community at high altitudes and latitudes, treatment has also probably played an important part in the scarcity of this species towards north-west Scotland where much of the good timber has been removed over the centuries (McVean & Ratcliffe 1962, McVean 1964a). There has also been a progressive run-down of many stands of the *Rhytidiadelphus* sub-community in eastern Scotland, where birch-dominated thickets are prominent (Birse 1984). Preferential planting of *Q. robur* in this region may also account for the scarcity of *Q. petraea* in such stands as are still oak-dominated (Jones 1959, Gardiner 1974), though other local occurrences of *Q. robur* in other sub-communities are not so easily explained. The striking prominence of this species in virtually pure populations in places like Wistman's Wood may be due in part to patterns of early migration in the post-Glacial and partly to geological preferences: in Devon, there is a rather well defined association between the occurrence of *Q. robur* on granite and *Q. petraea* on other rocks like the Culm Measures (Wigston 1974).

Zonation and succession

Zonations between the different types of *Quercus-Betula-Dicranum* woodland and from this community to other vegetation types are most commonly related to differences in microclimate, topography and soils and grazing and, as explained above, all these factors can interact in complex fashion, often in an inter-dependent way, so it is quite rare to find simply-explicable transitions.

The edaphic trend within the community is essentially one of soil maturity, running from lithomorphic rankers, such as are typical of the *Isothecium-Diplophyllum* sub-community, to the deeper, more stable and mature brown podzolic soils and podzols characteristic of the other sub-communities. To the west, such a trend is often visible in terms of a zonation from the former over the rocky ground of crags and screes to either the Typical sub-community (in less heavily grazed woods) or the *Anthoxanthum-Agrostis* sub-community (in heavily-grazed woods) on downwash soils over the gentler slopes below. And, on a smaller scale, the pattern may be repeated over the variably-sloping surfaces of every large boulder in differences in the stability and composition of the bryophyte mat. Between the two zones, there can be very complex mosaics as stretches of deeper soil run up between boulders and become fragmented into pockets between and atop the rocks; and, in the other direction, as boulders become progressively more isolated and sunken into the body of the creep soils. And such zonations are frequently complicated by climatically-related responses in the vegetation, so that shifts towards a more southerly, sunnier aspect may increase the proportion of ground occupied by either the Typical or *Anthoxanthum-Agrostis* sub-communities (especially perhaps the latter as lack of cool shade may depress *Vaccinium* and enhance the grass cover); or, conversely, a move towards more northerly aspects may produce a greater contribution from the *Isothecium-Diplophyllum* sub-community. Especially complex patterns may be found in landscapes cleft by repeated ravines, where alternating zones may lie side by side along a slope and where, within each gorge, there is a small-scale and almost unfathomable jumble of vegetation disposed over cliffs, ledges, tumbled blocks and slips of soil.

Quite commonly in the situations where the *Quercus-Betula-Dicranum* woodland is found, edaphic zonations continue to less base-poor profiles, usually strongly-leached brown earths. Such a transition can be related to a change in the underlying bedrock from more coarse-grained arenaceous sedimentaries or igneous or metamorphic rocks to more fine-grained ones which weather to a less readily leached mantle: this kind of switch is a fairly common feature in Silurian and Ordovician sequences in Wales and the Lake District and among volcanic rocks in the Lakes and parts of Scotland (e.g. Pearsall 1968, Pearsall & Pennington 1973). In such cases, the associated edaphic and vegetational changes can be sudden and clearly marked but, very often, it is the presence of superficials, like glacial drift or head, or the effect of flushing, that shifts the direction of soil development by masking the effect of the underlying acidic rocks and then the edaphic and vegetational gradient is much more gradual and ill defined. Typically, though, in all these situations, the *Quercus-Betula-Dicranum* woodland gives way, with varying degrees of sharpness, to the *Quercus-Betula-Oxalis* woodland, with a decline in ericoids and more calcifugous herbs and bryophytes and an increase in *Hyacinthoides, Rubus* and *Holcus mollis*. But uniformly intense grazing over these boundaries may soften the transition and, even where there is no grazing, very much the same kind of tree canopy extends over both communities with even *Corylus* providing no sure guide as to the move from one to the other (Figure 21).

The juxtaposition of highly acidic and markedly basic rocks, or the presence of very calcareous drift over siliceous substrates, or strong flushing with very base-rich waters, all of which can occur locally within the range of the *Quercus-Betula-Dicranum* woodland, result in a sharper shift from the community to either the *Fraxinus-Sorbus-Mercurialis* woodland or, in very wet situations, the *Alnus-Fraxinus-Lysimachia* woodland.

Then there is often a much more obvious floristic difference among all elements of the vegetation with calcicolous species replacing the calcifugous. In some areas though, extreme climatic conditions produce a convergence among certain components: this is well seen in the Suardal area of Skye where stands of all three of these communities, and of the *Quercus-Betula-Oxalis* woodland on intermediate soils, occur under scrubby canopies of *B. pubescens, S. aucuparia* and *Corylus* and with a prominent western element among the bryophytes (Birks 1973).

As well as confounding the simplicity of these kinds of zonations, grazing and browsing also produce transitions of their own which are essentially a reflection of the successional development of the community or, very commonly, its reversal. Burning is also frequently involved as a factor in such zonations. Seral progressions to the *Quercus-Betula-Dicranum* woodland have not been followed (they are rare, of course, and now often seen only in fenced sections of reserves or in forestry or reservoir enclosures) but the most obvious natural precursors to the community are probably the *Luzula sylvatica-Vaccinium myrtillus* tall-herb vegetation and certain kinds of *Calluna-Vaccinium* heath. Both of these show considerable overlap with the *Quercus-Betula-Dicranum* woodland in their edaphic and climatic requirements and essentially reproduce elements of its field and ground layers: McVean & Ratcliffe (1962) included stands of the *Luzula-Vaccinium* community as a 'treeless facies' of the woodland. Small stands of these rich vegetation types can sometimes be found in close spatial association with the community and, like it, survive as fragments within much-modified landscapes. They also extend the typical sub-shrub, herb and bryophyte cover characteristic of the community to considerably higher altitudes than those at which the woodland itself is found.

In fact, much more widespread are zonations to more species-poor kinds of *Calluna-Vaccinium* heath (in which bryophytes in particular are ill-represented), various types of calcifugous grasslands (notably less-improved *Festuca-Agrostis-Galium* swards) and stands of the *Pteridium-Galium saxatile* community, in landscapes which bear obvious signs of a long history of woodland clearance, grazing, burning and agricultural neglect. Quite often, the sequence of such vegetation types reflects the kind of underlying edaphic variation, running down from crags and screes over thin soils to colluvium, that can be seen picked out in the surviving woodland by the spatial sorting of *Vaccinium myrtillus*, grasses and bracken, though the proportional contribution of the last is often much greater with freedom from canopy shade on the open slopes. Fragments of the bryophyte-richness of the *Quercus-Betula-Dicranum* woodland may persist in derived *Calluna-Vaccinium* heath, though burning is very deleterious, and, among the sward of the plagioclimax *Festuca-Agrostis-Galium* grassland, continuous grazing and increased exposure to sun and wind can eliminate all but the more robust mosses; less intractable slopes may also have been improved by fertiliser application. Dense bracken stands, too, are usually impoverished: ericoids and herbs may persist patchily but the bryophytes are rapidly reduced by the combination of deep shade and thick litter. Finally, one very common additional element of these altered landscapes is coniferous forest which has been widely planted in the north-western uplands on soils which could otherwise support this community. Such woodlands have not been extensively sampled, though more open stands could probably be incorporated here as impoverished variants.

In some parts of Scotland, there is the further problem of the relationship between the *Quercus-Betula-Dicranum* woodland and what seems, from palynological evidence, to be the natural forest cover of the region, the *Pinus-Hylocomium* woodland (e.g. McVean & Ratcliffe 1962, Birks 1970, Gunson 1975, O'Sullivan 1977). There is a considerable floristic overlap between the associated species of these two woodland types and what may have happened here, with the destruction of the original, and apparently fairly mixed, woody cover of the pine forest is that birch and oak, especially the former, have now come to dominate in the *Quercus-Betula-Dicranum* woodland on the more acidic mineral soils that could naturally support pine.

Distribution
The community is widely distributed through the upland fringes of western and northern Britain from Dartmoor to Sutherland with especially good stands in mid and north Wales, the Lake District and along the north-western seaboard of Scotland. The sub-communities show a fairly clear climatically-related pattern of occurrence with the *Isothecium-Diplophyllum* type in the most oceanic areas, the Typical and *Anthoxanthum-Agrostis* sub-communities in somewhat less extreme situations and the *Rhytidiadelphus* sub-community in the more continental parts of eastern Scotland.

Affinities
The *Quercus-Betula-Dicranum* woodland unites more calcifugous and heathy woodlands which have sometimes been described within a general *Quercetum petraeae* where the stress has been on oak-dominance (e.g. Tansley 1939), or diagnosed as birch-dominated communities from higher latitudes (e.g. McVean & Ratcliffe 1962, Birks 1973, Birse 1982). Although the proportion of oak here does have some climatic basis, it is, in part, a treatment-related phenomenon which has little effect on the associated flora (e.g. McVean & Ratcliffe 1962, McVean 1964a, Tittensor & Steele 1971, Birks 1973).

Neither is it fruitful to make any distinction according to which of the oaks or birches is predominant: early (e.g. Tansley 1939) and later (e.g. Peterken 1981) schemes have sometimes used the oak species as diagnostic in this respect but, again, there are no directly-related differences in the other layers of the vegetation. Furthermore, the ecological significance of the representation of the two species varies according to the locality: sometimes it appears to have a geological basis (as in Dartmoor), in other cases preferential planting seems to have been largely responsible (in eastern Scotland).

Uniting all such woodlands within a single community has the advantage that it is possible to make some integrated assessment of national variation. The most obvious trend is the increasing contribution from the bryophytes, in particular a more Atlantic element, in moving from the comparatively drier east and south to the very wet west and north. We can thus now see the context of more extreme versions of this community described from Scotland (McVean & Ratcliffe 1962, Klötzli 1970, Birks 1973, Birse 1980, 1982, 1984) and generally related to the *Blechno-Quercetum*, the association in which Braun-Blanquet & Tüxen (1952) placed the very distinctive Irish oakwoods first described from Killarney (e.g. Rübel 1912, Richards 1938, Turner & Watt 1939, Tansley 1939). Reanalysis of old and new data from this kind of woodland in Ireland (Kelly 1981; see also Kelly & Moore 1975) has confirmed Braun-Blanquet & Tüxen's original recognition of two subtypes, the *scapanietosum* (originally called *isothecietosum*), significantly associated with wetter regions and rockier situations and very similar to the *Isothecium-Diplophyllum* sub-community here, and a *typicum*, including vegetation here split into the Typical and *Anthoxanthum-Agrostis* sub-communities. Kelly & Moore's (1975) *coryletosum*, significantly correlated with richer and less humic soils of higher pH, would here be placed in the oceanic *Blechnum* sub-community of the *Quercus-Betula-Oxalis* woodland.

The *Quercus-Betula-Dicranum* woodland thus incorporates some of the most oceanic Quercion forest in Western Europe, linking through the less extreme stands of the Typical and *Anthoxanthum-Agrostis* sub-communities with more mainstream associations like the *Vaccinio-Quercetum sessiliflorae* Clement et al. 1975, and through the *Isothecium-Diplophyllum* sub-community with the *Blechno-Quercetum scapanietosum* Kelly & Moore 1975 of western Ireland (Géhu 1975a). The *Rhytidiadelphus* sub-community of eastern Scotland provides a floristic connection in a different direction, to the Scandinavian calcifuge birchwoods in such associations as the *Betuletum myrtillo-hylocomiosum* (Nordhagen 1928, 1943), where species such as *Trientalis europaea*, *Cornus suecica* and *Linnaea borealis* become much more frequent.

Floristic table W17

	a	b	c	d	17
Quercus petraea	V (1–10)	V (3–10)	V (4–10)	I (8–9)	IV (1–10)
Betula pubescens	III (1–9)	III (1–7)	III (3–7)	IV (1–9)	III (1–9)
Sorbus aucuparia	III (1–5)	I (1–2)	II (1–5)	III (1–4)	II (1–5)
Fraxinus excelsior	I (2–6)	I (1–3)	I (1–7)	I (4)	I (1–7)
Acer pseudoplatanus	I (1–4)	I (1–2)	I (2–5)	I (1)	I (1–5)
Ilex aquifolium	I (1–4)			I (1–2)	I (1–4)
Fagus sylvatica		I (1–2)	I (1–5)	I (3–4)	I (1–5)
Betula pendula	I (2–5)	I (2–4)	I (1–3)	II (2–8)	I (1–8)
Quercus robur		I (6)	I (5–10)	II (2–8)	I (2–10)
Quercus hybrids				I (4–9)	I (4–9)
Corylus avellana	II (1–7)	III (1–7)	III (1–6)	I (1–5)	II (1–7)
Sorbus aucuparia	I (1–5)	III (1–4)	I (1–4)		II (1–5)
Betula pubescens sapling	I (1–4)	II (1–4)	I (2–4)		I (1–4)
Quercus petraea sapling	I (1–2)	II (1–4)	I (1–2)		I (1–4)
Ilex aquifolium	I (1–2)	II (1–5)	I (1–3)		I (1–5)
Crataegus monogyna	I (1–4)	I (1–2)	II (1–5)		I (1–5)
Fraxinus excelsior sapling	I (1)	I (1–3)	I (1–2)		I (1–3)
Fagus sylvatica sapling	I (1–4)	I (1–2)	I (4)		I (1–4)
Acer pseudoplatanus sapling	I (2)	I (1)	I (2)		I (1–2)
Rhododendron ponticum	I (1)	I (1–4)	I (1)		I (1–4)
Betula pendula sapling	I (2)			I (1)	I (1–2)
Salix caprea		I (2)		I (2–3)	I (2–3)
Juniperus communis communis				I (1–5)	I (1–5)
Deschampsia flexuosa	IV (2–7)	V (1–8)	III (1–8)	V (1–9)	V (1–9)
Rhytidiadelphus loreus	V (1–9)	IV (1–6)	V (1–7)	III (1–6)	IV (1–9)
Polytrichum formosum	V (1–5)	IV (1–6)	V (1–5)	III (1–4)	V (1–6)
Dicranum majus	V (1–6)	V (1–7)	III (1–5)	III (1–6)	IV (1–7)
Hylocomium splendens	IV (1–6)	III (2–5)	III (1–5)	V (1–9)	IV (1–9)
Pleurozium schreberi	IV (1–5)	IV (1–6)	II (1–5)	V (1–8)	IV (1–8)
Vaccinium myrtillus	III (1–7)	IV (1–9)	I (1–5)	V (1–9)	IV (1–9)
Plagiothecium undulatum	IV (1–5)	III (1–5)	IV (1–5)	II (1–4)	IV (1–5)

Species					
Isothecium myosuroides	IV (1–6)	II (1–4)	II (1–6)	I (1–2)	II (1–6)
Diplophyllum albicans	IV (1–4)	II (1–4)	I (1–4)		II (1–4)
Hypnum cupressiforme	III (1–5)	II (1–5)	I (1–3)	I (1–4)	II (1–5)
Blechnum spicant	III (1–5)	II (1–5)	I (1–2)	II (1–5)	II (1–5)
Lepidozia reptans	III (1–4)	I (1–4)	I (2–3)	I (1–2)	I (1–4)
Thuidium delicatulum	III (1–5)	I (1)	I (1)	I (3)	I (1–5)
Leucobryum glaucum	III (1–6)	I (1–6)	I (1)	I (2)	I (1–6)
Campylopus paradoxus	III (1–4)	I (1–2)	I (1–3)	I (1)	I (1–4)
Plagiochila spinulosa	III (1–4)	I (1–2)	I (2–3)		I (1–4)
Scapania gracilis	III (1–5)	I (1–3)	I (1–2)		I (1–5)
Bazzania trilobata	III (1–6)	I (1–5)	I (3)		I (1–6)
Molinia caerulea	II (1–7)	I (1–5)	I (1–6)	I (5–8)	I (1–8)
Dicranodontium denudatum	II (1–5)	I (1–3)	I (1–2)		I (1–5)
Saccogyna viticulosa	II (1–5)	I (1–5)	I (1–2)		I (1–5)
Hylocomium umbratum	II (1–3)		I (4)	I (4)	I (1–4)
Isopterygium elegans	II (1–4)	I (1–4)	I (1–4)		I (1–4)
Parmelia saxatilis	II (1–4)	I (1)		I (2)	I (1–4)
Hymenophyllum wilsonii	II (1–5)		I (1)		I (1–5)
Heterocladium heteropterum	II (1–4)		I (1–2)		I (1–4)
Racomitrium heterostichum	II (1–3)	I (2–4)			I (1–4)
Lophozia ventricosa	I (1–2)		I (1)	I (1)	I (1–2)
Marsupella emarginata	I (1–3)				I (1–3)
Barbilophozia attenuata	I (1–3)	I (1–2)			I (1–3)
Jamesoniella autumnalis	I (1–4)	I (1)			I (1–4)
Hypnum callichroum	I (1–4)	I (1)			I (1–4)
Racomitrium fasciculare	I (1–4)	I (1–2)			I (1–4)
Racomitrium lanuginosum	I (1–3)	I (4)			I (1–4)
Plagiochila asplenoides major	I (1–3)	I (1)			I (1–3)
Scapania nemorosa	I (1–2)				I (1–2)
Blepharostoma trichophyllum	I (1–2)				I (1–2)
Lejeunea ulicina	I (1–2)				I (1–2)
Plagiochila punctata	I (1–3)				I (1–3)
Sematophyllum micans	I (2–7)				I (2–7)
Hyocomium armoricum	I (1–3)				I (1–3)
Scapania umbrosa	I (1–2)				I (1–2)
Andreaea rupestris	I (1–2)				I (1–2)

Floristic table W17 (*cont.*)

	a	b	c	d	17
Tritomaria quinquedentata	I (1–2)				I (1–2)
Jungermannia gracilima	I (1)				I (1)
Adelanthus decipiens	I (1–4)				I (1–4)
Sematophyllum demissum	I (1–2)				I (1–2)
Hymenophyllum tunbrigense	I (1–2)				I (1–2)
Tritomaria exsecta	I (1–2)				I (1–2)
Nowellia curvifolia	I (1–3)				I (1–3)
Plagiochila killarniensis	I (1–5)				I (1–5)
Sphaerophorus fragilis	I (1–3)				I (1–3)
Sphagnum fimbriatum	I (1–3)				I (1–3)
Dicranum scottianum	I (1–2)				I (1–2)
Plagiochila atlantica	I (1–3)				I (1–3)
Primula vulgaris	I (1–2)				I (1–2)
Dryopteris aemula	I (1–4)				I (1–4)
Anastrepta orcadensis	I (1–2)				I (1–2)
Cladonia subcervicornis	I (1–2)				I (1–2)
Harpanthus scutatus	I (2)				I (2)
Plagiochila corniculata	I (1)				I (1)
Dryopteris dilatata	I (1–6)	II (1–4)	I (1–4)	I (1–5)	I (1–6)
Quercus petraea seedling	I (1–3)	II (1–2)	I (1–3)		I (1–3)
Cladonia squamosa	I (1–3)	II (1–2)	I (1–2)		I (1–3)
Quercus sp. seedling		I (1–2)			I (1–2)
Cladonia digitata		I (1–3)			I (1–3)
Eurhynchium striatum		I (1–2)			I (1–2)
Hypericum pulchrum		I (1–2)			I (1–2)
Galium saxatile	II (1–5)	II (1–4)	IV (1–6)	IV (1–4)	III (1–6)
Anthoxanthum odoratum	II (1–6)	II (1–4)	IV (1–8)	II (1–4)	II (1–8)
Agrostis capillaris	II (1–6)	I (1–5)	IV (1–7)	I (1–4)	II (1–7)
Holcus mollis	II (1–6)	I (1–5)	III (2–8)	I (1–3)	II (1–8)
Rubus fruticosus agg.	I (1–5)	I (1–7)	II (1–9)	I (5)	I (1–9)
Eurhynchium praelongum	I (1–2)	I (4)	II (1–3)	I (1)	I (1–4)
Dicranella heteromalla	I (1–2)	I (1–3)	II (1–4)	I (1)	I (1–4)

Digitalis purpurea			I (1)	I (1-5)
Dactylis glomerata				I (2-3)
Rumex acetosa				I (1-3)
Poa nemoralis				I (1-3)
Poa pratensis				I (2-3)
Calluna vulgaris	II (1-6)	III (1-6)	I (1)	III (1-9)
Pseudoscleropodium purum	I (1-4)	I (1-4)	II (1-9)	II (1-4)
Rhytidiadelphus triquetrus	I (1-5)	I (1-2)	I (1-4)	I (1-5)
Luzula pilosa	I (1-3)	I (1-4)	I (1-3)	I (1-4)
Trientalis europaea			II (1-5)	I (1-5)
Goodyera repens			I (1)	I (1)
Cornus suecica			I (2)	I (2)
Dicranum scoparium	IV (1-5)	III (1-6)	III (1-4)	III (1-6)
Mnium hornum	IV (1-4)	IV (1-4)	III (1-5)	III (1-5)
Pteridium aquilinum	III (1-7)	III (1-7)	IV (1-8)	III (1-8)
Thuidium tamariscinum	III (1-7)	III (1-5)	III (1-7)	III (1-8)
Oxalis acetosella	III (1-5)	I (1-4)	IV (1-7)	III (1-7)
Lophocolea bidentata s.l.	I (1-2)	III (1-4)	II (1-3)	II (1-6)
Agrostis canina montana	II (1-6)	I (1-5)	II (1-6)	II (1-6)
Festuca ovina	III (1-5)	I (2-3)	II (2-6)	II (1-6)
Hypnum jutlandicum	I (1-4)	III (2-5)	II (1-5)	II (1-6)
Rhytidiadelphus squarrosus	I (1-4)	I (1-4)	II (1-7)	II (1-7)
Lonicera periclymenum	I (1-5)	II (1-6)	I (1-3)	I (1-6)
Sorbus aucuparia seedling	I (1-4)	II (1-3)	I (1-3)	I (1-4)
Sphagnum quinquefarium	II (1-7)	II (1-7)	I (3-4)	I (1-7)
Potentilla erecta	II (1-7)	I (1-3)	I (1-4)	I (1-5)
Melampyrum pratense	II (1-7)	I (3-8)	II (2-5)	II (1-8)
Dryopteris filix-mas	I (1-4)	I (4)	I (1-5)	I (1-5)
Dryopteris borreri	I (1-2)	I (1-2)	I (1-5)	I (1-5)
Ilex aquifolium seedling	I (1-3)	I (1-2)	I (1-3)	I (1-3)
Viola riviniana	I (1-4)	I (1-3)	I (2-3)	I (1-5)
Holcus lanatus	I (1-4)	I (1)	I (1-4)	I (1-4)
Rubus idaeus	I (3-4)	I (1)	I (2-3)	I (1-4)
Thelypteris limbosperma	I (1-4)	I (2)	I (1-6)	I (1-6)
Calypogeia fissa	I (1-2)	I (1-3)	I (1)	I (1-3)

Floristic table W17 (*cont.*)

	a	b	c	d	17
Betula pubescens seedling	I(1–3)	I(1–2)	I(1–3)	I(1–4)	I(1–4)
Tetraphis pellucida	I(1–3)	I(1–2)	I(1–3)	I(1)	I(1–3)
Teucrium scorodonia	I(2–3)	I(1–4)	I(1–4)	I(5)	I(1–5)
Luzula sylvatica	I(1–7)	I(4)	I(4–9)	I(1–9)	I(1–9)
Hyacinthoides non-scripta	I(1–5)	I(1–2)	I(2–5)	I(1–3)	I(1–5)
Sphagnum palustre	I(2–8)	I(2–6)	I(2–6)	I(6)	I(2–8)
Carex pilulifera	I(2–3)	I(3)	I(1)	I(1)	I(1–3)
Erica cinerea	I(4)	I(1–5)	I(1)	I(1–7)	I(1–7)
Anemone nemorosa	I(1–2)	I(1–2)	I(1)	I(1)	I(1–2)
Athyrium filix-femina	I(1–4)		I(1–5)	I(1)	I(1–5)
Cladonia impexa	I(1–2)	I(1–5)		I(1)	I(1–5)
Succisa pratensis	I(1–4)	I(1)		I(1–2)	I(1–4)
Quercus hybrids sapling	I(1)	I(1)		I(1–3)	I(1–3)
Lophocolea cuspidata	I(1–2)	I(3)	I(1)		I(1–3)
Rhizomnium punctatum	I(1)	I(1)	I(1–2)		I(1–2)
Calypogeia muelleriana	I(1–2)	I(1)	I(1)		I(1–2)
Cladonia polydactyla	I(1–2)	I(1)	I(1–3)		I(1–3)
Barbilophozia floerkei	I(1–3)	I(1–2)	I(1–2)		I(1–3)
Cephalozia media	I(1)	I(1)	I(1)		I(1)
Bazzania tricrenata	I(1–3)	I(2)	I(2)		I(1–3)
Hedera helix	I(1–4)	I(1–4)	I(1–4)		I(1–4)
Lepidozia pearsonii	I(1–3)	I(1–3)	I(1–3)		I(1–3)
Solidago virgaurea	I(1–2)	I(1)	I(1)		I(1–2)
Deschampsia cespitosa	I(5)	I(4)	I(1–5)		I(1–5)
Ptilidium ciliare	I(1)	I(1)		I(1)	I(1)
Cladonia chlorophaea	I(1)	I(1–2)		I(1)	I(1–2)
Polytrichum commune	I(2–4)	I(1–6)		I(1–8)	I(1–8)
Dicranum fuscescens		I(2–3)	I(1)	I(1–3)	I(1–3)
Plagiochila asplenoides		I(1)	I(1–2)	I(2–4)	I(1–4)
Luzula multiflora		I(2)	I(1)	I(1–3)	I(1–3)
Mylia taylori	I(1–3)	I(3)			I(1–3)
Sphaerophorus globosus	I(1–2)	I(1–2)			I(1–2)
Cladonia arbuscula	I(1–3)	I(1–4)			I(1–4)

	a	b	c	d	17
Sphagnum russowii	I (2–3)	I (3)			I (2–3)
Cladonia furcata	I (1–2)	I (1–3)			I (1–3)
Ptychomitrium polyphyllum	I (1)	I (2)			I (1–2)
Cephalozia bicuspidata	I (1–2)		I (1–2)		I (1–2)
Fraxinus excelsior seedling	I (1–3)		I (1–2)		I (1–3)
Cladonia squamules	I (1–2)		I (1)		I (1–2)
Frullania tamarisci	I (1–3)		I (2)		I (1–3)
Lysimachia nemorum	I (1–3)		I (3)		I (1–3)
Hylocomium brevirostre	I (1–4)		I (1–4)		I (1–4)
Sphagnum subnitens	I (2)			I (1–4)	I (1–4)
Ptilium crista-castrensis	I (1–4)			I (2–4)	I (1–4)
Plagiothecium denticulatum	I (1–5)			I (1–3)	I (1–5)
Polypodium vulgare	I (2–3)			I (2)	I (2–3)
Erica tetralix	I (4)			I (1–4)	I (1–4)
Agrostis stolonifera			I (1–4)	I (2)	I (1–4)
Lophocolea heterophylla			I (1–2)	I (1)	I (1–2)
Quercus robur seedling			I (1–2)	I (1)	I (1–2)
Number of samples	131	48	69	55	303
Number of species/sample	32 (15–55)	29 (17–48)	23 (15–50)	22 (13–31)	28 (13–55)
Tree height (m)	12 (7–22)	14 (5–24)	16 (8–26)	14 (6–30)	14 (5–30)
Tree cover (%)	74 (10–100)	87 (60–100)	86 (70–100)	56 (15–90)	76 (10–100)
Shrub height (m)	3 (1–5)	3 (1–5)	3 (1–5)	3 (2–6)	3 (1–6)
Shrub cover (%)	19 (0–90)	12 (0–60)	12 (0–50)	1 (0–4)	13 (0–90)
Herb height (cm)	30 (4–120)	28 (10–80)	26 (3–62)	31 (10–70)	29 (3–120)
Herb cover (%)	42 (2–90)	53 (4–100)	60 (5–100)	80 (40–100)	55 (4–100)
Ground height (mm)	44 (10–100)	40 (10–100)	30 (10–90)	35 (30–50)	39 (10–100)
Ground cover (%)	62 (10–100)	54 (10–90)	47 (7–90)	51 (2–95)	55 (2–100)
Altitude (m)	144 (12–280)	160 (30–290)	175 (75–300)	159 (40–519)	162 (12–519)
Slope (°)	22 (2–80)	29 (0–80)	25 (4–50)	10 (0–28)	22 (0–80)

a   *Isothecium myosuroides-Diplophyllum albicans* sub-community

b   Typical sub-community

c   *Anthoxanthum odoratum-Agrostis capillaris* sub-community

d   *Rhytidiadelphus triquetrus* sub-community

17 *Quercus petraea-Betula pubescens-Dicranum majus* woodland (total)

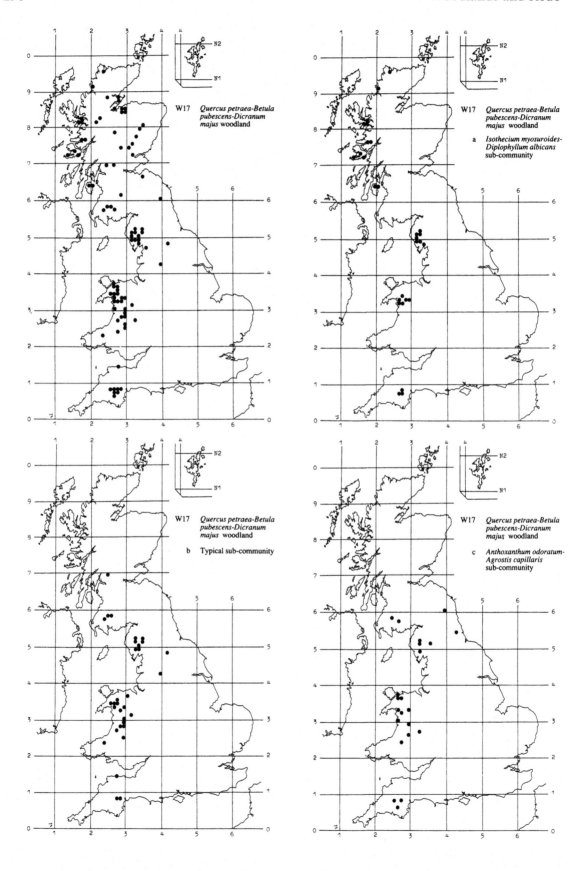

W17   *Quercus petraea-Betula pubescens-Dicranum majus* woodland

W17   *Quercus petraea-Betula pubescens-Dicranum majus* woodland

a   *Isothecium myosuroides-Diplophyllum albicans* sub-community

W17   *Quercus petraea-Betula pubescens-Dicranum majus* woodland

b   Typical sub-community

W17   *Quercus petraea-Betula pubescens-Dicranum majus* woodland

c   *Anthoxanthum odoratum-Agrostis capillaris* sub-community

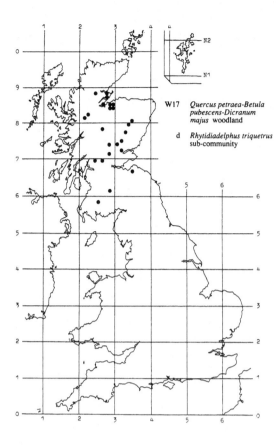

W17   *Quercus petraea-Betula pubescens-Dicranum majus* woodland

d   *Rhytidiadelphus triquetrus* sub-community

W18
Pinus sylvestris-Hylocomium splendens woodland

Synonymy

Scottish *Pinetum sylvestris* Tansley 1911; Highland Pine Forest Tansley 1939; Pinewood communities Steven & Carlisle 1959 *p.p.*; *Pinetum Hylocomieto-Vaccinietum* McVean & Ratcliffe 1962; Pinewood *Vaccinium*-moss association McVean 1964a; Pinewood *Vaccinium-Calluna* association McVean 1964a; Pinewood plot types 1–8 Bunce 1977; *Erica cinerea-Pinus sylvestris* Plantation Birse & Robertson 1976 *emend.* Birse 1980; Pine woodland Peterken 1981; *Pinetum scoticae* (Steven & Carlisle 1959) Birse 1980, 1982; Woodland plot type 28 Bunce 1982.

Constant species

Pinus sylvestris, Calluna vulgaris, Deschampsia flexuosa, Vaccinium myrtillus, V. vitis-idaea, Dicranum scoparium, Hylocomium splendens, Plagiothecium undulatum, Pleurozium schreberi, Rhytidiadelphus loreus.

Rare species

Arctostaphylos uva-ursi, Goodyera repens, Linnaea borealis, Moneses uniflora, Orthilia secunda, Pyrola media, P. rotundifolia, Mastigophora woodsii.

Physiognomy

The *Pinus sylvestris-Hylocomium splendens* woodland always has *Pinus sylvestris* as the most abundant tree. Indigenous Scots pine is generally referred to *P. sylvestris* var. *scotica* (Willd.) Schott (Gaussen *et al.* 1964), though populations show considerable and continuous variation in characters like crown shape, foliage colour and bark textures (Carlisle 1958, Steven & Carlisle 1959) and many of the distinctive morphological traits recognised in Scotland occur in populations elsewhere in the range of the species (Carlisle & Brown 1968). Moreover, up until about 1950, there had been extensive introduction of *P. sylvestris* of unknown or uncertain origin into or adjacent to stands of the native tree (Carlisle 1977, Faulkner 1977). Some earlier schemes have concentrated on the classification of apparently native woodlands (e.g. Steven & Carlisle 1959) and others have recognised distinct communities from either semi-natural or planted stands (e.g. Birse & Robertson 1976, Birse 1980, 1982, 1984). As defined here, the *Pinus-Hylocomium* woodland includes more natural pine forest, modified stands and plantations (as in McVean & Ratcliffe 1962 and McVean 1964a), though treatment is recognised as one of the factors important in controlling variation in the associated flora.

Although pine predominates here, the stocking density of stands is very variable. In many tracts, the canopy is very open, creating the impression of heath with trees, a kind of landscape that has long had great aesthetic appeal, organised on a large scale, allowing the trees to be seen in their often great individuality, and permitting glimpses of the surrounding scenery, which is frequently grand (e.g. Ratcliffe 1974, Thom 1977). A sensible lower limit of pine cover for defining this kind of woodland (in the absence of other canopy trees) is perhaps 25%, which may amount to no more than two large or several small specimens in a 50 × 50 m quadrat, but which makes an effective separation from more open ericoid vegetation with very widely-scattered pines which, in this scheme, is treated separately (cf. McVean & Ratcliffe (1962) where a treeless facies is incorporated into the community). At the other extreme are fully-stocked stands with a virtually complete canopy. Between the two, there is a continuous range of tree covers, though most semi-natural stands have a canopy of less than 70%. There is also a distinct tendency for a geographical separation with denser covers prevailing in eastern Scotland, more open ones being commoner in the west (Steven & Carlisle 1959, McVean & Ratcliffe 1962, Goodier & Bunce 1977); and there is an obvious association between this pattern of canopy structure and the distribution of the sub-communities, though not all their preferential features are directly related to the degree of closure.

Although stands with scattered trees are quite common here, *P. sylvestris* is essentially a gregarious species (Carlisle & Brown 1968) and it is a very characteristic feature of stretches of this kind of woodland that they are made up of a mosaic of quite well segregated age-classes, rather than being an intimate mixture of trees of all ages as in other kinds of semi-natural high forest. McVean & Ratcliffe (1962) noted three major arrangements at the time of their survey: more or less even-aged stands, 80–150 years old, sometimes with suppressed individuals of the same age; two-generation mixtures of pioneers, 150–200 years old, embedded in a matrix of straight-stemmed offspring, 80–100 years old; and pine-heath stands of varying densities but composed exclusively of broad-crowned pioneers, 150–200 years old. The proportions of these different kinds of groupings varies within the individual tracts of pine forest (Steven & Carlisle 1959), but a recent survey (Goodier & Bunce 1977) has confirmed the great predominance of older trees in semi-natural stands. Dead specimens, however, seem to be rather rare: in Scotland the maximum age of *P. sylvestris* is probably in excess of 300 years (Steven & Carlisle 1959). The implications of these types of structural arrangements for the natural regeneration of the community are taken up below but, again, it should be noted that there is some relationship between the canopy age-structure, in both these semi-natural stands and in younger even-aged plantations included here, and the floristics of the field and ground layers.

The age and size of the pine in this community are not always closely correlated and small, stunted trees towards the upper limit of woods can exceed 200 years of age (Steven & Carlisle 1959). Even in more favourable situations, the canopy is usually only 13–15 m high, exceptionally reaching more than 20 m on better soils in sheltered sites, though it should always be remembered that bigger specimens may have been extracted, especially from stands where growth is generally better and where access is easier. Despite this generally small stature and the often low cover of the canopy, *P. sylvestris* always dominates here in terms of proportional abundance and height: within stands of the community, associated trees and shrubs are rare and typically shorter than the pine. Birch is the commonest companion, with *B. pubescens* (often ssp. *carpatica*) more characteristic of western stands, *B. pendula* of those towards the east, and *Sorbus aucuparia* occurs occasionally. These species can be found either as scattered individuals under moderately close pine canopies or in thicker patches where the cover is more open. Strictly speaking, however, denser groups with a very low proportion of pine would, in this scheme, be considered as part of mosaics between the *Pinus-Hylocomium* woodland and either the *Quercus-Betula-Dicranum* woodland or, on somewhat more fertile soils, the *Quercus-Betula-Oxalis* woodland. Such mosaics are a very common feature of Scottish pine forest and they usually involve birch- (occasionally rowan-) dominated stands of these other communities, though oak can figure at lower altitudes, *Quercus petraea* being typical to the west, *Q. robur* and hybrids becoming more frequent in the east. It is in this kind of situation that *Ilex aquifolium* is found in close association with pine, especially in the milder west, though it is often nibbled to a low undershrub by sheep and deer. In other cases, *Juniperus communis* can be found as scattered bushes or in small patches within the *Pinus-Hylocomium* woodland but, again, where it thickens up in the local absence of pine, such stands are best seen as mosaics with the *Juniperus-Oxalis* woodland, a characteristic feature of more high-altitude situations in eastern Scotland. Quite commonly, this kind of variegation in the vegetation cover within pine forests also takes in stands of wetter woodland on valley and basin mires: this is why such species as *Alnus glutinosa*, *Salix cinerea*, *S. aurita*, *S. pentandra* and *Populus tremula* sometimes figure in descriptions of *P. sylvestris* woodland which have a broader basis of definition (e.g. Steven & Carlisle 1959). Finally, there may be occasional records for other exotic conifers seeding in from nearby plantations.

The associated flora of the *Pinus-Hylocomium* woodland has three major elements, *Deschampsia flexuosa*, ericoid sub-shrubs and bryophytes. As in other of our more calcifugous woodlands, *D. flexuosa* forms a fairly consistent grassy background to the field layer here and it can be especially prominent in more heavily grazed stands or under denser shade where the ericoids are reduced: it may dominate, for example, in pine plantations at the thicket stage (Birse 1980). Generally, though, it is variation among the two other elements that is more important here, both in general visual terms and in the particular definition of the different sub-communities. Among the ericoids, *Vaccinium myrtillus*, *V. vitis-idaea* and *Calluna vulgaris* are all more consistently frequent here than in any other kind of woodland, even the most floristically-similar *Juniperus-Oxalis* woodland where the same combination of highly acid soils and open canopies can occur. But the abundance of these species is very variable, being affected not only by canopy shade and grazing but also by such factors as their relative speed of colonisation when there is some change in the intensity of these variables or after burning (a frequent occurrence here) and inter-specific competition. *Calluna*, especially, is very sensitive to shade, and is much more prominent in sub-communities with a more open canopy, but *V. vitis-idaea* can suffer against *V. myrtillus* where the latter is growing very vigorously or where the trees are more closely set and, in very dense pine, all three can be eliminated, leaving bryophytes and *D. flexuosa* as effective dominants. When the canopy is opened up, a change which can precipitate a seral development from one sub-community to another, all the ericoids can expand or colonise, though the eventual prominence of the Vaccinia is very dependent on how much of them survived the period of denser shade, because *Calluna* is the most rapid invader and it may come to dominate, restricting the growth of the other species with a cover of leggy bushes more than half a metre high (McVean & Ratcliffe 1962). Grazing and browsing can also effect a marked reduction in the abundance of *V. myrtillus* and *Calluna*, both of which are readily eaten by stock and deer (the latter also by grouse), or alter the balance between them and favour a prominence of *V. vitis-idaea*, which is less palatable (Ritchie 1955).

Other sub-shrubs which can play a prominent, though less consistent, role here are *Empetrum nigrum* ssp. *nigrum, Erica tetralix*, which is very characteristic of moister, peatier soils, especially in the wetter west and in local transitions to mires, and *E. cinerea*, particularly distinctive of earlier and thinned stages in plantations though also occurring occasionally in more open, semi-natural stands. *E. nigrum* ssp. *hermaphroditum* and *V. uliginosum*, both of which are important species in certain types of Scandinavian pine associations (e.g. Aune 1977), are rare, as is *Arctostaphylos uva-ursi*.

The third important element of the flora here is the bryophytes, particularly bulkier mosses. Some of the species occur at consistently high frequencies throughout, being found even in denser stands of pine, where, with the reduction in ericoids, they can comprise the most prominent component of the ground vegetation: these species are *Hylocomium splendens, Dicranum scoparium* and *Pleurozium schreberi*. Two further species, *Plagiothecium undulatum* and *Rhytidiadelphus loreus*, are much less common under more closed canopies but they are so frequent elsewhere in the community that they attain constancy overall. This suite of more calcifugous mosses is very characteristic of a number of north-western woodland types on strongly acidic soils but one further species, not quite constant but very frequent under all but the densest canopies, is particularly distinctive of the *Pinus-Hylocomium* woodland and that is *Ptilium crista-castrensis*, a moss which maintains this important role in Scandinavian pine forests (e.g. Aune 1977). Other bryophytes occurring commonly throughout are *Hypnum jutlandicum* and *Lophocolea bidentata* s.l. and, more occasionally, *Hypnum cupressiforme s.l.*, *Polytrichum commune* (less frequent here than in calcifuge oak-birch woodlands of the north-west), *P. formosum*, *P. juniperinum, Campylopus paradoxus* and *Aulacomnium palustre*.

Then, there are further species which, though not common throughout, are preferentially frequent in different groups of sub-communities. *Rhytidiadelphus triquetrus, Pseudoscleropodium purum* and *Dicranum*

fuscescens provide an important part of the definition of *Pinus-Hylocomium* woodlands with more closed and younger canopies, especially in the drier east. By contrast, *Sphagnum capillifolium/quinquefarium* (not consistently distinguished in the data), *Dicranum majus* and *Scapania gracilis* are more characteristic of stands with more open and older covers of pine, particularly in the wetter west, where they may be accompanied by various less frequent species including other Sphagna of the Acutifolia group, like *S. russowii* and *S. girgensohnii*, *S. palustre, Leucobryum glaucum, Calypogeia trichomanis/muellerana* and some western or more strictly Atlantic bryophytes. The floristic distinction between these two groups is essentially that used by McVean & Ratcliffe (1962) in their definition of two pine associations and by Birse (1984) in his recognition of eastern and western races among semi-natural stands.

Among the bryophyte mat, there may be some sparsely-scattered *Cladonia* spp., including *C. cornuta, C. pyxidata, C. digitata, C. macilenta, C. impexa* and *C. arbuscula*, but these never attain the high frequencies and covers which mark many Scandinavian pine woods (Aune 1977).

Generally speaking, additional elements of the flora of the *Pinus-Hylocomium* woodland are either infrequent throughout or rather unevenly represented in the different sub-communities. *Pteridium aquilinum*, despite the often open cover of trees, is nothing like so common here as in the more calcifuge oak-birch woods of the region and, though *Blechnum spicant* becomes frequent in certain kinds of *Pinus-Hylocomium* woodland, ferns as a group are poorly represented. Other grasses, apart from *Deschampsia flexuosa*, are few in number, too. *Molinia caerulea* becomes more common towards the west and it can figure locally elsewhere in transitions to mires and, on somewhat less extreme soils or where there is consistent grazing, *Agrostis capillaris, A. canina* ssp. *montana, Anthoxanthum odoratum* and *Festuca ovina* can become prominent. Apart from *Melampyrum pratense*, which is frequent in all but the densest stands of pine, other herbaceous associates are usually sparse. There is sometimes a little *Potentilla erecta* and, in eastern Scotland, *Trientalis europaea* can be recorded; then, on rather more fertile soils, *Luzula pilosa, Oxalis acetosella* and *Galium saxatile* often appear together, a development which is sometimes accompanied by a rather more prolific representation of birch and rowan or, to the east, juniper, and which may mark transitions to woodlands dominated by these trees.

Finally, there is a group of herbs which, though generally infrequent, are especially characterisitic of the *Pinus-Hylocomium* woodland. The most uniformly distributed of these is *Listera cordata* which Steven & Carlisle (1959) found to be best represented in the south-eastern woodlands from Rannoch to Speyside, usually growing among hypnaceous mosses under tall *Calluna* beneath rather irregular pine covers; but other species occasionally found are the various wintergreens, *Pyrola minor* and the nationally rare *P. media*, *P. rotundifolia*, *Moneses uniflora* and *Orthilia secunda*, all of them plants with a strong Continental Northern distribution through Europe. The Northern Montane *Linnaea borealis* also occurs at a few sites. More common than any of these, though strongly associated with denser covers of pine, especially in eastern Scotland, and often found in plantations, is *Goodyera repens*.

Sub-communities

Erica cinerea-Goodyera repens **sub-community:** Pinewood community 1   Steven & Carlisle 1959; *Pinetum Hylocomieto-Vaccinietum triquetrosum* McVean & Ratcliffe 1962; Pinewood *Vaccinium*-moss association, *Hylocomium-Rhytidiadelphus* facies   McVean 1964a; *Erica cinerea-Pinus sylvestris* Plantation, Typical subcommunity, typical variant   Birse & Robertson 1976 emend. Birse 1980. This sub-community is characteristic of stands with a closer, even-aged and younger cover of pine and it is often found in plantations where there may be some admixtures of other conifers (like *Larix* and *Picea* spp.) and patchy colonisation, in the early stages or after thinning, by birch (generally *B. pendula*), rowan or even oak (Steven & Carlisle 1959, McVean & Ratcliffe 1962, Birse 1980).

The most obvious feature of the field layer here is generally the scarcity and low cover of the characteristic sub-shrubs of the community, particularly in the period after canopy closure and before thinning. Indeed, this feature may persist for some considerable time after the tree cover has been opened up, though *Calluna* is a fairly early invader (McVean & Ratcliffe 1962) and it is often accompanied here by *Erica cinerea*, so high frequencies (but usually low covers) of these two ericoids are characteristic. Vaccinia, by contrast, are very sparse and *Empetrum nigrum* ssp. *nigrum* is absent: a little *V. myrtillus* may persist through the period of deeper shade but it is slow to re-invade after thinning. In the absence of the sub-shrubs, the most prominent vascular plant is usually *Deschampsia flexuosa* and, in some stands, this may dominate the field layer, though it, too, can become very thin and is quite often eliminated completely as the canopy closes. Most of the other herbaceous associates of the community are also very sparse but a very good preferential for this kind of *Pinus-Hylocomium* woodland is *Goodyera repens*. This orchid is very frequent here, though it is generally found as scattered individuals or in small clumps, its rhizomes and runners ramifying the moss carpet. It can persist in quite dense shade, probably being strongly dependent on a mycorr-

hizal fungus, but it flowers best after the canopy has been thinned.

By and large, with this poor representation of vascular plants, the most prominent element of the flora here consists of bryophytes. Some of the characteristic mosses of the community are rather sparse here, notably *Rhytidiadelphus loreus*, *Ptilium crista-castrensis* and, a little less so, *Plagiothecium undulatum*, but others are very common: *Hylocomium splendens*, *Pleurozium schreberi*, *Dicranum scoparium*, *Hypnum jutlandicum* and *Lophocolea bidentata s.l.* And, in addition, *Rhytidiadelphus triquetrus* and *Pseudoscleropodium purum* are frequent. Among these, *R. triquetrus* and *H. splendens* are usually the most abundant, forming the bulk of a bryophyte carpet that can cover most of the ground or, in the very deepest shade, be reduced to patches amongst a covering of pine needles and areas of exposed mor humus.

Vaccinium myrtillus-V. vitis-idaea **sub-community:** Pinewood community 3   Steven & Carlisle 1959 p.p.; *Pinetum Hylocomieto-Vaccinietum myrtillosum* McVean & Ratcliffe 1962; Pinewood *Vaccinium*-moss association, *Vaccinium* facies   McVean 1964a; Pinewood plot types 1 & 2   Bunce 1977; *Erica cinerea-Pinus sylvestris* Plantation, *Polytrichum commune* sub-community   Birse & Robertson 1976 emend. Birse 1980; *Pinetum scoticae*, Typical subassociation, typical variant, eastern race   Birse 1984. The pine canopy here is typically a little less dense than in the *Erica-Goodyera* sub-community, though there is a considerable overlap in the ranges of tree cover and these two kinds of *Pinus-Hylocomium* woodland are often found in close spatial association in mosaics.

The essential difference between the two sub-communities lies in the relative prominence of the sub-shrubs and the bryophytes. Here the former are more varied and abundant, frequently covering up to 75% of the ground. *Vaccinium myrtillus* and *V. vitis-idaea* are generally the most extensive species, dominating in various proportions, but there is usually some *Calluna* and, quite often, some *Empetrum*. *Erica cinerea*, by contrast, is now infrequent. In more open places among the bushes, *Deschampsia flexuosa* can form a patchy carpet, sometimes, in more westerly stands, with a little *Molinia caerulea*, and, in the less intense shade of the canopy, some herbs flourish: *Melampyrum pratense* occurs frequently and there can be scattered plants of *Potentilla erecta* and *Galium saxatile*. *Goodyera* is occasionally found.

The bryophyte carpet is still extensive here, though usually not so well seen with the abundance of sub-shrubs. Indeed, it is a little more varied than in the *Erica-Goodyera* sub-community with *Plagiothecium undulatum*, *Rhytidiadelphus loreus* and *Ptilium crista-castrensis* becoming constant and *Dicranum fuscescens* joining *R. triquetrus* and *Pseudoscleropodium purum* as a preferential. Each of these can be abundant in particular stands, though *Hylocomium splendens* is the most consistent dominant. In more westerly stands, *Sphagnum capillifolium/quinquefarium* or *Dicranum majus* may make an occasional appearance but typically these mosses are not represented here.

Luzula pilosa sub-community: Pinewood communities 3, 4 & 5 Steven & Carlisle 1959; Pinewood plot types 3 & 4 Bunce 1977; *Erica cinerea-Pinus sylvestris* Plantation, Typical and *Polytrichum commune* sub-communities, *Oxalis acetosella* variants Birse & Robertson 1976 *emend.* Birse 1980; *Pinetum scoticae*, Typical subassociation, *Oxalis acetosella* variant Birse 1984. The general features of the vegetation here are very much as in the *Vaccinium* sub-community with a moderately close cover of pine, a field layer dominated by the Vaccinia with a little *Calluna* (usually only sparse *Empetrum*) and a rich and extensive bryophyte carpet in which the trio *R. triquetrus*, *P. purum* and *D. fuscescens* continue to be well represented in the general absence of Sphagna. But, among the herbs, there are some clear preferentials with *Luzula pilosa*, *Oxalis acetosella* and *Galium saxatile* all becoming frequent and grasses such as *Anthoxanthum odoratum*, *Agrostis capillaris*, *A. canina* ssp. *montana* and *Festuca ovina* occurring occasionally. Among eastern stands represented here, *Blechnum spicant* is a little commoner than usual in the region.

Sphagnum capillifolium/quinquefarium-Erica tetralix sub-community: Pinewood communities 2, 6, 8 & 9 Steven & Carlisle 1959; *Pinetum Vaccineto-Callunetum* McVean & Ratcliffe 1962 *p.p.*; Pinewood *Vaccinium-Calluna* association McVean 1964a *p.p.*; Pinewood plot types 5–8 Bunce 1977 *p.p.*; Woodland plot type 28 Bunce 1982; *Pinetum scoticae*, Typical subassociation, *Molinia* variant Birse 1984. In this sub-community, there is a distinct shift towards a more consistently open cover of pine, with denser plantations rarely represented and many stands having the kind of open heath structure or two-generation mixes that McVean & Ratcliffe (1962) detected. Among this less dense canopy, there can be some birch (usually *B. pubescens* in this predominantly western sub-community) and *S. aucuparia* and, in more southerly stands, a little *Juniperus*.

Sub-shrubs continue to be an important element in the field layer here, though there is a rather different balance among the common species of the community. *Calluna* is very frequent and is now the usual dominant with *V. myrtillus* often playing a subsidiary role and *V. vitis-idaea* and *Empetrum nigrum* ssp. *nigrum* very common but typically of low cover. *Erica cinerea* is

sometimes found but much more characteristic here are small amounts of *E. tetralix*.

Deschampsia flexuosa maintains its high frequency in this sub-community but generally it is not abundant and, quite often, a more prominent grass is *Molinia caerulea*. In drier places, there is occasionally some *Pteridium*, though its cover is usually low, and there can be scattered plants of *Blechnum spicant*, *Melampyrum pratense* and *Listera cordata*.

All the characteristic bryophytes of the community are well represented here with *H. splendens* and *P. cristacastrensis* often being especially abundant, but the really distinctive feature of the ground layer is the prominence of deep tussocks of Acutifolia Sphagna, especially *S. capillifolium* and *S. quinquefarium*. *Dicranum majus* also becomes frequent while *R. triquetrus*, *P. purum* and *D. fuscescens* are only rarely recorded. Other bryophytes sometimes found here include *Leucobryum glaucum*, *Calypogeia trichomanis/muellerana*, *Aulacomnium palustre*, *Barbilophozia floerkii*, the Sub-Atlantic *Scapania gracilis*, the Western British *Bazzania trilobata* and *Hylocomium umbratum* and the Northern Atlantic *Mastigophora woodsii* and *Herberta adunca* (Ratcliffe 1968).

Quite frequently, these various components are disposed over very uneven topography in this sub-community with the ground thrown into hummocks by bryophyte-covered pine stumps and boulders and with hollows where the *Pinus-Hylocomium* woodland gives way to mire forest or herbaceous bog vegetation.

Scapania gracilis sub-community: Pinewood community 2 Steven & Carlisle 1959 *p.p.*; *Pinetum Vaccinieto-Callunetum* McVean & Ratcliffe 1962 *p.p.*; Pinewood *Vaccinium-Calluna* association McVean 1964a *p.p.*; Pinewood plot types 5–8 Bunce 1977 *p.p.*; *Pinetum scoticae*, Typical subassociation, typical variant, western race Birse 1984. The general characteristics of the vegetation here are very similar to those of the *Sphagnum-Erica* sub-community with a more open pine canopy, a mixed cover of sub-shrubs and a hummocky bryophyte carpet in which Sphagna continue to play a prominent part. Here, however, *Erica tetralix* and *Molinia caerulea* are usually absent and there is a further enrichment of the bryophyte element. *Thuidium tamariscinum*, generally speaking a rare moss in the *Pinus-Hylocomium* woodland, becomes frequent but more striking is the common occurrence of *Scapania gracilis* and *Diplophyllum albicans* and the occasional appearance of the Sub-Atlantic *Anastrepta orcadensis*, a hepatic usually associated with bryophyte mats in vegetation above the forest zone (Ratcliffe 1968).

Habitat

The *Pinus-Hylocomium* woodland is characteristic of strongly-leached soils in the cooler parts of the north-western sub-montane zone in Britain. Floristic variation within the community is closely related to the density and age of the pine canopy but other factors, such as climatic and edaphic differences and the incidence of grazing and browsing and burning, also play an important role and make it difficult to attribute the character of the different sub-communities unequivocally to the antiquity or naturalness of the vegetation.

Two climatic variables, temperature and rainfall, have a major influence here. By and large, the *Pinus-Hylocomium* woodland is confined to those parts of Britain where the mean annual maximum temperature is 23 °C or less (Conolly & Dahl 1970), the annual accumulated temperature between 280 and 550 °C (500–1000°F: Page 1982), with most stands occurring in the central and north-west Highlands of Scotland. And the prevailing cool character of the climate is reflected in the occurrence in the community of Northern Montane plants like *Goodyera repens, Listera cordata, Trientalis europaea* and *Linnaea borealis* and the Continental Northern wintergreens. Apart from *Goodyera*, none of these is very common here but most of them are preferentially frequent and their British distributions roughly coincide with or are included within that of the *Pinus-Hylocomium* woodland. *Empetrum nigrum* ssp. *nigrum*, which has a wide range in Britain but has an Arctic-Alpine distribution through Europe as a whole, is also much better represented here than in any other calcifugous north-western woodland. The only major absentees among this group are the two other Arctic-Alpine sub-shrubs, *E. nigrum* ssp. *hermaphroditum* and *Vaccinium uliginosum*, which are important species in some Scandinavian pinewoods but very scarce here. As always in this community, it is possible that the great reduction in its extent has resulted in a floristic impoverishment, though the rarity of these two quite widely distributed species is very consistent throughout the range of the woodland.

Within this cooler zone, the community can be found from almost sea-level up to more than 600 m, extending noticeably higher in the east where, on Creag Fhiaclach in the north-west Cairngorms, it probably attains a natural altitudinal limit, with the pines becoming increasingly stunted and sparse beyond 640 m (McVean & Ratcliffe 1962: Plate 6, Ratcliffe 1977). In such situations, cold and exposure to wind probably restrict growth and reproduction, though this is not known for certain (Carlisle & Brown 1968). What is clear is that, though *P. sylvestris* can ascend to higher levels than oak, the *Pinus-Hylocomium* woodland is not an altitudinal replacement for the *Quercus-Betula-Dicranum* woodland: birch-dominated stands of the latter extend to roughly the same altitudes as the former and zonations from oak- to pine-dominance (they are actually rather uncommon despite Tansley's (1939) supposition) gener-

ally reflect edaphic transitions (McVean 1964a: see below). More accurately, the remaining distribution of the *Pinus-Hylocomium* woodland reflects a regional prominence of pine in the cooler parts of Scotland that is probably a direct inheritance of its prevalence there through the post-Glacial period (McVean & Ratcliffe 1962, O'Sullivan 1977). However, within this zone, its natural extent has been much reduced with widespread replacement by planted stands of pine and by sub-spontaneous development of the *Quercus-Betula-Dicranum* woodland; and outside this zone, planting of pine has extended the range of the *Pinus-Hylocomium* woodland into areas where the *Quercus-Betula-Dicranum* woodland might be expected as the natural climax forest. Many stands of the latter kind can be found around the Moray Firth and in central Scotland (e.g. Birse & Robertson 1976, Birse 1980, 1982, 1984, Blaxter 1983) and there may be others beyond these regions where acid soils under pine experience sufficiently high rainfall to acquire the distinctive bryophyte flora of this community.

It is this related combination of a generally wet climate and high soil acidity that gives the *Pinus-Hylocomium* woodland much of its remaining floristic character. In fact, *P. sylvestris* will grow on quite calcareous soils: calcicolous pinewoods are a feature of various parts of Europe (e.g. Aune 1977, Ellenberg 1978) and they may have occurred in Britain in the past. But, at the present time, our pine woodlands are found, for the most part, on very strongly leached, lime-free profiles and the characteristic soil type under the *Pinus-Hylocomium* woodland is the podzol (e.g. Steven & Carlisle 1959, Manley 1961, McVean & Ratcliffe 1962, Carlisle & Brown 1968, Fitzpatrick 1977). Typically, raw litter overlies a mat of mor humus, with slow integration of organic matter through fungal decay and arthropod activity. The surface pH is consistently low, generally 3.5–4.5, and there is marked leaching with a prominent E_b and deposition horizons with humus and iron. The texture throughout the mineral layers is generally sandy and the structure very poor, often single-grain with but minimal deposition of any clay coatings. In Scotland, such soils are very widespread over a variety of pervious siliceous substrates: especially important here are the extensive sandy and gravelly fluvioglacial deposits laid down over Dalradian and Moine schists and granites in the east and south, where the community occupies a very distinctive habitat over undulating terraces; elsewhere, podzols are developed directly from quartzites and sedimentaries like the Torridonian sandstone and Devonian Old Red Sandstone.

The highly acidic and impoverished soil conditions here are reflected in the *Pinus-Hylocomium* woodland as a whole by the great predominance of calcifuges in its flora, among the sub-shrubs, the herbs and the bryophytes, many of which it shares with other acidophilous woodlands of the north-western part of Britain. Some of these, of course, help maintain the mor humus regime and encourage podzolisation, by the provision of acidic litter. Pine itself produces large quantities of litter (Bray & Gorham 1964, Carlisle & Brown 1968), mainly leaves, small branches and bark, which decays very slowly (Kendrick & Burges 1962), but the sub-shrubs are also important, particularly *Calluna*, which can also acidify its root environment (e.g. Gimingham 1972). This is a more frequent and abundant plant in this community than under the often closed canopies of the *Quercus-Betula-Dicranum* woodland and it is a noticeable feature of the soils here that they are most strongly podzolised, with the development of a thin iron pan, in the open heathy vegetation of the *Sphagnum-Erica* and *Scapania* sub-communities where *Calluna* is frequently the field-layer dominant (Fitzpatrick 1977).

These two kinds of *Pinus-Hylocomium* woodland are also characteristic of the wetter areas in which the community is found, at lower altitudes towards the west and at higher altitudes in the east, where the annual rainfall is usually more than 1600 mm (*Climatological Atlas* 1952) with at least 180 wet days yr^{-1} (Ratcliffe 1968), in some places rising to 2000 mm and 200 wet days or beyond. This index of climate provides quite a good separation between the two groups of sub-communities, the *Sphagnum-Erica* and *Scapania* types on the one hand and the *Erica-Goodyera*, *Vaccinium* and *Luzula* types on the other, with acutifolia Sphagna and *Dicranum majus* diagnostic of the former, *Rhytidiadelphus triquetrus*, *Pseudoscleropodium purum* and *Dicranum fuscescens* preferential for the latter. Not only do the soils of the former show a somewhat stronger tendency towards podzolisation but they are generally noticeably moister, with thicker accumulations of litter, sometimes reaching 30 cm and forming what is in effect a layer of peat, constantly replenished by the decay of the *Sphagnum* tussocks. In the hummocky topography characteristic of the *Sphagnum-Erica* and *Scapania* sub-communities, there may also be some local impedence of drainage in the lower horizons. Here, then, the vegetation takes on some of the character of a mire forest with *Erica tetralix* and *Molinia caerulea* becoming prominent on profiles which approach peaty podzols or gley-podzols though, with very marked waterlogging, pine growth becomes severely hindered (Carlisle & Brown 1968) and there is a transition to other kinds of woodlands in very wet hollows (see below).

The increased surface moisture of the soils in the *Sphagnum-Erica* and *Scapania* sub-communities is also part of the complex of factors that imparts a more oceanic feel to their flora. The floristic element involved is not a large one and the two main species (*Sphagnum quinquefarium* and *Scapania gracilis*) are only of rather broadly Atlantic nature (Ratcliffe 1968) but, where they become especially prominent in the *Scapania* sub-community, with *Diplophyllum albicans* and *Anastrepta orcadensis*, in Wester Ross pinewoods with an annual rainfall over 3000 mm (220 wet days yr^{-1}), the vegetation begins to take on some of the character of the strongly oceanic *Quercus-Betula-Dicranum* woodlands of that region.

The other soil-related trend visible in the community has to do with fertility. By and large, the podzols under the *Pinus-Hylocomium* woodland are very infertile and perhaps only able to maintain a forest cover by the constant, though slow, cycling of nutrients derived from the litter (Carlisle & Brown 1968, Fitzpatrick 1977). But, in some situations, conditions are not so extreme, where, for example, there is some heterogeneity among the parent materials, with less siliceous strata interposed or a variety of superficials admixed, or where there is some flushing with less base-poor waters, a feature well seen in the Loch Maree woodlands where there is seepage from calcareous mudstones (Ratcliffe 1977). Then the effects of leaching are offset a little and the profile tends towards a brown podzolic soil or even a brown earth with a somewhat milder humus regime (Fitzpatrick 1977, Birse 1980, 1984). Among the different kinds of *Pinus-Hylocomium* woodland, there is a clear association between such conditions and the *Luzula* sub-community. This is quite a widely distributed vegetation type, being related to local edaphic variation rather than any regional pattern of soil differences, but it seems to be more frequent in the drier east, so its general affiliations are with the *Erica-Goodyera* and *Vaccinium* sub-communities.

As noted above, the tendency towards more pronounced leaching in soils under the *Pinus-Hylocomium* woodland is associated with and perhaps favoured by the openness of the pine canopy. The density of the tree cover certainly has a marked effect on the field and ground layers of the community by altering the balance between the more shade-tolerant bryophytes and the more light-demanding sub-shrubs. This can be seen within individual sub-communities, as when *Erica-Goodyera* plantations close over and are later thinned (Birse 1980, 1984), and in the general contrast between this sub-community and the *Vaccinium* type where canopies are usually rather more open (McVean & Ratcliffe 1962). And, within the community as a whole, there is a broad shift towards a prominence of sub-shrubs under the more consistently open canopies of the *Sphagnum-Erica* and *Scapania* sub-communities. But, apart from the effect on *Calluna*, the coincidence between the major contrast in canopy density and the most obvious difference in soil conditions within the *Pinus-Hylocomium* woodlands may be largely accidental. Vagaries of treatment history could have left us with the predominance of more open stands of pine that we now see in the wetter regions; extensive planting in the drier areas has certainly increased the proportion of denser stands there. By its effect on sub-shrub abundance, variation in canopy closure thus reinforces differences between the sub-communities, seen best among the species of bryophytes, that are directly related to regional variations in soil development.

The potential dominance of sub-shrubs in the field layer which increased penetration of light permits is often moderated here by the influence of herbivores; and these play an important role in controlling the ultimate density of trees by their effect on regeneration. Many stands of the community, particularly the more semi-natural remnants, occur within tracts of unenclosed upland, with the vegetation freely open to stock and deer. Before the mid-eighteenth century, mixed pasturing of these woodlands seems to have been common with cattle, sheep, goats and ponies all involved, though apparently all in fairly small numbers (Steven & Carlisle 1959). After that time, there was a switch to sheep and an increase in stock densities with the formation of the great Highland runs. Red deer are also plentiful in many areas and, with the creation of deer forests to provide shooting in the middle of the last century, they were often encouraged locally. There are also small numbers of roe and, in some northern pinewoods, feral sika (Steven & Carlisle 1959). The effects of all these grazers and browsers is especially great in winter, when harsh weather at higher levels drives them into the shelter of the tree cover where some snow-free herbage is still available. At such times, too, drifting of snow over fences can give animals easy access to plantation stands.

By their browsing of *Vaccinium myrtillus* and *Calluna* (the latter also being eaten by grouse), stock and deer can prevent these sub-shrubs attaining dominance under more open canopies, thus favouring less palatable species (notably *V. vitis-idaea*) or completely counteracting the effect of increased light penetration, shifting the balance of dominance back towards bryophytes and *Deschampsia flexuosa*: the *Erica-Goodyera* sub-community can contain some heavily grazed stands of this kind where there are quite open covers of pine. The other impact of the herbivores is on the pine itself, because they can consume such seedlings as have managed to get a hold in more open situations and bite out the leaders of saplings that have breached the sub-shrub canopy (a process that may take 20 years) but which have not grown tall enough to be out of reach. In fact, *P. sylvestris* is fairly resistant to repeated topping and can readily replace damaged leaders: Carlisle & Brown (1968) reported good regrowth of native pine after it had been held in check by browsing for ten years. But the cumulative effect of sustained herbivore predation can severely hinder regeneration where conditions

are otherwise suitable (Steven & Carlisle 1959, McVean 1961, Carlisle 1977, Lowe 1977, Booth 1977, Millar 1977).

Ground fires can be similarly deleterious, destroying with speed the result of decades of natural regeneration (e.g. Booth 1977) and sometimes killing mature trees, especially those with thin plated bark (usually older specimens: Carlisle & Brown 1968). Fires may occur accidentally, through careless behaviour by visitors or, in the days of steam trains, by showering of hot coals, or by spread from adjacent muirburns; in the extension of sheep runs, some stands were deliberately destroyed by fire. That fires have been widespread in the history of the *Pinus-Hylocomium* woodland is abundantly clear in the frequent presence of carbon or recognisable fragments of charcoal in the A horizon of the soils (Manley 1961, McVean & Ratcliffe 1962, Carlisle & Brown 1968, Fitzpatrick 1977).

But fire is probably of considerable importance in creating suitable conditions for regeneration (e.g. Carlisle & Brown 1968, Carlisle 1977). *P. sylvestris* can produce large quantities of seed, though it is a sporadic fruiter and a fairly precise set of conditions must coincide with a good crop of seed for germination to approach an optimum (Carlisle & Brown 1968, Bunce & Jeffers 1977, where extensive bibliographies of the popular but controversial subject of pine regeneration are given). Generally speaking, natural regeneration seems to be more likely under more open covers of trees, where there is a patchy cover of mor or peat over freely-draining mineral horizons, where sub-shrubs are not luxuriant and where grazing is not too heavy. Pine seed is light and can be wind-blown a considerable way, but Carlisle & Brown (1968) gave 90 m as a crude limiting distance for seed-parents. Ground fires, which burn off the competing vegetation, partially destroy the felted mat of mor and also provide a valuable release of nutrients from the ash, may be the best single producer of such conditions, perhaps to such an extent that *P. sylvestris* is effectively a fire-dependent species (Carlisle & Brown 1968, Carlisle 1977). But other forms of disturbance, such as upheaving of the soil by wind-thrown big trees, or even the scuffing of the ground surface by herbivores, may also assist seedlings in getting a hold. Certainly, the impression gained from the study of individual sites (Steven & Carlisle 1959) is that some form of disturbance, often man-related, is advantageous for initiating natural regeneration, though many other factors may affect the survival of the seedlings and saplings. At the present time, regeneration is very patchy: some stands, particularly in the east, are regenerating well locally; elsewhere the characteristic prevalence of older trees is still very noticeable and often inexplicable.

Grazing and burning are just two elements in the long

history of human influence to which our native pine-woods have been subject. At the present time, the *Pinus-Hylocomium* woodland comprises fragments of semi-natural forest, much reduced in extent by man and widely affected by timber removal, and younger stands of planted origin, sometimes interposed within the more ancient framework, sometimes on ground that seems unlikely to have been occupied by pine for considerable periods of time, if ever. There is good palynological evidence (e.g. Birks 1970, O'Sullivan 1977, Carlisle 1977) to suggest that, in its more semi-natural stands, the *Pinus-Hylocomium* woodland represents a vegetation type of great antiquity, providing continuity over some 8000 years with the original post-Glacial forest cover that developed in many parts of the Scottish Highlands. Subsequent climatic change seems to have markedly affected its abundance towards the west, where it succumbed widely to blanket mire extension but, in the drier east, it appears to have remained a more extensive forest type until the period of human expansion. Present climatic differences in these two regions are still reflected in the floristics of the community (and perhaps, too, in the regenerative ability of *P. sylvestris*, though systematic data are very patchy) but it is not very easy to say which kinds of *Pinus-Hylocomium* woodland represent more closely the original or more natural pine forest (or forests). Certainly the *Sphagnum-Erica* and *Scapania* sub-communities are typically found beneath older and more open stands of pioneer trees, and plantations are much more common in the *Erica-Goodyera*, *Vaccinium* and *Luzula* types but this latter group, too, includes stands which are probably semi-natural. And, among all these kinds of *Pinus-Hylocomium* woodland, there seems to be a much greater purity in the canopy than was the case in the original forest cover. The cumulative effect of the various kinds of interference and treatment has been gradually to select for pine as the dominant here, so that it is now easier to characterise distinct communities on the basis of the pre-eminence of pine (this community), birch (the *Quercus-Betula-Dicranum* woodland) or juniper (the *Juniperus-Oxalis* woodland), a situation especially well seen in eastern Scotland, where plants like *Trientalis europaea* and *Rhytidiadelphus triquetrus* provide a Continental element in all three. What seems to have been originally the case is that all these trees contributed to a much more mixed kind of forest with spatial patchworks and probably temporal shifts in dominance.

Zonation and succession

The *Pinus-Hylocomium* woodland on Creag Fhiaclach in the Cairngorms is probably one of the very few stands of forest vegetation in Britain where trees can be seen growing at a natural, climatically-related altitudinal limit. Here, the community gives way, through a gradual loss of its pine cover, to the more calcifugous kind of *Juniperus-Oxalis* woodland, the characteristic montane scrub vegetation in the east-central Highlands of Scotland.

Below this limit, and almost down to sea-level, the community survives as scattered stands in a number of geographical groupings, within some of which it constitutes a major element of the vegetation cover. In considering zonations, it is important to distinguish between such expanses of 'pine forest', defined on broad landscape criteria and containing a wide variety of vegetation types, including many non-wooded ones, and the stands of the different kinds of *Pinus-Hylocomium* woodland which form part of the mosaic.

Zonations within stands of the *Pinus-Hylocomium* woodland, between one sub-community and another, are generally related to edaphic variation and/or the density of the tree cover. In some places, notably the Speyside and Deeside forests, transitions from the *Vaccinium* sub-community to the *Sphagnum-Erica* sub-community can be seen as one moves from drier podzols to peaty podzols, a change that is often related on a broad scale to a shift to higher altitudes with their wetter climate. And, here and in the south-western forests around the Great Glen, there can be zonations from these sub-communities to the *Luzula* type where the soils become moderately enriched over less extreme substrates or because of gentle flushing.

Gradations from the *Vaccinium* or *Luzula* sub-communities to the *Sphagnum-Erica* sub-community are also typically associated with a decrease in the density of the pine canopy and the effect of this can be seen, too, over more uniformly dry soils, where the *Erica-Goodyera* sub-community passes to the *Vaccinium* sub-community as the canopy thins out. Indeed, such changes can be precipitated by thinning, and, in semi-natural stands, may represent a seral progression (see below).

Where a general uniformity in soil conditions is maintained, continued reduction in the pine cover usually marks a gradual zonation to some kind of heath, typically various sub-communities of the *Calluna-Vaccinium* heath (the floristic composition of which is essentially a *Pinus-Hylocomium* woodland without the pine) or the *Calluna-Arctostaphylos uva-ursi* heath, quite often a mosaic of both where burning patterns have been complex (McVean & Ratcliffe 1962). Where the ground conditions are moister, a frequent feature among stands of the *Sphagnum-Erica* sub-community towards the wetter west, these dry heaths are replaced in this kind of sequence by the *Scirpus-Erica* wet heath, a very widespread community of thin ombrogenous peats and the fringes of valley mires in western Britain. On such wetter ground, growth of *P. sylvestris* is much poorer (Carlisle & Brown 1968), so the zonation is essentially an edaphic one rather than a seral one, though drainage and fertilising has often artificially extended the range of pine on to such soils in Scotland (and, of course, elsewhere outside the natural range of the tree). Where such wet heath becomes wooded, *Betula pubescens* is usually the leading coloniser (often the only one) and extension of its canopy can lead to the development of stands of the *Betula-Molinia* woodland among the *Pinus-Hylocomium* woodland, a situation marked by a strong continuity in the field layer with species such as *Molinia caerulea*, *Erica tetralix* and Sphagna remaining prominent in both communities.

Transitions to even wetter ground in valley and basin mires, which are quite a common feature among the complex undulating topography of fluvioglacial deposits, involve the occurrence of small stands of base-poor communities like the *Carex echinata-Sphagnum* or *Carex rostrata-Sphagnum* mires, or, where these have been colonised by trees, the *Betula-Molinia* or *Salix-Carex* woodlands. Where there is a local influence of more base-rich waters, the *Pinguiculo-Caricetum* and *Carici-Saxifragetum* mires can be found, or their wooded equivalent the *Alnus-Fraxinus-Lysimachia* woodland. Within the areas where the *Pinus-Hylocomium* woodland is represented, all these different kinds of soil-related zonations can often be found within a relatively small compass, a feature which adds greatly to the overall diversity of the vegetation and contributes to the unique assemblage of communities which we include within the broader landscape definition of pine forest.

Over the range of soil types represented within what seems to be the natural range of *P. sylvestris* in Britain, it is probably able to colonise and attain dominance on a fairly broad spectrum of more freely draining brown earths, brown podzolic soils and podzols and, in such situations, its natural precursors are probably the various dry and wet heaths with which it can be found associated today. Where invasion has been observed, an initial slow start within the sub-shrub canopy, followed by some period of check through herbivore damage, has been succeeded by fairly rapid growth to produce vegetation resembling the *Erica-Goodyera* or *Vaccinium* sub-communities or, where the initial density of seedlings was more sparse, the *Sphagnum-Erica* type (McVean & Ratcliffe 1962). In some places, it is possible that the latter sub-community can develop from the former in a seral progression (McVean & Ratcliffe 1962), though it seems likely that there is some edaphic and climatic check on which direction succession takes: there may be two types of climax pinewood, one in the drier low-altitude eastern region, the other at higher altitudes there and in the wetter west.

What does seem certain is that, in the past, progressions to pine woodland involved a greater contribution from other tree species, notably birch and juniper, which continued to be well represented in mosaics of forest, at least in the eastern Highlands of Scotland. Now, we see some greater separation in the roles which these species play. Birch-dominated calcifugous forest (often with some oak, derived in part from planted stock) of the *Quercus-Betula-Dicranum* woodland, survives as distinct stands within the areas of pine forest and finds fragmentary representation in gaps, but it also seems to have sprung up widely on ground that mixed pine forest previously occupied. And the *Juniperus-Oxalis* woodland, though also represented among stands of the *Pinus-Hylocomium* woodland, now survives largely in isolation at higher altitudes. The original patchworks of mixed dominance may have been of some considerable importance in tempering the tendency towards podzolisation, birch and juniper being less active encouragers of mor than pine, and the greater extent of the forest cover must certainly have restricted the spread of *Calluna*, the abundance of which among the heaths of eastern Scotland may be very deleterious for any future expansion of pine.

Distribution

The *Pinus-Hylocomium* woodland is confined to Scotland and best represented in the central and north-western Highlands. Stands are fragmented into some major geographical groupings which also have some vegetational significance because they experience rather different climatic conditions and show different edaphic trends, features represented in the distribution of the various sub-communities. The *Erica-Goodyera*, *Vaccinium* and *Luzula* sub-communities are essentially types of the drier east, being well represented in the Speyside and Deeside forests, more sporadically on drier ground in the south-western forests of the Great Glen and around Rannoch. The *Erica-Goodyera* sub-community is also very common in young plantations around the Moray Firth. The *Sphagnum-Erica* sub-community, by contrast, is largely a type of wetter areas, being widespread to the south-west and in the Wester Ross forests, but also occurring at higher altitudes in Speyside and Deeside. The *Scapania* sub-community seems to be confined to the very wet parts of Wester Ross.

Pine plantations outside the native range of *P. sylvestris* are, in this scheme, considered as replacements of other woodland types, notably in southern England, the *Quercus-Betula-Deschampsia* woodland. In areas of higher rainfall, as in parts of Surrey, the floristic similarity between such stands and the drier types of *Pinus-Hylocomium* woodland is considerable, though the distinctive low-frequency preferentials of the community are absent. *Goodyera repens* and *Ptilium crista-castrensis* occur in pine plantations in East Anglia but are thought to have been introduced with pine stock from Scotland (Petch & Swann 1968).

Affinities

The *Pinus-Hylocomium* woodland unites the two kinds
of pine woodland described by McVean & Ratcliffe
(1962) and, though adding further data from plantations (largely in the *Erica-Goodyera* sub-community)
and from more enriched soils (the *Luzula* sub-community), mostly originating from the surveys of Birse (1980,
1982, 1984; see also Birse & Robertson 1976), the
scheme preserves the major distinction which they
recognised in the contrasts between the two groups of
sub-communities. Although there are clear floristic
differences between these two groups (and related environmental, and perhaps seral, variations), the preferential species are rather few in number and it seems best to
retain them within a single community. Certainly, it is
not possible to separate the sub-communities unequivocally into more and less natural types (cf. Birse 1980,
1984). Despite the great deal of recent interest in this
kind of woodland (e.g. Bunce & Jeffers 1977), detailed
understanding of their floristics has not progressed far
beyond the surveys of McVean & Ratcliffe (1962) and
Birse (1980 *et seq.*) and the account of the individual
forests and their geographical groupings still remains
largely as Steven & Carlisle (1959) left it.

The *Pinus-Hylocomium* woodland belongs, together
with the *Juniperus-Oxalis* woodland, to the Vaccinio-
Picetea and is best placed within the Dicrano-Pinion
alliance. Its nearest equivalent in Europe is to be found
in the pine forests of western Norway where, in Scandinavian terms, the climate is relatively oceanic (Aune
1977).

Floristic table W18

	a	b	c	d	e	18
Pinus sylvestris	V (6–8)	V (6–8)	V (5–8)	V (5–7)	V (6–7)	V (5–8)
Betula pendula	II (1–5)		I (1)			I (1–5)
Picea sitchensis	I (4–7)					I (4–7)
Larix spp.	I (3)					I (3)
Picea abies	I (1)			I (1)		I (1)
Juniperus communis communis	I (1)		I (2–6)			I (1–6)
Sorbus aucuparia	I (1)			I (1–2)	I (5)	I (1–5)
Betula pubescens	I (1)			I (1)	II (1–5)	I (1–5)
Hylocomium splendens	V (5–9)	V (7–9)	V (2–9)	V (1–9)	V (5–8)	V (1–9)
Pleurozium schreberi	V (1–7)	V (1–7)	V (1–6)	V (1–5)	V (3–5)	V (1–7)
Dicranum scoparium	V (1–6)	V (1–5)	V (1–4)	IV (1–3)	V (1–3)	V (1–6)
Calluna vulgaris	IV (1–8)	V (2–7)	V (1–8)	V (1–8)	V (2–8)	V (1–8)
Deschampsia flexuosa	III (1–5)	V (1–5)	V (3–7)	IV (1–3)	V (2–5)	V (1–7)
Plagiothecium undulatum	II (1–6)	IV (1–5)	IV (1–5)	IV (1–3)	IV (1–3)	IV (1–6)
Vaccinium myrtillus	I (1–3)	V (5–8)	V (3–9)	V (4–8)	V (6–8)	IV (1–9)
Rhytidiadelphus loreus	I (1)	V (1–5)	IV (1–8)	IV (1–5)	V (4–7)	IV (1–8)
Vaccinium vitis-idaea		V (5–9)	V (2–9)	IV (3–8)	IV (3–5)	IV (2–9)
Rhytidiadelphus triquetrus	IV (1–9)	V (1–7)	V (1–7)	I (4)		III (1–9)
Pseudoscleropodium purum	III (3–5)	III (1–4)	III (1–4)	I (2–4)	I (2–4)	III (1–5)
Dicranum fuscescens		III (1–5)	III (1–3)		I (1)	II (1–5)
Erica cinerea	IV (1–8)	I (2–4)	I (4)	II (1–4)	II (1–2)	II (1–8)
Goodyera repens	V (1–4)	II (1–3)	II (1–2)			II (1–4)
Pinus sylvestris seedling	II (1)	I (1)	I (1)	I (1–5)		I (1–5)
Lepidozia reptans	II (1–2)		I (1)			I (1–2)
Nardus stricta	I (1–3)		I (1)			I (1–3)
Luzula pilosa	I (1)		V (1–4)			II (1–4)
Galium saxatile	I (1)	I (1)	III (1–6)		I (1–2)	I (1–6)
Oxalis acetosella	I (1)		III (1–4)		II (2–4)	I (1–4)

Floristic table W18 *(cont.)*

	a	b	c	d	e	18
Sphagnum capillifolium/quinquefarium		II (1–4)	I (1)	V (1–8)	V (4–8)	III (1–8)
Dicranum majus	I (3)	II (1–3)	I (3–4)	III (1–4)	V (1–5)	III (1–5)
Pteridium aquilinum	I (1–3)		I (1–2)	II (1–3)	II (1–4)	I (1–4)
Erica tetralix	I (1)	I (3)	II (1–2)	III (1–6)		II (1–6)
Molinia caerulea		I (3–7)		II (1–8)		I (1–8)
Sphagnum girgensohnii				I (1–5)		I (1–5)
Leucobryum glaucum				I (1–4)		I (1–4)
Calypogeia trichomanis				I (2–3)		I (2–3)
Bazzania trilobata				I (1–3)		I (1–3)
Barbilophozia floerkei				I (2–3)		I (2–3)
Calypogeia muellerana				I (1–3)		I (1–3)
Sphagnum russowii				I (8)		I (8)
Scapania gracilis				II (1–3)	IV (1–2)	II (1–3)
Thuidium tamariscinum			I (1)	I (1–2)	III (1–3)	I (1–3)
Diplophyllum albicans					III (1–2)	I (1–2)
Anastrepta orcadensis				I (3)	II (1–3)	I (1–3)
Ptilium crista-castrensis	I (1)	IV (1–7)	III (1–5)	III (1–9)	III (2–7)	III (1–9)
Lophocolea bidentata s.l.	V (1–6)	III (1–3)	III (1–4)	III (1–3)	I (1)	III (1–6)
Hypnum jutlandicum	V (1–4)	II (1–3)	II (1–4)	II (1–4)	IV (2–3)	III (1–4)
Melampyrum pratense		III (1–4)	IV (1–5)	II (1–2)	III (1–3)	III (1–5)
Empetrum nigrum nigrum		IV (1–4)	II (1–2)	IV (1–9)	II (1–3)	III (1–9)
Blechnum spicant		I (1)	III (1–4)	I (1–2)	III (1–3)	II (1–4)
Listera cordata	I (2)	I (1)	I (1)	I (1–2)	I (1)	I (1–2)
Polytrichum commune	I (1–4)	I (1–4)	I (1–4)	I (1–4)		I (1–4)
Betula pubescens seedling	I (1)	I (1–4)	I (1)	I (1–2)		I (1–4)
Festuca ovina	I (2)		I (4)	I (1–3)		I (1–4)
Aulacomnium palustre	I (2)	I (1)		I (1–4)		I (1–4)
Cladonia cornuta	I (1)	I (1)		I (2)		I (1–2)
Agrostis canina montana	I (1)	I (1–5)	I (4)			I (1–5)
Potentilla erecta		I (1–3)	I (1–5)		I (2)	I (1–5)
Cladonia macilenta	I (1)	I (1)				I (1)

	a	b	c	d	e	18
Polytrichum juniperinum	I (1)	I (1)				I (1)
Cladonia digitata	I (1)		I (1)			I (1)
Trientalis europaea	I (1)		I (1)			I (1)
Agrostis capillaris	I (1-3)		I (1-4)			I (1-4)
Cladonia arbuscula	I (2)			I (1-2)		I (1-2)
Hypnum cupressiforme	I (1)	I (1)		I (1-2)		I (1-2)
Cladonia pyxidata	I (1)	I (1)		I (1)		I (1)
Cladonia impexa	I (1)	I (1)		I (1)		I (1)
Luzula multiflora			I (1)	I (1-2)		I (1-2)
Polytrichum formosum			I (5)	I (2-5)		I (2-5)
Campylopus paradoxus			I (2)	I (1-3)		I (1-3)
Number of samples	12	25	12	18	10	77
Number of species/sample	16 (12-21)	18 (14-20)	20 (16-29)	18 (12-29)	21 (18-24)	18 (12-29)
Tree height (m)	15 (9-21)	14 (9-20)	13 (6-18)	no data	15 (12-18)	14 (6-21)
Tree cover (%)	50 (40-60)	48 (30-70)	49 (25-60)	no data	37 (30-45)	47 (25-70)
Herb height (cm)	18 (3-46)	37 (12-70)	25 (10-60)	55 (30-65)	49 (10-100)	38 (3-100)
Herb cover (%)	25 (1-85)	75 (60-90)	73 (40-95)	no data	67 (45-75)	63 (1-95)
Ground height (mm)	no data	no data	no data	no data	no data	no data
Ground cover (%)	94 (80-100)	80 (60-90)	73 (10-90)	no data	79 (60-90)	81 (10-100)
Altitude (m)	93 (38-160)	342 (274-411)	376 (229-465)	180 (16-390)	159 (30-250)	260 (16-465)
Slope (°)	5 (0-16)	10 (0-35)	12 (2-25)	5 (0-20)	21 (8-33)	10 (0-35)

a   *Erica cinerea-Goodyera repens* sub-community

b   *Vaccinium myrtillus-Vaccinium vitis-idaea* sub-community

c   *Luzula pilosa* sub-community

d   *Sphagnum capillifolium/quinquefarium-Erica tetralix* sub-community

e   *Scapania gracilis* sub-community

18   *Pinus sylvestris-Hylocomium splendens* woodland (total)

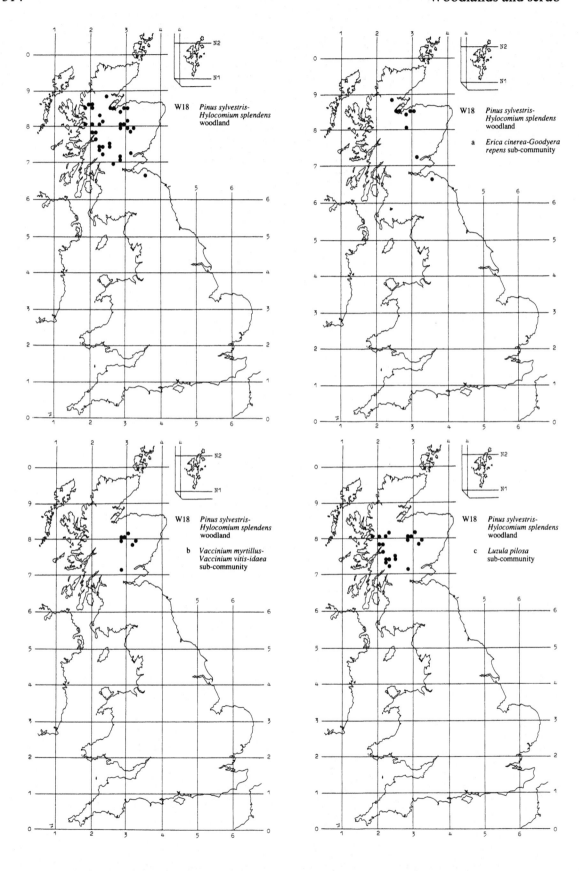

W18   *Pinus sylvestris-Hylocomium splendens* woodland

W18   *Pinus sylvestris-Hylocomium splendens* woodland

a   *Erica cinerea-Goodyera repens* sub-community

W18   *Pinus sylvestris-Hylocomium splendens* woodland

b   *Vaccinium myrtillus-Vaccinium vitis-idaea* sub-community

W18   *Pinus sylvestris-Hylocomium splendens* woodland

c   *Luzula pilosa* sub-community

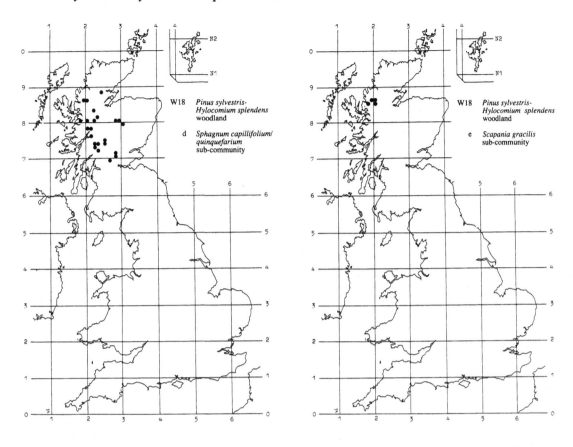

W18 *Pinus sylvestris-Hylocomium splendens* woodland

d *Sphagnum capillifolium/ quinquefarium* sub-community

W18 *Pinus sylvestris-Hylocomium splendens* woodland

e *Scapania gracilis* sub-community

W19
Juniperus communis ssp. *communis-Oxalis acetosella* woodland

Synonymy
Birch-juniper wood   Tansley 1939; Juniper heath   Pigott 1956a; Juniper scrub   Poore & McVean 1957; *Juniperus-Thelypteris* nodum   McVean & Ratcliffe 1962; Fern-rich juniper scrub   McVean 1964a; *Thelypterido-Juniperetum* Graham 1971; *Juniperus-Vaccinium* nodum   Huntley & Birks 1979a; *Juniperus-Campanula* nodum   Huntley & Birks 1979a; *Trientali-Juniperetum* Birse 1980; *Trientali-Betuletum pendulae* Birse 1982 *p.p.*

Constant species
Juniperus communis ssp. *communis, Agrostis canina* ssp. *montana, A. capillaris, Galium saxatile, Luzula pilosa, Oxalis acetosella, Vaccinium myrtillus, Hylocomium splendens, Thuidium tamariscinum.*

Rare species
Linnaea borealis, Orthilia secunda, Potentilla crantzii, Pyrola media.

Physiognomy
Juniperus communis ssp. *communis* is always the most abundant woody species in this community, though some stands have an open over-canopy of birch, almost invariably *Betula pubescens* and usually ssp. *carpatica,* with multi-stemmed trees typically less than 10 m high and showing bushy, contorted growth (Forbes & Kenworthy 1973, Huntley & Birks 1979a; Birse 1980, 1982). Increased amounts of birch, over juniper and the associated flora here, generally mark transitions to the *Quercus-Betula-Dicranum* woodland or, on somewhat less acidic and more fertile soils, the *Quercus-Betula-Oxalis* woodland. Such transitions often take the form of ill-defined mosaics (a feature especially well seen in the east-central Highlands of Scotland: Ratcliffe 1977, Huntley & Birks 1979a, b) and then separation of the different communities may be a question of the proportions of the two woody species. Similar problems can arise with *Pinus sylvestris,* which can occur here as isolated trees up to about 15 m tall, but which thickens up considerably in zonations to the *Pinus-Hylocomium* woodland.

The cover and physiognomy of the *Juniperus* itself here are rather variable. Very widely scattered bushes within a heathy or grassy context would not qualify for inclusion in this woodland type, but many stands have less than 60% cover of juniper and extensive stretches of a closed canopy are rather exceptional: the usual picture is of a patchy cover with some more open areas and others where the bushes form a virtually impenetrable thicket. The individual bushes themselves are also of very diverse form: in some stands, low plants with decumbent branches reaching little more than a metre predominate; in others, pyramidal, conical or narrow, cylindrical forms are in the majority, occasionally attaining 5 m in height. Some of this variation has a genetic basis, the distinctive characters persisting in cultivation, but bushes also change shape with age, innermost branches often being killed by self-thinning and older bushes tending to fall open. Exposure also has some effect on growth-form and this can sometimes be seen in relation to altitude: in both Scottish sites and in Teesdale, bushes at lower levels are taller, those in the highest situations sometimes as low as 50 cm (Ratcliffe 1977, Gilbert 1980). And grazing and browsing by stock and deer can also affect the canopy physiognomy, restricting the height of the bushes and opening up denser covers, in extreme cases leaving the juniper looking very straggly and moribund (Huntley & Birks 1979a, Gilbert 1980). Fungal attack has also been noted on sickly or dying bushes in Teesdale, with needle cast (*Lophodermium juniperinum*) and purple-shoot fungus (*Phomopsis* sp.) especially prevalent (Gilbert 1980). Where juniper populations in this community have been aged (a difficult procedure because of the often numerous, eccentric boles, which can rot with age), both even- and uneven-aged stands have been detected and old individuals, exceeding 100 years, encountered (Malins-Smith 1935, Kerr 1968, Gilbert 1980). In some stands, young junipers are decidedly scarce (Pigott 1956*a*, Gilbert 1980).

The other major elements in this kind of woodland are ericoids, ferns, herbs and bryophytes. *Vaccinium myrtillus* is the commonest sub-shrub with *V. vitis-idaea* and *Calluna vulgaris* somewhat less frequent, but the prominence of all three is very much affected by the evergreen shade of the juniper and by soil differences and grazing. *Calluna* especially is strongly limited by shade, *V. vitis-idaea* somewhat less so, *V. myrtillus* least of all, but the fairly dense growth of even isolated, younger juniper bushes generally confines all of them to the areas between, where they occur in various mixtures in a sub-shrub canopy that can attain half a metre or more in height. All of these species are more common and abundant on the more acidic soils of the *Vaccinium-Deschampsia* sub-community but, throughout, the prominence of *V. myrtillus* and *Calluna* can be greatly reduced by herbivores, shifting the balance of dominance in this layer of the vegetation.

Very often, the other physiognomically prominent element in this kind of woodland comprises ferns. None of these is itself constant but usually two or more are well represented and together they give the vegetation a very striking character. Commonest among them is *Blechnum spicant* but more distinctive, and often more abundant, is *Gymnocarpium dryopteris* and, somewhat less frequent, *Thelypteris phegopteris*. *T. limbosperma* can also be found and there are occasional records for *Pteridium aquilinum*, *Dryopteris dilatata* and *D. filix-mas*. Some of these species are demanding of a certain amount of shelter and shade and all of them seem to thrive best here among fairly close set, but not too densely crowded, juniper where there is probably also some protection from herbivores.

Between these bulkier components of the vegetation, there is distributed a fairly rich assemblage of herbs, forming a discontinuous but, in less heavily grazed stands, quite luxuriant sward. Among the grasses, *Agrostis capillaris* and *A. canina* ssp. *montana* are constant and *Anthoxanthum odoratum* is occasional to frequent, and each or all of these can be abundant in particular stands. Also common, but preferential for different kinds of *Juniperus-Oxalis* woodland, are *Deschampsia flexuosa* (more frequent on the more acidic soils of the *Vaccinium-Deschampsia* sub-community) and *Festuca ovina*, *F. rubra* and *Holcus mollis* (all encountered more often on the more fertile soils of the *Viola-Anemone* sub-community). In moister situations, there can be a little *Deschampsia cespitosa*. Another very common monocotyledon in the community, though generally not very abundant, is *Luzula pilosa*, and *L. multiflora* can also sometimes be found. *Carex pilulifera* and *C. binervis* occur occasionally and, where there is some base-enrichment, *C. flacca*.

The most frequent dicotyledons of this kind of woodland are *Oxalis acetosella*, *Galium saxatile*, *Potentilla erecta* and the Northern Montane *Trientalis europaea*, especially common in the centre of the range of the community in eastern Scotland but some of whose more far-flung localities are in stands of *Juniperus-Oxalis* woodland. *Veronica officinalis* occurs occasionally and *Melampyrum pratense* is sometimes found and, then, on the more fertile soils of the *Viola-Anemone* sub-community, there is a consistent enrichment of this element with an increased frequency of *Viola riviniana*, *Anemone nemorosa*, *Campanula rotundifolia*, *Cardamine flexuosa* and numerous less common preferentials and differentials. Other species of more restricted distribution in Britain which have been recorded in the community include the Northern Montane *Listera cordata*, *Rubus saxatilis* and *Linnaea borealis* and the Continental Northern *Pyrola minor*, *P. media* and *Orthilia secunda*.

Bryophytes almost always make a prominent contribution to the vegetation, often forming a patchy carpet among the herbs and extending into the centres of the more open ericoids and junipers. The most frequent species are *Hylocomium splendens* and *Thuidium tamariscinum* with *Pseudoscleropodium purum*, *Rhytidiadelphus squarrosus*, *Lophocolea bidentata* s.l., *Dicranum scoparium* and *Plagiochila asplenoides* (including var. *major*) also very common, *Plagiomnium rostratum*, *Mnium hornum*, *Polytrichum formosum*, *P. commune*, *Plagiothecium denticulatum*, *Hypnum jutlandicum* and *Ptilium crista-castrensis* occasional. Then, there are distinctive suites of preferential species in each of the sub-communities, with more marked calcifuges like *Plagiothecium undulatum*, *Rhytidiadelphus loreus*, *Dicranum majus* and *Pleurozium schreberi* becoming more frequent in the *Vaccinium-Deschampsia* sub-community, *Rhytidiadelphus triquetrus*, *Plagiomnium undulatum* and *P. affine* occurring more commonly in the *Viola-Anemone* sub-community.

Sub-communities

Vaccinium vitis-idaea-Deschampsia flexuosa **sub-community:** Juniper heath Pigott 1956*a*; Juniper scrub Poore & McVean 1957; *Juniperus-Thelypteris* nodum McVean & Ratcliffe 1962; Fern-rich juniper scrub McVean 1964*a*; *Thelypterido-Juniperetum*, Sub-association of *Lophozia* and *Campylopus* Graham 1971; *Juniperus-Vaccinium* nodum Huntley & Birks 1979*a*; *Trientali-Juniperetum*, Typical sub-community Birse 1980; *Trientali-Betuletum*, *Vaccinium vitis-idaea* Subassociation Birse 1984 *p.p.* This sub-community includes the more heathy and calcifugous stands of the *Juniperus-Oxalis* woodland in which there is often a rather open cover of juniper bushes, an abundance of ericoid sub-shrubs and an enriched suite of bryophytes growing luxuriantly over the accumulations of mor. *Vaccinium myrtillus*, though a constant throughout the community, is more frequent here than in the *Viola-Anemone* sub-community and often markedly more abundant, but better preferentials at most sites are *V. vitis-idaea* (not in Teesdale) and *Calluna vulgaris*, the former especially often rivalling or exceeding *V. myrtillus* in cover. Ferns can be prominent here among the woody species and the constant and frequent herbs of the community are all well represented. Indeed, among the vascular plants, only *Deschampsia flexuosa* is preferential to this sub-community: it becomes constant here and is often an abundant element in the grassy sward. A distinctive feature of this kind of *Juniperus-Oxalis* woodland in Morrone Birkwoods is a spatial patterning built around a hummock-hollow topography, with the ericoids largely confined to the tops of the hummocks which have a highly organic soil, the herbs growing on the less humic profiles of the hollows (Huntley & Birks 1979a). The cause of this physiography is unknown: it has been suggested that it might have developed from the wind-throw of trees which have now disappeared, though Huntley & Birks (1979a) noted that at least some of the hummocks were former nests of the wood-ant, *Formica lugubris*. Perhaps a more likely explanation in view of the harshness of the climate is that the hummocks are the product of frost-heaving (Cotton 1968).

Apart from the prominence of ericoids, it is bryophytes which give this sub-community its distinctive character. *Dicranum scoparium* increases in frequency here but much stronger preferentials are *Plagiothecium undulatum*, *Rhytidiadelphus loreus*, *Pleurozium schreberi*, *Hypnum cupressiforme* (including *H. jutlandicum*) and *Dicranum majus*. *Lophozia ventricosa* and *Scapania gracilis* occur occasionally and, growing epiphytically over decumbent, litter-clothed trunks, Graham (1971) reported an abundance of *Campylopus paradoxus* and *Barbilophozia floerkii*.

Viola riviniana-Anemone nemorosa sub-community: *Thelypterido-Juniperetum*, Typical sub-association Graham 1971; *Juniperus-Campanula* nodum Huntley & Birks 1979a; *Trientali-Juniperetum*, Anemone sub-community Birse 1980. The juniper canopy in this sub-community tends to be consistently denser than above, though a sparser cover of ericoid sub-shrubs between the bushes means that the vegetation can look more open. In fact, of the ericoids, *V. myrtillus* remains very frequent here, though its cover is usually very low, with just a few sparse shoots or widely-scattered bushes; *V. vitis-idaea* and *Calluna* are reduced in both frequency and abundance, occurring only occasionally and then typically in small amounts.

Grasses figure prominently in the sward between the

juniper bushes and, along with the community species, *Agrostis capillaris*, *A. canina* ssp. *montana* and *Anthoxanthum odoratum*, three good preferentials become frequent: *Festuca ovina*, *F. rubra* and, a little less common, *Holcus mollis*. *Deschampsia flexuosa*, by contrast, is scarce and only rarely abundant. The shift towards a flora of less markedly acidic, dry and impoverished soils can be seen, too, among the dicotyledons. Here *Viola riviniana* and *Anemone nemorosa* are especially good markers: these species are, at most, occasional in the *Vaccinium-Deschampsia* sub-community while here they are a constant, and in the case of *Anemone* sometimes abundant, element in the field layer. Less frequent, though also strongly preferential, are *Campanula rotundifolia*, *Cardamine flexuosa*, *Cerastium fontanum*, *Urtica dioica*, *Stellaria holostea*, *Fragaria vesca*, *Galium boreale* and *Adoxa moschatellina*. Other species recorded more occasionally include *Prunella vulgaris*, *Galium verum*, *Potentilla sterilis*, *Epilobium montanum* and *Lathyrus montanus* and, very exceptionally, where there is more marked base-enrichment, there can even be some *Mercurialis perennis* or *Carex flacca*. In more heavily grazed stands, many of these plants only attain any degree of luxuriance where there is some local protection against herbivores: with the ferns of the community, they tend therefore to be most marked within rings of juniper bushes or in the centre of older individuals that have begun to open out. In such situations on moister soils at Morrone, Huntley & Birks (1979a) also noted tall herbs such as *Cirsium helenioides*, *Geranium sylvaticum*, *Filipendula ulmaria* and *Trollius europaeus*.

Bryophyte cover remains high here, though there are some marked changes in the species represented. The mosses and hepatics typical of the community all remain frequent but more calcifugous species characteristic of the *Vaccinium-Deschampsia* sub-community become, at most, occasional. On the positive side, there is a marked increase in the occurrence of *Rhytidiadelphus triquetrus* and *Plagiomnium undulatum* and occasional records for *P. affine*, *Cirriphyllum piliferum*, *Drepanocladus uncinatus*, *Barbilophozia barbata* and *Isothecium myurum*.

Habitat

The *Juniperus-Oxalis* woodland is a community of high altitudes, mostly within the colder and relatively dry parts of north-west Britain. It occurs on quite a wide variety of soils and edaphic differences, together with grazing and browsing, have important influences on the floristics of the vegetation. Although the community has some claim to represent climax montane scrub, its present prominence at some sites may be due to the destruction of a tree canopy in which pine and birch probably figured abundantly.

The centre of distribution of the *Juniperus-Oxalis* woodland lies within the east-central Highlands of Scotland where the climate has a distinctly continental character (Ratcliffe 1968, Green 1974). Rainfall is low for these latitudes with only 800–1200 mm yr^{-1} (*Climatological Atlas* 1952) or about 160 wet days yr^{-1} (Ratcliffe 1968) and annual accumulated temperatures are within the range 280–550 °C (Page 1984) with the lowest February minima of any part of the country. Snowfall totals are moderate but morning snow-lie occurs on more than 60 days yr^{-1}, late frosts are very frequent and the mean annual maximum temperature is below 23 °C (Conolly & Dahl 1970, Huntley & Birks 1979*a*). None of these variables seems to be of direct importance in the actual favouring of the prominence of juniper itself, but the cold, dry nature of the climate has a clear influence on the associated flora: this is seen in the presence of Northern Montane plants, especially *Trientalis europaea* but also the less frequently recorded *Rubus saxatilis*, *Trollius europaeus* and the very rare *Linnaea borealis*, the Continental Northern wintergreens, *Galium boreale* and *Cirsium helenioides* and the Arctic-Alpine *Polygonum viviparum* and *Potentilla crantzii*. None of these species is confined to this community: *Trientalis* in particular is characteristic of eastern Scottish stands of a number of woodland types and the less basiphilous of the very rare plants among them are more frequent in the *Pinus-Hylocomium* woodland. But, taken together, they give our northern stands of juniper a very different look from the vegetation in which this tree dominates in the south-eastern lowlands; moreover, some of the more southerly stations of certain of these species, in Northumberland and the Pennines, occur in association with juniper in stands of this community (e.g. Pigott 1956*a*).

The *Juniperus-Oxalis* woodland has been recorded in the available samples at altitudes between 300 and 650 m. Towards the upper part of this range, harsh climatic conditions may also play a part in excluding some of the other woody dominants which, at lower levels, are commonly found over the associated flora characteristic here. This would certainly be true of oak (which is, in any case, rather uncommon in semi-natural woods in eastern Scotland: McVean & Ratcliffe 1962) and probably of *Betula pendula* and perhaps also *B. pubescens* ssp. *pubescens* (Forbes & Kenworthy 1973); and, at the very highest altitudes, as on Creag Fhiaclach in the Cairngorms, it may be true also of *Pinus sylvestris* (McVean & Ratcliffe 1962, Carlisle & Brown 1968). In extreme situations, then, the *Juniperus-Oxalis* woodland may be a climatic climax community; at lower levels, it is probably a seral scrub which, for the most part, does not progress to mature forest for quite other reasons.

Throughout this altitudinal range, the *Juniperus-Oxalis* woodland can be found on a wide variety of soils. What the profiles have in common is that they are free-draining, being derived in most cases from pervious bedrocks or coarser-textured superficials. However,

though the community typically avoids gleyed soils, the profiles are never drought-prone with the moderately high rainfall they receive and are often kept quite moist by a moderate amount of flushing or by virtue of a northerly aspect (McVean & Ratcliffe 1962, Huntley & Birks 1979*a*): a striking contrast to the frequently parched conditions which juniper favours in its southern lowland localities. Freedom from dryness plus good aeration, combined with the effect of the only moderately acid litter of juniper, all help maintain a humus regime which is not of the extreme mor type and in which there is probably quite an active and speedy incorporation and nutrient turnover. This is reflected in the floristics of the community as a whole in features like the constancy of *Oxalis acetosella* (Packham 1978), the prominence of ferns like *Gymnocarpium dryopteris* and *Thelypteris phegopteris* (Page 1982) and the frequency throughout of less demanding calcifuge herbs such as *Agrostis capillaris* and *Luzula pilosa* and bryophytes like *Hylocomium splendens*, *Thuidium tamariscinum*, *Rhytidiadelphus squarrosus*, *Pseudoscleropodium purum*, *Plagiomnium rostratum* and *Plagiochila asplenoides*.

Nonetheless, there is always a tendency to surface eluviation of the soils here and more demanding species such as *Vaccinium myrtillus*, *Galium saxatile* and *Agrostis canina* ssp. *montana* retain a frequent representation throughout. And, where the parent materials are more siliceous, as over quartzites, gneisses or rhyolites or on sandy superficials, or where there is no influence of flushing with more base-rich waters, this potential is fully realised in the development of strongly podzolised profiles. On such soils, the *Vaccinium-Deschampsia* subcommunity is characteristic with its much more obvious calcifuge elements: an abundance of ericoid sub-shrubs, frequent records for *Deschampsia flexuosa* and a prominent contribution from more exacting bryophytes. In such stands, the flora of the *Juniperus-Oxalis* woodland takes on much of the character of the *Pinus-Hylocomium* or *Quercus-Betula-Dicranum* woodlands and, as there, more open covers of trees may favour podzolisation by permitting the spread of *Calluna* and the accumulation of substantial amounts of mor.

In other situations, various factors can offset the tendency towards surface leaching, most notably the local occurrence of more calcareous bedrocks or superficials, or seepage from them of base-rich waters. The effects of such variation are well seen in the east central Highlands where schists and limestones occur within the Dalradian and Moine meta-sediments (Ratcliffe 1977, Huntley & Birks 1979*a*), in the Lake District where andesites are found among the Borrowdale Volcanics (Pearsall & Pennington 1973), in Teesdale where there is seepage from Carboniferous Limestone into soils derived from Whin Sill dolerite and its drift (Pigott 1978*b*) and in scattered localities on Silurian and Carboniferous shales where limy partings are interbedded. Where the effects of such variation prevail, the profiles tend towards the brown earth type with much better incorporation of mull-like humus and a rise in surface pH (from about pH4 to pH6, limited data from Morrone suggest: Huntley & Birks 1979b). On such soils, the *Viola-Anemone* sub-community is the characteristic type of *Juniperus-Oxalis* woodland, with a fading contribution from strongly calcifuge sub-shrubs, *D. flexuosa* and bryophytes, and a preferential rise in frequency of species indicative of less acidic and more fertile soils: *Viola riviniana* and *Anemone nemorosa* themselves, grasses like *Festuca rubra* and *Holcus mollis*, dicotyledons such as *Campanula rotundifolia*, *Cerastium fontanum*, *Cardamine flexuosa* and even *Urtica dioica*, and bryophytes like *Plagiomnium undulatum*, *P. affine* and *Cirriphyllum piliferum*. Then the vegetation comes to resemble closely the flora found occasionally on less extreme soils in the *Pinus-Hylocomium* and *Quercus-Betula-Dicranum* woodlands but much more consistently associated with the *Quercus-Betula-Oxalis* woodland, the major oak- and birch-dominated forest of brown earths in the north-western parts of Britain. And when plants such as *Adoxa moschatellina*, *Mercurialis perennis*, *Cirsium helenioides* and *Geranium sylvaticum* appear in sites where there is a more pronounced rise in the base-richness and calcareous character of the soils within the *Viola-Anemone* sub-community the field layer comes close to that found under mixtures of *Fraxinus excelsior*, *Betula pubescens* and *Corylus avellana* in the *Fraxinus-Sorbus-Mercurialis* woodland.

Edaphic variation on a combined axis of base-richness and fertility can account for most of the floristic differences between the sub-communities but its effects are overlain and confused by the influence of grazing and browsing. Most of the stands of the *Juniperus-Oxalis* woodland appear to be open to large herbivores, occurring as part of unenclosed upland, within areas used as sheep ranges and with unhindered access for deer; at lower altitudes, cattle are sometimes pastured and, locally, horses (Ratcliffe 1977, Huntley & Birks 1979a, b; Gilbert 1980). And, as in the south-eastern lowlands (e.g. Ward 1973, Fitter & Jennings 1975), rabbits could be important in some sites.

As in other kinds of north-western woodlands, the general effects of grazing and browsing of the field layer are to reduce the cover of the palatable ericoids, *Vaccinium myrtillus* and *Calluna*, and of many of the ferns and dicotyledons and to favour the spread of tillering grasses and of the bryophytes which benefit indirectly by the reduction of competition. The impact of such developments can be seen in both sub-communities in a transformation of lush and varied scrub into what is, in effect, a grassland with juniper bushes; and, because grazing and browsing tend to have their most

obvious effects on the preferentials of the two sub-communities, herbivore predation tends to favour a floristic convergence of the two kinds of *Juniperus-Oxalis* woodland. In the *Vaccinium-Deschampsia* sub-community, the less palatable *V. vitis-idaea* may persist in some abundance to give an obvious clue as to the affinities of the vegetation but, in some cases, careful scrutiny of the closely-cropped sward and its bryophyte mat is required to distinguish which species remain.

Compared with, say, *V. myrtillus*, juniper itself is not a very palatable plant and, indeed, can confer some measure of protection on more favoured species growing within rings of bushes or in the centre of older individuals that have begun to open out. But browsing by deer and, when they are sufficiently abundant, by sheep or cattle, has a clear effect on the physiognomy of bushes that are within reach: in more open stands, especially where the juniper is of lower stature, all the bushes may be accessible; in denser stands or where the juniper is taller, only the marginal individuals or lower branches can be browsed (e.g. Huntley & Birks 1979a, Gilbert 1980). In extreme cases, it seems likely that deer and stock can actually open up thicker stands and perhaps even eliminate the juniper; and, where horses browse, a rare occurrence but recorded in Teesdale (Gilbert 1980), bushes can be killed by gnawing of the bark.

Furthermore, by grazing out seedlings and young plants of juniper, herbivores can probably play a large part in controlling regeneration and help determine the eventual number and density of the bushes. The scarcity of established young junipers in stands of this community was commented on by Pigott (1956a) and Gilbert (1980) and, though the latter noted some losses from erosion, burning, fungal attack and the exceptional drought of 1976, monitoring of individual plants and enclosure showed conclusively that herbivores were a very important cause of destruction. These observations were confined to Teesdale and, as Huntley & Birks (1979b) stated, we are in urgent need of data on the exact impact of grazing and browsing on this vegetation in the centre of its distribution in Scotland. For the moment, the work of Gilbert (1980) and studies on southern juniper scrub (Ward 1973, Fitter & Jennings 1975) suggest that simple reduction or withdrawal of grazing may not itself be sufficient to ensure regeneration. Juniper seed is quite widely dispersed by birds, though Gilbert (1980) noted that excreted seeds were often eaten, probably by the woodmouse *Apodemus sylvaticus*. And germination seems to be achieved only with some difficulty (Miles & Kinnaird 1979).

Also of great importance is that establishment requires either bare ground or a short sward, where there is freedom from shade, conditions which are maintained by the disturbance and cropping that herbivores themselves provide (Ward 1973, Fitter & Jennings 1975). As with birch then (Pigott 1983), regeneration may thus depend more on the sudden relaxation of grazing than on the maintenance of continuously low herbivore numbers. This may account for the occurrence of markedly even-aged stands of juniper in this community and the persistence of many of the Teesdale stands around farms and on the boundaries between the in-by and out-by land where fluctuations in grazing intensities would be most strongly felt. Indeed, Gilbert (1980) went further than this observation with the interesting suggestion that the survival of the *Juniperus-Oxalis* woodland in Teesdale and parts of Northumberland, and perhaps also in the Lake District, was related to the dual economy of farming and mining over the past few centuries, with its erratic variation in the rigour with which the land was pastured. Miles & Kinnaird (1979) also pointed out that, like the *Pinus-Hylocomium* woodland, this community may be partly dependent on the occurrence of burning for the regeneration of its juniper cover: at Tynron in Dumfries, an accidental fire stimulated the first regeneration seen for many years, but subsequent attempts to encourage regeneration using various other techniques failed (Kerr 1968). Patchy burning of juniper was used by farmers as a means of limiting its encroachment on upland grazing land and, in the past, this may actually have ensured its perpetuation.

In Teesdale, juniper was cut until the early part of this century to provide firewood and large boughs were used as a base for making haystacks (Gilbert 1980). Such use may have been widespread and account for some of the present restriction of this community to more scattered sites towards the south of its range. But perhaps of more obvious importance to the present appearance of this vegetation is that, at lower altitudes at least, it may have lost a cover of other tree species by a combination of timber removal, burning and grazing. Part of the evidence for this is circumstantial: it is that the *Juniperus-Oxalis* woodland survives over a wide range of soil types which one would expect, in the more hospitable situations, to carry mixtures of pine and birch with rowan and, in more fertile sites at the lowest altitudes, perhaps a little ash and oak. But such a view receives strong support from palynological evidence, at least in the east-central Highlands, where just such a mixed pine-birch-juniper forest seems to have persisted from Boreal times until a few centuries ago (e.g. Birks 1970, O'Sullivan 1977). Juniper still figures in this region as an understorey component in other woodland types but, for the most part, the original mixed forest cover has been segregated out into remnants of *Pinus-Hylocomium* woodland and birch-dominated *Quercus-Betula-Dicranum* and *Quercus-Betula-Oxalis* woodlands, many stands of which seem to be fairly young sub-spontaneous developments on neglected land. In other places, juniper remains as the woody dominant in *Juniperus-Oxalis* woodland, in effect a relic understorey, stabilised by grazing and into which there has been no sustained re-invasion of trees.

Zonation and succession

Zonations between the two sub-communities of the *Juniperus-Oxalis* woodland are most frequently a reflection of edaphic variation. Soil differences are involved, too, in transitions from the community to other vegetation types but, throughout, grazing, burning and timber removal have confused these patterns. Complex treatment histories also make it difficult to see how the community relates to other vegetation types in seral sequences and in natural altitudinal zonations of climax communities.

Soil variations within stands of the *Juniperus-Oxalis* woodland are generally related to differences in parent materials or the degree of flushing, with the *Vaccinium-Deschampsia* sub-community typically giving way to the *Viola-Anemone* sub-community wherever there is an increase in base-richness and fertility. Sometimes, as where there is an abrupt and marked shift in bedrock type, the edaphic and vegetational contrasts are sharp; often, especially where the community occurs over heterogenous superficials or experiences diffuse flushing, the change is a much more gradual one, manifest in ill-defined mosaics rather than clear zonations, a feature very well seen over the complex geology of the Morrone Birkwoods (Huntley & Birks 1979a, b).

Often, now, the landscape context of stands of the *Juniperus-Oxalis* woodland is one dominated by various types of heath or grassland and, at lower altitudes, it is clear that the vegetation pattern is strongly related to biotic factors, partly deer-grazing but also various kinds of human interference. Nonetheless, zonations often preserve a measure of edaphic control. For example, transitions to *Vaccinium-Deschampsia* heath sometimes reflect an obvious move to drier and more strongly podzolised profiles, a switch which is sometimes related to the occurrence of highly siliceous parent materials or a complete waning of any impact of flushing with base-rich waters (e.g. Huntley & Birks 1979a). In other cases, this type of soil and vegetation zonation is a reflection of slope and aspect: in the east-central Highlands, the *Juniperus-Oxalis* woodland sometimes marks out damp hollows and north-facing slopes in tracts of heather moorland (McVean & Ratcliffe 1962). In the opposite direction, on free-draining soils which are increasingly base-rich and calcareous, the community typically passes to some kind of calcicolous sward, often preferentially grazed by deer and stock because of its nutritious bite. On limestone or schist outcrops in the Dalradian and Moine rocks of eastern Scotland, such zonations usually involve the *Festuca-Agrostis-Thymus* grassland (e.g. Huntley & Birks 1979a); in Teesdale, with the switch from intruded Whin Sill dolerite to Carboniferous Limestone, the *Sesleria-Galium* grassland is represented (e.g. Pigott 1956a, Graham 1971, Gilbert 1980).

In situations such as these, the *Juniperus-Oxalis* woodland occupies the moister sites, where aspect or seepage prevent the development of excessively-draining soils. In other cases, as where heavy-textured superficials dominate among the parent materials, it may mark out better-drained areas, passing to various kinds of mires in hollows where there is ground-water gleying or a strong influence of soligenous waters. At Morrone, for example, a complex of flush communities occurs among the *Juniperus-Oxalis* woodlands with *Pinguiculo-Caricetum* mire figuring prominently in more base-rich situations, the *Scirpus-Erica* wet heath in the more base-poor (Huntley & Birks 1979a).

In the absence of enclosure experiments or any sustained observations on the effects of burning within such sequences as these, it is difficult to know how far the limits of the *Juniperus-Oxalis* woodland are natural. However, waterlogging is strongly inimical to juniper, so the absence of the community from more hydromorphic soils probably represents a real edaphic boundary though, if there were no grazing, one might expect the *Juniperus-Oxalis* woodland to give way to mire scrub or forest in such situations, perhaps the *Alnus-Fraxinus-Lysimachia* woodland in more base-rich places or the *Salix-Carex* woodland in the less. On drier soils, grazing and burning have almost certainly reduced the extent of the community, the former favouring an extension of grasslands, the latter, especially on more base-poor profiles, encouraging a spread of heaths, and especially the dominance of *Calluna*. In many sites, the composition of the *Vaccinium-Deschampsia* heath adjacent to the community is very similar (essentially it is often *Juniperus-Oxalis* woodland without the juniper) and the balance between ericoids and juniper in the invasion process after burning may be quite a fine one.

Information on the precise effects of burning and herbivore predation is urgently needed for the sensitive conservation of this community. It would also enable us to see something of the seral processes in which juniper is involved and to assess the successional status of the community. To the north of its British range, juniper obviously colonises and maintains itself on a much broader spectrum of soil types than is the case in the southern lowlands and, in more natural circumstances, the community may develop along a number of seral pathways. Early invasion of very open habitats with fragmentary soils could be the normal mode of establishment for this light-demanding tree, in which case its colonisation of existing ungrazed herbaceous vegetation might depend on erosion on steeper slopes or clearance of the ground by spontaneous fires (e.g. Fitter & Jen-

nings 1975, Miles & Kinnaird 1979). Over rocky ground and on ledges, juniper can also invade coincidentally with ferns, ericoids and tall herbs, occasionally finding a place in vegetation types like the *Luzula-Vaccinium* and *Luzula-Geum* communities (McVean & Ratcliffe 1962, Huntley & Birks 1979a). These ungrazed communities show a considerable floristic overlap with the *Juniperus-Oxalis* woodland, the former, characteristic of more base-poor rocks, resembling the *Vaccinium-Deschampsia* sub-community, the latter, typical of more base-rich situations, sharing many species with the *Viola-Anemone* sub-community. Where juniper is able to overtop the associates, it essentially converts the vegetation to the *Juniperus-Oxalis* woodland, though stands on ledges are very fragmentary.

Subsequent developments probably depend on a number of factors. In theory, one would expect stands at lower altitudes to progress to forest dominated by pine and/or birch, with the *Vaccinium-Deschampsia* sub-community on more acidic, impoverished soils moving to the *Quercus-Betula-Dicranum* or *Pinus-Hylocomium* woodlands, the *Viola-Anemone* sub-community on more base-rich and fertile soils to the *Quercus-Betula-Oxalis* woodland or the *Luzula* sub-community of the *Pinus-Hylocomium* woodland. In some places, the different kinds of *Juniperus-Oxalis* woodland can be seen in gaps within these other woodland types, the field layers showing considerable continuity. In fact, re-invasion of junipers and ericoid-dominated covers is very difficult for both pine and birch, because they are light-demanding, and with pine there is the additional problem of rather erratic fruiting and a less efficient wind-dispersal of seed than is the case with birch (Carlisle & Brown 1968, Carlisle 1977). Grazing can open up the ground but destroys seedlings; burning without grazing may be more effective but tends to allow *Calluna* to invade first; and widespread removal of timber, especially common and more critical with pine, has reduced the number of seed-parents. Just such a combination of factors seems to have converted the original patchwork of pine, birch and juniper forests in the east-central Highlands to the pattern of woodland fragments among extensive tracts of heath and grassland that we see today (e.g. McVean & Ratcliffe 1962, O'Sullivan 1977).

Widespread forest destruction has also left very few places where the *Juniperus-Oxalis* woodland can be seen as a convincing altitudinal replacement for sub-montane forest: the best site is probably Creag Fhiaclach where the community forms a scrubby fringe between the *Pinus-Hylocomium* woodland and montane heath (McVean & Ratcliffe 1962). Elsewhere, though, the *Juniperus-Oxalis* woodland does persist as a fragmentary belt above a wide zone now largely converted to heath and grassland (Watt & Jones 1948, Poore & McVean 1957, McVean & Ratcliffe 1962). In such situations as these, the community occupies an analogous position to the juniper scrub of the 'lower-alpine' zone in Scandinavia (Du Rietz & Du Rietz 1925, Nordhagen 1943). As there, though in much more fragmentary fashion, it shares this zone with scrub dominated by Arctic-Alpine willows like *Salix lapponum*, *S. myrsinites* and *S. reticulata* which replaces the *Juniperus-Oxalis* woodland on wetter soils but has quite a strong floristic overlap with the *Viola-Anemone* sub-community. On the highly acidic rocks of the north-western parts of Scotland, this altitudinal zone is characterised by a scrubby heath in which *J. communis* ssp. *nana* plays a major role.

Distribution

The *Juniperus-Oxalis* woodland is largely confined to the east-central Highlands of Scotland, particularly the hills of the Cairngorm and Monadhliath ranges. More isolated stands have been recorded from southern Scotland, Northumberland, the Pennines and the Lake District.

Affinities

This community unites a variety of previously-described kinds of scrub or woodland dominated by *J. communis*

ssp. *communis*, with or without small amounts of birch (Pigott 1956a, McVean & Ratcliffe 1962, Graham 1971, Huntley & Birks 1979a, Birse 1980, 1982), and preserves the major floristic distinction between more and less calcifugous types recognised in Teesdale (Graham 1971), at Morrone (Huntley & Birks 1979a, b) and more widely in the east-central Highlands (Birse 1980, 1982). The close floristic affinities with the *Pinus-Hylocomium* woodland and, over much of the altitudinal range, the seral relationship to that community, argue for placing the *Juniperus-Oxalis* woodland in the Vaccinio-Picetea, perhaps in the alliance Vaccinio-Juniperion Passarge 1968 or within the Dicrano-Pinion (Libbert 1933) Matuszkiewicz 1962. If the climatic climax character of high-altitude stands is stressed, the most obvious floristic affinities are with Scandinavian sub-alpine juniper scrubs like the *Junipereto-Betuletum nanae myrtilletosum* (Nordhagen 1928, 1943) which Dahl (1956) placed with the montane *Nardus stricta* and *Carex bigelowii* communities. In these scrubs, however, the juniper is *J. communis* ssp. *nana* and there is a good representation of *Betula nana*: in Scotland, there is no evidence of an association between juniper and *B. nana* (Poore & McVean 1957).

Floristic table W19

	a	b	19
Betula pubescens	II (5–7)	I (6)	I (5–7)
Sorbus aucuparia		I (1)	I (1)
Juniperus communis communis	V (5–10)	V (7–10)	V (5–10)
Oxalis acetosella	V (1–7)	V (3–7)	V (1–7)
Hylocomium splendens	V (1–8)	IV (1–8)	V (1–8)
Vaccinium myrtillus	V (1–8)	IV (1–7)	V (1–8)
Galium saxatile	V (1–5)	IV (1–6)	V (1–6)
Luzula pilosa	IV (1–5)	IV (1–5)	IV (1–5)
Agrostis capillaris	IV (1–6)	IV (1–7)	IV (1–7)
Agrostis canina montana	IV (1–5)	IV (1–5)	IV (1–5)
Thuidium tamariscinum	IV (1–5)	IV (1–9)	IV (1–9)
Dicranum scoparium	V (1–4)	III (1–4)	III (1–4)
Plagiothecium undulatum	V (1–5)	II (1–4)	III (1–5)
Deschampsia flexuosa	V (2–6)	II (1–7)	III (1–7)
Vaccinium vitis-idaea	IV (1–8)	II (1–5)	III (1–8)
Rhytidiadelphus loreus	IV (1–4)	II (1–3)	III (1–4)
Pleurozium schreberi	IV (1–4)	II (1–4)	III (1–4)
Calluna vulgaris	IV (1–4)	II (1–4)	III (1–4)
Hypnum cupressiforme	IV (1–5)	II (1–5)	III (1–5)
Dicranum majus	III (1–5)	I (1–3)	II (1–5)
Lophozia ventricosa	I (1–3)		I (1–3)
Scapania gracilis	I (1)		I (1)

Floristic table W19 *(cont.)*

	a	b	19
Isothecium myosuroides	I (1–2)		I (1–2)
Barbilophozia hatcheri	I (1–6)		I (1–6)
Listera cordata	I (1)		I (1)
Pyrola minor	I (1)		I (1)
Cladonia pyxidata	I (1–2)		I (1–2)
Viola riviniana	II (1–3)	V (1–5)	III (1–5)
Anemone nemorosa	II (2–3)	IV (1–6)	III (1–6)
Rhytidiadelphus triquetrus	II (1–7)	IV (1–8)	III (1–8)
Festuca ovina	II (2–5)	III (1–5)	III (1–5)
Plagiomnium undulatum	I (1–2)	III (1–6)	II (1–6)
Festuca rubra	I (1–2)	III (1–5)	II (1–5)
Campanula rotundifolia	I (1–3)	III (1–4)	II (1–4)
Cardamine flexuosa	I (1–3)	III (1–4)	II (1–4)
Urtica dioica	I (2)	II (1–5)	I (1–5)
Holcus mollis	I (3–5)	II (1–8)	I (1–8)
Cerastium fontanum	I (1–2)	II (1–3)	I (1–3)
Barbilophozia barbata	I (1)	II (1–6)	I (1–6)
Drepanocladus uncinatus	I (1)	II (1–5)	I (1–5)
Rubus idaeus	I (1)	II (1–6)	I (1–6)
Cirriphyllum piliferum	I (1)	II (1–4)	I (1–4)
Stellaria holostea		II (1–4)	I (1–4)
Fragaria vesca		II (1–4)	I (1–4)
Plagiomnium affine		II (2–4)	I (2–4)
Adoxa moschatellina		II (1–4)	I (1–4)
Galium boreale		II (2–4)	I (2–4)
Mercurialis perennis		I (2–7)	I (2–7)
Galium verum		I (1–2)	I (1–2)
Prunella vulgaris		I (2)	I (2)
Carex flacca		I (2–3)	I (2–3)
Geum rivale		I (3)	I (3)
Potentilla sterilis		I (2–3)	I (2–3)
Epilobium montanum		I (1–5)	I (1–5)
Isothecium myurum		I (1–4)	I (1–4)
Lathyrus montanus		I (1–2)	I (1–2)
Polygonum viviparum		I (1–2)	I (1–2)
Pyrola media		I (1)	I (1)
Ranunculus acris		I (2–3)	I (2–3)
Poa trivialis		I (1–3)	I (1–3)
Rumex acetosa		I (1–2)	I (1–2)
Blechnum spicant	III (1–4)	III (1–5)	III (1–5)
Lophocolea bidentata s.l.	III (1–4)	III (1–5)	III (1–5)
Pseudoscleropodium purum	III (1–6)	III (1–4)	III (1–6)
Potentilla erecta	III (1–4)	III (1–4)	III (1–4)
Trientalis europaea	III (1–5)	III (1–3)	III (1–5)
Rhytidiadelphus squarrosus	III (1–5)	III (1–5)	III (1–5)
Plagiochila asplenoides	III (1–4)	III (1–5)	III (1–5)

Anthoxanthum odoratum	II (1–5)	III (1–6)	III (1–6)
Veronica officinalis	II (1–2)	II (1–5)	II (1–5)
Plagiomnium rostratum	II (1–5)	II (1–4)	II (1–5)
Gymnocarpium dryopteris	II (1–8)	II (3–7)	II (1–8)
Mnium hornum	II (1–4)	II (1–5)	II (1–5)
Polytrichum formosum	II (1–2)	II (1–2)	II (1–2)
Pteridium aquilinum	II (1–5)	II (1–5)	II (1–5)
Plagiothecium denticulatum	II (1–4)	II (1–4)	II (1–4)
Dryopteris dilatata	II (1–4)	I (2–4)	I (1–4)
Hypnum jutlandicum	II (2–5)	I (1–4)	I (1–5)
Ptilium crista-castrensis	II (1–6)	I (2–4)	I (1–6)
Carex pilulifera	I (1)	I (1–3)	I (1–3)
Deschampsia cespitosa	I (4)	I (1–4)	I (1–4)
Dryopteris filix-mas	I (1)	I (1)	I (1)
Thelypteris phegopteris	I (1–3)	I (1–2)	I (1–3)
Atrichum undulatum	I (2)	I (1–2)	I (1–2)
Eurhynchium praelongum	I (1–2)	I (2–7)	I (1–7)
Plagiothecium succulentum	I (1)	I (1–4)	I (1–4)
Polytrichum commune	I (1–5)	I (1–5)	I (1–5)
Thelypteris limbosperma	I (4–6)	I (1–3)	I (1–6)
Melampyrum pratense	I (2)	I (2)	I (2)
Luzula multiflora	I (1–3)	I (1–3)	I (1–3)
Carex binervis	I (1)	I (1)	I (1)
Rubus saxatilis	I (2)	I (1–2)	I (1–2)
Geranium sylvaticum	I (1)	I (2)	I (1–2)
Hylocomium umbratum	I (1)	I (1)	I (1)
Number of samples	31	38	69
Number of species/sample	28 (24–34)	32 (22–46)	30 (22–46)
Tree height (m)	7 (7–8)	9 (7–10)	8 (7–10)
Tree cover (%)	5 (0–50)	2 (0–30)	3 (0–50)
Shrub height (m)	1 (0.8–2.1)	1 (0.4–2.0)	1 (0.4–2.1)
Shrub cover (%)	62 (45–90)	78 (40–90)	73 (40–90)
Herb height (cm)	18 (5–30)	18 (5–35)	18 (5–35)
Herb cover (%)	65 (50–75)	59 (30–80)	61 (30–80)
Ground height (mm)	no data	no data	no data
Ground cover (%)	67 (50–80)	65 (25–85)	66 (25–85)
Altitude (m)	445 (365–556)	448 (366–663)	447 (365–663)
Slope (°)	10 (4–22)	14 (4–30)	12 (4–30)

a   *Vaccinium vitis-idaea-Deschampsia flexuosa* sub-community

b   *Viola riviniana-Anemone nemorasa* sub-community

19 *Juniperus communis communis-Oxalis acetosella* woodland (total)

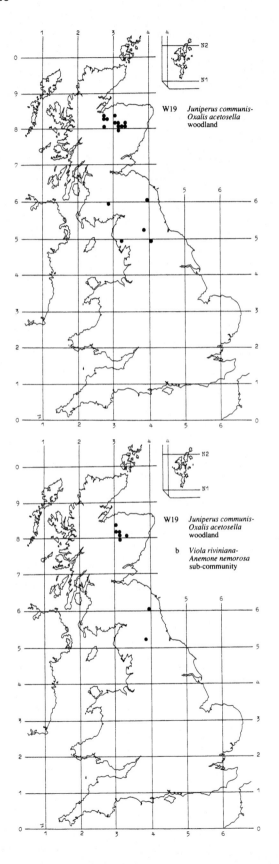

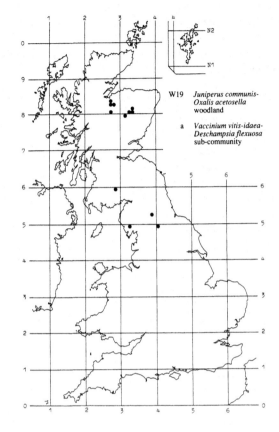

Salix lapponum-Luzula sylvatica scrub

W20

Synonymy

Willow scrub   Poore & McVean 1957 *p.p.*; *Salix lappo-num-Luzula sylvatica* nodum McVean & Ratcliffe 1962; Montane willow scrub association McVean 1964; *Salix lapponum-Vaccinium myrtillus* nodum Huntley 1979.

Constant species

Salix lapponum, Deschampsia cespitosa, D. flexuosa, Luzula sylvatica, Vaccinium myrtillus, Dicranum scoparium, Hylocomium splendens, Rhytidiadelphus loreus.

Rare species

Carex atrata, Polystichum lonchitis, Salix arbuscula, S. lanata, S. lapponum, S. myrsinites, S. reticulata.

Physiognomy

Isolated bushes of Arctic-Alpine and Arctic-Subarctic willows figure occasionally in higher-altitude stands of the moderately calcicolous kinds of ungrazed vegetation included in the *Luzula-Vaccinium* and *Luzula-Geum* communities but in some situations these willows are so abundant as to constitute a low bushy canopy to this *Salix lapponum-Luzula sylvatica* scrub with associated changes in the accompanying flora. Of the various willows represented here, *Salix lapponum* is the commonest and most widely distributed and usually it dominates, its much-branched bushes forming a patchy cover up to a metre or so high. But it can be accompanied or sometimes replaced by the generally smaller *S. myrsinites* or by *S. lanata*, especially prominent in the Clova-Caenlochan area (e.g. Huntley 1979), or *S. arbuscula*, particularly distinctive around Breadalbane (McVean & Ratcliffe 1962, Ratcliffe 1977). In some sites, too, drawn-up shoots of the normally diminutive *S. reticulata* make a contribution to the cover and very occasionally the canopy is further enriched by bushes of more widely distributed willows like *S. cinerea* and *S. phylicifolia*. Hybrid willows are sometimes found but sexual reproduction even within the rarer species may be very

infrequent: in many areas, the number of bushes is small and the sexes often widely separated (McVean 1964*a*).

There are usually some sub-shrubs growing among the willows. The commonest of these is *Vaccinium myrtillus*, which can be co-dominant, but *V. vitis-idaea*, *Empetrum nigrum* ssp. *hermaphroditum* and *Calluna vulgaris* all occur frequently and *V. uliginosum* occasionally, though generally their cover is low. Often more prominent is a strong contingent of grazing-sensitive herbs, the luxuriant growth of which among the shrubby canopy gives this kind of scrub a very distinctive look. *Luzula sylvatica* is the most frequent member of this group, its tussocky mats sometimes forming an extensive cover and hanging down from ledges in festoons, but other species are *Alchemilla glabra, Geum rivale, Rumex acetosa, Angelica sylvestris, Galium boreale* and the Arctic-Alpine *Rhodiola rosea, Oxyria digyna* and *Saussurea alpina*. Somewhat less frequently, there are records for *Solidago virgaurea, Succisa pratensis, Filipendula ulmaria, Valeriana officinalis, Coeloglossum viride* and Hieracia (not recorded to the species, but presumably mostly of the Alpina, Subalpina and Cerinthoidea sections: see Raven & Walters 1956); and occasionally there is some *Dryopteris dilatata* or *Thelypteris phegopteris* with *Polystichum lonchitis* and *Blechnum spicant* occurring rarely.

Among these is an equally rich and diverse assemblage of herbs of somewhat smaller stature. Continuing the Arctic-Alpine contribution are *Alchemilla alpina, Thalictrum alpinum, Polygonum viviparum* (all very frequent), *Saxifraga oppositifolia* (occasional) and *S. stellaris, S. aizoides, Epilobium anagallidifolium* (all scarce). But also very common are *Galium saxatile, Viola riviniana, Campanula rotundifolia, Oxalis acetosella* with, somewhat less frequently, *Huperzia selago, Ranunculus acris, Philonotis fontana, Selaginella selaginoides* and, very occasionally, *Caltha palustris, Euphrasia officinalis* agg., *Rhinanthus minor, Anemone nemorosa* and *Thymus praecox*.

Then, there is usually some contribution from grasses,

typically growing as discrete tussocks. *Deschampsia cespitosa* (sometimes viviparous) and *D. flexuosa* are the commonest species and generally the most abundant, but *Festuca ovina* (and *F. vivipara* when recorded separately), *Agrostis canina*, *A. capillaris*, *Anthoxanthum odoratum* all occur quite frequently and *Festuca rubra* and *Nardus stricta* more occasionally. *Carex bigelowii*, *C. binervis* and, more rarely, *C. flacca* and the Arctic-Alpine *C. atrata* have also been recorded here. The total vascular flora of this kind of vegetation is thus very large and individual stands are characteristically rich, though the contribution from the various components and species is typically diverse and, in any single locality, the frequencies of the occasional associates can show considerable variation.

The same feature is true of the bryophyte element. Very common throughout the community and often forming a thick, luxuriant carpet are *Hylocomium splendens*, *Rhytidiadelphus loreus*, *Dicranum scoparium*, *D. majus*, *Mnium hornum*, *Ptilidium ciliare*, *Pleurozium schreberi*, *Thuidium tamariscinum* and *Sphagnum subnitens*. Rather less frequent are *Rhytidiadelphus triquetrus*, *R. squarrosus*, *Polytrichum alpinum*, *Rhizomnium punctatus*, *Plagiomnium undulatum*, *Plagiothecium undulatum*, *Hypnum cupressiforme* and a variety of Sphagna, *S. girgensohnii*, *S. russowii*, *S. capillifolium*, *S. palustre*, *S. recurvum*, *S. teres* and *S. squarrosum*. Other scarce bryophytes include *Hypnum callichroum*, *Ptilium crista-castrensis*, *Drepanocladus revolvens*, *D. uncinatus*, *Ctenidium molluscum*, *Fissidens adianthoides* and *Cratoneuron commutatum*.

Lichens are generally sparse but *Peltigera canina* may be conspicuous and there can be some *Cladonia squamosa* or *C. pyxidata*.

Habitat

The *Salix-Luzula* scrub is a community of ungrazed, high-altitude rocky slopes and ledges with wet, mesotrophic and base-rich soils. It probably represents the sub-alpine climax vegetation in such situations, though herbivore predation has reduced it to the status of a relic community of generally small, isolated stands, a fragmentation which affects the composition and structure of the community.

The *Salix-Luzula* scrub is the most high-level kind of tree- or shrub-dominated vegetation in Britain, the altitudinal range of available samples being from 630 m to over 900 m: it overlaps a little with the *Juniperus-Oxalis* woodland where *Juniperus communis* ssp. *communis* and *Betula pubescens* ssp. *carpatica* can dominate on drier, more acidic soils but the mean altitude of stands here is over 300 m higher. Apart from some fragmentary stands in the Moffat Hills in Dumfries (Ratcliffe 1959*b*, 1977), the community is confined to the Scottish Highlands where there is a harsh montane climate with long, bitter winters and short, cool summers. The annual accumulated temperature lies within the range 277–556 °C (500–1000°F) (Gregory 1954) and the mean annual maximum temperature is less than 21 °C (Conolly & Dahl 1970), conditions which are strongly reflected in the vegetation by the large contingent of Arctic-Alpine species, not least the willows themselves. Annual precipitation is generally more than 1600 mm (*Climatological Atlas* 1952), with at least 180 wet days yr^{-1} (Ratcliffe 1968) but, at these altitudes, much of this falls as snow in the winter months and, with the lateness of spring, this tends to be long on the ground.

Late snow-lie is especially marked in the east-central and southern Highlands where the *Salix-Luzula* scrub has its centre of distribution but, throughout its range, there is a tendency for stands to occupy sites with a north to east aspect which would afford some protection against early melt. McVean & Ratcliffe (1962) suggested that a snow cover might be important in localising the occurrence of the willows by giving some shelter from air frosts, very frequent and occurring into late spring at these altitudes. Whether this is always the case has been questioned (Huntley 1979), though comparable communities in Scandinavia are clearly associated with long snow-lie (e.g. Nordhagen 1928, 1943, Dahl 1956).

Within this climatic zone, the *Salix-Luzula* scrub is typically found on soils with a degree of base-, and probably nutrient-, enrichment. It is distinctly associated with more calcareous rocks, notably among the Dalradian meta-sediments of the central and southern Highlands, where it occurs on banks and ledges in limestones, mica-schists and epidiorite intrusions in Breadalbane and the Clova-Caenlochan hills (Ratcliffe 1977, Huntley 1979), and, more locally, on the Moine Assemblage, as on Creag Meagaidh in Inverness and on Ben Hope in Sutherland (Ratcliffe 1977). In the Moffat Hills, fragmentary stands are found on calcitic crush-zones (Ratcliffe 1959*b*). However, the soils are not the kind of alpine rendzinas typical of the *Dryas-Silene* community. Characteristically, seepage of ground water is strong and the profiles are permanently wet and usually unstructured accumulations of silt and rock fragments, often unstable and, on the steepest slopes, retained only by the mat of vegetation rooted into the rocks. Very few analytical data are available from these soils but McVean & Ratcliffe (1962) recorded pHs of 5.7 and 6.9 under two stands with similar amounts of calcium as in the *Dryas-Silene* community and the more calcicolous kinds of *Festuca-Agrostis-Thymus* grassland. Litter incorporation and turnover of such nutrients as there are in the system are probably quite rapid with the abundance of *Luzula sylvatica* in particular tending to favour the development of moder humus.

The floristic response to these edaphic conditions is

quite varied. A strictly calcicolous element in the vegetation is not well defined, though, among the willows themselves, *S. lanata*, *S. myrsinites*, *S. arbuscula* and *S. reticulata* are fairly exacting species (McVean & Ratcliffe 1962) and some plants may be largely or totally excluded by competition from the tall herbage (e.g. *Saxifraga oppositifolia*, *Thymus praecox*, *Carex flacca*, *C. pulicaris*) or by the wetness of the soils (e.g. *Polystichum lonchitis*). Much more obvious is a group of species indicative of mesotrophic conditions, often associated in the uplands with base-enrichment: among these would figure *S. lapponum*, most of the tall herbs, *Viola riviniana* and bryophytes like *Thuidium tamariscinum* and *Plagiomnium undulatum*. But calcifuges are quite strongly represented here with *Vaccinium myrtillus* and other sub-shrubs and a number of more exacting bryophytes: this may be a measure of soil heterogeneity across larger ledges or a reflection of some vertical differentiation in the profile with a patchy surface mat of litter, humus and bryophytes insulated somewhat from seepage through the mineral layer beneath.

Even where these climatic and edaphic requirements are met, the *Salix-Luzula* scrub is found only over very rocky slopes and on ledges whose inaccessibility affords protection from grazing and browsing by deer and sheep (and, in some areas in the past, cattle, as at Caenlochan: Huntley 1979). One very characteristic type of site which combines such features is the complex of fairly steep slopes below series of low cliffs associated with sequential waterfalls, where the severe topography prevents access to herbivores from above and below. The community is not, however, invariably present on all such suitable sites, being often replaced by various kinds of *Luzula-Geum* or *Luzula-Vaccinium* communities on identical soils. Variation in exposure and snow-lie may play some part in such differentiation but, often, the absence of willows may be simply a question of loss through death of existing bushes and failure to re-invade. Individuals of some of these species may not be long-lived, despite their sometimes gnarled and venerable appearance (Meikle 1984) and, with the progressive isolation of populations with centuries of pastoral exploitation of the uplands and often wide separation of the sexes (Poore & McVean 1957), continued colonisation may be very difficult. Where the community does persist, the physical configuration of the rock exposures and ledges affects the physiognomy and floristics of the vegetation by the simple limitation of space and the presentation of surfaces of different shape and slope.

Zonation and succession

Most commonly, zonations from the *Salix-Luzula* scrub to other vegetation types are a reflection of variations in grazing pressure though, since these typically relate to differences in topography, edaphic factors sometimes play a part.

Usually, the community occurs as small stands isolated on rocky knolls and ledges within a montane landscape largely transformed by grazing to various kinds of close-cropped grassy or herb-dominated swards. In areas where calcareous rocks and soils are more common, as over the Dalradian schists and limestones of Breadalbane and in the Clova-Caenlochan district, the *Salix-Luzula* scrub overlaps altitudinally with more high-level stands of the *Festuca-Agrostis-Thymus* grassland and the *Festuca-Agrostis-Alchemilla* grass-heath. More flushed types of these two communities show strong floristic affinities with the *Salix-Luzula* scrub and have probably been derived from it by elimination of willows and tall herbs and a favouring of grasses and grazing-resistant dicotyledons. At higher altitudes, the community can also be found in association with the *Festuca-Alchemilla-Silene* dwarf-herb community in which grazing-tolerant Arctic-Alpines are well represented and this kind of vegetation, too, may be a derivative of *Salix-Luzula* scrub, though it is characteristic of more exposed situations where solifluction and cryoturbation are important, so this transition may be partly an edaphic one. Floristically and environmentally, the *Dryas-Silene* community can be seen as intermediate between the *Festuca-Alchemilla-Silene* community and the *Salix-Luzula* scrub. Typically, it is much less heavily grazed than the former, though often in more unstable situations than the latter and it can occasionally have some of the larger Arctic-Alpine willows. Good transitions from the *Dryas-Silene* community to the *Salix-Luzula* scrub can be seen on Meall na Samhna, Carn Gorm and Beinn Dearg (Ratcliffe 1977), the willows increasing their cover in the latter vegetation and shading out many species of the former. Where calcareous rocks and soils form a much more local intrusion into landscapes dominated by acidic substrates, as among the Moinian and Lewisian rocks of the Cairngorms and the north-west Highlands, floristic transitions are usually much sharper: here ledges with the *Salix-Luzula* scrub may be surrounded by a small zone of more calcicolous grassland on flushed soils but often there is a fairly quick zonation to calcifuge grasslands and heaths on base-poor brown earths, gleys and peaty soils.

In both these kinds of situations, the community typically occupies only some of the available ledges, others supporting very similar *Luzula-Geum* vegetation or, where flushing is with somewhat less base-rich waters, the *Luzula-Vaccinium* community. How far the willows could spread into these other vegetation types or into flushed, calcicolous swards, if grazing were to be reduced, is unknown. Reconstitution of the *Salix-Luzula* scrub would probably be very difficult, though enclosure of the surrounds of some more vigorous and mixed-sex willow populations would be an instructive exercise. On the Durness Limestone above Inchnadamph, Poore & McVean (1957) sampled a low-altitude scrub (210–274 m) where *Salix myrsinites* seemed to have re-invaded *Festuca-Agrostis-Thymus* grassland and such expansion might be possible at higher levels.

By and large, however, the *Salix-Luzula* scrub persists now as remnants isolated well above any other kind of woody vegetation. It seems reasonable to suppose, comparing the Scottish situation with Scandinavia, where similar vegetation is widespread, that the community was at one time much more common, replacing scrubby hazel-, rowan- and birch-dominated *Fraxinus-Sorbus-Mercurialis* woodland at high altitudes on wet, calcareous soils. Nowhere, now, do such zonations persist intact.

Distribution

The community is widespread but local through the southern and central Highlands of Scotland, being especially well-developed in Breadalbane and around Clova-Caenlochan, with more isolated stands in the north-west Highlands and, far to the south, in the Moffat Hills. *Salix lapponum* survives at high altitudes on Helvellyn in the Lake District, though it occurs there on rather dry rocks without the luxuriant assemblage typical of the *Salix-Luzula* community.

Affinities

The definition of the *Salix-Luzula* scrub is based on the data of McVean & Ratcliffe (1962) and Huntley (1979) with no further sampling. The nearest equivalents to the community in Europe are the various kinds of sub-alpine willow scrub described from Scandinavia by Nordhagen (1928, 1943) and Dahl (1956), particularly the *Salicetum geraniosum alpicolum* from Sikilsdalen (Nordhagen 1943) and the *Rumiceto-Salicetum lapponae* from the Rondane area (Dahl 1956). However, two features are notable in comparing these vegetation types with our own montane willow scrub: first, the former are generally much richer in tall herbs, having a field layer more like that of our wetter *Fraxinus-Sorbus-Mercurialis* woodland (*Crepis* sub-community); and, second, in Scandinavia, the Arctic-Alpine willows also extend into mire vegetation like that of the *Carex saxatilis* and *Carex rostrata-Sphagnum warnstorfii* mires. The *Salix-Luzula* scrub clearly belongs among the sub-alpine and alpine tall-herb communities of the Betulo-Adenostyletea, in which Ellenberg (1978) has distinguished a Salicion arbusculae with prominent dwarf willows.

Floristic table W20

Salix lapponum	V (1–9)		*Galium saxatile*	III (1–3)
Salix lanata	II (4–7)		*Viola riviniana*	III (1–3)
Salix myrsinites	II (4–8)		*Pleurozium schreberi*	III (1–4)
Salix reticulata	II (1–5)		*Sphagnum subnitens*	III (1–8)
Salix arbuscula	I (6)		*Thuidium tamariscinum*	III (1–5)
Salix phylicifolia	I (5)		*Vaccinium vitis-idaea*	III (2–3)
Salix cinerea	I (2–3)		*Festuca ovina*	III (3–4)
Vaccinium myrtillus	V (1–7)		*Angelica sylvestris*	III (1–5)
Hylocomium splendens	V (2–8)		*Thalictrum alpinum*	III (2–3)
Rhytidiadelphus loreus	V (1–7)		*Polygonum viviparum*	III (1–3)
Deschampsia flexuosa	IV (2–5)		*Agrostis canina*	II (2–3)
Luzula sylvatica	IV (1–8)		*Campanula rotundifolia*	II (1–3)
Deschampsia cespitosa	IV (2–5)		*Galium boreale*	II (1–3)
Dicranum scoparium	IV (1–5)		*Rhizomnium punctatum*	II (1–5)
Alchemilla alpina	III (2–5)		*Rhytidiadelphus triquetrus*	II (1–5)
Alchemilla glabra	III (1–6)		*Sphagnum girgensohnii*	II (2–6)
Geum rivale	III (1–6)		*Plagiochila asplenoides*	II (1–2)
Rumex acetosa	III (1–4)		*Dryopteris dilatata*	II (1–3)
Rhodiola rosea	III (1–4)		*Oxalis acetosella*	II (1–3)
Dicranum majus	III (1–6)		*Plagiothecium undulatum*	II (2–3)
Mnium hornum	III (1–3)		*Polytrichum alpinum*	II (1–2)
Ptilidium ciliare	III (1–3)		*Selaginella selaginoides*	II (2)
Empetrum nigrum hermaphroditum	III (2–5)		*Rhytidiadelphus squarrosus*	II (1–2)
			Sphagnum palustre	II (1–3)

Agrostis capillaris	II (2–3)	*Saxifraga stellaris*	I (1)
Carex bigelowii	II (1–2)	*Calliergon cuspidatum*	I (2)
Oxyria digyna	II (1–3)	*Dicranum bonjeanii*	I (1–2)
Saussurea alpina	II (1–5)	*Drepanocladus uncinatus*	I (2–3)
Plagiomnium undulatum	II (2–3)	*Sphagnum recurvum*	I (1–2)
Calluna vulgaris	II (2–3)	*Sphagnum squarrosum*	I (2–4)
Anthoxanthum odoratum	II (2–5)	*Atrichum undulatum*	I (1–2)
Solidago virgaurea	II (1–2)	*Sphagnum capillifolium*	I (1–2)
Saxifraga oppositifolia	II (1–3)	*Sphagnum russowii*	I (1)
Hypnum cupressiforme	II (2–4)	*Thymus praecox*	I (2–3)
Huperzia selago	II (1–3)	*Nardus stricta*	I (3)
Diplophyllum albicans	II (2)	*Euphrasia officinalis* agg.	I (2–3)
Lophocolea bidentata s.l.	I (1–3)	*Rhinanthus minor*	I (1–2)
Festuca rubra	I (2–3)	*Valeriana officinalis*	I (3–4)
Epilobium anagallidifolium	I (1–2)	*Aulacomnium palustre*	I (2–3)
Sphagnum teres	I (1–3)	*Coeloglossum viride*	I (1)
Chiloscyphus polyanthos	I (1–2)	*Saxifraga hypnoides*	I (2–3)
Vaccinium uliginosum	I (1–6)	*Carex atrata*	I (1–2)
Hypnum callichroum	I (1–2)	*Fissidens adianthoides*	I (2)
Ranunculus acris	I (2–3)	*Ctenidium molluscum*	I (3)
Ptilium crista-castrensis	I (2–4)	*Carex flacca*	I (2)
Succisa pratensis	I (1–4)	*Scapania undulata*	I (2–3)
Philonotis fontana	I (1–3)	*Cratoneuron commutatum*	I (2–3)
Peltigera canina	I (1–4)	*Anemone nemorosa*	I (2–4)
Carex binervis	I (1–4)	*Barbilophozia barbata*	I (2–3)
Filipendula ulmaria	I (2–4)	*Luzula multiflora*	I (2)
Racomitrium lanuginosum	I (2–4)		
Saxifraga aizoides	I (2–3)	Number of samples	19
Drepanocladus revolvens	I (2)	Number of species/sample	40 (29–58)
Plagiothecium denticulatum	I (2)		
Hieracium sp.	I (1–2)	Vegetation height (cm)	58 (15–90)
Cladonia squamosa	I (2–3)	Vegetation cover (%)	94 (70–100)
Thelypteris phegopteris	I (2)		
Caltha palustris	I (3–4)	Altitude (m)	787 (630–914)
		Slope (°)	40 (10–90)

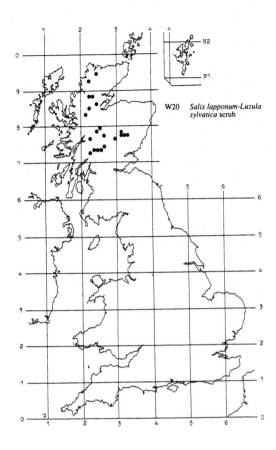

W20   *Salix lapponum-Luzula
        sylvatica* scrub

W21
Crataegus monogyna-Hedera helix scrub

Synonymy

Scrub and Chalk scrub *auct. angl.*; Scrub associations
Tansley 1911; Progressive scrub Moss 1913; Retrogressive scrub Moss 1913; Thicket scrub Salisbury
1918*b*; Woodland scrub Salisbury 1918*b*; *Fruticetum*
Tansley 1939 *p.p.*

Constant species

Crataegus monogyna, Rubus fruticosus agg., *Hedera
helix.*

Rare species

*Himantoglossum hircinum, Orchis militaris, O. purpurea,
O. simia, Silene nutans, Seseli libanotis.*

Physiognomy

The *Crataegus monogyna-Hedera helix* scrub is a compendious community which includes most of the seral
thorn scrub and many hedges in the British Isles. The
vegetation is always dominated by various mixtures of
smaller trees and shrubs, undershrubs and woody
climbers and sprawlers but physiognomically it is quite
diverse and sometimes difficult to separate from more
open herbaceous vegetation with scattered woody
plants on the one hand and woodland on the other.
Typically, however, the woody cover of the community
as defined here is dense, often closed or almost so, such
that half-and-half mixtures of grassland and scrub
would be considered as mosaics of this kind of vegetation and others. The canopy can, however, be quite
low, sometimes little more than a metre high and only
rarely more than 5 m. But, although stands can be very
uneven-topped, where once discrete groups of shrubs,
trees and undershrubs are coalescing, the woody cover is
characteristically unstratified. Saplings of some species
of taller trees are common in the community and
occasional specimens may protrude a little but they
never form an overtopping canopy.

In floristic terms, the woody component of this vegetation is quite varied, being influenced not only by
edaphic differences but also by the availability of seed-
parents and the vagaries of dispersal and establishment;
and, once the canopy has begun to close, there is some
competitive interplay between certain of the species.
However, a strong common element is provided by
various spinose plants. *Crataegus monogyna* is the most
frequent of these overall and it is often the most abundant tree: it is usually among the first invaders of the
various kinds of neglected herbaceous vegetation from
which the community often develops and, except on
shallower soils (usually rendzinas here), it can be very
abundant from the start. It is also the most widely
planted species of hedges, and in younger stretches, it
may be the sole dominant. *C. laevigata*, by contrast, is
very rare, though it may appear in stands developing
within or close to long-established woodlands in southern Britain (Ross in Tansley 1939, Bradshaw 1953). The
other common thorny tree here is *Prunus spinosa*: it is
somewhat less frequent overall than *C. monogyna* but it
is often prominent in situations where established
bushes can sucker in to developing scrub and it probably
has an edge over hawthorn on heavier, moister soils. It is
also more resistant to salt-spray than *C. monogyna* and
tends to replace it as the common dominant on exposed
sea-cliffs, although much coastal blackthorn scrub is
better placed in the *Prunus-Rubus* community. Feral *Prunus domestica* (often recognisably ssp. *insititia*) is sometimes found in hedges or other stands of the *Crataegus-
Hedera* scrub, not always close to settlements.

After *C. monogyna*, the commonest member of the
community is *Rubus fruticosus* agg. which is often very
abundant here on all but the driest soils, forming clumps
of decumbent, arching or erect shoots which can fuse
into an impenetrable tangle among the taller trees and
sprawl over their lower branches. Where scrub is developing from established woodland or actively spreading,
the Rubi often occur as an advancing front over which
other trees can leapfrog to colonise the open ground
beyond, so that a complex patchwork of young trees and
bramble patches develops (e.g. Tansley 1922). Thick

Rubus is densely shading (many taxa are also partly evergreen) and may hinder the establishment of tree seedlings, dense clumps persisting as enclaves among the taller areas of scrub; though, where the canopy of existing trees closes over, the brambles are themselves shaded out to a sparse cover or totally eliminated. Where the scrub has stabilised, *Rubus* may remain dominant in what is essentially a distinct marginal band of the *Rubus-Holcus* underscrub, as along uncut hedgebanks. On drier soils where the community develops, especially more calcareous ones, the procumbent *Rubus caesius* can be found but *R. idaeus* is rather rare here.

Roses, too, can be a prominent feature of the *Crataegus-Hedera* scrub, readily colonising even quite rank herbaceous vegetation. *Rosa canina* agg. is the commonest taxon (*R. dumalis* has been distinguished in some more northerly stands) and it can form large irregular bushes with arching stems up to 3 m tall. Less frequent among the shrubby roses, but rather characteristic of early stages in the invasion of chalky soils, are *R. rubiginosa* and, much more local, *R. agrestis*. Another Rubiginosae rose, *R. micrantha*, has a less marked preference for calcareous soils. Of different habit is *R. arvensis*, which occurs occasionally throughout, as a decumbent bush or, much more conspicuously, as a sprawler, its long shoots reaching several metres in height in the crowns of shrubs and trees. *R. pimpinellifolia* occurs locally, though sometimes in considerable abundance, over sandy soils and limestones.

Few other small trees or shrubs are found with any frequency throughout the *Crataegus-Hedera* scrub. *Corylus avellana*, for example, is rather uncommon. It prefers deeper and moister profiles, so does best in situations where there are pockets of soil among barer rocky ground, where it can be locally abundant, as over limestone talus admixed with soil (e.g. Merton 1970), on pavements with downwash into grikes, or over spoil and made slopes, as on railway embankments (e.g. Sargent 1984). It can also figure prominently in stands developed from degenerating woodland (as in Moss's (1913) 'retrogressive scrub'), clear-felled and abandoned woodland or mis-managed coppice. *Ilex aquifolium* is even scarcer: it can invade closed ungrazed swards, but is commoner on more base-poor soils than are usual here and most of the scrubby vegetation in which it is prominent (like the New Forest holms: Peterken & Tubbs 1965) is best considered as immature stands of less calcicolous oak-birch or beech woodland. The peculiar holly scrub of Dungeness is again developed in association with heath (Scott 1965, Peterken & Lloyd 1967). However, *Ilex* can be locally prominent here and is especially abundant in some areas in hedgerow stands, where there are sometimes suspicions that it has been planted (Peterken & Lloyd 1967). One further infrequent associate is *Salix cinerea* which is occasionally encountered on heavier, moister soils.

Some other species are common but very much confined to particular kinds of *Crataegus-Hedera* scrub. On more naturally eutrophic or disturbed and enriched soils, *Sambucus nigra* is very frequent and locally abundant. It is characteristic here of two sub-communities on mesotrophic mull profiles and on made or fragmentary soils in industrial landscapes and derelict land (in the latter situations sometimes being accompanied by prominent bushes of the garden-escape *Buddleja davidii*); but it also marks out sites of local enrichment on more impoverished calcareous soils, being especially associated with rabbit warrens on the Chalk. Much more distinctive of the more calcicolous kinds of scrub included here is a group of species of which *Viburnum lantana*, *Cornus sanguinea* and *Ligustrum vulgare* are the strongest preferentials with *Rhamnus catharticus* and *Euonymus europaeus* less markedly confined and *Juniperus communis* ssp. *communis* as a locally abundant member.

This type of vegetation is also richer in climbers than are other kinds of *Crataegus-Hedera* scrub with *Tamus communis* and *Clematis vitalba* preferentially frequent. *Lonicera periclymenum* occurs more evenly throughout the community and *Hedera helix* is sometimes seen among the crowns of the trees, though it is more characteristic as a ground-growing plant here. Towards the south-west, *Rubia peregrina* can be found.

The other common group of woody plants in the *Crataegus-Hedera* scrub comprises saplings of certain woodland canopy trees. Frequently, such species invade together with or even in advance of the smaller trees and shrubs, coexisting for some time in a single irregular stratum before they overtop them. The commonest overall is *Fraxinus excelsior* with *Acer pseudoplatanus* more occasional and more obviously confined to moister profiles. Also preferential to heavier soils is *Quercus robur*, much better able to colonise unwooded ground than it is to regenerate under more closed canopies and, with its large food supply in the seed, able to grow up through quite rank swards (Jones 1959). *Acer campestre* is also a little more common on moister soils and to the west and north *Ulmus glabra*. On more eutrophic moist soils, suckering elms of the *procera* and *carpinifolia* sections can also be very prominent here, often spreading from planted trees around settlements and in hedgerows over neglected farmland and abandoned gardens, their dense clones showing the characteristic rounded or semi-circular plan (e.g. Rackham 1975, 1980, Peterken 1981). Saplings of *Fagus sylvatica* can also be found, but they tend to occur only in fairly close proximity to seed-parents and their abundance is very much controlled by the sporadic pattern of masting. Where good mast years have coincided with agricultural neglect near to beech woodland (as in the situations recorded by Watt 1924,

1934*a, b*), *Fagus* can figure prominently in this community on both deeper and moister (though not strongly-gleyed) profiles and dry soils. Other trees occasionally found on free-draining, more calcareous soils are young *Taxus baccata* and *Sorbus aria*.

The field layer beneath the more or less continuous woody covers included in the *Crataegus-Hedera* scrub is typically species-poor. The most characteristic feature throughout is a ground carpet of *Hedera helix* which is often very abundant here, frequently accounting for the bulk of the cover and a good indicator of the uninterrupted canopy expansion and slowly-deepening shade that distinguishes many stands. Also typical, though clearly preferential to more eutrophic and disturbed soils (that is, those situations where *Sambucus* tends to occur) are *Urtica dioica* and *Galium aparine* though, throughout, these species are only abundant where the canopy is thinner, becoming more confined to enclaves and margins as the shrub and tree cover closes. On lighter, more calcareous and impoverished soils, *Urtica* and *Galium* continue to be represented sporadically (often around areas of local enrichment like rabbit burrows) but a more frequent indicator of disturbance is *Brachypodium sylvaticum*. This grass is a characteristic plant of scrub that has developed over ploughed and abandoned sites over limestones (commonplace, for example, after wartime ploughing campaigns on marginal land) but is also widely found where the community has sprung up on disturbed, base-rich soils over limestone screes and spoil. It can also mark out areas of scrub which have developed by run-down of a pre-existing woodland cover, though it is by no means an infallible indicator of this kind of origin.

Two other groups of herbs characterise the community. The first comprises species typical of more calcicolous Carpinion and Fagion woodlands and these are best represented in more long-established stands well on their way to maturity and in scrub that has developed adjacent to existing woodland of this kind or by its degeneration. *Mercurialis perennis* is the commonest and most distinctive of this suite and, like *Hyacinthoides non-scripta* and *Anemone nemorosa*, which are much less frequent and more obviously preferential to moister soils over which the community occurs, it is slow to appear in or spread into newly-developing stands. Other members of this group include *Arum maculatum*, *Geranium robertianum*, *Phyllitis scolopendrium*, *Stachys sylvatica* and *Circaea lutetiana* with, on moister soils, *Poa trivialis* and *Glechoma hederacea* and, towards the south, *Iris foetidissima*, all of which are encountered occasionally. Plants of less base-rich woodlands are generally more scarce here, though *Silene dioica* and *Stellaria holostea* occur in some stands. *Holcus mollis* is very patchy and *Pteridium aquilinum* infrequent and usually sparse, though it can show local abundance

among more open patches with brambles. On strongly-gleyed soils, *Juncus effusus* and *Deschampsia cespitosa* can be found.

The other group of species occurring here consists of plants which have either survived from the preceding herbaceous vegetation or which have appeared among degenerating scrub, usually the former. By and large, they are shade-intolerant and thus very much confined to more open enclaves and margins, so their frequency among these generally closed canopies is low. Nonetheless, such species can give this kind of scrub considerable floristic diversity and where some grazing or mowing has stabilised the vegetation, they may become a more or less permanent feature among or fronting the trees and shrubs. Grassland assemblages are most frequently encountered with, on less calcareous soils, Arrhenatherion plants such as *Arrhenatherum elatius*, *Holcus lanatus* and *Heracleum sphondylium* and, on more calcareous ones, Mesobromion species like *Bromus erectus*, *Brachypodium pinnatum* and *Sanguisorba minor*. Where the community occurs on more disturbed ground, weedy plants such as *Lamium purpureum*, *Stellaria media*, *Cirsium arvense*, *Arctium minus* agg. and *Bromus sterilis* can occur or, on more calcareous soils, *Erigeron acer*, *Inula conyza*, *Hypericum perforatum* and *H. montanum*. Some of the orchids which can be found in rather more open stands of the community, sometimes in local abundance, may also benefit from the ground disturbance with which the development of this scrub is often associated or from the structural changes that invasion of trees and shrubs involves. Among the species encountered here are *Ophrys insectifera*, *Plantanthera bifolia*, *P. chlorantha* and the rarer *Himantoglossum hircinum* (Good 1936), *Orchis simia*, *O. purpurea* and *O. militaris* (Farrell 1985).

Bryophytes are generally sparse in this vegetation and few species occur more than occasionally. *Eurhynchium praelongum* is the commonest with *Brachythecium rutabulum*, *Plagiomnium undulatum* and *Eurhynchium confertum*.

Sub-communities

Hedera helix-Urtica dioica **sub-community.** These are species-poor scrubs with a dense canopy that is most often dominated by *Crataegus* or various mixtures of hawthorn, *Prunus* and *Sambucus* with patches and/or a marginal fringe of *Rubus* and *Rosa canina* agg. *Corylus* and saplings of *Fraxinus* occur occasionally but other shrubs and trees are generally scarce, though to the west and north, young *Acer pseudoplatanus* can be locally prominent. Other stands, especially on the heavy clays of the Midlands and East Anglia, are dominated by young suckers of *procera* or *carpinifolia* elms, both in hedges, which are common in this sub-community, and in field scrubs produced by spread from existing trees. On wastelands, *Buddleja* sometimes figures prominently and other garden-escapes like *Lupinus arboreus* and *Laburnum anagyroides* can be locally abundant.

Under the typically densely-shading canopy, the field layer is usually impoverished and sometimes very sparse, though *Hedera* can be extensive as a ground carpet. The only other common plants are *Urtica* and *Galium aparine* and even these may be reduced to sparse and puny individuals beneath thicker covers. *Poa trivialis* and *Glechoma hederacea* sometimes occur on moister soils or, in drier situations, *Brachypodium sylvaticum* but these are generally very infrequent. Somewhat more common and slightly preferential to this sub-community are *Silene dioica* and species of rank and more weedy Arrhenatherion swards, especially *Arrhenatherum* itself, *Heracleum sphondylium, Holcus lanatus* and also *Elymus repens, Calystegia sepium, Bromus sterilis, Cirsium arvense* and *Arctium minus* agg. *Pteridium* also occurs very occasionally, sometimes in abundance. Often, mixtures of these species occur patchily throughout the scrub and may represent survivors of herbaceous vegetation that is being gradually eclipsed but, along narrow verges, they often form a permanent fronting fringe to hedgerow scrub, being stabilised by mowing.

***Mercurialis perennis* sub-community.** In both canopy and field layer, this sub-community is a little richer than the last, having a composition that approximates to that of young calcicolous Carpinion woodland, particularly that of moister, eutrophic soils. *Crataegus, Prunus* and *Sambucus* are still very common here and mixtures of these species often dominate; *Rubus* and *Rosa* spp. (including rather more *R. arvensis* here) are also usually prominent within and around stands. But with some *Corylus*, frequent saplings of *Fraxinus* and *Acer pseudoplatanus*, occasional preferential occurrences of young *Quercus robur* and *Acer campestre* and sparse records for *Euonymus* and *Cornus*, the floristics of the woody cover are often qualitatively indistinguishable from the *Fraxinus-Acer-Mercurialis* woodland. Quantitatively, however, the balance of the various components is different and structurally the vegetation is dominated by smaller woody plants, having a canopy usually around 5 m high with occasional taller emergent trees. In some stands fairly close to mature *Fagus*, beech saplings may be locally abundant and other tracts are dominated by young elm suckers.

Such stands may be very dense but usually here the canopy is a little more open and the presence of some less heavily shading trees makes for greater lighter penetration. This, together with the probably greater age of these scrubs, means that a richer and more extensive field layer can develop. *Hedera* is still often extensive as a ground carpet and patches of *Urtica* and *Galium aparine* are very

common but *Mercurialis* is now much increased in frequency, though it usually occurs as discrete clonal patches rather than as a continuous sheet. Also preferential here are *Arum maculatum, Poa trivialis, Glechoma hederacea, Hyacinthoides non-scripta, Allium ursinum* and, on spring-waterlogged soils, *Anemone nemorosa*.

Bryophytes are more frequent here than in the last sub-community, though their cover is typically low with *Eurhynchium praelongum* and *Brachythecium rutabulum* forming sparse wefts.

***Brachypodium sylvaticum* sub-community.** The composition of the woody cover here is very much as in the *Hedera-Urtica* sub-community except that *Sambucus* is much reduced in frequency and there are occasional records for more calcicolous species like *Ligustrum, Cornus, Rhamnus* and *Viburnum lantana*. *Juniperus* and young *Taxus* can also be found in some stands but none of these species is as common as in the *Viburnum* sub-community. *Fagus* saplings can be locally prominent and some stands are dominated by elm suckers.

In the field layer, *Hedera* remains frequent and often abundant but *Urtica* and *Galium aparine* are rather uncommon and usually of low cover. Much more obvious here is *Brachypodium sylvaticum* with occasional *Fragaria vesca* and *Viola riviniana*. Other species can be quite numerous, though no other plants are preferential: rather, there are scattered individuals of a wide variety of species occurring throughout the community or more frequently in other sub-communities, including *Silene dioica, Holcus mollis, H. lanatus, Cirsium arvense, Pteridium aquilinum* (on somewhat less base-rich soils), *Mercurialis perennis, Arum maculatum, Geranium robertianum* and *Phyllitis scolopendrium* (on rather more base-rich soils). Bryophytes are poorly represented with only very occasional *Eurhynchium praelongum, Brachythecium rutabulum* and *Plagiomnium undulatum*.

***Viburnum lantana* sub-community:** Chalk scrub association Tansley & Rankin 1911: Limestone scrub Moss 1911; Progressive scrub Moss 1913 *p.p.*; Retrogressive scrub Moss 1913; Chalk scrub Tansley 1925, Smith 1980; Juniper scrub Watt 1934a; Hawthorn scrub Watt 1934a *p.p.*; *Clematito-Prunetum* Shimwell 1968a; *Crataegus-Rosa pimpinellifolia* community Shimwell 1968a; *Geranio-Coryletum* Shimwell 1968a; Southern mixed shrub communities Duffey *et al.* 1974. In terms of its woody component, this is by far the most distinctive kind of *Crataegus-Hedera* scrub. In the first place, various of the general community species show a reduction in frequency or abundance here, tending to be better represented on deeper soils which are in the minority. *Prunus spinosa*, for example, is less common than usual and *Crataegus* and *Rubus fruticosus* agg., though still frequent, often have lower cover than elsewhere. *Corylus*, too, favours deeper soils though it performs well on rubbly slopes with admixed talus and soil. *Sambucus* is local and distinctly associated with disturbed and enriched sites but, since these can be widespread, it figures fairly frequently. But, more obviously, various other species are strongly preferential to this sub-community, *Viburnum lantana*, *Cornus* and *Ligustrum* becoming constant and *Juniperus* occurring more frequently than elsewhere. Roses are also often conspicuous with *R. rubiginosa*, *R. micrantha* and, more locally, *R. agrestis* and *R. pimpinellifolia* being recorded along with *R. canina* agg. and *R. arvensis*. *Rubus caesius* can occur in the early stages of colonisation. Among the woodland canopy trees represented by saplings, there are also some distinctive features. Young *Fraxinus* and, on deeper soils to the west and north, *Acer pseudoplatanus* are common and there can be some *Fagus sylvatica* near to established seed-parents, but more frequent than usual are *Taxus* and *Sorbus aria*. *Quercus robur* saplings, on the other hand, are very scarce. The other prominent component comprises climbers. *Hedera* and *Lonicera* are occasionally seen among the tree and shrub canopies but much more characteristic here are *Tamus* and *Clematis*.

When rich mixtures of these species are present, this kind of scrub can present a splendid sight, especially when the deciduous species get their autumn colours and the plants are in full fruit. As always, there is considerable physiognomic variation between stands, with some more open and others almost closed, though, in line with the general definition of the community, less dense woody covers should be considered as scrub/grassland mosaics. There are also differences in the canopy composition in relation to regional and local climate and soils. Many of the preferentials, for example, have a distinctly Continental distribution in Britain, being best represented on the Chalk of the south and east and becoming increasingly scarce in moving to the north-western limit of the *Crataegus-Hedera* scrub (e.g. Pigott & Huntley 1978). And, among these, *Cornus* seems especially well adapted to the colonisation of more impoverished soils by rapid and prolific seed germination (Lloyd & Pigott 1967), after which it can sucker profusely and become locally dominant. Then, there is the tendency for *Juniperus* (and sapling *Taxus*) to be associated with shallower, drier soils on steeper, more exposed slopes, a preference which, in contrast to the quantitative prominence of *Crataegus* on deeper, moister soils on gentler, sheltered slopes here, led Watt (1934*a*) to propose his classic two-sere view of scrub and beechwood development on the Chilterns. *Crataegus-Hedera* scrubs with *Juniperus* have not been widely sampled but available data suggest that both types of Watt's scrubs can be accommodated in this single sub-community.

The field layer beneath denser woody covers here is not especially distinctive. *Hedera* and *Brachypodium sylvaticum* are the commonest species and the occasional presence of plants such as *Mercurialis perennis*, *Arum maculatum* and *Geranium robertianum* gives the vegetation something of the appearance of the *Mercurialis* sub-community, though associates of moister soils like *Poa trivialis* and *Glechoma hederacea* are very scarce. In more open places, there can be greater enrichment. On shallower, rocky soils, *Teucrium scorodonia* and *Origanum vulgare* are very characteristic of this kind of scrub and there can be some *Hypericum perforatum*, *H. montanum* and more ephemeral species such as *Erigeron acer* and *Inula conyza*. In places with local soil enrichment, as where this scrub is colonising around rabbit burrows, *Urtica* and *Galium aparine* can be prominent together with plants like *Verbascum thapsus*, *Atropa belladonna* and *Solanum dulcamara* among an abundance of *Sambucus*. Then, where invasion is occurring over more intact soil covers, there may be survivors of the preceding herbaceous vegetation, an often extremely rich calcicolous grassland element in the more open mosaics characteristic of the early stages of scrub development but, with increasing canopy closure, generally limited to ranker grasses like *Bromus erectus* or *Brachypodium pinnatum* with dicotyledons such as *Sanguisorba minor* and *Helianthemum nummularium* which are able to grow up through the tall sward. Some of the rarer orchids noted above are particularly associated with these kinds of transition: some of the stations of *Orchis simia*, *O. purpurea* and *O. militaris* (Farrell 1985) occur in more open stands of this sub-community and much of the spread of *Himantoglossum hircinum* early in this century has been around this scrub (Good 1936). Another rarity which seems to be closely related to rank grassy scrubs of this type is *Seseli libanotis* (Dony 1953).

Also best considered here is the distinctive kind of vegetation described from Derbyshire by Moss (1913) as 'retrogressive *Corylus* scrub' and defined by Shimwell (1968*a*) as the *Geranio-Coryletum*. This open scrub would, strictly speaking, be considered in this scheme as a complex mosaic of the *Viburnum* sub-community with the local calcicolous grassland, but it has some other peculiar floristic features. Although the woody cover is generally orthodox, *V. lantana* is already scarce this far north and *Corylus* is unusually abundant. Furthermore, although *Brachypodium sylvaticum*, *Teucrium* and *Origanum* are frequent, together with some common woodland herbs like *Mercurialis perennis*, *Viola riviniana* and *Melica uniflora*, there is a very striking mixture of species with strong northern affinities, such as *M. nutans*, *Trollius europaeus* and *Rubus saxatilis* with others like *Geranium sanguineum*, *Convallaria majalis*, *Silene nutans* and *Aquilegia vulgaris* that seems to indicate rather specialised environmental and historical relationships.

Bryophytes are usually not numerous or abundant in more closed stands of the *Viburnum* sub-community but the usual *Eurhynchium praelongum* and *Brachythecium rutabulum* are sometimes accompanied by *Ctenidium molluscum*, *Thuidium tamariscinum* or *Fissidens cristatus*.

Habitat

The *Crataegus-Hedera* scrub is the typical sub-climax woody community of circumneutral to base-rich soils throughout the British lowlands. It usually develops by the invasion of neglected bare ground or untreated herbaceous vegetation or where woodland has been degraded and its floristics and physiognomy reflect its transitional and unstable character as well as being related to edaphic and climatic variation. Hedgerow stands are often of planted origin but, even where the establishment of a woody cover is more natural, human influence affects colonisation through the previous treatment of the ground and the availability of seed-parents; and re-imposition of grazing, mowing or burning can halt or reverse the successional process at any stage.

The most important factor governing the development of the community in the intensively-used British landscape is the disruption of the stability of the existing ground cover. Sometimes, this happens naturally and exposes fresh bare surfaces, as where the woody species of this scrub have invaded directly or early in the colonisation of landslips in softer deposits (splendidly seen on the Axmouth–Lyme Regis Undercliffs: Ratcliffe 1977) or of fresh talus and rock falls beneath harder cliffs. But, very often here, such disruption is to some extent artificial. The *Crataegus-Hedera* scrub has developed widely, for example, on many kinds of neglected made ground (on derelict land, over spoilheaps and on verges and embankments: e.g. Sargent 1984) and on cultivated land that has been abandoned (in gardens, allotments and arable fields: e.g. Brenchley & Adam 1915, Salisbury 1918*b*, Tansley 1939, Lloyd & Pigott 1967). It is also extremely common where grasslands, previously maintained as plagioclimax vegetation, have been subject to a relaxation of grazing and/or mowing, with stands widely distributed over neglected pastures, meadows, verges, commons and graveyards (e.g. Watt 1924, 1934*a*, Tansley 1939) and on land that has experienced some decline in natural herbivore populations, notably rabbits following myxomatosis (e.g. Thomas 1960, 1963, Wells 1969).

The *Crataegus-Hedera* scrub includes most of the more well-established woody vegetation that develops in such varied situations as these except where the soils are either markedly acid or strongly waterlogged. In the former case, it is generally replaced by the *Ulex-Rubus* scrub or various kinds of ericoid heath which are often colonised by birch (and, seeding in from plantations,

pine); on wetter ground, by young woodlands with *Salix cinerea*, *Betula pubescens* and *Alnus*. Apart from brambles, none of these species is common here, though pine can figure locally, even among *Juniperus* (Ward 1973) if there is an abundant seed source nearby. But, within these very broad limits, the soils here are very variable, ranging from man-made raw soils on rock waste and demolished buildings, through man-made soils on restored ground, new verges and hedgebanks, to more natural profiles of different degrees of maturity. In terms of base-status, the soils run from quite base-poor brown earths to very base-rich and calcareous rendzinas; and the drainage of the profiles varies from impeded (surface-water gleys being especially important) to excessive; some soils are highly eutrophic, others are very poor in nutrients.

Some of the woody colonisers characteristic of the *Crataegus-Hedera* scrub are able to invade and establish over virtually the whole spectrum of these soils and it is these more catholic species that provide the core of the floristic definition of the community. *Crataegus*, *Prunus*, *Rubus fruticosus* agg. and *Rosa canina* agg. are well represented throughout and are often among the first colonisers, fading in quantitative importance only on the most shallow and dry profiles. Among the woodland trees that invade early, both *Fraxinus* and *Acer pseudoplatanus* are also relatively undemanding as far as soil conditions are concerned.

Below this general level, there are some clear edaphic preferences among other woody species which help to define the different kinds of *Crataegus-Hedera* scrub. *Sambucus*, for example, is distinctly associated with more eutrophic soils, though the source of the enrichment is very varied. Some of the profiles here are naturally richer than others and the preferential frequency of *Sambucus* in the *Hedera-Urtica* and *Mercurialis* sub-communities is partly a reflection of the fact that many soils under these kinds of scrub are mulls, rather than more oligotrophic rendzinas which predominate in the *Brachypodium* and *Viburnum* sub-communities. The former sub-communities are also common on abandoned arable land which has received some additions of fertiliser in the past and in a wide variety of situations where there has been or continues to be some kind of ground disturbance. In the *Brachypodium* and particularly the *Viburnum* sub-community, *Sambucus* often marks areas of more local enrichment, as around rabbit burrows, where its success is enhanced because of the unpalatable nature of its bark. Suckering elms of the *procera* and *carpinifolia* groups also tend to be commoner on the more eutrophic soils of the *Hedera-Urtica* and *Mercurialis* sub-communities, though it is not clear whether they are responding to or helping produce enrichment (Martin & Pigott 1975, Rackham 1975, 1980).

Among the more eutrophic kinds of *Crataegus-*

Hedera scrub, there are further differences among the woody species related to soil moisture. *Quercus robur* and, especially in more southerly stands, *Acer campestre* and suckering elms are all preferentially frequent invaders in the scrub of the *Mercurialis* sub-community which is particularly characteristic of ill-draining clays and shales with stagnogleys and pelosols in abandoned arable land and neglected pastures in the Midlands and East Anglia (e.g. Brenchley & Adam 1915, Ross in Tansley 1939).

Generally, the soils of the other types of *Crataegus-Hedera* scrub are more free-draining and, among these, the major edaphic influence is related to base-status. The *Hedera-Urtica* sub-community is typical of less base-rich profiles, many of them fragmentary soils over non-calcareous waste or made soils on verges and hedge-banks, others more natural brown earths on neglected agricultural land. By contrast, the soils of the *Brachypodium* and particularly the *Viburnum* sub-community are characteristically base-rich and calcareous, in the latter type of *Crataegus-Hedera* scrub often classic rendzinas where extreme base-richness is combined with excessive drainage and impoverishment. Much of the very distinctive richness of the canopy of the *Viburnum* sub-community comes from the preferential frequency of more calcicolous species: *Viburnum lantana*, *Cornus*, *Ligustrum* and, to a lesser extent, *Rhamnus* and *Euonymus* are all best represented here, together with a number of less common roses, like *R. rubiginosa*, *R. agrestis* and *R. pimpinellifolia*, and *Clematis*. Most of the southern British localities of *Juniperus* are also on the more shallow and calcareous of these rendzinas (Watt 1934a), though the differential survival of this shrub on more marginal land may give us only a partial view of its edaphic preferences in this part of the country (Ward 1973). Among invading woodland trees, the more calcicolous *Taxus* and *Sorbus aria* are preferential here and, where seed-parents are near, *Fagus* can figure prominently, out-performing *Q. robur* on these shallower soils and only giving way to *Taxus* in more extreme conditions (Watt 1926, 1934a).

The *Viburnum* sub-community is by no means confined to rendzinas over Chalk: it occurs on calcareous, lithomorphic soils over other limestone bedrocks and waste as far north as the southern Lake District. But it is best represented to the south and east, where Chalk is the predominant limestone, because many of its distinctive calcicolous woody species happen to have strongly Continental affinities, becoming increasingly scarce towards the north and west where summers are cooler and shorter. No single isotherm provides a precise limit for this kind of *Crataegus-Hedera* scrub and, though a mean annual maximum temperature of 26 °C (Conolly & Dahl 1970) marks a crude final boundary for the sub-community as a whole, some species (*V. lantana* itself, for example) have a distribution more strictly confined to the warmest parts of the country. Others, like *Juniperus*, occur widely further north (though on different kinds of soils) and yet are rather local in their distribution to the south. Among the colonising woodland trees, *Taxus* is likewise local and *Fagus*, though of Continental range through Europe, is of restricted natural occurrence in Britain, though it often colonises in this scrub from planted stock far to the north of its limit.

A number of factors are of very considerable importance in limiting the ability of the woody species of the community to colonise the various climatic and edaphic situations they prefer, and this means that stands, especially younger ones, can show great diversity, even within individual sub-communities. The physical character of the existing ground cover is one: whether it is bare (and then, whether it is stable or not) or already occupied by herbaceous vegetation (and then, how rank this is). *Fraxinus*, for example, is able to get a hold on quite mobile talus, *Corylus* favouring more stable and finer material admixed with soil, *Acer pseudoplatanus* faring especially well on grossly-disturbed and abandoned sites like old spoil heaps. In Cressbrook Dale in Derbyshire, Scurfield (1959) and Merton (1970) demonstrated how much of the scrub and young woodland springing up on barer ground (and belonging mostly to the *Brachypodium* and *Viburnum* sub-communities) was dominated by variation in the proportions of these species according to their substrate preferences. One other species which can rapidly colonise open, shallow and very impoverished limestone soils here is *Cornus*: it has the advantage of a berry containing two seeds which germinate very quickly provided there has been a cold pre-treatment and it can become very prominent in stands of the *Brachypodium* and especially the *Viburnum* sub-community which have grown up on marginal Chalkland, ploughed once and then abandoned (Lloyd & Pigott 1967: see also Tansley 1922). On enriched bare soils, where the *Hedera-Urtica* sub-community often develops, *Sambucus* (and the garden-escape *Buddleja*) often get an early hold, sometimes in more unorthodox situations, as where such scrub develops up the walls of half-demolished buildings.

Other species are able to colonise closed swards, even quite rank ones. *Crataegus*, *Prunus*, *Rubus fruticosus* agg. and *Rosa canina* agg. are all plants which favour somewhat deeper and moister soils than are characteristic of the above situations, but they can also grow through a herbaceous cover and commonly come to dominate in all kinds of *Crataegus-Hedera* scrub derived from grasslands where grazing or mowing have been relaxed. On heavier, moister soils, *Q. robur* is also a characteristically early invader, well adapted, with its big seeds that benefit from a cover of litter, to growing through ranker herbage (Jones 1959). *Procera* and *carpinifolia* elms, as well as *Prunus* and in some areas *Populus tremula*, can circumvent competition from taller plants by suckering under the vegetation, and these species may dominate locally where established trees have spread into overgrown woodland rides or fields adjacent to hedges.

A second factor influencing colonisation is the supply of seed of the woody species, more particularly the location of seed-parents and dispersal from them. Some plants here are extremely widely distributed, notably *Crataegus* which has been universally planted in lowland hedges, and *Rubus fruticosus* agg. which is so catholic as to figure, as one taxon or another, in a very wide variety of vegetation types; this undoubtedly gives these species some advantage in the invasion of newly-available ground. Others benefit from having been planted nearby too, like *Fagus*, which can seed into this kind of scrub well beyond the area where it is a natural forest dominant. On the other hand, there are species like *Juniperus* which are rather localised and which may have declined to such critically low levels in some areas as to be able to re-expand only with difficulty, even were conditions for establishment to become suitable (Ward 1973).

As far as dispersal is concerned, it is noteworthy that many of the most successful species of the *Crataegus-Hedera* scrub, both generally and among the preferentials of more calcareous soils, have brightly-coloured and fleshy fruits attractive to birds, among which larger members of the thrush family seem to be particularly important as dispersal agents, flocking in large numbers in winter and often totally depleting the crop (Fuller 1982). These species show diet preferences, though this varies with availability (Hartley 1954) and no systematic work seems to have been done on the effectiveness of these birds in dispersing particular trees or shrubs. Heavier fruits like acorns and hazel nuts are often dispersed by small mammals, and the former by larger birds, especially jays, sometimes over considerable distances: acorns, for example, can be carried up to 200 m from the parent tree (Mellanby 1968) and this can give *Q. robur* some advantage in this community over *Fagus*, whose fruits often just drop from the tree (Watt 1923, 1925). Other important invaders such as *Fraxinus* and *Acer pseudoplatanus* have fruits which can be wind-dispersed and prevailing wind direction may be a controlling factor in their prominence. In considering the invasion of pastures where grazing has been relaxed, it should also be remembered that many of the commonest woody species of this scrub are already present in many swards as very small individuals, a few centimetres high and growing very slowly, but able to get away quickly if unchecked.

Once established in such situations, those species which are well armed with thorns or prickles are protected to some extent against renewed herbivore predation and can offer some shelter to new woody invaders that may not be so resistant. Other species benefit by being unpalatable: *Sambucus*, for example, once established around rabbit burrows, is largely immune from barking by these animals and *Juniperus* can play a crucial role in protecting young *Fagus* and *Taxus* which grow up in the *Viburnum* sub-community (Watt 1926, 1934a). Tighter canopies of early invaders can also offer shelter from wind in exposed situations. However, as soon as the cover of trees and shrubs begins to close, the influence of shade begins to outweigh any protective effect for those species more sensitive to reduced light penetration. Rubi and roses in particular tend to become more confined to the margins of all the different kinds of *Crataegus-Hedera* scrub or to enclaves where they established themselves early as a thick, smothering cover that precludes invasion of taller species; and young saplings of many of the trees, notably *Fraxinus* and *Acer pseudoplatanus*, can be overtaken and shaded out. Where *Juniperus* figures in the *Viburnum* sub-community, it too is eventually killed by its crowding neighbours, often by the shade-tolerant *Taxus* which it nurses but which eventually overtops it (Watt 1926).

The increasing canopy shade, and probably also intense root-competition, are the major factors responsible for the characteristically impoverished field layer of the *Crataegus-Hedera* scrub. As defined here, the community includes only denser stands of scrub within which there is usually but scanty survival of any pre-existing herbaceous vegetation: such elements can be a continuing source of enrichment in younger, more open mosaics or stabilised zonations (see below) but, with the advance of scrub development, they become progressively confined to margins and any remaining uncolonised areas. Those plants which do persist vary considerably, particularly in more disturbed situations where there can be many low-frequency adventives, but the recognisable assemblages described above usually differ according to soil conditions and the particular pattern of treatment and neglect of the vegetation in the period before the scrub developed. Where grasslands have been colonised, the survivors are typically important species of ranker swards (unless grazing or mowing have been re-imposed around the scrub) with Arrhenatherion assemblages on the less base-rich soils, represented here in the *Hedera-Urtica* sub-community, Mesobromion species on the more base-rich, as in the *Viburnum* sub-community. Where waste ground or abandoned arable land has been invaded, Arrhenatherion plants may figure again where there has been time for the establishment of an intervening grassy phase before the woody species have colonised. On more calcareous soils, *Brachypodium sylvaticum* is a more persistent survivor from abandoned ploughland, being more shade-tolerant than most of the other grassland herbs: both the *Brachypodium* and *Viburnum* sub-communities probably include stands which have developed in this kind of situation. Where trees and shrubs colonise more open ground directly, weeds often remain prominent for some time with coarser, eutrophic species on richer soils; on shallow rendzinas that have been ploughed and abandoned, there may be remnants of the distinctive dicotyledon-dominated calcicolous floras characteristic of impoverished, base-rich soils, often preferentially invaded by *Cornus* to produce scrub of the *Viburnum* sub-community (e.g. Lloyd & Pigott 1967).

On the positive side, the single most obvious response to the deepening shade, the great spread in all kinds of *Crataegus-Hedera* scrub of the ground carpet of ivy, marks a floristic convergence in the middle years of the community, as the variety of the preceding herbaceous floras is finally extinguished and before there is any obvious development of a rich woodland element. The prominence of the other more or less common component of the community, *Urtica* and *Galium aparine* may also be partly an indirect response to shade. On more disturbed, enriched soils, these species are often present before the scrub develops but, where closed swards are being invaded, they typically appear with the trees and shrubs, perhaps responding to the appearance of patches of enriched bare ground where other herbs have succumbed to the shade and decayed, releasing a flush of nutrients. In some cases, the use of scrub as bird-roosts may play a part in soil eutrophication.

The *Crataegus-Hedera* scrub can remain in this very dense and dark phase for considerable periods of time with often little more than a sparse carpet of ivy beneath or sometimes total extinction of the field layer. But, where succession progresses, a new phase of enrichment with the appearance of woodland herbs ensues and this community includes such transitional vegetation prior to the development of the distinctive woodland canopy. The initiation of this stage is partly a function of the opening up of the woody cover, as overtopped small trees and shrubs thin out to leave some gaps. Some existing species, like *Hedera*, *Urtica*, *Galium aparine* and *Brachypodium sylvaticum*, can show a renewed expansion, providing some floristic continuity and *Rubus* may spread again as an underscrub, but distinctive woodland species now begin to appear too. However, since some of the most important of these, notably *Mercurialis*, *Hyacinthoides* and, on moister soils, *Anemone*, are very slow to spread, there may be a very considerable time-lag before anything like a full complement of herbs develops, a feature very well seen in the colonisation of Geescroft Wilderness at Rothamsted (Brenchley & Adam 1915, Tansley 1939, Pigott 1977). Although no data on the age of the stands included here were available, it seems likely that the most long-established scrubs are those of the *Mercurialis* sub-community where the elements of a richer kind of Carpinion or

Fagion woodland are well represented. Some stands in the *Hedera-Urtica* sub-community have a field layer similar to that of less calcicolous Carpinion woodland.

Zonation and succession

Zonations between the *Crataegus-Hedera* scrub and other vegetation types are very varied but they usually represent stages in diverse successions between open ground or herbaceous vegetation on the one hand and woodland on the other, sometimes progressive, sometimes stabilised, sometimes moving in reverse and often represented by only part of the full sequence of communities. Surprisingly, in view of the very widespread occurrence of these successions, very few systematic studies of their operation have been undertaken. Potentially valuable early studies (e.g. Salisbury 1918b, Tansley 1922, Tansley & Adamson 1915, Brenchley & Adam 1915, Ross in Tansley 1939, Hope-Simpson 1940b, 1941b) were, for one reason or another, not followed through in detail and what other knowledge we have usually comes from comparison of different stands at different stages of development (e.g. Watt 1926, 1934a, Scurfield 1959, Merton 1970, Duffey et al. 1974) or more generalised observations (e.g. Thomas 1960, 1963). We are therefore much in need of long-term investigations of the progression of even the most ordinary processes here.

Active and direct scrub colonisation of recently- or grossly-disturbed substrates is more common in artificial habitats like urban and industrial wasteland than in naturally unstable situations which are quite rare in the subdued landscape of lowland Britain. On neglected derelict land, the *Crataegus-Hedera* scrub is usually represented by the *Hedera-Urtica* sub-community springing up among mosaics of weed communities or weedy Arrhenatherion swards, together with *Rubus-Holcus* underscrub. Patterning is often rather chaotic and progression to woodland uncommon, partly because of the frequently recent and continuing disturbance and partly because, in more extensive sites, seed-parents of canopy trees can be quite rare. But *Acer pseudoplatanus* sometimes becomes very prominent and less isolated sites may have patchy stands of the *Holcus lanatus* sub-community of the *Quercus-Pteridium-Rubus* woodland, dominated by sycamore with a little ash and *Q. robur*, which have perhaps developed from the scrub.

Similar mixtures of weedy vegetation with scrub of the *Hedera-Urtica* sub-community can characterise abandoned ploughland on less base-rich brown earths. The frequency of this kind of site has varied according to the vagaries of the agricultural economy, fields sometimes falling into neglect when arable cultivation became unprofitable or in periods when incentives for ploughing pushed such cultivation on to marginal land that proved too intractable. Colonisation of such abandoned ground by woody species can follow diverse courses but with generally abundant supplies of seed in hedges and woodlands nearby, it is often more rapid and orthodox than on derelict land and some of the classic descriptions of the development of the *Crataegus-Hedera* scrub relate to such situations (Salisbury 1918*b* Brenchley & Adam 1915, Ross in Tansley 1939). As before, it seems likely that the *Hedera-Urtica* sub-community develops into some kind of *Quercus-Pteridium-Rubus* woodland on more base-poor soils or, where *Fagus* figures prominently among the invading trees, into the *Fagus-Rubus* woodland (e.g. Watt 1924, 1934*b*). On more base-rich but heavy soils, very common over the claylands of the Midlands and East Anglia, succession is more likely to lead to the *Fraxinus-Acer-Mercurialis* woodland with the *Mercurialis* sub-community perhaps supervening between the *Hedera-Urtica* sub-community and richer forms of this forest; or the latter kind of *Crataegus-Hedera* scrub may develop directly to the *Hedera* sub-community of the *Fraxinus-Acer-Mercurialis* woodland, as seems to have happened on Broadbalk Wilderness at Rothamsted (Brenchley & Adam 1915, Tansley 1939) and in the Hayley Triangle (Rackham 1975).

Zonations involving the *Hedera-Urtica* and *Mercurialis* sub-communities are also common on less and more base-rich brown earths in lowland agricultural landscapes where scrub development has been precipitated by a decline in grazing by stock or wild herbivores, notably rabbits, or by neglect of mowing in field corners or more inaccessible parts of meadows. Usually, in such cases, weedy elements figure little among the associated herbaceous vegetation, grasslands being more prominent. Typically, on these often quite enriched soils, these are mesotrophic swards like the *Lolio-Cynosuretum* or the *Centaureo-Cynosuretum* or, more often in direct contact with the scrub, rank *Arrhenatheretum* or, on moister soils, the *Deschampsia-Holcus* grassland. *Rubus-Holcus* underscrub again frequently forms a fringe between the grassland and the scrub. On sea-cliffs, where the *Crataegus-Hedera* scrub can develop in more sheltered situations, these grasslands are replaced by sequences of more maritime swards like the *Festuca-Holcus* or *Festuca-Daucus* grasslands or, where there is some grazing, the *Festuca-Plantago* grassland. In such situations, progressions to woodland are rare, but on neglected inland pastures and meadows the *Crataegus-Hedera* scrub seems to follow the same sequence as in successions on open ground, to *Quercus-Pteridium-Rubus* or *Fagus-Rubus* woodland on less base-rich soils or to *Fraxinus-Acer-Mercurialis* woodland on the more base-rich but moist soils.

On base-rich but more free-draining rendzinas, zonations typically involve the *Brachypodium* or *Viburnum* sub-communities and on freely-weathering limestone

slopes free of heavy grazing, a situation seen especially well on Carboniferous Limestone exposures following myxomatosis, complete sequences from open fern-dominated vegetation, through tussocky *Arrhenatherum* to these kinds of *Crataegus-Hedera* scrub and woodland, can be found (e.g. Merton 1970). Usually, on Carboniferous Limestone, most of which lies beyond the natural limit of *Fagus*, different kinds of *Fraxinus-Acer-Mercurialis* woodland represent the climax forest, with the *Geranium* sub-community predominating on steeper screes that have acquired a woodland cover fairly recently (Pigott 1960, 1969, Merton 1970). Longer-established *Fraxinus-Acer-Mercurialis* woodland, among which open areas of *Viburnum* sub-community scrub can be found on rocky knolls, is very locally of the *Teucrium* sub-community (Moss 1913, Pigott 1960, 1969, Shimwell 1968*a*, *b*).

On the gentler landscapes of the more southerly limestones like the Chalk and Oolite, zonations between the *Brachypodium* and *Viburnum* sub-communities and more open vegetation are often associated with artificially-disturbed sites like spoil heaps and quarry floors (e.g. Tansley 1922, Tansley & Adamson 1925, Hope-Simpson 1940*a*, 1941*a*) and rabbit warrens (Tansley 1939). In such places, these kinds of *Crataegus-Hedera* scrub can be found in mosaics with the calcicolous weedy vegetation of the Atropion alliance of the Epilobietea. On more impoverished and disturbed limestone soils, such as are exposed by ploughing and abandonment of very shallow rendzinas, the *Viburnum* sub-community can be seen developing among open swards of the *Festuca-Thymus-Hieracium* grassland (Lloyd & Pigott 1967).

Where scrub develops at a later stage in successions from open ground over limestones or invades long-established grasslands where grazing has been relaxed, the *Brachypodium* and *Viburnum* sub-communities are typically found in mosaics with closed Mesobromion swards. What these are depends partly on the regional and local climate, partly on the soil conditions and partly on the grazing regime before scrub invasion. By and large, the distribution of these two kinds of *Crataegus-Hedera* scrub coincides with the range of the major lowland plagioclimax calcicolous grassland, the *Festuca-Avenula* sward, and different types of this community usually constitute the starting point for successions; towards the north-west, there is a small overlap with *Sesleria albicans* swards, the *Sesleria-Scabiosa* and *Sesleria-Galium* grasslands; along the oceanic seaboard of the west and south, the *Festuca-Carlina* grassland can show sporadic progression to scrub, though the edaphic and climatic conditions characteristic of this sward are usually too extreme to support any growth of woody species.

Most of the trees and shrubs of the *Brachypodium* and

Viburnum sub-communities can invade (or, if suppressed saplings are already present, get away in) these grasslands once grazing is relaxed but, quite often, a coincidental spread of ranker grasses means that scrub development is occurring among mosaics of these swards and grasslands dominated by *Bromus erectus* and/or *Brachypodium pinnatum* (especially to the south) or *B. sylvaticum* or, on somewhat deeper soils, *Festuca rubra, Avenula pubescens* or *Arrhenatherum elatius*. And, among these latter communities, the advance of scrub may be accompanied by a fringe of *Rubus-Holcus* underscrub, a feature well seen in Tansley's (1922) account of invasion at Downley Bottom on the South Downs. After the demise of rabbits with myxomatosis in the mid-1950s, mixtures of all these vegetation types have become extremely common, particularly on the southern Chalklands (e.g. Thomas 1960, 1963, Wells 1969, Ratcliffe 1977). Although advanced scrub development in this kind of landscape usually results in a great floristic impoverishment, more open mosaics of short swards, rank grasslands and scrub on limestones represent one of the richest complexes of vegetation types that are found in lowland Britain, with a structural variety that can be of importance to the diversity of invertebrate and bird populations (e.g. Duffey *et al.* 1974, Smith 1980, Fuller 1982).

Post-myxomatosis successions are too young for any extensive progression to mature forest yet to have occurred but, over much of the southern Chalk, the *Fagus-Mercurialis* woodland is probably the natural climax community on moderately free-draining rendzinas, with the *Taxus* woodland replacing it locally on slopes with a warmer topoclimate (Watt 1926, 1934a). Beyond the natural limit of beech, such successions probably terminate in the *Fraxinus-Acer-Mercurialis* woodland with *Fraxinus* and *Acer pseudoplatanus* very well represented in younger, secondary canopies.

In all these different kinds of succession, scrub advance can be halted by a re-imposition of grazing or mowing which typically results in sharp boundaries between the grassland and scrub quite unlike the gradual and uneven transitions of active colonisation. Such abrupt zonations are the norm on hedgebanks where linear stands of *Crataegus-Hedera* scrub, often of planted origin though sometimes the remnants of woodlands adjacent to the path or road, are fronted by mesotrophic or calcicolous grasslands artificially maintained, usually by mowing, with perhaps a very condensed zone of *Rubus-Holcus* underscrub between. All the sub-communities can be found in such zonations, the particular arrangement of the scrub, underscrub and grassland components varying according to the verge and hedgebank structure, something which often shows marked regional differences along older routes. Similar condensed transitions can be seen where the *Crataegus-Hedera* scrub occurs as a marginal fringe to mature woodland.

Very heavy grazing or cutting or burning of scrub followed by grazing and mowing can reverse successions to the *Crataegus-Hedera* scrub though it is probably very difficult to restore the richer, more long-established swards from which the seral progressions often start, especially where scrub invasion is well advanced and destruction of the trees and shrubs involves great disturbance. Then, the scrub is likely to be replaced by weedy vegetation or rank Arrhenatherion grasslands. Gradual scrub destruction by long-continued grazing is probably more successful in restoring richer herbaceous communities and some stands of the community may represent an intermediate stage on a retrogressive sere from woodland back to grassland, initiated and maintained by grazing (Moss 1913).

Distribution

The *Crataegus-Hedera* scrub is widely distributed through the British lowlands. The *Brachypodium* and *Viburnum* sub-communities are generally confined to areas with more free-draining calcareous soils so their ranges reflect the occurrence of drift-free limestones; and the latter type is much better represented in the warmer south and east, having its centre of distribution on the Chalk. The *Mercurialis* sub-community is concentrated on more heavy-textured base-rich soils, being especially common in areas with clays and shales or ill-draining superficials.

Affinities

The community brings together a wide variety of previously described vegetation types often simply termed 'scrub', or 'Chalk scrub' in the case of the *Viburnum* sub-community, and only rarely (e.g. Shimwell 1968 *a, b*) subjected to more detailed phytosociological analysis. Available data are relatively few and, pending further investigation, it seems best to retain the types distinguished here within the same community. The *Crataegus-Hedera* scrub thus contains most of the more well-established British woody vegetation of the order Prunetalia, often placed in a distinct class of scrubs and underscrubs, the Rhamno-Prunetea, though incorporated by Ellenberg (1978) into the Querco-Fagetea. The community is best accommodated among the more calcicolous scrubs of the Berberidion alliance, though the *Hedera-Urtica* sub-community has affinities with the Rubion subatlanticum. Similar vegetation types have been described from Germany (Tüxen 1952, Ellenberg 1978) and The Netherlands (Westhoff & den Held 1969).

Floristic table W21

	a	b	c	d	21
Crataegus monogyna	V (1–10)	IV (2–8)	IV (1–8)	V (2–8)	V (1–10)
Rubus fruticosus agg.	V (1–10)	III (2–7)	IV (2–6)	V (2–8)	IV (1–10)
Prunus spinosa	III (1–9)	II (2–9)	III (1–10)	II (3–6)	III (1–10)
Fraxinus excelsior sapling	II (1–4)	III (3–7)	II (1–10)	III (3–4)	III (1–10)
Rosa canina agg.	III (1–7)	II (1–5)	II (4–6)	III (3–7)	III (1–7)
Corylus avellana	II (1–7)	II (4–8)	II (3–5)	II (4–6)	II (1–8)
Rosa arvensis	I (2)	II (2–5)	I (1)	I (1–4)	I (1–5)
Acer pseudoplatanus sapling	I (2–6)	II (3–7)	I (3)	II (3–5)	I (2–7)
Lonicera periclymenum	I (2–5)	I (3–4)	I (1–4)	I (3–5)	I (1–5)
Euonymus europaeus	I (1–3)	I (3–4)	I (4)	I (3)	I (1–4)
Ilex aquifolium	I (1–2)	I (1–9)	I (1–2)	I (4)	I (1–9)
Salix cinerea	I (7)		I (1–8)	I (5)	I (1–8)
Rhamnus catharticus	I (4)		I (4–5)	I (4–5)	I (4–5)
Betula pendula sapling	I (3)	I (3)	I (7)		I (3–7)
Fagus sylvatica sapling		I (4–7)	I (1–4)	I (1–4)	I (1–7)
Sambuca nigra	III (1–6)	III (1–6)	I (4)	II (1–4)	II (1–6)
Quercus robur sapling	I (1)	II (3–9)	I (5)	I (3)	I (1–9)
Acer campestre	I (1–4)	II (3–5)	I (3)	I (1–4)	I (1–5)
Ulmus procera/carpinifolia suckers	I (1–8)	II (4–10)	I (9–10)		I (1–10)
Malus sylvestris	I (1–2)	I (1–4)			I (1–4)
Ulmus glabra sapling	I (1)	I (1–6)			I (1–6)
Crataegus laevigata		I (3)			I (3)
Ligustrum vulgare	I (1–5)	I (1–6)	II (2–5)	V (1–5)	II (1–6)
Viburnum lantana	I (4)	I (4)	I (4–8)	V (4–8)	II (4–8)
Cornus sanguinea	I (5)	I (3–6)	I (3–6)	IV (3–6)	I (3–6)
Tamus communis	I (3)	I (1–4)	I (1–4)	III (1–4)	I (1–4)
Clematis vitalba		I (3)	I (1–6)	III (1–4)	I (1–6)
Taxus baccata sapling	I (8)	I (7)	I (3–8)	II (3)	I (3–8)
Juniperus communis communis			I (4–8)	II (4–7)	I (4–8)
Sorbus aria sapling				I (4–5)	I (4–5)
Hedera helix	IV (2–10)	IV (3–10)	IV (1–10)	IV (6–10)	IV (1–10)

Urtica dioica	IV (1–8)	IV (1–10)	II (3–4)	II (3–4)	III (1–10)
Galium aparine	IV (1–7)	IV (2–8)	II (3–5)	III (1–4)	III (1–8)
Silene dioica	II (2–5)	I (4)	I (1–5)	I (5)	I (1–5)
Heracleum sphondylium	II (1–4)	I (1–3)	I (1–4)	I (4)	I (1–4)
Holcus lanatus	II (3–5)	I (2)	I (4–6)		I (2–6)
Arrhenatherum elatius	II (1–6)		I (4)		I (1–6)
Elymus repens	I (3–6)				I (3–6)
Calystegia sepium	I (1–5)				I (1–5)
Mercurialis perennis	I (1–6)	IV (2–7)	I (2–3)	II (2–7)	II (1–7)
Eurhynchium praelongum	I (1)	III (1–8)	II (1–6)	II (2–6)	II (1–8)
Arum maculatum	I (1–5)	III (2–5)	I (1–4)	II (2–4)	II (1–5)
Poa trivialis	I (3–7)	II (2–7)	I (1–2)	I (1)	I (1–7)
Glechoma hederacea	I (1–5)	II (2–6)	I (3)	I (2)	I (1–6)
Brachythecium rutabulum	I (1)	II (1–6)	I (4–5)	I (5)	I (1–6)
Hyacinthoides non-scripta		II (2–6)			I (2–6)
Allium ursinum	I (5)	I (3–8)			I (3–8)
Anemone nemorosa		I (2–7)			I (2–7)
Brachypodium sylvaticum	I (1–6)	I (1–4)	IV (1–9)	IV (3–6)	III (1–9)
Fragaria vesca			II (2–4)	II (2–3)	I (2–4)
Viola riviniana			II (1–7)		I (1–7)
Teucrium scorodonia			I (3–6)	II (1–4)	I (1–6)
Origanum vulgare				II (1–3)	I (1–3)
Bromus erectus				II (2–5)	I (2–5)
Brachypodium pinnatum				II (1–5)	I (1–5)
Sanguisorba minor				II (1–3)	I (1–3)
Stachys sylvatica	I (1–3)	I (2–5)	I (4)	I (2)	I (1–5)
Geranium robertianum	I (1–2)	I (3–5)	I (1–6)	I (1–6)	I (1–6)
Phyllitis scolopendrium	I (1–5)	I (4–6)	I (2–7)	I (1–6)	I (1–7)
Iris foetidissima	I (3)	I (3–4)	I (4)	I (3–4)	I (3–4)
Pteridium aquilinum	I (2–8)	I (1–2)	I (1–3)	I (1–8)	I (1–8)
Circaea lutetiana	I (4)	I (2–3)	I (2)		I (2–4)
Stellaria holostea	I (1–3)	I (2)	I (4)		I (1–4)
Dryopteris filix-mas	I (1)	I (1–3)	I (1)		I (1–3)
Dactylis glomerata	I (1–3)	I (1–2)			I (1–3)

Floristic table W21 (*cont.*)

	a	b	c	d	21
Rumex sanguineus	I (1–4)	I (3)			I (1–4)
Moehringia trinervia	I (1–2)	I (3)			I (1–3)
Torilis japonica	I (1–2)	I (3)			I (1–3)
Lamium purpureum	I (3)	I (3)			I (3)
Stellaria graminea	I (4)	I (2)			I (2–4)
Stellaria media	I (1)	I (3)			I (1–3)
Arctium minus agg.	I (5)	I (3)			I (3–5)
Holcus mollis	I (2)		I (3–6)		I (2–6)
Cirsium arvense	I (4)		I (1–3)		I (1–4)
Ranunculus repens	I (1–2)		I (2)		I (1–2)
Bromus sterilis	I (3–4)		I (1)		I (1–4)
Potentilla sterilis	I (1)		I (2–3)		I (1–3)
Plagiomnium undulatum		I (3–4)	I (4–7)		I (3–7)
Juncus effusus		I (3)	I (4)		I (3–4)
Geum urbanum		I (2)	I (1)		I (1–2)
Eurhynchium confertum		I (3)	I (3)		I (3)
Rubia peregrina			I (3–5)	I (5)	I (3–5)
Number of samples	30	34	20	31	115
Number of species/sample	11 (4–24)	15 (6–27)	16 (7–29)	14 (8–18)	14 (4–29)
Shrub height (m)	3 (1–8)	5 (1–15)	2 (1–5)	3 (2–3)	3 (1–15)
Shrub cover (%)	97 (80–100)	87 (60–100)	91 (25–100)	98 (90–100)	92 (25–100)
Herb height (cm)	78 (10–150)	46 (15–150)	35 (10–150)	42 (20–150)	46 (10–150)
Herb cover (%)	73 (5–100)	91 (50–100)	79 (5–100)	77 (10–100)	81 (5–100)
Ground height (mm)	10	18 (10–40)	20 (10–40)	10	15 (10–40)
Ground cover (%)	1 (0–50)	11 (0–85)	16 (0–100)	17 (0–85)	9 (0–100)
Altitude (m)	87 (4–270)	73 (2–150)	82 (5–130)	60 (10–130)	66 (2–270)
Slope (°)	2 (0–40)	4 (0–50)	7 (0–40)	7 (0–30)	4 (0–50)

a *Hedera helix-Urtica dioica* sub-community
b *Mercurialis perennis* sub-community
c *Brachypodium sylvaticum* sub-community
d *Viburnum lantana* sub-community
21 *Crataegus monogyna-Hedera helix* shrub (total)

W22
Prunus spinosa-Rubus fruticosus scrub

Synonymy
Scrub *auct. angl. p.p.*; Scrub associations Tansley 1911; Cliff Scrub Nodum   Malloch 1970 *p.p.*; *Primula vulgaris-Prunus spinosa* Association (R. Tx 1952) Birse 1980.

Constant species
Prunus spinosa.

Physiognomy
Prunus spinosa is a frequent and locally abundant species in the *Crataegus-Hedera* scrub but, in this community, it is the sole woody constant and almost always an overwhelming dominant in a consistently more species-poor canopy. Indeed, it is usually the only tree or shrub present, though *Ulex europaeus* is occasional on more base-poor soils and *Corylus avellana* and *Ligustrum vulgare* can occur on the more base-rich. The height of the canopy is rather variable, but generally quite low, well-grown *Prunus* in sheltered situations typically attaining no more than 4 m and scrub in very exposed places, as on some sea-cliffs, having a cover often less than 1 m. Usually the canopy is closed or almost so, being especially dense where the trees are wind-pruned.

Undershrubs are not numerous, though there is often some *Rubus fruticosus* agg., growing sparsely beneath the *Prunus* or, more obviously, forming a thick tangled fringe. *R. idaeus* also occurs occasionally and can be locally abundant and there can be some roses, usually *Rosa canina* with *R. dumalis, R. tomentosa* and *R. sherardii* recorded more rarely (Birse 1980). Woody sprawlers and climbers are uncommon, but *Lonicera periclymenum* is occasional and, to the south-west, *Rubia peregrina* may be found.

The field layer of the *Prunus* scrub is characteristically species-poor and often rather sparse, especially in denser stands, where it is limited to a patchy cover beneath the trees themselves and a marginal belt. *Pteridium aquilinum* is the commonest species throughout, though it is only locally abundant and often reduced to scattered fronds. *Urtica dioica* and *Galium aparine* also occur frequently, the latter often sprawling in some abundance up the scrub fringes. Then, there are often some grasses, though the species represented vary somewhat in the different sub-communities. *Poa trivialis* and *Holcus mollis* are among the commonest but *H. lanatus* and *Agrostis capillaris* occur throughout and there can also be some *Brachypodium sylvaticum, Festuca rubra* and *Dactylis glomerata*. A further quite frequent component, though less universally prominent here than in the *Crataegus-Hedera* scrub, is a ground carpet of *Hedera helix*.

Herbaceous dicotyledons can be quite numerous, though none is more than occasional overall. *Silene dioica* is fairly common, though less frequent on more base-rich soils; *Viola riviniana* is likewise represented throughout, though it shows the reverse edaphic trend. Other species encountered at low frequencies include *Moehringia trinervia, Digitalis purpurea, Teucrium scorodonia* and *Stachys sylvatica* with, showing some obvious preference to the different sub-communities, *Hyacinthoides non-scripta, Mercurialis perennis, Veronica chamaedrys*, some maritime herbs and a variety of low-frequency associates detailed below. Ferns can be found in some stands with *Dryopteris filix-mas, D. borreri* and *Phyllitis scolopendrium* recorded occasionally.

Bryophytes are few, though they can have high local abundance. *Eurhynchium praelongum, Plagiomnium undulatum* and *Brachythecium rutabulum* are the commonest species with *Eurhynchium striatum, Thuidium tamariscinum, Atrichum undulatum* and *Lophocolea bidentata s.l.* occurring more sporadically.

Sub-communities

Hedera helix-Silene dioica **sub-community:** Cliff Scrub Nodum   Malloch 1970 *p.p.*; *Primula vulgaris-Prunus spinosa* Association (R. Tx. 1952) Birse 1980 *p.p.* Dense and quite low covers of *Prunus* are characteristic here with occasional *Ulex*. *Rubus fruticosus* agg. occurs frequently and it can be locally abundant sprawling up and over the scrub margins and sometimes forming a sparse underscrub. *Lonicera* is occasional.

The most obvious feature of the field layer is usually a ground carpet of *Hedera* and this can extend, as a thin cover, beneath quite dense canopies, together with a patchy mat of *Poa trivialis* and *Holcus mollis* and scattered clumps of *Urtica*. *Hyacinthoides*, though only occasional, is preferential to this sub-community and, when present in abundance, can give stands a distinct vernal aspect. In more open places and around the fringes of the scrub, the herbaceous component thickens up considerably and, among the patches of *Pteridium*, *Urtica* and *Galium aparine*, there is frequently some *Silene dioica* along with scattered plants of community occasionals like *Digitalis* and *Teucrium scorodonia*. Bryophytes are usually few and represented by the common community species.

***Viola riviniana-Veronica chamaedrys* sub-community:** *Primula vulgaris-Prunus spinosa* Association (R. Tx. 1952) Birse 1980 *p.p.* The woody cover of this sub-community is generally taller and a little more open than usual, though still invariably dominated by *Prunus* with scarce *Corylus*. *Rubus fruticosus* agg. is only occasional here but *Rosa canina* agg. is weakly preferential.

However, the major distinctive features are to be seen in the field layer. *Pteridium*, *Urtica* and *Galium aparine* are all still common but *Hedera* is much reduced in frequency and there is a variety of preferentials characteristic of somewhat moister and base-rich soils. The most obvious of these are *Viola riviniana* and *Veronica chamaedrys* but *Oxalis acetosella*, *Geranium robertianum*, *Primula vulgaris*, *Filipendula ulmaria* and *Geum urbanum* occur occasionally and, in some stands, there are prominent patches of *Mercurialis perennis*. There are also some weakly preferential bryophytes, *Eurhynchium striatum* and *Thuidium tamariscinum* occurring more commonly here than elsewhere.

***Dactylis glomerata* sub-community:** Cliff Scrub Nodum Malloch 1970 *p.p.* *Prunus*, occasionally with some *Ulex*, typically forms a rather open, patchy canopy here which, in the exposed situations characteristic of this sub-community, is low and wind-pruned. In extreme situations, the *Prunus* cover can be less than 50 cm high with herbs growing through it: sparse *Pteridium* fronds can emerge from the canopy and Malloch (1970) reported some peculiar examples of coastal scrub which had tussocks of *Armeria maritima* resting on top of the *Prunus* branches and rooted in the soil below.

Among the bushes, grasses are often conspicuous, with *Dactylis glomerata* constant and *Brachypodium sylvaticum*, *Festuca rubra*, *Holcus lanatus* and *Agrostis capillaris* occuring occasionally. *Pteridium* and *Galium*

aparine (though not *Urtica*) can form patches of more luxuriant vegetation with scattered *Silene dioica* and, preferential here, *Rumex acetosa* and *Plantago lanceolata*. On sea-cliffs, there can also be some obviously maritime plants like *Armeria maritima* and *Silene vulgaris* ssp. *maritima*.

Habitat
The *Prunus* scrub is most characteristic of mesotrophic mull soils of moderate base-status in the lowland parts of Britain. Typically, it develops from grasslands where grazing has been relaxed and theoretically it is a sub-climax vegetation, seral to certain types of high forest. In exposed sites, however, it may represent a local climatic climax and it can persist, too, along woodland margins as a more or less permanent fringing vegetation.

The environmental conditions here overlap quite considerably with those of the *Crataegus-Hedera* scrub but there are two particular features of the habitat which may favour the overwhelming predominance of *Prunus spinosa* over what is a fairly similar range of field layers. One is edaphic, because this tree seems to fare best on soils that are deep, moist and fairly nutrient-rich. In so far as the profiles have been investigated here, mull brown earths seem to be the characteristic type (Coombe & Frost 1956, Géhu 1964, Malloch 1970, Birse 1980) and, though some of the soils are quite base-rich and calcareous, this community is typically absent from oligotrophic and excessively-draining rendzinas, on to which the *Crataegus-Hedera* scrub can extend as the *Viburnum* sub-community, where *Prunus* itself is noticeably scarce. On the other hand, the *Prunus* scrub does not extend far on to markedly acid soils: *Ulex europaeus* occurs here occasionally but usually the community is replaced on such profiles by the *Ulex-Rubus* scrub where, again, *Prunus* is rare. Few field-layer species are common enough to be said to match *Prunus* in its edaphic responses though the fairly frequent occurrence of *Pteridium* is another good indicator of the combination of appreciable soil depth and moderate moisture content. Topographically, the soil preferences of the community are often marked by the fact that it picks out accumulations of colluvium or more stabilised areas of softer deposits, like drift or head, on steep slopes.

The other environmental situation where *Prunus* may have an edge on some other common scrub dominants like *Crataegus monogyna* or *Sambucus nigra* and from which saplings of woodland canopy trees are probably excluded, is in more exposed places. On sea-cliffs, for example, though the community is scarce and local in very wind-blown situations, it is very common and abundant where there is a modicum of shelter but perhaps sufficient exposure to prevent vigorous growth of other woody species (Malloch 1970). In such places, the *Prunus* canopy looks quite healthy, though it is here that the shortest and most wind-pruned covers are seen.

As with the *Crataegus-Hedera* scrub, this community develops only where there is an opportunity for natural succession to proceed. It can colonise open ground directly, provided the edaphic conditions are suitable, whether these are natural or artificial. However, though it can appear on newly-exposed soft materials like slumping clay or drift, it very often develops from existing herbaceous vegetation where grazing has been relaxed. In these situations, its eventual dominance may depend partly on the chance availability of seed but the great ability of *Prunus* to sucker once established probably plays an important role in the development of the characteristically impoverished and dense canopy here.

The heavy shade cast by the woody cover is undoubtedly responsible in large measure for the often sparse or localised and species-poor nature of the field layer in the community, which is rarely continuous and well developed only in more open places and around ungrazed fringes. More specifically, shade restricts the abundance of *Pteridium* and a *Rubus* underscrub here, both of which could be potentially very abundant on these soils, and limits the expression of any woodland flora until the canopy begins to open up as the *Prunus* is shaded out by any overtopping trees: these may never colonise or emerge, of course, in very exposed situations.

The differences between the various sub-communities are partly related to the maturity of the scrub. On sea-cliffs, the *Dactylis* sub-community, with its more discontinuous and wind-pruned cover of *Prunus* and its prominent remnants of a maritime grassland flora, probably represents the seaward limit of scrub development, where salt-spray influence is low but where exposure to wind curtails canopy closure. In more sheltered situations, the *Hedera-Silene* sub-community represents an advancement with canopy closure and the ground carpet of ivy typical of a more or less continuous shrub cover and gloomy interior. The *Viola-Veronica* sub-community may be more mature still: here the canopy is a little more open and there are elements of a woodland field layer.

However, soil differences probably also play some part in floristic variation here, with the *Hedera-Silene* sub-community being characteristic of drier and perhaps marginally acid profiles, the *Viola-Veronica* sub-community occurring on more base-rich and sometimes heavier soils and the *Dactylis* sub-community typical of fairly moist mulls.

Zonation and succession

The *Prunus-Rubus* scrub is typically found in zonations and mosaics with grasslands, underscrubs and woodlands, sometimes representing active successions on wasteland, in neglected farmland or in coppice plots and clearings, in other cases in more stabilised sequences in exposed situations or along wood margins and rides, by hedges and in field corners.

Most often, the associated grasslands are ranker, mesotrophic swards, among which various kinds of *Arrhenatheretum* figure prominently or, on somewhat moister soils, the *Holco-Juncetum* or *Holcus-Deschampsia* grassland. The former are very common where the *Prunus-Rubus* scrub is colonising drier pasture or where it constitutes the core of hedges, where the *Hedera-Silene* sub-community is typical; the latter occur with the community in ill-drained pastures or where scrub is spreading into rides and clearings within woodlands on heavy and especially trampled soils and here the *Viola-Veronica* sub-community is more frequent. Less commonly, the *Prunus-Rubus* scrub occurs among somewhat more calcifugous grasslands and heaths, as where the community is invading neglected commons. Frequently, the transition from the scrub to the grassland is marked by an untidy fringe of the *Rubus-Holcus* or *Pteridium-Rubus* underscrubs.

The kinds of woodland most frequently associated with the *Prunus-Rubus* scrub are the *Quercus-Pteridium-Rubus* woodland on more base-poor soils and the *Fraxinus-Acer-Mercurialis* on heavier, more base-rich soils and the community grades floristically to these forest types through their *Hedera* sub-communities. More locally, towards the south, the *Prunus-Rubus* scrub can be found with the *Fagus-Rubus* woodland and, around flushes, it can pass to the *Alnus-Fraxinus-Lysimachia* woodland. Although successions have not been followed, it seems likely that the community can develop into any of these woodland types where saplings of the canopy trees are able to invade and overtop the *Prunus*: although forming a dense canopy, *Prunus* is readily shaded out once trees have begun to emerge. The most usual climax woodlands would seem to be the *Quercus-Pteridium-Rubus* and *Fraxinus-Acer-Mercurialis* woodlands and stands of the *Prunus-Rubus* scrub can quickly spring up in stretches of these communities where rides are neglected or areas cleared and left.

However, although complete zonations from grasslands, through underscrub and the *Prunus-Rubus* scrub to woodland can be seen, the community often occurs in broken or abbreviated zonations where succession has been halted for one reason or another. It is common, for example, as a narrow static fringe along woodland rides and around wood margins. Many hedgerow stands have a similar appearance, with a linear stand of the scrub and a much-compressed zone of underscrub and grassland maintained by mowing or grazing.

On sea-cliffs with more mesotrophic soils, the *Prunus-Rubus* scrub often terminates the sequence of vegetation types on the unenclosed cliff top, the *Hedera-Silene* sub-community occurring in more sheltered situations, the *Dactylis* sub-community in more exposed places, and giving way below to maritime grassland, usually some type of *Festuca-Holcus* grassland. Sometimes there is an intervening zone of the *Pteridium-Rubus* underscrub and, on moister soils, the *Festuca-Hyacinthoides* grassland beyond this, *Prunus* and then *Rubus* and *Pteridium* petering out as one moves seawards. In other places, these communities can be replaced by the *Arrhenatheretum* in this transitional zone.

Distribution

The *Prunus-Rubus* scrub is of widespread distribution through the British lowlands.

Affinities

Although the community to some extent replicates the field-layer variation found within the *Crataegus-Hedera* scrub, it seems best to retain it as a separate unit characterised by the distinctly impoverished canopy. Similar scrubs have been described from northern France (Géhu 1964), from The Netherlands (Doing 1962, Westhoff & den Held 1969) and from Germany (Ellenberg 1978) and are usually placed in the alliance Rubion subatlanticum in the Prunetalia.

Floristic table W22

	a	b	c	22
Prunus spinosa	V (7–10)	V (8–9)	V (5–10)	V (5–10)
Rubus fruticosus agg.	IV (1–5)	II (1–5)	II (2–4)	III (1–5)
Rubus idaeus	I (3–6)	I (1–5)		I (1–6)
Corylus avellana	I (4)	I (1–4)		I (1–4)
Ulex europaeus	II (1–7)		II (1–4)	II (1–7)
Lonicera periclymenum	II (2–3)		I (4)	I (2–4)
Rosa canina agg.		I (1–4)		I (1–4)
Hedera helix	III (4–10)	I (8)	I (1–4)	II (1–10)
Silene dioica	III (1–6)	I (1–3)	II (2–3)	II (1–6)
Hyacinthoides non-scripta	II (1–8)	I (1–5)		I (1–8)
Stellaria media	II (2–4)			I (2–4)
Viola riviniana	I (1–4)	III (1–4)	I (3)	I (1–4)
Veronica chamaedrys	I (1)	III (1–2)		I (1–2)
Mercurialis perennis	I (1–2)	II (1–8)		I (1–8)
Eurhynchium striatum	I (5)	II (2–6)		I (2–6)
Oxalis acetosella	I (1–2)	II (1–8)		I (1–8)
Thuidium tamariscinum	I (1)	II (1–8)		I (1–8)
Geum urbanum	I (4)	II (1–6)		I (1–6)
Filipendula ulmaria	I (3)	II (1–5)		I (1–5)
Geranium robertianum		II (1–3)	I (1)	I (1–3)
Primula vulgaris		II (1–5)		I (1–5)
Dactylis glomerata			V (2–5)	II (2–5)
Brachypodium sylvaticum		I (1)	II (2–4)	I (1–4)
Festuca rubra		I (1)	II (2–6)	I (1–6)
Rumex acetosa	I (1)		II (1–4)	I (1–4)
Agrostis capillaris	I (1–3)	I (4)	II (2–3)	I (1–4)
Holcus lanatus	I (4)	I (1–3)	II (2–3)	I (1–4)
Plantago lanceolata			II (1–4)	I (1–4)
Silene vulgaris maritima			II (2–4)	I (2–4)
Armeria maritima			I (2–5)	I (2–5)
Achillea millefolium			I (2–3)	I (2–3)

	a	b	c	22
Pteridium aquilinum	III (1–6)	III (1–4)	III (1–4)	III (1–6)
Galium aparine	III (1–7)	III (1–6)	II (2–4)	III (1–7)
Eurhynchium praelongum	II (2–6)	II (1–7)		II (1–7)
Plagiomnium undulatum	II (2–7)	II (1–6)		II (1–7)
Poa trivialis	II (1–4)	II (1–6)		II (1–6)
Holcus mollis	II (1–7)	II (1–6)		II (1–7)
Urtica dioica	II (2–5)	II (3–5)		II (2–5)
Moehringia trinervia	II (1–3)	II (2–3)		I (1–3)
Brachythecium rutabulum	II (1–6)	II (1–5)		I (1–6)
Digitalis purpurea	I (1–3)	I (1)	I (1)	I (1–3)
Teucrium scorodonia	I (2–4)		I (5)	I (2–5)
Phyllitis scolopendrium	I (1–4)	I (4)	I (2–4)	I (1–4)
Dryopteris filix-mas	I (1–4)	I (2–3)		I (1–4)
Stachys sylvatica	I (1–2)	I (1–4)		I (1–4)
Dryopteris borreri	I (1)	I (1–2)		I (1–2)
Cardamine flexuosa	I (1–2)	I (1–2)		I (1–2)
Atrichum undulatum	I (1–4)	I (1–2)		I (1–4)
Lophocolea bidentata s.l.	I (1–4)	I (1–2)		I (1–4)
Number of samples	22	8	19	49
Number of species/sample	12 (4–31)	26 (11–38)	10 (7–14)	14 (4–38)
Shrub height (m)	2 (1–4)	4 (3–5)	1 (0.2–2)	2 (0.2–5)
Shrub cover (%)	92 (70–100)	85 (75–95)	78 (25–90)	85 (70–100)
Herb height (cm)	63 (10–150)	45 (10–75)	no data	58 (10–150)
Herb cover (%)	56 (30–100)	59 (20–80)	no data	57 (20–100)
Altitude (m)	51 (4–244)	145 (91–213)	28 (4–54)	57 (4–244)
Slope (°)	15 (0–75)	11 (1–20)	18 (0–45)	16 (0–75)

a   *Hedera helix-Silene dioica* sub-community
b   *Viola riviniana-Veronica chamaedrys* sub-community
c   *Dactylis glomerata* sub-community
22 *Prunus spinosa-Rubus fruticosus* scrub (total)

W23
Ulex europaeus-Rubus fruticosus scrub

Synonymy

Scrub associations Tansley 1911 *p.p.*; *Ulicetum* Tansley 1939 *p.p.*; *Ulex europaeus* scrub Grubb *et al.* 1969 *p.p.*; Cliff Scrub Nodum Malloch 1970 *p.p.*; *Pteridium aquilinum-Ulex europaeus* Association (Birse & Robertson 1976) Birse 1984.

Constant species

Rubus fruticosus agg., *Ulex europaeus*, *Agrostis capillaris*.

Physiognomy

The *Ulex europaeus-Rubus fruticosus* agg. scrub has a fairly low woody cover, usually between 1 and 2 m high, in which *U. europaeus* is generally the dominant plant. Its physiognomy is very variable: where the scrub is browsed (the young gorse shoots providing a palatable bite), the bushes can be kept trimmed to rounded hummocks, sometimes with a central tuft of inaccessible branches; where the gorse grows free of browsing or has not been recently burned, the bushes grow tall and leggy, with their foliage out of reach of any herbivores that get access subsequently; where there is regeneration after burning from basal or buried stems, there can be a low sward of young shoots. Quite often, gorse is accompanied here by *Cytisus scoparius*, particularly on the more acid soils, and in some stands this can be the sole dominant.

U. *europaeus* figures widely as a component of various kinds of heath in Britain, where it can accompany *U. gallii* to the west and/or *U. minor* in the south, as well as *Calluna vulgaris* and *Erica cinerea*. Apart from *Calluna*, which makes a very occasional contribution, all these species are typically absent here. Usually, the only other members of the scrubby cover are Rubi: *Rubus fruticosus* agg. is very frequent, though its cover is generally low, and *R. idaeus* occurs occasionally.

In dense *Ulex-Rubus* scrub, there is next to no vegetation beneath the bushes, the ground being bare or covered with a layer of the cast spiny shoots, and herbaceous plants being limited to areas between the gorse; where the growth is more open and leggy, the herbage can be more or less continuous, especially where animals graze beneath the bushes. Strictly speaking, the associated plants can hardly be called a field layer: by and large, the vegetation presents the appearance of a grassland with gorse and broom, and the affinities of the herbaceous element are clearly with the more mesotrophic forms of *Festuca-Agrostis-Galium* grassland. *Agrostis capillaris* is usually very common and *Festuca rubra* frequent with *Holcus lanatus* and *Dactylis glomerata* represented a little more unevenly. Other grasses found occasionally include *Deschampsia flexuosa*, *Holcus mollis*, *Poa pratensis* and, on somewhat less acid soils, *Arrhenatherum elatius* and *Brachypodium sylvaticum*. The commonest dicotyledons in the community are *Galium saxatile*, *Potentilla erecta*, *Rumex acetosa*, *Silene dioica* and *Digitalis purpurea*. *Viola riviniana*, *Cerastium fontanum* and *Achillea millefolium* are recorded very occasionally but there are none of the intimate mixtures of calcicoles characteristic of the occurrences of *U. europaeus* in what has generally been called 'Chalk-' or 'Limestone-heath'. Neither are markedly eutrophic plants like *Urtica dioica* or *Galium aparine* represented, although disturbed or burned areas within the *Ulex-Rubus* scrub may have dense patches of *Epilobium angustifolium*. *Pteridium aquilinum* occurs quite frequently but only exceptionally does it have high cover, being found more usually as sparse scattered fronds.

In the grassy sward, bryophytes can be quite abundant but the species involved are few. *Rhytidiadelphus squarrosus* is the most frequent and characteristic but *Eurhynchium praelongum* and *Pseudoscleropodium purum* also occur occasionally.

Sub-communities

***Anthoxanthum odoratum* sub-community:** *Pteridium aquilinum-Ulex europaeus* Association (Birse & Robertson 1976) Birse 1984. *U. europaeus*, sometimes with or occasionally replaced by *Cytisus*, dominates here in rather open and sometimes tall covers with patchy *R. fruticosus* agg. and infrequent *R. idaeus*. The herbaceous component is usually extensive and quite rich with *Agrostis capillaris*, *Festuca rubra* and *Holcus lanatus* being especially prominent among the grasses and here frequently accompanied by small amounts of *Anthoxanthum odoratum* and *Poa pratensis* and, more occasionally, by *Deschampsia flexuosa*. Among the dicotyledons, *Potentilla erecta* is preferentially common, occurring as scattered plants with *Galium saxatile*, *Cerastium fontanum*, *Viola riviniana* and *Rumex acetosa*. Very occasionally, there can be a little *Calluna*.

***Rumex acetosella* sub-community:** Cliff Scrub Nodum Malloch 1970 *p.p.*; *Pteridium aquilinum-Ulex europaeus* Association (Birse & Robertson 1976) Birse 1984 *p.p.* The general features of the vegetation here are much as in the last sub-community with rather patchy covers of gorse and/or broom, a little *Rubus fruticosus* agg. and a grassy ground cover in which *Festuca-Agrostis* grassland species figure prominently, *Agrostis capillaris* being especially frequent and abundant. However, *Anthoxanthum odoratum* and *Potentilla erecta* are noticeably less common and there is a marked enrichment with species characteristic of light and/or disturbed soils. *Rumex acetosella* and *Hypochoeris radicata* are the most frequent among these but *Senecio jacobaea*, *Crepis capillaris*, *Jasione montana* and *Aira praecox* also occur, sometimes as scattered plants in little breaks in the swards, often on more extensive patches of open ground in disturbed places, on ant-hills or around rock exposures.

***Teucrium scorodonia* sub-community:** Cliff Scrub Nodum Malloch 1970 *p.p.* Fairly dense covers of *U. europaeus* and *Rubus* provide the usual canopy here and there is a marked reduction in the cover and variety of the grassy element. *Festuca rubra* and *Agrostis capillaris* both still occur occasionally and *Dactylis glomerata* and *Brachypodium sylvaticum* are preferential at low frequencies but, generally, these occur as scattered tussocks among the gorse and bramble cover. More characteristic, in open places, is *Teucrium scorodonia* along with a little *Silene dioica*, *Digitalis purpurea* and *Pteridium*. A patchy cover of *Hedera helix* can extend beneath the gorse.

Habitat

The *Ulex-Rubus* scrub is characteristic of moderately to strongly acid brown soils, free-draining though not always dry and not markedly oligotrophic. It is probably a fairly natural colonising vegetation on such profiles throughout the British lowlands and in the upland fringes, though its establishment and spread are much encouraged by disturbance and agricultural neglect. It

can be a seral precursor to or, in coppice plots and clearings, a temporary replacement for more calcifugous woodlands but many stands seem to be stable or, in pasture, to show a cyclical pattern of development and run-down.

The community seems most consistently associated with base-poor brown earths of pH 4–6, without any drainage impedence, though often quite deep and relatively moist, at least when compared with those over which *U. europaeus* occurs with ericoid sub-shrubs. Such soils have developed naturally over wide areas of lowland Britain, being especially associated with more pervious arenaceous bedrocks, like sands and sandstones, and lighter-textured superficials including some aeolian deposits, more free-draining material among glacial and peri-glacial deposits, colluvium and head. Very frequently, the profiles have been subject to some measure of improvement in association with low-intensity and often sporadic arable cultivation or pastoral agriculture. And, in some places, such activities have brought more extreme podzolised profiles into the more fertile state characteristic of low base-status brown earths, as around agricultural enclaves within lowland heaths and along the fringes of the uplands.

The *Ulex-Rubus* scrub is most often found on such soils in places which are of marginal value in the agricultural landscape: where tractable slopes are broken by banks and rocky areas, around the limits of enclosures and near settlements and farms and along pathways and hedgebanks. With this kind of association, *U. europaeus* is very much a more calcifugous equivalent of *Juniperus communis* ssp. *communis*.

In less intensive agricultural economies, gorse has played some part as a provider of fodder. Its young shoots are soft and nutritious and, in the past, older, harder shoots were cut and ground for feeding to cattle (Tansley 1939). It has been maintained (Roberts in Tansley 1939) that gorse was deliberately introduced into some parts of Wales for cutting as feed and it has certainly been planted to provide fox cover in central and southern England and perhaps elsewhere as hedges. Then, where it has been an impediment to agriculture, it has frequently been burned back, though it sprouts readily from undamaged basal or buried shoots and fire stimulates germination of its seed (and that of *Cytisus*).

U. europaeus is well adapted to maintain itself in such situations where exploitation comes and goes. As well as an explosive mechanism of seed-dispersal, the pods ripening and bursting in mid-summer, it is probably spread along paths by ants which carry the seeds away to devour the oily caruncle. Once established, it is very resistant to grazing, persisting as low hummocks in mixed swards and getting away to form leggy bushes if there is any relaxation of herbivore pressure: the demise of rabbits in myxomatosis was widely followed by upgrowth of *U. europaeus*, as at Lullington Heath in Sussex, where stands of this kind of scrub developed from *Calluna-Erica* heath (Grubb *et al.* 1969). Rabbits can undermine and kill *U. europaeus* where bushes provide shelter for warren entrances in open grassland, but the disturbance which the animals create itself provides further ground for re-establishment (Duffey *et al.* 1974).

Two further features of *U. europaeus* may be of particular importance in certain situations. First, although it seems to favour soils which are initially not very impoverished, it can itself enrich its environment through the activity of root-nodule nitrogen-fixing bacteria, an activity whose consequences are often most marked after gorse is cleared when eutrophic weeds frequently spring up on the disturbed ground. *Cytisus* also has such root nodules. Second, as Grubb *et al.* (1969; see also Grubb & Suter 1971) showed, *U. europaeus* can acidify its root environment, probably not by the accumulation of mor and enhancement of podzolisation, as under *Calluna*, but by the immobilisation of bases in the standing crop and litter. This can enable it to thrive on initially quite base-rich and calcareous soils, such as develop from more limy superficials, thus giving the plant an advantage when grazing is relaxed, when it can spread and extinguish many elements of the previous sward.

There is one other particular situation where the clear association of the community with moderately acid and moderately enriched soils and disturbance is very evident, and that is within woodland coppice plots or clearings. The *Ulex-Rubus* scrub frequently develops after underwood or canopy have been removed from stands of the *Quercus-Pteridium-Rubus* woodland on more free-draining soils or from the *Quercus-Betula-Deschampsia* woodland on profiles that are not too strongly podzolised and impoverished. Such occurrences give some clue as to the possible seral successions in which the community is involved (see below).

Even within a woodland context, the *Ulex-Rubus* scrub has its characteristic suite of calcifuges and more mesophytic herbs, strongly indicative of the typical balance of edaphic conditions here. But it is where the community develops among a pastoral or heathy landscape that this element is most prominent. Usually, in such situations, it is represented by either the *Anthoxanthum* or *Rumex* sub-communities. The former is more characteristic of undisturbed places where the scrub is well developed among pasture or along wood margins or hedges. The latter very typically has some measure of disruption of the ground surface, either through the scuffing of grazing animals or the activities of mound-building ants, or where rock crops out; this kind of *Ulex-Rubus* scrub also developed at Lullington in places where gorse had been cleared and mowing imposed

(Grubb *et al.* 1969). The *Teucrium* sub-community sometimes occurs as a dense fringe to mature woodland but it is also common on some sea-cliffs, where it occupies an edaphically intermediate position between the *Prunus-Rubus* scrub of deeper, moister soils and the maritime heaths of shallow, drier profiles.

Zonation and succession

Most commonly, the *Ulex-Rubus* scrub occurs as small stands in mosaics with grasslands, heaths, underscrubs and other kinds of scrub on marginal agricultural land. It is also widespread as a fringe to certain kinds of woodland and can occur in hedgerow sequences.

The former kinds of pattern are very diverse and often related to complex histories of land use but the range of communities involved is fairly small and distinctive. The grasslands most frequently encountered with the *Ulex-rubus* scrub, and forming the ground where the community occurs in pasture, are the more mesotrophic forms of the *Festuca-Agrostis-Galium* grassland and the more calcifugous types of improved swards in the *Lolio-Cynosuretum* and *Centaureo-Cynosuretum*. In all these communities, the characteristic herbs of the *Ulex-Rubus* scrub figure prominently and, in zonations, there is often a strong floristic continuity between the open grassland and the vegetation beneath and among the gorse. In sea-cliff sequences, these communities are replaced by their maritime equivalents in the *Festuca-Holcus* and *Festuca-Plantago* grasslands. Where grazing is less intense, transitions from scrub to grassland may be marked by a zone with bracken, either the *Pteridium-Galium* community or the *Pteridium-Rubus* underscrub, invariably of the *Teucrium* sub-community on these soils.

Where there is some diversity in the soils, the *Ulex-Rubus* scrub can be found colonising alongside other kinds of scrub or with heath. On less base-poor soils, which are sometimes moister than the profiles characteristic here, the community may give way to the *Crataegus-Hedera* scrub or, commonly on exposed sea-cliffs, to the *Prunus-Rubus* scrub, where mixtures of *U. europaeus* and *Prunus spinosa* can sometimes be found over a rather similar field layer to that typical of this community. Where there is a switch to more acidic and impoverished profiles, the *Ulex-Rubus* scrub can be found among heaths: on commons, the community often forms a zone around enclosures and settlements and linear strips running along tracks through the heath. The particular heath communities involved vary according to the geographic region: to the south of England, the *Calluna-Ulex minor*, *Ulex minor-Agrostis curtisii* or *Ulex gallii-Agrostis curtisii* heaths usually provide the context, to the west the *Calluna-Ulex gallii* or *Calluna-Erica cinerea* heaths, and on the more maritime parts of sea cliffs, the *Calluna-Scilla verna* heath. *U. europaeus* continues to be a local dominant in some of these communities and there is often a considerable overlap in the associated herbaceous floras of both the heath and the scrub.

Both *U. europaeus* and *Cytisus* are naturally rather short-lived species (individuals lasting ten years or so) and imposition of grazing or periodic burning may prevent any successional developments beyond the establishment of *Ulex-Rubus* scrub so that, in many cases, there may be a cyclical alternation of the community with grassland or heath, perhaps with a phase of dominance by eutrophic tall herbs where burning and disturbance expose rich soils. In exposed places, too, as on sea-cliffs, the *Ulex-Rubus* scrub may represent an end point in the development of woody vegetation. But where there is shelter from winds and some freedom from grazing and burning, trees can invade more open stands of the community: like juniper, gorse itself may provide some measure of protection from herbivores, saplings growing up inside more leggy bushes or within enclosed enclaves of herbaceous vegetation.

The commonest invaders are birch and oak, *Betula pendula* and *Quercus robur* predominating but *B. pubescens* and *Q. petraea* becoming more frequent to the north and west and *Q. petraea* locally frequent even in the south-east. *Pinus sylvestris* is also a prominent coloniser where it is able to seed in from nearby plantations and there can be some *Sorbus aucuparia* and *Ilex aquifolium*. Scattered saplings of all these species and locally dense stands of young birch among patches of the *Ulex-Rubus* scrub with heaths, open grassland areas and tracts of bracken have become a characteristic feature of many lowland commons with the decline in their traditional uses as a source of grazing and mown crops.

Successions in such situations as these have never been followed in detail but it seems likely that the usual replacements for the *Ulex-Rubus* scrub in uninterrupted seres are the *Quercus-Pteridium-Rubus* woodland on more fertile brown earths (perhaps the *Holcus* sub-community with its grassy field layer lacking slow-spreading herbs like *Hyacinthoides non-scripta*), itself perhaps succeeded within the natural British range of beech by the *Fagus-Rubus* woodland, and the *Quercus-Betula-Deschampsia* woodland on somewhat more acidic and oligotrophic soils. Where this kind of scrub extends into the upland fringes, the oak-birch climax forests may be represented by the north-western *Quercus-Betula-Oxalis* and *Quercus-Betula-Dicranum* woodlands.

The *Ulex-Rubus* scrub is also found in close spatial association with woodlands, generally the *Quercus-Pteridium-Rubus* woodland, where underwood or canopy has been cut and, in the absence of cleaning or careful attention to planted saplings, it may herald a run-down of the woodland. The community also occurs widely as a more or less permanent fringe to this kind of woodland, kept in check by intensive use of the neighbouring land.

Distribution

The *Ulex-Rubus* scrub has a widespread distribution on marginal land throughout the lowlands and upland fringes.

Affinities

The community includes most of the vegetation dominated by *U. europaeus* (or *Cytisus*) in the absence of other species of *Ulex* and ericoid sub-shrubs and it is therefore somewhat narrower than the *Ulicetum* of early British accounts (e.g. Tansley 1939), forming part of the general scrub associations described from more acidic soils (e.g. Tansley 1911). Similar communities have been described from France (Géhu 1964), and The Netherlands (Doing 1962, Westhoff & den Held 1969): the former preferred placing the community within the Nardo-Callunetea because of the heathy affinities of the associated flora, the latter favoured the erection of a Ulici-Sarothamnion alliance within the Quercetea to contain stands in which gorse or broom were dominant with bramble and bracken. This view echoes Tansley's (1939) remark that *U. europaeus* cannot be considered a proper member of the heath formation.

Floristic table W23

	a	b	c	23
Ulex europaeus	V (5–9)	V (2–9)	V (5–10)	V (2–10)
Rubus fruticosus agg.	V (1–6)	IV (1–6)	V (3–9)	V (1–9)
Cytisus scoparius	II (1–10)	II (1–6)		II (1–10)
Rubus idaeus	II (1–3)	I (2–3)		II (1–3)
Agrostis capillaris	V (1–6)	V (2–8)	II (3–4)	IV (1–8)
Holcus lanatus	IV (1–4)	IV (1–3)	I (3)	III (1–4)
Galium saxatile	III (1–3)	III (1–3)		III (1–3)

Floristic table W23 *(cont.)*

	a	b	c	23
Rhytidiadelphus squarrosus	III (1–5)	III (1–5)		III (1–5)
Holcus mollis	II (2–8)	II (1–6)		II (1–8)
Eurhynchium praelongum	II (1–7)	II (1–4)		II (1–7)
Cerastium fontanum	II (1)	II (1–2)		I (1–2)
Viola riviniana	II (1–6)	II (3)		I (1–6)
Pseudoscleropodium purum	II (1–5)	II (1–4)		I (1–5)
Anthoxanthum odoratum	IV (1–4)	I (1–5)	I (2–3)	III (1–5)
Potentilla erecta	III (1–2)	I (2–4)		II (1–4)
Poa pratensis	III (2–4)	I (2–4)		II (2–4)
Deschampsia flexuosa	II (2–5)	I (4)		I (2–5)
Calluna vulgaris	II (2)	I (1)		I (1–2)
Rumex acetosella	I (1)	IV (1–3)		II (1–3)
Hypochoeris radicata		IV (1–3)	I (1)	II (1–3)
Senecio jacobaea		III (1–2)		I (1–2)
Plantago lanceolata		III (1–3)		I (1–3)
Crepis capillaris		II (1–2)		I (1–2)
Jasione montana		II (1–2)		I (1–2)
Aira praecox		I (1)		I (1)
Teucrium scorodonia		I (3–5)	V (2–5)	II (2–5)
Hedera helix			II (2–6)	I (2–6)
Brachypodium sylvaticum			II (2–4)	I (2–4)
Pteridium aquilinum	III (2–5)	III (1–7)	III (1–4)	III (1–7)
Festuca rubra	III (1–4)	II (2–4)	II (3–4)	II (1–4)
Dactylis glomerata	I (1)	II (1–3)	II (2–4)	II (1–4)
Rumex acetosa	II (1–3)	I (1)	I (3)	I (1–3)
Achillea millefolium	I (1)	I (1)	I (2)	I (1–2)
Silene dioica	I (1–3)	I (1–2)	I (3–4)	I (1–4)
Digitalis purpurea	I (1)	I (1)	I (1–3)	I (1–3)
Campanula rotundifolia	I (1–2)	I (1)		I (1–2)
Rhytidiadelphus triquetrus	I (1–8)	I (1)		I (1–8)
Veronica officinalis	I (1)	I (1–2)		I (1–2)
Veronica chamaedrys	I (1–4)	I (3)		I (3–4)
Arrhenatherum elatius		I (1–2)	I (3–4)	I (1–4)
Number of samples	14	9	9	32
Number of species/sample	18 (15–23)	23 (17–30)	9 (4–14)	16 (4–30)
Vegetation height (cm)	177 (30–250)	140 (80–250)	108 (60–220)	143 (30–250)
Vegetation cover (%)	98 (92–100)	97 (90–100)	100	99 (90–100)
Altitude (m)	34 (6–50)	62 (32–105)	33 (3–60)	41 (3–105)
Slope (°)	8 (0–16)	10 (3–16)	19 (4–45)	13 (0–45)

a   *Anthoxanthum odoratum* sub-community

b   *Rumex acetosella* sub-community

c   *Teucrium scorodonia* sub-community

23   *Ulex europaeus-Rubus fruticosus* scrub (total)

W24
Rubus fruticosus-Holcus lanatus underscrub

Synonymy
Marginal Society   Salisbury 1916, 1918*a p.p.*; Marginal flora   Tansley 1939 *p.p.*

Constant species
Rubus fruticosus agg., *Holcus lanatus*.

Physiognomy

The *Rubus fruticosus* agg.-*Holcus lanatus* underscrub is typically dominated by mixtures of brambles, rank grasses and tall dicotyledons, forming an untidy cover of rather variable height, but usually less than 1 m. Although it is very commonly found in close association with taller woody vegetation, in active successions and in stabilised zonations around scrub and woodland margins, trees and shrubs are characteristically sparse within the community itself. There are sometimes scattered *Crataegus monogyna*, *Prunus spinosa*, *Sambucus nigra* and saplings of *Fraxinus excelsior*, *Acer pseudoplatanus*, *Fagus sylvatica* or *Quercus robur*, but their total cover is generally low.

R. *fruticosus* agg. is a constant component of the vegetation but its abundance is rather variable. In many stands it is very plentiful, forming dense clumps of tangled arching shoots and, in such cases, most of the other plants are confined to the margins of the bushes or areas between them. In other stands, brambles occur as more widely scattered bushes or as sparse shoots throughout, when the proportion of herbs is consequently greater. Where the community occurs on hedgebanks, where it typically forms a narrow zone between hedge and verge, the bramble cover may be occasionally cut back, but this vegetation is not regularly mown. Other undershrubs are relatively infrequent but *Rosa canina* agg. or *R. arvensis* are sometimes found and, on more acid soils, there can be some *Ulex europaeus*.

A rank growth of grasses is usually a prominent feature of the community. The commonest species throughout are *Holcus lanatus* and *Dactylis glomerata*, but *Arrhenatherum elatius* also occurs quite frequently and, with *Festuca rubra*, is especially characteristic of one kind of *Rubus-Holcus* underscrub. *Brachypodium sylvaticum* is occasionally found on more base-rich soils and, where drainage is impeded, *Deschampsia cespitosa* can be abundant. Where there is little or no grazing, the usual state of affairs, these species grow as bulky tussocks between or fronting the bramble; where herbivores have gained access after the bramble cover has developed, the grasses may continue to bulk large in the cropped sward extending between the stabilised or regressing bush cover. Smaller grasses, too, can make some contribution to the vegetation. *Poa trivialis* occurs occasionally and can extend as a patchy mat beneath quite dense bramble and, in more open, weedy vegetation, *Agrostis stolonifera* is common. On more acidic soils, species such as *Agrostis capillaris* and *Anthoxanthum odoratum* can be found, though an increase in such grasses, and in the abundance of *Ulex europaeus*, usually marks a transition to the *Ulex-Rubus* scrub.

In this coarse, grassy ground, taller dicotyledons occur as scattered plants or with patchy local abundance. *Urtica dioica*, often accompanied by scrambling *Galium aparine*, is common throughout but most of the remaining species segregate into two groups, each characteristic of the different kinds of *Rubus-Holcus* underscrub. In the *Arrhenatherum-Heracleum* subcommunity, plants of unmown and ungrazed grasslands predominate with umbellifers like *Heracleum sphondylium*, *Anthriscus sylvestris* and *Chaerophyllum temulentum* figuring frequently; in the *Cirsium* sub-community, such plants occur occasionally, but the vegetation has a more marked weedy element with *Cirsium arvense*, *C. vulgare* and *Epilobium angustifolium* occurring commonly. In contrast to the *Pteridium-Rubus* underscrub, where some of these species are also represented, along with brambles and certain of the grasses, *Pteridium aquilinum* is rare here.

Smaller dicotyledons are often overwhelmed by the dense brambles and bulky herbage but a variety of species occur at low frequencies throughout. Some of these, like *Ranunculus acris*, *Equisetum arvense*, *Trifolium repens*, *Hypochoeris radicata* and *Lotus corniculatus* are survivors from the previous herbaceous vegetation which persist in more open places. Others can tolerate considerable shade and form a patchy understorey to less dense bramble covers, being especially well developed where the community forms a stabilised fringe to hedges or woodland. In such situations, a ground carpet of *Hedera helix* is characteristic and there can be scattered plants of *Geranium robertianum*, *Geum urbanum*, *Veronica chamaedrys*, *Viola riviniana*, *Arum maculatum* and, where the community abuts on to older woods and hedges, *Mercurialis perennis*.

Bryophytes are generally sparse but *Eurhynchium praelongum* occurs occasionally and there are infrequent records for *Pseudoscleropodium purum*, *Hypnum cupressiforme* and *Brachythecium rutabulum*.

Sub-communities

***Cirsium arvense-Cirsium vulgare* sub-community.** The general feature of complementary proportions of brambles and herbs in a patchy mosaic is preserved here but, among the latter, bulkier grasses, apart from *Holcus lanatus*, tend to be less prominent: *Dactylis* and *Arrhenatherum* are fairly infrequent and *Festuca rubra* absent. *Agrostis stolonifera*, on the other hand, is preferentially frequent and sometimes abundant as a ground carpet between the bramble with occasional *Ranunculus repens*, *Glechoma hederacea*, *Prunella vulgaris* and *Fragaria vesca*. But, more obviously here, there are preferentially frequent records for taller weedy species: along with *Urtica*, *Cirsium arvense* and *C. vulgare* are common and there can be prominent clumps of *Epilobium angustifolium*. On drier, more base-poor soils, *Digitalis purpurea* and *Silene dioica* can be found and there may be some *Ulex europaeus*. On wetter ground, as where this subcommunity develops near to areas of open water or on ill-drained land where pools form in hollows, plants such as *Epilobium hirsutum*, *Phalaris arundinacea* or *Phragmites australis* can be locally abundant and there may be some *Solanum dulcamara*.

***Arrhenatherum elatius-Heracleum sphondylium* sub-community.** This kind of *Rubus-Holcus* underscrub is less diverse than the above but consistently enriched by a very distinctive group of preferentials, whose general affinities are clearly with Arrhenatherion‧ grasslands. *Rubus* remains very frequent and can be abundant but, among the grasses, *Dactylis* increases in frequency somewhat and *Arrhenatherum* and *Festuca rubra* become constant; *Poa pratensis* is also recorded occasionally. Along with the frequent *Holcus lanatus*, the total cover of these species is generally high. Then, *Urtica* is joined here by frequent *Heracleum sphondylium*, *Anthriscus sylvestris* and, especially characteristic of this kind

of vegetation, *Chaerophyllum temulentum*, the sequential flowering of these umbellifers being very prominent through the early summer months. Other frequent herbs here, of slightly shorter stature, are *Cruciata laevipes*, *Achillea millefolium*, *Stellaria holostea*, *Lapsana communis* and, more occasional but very conspicuous as a front to recently-disturbed hedgerow stands, *Alliaria petiolata*. In more open places, *Taraxacum officinale* agg. may be prominent: it can be a marked feature when flowering in spring along the trampled or mown margins of the *Rubus-Holcus* underscrub on verges. A further element is provided by scrambling or climbing herbs: *Galium aparine* is constant and occasionally accompanied by *Lathyrus pratensis*, *Vicia sepium* or *V. sativa* ssp. *nigra*. Where the soil is moist, as where this vegetation grows over roadside ditches, *Filipendula ulmaria* can be locally abundant.

Habitat

The *Rubus-Holcus* underscrub is a very typical community of abandoned and neglected ground in the British lowlands where it can be found on a wide variety of circumneutral and less oligotrophic soils. It is extremely common on derelict land, in run-down arable, pasture and meadow and over disused gardens, allotments and graveyards and here it can represent an early stage in successions to mixed deciduous or less calcifugous oak-birch woodlands. But many stands appear to be static and the community is very frequent as a component of stabilised zonations around wood margins and along hedgerows. It can also figure as a temporary or persistent vegetation type where woodland has been coppiced or cleared.

R. fruticosus agg. is a very ready and early invader of a wide range of soil types and can quickly establish itself as a dominant. The *Rubus-Holcus* underscrub includes the kind of vegetation where brambles, along with a variety of mesophytic herbs, become prominent on profiles which are less extreme in their base-status and moisture content. The community is best developed on fairly deep, moist, circumneutral brown earths, the vigorous growth of grasses such as *Holcus lanatus*, *Dactylis* and *Arrhenatherum* and dicotyledons like the *Cirsium* spp. and the umbellifers being a good indication of these edaphic conditions. Although the soils may be patchily wet, as where the *Rubus-Holcus* underscrub develops over unevenly-draining ground or spreads over ditches, uniform or prolonged waterlogging is not characteristic: in such situations, the community is generally replaced by immature kinds of *Alnus-Urtica* woodland, often dominated by osiers. Nonetheless, soil moisture may sometimes be too excessive here to allow the spread of *Pteridium*: the rarity of bracken is one good criterion for separating this community from the *Pteridium-Rubus* underscrub.

Towards the other extreme, the *Rubus-Holcus* underscrub does not extend far on to sharply-draining soils, whether these are base-rich or base-poor. On the former profiles, it is usually replaced by immature stands of calcicolous *Crataegus-Hedera* scrub, on the latter by *Ulex-Rubus* scrub, in both of which *R. fruticosus* agg. can be represented (though often not very vigorously) but where mesophytic associates are scarce. Typically, such soils as these are oligotrophic, whereas the profiles of the *Rubus-Holcus* underscrub are either naturally quite rich mulls or soils where disturbance or manuring have enhanced fertility: the prominence of such species as *Urtica* and *Galium aparine* in this vegetation is a good indication of this.

Circumneutral and mesotrophic soils of the kind likely to be invaded by the *Rubus-Holcus* underscrub are very widespread through the British lowlands, developing naturally from a diverse range of parent materials. Similar edaphic conditions are also found on a variety of man-made soils on road verges and restored land and the community will also develop on man-made raw soils as over demolished buildings and quarry waste with some admixture of fine material. In some of these situations, the community develops directly on open ground, either naturally exposed, as in landslips in soft cliffs (locally common around the south and much more extensive along the east coasts of England) or artificially cleared, as on derelict land and abandoned arable fields and gardens. Very often, though, the *Rubus-Holcus* underscrub grows up subsequent to the occupation of the ground by some kind of herbaceous vegetation, either an earlier stage in natural colonisation or grassland previously maintained as a plagioclimax by grazing or mowing. The differences between the two sub-communities can be largely understood in terms of a complex of factors related to these various modes of development and to the nature of the soils associated with each.

The *Cirsium* sub-community is characteristic of a wide variety of situations where brambles are fairly early invaders of more open ground, often with recent disturbance. It is very characteristic of waste-ground and tips of soil, but also develops on abandoned arable land and in coppice plots and cleared woodland where the soil has been greatly churned up. In such cases, the pre-existing vegetation, if there is any, or the associated flora invading with the bramble, is marked by a prominent weedy element (*Cirsium* spp.) and, since the ground conditions are very heterogenous, by a patchy development of species favouring locally wet conditions (*Epilobium hirsutum*, *Phalaris*, *Phragmites*) or picking out sites of enrichment, as around bonfires (*Urtica*, *Galium aparine*). Disturbance of more acidic soils, as in calcifugous woodlands or in heaths, allows plants like *Digitalis* and *Silene dioica* to flourish in some stands.

The *Arrhenatherum-Heracleum* sub-community, by contrast, is much more typical of situations where the *Rubus-Holcus* underscrub has developed within established grassland, either later in successional sequences or, more usually, where grazing and mowing have been abandoned in agricultural land, grassy open spaces or on verges. Natural decline in herbivore populations, as after myxomatosis, can also precipitate the development of this sub-community. In many of the places where this kind of *Rubus-Holcus* underscrub appears, the soils are more mature and of greater fertility, frequently because of the past addition of fertilisers. The characteristic associates of the bramble here are Arrhenatherion species typical of ungrazed and unmown swards on such profiles and, being often of bulky physiognomy or of climbing or scrambling habit, they can maintain themselves consistently among the bramble cover. Very often, too, this sub-community occurs in a stabilised form as a narrow fringe between woodlands, scrub or hedges to the one side and pasture, meadow, mown verge or arable land to the other.

Where the *Rubus-Holcus* underscrub develops in coppice plots or on the site of cleared woodland, some woodland herbs can persist as a shade-tolerant field layer beneath the developing bramble but such plants may also migrate in if the community occupies ground adjacent to established woodland for some considerable time. Although some species found in the open grassy areas can grow in the shade of the bramble (e.g. *Brachypodium sylvaticum*, *Poa trivialis*, *Ranunculus repens*), there is often a fairly sharp disjunction in extensive stands between the vegetation within and outside the *Rubus* cover. In much-compressed hedgebank sequences, where there is shade cast from above but light admitted from the side, much more intimate mixtures of woodland and grassland herbs occur in stands of the community.

Zonation and succession

The *Rubus-Holcus* underscrub is found in mosaics and zonations with a wide variety of herbaceous communities but a relatively small number of scrubs and woodland types. Often, these patterns are a clear expression of seral sequences in various stages of active development but many are incomplete or, very commonly, stabilised, either by their coming to a natural halt or by being subject to some form of treatment along their margins.

In areas of more open ground, the community, usually represented by the *Cirsium* sub-community, typically occurs with weedy vegetation in which *Chenopodium* spp., *Artemisia vulgaris*, Rumices, *Epilobium angustifolium*, *Urtica dioica* or, along pathways, trampling-resistant plants like *Plantago major* and *Poa annua*, occur: mosaics of communities dominated by such species, with patchy *Rubus-Holcus* underscrub among them, are very characteristic of derelict land. Continuing disturbance (often including burning) and the shortage of seed-parents often limits colonisation by shrubs and trees but zonations sometimes continue to *Crataegus-Hedera* scrub and even to woodland, usually some form of *Quercus-Pteridium-Rubus* woodland. Less frequently, the *Cirsium* sub-community can be found among various kinds of heath on disturbed commons, though here it usually marks out small enclaves of more enriched soils among stretches of *Ulex-Rubus* scrub.

The usual context for the *Arrhenatherum-Heracleum*

Figure 25. Patterns of grassland, underscrub and scrub in neglected pasture and on managed verge.
W21   *Crataegus-Hedera* scrub
W24   *Rubus-Holcus* underscrub
MG1   *Arrhenatherum* grassland
MG6   *Lolium-Cynosurus* grassland

sub-community is an agricultural landscape where this type of *Rubus-Holcus* underscrub generally forms a part of patterns involving pastures like the *Lolio-Cynosuretum*, grazed and mown grasslands like the *Centaureo-Cynosuretum* or periodically-cut swards like the *Arrhenatheretum*. Transitions through to scrub and woodland are more common here where neglect has been of long standing, and usually involve the *Crataegus-Hedera* or *Prunus-Rubus* scrubs and the *Quercus-Pteridium-Rubus* woodland or its beech analogue, the *Fagus-Rubus* woodland. On somewhat more base-rich brown earths, the *Fraxinus-Acer-Mercurialis* woodland may figure. The *Arrhenatherum-Heracleum* sub-community is also very widespread in zonations around wood margins and along hedgebanks, where sequences of these grassland, underscrub, scrub and woodland communities can be compressed into the space of a few metres (Figure 25).

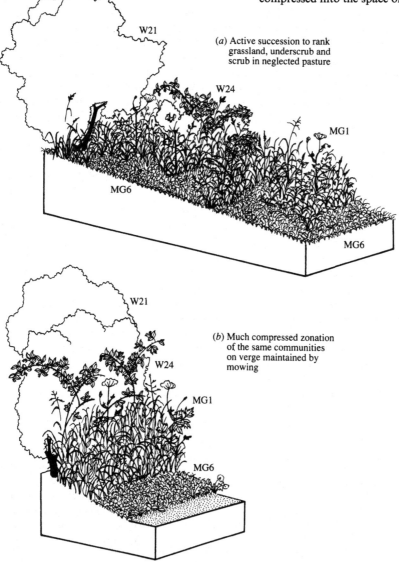

(a) Active succession to rank grassland, underscrub and scrub in neglected pasture

(b) Much compressed zonation of the same communities on verge maintained by mowing

In such situations and in more extensive stands where dense growth of the brambles may hinder invasion by shrubs or trees, the *Rubus-Holcus* underscrub can attain some degree of stability. Often, however, taller woody species invade with the *Rubus* and overtake it in a progression to scrub and woodland. In most cases, the *Quercus-Pteridium-Rubus* woodland appears to be the natural climax on the kinds of soils typical of the *Rubus-Holcus* underscrub or, more locally, the *Fagus-Rubus* woodland. In both of these communities, brambles play an important role in the field layer and, in younger stands of the *Quercus-Pteridium-Rubus* woodland, *Holcus lanatus* and various of the other mesophytic or weedy species important here, continue to be well represented. It is in this kind of woodland, too, that the *Rubus-Holcus* underscrub shows its most prominent resurgence after coppicing or clear-felling and dense

stands of the community may hinder regrowth of stools or establishment of any new planted saplings.

Distribution
The community is ubiquitous on suitable soils throughout the British lowlands.

Affinities
This very common vegetation type has frequently been referred to in the British literature, though usually included within ill-defined scrub communities or marginal vegetation (e.g. Salisbury 1916, 1918a, Tansley 1939). Similar vegetation in mainland Europe has generally been placed in the Rubion subatlanticum alliance in the Prunetalia (e.g. Doing 1962, Westhoff & den Held 1969, Ellenberg 1978).

Floristic table W24

	a	b	24
Fraxinus excelsior sapling	I (4)	I (7)	I (4–7)
Acer pseudoplatanus sapling	I (3–4)	I (1)	I (1–4)
Fagus sylvatica sapling	I (1–5)	I (5)	I (1–5)
Crataegus monogyna	II (1–5)	I (5)	II (1–5)
Sambucus nigra	I (1–4)		I (1–4)
Quercus robur sapling	I (2–3)		I (2–3)
Prunus spinosa	I (1–3)		I (1–3)
Corylus avellana	I (4)		I (4)
Betula pubescens sapling	I (5)		I (5)
Rubus fruticosus agg.	V (1–8)	IV (2–6)	V (1–8)
Holcus lanatus	IV (2–8)	III (3–4)	IV (2–8)
Cirsium arvense	III (1–4)	I (1–4)	II (1–4)
Agrostis stolonifera	III (1–8)	I (3–4)	II (1–8)
Cirsium vulgare	III (1–4)	I (1)	II (1–4)
Epilobium angustifolium	II (2–8)	I (1)	II (1–8)
Glechoma hederacea	II (2–4)	I (1–2)	II (1–4)
Ranunculus repens	II (1–4)	I (3)	I (1–4)
Prunella vulgaris	II (1–5)	I (1)	I (1–5)
Silene dioica	II (2–5)	I (3–4)	I (2–5)
Digitalis purpurea	II (3–6)	I (1)	I (1–6)
Brachythecium rutabulum	II (3–4)	I (1)	I (1–4)
Fragaria vesca	II (2–5)		I (2–5)
Epilobium hirsutum	II (3–5)		I (3–5)
Solanum dulcamara	II (1–4)		I (1–4)
Ulex europaeus	II (2–5)		I (2–5)
Senecio jacobaea	I (1–3)		I (1–3)
Phalaris arundinacea	I (3–4)		I (3–4)
Phragmites australis	I (4–6)		I (4–6)
Cirsium palustre	I (1–3)		I (1–3)
Cerastium fontanum	I (2–3)		I (2–3)

Floristic table W24 *(cont.)*

	a	b	24
Dactylis glomerata	III (1–7)	V (1–4)	III (1–7)
Urtica dioica	III (2–9)	IV (1–4)	III (1–9)
Galium aparine	II (2–6)	IV (1–4)	III (1–6)
Arrhenatherum elatius	II (1–5)	IV (2–7)	III (1–7)
Heracleum sphondylium	II (1–6)	IV (1–6)	III (1–6)
Taraxacum officinale agg.	I (1–4)	IV (1–4)	II (1–4)
Festuca rubra		IV (1–7)	II (1–7)
Anthriscus sylvestris	I (3)	III (2–5)	II (2–5)
Achillea millefolium	I (1)	III (1–3)	II (1–3)
Chaerophyllum temulentum		III (2–5)	II (2–5)
Cruciata laevipes		III (1–4)	II (1–4)
Poa pratensis	I (4)	II (1–7)	I (1–7)
Elymus repens	I (3)	II (2–3)	I (2–3)
Potentilla sterilis	I (4)	II (1–4)	I (1–4)
Dryopteris filix-mas	I (1)	II (1–3)	I (1–3)
Vicia sepium		II (2)	I (2)
Vicia sativa nigra		II (1–4)	I (1–4)
Lathyrus pratensis		II (2–6)	I (2–6)
Filipendula ulmaria		II (2–5)	I (2–5)
Stellaria holostea		II (4–7)	I (4–7)
Lapsana communis		II (1–2)	I (1–2)
Ranunculus ficaria		I (5–6)	I (5–6)
Alliaria petiolata		I (1–4)	I (1–4)
Hedera helix	II (2–3)	II (1–4)	II (1–4)
Poa trivialis	II (2–6)	II (1–6)	II (1–6)
Brachypodium sylvaticum	II (3–7)	II (1–7)	II (1–7)
Eurhynchium praelongum	II (2–3)	II (1–4)	II (1–4)
Geranium robertianum	II (1–2)	II (1–3)	II (1–3)
Geum urbanum	II (1–3)	II (1–4)	II (1–4)
Mercurialis perennis	II (2–3)	II (2–4)	II (2–4)
Veronica chamaedrys	I (3–4)	I (2–3)	I (2–4)
Ranunculus acris	I (1–4)	I (1–2)	I (1–4)
Viola riviniana	I (1–3)	I (4)	I (1–4)
Equisetum arvense	I (1–2)	I (2)	I (1–2)
Hypochoeris radicata	I (1–4)	I (1)	I (1–4)
Anthoxanthum odoratum	I (4)	I (3)	I (3–4)
Lolium perenne	I (1–4)	I (3)	I (1–4)
Rosa canina agg.	I (2–4)	I (2)	I (2–4)
Bromus sterilis	I (4–7)	I (2)	I (2–7)
Stachys sylvatica	I (2–3)	I (3)	I (2–3)
Deschampsia cespitosa	I (2–5)	I (3)	I (2–5)
Agrostis capillaris	I (1–4)	I (4)	I (1–4)
Trisetum flavescens	I (4)	I (1–3)	I (1–4)
Centaurea nigra	I (1–2)	I (4)	I (1–4)
Trifolium repens	I (3–4)	I (2)	I (2–4)
Arum maculatum	I (4)	I (4)	I (4)

Solidago virgaurea	I (3)	I (5)	I (3–5)
Pseudoscleropodium purum	I (4)	I (1)	I (1–4)
Epilobium montanum	I (1–3)	I (1)	I (1–3)
Hypnum cupressiforme	I (5)	I (1)	I (1–5)
Rumex sanguineus	I (4)	I (1)	I (1–4)
Crataegus monogyna seedling	I (1)	I (2)	I (1–2)
Lotus corniculatus	I (4)	I (1)	I (1–4)
Number of samples	28	11	39
Number of species/sample	17 (8–39)	22 (16–33)	18 (8–39)
Shrub height (m)	4 (1–9)	3 (1–9)	4 (1–9)
Shrub cover (%)	29 (0–50)	40 (0–60)	35 (0–60)
Herb height (cm)	62 (10–150)	45 (15–150)	57 (10–150)
Herb cover (%)	86 (20–100)	88 (50–100)	87 (20–100)
Ground height (mm)	16 (10–50)	15 (10–20)	16 (10–50)
Ground cover (%)	8 (0–50)	4 (0–25)	7 (0–50)
Altitude (m)	83 (10–140)	124 (5–250)	98 (5–250)
Slope (°)	6 (0–60)	21 (0–70)	11 (0–70)

a   *Cirsium arvense-Cirsium vulgare* sub-community

b   *Arrhenatherum elatius-Heracleum sphondylium* sub-community

24   *Rubus fruticosus-Holcus lanatus* underscrub (total)

W25
Pteridium aquilinum-Rubus fruticosus underscrub

Synonymy

Marginal Society Salisbury 1981*a p.p.*; *Pteridietum auct. angl. p.p.*

Constant species

Pteridium aquilinum, Rubus fruticosus agg.

Physiognomy

The *Pteridium aquilinum-Rubus fruticosus* agg. underscrub brings together vegetation dominated by mixtures of bracken and bramble. As with the *Rubus-Holcus* underscrub, although this community is often found closely associated with taller woody vegetation, shrubs and trees generally make a negligible contribution to the cover: scattered *Crataegus monogyna*, *Sambucus nigra* and *Prunus spinosa* are sometimes found and there can be very occasional saplings of *Fraxinus excelsior*, *Acer pseudoplatanus*, *Quercus robur* or *Fagus sylvatica*.

Pteridium is generally the more abundant of the two constants and, by mid-summer when its fronds are fully unfurled, it can form a virtually complete canopy to the vegetation up to a metre or more in height. In other stands, brambles are more prominent, forming a thick tangle of arching shoots with patches of bracken between or scattered fronds throughout. The brambles, which generally retain some of their leaves through the winter, may become more conspicuous when the bracken has died back, though their shoots often hold the dead fronds upright until they decay. Other undershrubs are infrequent, though *Rubus idaeus*, *Rosa canina* agg. and *Rosa arvensis* have been recorded in one subcommunity and *Ulex europaeus* is a scarce associate in the other.

No other plants attain constancy throughout and many show a marked preference for one or other of the sub-communities. Also, with the often dense cover of the dominants, the abundance of these associates is sometimes low and some are very much confined to more open areas between the bracken and bramble. Of those species which can be found throughout, *Urtica dioica* and *Holcus lanatus* are the most common, with *Silene dioica*, *Rumex acetosa* and *Viola riviniana* occasional. *Hedera helix* sometimes forms a patchy ground carpet and there can be some scrambling *Lonicera periclymenum*. Weedy plants like *Epilobium angustifolium* and *Cirsium arvense* can be locally prominent and, in more open places, *Festuca rubra*, *Arrhenatherum elatius* and *Heracleum sphondylium* are sometimes found, though these are never so frequent here as in the *Rubus-Holcus* underscrub.

Bryophytes are strongly preferential to one of the sub-communities but, even there, are not very numerous or consistently abundant.

Sub-communities

***Hyacinthoides non-scripta* sub-community.** The vegetation is a little richer here than in the *Teucrium* sub-community with more frequent records for scattered shrubs and saplings, notably *Crataegus* and *Sambucus*, occasional *Rubus idaeus* and roses among the bracken and bramble and, more obviously, a larger contingent of herbaceous associates. Foremost among these is *Hyacinthoides non-scripta* which is often present in abundance and which, flowering and fading before the *Pteridium* canopy closes, gives this vegetation a distinct vernal aspect. In the rare cases where the *Pteridium-Rubus* underscrub extends on to soils with some spring waterlogging, *Hyacinthoides* may be replaced by *Anemone nemorosa*. Grasses are often conspicuous, *Holcus mollis* in particular becoming prominent as *Hyacinthoides* fades and sometimes being accompanied by *Dactylis glomerata* and *Poa trivialis* with occasional *Holcus lanatus*. Among the dicotyledons, *Urtica dioica* and *Galium aparine* are the most frequent and they can be patchily abundant. Then, together with occasional plants of the community species *Silene dioica*, *Viola riviniana* and *Rumex acetosa*, there may be some *Stellaria holostea*, *Glechoma hederacea* and *Dryopteris filix-mas*. Taken together, these species give stands of this sub-community the appearance of a disturbed *Quercus-Pteridium-Rubus* woodland without an intact cover of shrubs and trees. On somewhat more base-rich soils, elements of a more calcicolous woodland flora can be represented with records for *Mercurialis perennis*, *Arum maculatum* and *Geranium robertianum* but such plants are generally scarce.

Bryophytes are more frequent here than in the *Teucrium* sub-community with *Eurhynchium praelongum* frequent and *Brachythecium rutabulum* occasional, both sometimes with high cover.

***Teucrium scorodonia* sub-community.** Apart from occasional bushes of *Ulex europaeus*, the taller elements of this kind of *Pteridium-Rubus* underscrub are generally confined to mixtures of bracken and bramble and there is no marked vernal aspect to the vegetation, with *Hyacinthoides* being very scarce. Indeed, most of the herbs of the former sub-community are markedly uncommon with only *Urtica* attaining occasional frequency along with the community species *Viola riviniana*, *Rumex acetosa* and *Silene dioica*. But *Holcus lanatus* increases its representation and, with occasional *Anthoxanthum odoratum* and *Agrostis capillaris*, can give stands a markedly grassy appearance. More strongly preferential, however, is *Teucrium scorodonia* and this may occasionally be accompanied by *Digitalis purpurea* and the more diminutive *Potentilla erecta*, *Galium saxatile* and *Luzula multiflora*.

Habitat

The *Pteridium-Rubus* underscrub is characteristic of deeper and generally free-draining, circumneutral to moderately acid and fairly fertile soils in the British lowlands. It is most commonly found in close association with woodlands, less frequently among heaths, and often appears to have replaced them or spread from them as a result of changes in their treatment. Once established, the community may attain a measure of stability, preventing re-invasion of woody species unless disturbed.

Pteridium performs best on deeper soils which are free from waterlogging and it has become most widely established in Britain on more acidic profiles of this kind in the *Pteridium-Galium saxatile* community. In the *Pteridium-Rubus* underscrub, its dominance extends on to less oligotrophic brown earths, edaphic conditions well marked by the importance here of *Rubus fruticosus* agg., the occurrence of such herbs as *Holcus lanatus*, *Viola riviniana* and *Silene dioica* and the relative infrequency of calcifuges like *Galium saxatile*, *Potentilla erecta* and *Deschampsia flexuosa*. Although *Pteridium* can produce very large amounts of nutrient-poor and slowly-rotting litter, the humus regime of the soils beneath this underscrub is often of the mull type and the

frequency and patchy abundance of *Urtica dioica* and *Galium aparine* point to a turnover of nutrients that is, at least locally, brisk.

Soils suitable for the establishment or spread of the *Pteridium-Rubus* underscrub are widespread throughout the British lowlands, though the bulk of them are under intensive agriculture. Bracken can colonise neglected open ground by spore dispersal but this is probably of minor significance: generally, the *Pteridium-Rubus* underscrub seems to develop by vegetative expansion from existing bracken. Most often here this is in woodland where *Pteridium* is a frequent component of the field layer but one whose abundance and vigour are held strongly in check by canopy shade. If the woody cover is removed by coppicing or clearance, the luxuriance and abundance of the bracken increase greatly, so that it can become dominant over what is essentially a woodland field layer, often with some additional species indicative of the disturbance that often accompanies such treatments. This is exactly the composition of the *Hyacinthoides* sub-community which is most frequently encountered in rides, clearings and more extensive open areas within wood-pasture where such an origin seems most likely. This kind of *Pteridium-Rubus* underscrub also persists as a field-margin vegetation type in less intensive agricultural landscapes where clearance of woody vegetation has been less assiduous as, for example, on many coastal cliffs.

The *Teucrium* sub-community may also often develop in the same way, though the poorer representation of eutrophic species and the shift towards a mildly calcifugous flora suggest that it is characteristic of somewhat drier, more acidic and impoverished profiles. Perhaps of equal importance is the great scarcity of *Hyacinthoides*, a plant slow to spread or to re-establish itself once eliminated, and the fact that *Teucrium*, *Digitalis* and *Holcus lanatus* are all very typical of more grossly disrupted or younger woodland habitats on fairly base-poor brown earths, as in much-disturbed coppice compartments, thinned plantations or along the outer margins of woodland, all of them situations where this sub-community is very common. This kind of *Pteridium-Rubus* underscrub also extends the occurrence of the community on to less extreme heathland soils where a combination of sporadic disturbance, notably burning, and the abandonment of traditional treatments like bracken-cutting has allowed this vegetation to spread over abandoned settlements and along pathways.

Grazing and browsing may also play some part in the development of the community. Herbivores can greatly hinder coppice regrowth or the establishment of planted saplings and allow the *Pteridium-Rubus* underscrub to become established in coppices or young plantations. But they also strongly affect the proportions of bracken and bramble: both can increase greatly with release from shade and their eventual abundance here may be partly a reflection of the original amounts of each. Bramble, though, is readily browsed and can be totally eliminated where herbivores have access thus tilting the balance in favour of bracken-dominance.

With its annual pattern of frond development, *Pteridium* serves as an effective replacement in this community for a deciduous woody canopy and, except where it becomes very dense, does not pose any new threat with its shade to the vernal and shade-bearing elements of the original woodland field layer. The *Hyacinthoides* sub-community, where these elements are more obvious, may thus have a considerable degree of stability, especially since the bracken canopy may prevent the re-establishment of light-demanding woody species which could eventually overtop and shade out the bracken. The accumulation of bracken litter, which can overwhelm many smaller herbs, may also not be so great on the more mesotrophic soils of this kind of *Pteridium-Rubus* underscrub. The *Teucrium* sub-community, on the other hand, may be more dependent for its maintenance on the repeated opening of the bracken canopy by the physical disturbance of thinning operations in plantations or by fire on heaths.

Zonation and succession

The *Pteridium-Rubus* underscrub is commonly found in close association with woodlands or other replacements for them within predominantly agricultural or heath landscapes. Sometimes, the derivation from woodland is clear, as where the community occupies rides and clearings, extensive glades in wood-pasture, old coppice compartments or clear-felled sites; in other cases, the historical connection with long-lost woodland is more tenuous. In general, though, this is a vegetation type of regressive seres and the overwhelming dominance of bracken itself, together with the effects of grazing and burning, may prevent re-establishment of woodland.

The community shows a close floristic relationship to the *Quercus-Pteridium-Rubus* woodland, the main Carpinion forest of base-poor brown earths in the British lowlands, and, where it occurs within or adjacent to woodland, it is almost always of this type, the under-scrub running from glades and rides under the tree and shrub canopy with very little change apart from some attenuation of the bracken and bramble cover. The *Hyacinthoides* sub-community is very similar to the field layer of the Typical sub-community of the *Quercus-Pteridium-Rubus* woodland, the *Teucrium* sub-community to the *Holcus lanatus* sub-community of the woodland, a common forest type of plantations; and, in the woodland context, the particular kinds of vegetation involved may be largely a reflection of treatment.

Where the woodland has totally disappeared, stands of the community can remain isolated, marking out the original site and sharply delimited from the surrounding landscape of improved agricultural grasslands or arable. In less intensive landscapes, the *Pteridium-Rubus* under-scrub can persist as a field-border community and then it may form part of a fairly ordered sequence from grass-lands through to remnants of woodland or scrub. Par-ticularly striking zonations of this kind can be seen on the sea-cliffs of western Britain where the *Pteridium-Rubus* underscrub occurs interposed between maritime grasslands and scrub (Malloch 1970, 1971). In these situations, the different edaphic preferences of the two sub-communities are well seen, the *Hyacinthoides* type being found between *Festuca-Hyacinthoides* grassland and the *Prunus-Rubus* scrub on moister, mesotrophic soils, the *Teucrium* type occurring with the *Festuca-Holcus* grassland and the *Ulex-Rubus* scrub on more acidic profiles.

In more degraded heath landscapes, the *Pteridium-Rubus* underscrub is represented by the *Teucrium* sub-community which usually accounts for but a small proportion of the bracken-dominated vegetation, most of it clearly belonging to the more calcifugous *Pteridium-Galium* community. Vegetation patterns in such situations are usually complex mosaics but small stands of the *Ulex-Rubus* scrub can often be found with the *Teucrium* sub-community forming a patchwork over the less-impoverished soils. Although natural senescence of bracken may permit re-invasion of woody species, healthy *Pteridium-Rubus* underscrub is probably very resistant to progression back to climax forest. Birch or, on moister and more fertile soils, ash and sycamore are possible invaders here but none is well equipped to grow on thick bracken litter or in denser shade. *Quercus robur* can fare better, though it too may suffer under thicker bracken covers (Jones 1959). The proportion of surviving bramble in the underscrub may be important here, since it casts lighter shade and helps maintain the mull humus regime on which most of the eventual forest dominants can thrive, hindering the run-down to more impoverished conditions. Although less widespread than the *Pteridium-Galium* community, the persistence of the *Pteridium-Rubus* underscrub bears similar testi-mony to the long history of woodland clearance, grazing and improvement in the southern lowlands of Britain.

Distribution

The community is widespread on suitable soils through-out lowland Britain.

Affinities

The classification of bracken-dominated communities has always posed a problem and the usual solution in Britain has been to recognise a single compendious

Pteridietum (e.g. Tansley 1939). The *Pteridium-Rubus* underscrub accounts for a fairly small proportion of such a vegetation type, the bulk of our stands belonging to the *Pteridium-Galium* community. The floristic affinities of that community argue for placing it in the Nardo-Callunetea whereas the *Pteridium-Rubus* underscrub is best retained with scrubs and bramble-dominated vegetation in the Prunetalia, probably in the Rubion subatlanticum.

Floristic table W25

	a	b	25
Crataegus monogyna	II (1–4)	I (4)	I (1–4)
Sambucus nigra	II (1–4)		I (1–4)
Prunus spinosa	I (1–6)		I (1–6)
Fraxinus excelsior sapling	I (1–6)		I (1–6)
Acer pseudoplatanus sapling	I (2–5)		I (2–5)
Quercus robur sapling	I (3–6)		I (3–6)
Fagus sylvatica sapling	I (4–5)		I (4–5)
Pteridium aquilinum	V (1–10)	V (6–10)	V (1–10)
Rubus fruticosus agg.	III (2–5)	IV (1–8)	IV (1–8)
Hyacinthoides non-scripta	IV (3–10)	I (1–9)	III (1–10)
Urtica dioica	III (1–6)	II (2–4)	III (1–6)
Galium aparine	III (1–5)	I (1–4)	II (1–5)
Eurhynchium praelongum	III (1–6)	I (3)	II (1–6)
Holcus mollis	III (2–10)	I (3)	II (2–10)
Glechoma hederacea	II (1–8)	I (3–4)	II (1–8)
Dactylis glomerata	II (1–6)	I (2–4)	I (1–6)
Geranium robertianum	II (1–4)		I (1–4)
Dryopteris filix-mas	II (1–6)		I (1–6)
Stellaria holostea	II (1–4)		I (1–4)
Brachythecium rutabulum	II (1–7)		I (1–7)
Conopodium majus	II (1–4)		I (1–4)
Rosa canina agg.	I (2–7)		I (2–7)
Rubus idaeus	I (1–3)		I (1–3)
Rosa arvensis	I (1–8)		I (1–8)
Mercurialis perennis	I (1–6)		I (1–6)
Arum maculatum	I (1–3)		I (1–3)
Poa trivialis	I (2–3)		I (2–3)
Angelica sylvestris	I (1–4)		I (1–4)
Circaea lutetiana	I (1–3)		I (1–3)
Anemone nemorosa	I (1–6)		I (1–6)
Teucrium scorodonia	I (1–5)	IV (2–4)	II (1–5)
Holcus lanatus	II (1–5)	III (1–9)	II (1–9)
Digitalis purpurea	I (4–5)	II (1–4)	II (1–5)
Agrostis capillaris	I (3–8)	II (1–7)	I (1–8)
Anthoxanthum odoratum		II (2–8)	I (2–8)
Galium saxatile		I (3–4)	I (3–4)
Potentilla erecta		I (1–4)	I (1–4)
Luzula multiflora		I (1–7)	I (1–7)
Ulex europaeus		I (1–6)	I (1–6)

Floristic table W25 *(cont.)*

	a	b	25
Viola riviniana	II (1–4)	II (1–4)	II (1–4)
Rumex acetosa	II (1–4)	II (1–4)	II (1–4)
Silene dioica	II (2–5)	II (3–4)	II (2–5)
Hedera helix	I (4–9)	I (3–10)	I (3–10)
Lonicera periclymenum	I (2–4)	I (2–6)	I (2–6)
Ranunculus ficaria	I (1–7)	I (3–4)	I (1–7)
Epilobium angustifolium	I (1–3)	I (3–4)	I (1–4)
Dryopteris dilatata	I (1–4)	I (2)	I (1–4)
Cirsium arvense	I (1–4)	I (2)	I (1–4)
Heracleum sphondylium	I (3–7)	I (1)	I (1–7)
Festuca rubra	I (3–4)	I (3–5)	I (3–5)
Galium mollugo	I (1–2)	I (1)	I (1–2)
Primula vulgaris	I (4–5)	I (4)	I (4–5)
Mnium hornum	I (1–3)	I (2)	I (1–3)
Arrhenatherum elatius	I (3–5)	I (4)	I (3–5)
Number of samples	32	22	54
Number of species/sample	18 (3–38)	no data	
Shrub height (m)	3 (1–6)	no data	
Shrub cover (%)	4 (0–35)	no data	
Herb height (cm)	63 (10–150)	no data	
Herb cover (%)	95 (25–100)	no data	
Ground height (mm)	18 (10–50)	no data	
Ground cover (%)	13 (0–100)	no data	
Altitude (m)	103 (3–250)	no data	
Slope (°)	9 (0–45)	no data	

a *Hyacinthoides non-scripta* sub-community
b *Teucrium scorodonia* sub-community
25 *Pteridium aquilinum-Rubus fruticosus* underscrub (total)

INDEX OF SYNOMYMS TO WOODLANDS AND SCRUB

The vegetation types are listed alphabetically, then by date of ascription of the name, with the code number of the equivalent NVC community thereafter. The NVC communities themselves are included in the list with a bold code.

Alder carr   Sinker 1962   W5
Alder stand type 7Aa   Peterken 1981   W6,W7
Alder stand type 7Ab   Peterken 1981   W6,W7
Alder stand type 7Ba   Peterken 1981   W5
Alder stand type 7Bb   Peterken 1981   W5
Alder stand type 7Bc   Peterken 1981   W7
Alder stand type 7D   Peterken 1981   W7,W9
Alder stand type 7E   Peterken 1981   W7
Alder stand type 7Eb   Peterken 1981   W7
Alder thicket   Rankin 1911*b*   W5
Alder wood   Rankin 1911*b*   W3
Alder wood   Clapham 1940   W5
Alder woodland   Peterken 1981   W5,W6,W7,W9
Alder woodland type 1b   McVean 1956*b*   W7
Alder woodland type 1c   McVean 1956*b*   W7
Alder woodland type 2a   McVean 1956*b*   W4
Alder woodland type 2b   McVean 1956*b*   W7
Alder woodland type 2c   McVean 1956*b*   W2,W5
Alder woodland type 3a   McVean 1956*b*   W4
Alder woodland type 3c   McVean 1956*b*   W5
Alnus glutinosa-Carex paniculata woodland   **W5**
Alnus glutinosa-Fraxinus excelsior-Lysimachia nemorum woodland   **W7**
Alnus glutinosa-Urtica dioica woodland   **W6**
Alnus-Salix-Betula woodland   Meres Report 1980   W2
Alnus-Salix-Betula woodland XXIi   Meres Report 1980   W6
Alnus-Salix-Betula woodland XXIii   Meres Report 1980   W2
Alnus-Salix woodland XXi   Meres Report 1980   W6
Alnus-Salix woodland XXii   Meres Report 1980   W6
Alnus-Salix woodland XXiii   Meres Report 1980   W5
Ash-lime stand type 4A   Peterken 1981   W10
Ash-lime stand type 4Ba   Peterken 1981   W8

Ash-lime stand type 4Bb   Peterken 1981   W8
Ash-lime stand type 4C   Peterken 1981   W8
Ash-lime woodland   Peterken 1981   W8,W10
Ash-maple-hazel woods   Rackham 1980   W8,W10
Ash-maple stand type 2A   Peterken 1981   W8
Ash-maple stand type 2B   Peterken 1981   W8
Ash-maple stand type 2C   Peterken 1981   W8
Ash-maple woodland   Peterken 1981   W8
Ash-oak wood   Tansley 1939   W8
Ash-oak wood, *Allium ursinum* society   Tansley 1939   W8
Ash-oakwood association   Moss *et al.* 1910   W8
Ash-oakwood Association   Tansley 1911   W8
Ashwood   McVean 1964*a*   W9
Ashwood association   Moss *et al.* 1910   W8
Ash-wych elm stand type 1Aa   Peterken 1981   W8
Ash-wych elm stand type 1Ab   Peterken 1981   W8,W9
Ash-wych elm stand type 1Ba   Peterken 1981   W8
Ash-wych elm stand type 1Bb   Peterken 1981   W8
Ash-wych elm stand type 1D   Peterken 1981   W8,W9
Ash-wych elm woodland   Peterken 1981   W8,W9
Beech-Ash-Oak Association type 7   McNeill 1961   W12
Beech-Ash-Yew Association   McNeill 1961   W12
Beech-Ash-Yew Assocation type 5a   McNeill 1961   W12
Beech-Ash-Yew Association type 5b   McNeill 1961   W12
Beech-Ash-Yew Association type 6   McNeill 1961   W12
Beech associes, sere 1   Watt 1924   W14
Beech associes, sere 2   Watt 1924   W14
Beech associes, sere 3   Watt 1924   W12
Beech associes, sere 4   Watt 1924   W12
Beech associes, sere A   Watt 1934*b*   W14

Hazel-ash stand type 3D   Peterken 1981   W10,W11
Hazel-ash woodland   Peterken
  1981   W8,W9,W10,W11
Hazel scrub   McVean & Ratcliffe 1962   W9
Heathy birchwood   Pigott 1956a   W11
Herb-rich birch and oakwood assciation   McVean
  1964a   W11
Herb-rich birchwood   Pigott 1956a   W9
Highland birchwoods   Tansley 1939   W11
Highland oakwoods   Tansley 1939   W11
Highland Pine Forest   Tansley 1939   W18
Hornbeam-ash woods   Rackham 1980   W8,W10
Hornbeam stand type 9Aa   Peterken 1981   W10
Hornbeam stand type 9Ab   Peterken 1981   W8
Hornbeam stand type 9Ba   Peterken 1981   W10
Hornbeam stand type 9Bb   Peterken 1981   W8
Hornbeam woodland   Peterken 1981   W8,W10
Hornbeam-woods   Rackham 1980   W8,W10
Hymenophyllum wilsonii-Isothecium myosuroides
  Association   Birks 1973   W17
Hyperico-Fraxinetum   Klötzli 1970   W8
Juniper scrub   Watt 1934a   W21
Juniper heath   Pigott 1956a   W19
Juniper scrub   Poore & McVean 1957   W19
Juniperus-Campanula nodum   Huntley & Birks
  1979a   W19
Juniperus communis-Oxalis acetosella woodland   **W19**
Juniperus-Thelypteris nodum   McVean & Ratcliffe
  1962   W19
Juniperus-Vaccinium nodum   Huntley & Birks
  1979a   W19
Lime-oak woodland   Rackham 1980   W10
Lime-ash woods   Rackham 1980   W8
Limestone scrub   Moss 1911   W21
Limewoods   Rackham 1980   W8,W10
Loch Lomond Community type 1   Tittensor & Steele
  1971   W17
Loch Lomond Community type 2   Tittensor & Steele
  1971   W11,W17
Loch Lomond Community type 3   Tittensor & Steele
  1971   W11
Loch Lomond Community type 4   Tittensor & Steele
  1971   W11
Loch Lomond Community type 6   Tittensor & Steele
  1971   W17
Loch Lomond Community type 7   Tittensor & Steele
  1971   W17
Loch Lomond Community type 8   Tittensor & Steele
  1971   W11
Loch Lomond Community type 9   Tittensor & Steele
  1971   W11
Loch Lomond Community type 10   Tittensor &
  Steele 1971   W11
Loch Lomond oakwoods   Tittensor & Steele
  1971   W11,W17

Lonicero-Quercetum (Birse & Robertson 1976) Birse
  1984   W10,W11
Lonicero-Quercetum, Endymion subassociation (Birse
  & Robertson 1976) Birse 1984   W10,W11
Lonicero-Quercetum, Typical subassociation (Birse &
  Robertson 1976) Birse 1984   W11
Lowland birchwoods   Rackham 1980   W10
Luzulo-Betuletum odoratae   Birse 1984   W11
Luzulo-Betuletum, Rubus saxatilis
  subassociation   Birse 1984   W11
Malham Tarn Birchwoods   Proctor 1974   W4
Malham Tarn Birchwoods   Adam *et al.* 1975   W4
Maple-hornbeam woods   Rackham 1980   W8
Marginal flora   Tansley 1939   W24
Marginal Society   Salisbury 1916   W24
Marginal Society   Salisbury 1918a   W24,W25
Mercurialis society   Adamson 1912   W8
Mercury beechwoods   Watt & Tansley 1930   W12
Mixed deciduous woodland   McVean & Ratcliffe
  1962   W9
Mixed Deciduous Woodland   Ferreira 1978   W9
Montane willow scrub association   McVean
  1964   W20
Oak-birch-heath association   Moss *et al.* 1910   W16
Oak-birch-heath association   Tansley 1911   W16
Oak-hazel woods   Moss 1907   W8
Oak-hazel woods   Salisbury 1916   W8
Oak-hornbeam stand type 9Aa   Peterken 1981   W10
Oak-hornbeam woodland   Rackham 1980   W10
Oak-hornbeam woods, marginal & path
  floras   Salisbury 1916   W8
Oak-hornbeam woods, marginal and path
  floras   Salisbury 1918   W8
Oak-lime stand type 5A   Peterken 1981   W10
Oak-lime stand type 5B   Peterken 1981   W11
Oak-lime woodland   Peterken 1981   W10,W11
Oakwoods   Pigott 1955   W16
Oakwoods   Pigott 1956b   W16
Oakwoods   Rackham 1980   W10,W16
Open carr   Pearsall 1918   W3
Osmundo-Alnetum glutinosae   Klötzli 1970   W2,W5
Osmundo-Alnetum glutinosae (Klötzli 1970) Wheeler
  1980c   W2,W5
Osmundo-Alnetum glutinosae chrysosplenietosum
  (Klötzli 1970) Wheeler 1980c   W5
Osmundo-Alnetum glutinosae
  hydrocotyletosum   Klötzli 1970   W5
Osmundo-Alnetum glutinosae lycopetosum   Klötzli
  1970   W5
Osmundo-Alnetum glutinosae typicum (Klötzli 1970)
  Wheeler 1980c   W5
Osmundo-Alnetum sphagnetosum   Wheeler
  1980c   W4
Oxalido-Betuletum   Graham 1971   W11
Oxalido-Betuletum typicum   Graham 1971   W11

Quercetum sessiliflorae, drier sub-association   Moss
  1911   W10

Quercetum sessiliflorae, drier sub-association   Moss
  1913   W10

Quercetum sessiliflorae, heathy sub-association   Moss
  1911   W16

Quercetum sessiliflorae, heathy sub-association   Moss
  1913   W16

Quercetum sessiliflorae, *Pteris-Holcus-Scilla*
  complementary society   Woodhead 1906   W10

Querco-Betuletum   Klötzli 1970   W10,W16

Querco-Fraxinetum   Klötzli 1970   W8

Querco-Fraxinetum filipendulietosum   Klötzli
  1970   W8

Querco-Fraxinetum typicum   Klötzli 1970   W8

Querco-Ulmetum glabrae   Birse & Robertson 1976
  emend. Birse 1984   W8,W9

Querco-Ulmetum glabrae, *Allium ursinum*
  subassociation   Birse 1984   W8

Querco-Ulmetum glabrae, Typical
  subassociation   Birse & Robertson 1976 *emend.*
  Birse 1984   W8,W9

Quercus petraea-Betula pubescens-Dicranum majus
  woodland   **W17**

Quercus petraea-Betula pubescens-Oxalis acetosella
  woodland   **W11**

Quercus robur-Carpinus woodland   Salisbury
  1916   W8,W10

Quercus robur-Carpinus woods, *Anemone*
  society   Salisbury 1916   W8,W10

Quercus robur-Carpinus woods, *Ficaria*
  society   Salisbury 1916   W8

Quercus robur-Carpinus woods, *Mercurialis*
  society   Salisbury 1916   W8

Quercus robur-Carpinus woods, *Pteris*
  society   Salisbury 1916   W10

Quercus robur-Carpinus woods, *Scilla*
  society   Salisbury 1916   W10

Quercus robur-Pteridium aquilinum-Rubus fruticosus
  woodland   **W10**

Quercus robur wood, *Holcus* society   Adamson
  1912   W10

Quercus robur wood, *Pteris-Holcus* society   Adamson
  1912   W10

Quercus sessiliflora-Carpinus woodland   Salisbury
  1918a   W8,W10

Quercus sessiliflora-Carpinus woods, *Anemone*
  society   Salisbury 1918a   W8

Quercus sessiliflora-Carpinus woods, *Ficaria*
  society   Salisbury 1918a   W8

Quercus sessiliflora-Carpinus woods, *Mercurialis*
  society   Salisbury 1918a   W8

Quercus sessiliflora-Carpinus woods, *Nepeta*
  society   Salisbury 1918a   W8

Quercus sessiliflora-Carpinus woods, *Pteris*

  society   Salisbury 1918a   W10

Quercus sessiliflora-Carpinus woods, *Rubus*
  society   Salisbury 1918a   W10

Quercus sessiliflora-Carpinus woods, *Scilla*
  society   Salisbury 1918a   W10

Quercus spp.-*Betula* spp.-*Deschampsia flexuosa*
  woodland   **W16**

Retrogressive scrub   Moss 1913   W21

Rhamnetum   Godwin & Tansley 1929   W2

Rhamnetum   Godwin 1936   W2

Rhamnetum   Godwin 1943a, b   W2

Rhamno-franguletum   Godwin & Tansley 1929   W2

Rhamno-franguletum   Godwin 1936   W2

Rhamno-franguletum   Godwin 1943a, b   W2

Rhamnus catharticus sociation   Wheeler 1980c   W2

Rubus fruticosus-Holcus lanatus
  underscrub   **W24**

Rusco-Fagetum   Géhu 1975b   W15

Salix carr   Ingram *et al.* 1959   W3

Salix carr   Willis & Jefferies 1959   W1

Salix cinerea carr   Wheeler 1980c   W1

Salix cinerea-Betula pubescens-Phragmites australis
  woodland   **W2**

Salix cinerea-Galium palustre woodland   **W1**

Salix lapponum-Luzula sylvatica nodum   McVean &
  Ratcliffe 1962   W20

Salix lapponum-Luzula sylvatica scrub   **W20**

Salix lapponum-Vaccinium myrtillus nodum   Huntley
  1979   W20

Salix pentandra-Carex rostrata woodland   **W3**

Sanicle beechwoods   Watt & Tansley 1930   W12

Scottish beechwoods   Watt 1931   W11

Scottish beechwoods, *Holcus* type   Watt 1931   W11

Scottish *Pinetum sylvestris*   Tansley 1911   W18

Scrub *auct. angl.*   W21,W22

Scrub associations   Tansley 1911   W21,W22,W23

Scutellaria galericulata-Alnus glutinosa
  Association   Birse 1982   W1,W5

Semi-swamp carr   Lambert 1951   W2,W5

Sorbo-Brachypodietum   Graham 1971   W9

Southern mixed shrub communities   Duffey *et al.*
  1974   W21

Southern Pennine oakwoods   Scurfield 1953   W16

Southern Pennine oakwoods, *Deschampsia-Vaccinium*
  type   Scurfield 1953   W16

Sphagno-Salicetum atrocinereae   Birse 1984   W3,W4

Sphagnum palustre-Betula pubescens
  Community   Birse 1982   W4

Spiraea-Deschampsia society   Adamson 1912   W8

Spiraea society   Adamson 1912   W8

Suckering elm woodland   Peterken 1981   W8

Swamp carr   Pallis 1911   W5

Swamp carr   Lambert 1951   W5

Sweet chestnut coppice   Ford & Newbould
  1970   W10

INDEX OF SPECIES IN WOODLANDS AND SCRUB

The species are listed alphabetically, with the code numbers of the NVC communities in which they occur thereafter. Bold codes indicate that a species is constant throughout the community, italic codes that a species is constant in one or more sub-communities.

Acer campestre   W8, W10, W12, W21
Acer campestre seedling   W8, W10, W12
Acer pseudoplatanus   W6, W7, *W8*, W9, W10, W12, W13, W14, W15, W16, W17
Acer pseudoplatanus sapling   W6, W7, W8, W9, W10, W12, W13, W14, W16, W17, W21, W24, W25
Acer pseudoplatanus seedling   W8, W9, W10, W12, W14, W15, W16
Achillea millefolium   W22, W23, W24
Achillea ptarmica   W7
Adelanthus decipiens   W17
Adoxa moschatellina   W8, W9, W19
Aegopodium podagraria   W10
Aesculus hippocastanum   W12
Agrostis canina canina   W1, W2, W4, W7, W20
Agrostis canina montana   W11, W17, W18, **W19**
Agrostis capillaris   W1, W4, W7, W9, W10, **W11**, W14, W15, W16, *W17*, W18, **W19**, W20, W22, **W23**, W24, W25
Agrostis stolonifera   W1, W2, W3, W4, W5, W6, W9, W14, W16, W17, W24
Aira praecox   W23
Ajuga reptans   W2, W3, *W5*, W7, W8, W9, W10, W11, W12
Alchemilla alpina   W20
Alchemilla glabra   W9, W20
Alchemilla xanthochlora   W9
Alliaria petiolata   W8, W12, W24
Allium ursinum   *W8*, W9, W12, W21
Alnus glutinosa   W1, W2, W3, W4, **W5**, **W6**, **W7**, W8, W9, W10
Alnus glutinosa sapling   W1, W2, W4, W5, W6, W7
Alnus glutinosa seedling   W3
Amblystegium riparium   W2
Amblystegium serpens   W2, W5, W8, W10
Anastrepta orcadensis   W17, W18

Andreaea rupestris   W17
Anemone nemorosa   W7, *W8*, W9, *W10*, *W11*, W12, W14, W17, *W19*, W20, W21, W25
Angelica sylvestris   W1, W2, **W3**, W4, W5, W6, W7, W8, W9, W11, W20, W25
Anthoxanthum odoratum   W4, W7, W9, W10, **W11**, W16, *W17*, W19, W20, *W23*, W24, W25
Anthriscus sylvestris   W8, W9, W10, W12, W24
Apium nodiflorum   W1, W6
Arctium minus   W8, W12, W21
Armeria maritima   W22
Arrhenatherum elatius   W6, W8, W9, W10, W12, W21, W23, *W24*, W25
Arum maculatum   W8, W9, W12, W13, W14, W21, W24, W25
Asplenium ruta-muraria   W8
Asplenium trichomanes   W8
Asplenium viride   W8
Athyrium filix-femina   W2, W3, W4, W5, W6, W7, W8, W9, W10, W11, W17
Atrichum undulatum   W2, W5, W7, W8, W9, W10, W11, W14, W19, W20, W22
Atropa belladonna   W13
Aulacomnium androgynum   W3
Aulacomnium palustre   W2, W4, W18, W20

Barbilophozia attenuata   W17
Barbilophozia barbata   W19, W20
Barbilophozia floerkii   W16, W17, W18
Barbilophozia hatcheri   W19
Bazzania tricrenata   W17
Bazzania trilobata   W17, W18
Berula erecta   W2, W5
Betula hybrids   W7, W11
Betula hybrids sapling   W7
Betula hybrids seedling   W11

Betula pendula W1, W4, W5, W6, W7, W8, W9,
   W10, W11, W12, W14, W15, **W16**, W17, W18
Betula pendula sapling W2, W8, W10, W11, W14,
   W15, W16, W17, W21
Betula pendula seedling W10, W16, W18
Betula pubescens W1, **W2**, W3, **W4**, W5, W6, W7,
   W8, W9, W10, *W11*, W12, W14, W15, W16, *W17*,
   W18, W19
Betula pubescens sapling W1, W4, W6, W7, W9,
   W10, W11, W12, W16, W17, W24
Betula pubescens seedling W4, W11, W16, W17, W18
Blepharostoma trichophyllum W17
Blechnum spicant W4, W9, W10, *W11*, W15, W16,
   W17, W18, W19
Brachypodium pinnatum W21
Brachypodium sylvaticum W7, *W8*, *W9*, W10, W11,
   W12, W13, W14, *W21*, W22, W23, W24
Brachythecium rivulare W7
Brachythecium rutabulum W1, *W2*, W3, W4, **W5**,
   W6, W7, *W8*, W9, W10, W12, W14, W21, W22,
   W24, W25
Brachythecium velutinum W13
Bromus erectus W21
Bromus ramosus W8, W9, W10, W12, W14
Bromus sterilis W12, W21, W24
Bryum pseudotriquetrum W5
Buxus sempervirens W12, W13

Calamagrostis canescens W2, W5
Calliergon cordifolium W1, W3, W5
Calliergon cuspidatum W1, W2, **W3**, W4, W5, W7,
   W20
Calliergon giganteum W2, W3
Calliergon stramineum W4
Callitriche stagnalis W1, W3, W5, W8
Calluna vulgaris W4, W11, *W15*, W16, *W17*, **W18**,
   W19, W20, W23
Caltha palustris W1, W2, **W3**, W5, W6, W7, W20
Calypogeia fissa W2, W4, W15, W16, W17
Calypogeia muellerana W17, W18
Calypogeia trichomanis W16, W18
Calystegia sepium W2, W5, W21
Campanula latifolia W8
Campanula rotundifolia W11, W19, W20, W23
Campanula trachelium W8, W12
Campylium stellatum W5
Campylopus paradoxus W2, W4, W16, W17, W18
Campylopus pyriformis W16
Cardamine amara W5, W7
Cardamine flexuosa W1, W3, W6, W7, W8, W9,
   W19, W22
Cardamine impatiens W8
Cardamine pratensis **W3**, *W5*, W7
Carex acuta W2
Carex acutiformis W2, W5, W6, W8

Carex appropinquata W3, W5
Carex atrata W20
Carex bigelowii W20
Carex binervis W19, W20
Carex digitata W8
Carex echinata W3
Carex elata W5
Carex elongata W5
Carex flacca W12, W19, W20
Carex laevigata W3, W7
Carex nigra W1, W2, W3, W4, W7
Carex pallescens W9
Carex paniculata W1, W2, W3, **W5**, W6
Carex pendula W7, W8
Carex pilulifera W11, W15, W17, W19
Carex pseudocyperus W5
Carex remota W1, W2, W5, W7, W8, W14
Carex riparia W1, W5, W6
Carex rostrata **W3**, W4
Carex strigosa W8
Carex sylvatica W8, W9, W10, W12, W14
Carex vesicaria W1, W2, W3
Carpinus betulus W8, W10, W12
Carpinus betulus sapling W10
Castanea sativa W8, W10, W16
Castanea sativa sapling W8
Centaurea nigra W24
Cephalanthera damasonium W12
Cephalozia biscuspidata W3, W16, W17
Cephalozia media W17
Cerastium fontanum W11, W19, W23, W24
Ceratodon purpureus W10
Chaerophyllum temulentum W24
Chiloscyphus pallescens W3
Chiloscyphus polyanthos W1, W3, W4, W7, W20
Chrysosplenium oppositifolium W3, *W5*, W6, *W7*,
   W8, W9
Cicuta virosa W5
Circaea intermedia W9, W14
Circaea lutetiana W6, W7, W8, W9, W10, W12,
   W14, W21, W25
Cirriphyllum piliferum W7, W8, W9, W11, W19
Cirsium arvense W21, W24, W25
Cirsium helenioides W9
Cirsium palustre W1, W2, W3, W4, *W5*, W6, W7,
   W9, W24
Cirsium vulgare W24
Cladium mariscus W2
Cladonia arbuscula W17, W18
Cladonia chlorophaea W17
Cladonia coccifera W16
Cladonia coniocraea W16
Cladonia cornuta W18
Cladonia deformis W18
Cladonia digitata W17, W18

Eurhynchium striatum (cont.)
W17, W22
Eurhynchium swartzii   W7, W8, W9

Fagus sylvatica   W4, W7, W8, W9, W10, W11, **W12**, W13, **W14**, **W15**, W16, W17
Fagus sylvatica sapling   W8, W10, W12, W14, W15, W16, W17, W21, W24, W25
Fagus sylvatica seedling   W4, W8, W10, W12, **W14**, W15, W16
Festuca gigantea   W8, W9, W14
Festuca ovina   W10, W11, W16, W17, W18, W19, W20
Festuca pratensis   W11
Festuca rubra   W11, W16, W19, W20, W22, W23, *W24*, W25
Filipendula ulmaria   W1, *W2*, **W3**, W4, W5, W6, **W7**, W8, *W9*, W20, W22, W24
Fissidens adianthoides   W3, W20
Fissidens bryoides   W8, W9
Fissidens taxifolius   W8, W9, W12, W14
Fragaria vesca   W7, W8, W9, W10, W12, W13, W19, W21, W24
Frangula alnus   W1, W2, W5, W16
Fraxinus excelsior   W2, W4, W5, W6, W7, **W8**, **W9**, W10, W11, *W12*, W13, W14, W17
Fraxinus excelsior sapling   W2, W6, W7, *W8*, W9, W10, W12, W13, W14, W17, W21, W24, W25
Fraxinus excelsior seedling   W2, W7, W8, W9, W10, W11, W12, W13, W14, W17
Frullania tamarisci   W17

Galeopsis tetrahit   W3, W6
Galium aparine   W1, W2, W3, W5, *W6*, W7, W8, W9, W10, W11, W12, *W21*, W22, *W24*, W25
Galium boreale   W19, W20
Galium mollugo   W12, W25
Galium odoratum   W8, W9, W10, W12, W14
Galium palustre   **W1**, W2, **W3**, W4, **W5**, W6, W7, W8
Galium saxatile   W4, W7, W9, W10, **W11**, W16, *W17*, W18, **W19**, W20, W23, W25
Galium uliginosum   W1, W3, W4, W6
Galium verum   W14, W19
Geranium robertianum   W2, W5, W6, W7, W8, W9, W10, W11, W12, W21, W22, W24, W25
Geranium sanguineum   W8
Geranium sylvaticum   W9, W19
Geum rivale   **W3**, W7, W8, W9, W19, W20
Geum urbanum   W6, W7, W8, W9, W12, W14, W21, W22, W24
Glechoma hederacea   W2, W6, W7, W8, W10, W12, W13, W21, W24, W25
Glyceria fluitans   W1, W7
Glyceria maxima   W2, W5
Goodyera repens   W17, *W18*

Gymnocarpium dryopteris   W19
Gymnocolea inflata   W15, W16

Harpanthus scutatus   W17
Hedera helix   W1, W2, W3, W4, W5, W6, W7, *W8*, W9, *W10*, *W12*, W13, W14, W15, W16, W17, **W21**, W22, W23, W24, W25
Heracleum sphondylium   W6, W8, W9, W10, W12, W21, *W24*, W25
Herbertus hutchinsiae   W18
Heterocladium heteropterum   W17
Hieracium sect. *Alpina*   W20
Hieracium sect. *Cerinthoidea*   W20
Hieracium sect. *Subalpina*   W20
Holosteum umbellatum   W8
Holcus lanatus   W1, W2, W3, W4, W5, W6, W7, W8, W9, *W10*, W11, W14, W15, W16, W17, W21, W22, *W23*, **W24**, W25
Holcus mollis   *W4*, W6, W7, W8, W9, *W10*, **W11**, W14, W15, W16, W17, W19, W21, W22, W23, W25
Homalothecium lutescens   W8
Humulus lupulus   W2, W5, W6
Huperzia selago   W20
Hyacinthoides non-scripta   W6, W7, *W8*, W9, *W10*, *W11*, W12, W14, W15, W16, W17, W21, W22, *W25*
Hydrocotyle vulgaris   W1, W2, W3, W4, W5
Hylocomium brevirostre   W9, W11, W17
Hylocomium splendens   W9, **W11**, W15, **W17**, **W18**, **W19**, **W20**
Hylocomium umbratum   W17, W18, W19
Hymenophyllum tunbrigense   W17
Hymenophyllum wilsonii   W17
Hyocomium armoricum   W17
Hypericum hirsutum   W8, W12
Hypericum perforatum   W8
Hypericum pulchrum   W11, W14, W17
Hypericum tetrapterum   W5
Hypnum callichroum   W17, W20
Hypnum cupressiforme   W2, W3, W4, W5, W8, W9, W10, W11, W14, W15, W16, W17, W18, *W19*, W20, W24
Hypnum jutlandicum   W4, W16, W17, *W18*, W19
Hypochoeris radicata   *W23*, W24

Ilex aquifolium   W4, W5, W6, W7, W8, W9, W10, W12, W13, **W14**, *W15*, W16, W17, W18, W21
Ilex aquifolium seedling   W8, W10, W11, W12, W14, *W15*, W16, W17
Impatiens capensis   W5
Impatiens glandulifera   W6
Inula conyza   W13
Iris foetidissima   W8, W12, W13, W21
Iris pseudacorus   W1, W5, W6, W7
Isopterygium elegans   W4, W7, W10, W13, W14, W15, W16, W17

Philonotis fontana   W20
Phragmites australis   W1, **W2**, W3, W5, W6, W24
Phyllitis scolopendrium   W8, W9, W12, W21, W22
Picea abies   W18
Picea sitchensis   W18
Pinus nigra var. *maritima*   W10
Pinus sylvestris   W4, W6, W9, W10, W14, W15, W16, **W18**
Pinus sylvestris seedling   W18
Plagiochila asplenoides   W3, W7, W8, W9, W11, W17, W19, W20
Plagiochila atlantica   W17
Plagiochila killarniensis   W17
Plagiochila spinulosa   W17
Plagiochila corniculata   W17
Plagiomnium affine   W3, W6, W8, W11, W19
Plagiomnium elatum   W3
Plagiomnium ellipticum   W3
Plagiomnium rostratum   W3, W5, W8, W9, W10, W19
Plagiomnium undulatum   W1, W2, W5, W6, W7, W8, **W9**, W10, *W11*, W12, W19, W20, W21, W22
Plagiothecium denticulatum   W2, W4, W6, W7, W8, W9, W10, W11, W14, W16, W17, W19, W20
Plagiothecium succulentum   W7, W9, W16, W19
Plagiothecium sylvaticum   W2, W8, W16
Plagiothecium undulatum   W10, W11, W14, W15, W16, **W17**, **W18**, *W19*, W20
Plantago lanceolata   W22, W23
Platanthera chlorantha   W8
Pleurozium schreberi   W4, *W11*, W16, **W17**, **W18**, *W19*, W20
Poa nemoralis   W8, W10, W12, W14, W15, W17
Poa pratensis   W11, W17, W23, W24
Poa trivialis   W2, W3, W4, W5, *W6*, W7, W8, W9, W10, W11, W12, W14, W19, W21, W22, W24, W25
Pohlia nutans   W2, W4, W16
Polygonatum multiflorum   W8
Polygonatum odoratum   W8
Polygonum viviparum   W19, W20
Polypodium vulgare   W8, W17
Polystichum aculeatum   W8
Polystichum lonchitis   W8
Polystichum setiferum   W8
Polytrichum commune   W4, W7, W11, W14, W17, W19
Polytrichum formosum   W4, W9, W10, *W11*, W14, W15, **W17**, W18, W19
Polytrichum alpinum   W20
Polytrichum juniperinum   W18
Polytrichum longisetum   W11, W18
Populus canescens   W5
Populus nigra   W5
Populus tremula   W5, W6, W8, W9, W10, W16
Potamogeton polygonifolius   W3

Potentilla erecta   W2, *W4*, W9, **W11**, W16, W17, W18, W19, W23, W25
Potentilla palustris   W1, W3, W4
Potentilla sterilis   W7, W8, W9, W10, W19, W21, W24
Primula elatior   W8
Primula vulgaris   W7, W8, W9, W10, W11, W12, W17, W22, W25
Primula vulgaris × *elatior*   W8
Prunella vulgaris   W6, W8, W9, W10, W11, W19, W24
Prunus padus   W7, W8, W9
Prunus avium   W8, W10, W12, W14
Prunus spinosa   W6, W7, W8, W10, W21, **W22**, W24, W25
Pseudoscleropodium purum   W4, W10, **W11**, W16, W17, W18, W19, W23, W24
Pseudotsuga menziesii   W10
Pteridium aquilinum   W4, W7, W8, W9, **W10**, **W11**, W14, *W15*, **W16**, *W17*, W18, W19, W21, W22, W23, **W25**
Ptilidium ciliare   W17, W20
Ptilium crista-castrensis   W17, W18, W19, W20
Ptychomitrium polyphyllum   W17
Pyrola media   W19
Pyrola minor   W11, W19
Pyrola rotundifolia   W3
Pyrus communis   W8

Quercus hybrids   W7, W8, W9, W10, W11, W15, W17
Quercus hybrids sapling   W4, W7, W10, W16
Quercus hybrids seedling   W11, W16, W17
Quercus petraea   W4, W7, W8, W9, W10, *W11*, W15, *W16*, **W17**
Quercus petraea sapling   W4, W7, W10, W16, W17
Quercus petraea seedling   W2, W15, W16, W17, W18
Quercus robur   W1, W2, W4, W5, W6, W7, *W8*, W9, **W10**, W11, W12, W13, W14, *W15*, W16, W17, W18
Quercus robur sapling   W1, W4, W7, W8, W10, W11, W12, W15, W16, W21, W24, W25
Quercus robur seedling   W4, W10, W11, W14, W15, W16, W17
Quercus sp. seedling   W17

Racomitrium fasciculare   W17
Racomitrium heterostichum   W17
Racomitrium lanuginosum   W17, W20
Ranunculus acris   W3, W5, W6, W7, W8, W9, W11, W19, W20, W24
Ranunculus auricomus   W8
Ranunculus bulbosus   W12
Ranunculus ficaria   W6, W7, *W8*, W9, W10, W12, W24, W25
Ranunculus flammula   W1, W3, W5, W7

Sphagnum teres   W20
Stachys sylvatica   W6, W7, W8, W9, W10, W12,
   W21, W22, W24
Stellaria alsine   W7
Stellaria graminea   W21
Stellaria holostea   W7, W8, W9, W10, W11, W19,
   W21, W24, W25
Stellaria media   W2, W6, W10, W14, W21, W22
Succisa pratensis   W3, W4, W7, W9, W11, W17, W20
Symphytum officinale   W2, W5

Tamus communis   W2, W8, W12, W13, W21
Taraxacum officinale   W6, W8, *W24*
Taxus baccata   W8, W10, *W11*, **W13**, W14, W15
Taxus baccata sapling   W8, W12, W13, W21
Taxus baccata seedling   W12, W13, W14
Tetraphis pellucida   W15, W16, W17
Teucrium scorodonia   W7, *W8*, W10, W11, W12,
   W16, W17, W21, W22, *W23*, *W25*
Thalictrum alpinum   W20
Thalictrum flavum   W5
Thamnobryum alopecurum   W8, W12
Thelypteris limbosperma   W2, W10, W11, W17, W19
Thelypteris palustris   W2, W5
Thelypteris phegopteris   W2, W19, W20
Thuidium delicatulum   W9, W17
Thuidium tamariscinum   W1, W7, W8, **W9**, W10,
   W11, W13, W14, W15, W16, W17, W18, **W19**,
   W20, W22
Thymus praecox   W20
Tilia cordata   W8, W10
Tilia platyphyllos   W8
Tilia × *vulgaris*   W10
Torilis japonica   W21
Tortella tortuosa   W8
Trientalis europaea   W11, W17, W18, W19
Trifolium repens   W24
Trisetum flavescens   W24
Tritomaria exsecta   W17

Tritomaria quinquedentata   W17
Trollius europaeus   W3, W9
Tussilago farfara   W7, W9

Ulex europaeus   W10, W16, W22, **W23**, W24, W25
Ulex gallii   W16
Ulmus glabra   W7, *W8*, W9, W10, W14
Ulmus glabra sapling   W7, W8, W9, W10, W12, W21
Ulmus minor   W8
Ulmus carpinifolia suckers   W8, W21
Ulmus procera   W8, W21
Ulmus sp. suckers   W8, W21
Urtica dioica   W2, W3, W5, W6, *W7*, W8, W9, W10,
   W12, W13, W14, W19, *W21*, W22, *W24*, W25

Vaccinium myrtillus   W4, W10, W11, *W15*, W16,
   W17, **W18**, **W19**, **W20**
Vaccinium oxycoccos   W4
Vaccinium uliginosum   W20
Vaccinium vitis-idaea   W11, **W18**, *W19*, W20
Valeriana dioica   **W3**, W5
Valeriana officinalis   W1, W2, W3, W4, W5, W7, W9,
   W20
Veronica beccabunga   W3
Veronica chamaedrys   W8, W9, W10, *W11*, W12,
   W22, W23, W24
Veronica montana   W6, W7, W8, W9
Veronica officinalis   W11, W19, W23
Veronica scutellata   W3
Viburnum lantana   W8, W10, W12, *W21*
Viburnum opulus   W2, W5, W6, W7, W8, W10, W12
Vicia sativa nigra   W24
Vicia sepium   W8, W9, W24
Viola odorata   W8
Viola palustris   W3, W4, W5, W7
Viola reichenbachiana   W8
Viola riviniana   W7, W8, **W9**, W10, **W11**, W14, W17,
   W19, W20, W21, W22, W23, W24, W25

BIBLIOGRAPHY

Abeywickrama, B.S. (1949). *A study of the variation in the field layer vegetation of two Cambridgeshire woods.* Cambridge University: PhD thesis.

Adam, P. (1976). *Plant sociology and habitat factors in British saltmarshes.* Cambridge University: PhD thesis.

Adam, P., Birks, H.J.B., Huntley, B. & Prentice, I.C. (1975). Phytosociological studies at Malham Tarn moss and fen, Yorkshire, England. *Vegetatio,* **30,** 117–32.

Adamson, R.S. (1912). An ecological study of a Cambridgeshire woodland. *Journal of the Linnean Society (Botany),* **40,** 339–87.

Adamson, R.S. (1921). The woodlands of Ditcham Park, Hampshire. *Journal of Ecology,* **9,** 114–219.

Adamson, R.S. (1932). Notes on the natural regeneration of woodland in Essex. *Journal of Ecology,* **20,** 152–6.

Anderson, M.L. (1950). *The Selection of Tree Species.* Edinburgh: Oliver & Boyd.

Anderson, M.L. (1967). *A History of Scottish Forestry.* London: Nelson.

Ash, J.E. & Barkham, J.P. (1976). Changes and variability in the field layer of a coppiced woodland in Norfolk, England. *Journal of Ecology,* **64,** 697–712.

Aune, E.I. (1977). Scandinavian pine forests and their relationships to the Scottish Pinewoods. In *Native Pinewoods of Scotland,* ed. R.G.H. Bunce & J.N.R. Jeffers, pp. 5–9. Cambridge: Institute of Terrestrial Ecology.

Avery, B.W. (1958). A sequence of beechwood soils on the Chiltern hills, England. *Journal of Soil Science,* **9,** 210–24.

Avery, B.W. (1964). *The Soils and Land Use of the District around Aylesbury and Hemel Hempstead.* Memoirs of the Soil Survey of Great Britain. London: HMSO.

Avery, B.W. (1980). *Soil Classification for England and Wales (Higher Categories). Soil Survey Technical Monograph No. 14.* Harpenden: Soil Survey of England and Wales.

Barker, S. (1985). *The Woodlands and Soils of the Coniston Basin, Cumbria.* Lancaster University: PhD thesis.

Barkham, J.P. & Norris, J.M. (1967). The changing ground flora of some Cotswold beechwoods. *Proceedings of the Cotteswold Naturalists' Field Club,* **35,** 107–11.

Barkham, J.P. & Norris, J.M. (1970). Multivariate procedures in an investigation of vegetation and soil relationships of two beech woodlands, Cotswold hills, England. *Ecology,* **51,** 630–9.

Barkman, J.J. (1958). *Phytosociology and Ecology of Cryptogamic Epiphytes.* Essen: Van Gorcum.

Bartley, D.D. (1960). Ecological studies at Rhosgoch Common, Radnorshire. *Journal of Ecology,* **48,** 205–13.

Bellamy, D.J. & Rose, F. (1961). The Waveney–Ouse valley fens of the Suffolk–Norfolk border. *Transactions of the Suffolk Naturalists' Society,* **11,** 367–85.

Birks, H.J.B. (1969). *The Late-Weichselian and Present Vegetation of the Isle of Skye.* Cambridge University: PhD thesis.

Birks, H.J.B. (1970). The Flandrian forest history of Scotland: a preliminary synthesis. In *British Quaternary Studies, Recent Advances,* ed. F.W. Shotton, pp. 119–35. Oxford: Clarendon Press.

Birks, H.J.B. (1973). *The Past and Present Vegetation of the Isle of Skye: a Palaeoecological Study.* Cambridge: Cambridge University Press.

Birse, E.L. (1980). *Plant Communities of Scotland: A Preliminary Phytocoenonia.* Aberdeen: Macaulay Institute for Soil Research.

Birse, E.L. (1982). The main types of woodland in North Scotland. *Phytocoenologia,* **10,** 9–55.

Birse, E.L. (1984). *The Phytocoenonia of Scotland: Additions and Revisions.* Aberdeen: Macaulay Institute for Soil Research.

Birse, E.L. & Dry, F.T. (1970). *Assessment of Climatic Conditions in Scotland. 1. Based on Accumulated Temperature and Potential Water Deficit* (Map). Aberdeen: Macaulay Institute for Soil Research.

Birse, E.L. & Robertson, J.S. (1976). *Plant Communities and Soils of the Lowland and Southern Upland Regions of Scotland.* Aberdeen: Macaulay Institute for Soil Research.

Blackman, G.E. & Rutter, A.J. (1954). Biological Flora of the British Isles: *Endymion nonscriptus* (L.) Garcke. *Journal of Ecology,* **42,** 629–38.

Blaxter, C.M. (1983). *Ecological Studies on the Braco Castle Pinewood.* University of Stirling: BSc thesis.

Böcher, T.W. (1954). Oceanic and continental vegetation complexes in south-west Greenland. *Meddelelser Grønland,* **148,** 82–418.

Booth, T.C. (1977). Pinewoods managed by the Forestry Commission. In *Native Pinewoods of Scotland,* ed. R.G.H. Bunce & J.N.R. Jeffers, pp. 112–5. Cambridge: Institute of Terrestrial Ecology.

Bradshaw, A.D. (1953). Human influence on hybridisation in *Crataegus.* In *The Changing Flora of Britain,* ed. J.E. Lousley, pp. 181–3. London: Botanical Society of the British Isles.

Bradshaw, M.E. (1962). The distribution and status of five species of the *Alchemilla vulgaris* L. aggregate in Upper Teesdale. *Journal of Ecology*, **50**, 681–706.

Bradshaw, M.E. & Jones, A.V. (1976). *Phytosociology in Upper Teesdale: Guide to the vegetation maps of Widdybank Fell*. With six accompanying maps. Durham: University of Durham, Department of Extra Mural Studies.

Braun-Blanquet, J. (1928). *Pflanzensoziologie. Grundzüge der Vegetationskunde*. Berlin: Springer.

Braun-Blanquet, J. (1932). *Plant Sociology*. New York: McGraw-Hill.

Braun-Blanquet, J. & Tüxen, R. (1952). Irische Pflanzengesellschaften. *Veröffentlichungen des Geobotanischen Institutes Rübel in Zürich*, **25**, 224–415.

Bray, J.R. & Gorham, E. (1964). Litter production in forests of the world. *Advances in Ecological Research*, **2**, 101–57.

Brenchley, W.E. & Adam, H. (1915). Recolonisation of cultivated land allowed to revert to natural conditions. *Journal of Ecology*, **3**, 193–210.

Bridgewater, P. (1970). *Phytosociology and community boundaries of the British heath formation*. Durham University: PhD thesis.

Brown, A.H.F. & Oosterhuis, L. (1981). The role of buried seeds in coppice woods. *Biological Conservation*, **21**, 19–38.

Brown, J.M.B. (1953). *Studies on British beechwoods*, Forestry Commission Bulletin No. 20. London: HMSO.

Brown, J.M.B. (1964). Forestry. In *The Soils and Land Use of the District around Aylesbury and Hemel Hempstead*, B.W. Avery, pp. 191–8. London: HMSO.

Bunce, R.G.H. (1977). The range of variation within the pinewoods. In *Native Pinewoods of Scotland*, ed. R.G.H. Bunce & J.N.R. Jeffers, pp. 10–25. Cambridge: Institute of Terrestrial Ecology.

Bunce, R.G.H. (1982). *A Field Key for Classifying British Woodland Vegetation. Part I.* Cambridge: Institute of Terrestrial Ecology.

Bunce, R.G.H. & Jeffers, J.N.R. (ed.) (1977). *Native Pinewoods of Scotland*. Cambridge: Institute of Terrestrial Ecology.

Carlisle, A. (1958). A guide to the named varieties of Scots pine (*Pinus sylvestris* Linnaeus). *Forestry*, **31**, 203–24.

Carlisle, A. (1977). The impact of man on the native pinewoods of Scotland. In *Native Pinewoods of Scotland*, ed. R.G.H. Bunce & J.N.R. Jeffers, pp. 70–7. Cambridge: Institute of Terrestrial Ecology.

Carlisle, A. & Brown, A.H.F. (1968). Biological Flora of the British Isles: *Pinus sylvestris* L. *Journal of Ecology*, **56**, 269–307.

Carroll, D.M., Hartnup, R. & Jarvis, R.A. (1979). *Soils of South and West Yorkshire*. Harpenden: Soil Survey of England & Wales.

Chandler, T.J. & Gregory, S. (ed.) (1976). *The Climate of the British Isles*. London: Longman.

Chippindale, H.G. & Milton, W.E.J. (1934). On the viable seeds present in soil beneath pastures. *Journal of Ecology*, **22**, 508–31.

Christy, M. (1897). *Primula elatior* in Britain: its distribution, hybrids and allies. *Journal of the Linnean Society (Botany)*, **33**, 172–201.

Christy, M. (1922). *Primula elatior* Jacquin: its distribution in Britain. *Journal of Ecology*, **10**, 200–10.

Christy, M. (1924). *Primula elatior* Jacquin: its distribution in Britain. *Journal of Ecology*, **12**, 314–6.

Christy, M. & Worth, R.H. (1922). The ancient dwarfed woods of Dartmoor. *Transactions of the Devonshire Association of Science, Literature and Art*, **54**, 291–342.

Clapham, A.R. (1940). The role of bryophytes in the calcareous fens of the Oxford district. *Journal of Ecology*, **38**, 71–80.

Clapham, A.R. & Nicholson, B.E. (1975). *The Oxford Book of Trees*. Oxford: Oxford University Press.

Clement, B., Gloaguen, J.-C., & Touffet, J. (1975). Contribution à l'étude phytosociologique des forêts de Bretagne. In *La Vegétation des Forêts Caducifoliées Acidiphiles*, ed. J.-M. Géhu, pp. 53–72. Leutershausen: Cramer.

Climatological Atlas of the British Isles (1952). London: Meteorological Office.

Coate, P.H. & Son (undated). *Your Guide to the Willow Industry*. Stoke St Gregory: Coate & Son.

Conolly, A.P. & Dahl, E. (1970). Maximum summer temperature in relation to the modern and Quaternary distributions of certain arctic-montane species in the British Isles. In *Studies in the Vegetational History of the British Isles*, ed. D. Walker & R.G. West, pp. 159–224. Cambridge: Cambridge University Press.

Conway, V.M. (1942). Biological Flora of the British Isles: *Cladium mariscus* (L.) R. Br. *Journal of Ecology*, **30**, 211–16.

Coombe, D.E. & Frost, L.C. (1956). The heaths of the Cornish Serpentine. *Journal of Ecology*, **44**, 226–56.

Corbett, W.M. (1973). *Breckland Forest Soils*. Harpenden: Soil Survey of England and Wales.

Corley, M.F.V. & Hill, M.O. (1981). *Distribution of Bryophytes in the British Isles*. Cardiff: British Bryological Society.

Cotton, D.E. (1968). *An Investigation of the Structure, Morphology and Ecology of some Pennine Soil Microtopographic Features*. Lancaster University: PhD thesis.

Cousens, J.E. (1965). The status of the pedunculate and sessile oaks in Britain. *Watsonia*, **6**, 161–76.

Crampton, C.B. (1911). *The Vegetation of Caithness considered in Relation to the Geology*. Committee for the Survey and Study of British Vegetation.

Cross, J.R. (1975). Biological Flora of the British Isles: *Rhododendron ponticum* L. *Journal of Ecology*, **63**, 345–64.

Curtiss, L.F., Courtney, F.M. & Trudgill, S. (1976). *Soils in the British Isles*. London: Longman.

Dahl, E. (1956). *Rondane: mountain vegetation in south Norway and its relation to the environment*. Oslo: Aschehoug.

Dahl, E. (1968). *Analytical Key to British Macrolichens, 2nd edition*. London: British Lichen Society.

Dahl, E. & Hadač, E. (1941). Strandgesellschaften der Insel Ostøy im Oslofjord. Eine pflanzensoziologische Studie. *Nytt Magasin for Naturvidenskapene B*, **82**, 251–312.

Daniels, R.E. (1978). Floristic analyses of British mires and mire communities. *Journal of Ecology*, **66**, 773–802.

David, R.W. (1978). The distribution of *Carex elongata* in Britain. *Watsonia*, **12**, 158–60.

Davy, A.J. (1980). Biological Flora of the British Isles:

Deschampsia caespitosa (L.) Beauv. *Journal of Ecology*, **68**, 1075–96.

Davy, A.J. & Taylor, K. (1974*a*). Water characteristics of contrasting soils in the Chiltern Hills and their significance for *Deschampsia caespitosa* (L.) Beauv. *Journal of Ecology*, **62**, 367–78.

Day, W.R. (1946). The pathology of beech on chalk soils. *Quarterly Journal of Forestry*, **40**, 72–82.

Delelis-Dusollier, A. & Géhu, J.-M. (1972). Aperçu phytosociologique sur les Fourres à *Taxus* de la basse Vallée de la Seine et comparaison avec ceux de l'Angleterre. *Documents Phytosociologiques*, **I**, 39–46.

Dengler, A. (1930). *Waldbau auf Ökologischer Grundlage*. Berlin.

Dethioux, M.-H. (1955). Aperçu sur la végétation de la forêt de Meerdael et des bois environnants. *Agricultura*, **3**, 261–92.

Dimbleby, G.W. (1962). *The Development of British Heathlands and their Soils*. Oxford: Oxford University Press.

Dimbleby, G.W. & Gill, J.M. (1955). The occurrence of podzols under deciduous woodlands in the New Forest. *Forestry*, **28**, 95–105.

Doing, H. (1962). *Systematische Ordnung und floristische Zusammensetzung niederländischer Wald- und Gebuschgesellschaften*. Wageningen University: PhD thesis.

Dony, J.G. (1953). *Flora of Bedfordshire*. Luton: The Corporation of Luton Museum and Art Gallery.

Duchaufour, Ph. (1950). Recherches sur l'évolution des sols calcaires en Lorraine. *Annales de l'Ecole Eaux Forestière*, **12**, 100–53.

Duchafour, Ph. (1956). *Pédologie: applications forestières et agricoles*. Nancy: Ecole Eaux Forestière.

Duffey, E. (1971). The management of Woodwalton Fen: the multi-disciplinary approach. In *The Scientific Management of Animal and Plant Communities for Conservation*, ed. E. Duffey & A.S. Watt, pp. 581–97. Oxford: Blackwell.

Duffey, E., Morris, M.G., Sheail, J., Ward, L.K., Wells, D.A. & Wells, T.C.E. (1974). *Grassland Ecology and Wildlife Management*. London: Chapman and Hall.

Dumont, J.-M. (1975). Les anciens taillis à écorce de la région du Plateau des Tailles (Haute Ardenne Belge). In *La Végétation des Forêts Caducifoliées Acidiphiles*, ed. J.-M. Géhu, pp. 89–106. Leutershausen: Cramer.

Du Rietz, G.E. & Du Rietz, G. (1925). Floristiska anteckningar fran Bleckinge skargard. *Botaniska Notiser 1925*, 66–76.

Durin, L., Géhu, J.-M., Noirfalise, A. & Sougnez, N. (1968). Les hêtraies atlantiques et leur essaim climatiques dans le nord-ouest de la France. *Bulletin de la Société de Botanique du nord de la France*, No. spéc. 20e anniv., 59–89.

Duvigneaud, J. (1975). Les chênaies acidiphiles de la région liégoise (Belgique). Les causes de leur dégradation. Leurs possibilités d'évolution. In *La Végétation des Forêts Caducifoliées Acidiphiles*, ed. J.-M. Géhu, pp. 107–116. Leutershausen: Cramer.

Edlin, H.L. (1958). *The Living Forest*. London: Thames and Hudson.

Edwards, M.E. & Birks, H.J.B. (1986). Vegetation and ecology of four western oakwoods (*Blechno-Quercetum petraeae* Br.-Bl. & Tx. 1952) in North Wales.

Phytocoenologia, **14**, 237–61.

Ellenberg, H. (1978). *Vegetation Mitteleuropas mit den Alpen*, 2 Auflage. Stuttgart: Ulmer.

Ellis, E.A. (1965). *The Broads*. London: Collins.

Elwes, H.J. & Henry, A.H. (1906). *The Trees of Great Britain and Ireland*. Edinburgh: privately printed.

Evans, E.M. (1954). *Studies in Bryophyte Ecology*. Sheffield University: PhD thesis.

Farrell, L. (1985). Biological Flora of the British Isles: *Orchis militaris* L. *Journal of Ecology*, **73**, 1041–54.

Farrow, E.P. (1915). On the ecology of the vegetation of Breckland. I. General description of Breckland and its vegetation. *Journal of Ecology*, **3**, 211–28.

Faulkner, R. (1977). The gene-pool of Caledonian Scots pine – its conservation and uses. In *Native Pinewoods of Scotland*, ed. R.G.H. Bunce & J.N.R. Jeffers, pp. 96–9. Cambridge: Institute of Terrestrial Ecology.

Ferreira, R.E.C. (1978). *A Preliminary Vegetation Survey of Selected Cleughs in Western Borders*. Edinburgh: Nature Conservancy Council.

Firbas, F. (1949). *Spät- und nacheiszeitliche Waldgeschichte Mitteleuropas nördlich der Alpen, I*. Jena: Fischer.

Fitter, A.H. & Jennings, R.D. (1975). The effects of sheep grazing on the growth and survival of seedling junipers (*Juniperus communis* L.). *Journal of Applied Ecology*, **12**, 637–42.

Fitter, A.H., Browne, J., Dixon, T. & Tucker, J.J. (1980). Ecological studies at Askham Bog Nature Reserve. I. Inter-relations of vegetation & environment. *Naturalist*, **105**, 89–101.

Fitzpatrick, E.A. (1977). Soils of the native pinewoods of Scotland. In *Native Pinewoods of Scotland*, ed. R.G.H. Bunce & J.N.R. Jeffers, pp. 35–41. Cambridge: Institute of Terrestrial Ecology.

Forbes, J.C. & Kenworthy, J.B. (1973). Distribution of two species of birch forming stands on Deeside, Aberdeenshire. *Transactions of the Botanical Society of Edinburgh*, **42**, 101–10.

Ford, E.D. & Newbould, P.J. (1970). Stand structure and dry weight production through the sweet chestnut (*Castanea sativa* Mill.) coppice cycle. *Journal of Ecology*, **58**, 275–96.

Ford, E.D. & Newbould, P.J. (1971). The leaf canopy of a coppiced deciduous woodland. I. Development and structure. *Journal of Ecology*, **59**, 843–62.

Fordham, S.J. & Green, R.D. (1980). *Soils of Kent*. Harpenden: Soil Survey of England and Wales.

Frileux, P.-N. (1975). Contribution à l'étude des forêts acidiphiles de Haute Normandie. In *La Végétation des Forêts Caducifoliées Acidiphiles*, ed. J.-M. Géhu pp. 287–300. Leutershausen: Cramer.

Fuller, R.J. (1982). *Bird Habitats in Britain*. Calton: T. & A.D. Poyser.

Furness, R.R. (1978). *Soils of Cheshire*. Harpenden: Soil Survey of England & Wales.

Gardiner, A.S. (1974). A history of the taxonomy and distribution of the native oak species. In *The British Oak*, ed. M.G. Morris & F.H. Perring, pp. 13–26. Faringdon: Botanical Society of the British Isles.

Gaussen, H., Heywood, V.H. & Chater, A.O. (1964). *Pinus* L. In *Flora Europaea I*, ed. T.G. Tutin *et al.*, pp. 32–5. Cambridge: Cambridge University Press.

Géhu, J.-M. (1964). Sur la végétation phanérogamique halophile des falaises bretonnes. *Revue générale Botanique*, **71**, 73–78.

Géhu, J.-M. (ed.) (1975a). *La Végétation des Forêts Caducifoliées Acidiphiles*. Leutershausen: Cramer.

Géhu, J.-M. (1975b). Aperçu sur les chênaies-hêtraies acidiphiles du Sud de l'Angleterre. L'exemple de la New Forest. In *La Végétation des Forêts Caducifoliées Acidiphiles*, ed. J.-M. Géhu, pp. 133–140. Leutershausen: Cramer.

Gilbert, O.L. (1970). Biological Flora of the British Isles: *Dryopteris villarii* (Bellardi) Woynar. *Journal of Ecology*, **58**, 301–13.

Gilbert, O.L. (1980). Juniper in Upper Teesdale. *Journal of Ecology*, **68**, 1013–24.

Giller, K.E. (1982). *Aspects of the Ecology of a Flood-plain Mire in Broadland*. University of Sheffield: PhD thesis.

Gimingham, C.H. (1972). *Ecology of Heathlands*. London: Chapman and Hall.

Godwin, H. (1936). Studies in the ecology of Wicken Fen. III. The establishment and development of Fen Scrub (Carr). *Journal of Ecology*, **24**, 82–116.

Godwin, H. (1941). Studies in the ecology of Wicken fen. IV. Crop-taking experiments. *Journal of Ecology*, **29**, 83–106.

Godwin, H. (1943a). Biological Flora of the British Isles: *Rhamnus cathartica* L. *Journal of Ecology*, **31**, 69–76.

Godwin, H. (1943b). Biological Flora of the British Isles: *Frangula alnus* Miller. *Journal of Ecology*, **31**, 77–92.

Godwin, H. (1975). *The History of the British Flora*. Cambridge: Cambridge University Press.

Godwin, H. (1978). *Fenland: its Ancient Past and Uncertain Future*. Cambridge: Cambridge University Press.

Godwin, H. & Bharucha, F.R. (1932). Studies in the ecology of Wicken Fen. II. The Fen water-table and its control of plant communities. *Journal of Ecology*, **20**, 157–91.

Godwin, H. & Clifford, M.H. (1938). Studies of the post-glacial history of British vegetation. I. Origin and stratigraphy of fenland deposits near Woodwalton, Hunts. II. Origin and stratigraphy of deposits in southern Fenland. *Philosophical Transactions of the Royal Society of London, series B*, **229**, 323–406.

Godwin, H. & Tansley, A.G. (1929). The Vegetation of Wicken Fen. In *The Natural History of Wicken Fen, Part V*, pp. 387–446. Cambridge: Bowes & Bowes.

Godwin, H. & Turner, J.S. (1933). Soil acidity in relation to vegetational succession in Calthorpe Broad, Norfolk. *Journal of Ecology*, **21**, 235–62.

Godwin, H., Clowes, D.R. & Huntley, B. (1974). Studies in the ecology of Wicken Fen. V. Development of fen carr. *Journal of Ecology*, **62**, 197–214.

Good, R. (1936). On the distribution of the lizard orchid (*Himantoglossum hircinum* Koch.). *New Phytologist*, **35**, 142–70.

Goodier, R. & Bunce, R.G.H. (1977). The native pinewoods of Scotland: the current state of the resource. In *Native Pinewoods of Scotland*, ed. R.G.H. Bunce & J.N.R. Jeffers, pp. 78–87. Cambridge: Institute of Terrestrial Ecology.

Graham, G.G. (1971). *Phytosociological Studies of Relict Woodlands in the North-east of England*. Durham University: MSc thesis.

Graham, G.G. (1988). *The Flora and Vegetation of County Durham*. Sunderland: The Durham Flora Committee and the Durham County Conservation Trust.

Green, F.H.W. (1974). Climate and weather. In *The Cairngorms – their Natural History and Scenery*, ed. D. Nethersole-Thompson & A. Watson, pp. 228–36. London: Collins.

Gregory, S. (1954). Accumulated temperature maps of the British Isles. *Transactions of the Institute of British Geographers*, **20**, 59–73.

Grime, J.P. & Lloyd, P.S. (1973). *An Ecological Atlas of Grassland Plants*. London: Edward Arnold.

Grubb, P.J. & Suter, M.B. (1971). The mechanism of acidification of soil by *Calluna* and *Ulex* and the significance for conservation. In *The Scientific Management of Animal and Plant Communities for Conservation*, ed. E. Duffey & A.S. Watt, pp. 115–33. Oxford: Blackwell.

Grubb, P.J., Green, H.E. & Merrifield, R.C.J. (1969). The ecology of chalk heath: its relevance to the calcicole-calcifuge and soil acidification problems. *Journal of Ecology*, **57**, 175–212.

Gunson, A.R. (1975). The vegetation history of North-east Scotland. In *Quaternary Studies in North-east Scotland*, ed. A.D.M. Gemmell, pp. 61–72. Aberdeen: University of Aberdeen.

Harley, J.L. (1937). Ecological observations on the mycorrhiza of beech. *Journal of Ecology*, **25**, 421–3.

Harley, J.L. (1949). Soil conditions and the growth of beech seedlings. *Journal of Ecology*, **37**, 28–37.

Harris, G.T. (1921). Ecological notes on Wistman's Wood and Black Tor Copse, Dartmoor. *Transactions of the Devonshire Association of Science, Literature and Art*, **53**, 232–45.

Hartley, P.H.T. (1954). Wild fruits in the diet of British thrushes. A study in the ecology of closely-allied species. *British Birds*, **17**, 97–107.

Hartmann, F.K., & Jahn, G. (1967). *Waldgesellschaften des mitteleuropäischen Gebirgeraumes nördlich der Alpen*. Stuttgart: Gustav Fischer Verlag.

Haslam, S.M. (1965). Ecological studies in Breck fens. I. Vegetation in relation to habitat. *Journal of Ecology*, **53**, 599–619.

Haslam, S.M. (1971a). Community regulation in *Phragmites communis* Trin. I. Monodominant stands. *Journal of Ecology*, **59**, 65–73.

Haslam, S.M. (1971b). Community regulation in *Phragmites communis* Trin. II. Mixed stands. *Journal of Ecology*, **59**, 75–88.

Haslam, S.M. (1972). Biological Flora of the British Isles: *Phragmites communis* Trin. *Journal of Ecology*, **60**, 585–610.

Hawksworth, D.L. (1972). The natural history of Slapton Ley Nature Reserve. IV. Lichens. *Field Studies*, **3**, 535–78.

Hey, R.W. & Perrin, R.M.S. (1960). *The Geology and Soils of Cambridgeshire*. Cambridge: Cambridge Natural History Society.

Hill, M.O. (1979). *TWINSPAN – a FORTRAN program for arranging multivariate data in an ordered two-way table by classification of the individuals and attributes*. New York: Cornell University.

Hill, M.O., Bunce, R.G.H. & Shaw, M.W. (1975). Indicator

Species Analysis, a divisive polythetic method of classification and its application to a survey of native pinewoods in Scotland. *Journal of Ecology*, **63**, 597–613.

Hodge, C.A.H. & Seale, R.S. (1966). *The Soils of the District around Cambridge*. Memoirs of the Soil Survey of Great Britain: England and Wales. Harpenden: Soil Survey.

Hodgson, J.M. (1967). *Soils of the West Sussex Coastal Plain*. Bulletin of the Soil Survey of Great Britain: England and Wales. Harpenden: Soil Survey.

Hooper, M.D. (1973). History. In *Monks Wood: a Nature Reserve Record*, ed. R.C. Steele & R.C. Welch, pp. 22–35. Huntingdon: Natural Environment Research Council.

Hope-Simpson, J.F. (1940*b*). The utilisation and improvement of chalk down pasture. *Journal of the Royal Agricultural Society of England*, **100**, 44–9.

Hope-Simpson, J.F. (1941*b*). Studies on the vegetation of the English Chalk. VIII. A second survey of the chalk grasslands of the South Downs. *Journal of Ecology*, **29**, 217–67.

Hope-Simpson, J.F. & Willis, A.J. (1955). Vegetation. In *Bristol and its Adjoining Counties*, ed. C.M. MacInnes & W.F. Whittard, pp. 91–109. Bristol: British Association for the Advancement of Science.

Hopkinson, J.W. (1927). Studies on the vegetation of Nottinghamshire. I. The ecology of the Bunter Sandstone. *Journal of Ecology*, **15**, 130–71.

Hoskins, W.G. & Stamp. L.D. (1963). *The Common Lands of England & Wales*. London: Collins.

Hulten, E. (1950). *Atlas of the Distribution of Vascular Plants in N.W. Europe*. Stockholm: Generalstabens Litografiska Anstalts Förlag.

Huntley, B. (1979). The past and present vegetation of the Caenlochan National Nature Reserve, Scotland. I. Present vegetation. *New Phytologist*, **83**, 215–83.

Huntley, B. & Birks, H.J.B. (1979*a*). The past and present vegetation of the Morrone Birkwoods National Nature Reserve, Scotland, I. A primary phytosociological survey. *Journal of Ecology*, **67**, 419–46.

Huntley, B. & Birks, H.J.B. (1979*b*). The past and present vegetation of the Morrone Birkwoods National Nature Reserve, Scotland. II. Woodland vegetation and soils. *Journal of Ecology*, **67**, 447–67.

Huntley, B. & Birks, H.J.B. (1983). *An atlas of past and present pollen maps for Europe: 0–13 000 years ago*. Cambridge: Cambridge University Press.

Huntley, B., Huntley, J.P. & Birks, H.J.B. (1981). PHYTOPAK: a suite of computer programs designed for the handling and analysis of phytosociological data. *Vegetatio*, **45**, 85–95.

Hutchings, M.J. & Barkham, J.P. (1976). An investigation of shoot interactions in *Mercurialis perennis* L., a rhizomatous perennial herb. *Journal of Ecology*, **64**, 723–43.

Ingram, H.A.P., Anderson, M.C., Andrews, S.M., Chinery, J.M., Evans, G.B. & Richards, C.M. (1959). Vegetation studies at Semerwater. *The Naturalist*, **871**, 113–27.

Issler, E. (1926). *Les associations végétales des Vosges méridionales et de la plaine rhénane avoisinante. Les forêts*. Colmar.

Iversen, J. (1944). *Viscum, Hedera* and *Ilex* as climatic indicators. *Geologiska föreningen Stockholm Förhandlingar*, **66**, 463ff.

Ivimey-Cook, R.B. & Proctor, M.C.F. (1966). The plant communities of the Burren, Co. Clare. *Proceedings of the Royal Irish Academy, Series B*, **64**, 211–301.

Ivimey-Cook, R.B., Proctor, M.C.F. & Rowland, D.M. (1975). Analysis of the plant communities of a heathland site: Aylesbeare Common, Devon, England. *Vegetatio*, **31**, 33–45.

James, P.W., Hawksworth, D.L. & Rose, F. (1977). Lichen Communities in the British Isles: A Preliminary Conspectus. In *Lichen Ecology*, ed. M.R.D. Seaward, pp. 295–413. London: Academic Press.

Jarvis, M.G. (1973). *Soils of the Wantage and Abingdon District*. Memoirs of the Soil Survey of Great Britain: England and Wales. Harpenden: Soil Survey.

Jarvis, M.S. (1960). *The influence of climatic factors on the distribution of some Derbyshire plants*. Sheffield University: PhD thesis.

Jarvis, P.G. (1964). Interference by *Deschampsia flexuosa* (L.) Trin. *Oikos*, **15**, 56–78.

Jarvis, R.A., Bendelow, V.C., Bradley, R.I., Carroll, D.M., Furness, R.R., Kilgour, I.N.L. & King, S.J. (1984). *Soils and their Use in Northern England*. Harpenden: Soil Survey of England & Wales.

Jennings, J.N. & Lambert, J. (1951). Alluvial stratigraphy and vegetation succession in the region of the Bure valley broads. I. Surface features and general stratigraphy. *Journal of Ecology*, **39**, 106–19.

Jermy, A.C., Arnold, H.R., Farrell, L. & Perring, F.H. (1978). *Atlas of Ferns of the British Isles*. London: Botanical Society of the British Isles and British Pteridological Society.

Jones, E.W. (1945). Biological Flora of the British Isles: *Acer* L. *Journal of Ecology*, **32**, 215–52.

Jones, E.W. (1959). Biological Flora of the British Isles: *Quercus* L. *Journal of Ecology*, **47**, 169–222.

Jones, E.W. (1968). The taxonomy of British species of *Quercus. Proceedings of the Botanical Society of the British Isles*, **7**, 183–4.

Jowett, G.H. & Scurfield, G. (1949). A statistical investigation into the distribution of *Holcus mollis* L. and *Deschampsia flexuosa* (L.) Trin. *Journal of Ecology*, **37**, 68–81.

Jowett, G.H. & Scurfield, G. (1952). Statistical investigations into the success of *Holcus mollis* L. and *Deschampsia flexuosa* (L.) Trin. *Journal of Ecology*, **40**, 393–404.

Kelly, D.L. (1981). The native forest vegetation of Killarney, south-west Ireland: an ecological account. *Journal of Ecology*, **69**, 437–72.

Kelly, D. & Moore, J.J. (1975). A preliminary sketch of the Irish acidophilous oakwoods. In *La Végétation des Forêts Caducifoliées Acidiphiles*, ed. J.-M. Géhu, pp. 375–87. Leutershausen: Cramer.

Kendrick, W.B. & Burgess, A. (1962). Biological aspects of the decay of *Pinus sylvestris* leaf litter. *Nova Hedwigia*, **4**, 313–42.

Kennedy, D. & Brown, I.R. (1983). The morphology of the hybrid *Betula pendula* Roth. × *B. pubescens* Ehrh. *Watsonia*, **14**, 329–36.

Kerr, A.J. (1968). *Tynron Juniper Wood National Nature Reserve: Management and Research 1957–68*. Nature

Conservancy Council: unpublished report.

Kinzel, W. (1926). *Neue Tabellen zu Frost und Licht als beeinflussende Kraefte bei der Samenkeimung.* Stuttgart.

Klötzli, F. (1970). Eichen-, Edellaub- und Bruchwälder der Britischen Inseln. *Schweizerischen Zeitschrift für Forstwesen,* **121,** 329–66.

Kubiena, W.L. (1953). *The Soils of Europe.* London: Murby.

Kühn, K. (1937). *Die Pflanzengesellschaften im Neckargebiet der Schwäbischen Alb.* Ohringen.

Lambert, J.M. (1946). The distribution and status of *Glyceria maxima* (Hartm.) Holmb. in the region of Surlingham and Rockland Broads, Norfolk. *Journal of Ecology,* **33,** 230–67.

Lambert, J.M. (1951). Alluvial stratigraphy and vegetational succession in the region of the Bure Valley Broads. III. Classification, status and distribution of communities. *Journal of Ecology,* **39,** 149–70.

Lambert, J.M. (1965). The Vegetation of Broadland. In *The Broads,* ed. E.A. Ellis, pp. 69–92. London: Collins.

Lambert, J.M. & Jennings, J.N. (1951). Alluvial stratigraphy and vegetational succession in the region of the Bure Valley Broads. II. Detailed vegetational-stratigraphical relationships. *Journal of Ecology,* **39,** 120–48.

Lambert, J.M. & Jennings, J.N. (1965). Appendix A: Maps. In *The Broads,* ed. E.A. Ellis, pp. 268–311. London: Collins.

Lambert, J.M., Jennings, J.N., Smith, C.T., Green, C. & Hutchinson, J.N. (1960). The making of the broads: a reconsideration of their origin in the light of new evidence. *Royal Geographical Society Research Memoirs,* **3.**

Lambert, J.M., Jennings, J.N. & Smith, C.T. (1965). The Origin of the Broads. In *The Broads,* ed. E.A. Ellis, pp. 37–65. London: Collins.

Leach, W. (1925). Two relict upland oakwoods in Cumberland. *Journal of Ecology,* **13,** 289–300.

LeBrun, J., Noirfalise, A., Heinemann, P. & vanden Berghen, C. (1949). Les Associations végétales de Belgique. *Centre de Recherches écologiques et phytosociologiques de Gembloux, Communication No.* **8,** 105–207.

LeBrun, J., Noirfalise, A. & Sougnez, N. (1955). Sur la flore et la végétation du territoire belge de la Basse-Meuse. *Bulletin de la Société Royale de Botanique de Belgique,* **87,** 157–94.

Lemée, G. (1937). Recherches écologiques sur la végétation du Perche. *Revue Générale de Botanique,* **49.**

Linnard, W. (1982). *Welsh Woods and Forests: History and Utilization.* Cardiff: National Museum of Wales.

Lloyd, P.S. & Pigott, C.D. (1967). The influence of soil conditions on the course of succession on the Chalk of southern England. *Journal of Ecology,* **55,** 137–46.

Lock, J.M. & Rodwell, J.S. (1981). Observations on the Vegetation of Crag Lough, Northumberland. A Report to the National Trust. Unpublished manuscript.

Loveday, J. (1962). Plateau deposits of the southern Chiltern hills. *Proceedings of the Geological Association of London,* **73,** 83–102.

Lowe, J. (1897). *The Yew Trees of Gt. Britain and Ireland.* London.

Lowe, V.P.W. (1977). Pinewoods as habitats for mammals. In *Native Pinewoods of Scotland,* ed. R.G.H. Bunce & J.N.R. Jeffers, pp. 103–11. Cambridge: Institute of Terrestrial Ecology.

Luck, K.E. (1964). *Studies in the autecology of* Calamagrostis epigejos *(L.) Roth and* C. canescens *(Weber) Roth.* Cambridge University: PhD thesis.

Lye, K.A. (1967). Studies in the growth and development of oceanic bryophyte communities. *Svensk botanisk tidskrift* **61,** 297–310.

Mackney, D. (1961). A podzol development sequence in oakwoods and heath in central England. *Journal of Soil Science,* **12,** 23–40.

Malcuit, G. (1929). Les associations végétales de la vallée de la Lanterne. *Archives de Botanique,* **II.**

Malins-Smith, A. (1935). Age and rate of juniper growth on Moughton Fell. *The Naturalist,* **60,** 121–30.

Malloch, A.J.C. (1970). *Analytical Studies of Cliff-top Vegetation in South-West England.* Cambridge University: PhD thesis.

Malloch, A.J.C. (1971). Vegetation of the maritime cliff-tops of the Lizard and Land's End peninsulas, West Cornwall. *New Phytologist,* **70,** 1155–97.

Malloch, A.J.C. (1988). VESPAN II. Lancaster: University of Lancaster.

Manil, G. (1956). Aspects dynamiques du profil pédologique. Transactions of the Sixth International Congress of Soil Science, E, 439–41.

Manley, D.J.R. (1961). *Pollen Distribution Studies in some Scottish Pinewood Soils.* Aberdeen University: MSc thesis.

Manley, G. (1936). The climate of the Northern Pennines. *Quarterly Journal of the Royal Meteorological Society,* **62,** 103–13.

Manley, G. (1940). Snowfall in Britain. *Meteorological Magazine,* **75,** 41ff.

Manley, G. (1945). The effective rate of altitudinal change in temperate Atlantic climates. *Geographical Review,* **35,** 408–17.

Mansfield, A. (1952). *Historical geography of the woodlands of the southern Chilterns.* University of London: MSc thesis.

Martin, M.H. (1968). Conditions affecting the distribution of *Mercurialis perennis* L. in certain Cambridgeshire woodlands. *Journal of Ecology,* **56,** 777–93.

Martin, M.H. & Pigott, C.D. (1975). Soils. In *Hayley Wood: its History and Ecology,* O. Rackham, pp. 61–71. Cambridge: Cambridgeshire & Isle of Ely Naturalists' Trust.

Matthews, J.R. (1955). *Origin and Distribution of the British Flora.* London: Hutchinson.

Matuszkiewicz, W. (1963). Zur systematischen Auffassung der oligotrophen Bruchwaldgesellschaften im Osten der Pommerschen Seenplatte. *Ebenda,* **10,** 149–55.

Matuszkiewicz, W. (1981). *Przewodnik do oznaczania zbiorowisk roslinnych Polski.* Warzawa: Panstwowe Wydawnictwo Naukowe.

McNeill, W.M. (1961). A key to beech (*Fagus sylvatica* L.) associations on chalk uplands in England, devised by Ray Bourne. *Forestry,* **34,** 116–8.

McVean, D.N. (1953). Biological Flora of the British Isles: *Alnus glutinosa* (L.) Gaertn. *Journal of Ecology,* **41,** 447–66.

McVean, D.N. (1956b). Ecology of *Alnus glutinosa* (L.) Gaertn. V. Notes on some British Alder populations. *Journal of Ecology,* **44,** 321–30.

McVean, D.N. (1958). Island vegetation of some west Highland fresh-water lochs. *Transactions of the Botanical Society of Edinburgh*, **37**, 200–8.

McVean, D.N. (1961). Flora and vegetation of the islands of St Kilda and North Rona in 1958. *Journal of Ecology*, **49**, 39–54.

McVean, D.N. (1964a). Woodland and Scrub. In *The Vegetation of Scotland*, ed. J.H. Burnett, pp. 144–65. Edinburgh: Oliver & Boyd.

McVean, D.N. & Ratcliffe, D.A. (1962). *Plant Communities of the Scottish Highlands*. London: HMSO.

Meikle, R.D. (1975). *Salix* L. In *Hybridisation and the Flora of the British Isles*, ed. C.A. Stace, pp. 304–38. London: Academic Press and the Botanical Society of the British Isles.

Meikle, R.D. (1984). *Willows and Poplars of Great Britain and Ireland*. London: Botanical Society of the British Isles.

Mellanby, K. (1968). The effect of some mammals and birds on regeneration of oak. *Journal of Applied Ecology*, **5**, 359–66.

Melville, R. (1975). *Ulmus* L. In *Hybridisation and the Flora of the British Isles*, ed. C.A. Stace, pp. 292–9. London: Academic Press and the Botanical Society of the British Isles.

Melville, R. (1978). On the discrimination of species in hybrid swarms with special reference to *Ulmus* and the nomenclature of *U. minor* Mill. and *U. carpinifolia* Gled. *Taxon*, 27. 345–51.

Meres Report (1980). *Survey of Shropshire, Cheshire and Staffordshire Meres*. Banbury: Nature Conservancy Council England Field Unit.

Merton, L.F.H. (1970). The history and status of the woodlands of the Derbyshire Limestone. *Journal of Ecology*, **58**, 723–44.

Miles, J. & Kinnaird, J.W. (1979). The establishment and regeneration of birch, juniper and scots pine in the Scottish Highlands. *Scottish Forestry*, **33**, 102–19.

Millar, C. (1977). Some gaps in our knowledge of the Scottish pinewood ecosystem. In *Native Pinewoods of Scotland*, ed. R.G.H. Bunce & J.N.R. Jeffers, pp. 56–9. Cambridge: Institute of Terrestrial Ecology.

Milton, W.E.J. (1936). Buried viable seeds of enclosed and unenclosed hill land. *Bulletin of the Welsh Plant Breeding Station, Series H*, **14**, 58–72.

Moor, M. (1952). Die Fagion-Gesellschaften im Schweizer Jura. *Ebenda*, **31**, 1–201.

Moore, J.J. (1962). The Braun-Blanquet System: a reassessment. *Journal of Ecology*, **50**, 761–9.

Moravec, J. (1979). Das *Violo reichenbachianae-Fagetum* – eine neue Buchenwaldassoziation. *Phytocoenologia*, **6**, 484–504.

Moss, C.E. (1907). *Geographical Distribution of Vegetation in Somerset: Bath and Bridgewater District*. London: Royal Geographical Society.

Moss, C.E. (1911). The plant formation of calcareous soils. A. The sub-formation of the Older Limestones. In *Types of British Vegetation*, ed. A.G. Tansley, pp. 146–61. Cambridge: Cambridge University Press.

Moss, C.E. (1913). *Vegetation of the Peak District*. Cambridge: Cambridge University Press.

Moss, C.E., Rankin, W.M. & Tansley, A.G. (1910). The woodlands of England. *New Phytologist*, **9**, 113–49.

Mukerji, S.K. (1936). Contributions to the autecology of *Mercurialis perennis* L. *Journal of Ecology*, **24**, 38–81.

Munault, A.V. (1959). Première contribution à l'étude palynologique des sols forestiers du district Picardo-Brabançon. *Bulletin de la Société Royale Forestière de Belgique*, **10**.

Neuhäusl, R. (1977). Comparative ecological study of European oak-hornbeam forests. *Naturaliste canadien*, **104**, 109–17.

Neuhäuslova-Novotna, Z. (1977). Beitrag zur Kenntnis des *Carici remotae-Fraxinetum* in der Tschechischen Sozialistischen Republik. *Folia Geobotanica et Phytotaxonomica*, **12**(3), 225–43.

Newbould, P.J. (1960). The ecology of Cranesmoor, a New Forest valley bog. *Journal of Ecology*, **48**, 361–83.

Noirfalise, A. (1952). La Frênaie à Carex (Cariceto remotae-Fraxinetum Koch 1926). *Mémoires d'Institut Royal des Sciences Naturelles de Belgique*, **122**.

Noirfalise, A. (1968). Le Carpinion dans l'Ouest de l'Europe. *Feddes Repertorium*, **79**, 69–85.

Noirfalise, A. (1969). La chênaie mélangée à Jacinte du domaine atlantique de l'Europe (Endymio-Carpinetum). *Vegetatio*, **17**, 131–50.

Noirfalise, A. & Sougnez, N. (1956). Les chênaies de l'Ardenne verviétoise. *Pédologie*, **6**, 119–43.

Noirfalise, A. & Sougnez, N. (1963). Les forêts du Bassin de Mons. *Pédologie*, **13**, 200–15.

Nordhagen, R. (1928). *Die vegetation und Flora des Sylenegebiets*. Oslo.

Nordhagen, R. (1943). *Sikilsdalen og Norges Fjellbeiter*. *Bergens Museums Skrifter 22*. Bergen: Griegs.

Oberdorfer, E. (1953). Der europäischer Auenwald. *Beiträge zur naturkundlichen Forschung in Südwestdeutschland*, **12**, 23–69.

Oberdorfer, E. (1957). Süddeutsche Pflanzengesellschaften. *Pflanzensoziologie*, **10**.

Okali, D.U.U. (1966). A comparative study of the ecologically related tree species *Acer pseudoplatanus* and *Fraxinus excelsior*. II. The analysis of adult tree distribution. *Journal of Ecology*, **54**, 419–25.

O'Sullivan, P.E. (1977). Vegetation history and the native pinewoods. In *Native Pinewoods of Scotland*, ed. R.G.H. Bunce & J.N.R. Jeffers, pp. 60–9. Cambridge: Institute of Terrestrial Ecology.

Ovington, J.D. (1953). A study of invasion by *Holcus mollis* L. *Journal of Ecology*, **41**, 35–52.

Packham, J.R. (1975). *The biology of two woodland herbs, Oxalis acetosella L. and Galeobdolon luteum Huds.* Wolverhampton Polytechnic: PhD thesis (CNNA).

Packham, J.R. (1978). Biological Flora of the British Isles: *Oxalis acetosella* L. *Journal of Ecology*, **66**, 669–93.

Packham, J.R. (1983). Biological Flora of the British Isles: *Lamiastrum galeobdolon* (L.) Ehrend. & Polatschek. *Journal of Ecology*, **71**, 975–97.

Packham, J.R. & Willis, A.J. (1967). Soil nutrients and changes in the vegetation of an uncultivated field on Keuper Marl at Studley, Warwickshire. *Proceedings of the Birmingham Natural History Society*, **21**, 2–21.

Packham, J.R. & Willis, A.J. (1977). The effects of shading on *Oxalis acetosella*. *Journal of Ecology*, **65**, 619–42.

Page, C.N. (1982). *The Ferns of Britain and Ireland*. Cambridge: Cambridge University Press.

Pallis, M. (1911). The River Valleys of East Norfolk: their Aquatic and Fen Formations. In *Types of British Vegetation*, ed. A.G. Tansley, pp. 214–45. Cambridge: Cambridge University Press.

Parker, E.V. (1974). *Beech Bark Disease*. London: HMSO.

Passarge, H. (1961). Zur soziologischen Gliederung der *Salix cinerea*-gebusche norddeutschlands. *Vegetatio*, 10, 209–28.

Paulson, R. (1926). The beechwood: its canopy and carpet. *Transactions of the South-eastern Union of Scientific Societies*, 24–37.

Pearsall, W.H. (1918). The aquatic and marsh vegetation of Esthwaite Water. *Journal of Ecology*, 5, 53–74.

Pearsall, W.H. (1968). *Mountains and Moorlands*. London: Collins.

Pearsall, W.H. & Pennington, W. (1973). *The Lake District*. London: Collins.

Penistan, M.J. (1974). Growing oak. In *The British Oak*, ed. M.G. Morris & F.H. Perring, pp. 98–112. Faringdon: Botanical Society of the British Isles.

Perring, F.H. (1968). *Critical Supplement to the Atlas of the British Flora*. London: Nelson.

Perring, F.H. & Walters, S.M. (1962). *Atlas of the British Flora*. London & Edinburgh: Nelson.

Petch, C.P. & Swann, E.L. (1968). *Flora of Norfolk*. Norwich: Jarrold & Sons.

Peterken, G.F. (1965). *The status and growth of holly (Ilex aquifolium L.) in the New Forest*. London University: PhD thesis.

Peterken, G.F. (1966). Mortality of holly (*Ilex aquifolium* L.) seedlings in relation to natural regeneration in the New Forest. *Journal of Ecology*, 54, 259–70.

Peterken, G.F. (1969). Development of vegetation in Staverton Park, Suffolk. *Field Studies*, 3, 1–39.

Peterken, G.F. (1972). Conservation coppicing and the coppice crafts. *Quarterly Journal of the Devon Trust for Nature Conservation*, 4, 157–64.

Peterken, G.F. (1974). A method for assessing woodland flora for conservation using indicator species. *Biological Conservation*, 6, 239–45.

Peterken, G.F. (1977). Habitat conservation priorities in British and European woodlands. *Biological Conservation*, 11, 223–36.

Peterken, G.F. (1981). *Woodland Conservation and Management*. London: Chapman & Hall.

Peterken, G.F. & Harding, P.T. (1975). Woodland conservation in eastern England: comparing the effects of changes in three study areas. *Biological Conservation*, 8 279–98.

Peterken, G.F. & Lloyd, P.S. (1967). Biological Flora of the British Isles: *Ilex aquifolium* L. *Journal of Ecology*, 55, 841–58.

Peterken, G.F. & Newbould. P.J. (1966). Dry matter production by *Ilex aquifolium* L. in the New Forest. *Journal of Ecology*, 54, 143–50.

Peterken, G.F. & Tubbs, C.R. (1965). Woodland regeneration in the New Forest, Hampshire, since 1650. *Journal of Applied Ecology*, 2, 159–70.

Philippi, G. (1965). Moosgesellschaften des morschen Holzes und des Rohhumus im Schwarzwald im der Rhön, im

Weserbergland und im Harz. *Nova Hedwigia*, 9, 185–232.

Phillips. E.N.M. (1982). *Viola riviniana* Rohb. and *V. reichenbachiana* Jord. ex Bor. *B.S.B.I. News*, 31, 19.

Pigott, C.D. (1955). Biological Flora of the British Isles: *Thymus* L. genus. *Journal of Ecology*, 43, 365–8.

Pigott, C.D. (1956a). The vegetation of Upper Teesdale in the North Pennines. *Journal of Ecology*, 44, 545–86.

Pigott, C.D. (1956b). Vegetation. In *Sheffield and its Region*, ed. D.L. Linton pp. 78–89. Sheffield: British Association.

Pigott, C.D. (1958). Biological Flora of the British Isles: *Polemonium caeruleum* L. *Journal of Ecology*, 46, 507–25.

Pigott, C.D. (1960). Natural History. In *Peak District: National Park Guide No. 3*. London: HMSO.

Pigott, C.D. (1969). The status of *Tilia cordata* and *T. platyphyllos* on the Derbyshire limestone. *Journal of Ecology*, 57, 491–504.

Pigott, C.D. (1970b). The response of plants to climate and climatic change. In *The Flora of a Changing Britain*, ed. F. Perring, pp. 32–44. Oxford: Botanical Society of the British Isles.

Pigott, C.D. (1977). The scientific basis of practical conservation: aims and methods of conservation. *Proceedings of the Royal Society of London, Series B*, 197, 59–68.

Pigott, C.D. (1978b). Soil Development. In *Upper Teesdale*, ed. A.R. Clapham, pp. 129–40. London: Collins.

Pigott, C.D. (1982). The experimental study of vegetation. *New Phytologist*, 90, 389–404.

Pigott, C.D. (1983). Regeneration of oak-birch woodland following exclusion of sheep. *Journal of Ecology*, 71, 629–46.

Pigott, C.D. (1984). The flora and vegetation of Britain: ecology and conservation. *New Phytologist*, 98, 119–28.

Pigott, C.D. & Huntley, J.P. (1978). Factors controlling the distribution of *Tilia cordata* at the northern limits of its geographical range. I. Distribution in north-west England. *New Phytologist*, 81, 429–41.

Pigott, C.D. & Huntley, J.P. (1980). Factors controlling the distribution of *Tilia cordata* at the northern limits of its geographical range. II. History in north-west England. *New Phytologist*, 84, 145–64.

Pigott, C.D. & Huntley, J.P. (1981). Factors controlling the distribution of *Tilia cordata* at the northern limits of its geographical range. III. Nature and causes of seed sterility. *New Phytologist*, 87, 817–39.

Pigott, C.D. & Taylor, K. (1964). The distribution of some woodland herbs in relation to the supply of nitrogen and phosphorus in the soil. *Journal of Ecology*, 52 (supplement), 175–86.

Pigott, C.D. & Walters, S.M. (1953). Is the box-tree a native of England? In *The Changing Flora of Britain*, ed. J.E. Lousley, pp. 184–7. Oxford: Botanical Society of the British Isles.

Pigott, C.D. & Wilson, J.H. (1978). The vegetation of North Fen at Esthwaite in 1967–69. *Proceedings of the Royal Society of London, Series A*, 200, 331–51.

Poore, M.E.D. (1955a). The use of phytosociological methods in ecological investigations. I. The Braun-Blanquet System. *Journal of Ecology*, 43, 226–44.

Poore, M.E.D. (1955b). The use of phytosociological methods in ecological investigations. II. Practical issues involved in an attempt to apply the Braun-Blanquet System. *Journal of Ecology*, **43**, 245–69.

Poore, M.E.D. (1955c). The use of phytosociological methods in ecological investigations. III. Practical application. *Journal of Ecology*, **43**, 606–51.

Poore, M.E.D. (1956b). The ecology of Woodwalton Fen. *Journal of Ecology*, **44**, 455–92.

Poore, M.E.D. & McVean, D.N. (1957). A new approach to Scottish mountain vegetation. *Journal of Ecology*, **45**, 401–39.

Proctor, M.C.F. (1960). Mosses and liverworts of the Malham district. *Field Studies*, **2**, 61–85.

Proctor, M.C.F. (1974). The vegetation of the Malham Tarn fens. *Field Studies*, **4**, 1–38.

Rackham, O. (1967). The history and effects of coppicing as a woodland practice. In *The Biotic Effects of Public Pressures on the Environment*, ed. E. Duffey, pp. 82–93. Monks Wood Experimental Station: Institute of Terrestrial Ecology.

Rackham, O. (1971). Historical studies and woodland conservation. In *The Scientific Management of Animal and Plant Communities for Conservation*, ed. E. Duffey & A.S. Watt, pp. 563–80. Oxford: Blackwell Scientific Publications.

Rackham, O. (1975). *Hayley Wood: its History and Ecology*. Cambridge: Cambridgeshire and Isle of Ely Naturalists' Trust Ltd.

Rackham, O. (1976). *Trees and Woodland in the British Landscape*. London: Dent.

Rackham, O. (1980). *Ancient Woodland*. London: Arnold.

Rankin, W.M. (1911b). The Valley Moors of the New Forest. In *Types of British Vegetation*, ed. A.G. Tansley, pp. 259–64. Cambridge: Cambridge University Press.

Ratcliffe, D.A. (1959b). The Mountain Plants of the Moffat Hills. *Transactions of the Botanical Society of Edinburgh*, **37**, 257–71.

Ratcliffe, D.A. (1968). An ecological account of Atlantic bryophytes in the British Isles. *New Phytologist*, **67**, 365–439.

Ratcliffe, D.A. (1974). The Vegetation. In *The Cairngorms: their Natural History and Scenery*, ed. D. Nethersole-Thompson & A. Watson, pp. 42–76. London: Collins.

Ratcliffe, D.A. (ed.) (1977). *A Nature Conservation Review*. Cambridge: Cambridge University Press.

Raven, J. & Walters, M. (1956). *Mountain Flowers*. London: Collins.

Richards, P.W. (1938). The bryophyte communities of a Killarney oakwood. *Annales Bryologiques*, **11**, 108–30.

Richens, R.H. (1983). *Elm*. Cambridge: Cambridge University Press.

Ritchie, J.C. (1955). Biological Flora of the British Isles: *Vaccinium vitis-idaea* L. *Journal of Ecology*, **43**, 701–8.

Roden, D. (1968). Woodland and its management in the medieval Chilterns. *Forestry*, **41**, 59–71.

Roisin, P. (1969). *Le domaine phytogéographique atlantique d'Europe*. Gembloux: Les Presses agronomiques.

Rose, F. (1950). The East Kent Fens. *Journal of Ecology*, **38**, 292–302.

Rose, F. (1974). The epiphytes of oak. In *The British Oak: its history and natural history*, ed. M.G. Morris & F.H. Perring, pp. 250–73. Faringdon: Botanical Society of the British Isles.

Rübel, E.A. (1912). The Killarney Woods. *New Phytologist*, **11**, 54–7.

Salisbury, E.J. (1916). The oak-hornbeam woods of Hertfordshire. I and II. *Journal of Ecology*, **4**, 83–117.

Salisbury, E.J. (1918a). The oak-hornbeam woods of Hertfordshire. III and IV. *Journal of Ecology*, **6**, 14–52.

Salisbury, E.J. (1918b). The ecology of scrub in Hertfordshire: a study in colonisation. *Transactions of the Hertfordshire Natural History Society*, **17**, 53–64.

Salisbury, E.J. (1924). The effects of coppicing as illustrated by the woods of Hertfordshire. *Transactions of the Hertfordshire Natural History Society*, **18**, 1–21.

Salisbury, E.J. (1964). *Weeds and Aliens*, 2nd edition. London: Collins.

Sargent, C. (1984). *Britain's railway vegetation*. Cambridge: Institute of Terrestrial Ecology.

Scheys, G., Dudal, R. & Baeyens, L. (1954). Une interprétation de la morphologie de podzols humoferriques. *Transactions of the Fifth International Congress of Soil Science*, **4**, 274–81.

Schwickerath, M. (1944). Das Hohe Venn und seine Randgebiete. *Pflanzensoziologie*, **6**.

Scott, G.A.M. (1965). The shingle succession at Dungeness. *Journal of Ecology*, **53**, 21–31.

Scurfield, G. (1953). Ecological observations in southern Pennine woodland. *Journal of Ecology*, **41**, 1–12.

Scurfield, G. (1959). The ashwoods of the Derbyshire Carboniferous Limestone: Monks Dale. *Journal of Ecology*, **47**, 357–69.

Seaward, M.R.D. & Hitch, C.J.B. (1982). *Atlas of the Lichens of the British Isles, Volume I*. Cambridge: Institute of Terrestrial Ecology.

Shaw, M.W. (1974). The reproductive characteristics of oak. In *The British Oak*, ed. M.G. Morris & F.H. Perring, pp. 162–81. Faringdon: Botanical Society of the British Isles.

Shimwell, D.W. (1968a). *The Phytosociology of Calcareous Grasslands in the British Isles*. University of Durham: PhD thesis.

Shimwell, D.W. (1968b). *The Vegetation of the Derbyshire Dales: A Report to the Nature Conservancy*. Attingham Park, Shrewsbury: Nature Conservancy Midland Region.

Shimwell, D.W. (1971c). *The Description and Classification of Vegetation*. London: Sidgwick & Jackson.

Sinker, C.A. (1962). The North Shropshire meres and mosses: a background for ecologists. *Field Studies*, **4**, 101–37.

Sjörs, H. (1954). Slettërangar: grangärde Finnmark. *Acta Phytogeographica Suecica*, **34**, 1–135.

Smith, C.J. (1980). *Ecology of the English Chalk*. London: Academic Press.

Soil Survey (1974). Soil Map of England and Wales. Southampton: Ordnance Survey.

Soil Survey (1983). 1:250,000 Soil Map of England and Wales: six sheets and legend. Harpenden: Soil Survey of England and Wales.

Sougnez, N. (1975). Les chênaies silicoles de Belgique. In *La Végétation des Forêts Caducifoliées Acidiphiles*, ed. J.-M. Géhu, pp. 183–250. Leutershausen: Cramer.

Spence, D.H.N. (1964). The macrophytic vegetation of freshwater lochs, swamps and associated fens. In *The Vegetation of Scotland*, ed. J.H. Burnett, pp. 306–425. Edinburgh: Oliver & Boyd.

Steven, H.M. & Carlisle, A. (1959). *The Native Pinewoods of Scotland*. Edinburgh: Oliver & Boyd.

Streeter, D.T. (1974). Ecological aspects of oak woodland conservation. In *The British Oak*, ed. M.G. Morris & F.H. Perring, pp. 274–97. Faringdon: Botanical Society of the British Isles.

Summerhayes, V.S. (1968). *Wild Orchids of Britain*, second edition. London: Collins.

Tansley, A.G. (ed.) (1911). *Types of British Vegetation*. Cambridge: Cambridge University Press.

Tansley, A.G. (1922). Studies on the vegetation of the English Chalk. II. Early stages of redevelopment of woody vegetation on chalk grassland. *Journal of Ecology*, **10**, 168–77.

Tansley, A.G. (1925). The Vegetation of the Southern English Chalk (Obere Kreide-Formation). *Veröffentlichungen des Geobotanischen Institutes Rübel in Zürich*, **3**, 406–30.

Tansley, A.G. (1939). *The British Islands and their Vegetation*. Cambridge: Cambridge University Press.

Tansley, A.G. & Adamson, R.S. (1925). Studies of the vegetation of the English Chalk. III. The chalk grasslands of the Hampshire–Sussex border. *Journal of Ecology*, **13**, 177–223.

Tansley, A.G. & Rankin, W.M. (1911). The plant formation of calcareous soils. B. the sub-formation of the Chalk. In *Types of British Vegetation*, ed. A.G. Tansley, pp. 161–86. Cambridge: Cambridge University Press.

Taylor, K. (1980). The growth of *Rubus vestitus* in a mixed deciduous woodland. *Journal of Ecology*, **68**, 51–62.

Thom, V.M. (1977). The appreciation of pinewoods in the countryside. In *Native Pinewoods of Scotland*, ed. R.G.H. Bunce & J.N.R. Jeffers, pp. 100–2. Cambridge: Institute of Terrestrial Ecology.

Thomas, A.S. (1960). Changes in vegetation since the advent of myxomatosis. *Journal of Ecology*, **48**, 287–306.

Thomas, A.S. (1963). Further changes in vegetation since the advent of myxomatosis. *Journal of Ecology*, **51**, 151–86.

Tittensor, R.M. (1970a). History of the Loch Lomond Oakwoods. I. Ecological history. *Scottish Forestry*, **24**, 100–10.

Tittensor, R.M. (1970b). History of the Loch Lomond Oakwoods. II. Period of intensive management. *Scottish Forestry*, **24**, 110–18.

Tittensor, R.M. & Steele, R.C. (1971). Plant communities of the Loch Lomond oakwoods. *Journal of Ecology*, **59**, 561–82.

Tombal, D. (1975). Diagnose phytocoenologique des forêts proclimaciques acidiphiles de la région de Paris. In *La Végétation des Forêts Caducifoliées Acidiphiles*, ed. J.-M. Géhu, pp. 301–10. Leutershausen: Cramer.

Troup, R.S. (1966). *Silvicultural Systems, 2nd edition*. Oxford: Oxford University Press.

Tubbs, C.R. (1964). Early encoppicements in the New Forest. *Forestry*, **37**, 95–105.

Tubbs, C.R. (1968). *The New Forest: An Ecological History*. Newton Abbot: David & Charles.

Turner, J.S. & Watt, A.S. (1939). The oakwoods (*Quercetum sessiliflorae*) of Killarney, Ireland. *Journal of Ecology*, **27**, 202–33.

Tutin, T.G., Heywood, V.H., Burges, N.A., Valentine, D.H., Walters, S.M. & Webb, D.A. (1964). *Flora Europaea, Volume 1*. Cambridge: Cambridge University Press.

Tutin, T.G., Heywood, V.H., Burges, N.A., Moore, D.M., Valentine, D.H., Walters, S.M. & Webb, D.A. (1968). *Flora Europaea, Volume 2*. Cambridge: Cambridge University Press.

Tutin, T.G., Heywood, V.H., Burges, N.A., Moore, D.M., Valentine, D.H., Walters, S.M. & Webb, D.A. (1972). *Flora Europaea, Volume 3*. Cambridge: Cambridge University Press.

Tutin, T.G., Heywood, V.H., Burges, N.A., Moore, D.M., Valentine, D.H., Walters, S.M. & Webb, D.A. (1976). *Flora Europaea, Volume 4*. Cambridge: Cambridge University Press.

Tutin, T.G., Heywood, V.H., Burges, N.A., Moore, D.M., Valentine, D.H., Walters, S.M. & Webb, D.A. (1980). *Flora Europaea, Volume 5*. Cambridge: Cambridge University Press.

Tüxen, R. (1937). Die Pflanzengesellschaften Nordwestdeutschlands. *Mitteilungen der Florist-soziologischen Arbeitsgemeinschaft*, **3**, 1–170.

Tüxen, R. (1952). Hecken und Gebüsche. *Mitteilungen der Geographischen Gesellschaft in Hamburg*, **50**, 85–117.

Tüxen, R. (1955). Das System der nordwestdeutschen Pflanzengesellschaften. *Mitteilungen der Florist-soziologischen Arbeitsgemeinschaft*, **NF5**, 155–76.

Valentine, D.H. (1947). Studies in British Primulas. I. Hybridization between Primrose and Oxlip (*Primula vulgaris* Huds. and *P. elatior* Schreb.). *New Phytologist*, **46**, 229–53.

Valentine, D.H. (1948). Studies in British Primulas. II. Ecology and taxonomy of primrose and oxlip (*Primula vulgaris* Huds. and *P. elatior* Schreb.). *New Phytologist*, **47**, 111–30.

Valentine, D.H. (1951). Studies in British Primulas. III. Hybridisation between *Primula elatior* (L.) Hill and *P. veris* L. *New Phytologist*, **50**, 383–99.

Vanden Berghen, C. (1953). Contribution à l'Etude des Groupements végétaux notés dans la vallée de l'Ourthe en amont de Laroche-en-Ardenne. *Bulletin de la Société Royale de Botanique de Belgique*, **85**, 195–277.

van Zeist, W. (1959). Studies on the Post-boreal vegetational history of south-eastern Dienthe (Netherlands). *Acta Botanica Neerlandica*, **8**, 156.

van Zeist, Q. (1964). A palaeobotanical study of some bogs in western Brittany (Finistère) France. *Palaeohistoria*, **10**, 157–80.

Walker, D. (1970). Direction and rate in some British Post-glacial hydroseres. In *Studies in the Vegetation History of the British Isles*, ed. D. Walker & R.G. West, pp. 117–40. Cambridge: Cambridge University Press.

Walters, S.M. (1964). *Betula* L. In *Flora Europaea I*, ed. T.G. Tutin *et al.*, pp. 57–8. Cambridge: Cambridge University Press.

Ward, L.K. (1973). The conservation of juniper. I. Present status of juniper in southern England. *Journal of Applied Ecology*, **10**, 165–88.

Wardle, P. (1959). The regeneration of *Fraxinus excelsior* in woods with a field layer of *Mercurialis perennis*. *Journal of Ecology*, **47**, 483–97.

Watt, A.S. (1919). On the causes of failure of natural regeneration in British oakwoods. *Journal of Ecology*, **7**, 173–203.

Watt, A.S. (1923). On the ecology of British beechwoods with special reference to their regeneration. I. The causes of failure of natural regeneration of the beech (*Fagus sylvatica* L.). *Journal of Ecology*, **11**, 1–48.

Watt, A.S. (1924). On the ecology of British beechwoods with special reference to their regeneration. II. The development and structure of the beech communities on the Sussex Downs. *Journal of Ecology*, **12**, 145–204.

Watt, A.S. (1925). On the ecology of British beechwoods with special reference to their regeneration. II (cont.). The development and structure of the beech communities on the Sussex Downs. *Journal of Ecology*, **13**, 27–73.

Watt, A.S. (1926). Yew communities of the South Downs. *Journal of Ecology*, **14**, 282–316.

Watt, A.S. (1931*a*). Preliminary observations on Scottish beechwoods. Introduction and Part I. *Journal of Ecology*, **19**, 137–57.

Watt, A.S. (1934*a*). The vegetation of the Chiltern Hills, with special reference to the beechwoods and their seral relationships. I. *Journal of Ecology*, **22**, 230–70.

Watt, A.S. (1934*b*). The vegetation of the Chiltern Hills, with special reference to the beechwoods and their seral relationships. II. *Journal of Ecology*, **22**, 445–507.

Watt, A.S. & Fraser, G.K. (1933). Tree roots and the field layer. *Journal of Ecology*, **21**, 404–14.

Watt, A.S. & Jones, E.W. (1948). The ecology of the Cairngorms. I. The environment and the altitudinal zonation of the vegetation. *Journal of Ecology*, **36**. 283–304.

Watt, A.S. & Tansley, A.G. (1930). British Beechwoods. In *Fifth International Botanical Congress Abstract of Communications*, pp. 105–14. Cambridge: Cambridge University Press.

Webb, J.A. & Moore, P.D. (1982). The late Devensian vegetation history of the Whitlaw Mosses, southeast Scotland. *New Phytologist*, **91**, 341–98.

Welch, R.C. (1972). *Windsor Forest Study: Wildlife Conservation Report*. London: Crown Estate Commission & Nature Conservancy.

Wells, T.C.E. (1969). Botanical aspects of conservation management of chalk grasslands. *Biological Conservation*, **2**, 36–44.

Wells, T.C.E. (1973). Botanical aspects of chalk grassland management. In *Chalk Grassland. Studies on its Conservation and Management in South-east England*, ed. A.C. Jermy & P.A. Stott, pp. 10–15. Maidstone: Kent Trust for Nature Conservation.

Westhoff, V. & den Held, A.J. (1969). *Plantengemeenschappen in Nederland*. Zutphen: Thieme.

Wheeler, B.D. (1975). *Phytosociological studies on Rich Fen Systems in England & Wales*. University of Durham: PhD thesis.

Wheeler, B.D. (1978). The wetland plant communities of the River Ant valley, Norfolk. *Transactions of the Norfolk & Norwich Naturalists' Society*, **24**, 153–87.

Wheeler, B.D. (1980*a*). Plant communities of rich-fen systems in England & Wales. I. Introduction. Tall sedge and reed communities. *Journal of Ecology*, **68**, 368–95.

Wheeler, B.D. (1980*c*). Plant communities of rich-fen systems in England & Wales. III. Fen meadow, fen grassland and fen woodland communities and contact communities. *Journal of Ecology*, **68**, 761–88.

Wheeler, B.D. (1983). A manuscript copy of a chapter 'British Fens: A Review' now published in *European Mires*, ed. P.D. Moore (1984), pp. 237–81. London: Academic Press.

White, G. (1788). *The Natural History of Selbourne*. London: Bensley.

White, J.M. (1932). The fens of North Armagh. *Proceedings of the Royal Irish Academy, series B*, **40**, 15.

Wigginton, M.J. & Graham, G.G. (1981). *Guide to the Identification of some Difficult Plant Groups*. Banbury: Nature Conservancy Council, England Field Unit.

Wigston, D.L. (1974). Cytology and genetics of oaks. In *The British Oak*, ed. M.G. Morris & F.H. Perring, pp. 27–50. Faringdon: Botanical Society of the British Isles.

Williamson, R. (1978). *The Great Yew Forest*. London: Macmillan.

Willis, A.J. & Jefferies, R.L. (1959). The plant ecology of the Gordano Valley. *Proceedings of the British Naturalists Society*, **31**, 297–304.

Wilson, J.F. (1968). *The control of density in some woodland plants*. Lancaster University: PhD thesis.

Wood, R.F. & Nimmo, M. (1962). *Chalk Downland Afforestation*, Forestry Commission Bulletin No. 34. London: HMSO.

Woodell, S.R.J. (1969). Natural hybridisation in Britain between *Primula vulgaris* and *P. elatior*. *Watsonia*, **7**, 115–27.

Woodhead, T.W. (1906). The ecology of woodland plants in the neighbourhood of Huddersfield. *Journal of the Linnean Society (Botany)*, **37**, 33–406.

Wooldridge, S.W. & Goldring, F. (1953). *The Weald*. London: Collins.

Yapp, W.B. (1953). The high level woodlands of the English Lake District. *North-west Naturalist*, NS1, 190–207 and 370–83.

Printed in the United States
By Bookmasters